ANNUAL REVIEW OF ASTRONOMY AND ASTROPHYSICS

EDITORIAL COMMITTEE (1980)

A. BOESGAARD
G. BURBIDGE
M. COHEN
H. GURSKY
D. LAYZER
T. OWEN
J. G. PHILLIPS
W. L. W. SARGENT

Responsible for organization of Volume 18
(Editorial Committee, 1978)

G. BURBIDGE
L. GOLDBERG
H. GURSKY
D. LAYZER
T. OWEN
A. A. PENZIAS
J. G. PHILLIPS
W. L. W. SARGENT
G. WALLERSTEIN
J. AARONS (Guest)
M. COHEN (Guest)
H. HUDSON (Guest)
H. E. SMITH (Guest)

Production Editor   E. P. BROWER
Indexing Coordinator   M. A. GLASS
Subject Indexer   D. GOLDSMITH

ANNUAL REVIEW OF ASTRONOMY AND ASTROPHYSICS

GEOFFREY BURBIDGE, *Editor*
Kitt Peak National Observatory

DAVID LAYZER, *Associate Editor*
Harvard College Observatory

JOHN G. PHILLIPS, *Associate Editor*
University of California, Berkeley

VOLUME 18

1980

ANNUAL REVIEWS INC.   4139 EL CAMINO WAY   PALO ALTO, CALIFORNIA 94306

ANNUAL REVIEWS INC.
Palo Alto, California, USA

COPYRIGHT © 1980 BY ANNUAL REVIEWS INC PALO ALTO CALIFORNIA USA. ALL RIGHTS RESERVED The appearance of the code at the bottom of the first page of an article in this serial indicates the copyright owner's consent that copies of the article may be made for personal or internal use, or for the personal or internal use of specific clients. This consent is given on the condition, however, that the copier pay the stated per-copy fee of $1.00 per article through the Copyright Clearance Center, Inc. (P.O. Box 765, Schenectady, NY 12301) for copying beyond that permitted by Sections 107 or 108 of the US Copyright Law. The per-copy fee of $1.00 per article also applies to the copying, under the stated conditions, of articles published in any Annual Review serial before January 1, 1978. Individual readers, and nonprofit libraries acting for them, are permitted to make a single copy of an article without charge for use in research or teaching. This consent does not extend to other kinds of copying, such as copying for general distribution, for advertising or promotional purposes, for creating new collective works, or for resale.

REPRINTS The conspicuous number aligned in the margin with the title of each article in this volume is a key for use in ordering reprints. Available reprints are priced at the uniform rate of $1.00 each postpaid. The minimum acceptable reprint order is 5 reprints and/or $5.00 prepaid. A quantity discount is available.

International Standard Serial Number : 0066-4146
International Standard Book Number : 0-8243-0918-9
Library of Congress Catalog Card Number : 63-8846

Annual Reviews Inc. and the Editors of its publications assume no responsibility for the statements expressed by the contributors to this Review.

FILMSET BY TYPESETTING SERVICES LTD, GLASGOW, SCOTLAND
PRINTED AND BOUND IN THE UNITED STATES OF AMERICA

PREFACE

This volume of the *Annual Review of Astronomy and Astrophysics* was planned at the meeting of the Editorial Committee held May 13, 1978 in San Francisco. Committee members present were: Editor Geoffrey Burbidge; Associate Editors David Layzer and John Phillips; Committee members Herbert Gursky, Tobias Owen, and George Wallerstein; and Production Editor Rosalie West. Guests who replaced absent committee members were: Jonathon Aarons, Marshall Cohen, Hugh Hudson, and H. E. Smith.

Once again I would like to thank the Associate Editors for carrying out the scientific editorial work very effectively and the Production Editor, Elizabeth Brower, for her excellent work.

THE EDITOR

SOME RELATED ARTICLES APPEARING
IN OTHER *ANNUAL REVIEWS*

From the *Annual Review of Earth and Planetary Sciences*, Volume 8 (1980)

Refractory Inclusions in the Allende Meteorite, Lawrence Grossman
Origin and Evolution of Planetary Atmospheres, James B. Pollack and Yuk L. Yung
Geomorphological Processes on Terrestrial Planetary Surfaces, Robert P. Sharp
The Moons of Mars, Joseph Veverka and Joseph A. Burns

From the *Annual Review of Nuclear and Particle Science*, Volume 30 (1980)

Particle Collisions Above 10 TeV as Seen by Cosmic Rays, T. K. Gaisser and G. B. Yodh

CONTENTS

ON SOME TRENDS IN THE DEVELOPMENT OF ASTROPHYSICS, *V. A. Ambartsumian*	1
THE MASSES OF CEPHEIDS, *Arthur N. Cox*	15
INFRARED SPECTROSCOPIC OBSERVATIONS OF THE OUTER PLANETS, THEIR SATELLITES, AND THE ASTEROIDS, *Harold P. Larson*	43
FORMATION OF THE TERRESTRIAL PLANETS, *George W. Wetherill*	77
STELLAR MASSES, *Daniel M. Popper*	115
THE STRUCTURE OF EXTENDED EXTRAGALACTIC RADIO SOURCES, *George Miley*	165
INTERSTELLAR SHOCK WAVES, *Christopher F. McKee and David J. Hollenbach*	219
ENVELOPES AROUND LATE-TYPE GIANT STARS, *B. Zuckerman*	263
COSMIC-RAY CONFINEMENT IN THE GALAXY, *Catherine J. Cesarsky*	289
OPTICAL AND INFRARED POLARIZATION OF ACTIVE EXTRAGALACTIC OBJECTS, *J. R. P. Angel and H. S. Stockman*	321
WHITE DWARF STARS, *James Liebert*	363
NUCLEAR ABUNDANCES AND EVOLUTION OF THE INTERSTELLAR MEDIUM, *P. G. Wannier*	399
STELLAR CHROMOSPHERES, *Jeffrey L. Linsky*	439
MEASUREMENTS OF THE COSMIC BACKGROUND RADIATION, *Rainer Weiss*	489
MICROWAVE BACKGROUND RADIATION AS A PROBE OF THE CONTEMPORARY STRUCTURE AND HISTORY OF THE UNIVERSE, *R. A. Sunyaev and Ya. B. Zel'dovich*	537
INDEXES	
AUTHOR INDEX	561
SUBJECT INDEX	577
CUMULATIVE INDEX OF CONTRIBUTING AUTHORS, VOLUMES 14–18	587
CUMULATIVE INDEX OF CHAPTER TITLES, VOLUMES 14–18	589

ANNUAL REVIEWS INC. is a nonprofit corporation established to promote the advancement of the sciences. Beginning in 1932 with the *Annual Review of Biochemistry*, the Company has pursued as its principal function the publication of high quality, reasonably priced Annual Review volumes. The volumes are organized by Editors and Editorial Committees who invite qualified authors to contribute critical articles reviewing significant developments within each major discipline.

Annual Reviews Inc. is administered by a Board of Directors whose members serve without compensation. The Board for 1980 is constituted as follows:

Dr. J. Murray Luck, Founder and Director Emeritus of Annual Reviews Inc.
  Professor Emeritus of Chemistry, Stanford University
Dr. Joshua Lederberg, President of Annual Reviews Inc.
  President, The Rockefeller University
Dr. James E. Howell, Vice President of Annual Reviews Inc.
  Professor of Economics, Stanford University
Dr. William O. Baker, *President, Bell Telephone Laboratories*
Dr. Robert W. Berliner, *Dean, Yale University School of Medicine*
Dr. Sidney D. Drell, *Deputy Director, Stanford Linear Accelerator Center*
Dr. Eugene Garfield, *President, Institute for Scientific Information*
Dr. William D. McElroy, *Chancellor, University of California, San Diego*
Dr. William F. Miller, *President, SRI International*
Dr. Colin S. Pittendrigh, *Director, Hopkins Marine Station*
Dr. Esmond E. Snell, *Professor of Microbiology and Chemistry,*
  University of Texas, Austin
Dr. Harriet A. Zuckerman, *Professor of Sociology, Columbia University*

The management of Annual Reviews Inc. is constituted as follows:
John S. McNeil, Chief Executive Officer and Secretary-Treasurer
William Kaufmann, Editor-in-Chief
Sharon E. Hawkes, Production Manager
Ruth E. Severance, Promotion Manager

Annual Reviews are published in the following sciences: Anthropology, Astronomy and Astrophysics, Biochemistry, Biophysics and Bioengineering, Earth and Planetary Sciences, Ecology and Systematics, Energy, Entomology, Fluid Mechanics, Genetics, Materials Science, Medicine, Microbiology, Neuroscience, Nuclear and Particle Science, Pharmacology and Toxicology, Physical Chemistry, Physiology, Phytopathology, Plant Physiology, Psychology, Public Health, and Sociology. The *Annual Review of Nutrition* will begin publication in 1981. In addition, five special volumes have been published by Annual Reviews Inc.: *History of Entomology* (1973), *The Excitement and Fascination of Science* (1965), *The Excitement and Fascination of Science, Volume Two* (1978), *Annual Reviews Reprints: Cell Membranes, 1975–1977* (published 1978), and *Annual Reviews Reprints: Immunology, 1977–1979* (published 1980). For the convenience of readers, a detachable order form/envelope is bound into the back of this volume.

1977, Erevan, USSR

ON SOME TRENDS IN THE DEVELOPMENT OF ASTROPHYSICS

✇2160

V. A. Ambartsumian

Academy of Sciences of Armenian SSR, Erevan, Armenia, USSR

INTRODUCTION

During my studies at the University of Leningrad (1925–1928) I paid chief attention to astronomical and mathematical courses. While I always felt the necessity of better knowledge of physics, at that time this discipline did not attract me very much. It is true that during my last two university years the logical beauty of quantum mechanics as well as of some aspects of statistical physics impressed me deeply. But even now I feel that my knowledge of physics was incomplete and insufficient for a theoretical astrophysicist.

Perhaps this circumstance, as well as a lack of physical intuition, were the reasons why during the fifty years of my scientific work I have concentrated mainly in directions where logical consistency is of greater importance than physical insight. At the same time, I have spent much time in the study of the data obtained by observers.

Modern astrophysics deals with an unusual diversity and richness of observational data, with a huge variety of cosmical bodies and systems. These bodies and systems are sometimes of different scale and properties. At the same time, one meets here a great diversity of roads of scientific investigations and ways of thinking.

Nevertheless, it happened that my personal scientific efforts have been almost completely devoted to three main directions of theoretical work: 1. invariance principles as applied to the theory of radiative transfer, 2. inverse problems of astrophysics, and 3. the empirical approach to the problems of the origin and the evolution of stars and galaxies.

In the following pages I'll give a short review of results received in each of these three directions.

1. INVARIANCE PRINCIPLES AND THE THEORY OF RADIATIVE TRANSFER

The problem of scattering and absorption of light in a medium that consists of plane-parallel layers was considered in the classical work of Schwarzschild, Shuster, Eddington, Milne, and Chandrasekhar. In essence, their method was connected with the consideration of the balance of radiative energy in all elementary volumes inside the medium. As a result the problem in each case can be reduced to some integral equation with the kernel Ei$(/\tau - t/)$ where Ei is the so-called integral logarithm

$$\text{Ei } y = \int_1^\infty e^{-yt} \frac{dt}{t}.$$

The case of isotropic and monochromatic scattering is comparatively simple. But the general problem of anisotropic scattering with some law of redistribution of frequencies (which is important for the theory of absorption lines) is connected with many complications and difficulties.

Early, as a university student, I tried to contribute to this field. My diploma work was devoted to the integral equation of radiative equilibrium. However, the first essential results were achieved only in 1932–1933 when I worked out a method of successive analysis of Lyman-continuum and L_α radiation fields in dealing with the radiative equilibrium of planetary nebulae. Before World War II I found also a simple way of considering the monochromatic scattering problem in deep layers of a medium (for example, in deep layers of the sea) with arbitrary indicator of scattering. But all this was within the framework of classical methods. Only in 1941 did I find that there are other possibilities.

Let us consider a medium consisting of plane-parallel layers occupying the half space $z < 0$ with the boundary plane $z = 0$. The parallel beam of light of the flux density πS falling on this boundary under the angle arc cos η to the normal will penetrate into the medium and suffer there innumerable elementary processes of scattering and absorption. As a result, some part of the initial beam will be scattered back into the half-space $z = 0$ in different directions. This phenomenon is called the "diffuse reflection" of light from the medium. The intensity $I(\xi)$ of the light diffusely reflected in the direction arc cos ξ will depend both on η and ξ

$$I(\xi) = Sr(\eta, \xi).$$

According to classical methods, in order to find the function $r(\eta, \xi)$ it is necessary to solve the above-mentioned integral equation for the different values of parameter η to find the radiation field for every η as a function of depth. However, we use then only the intensities at $z = 0$ which define $r(\eta, \xi)$.

In order to avoid the calculation of the data which describe the radiation field at layers $z < 0$, I decided to try to make use of the following circumstance. The function $r(\eta, \xi)$ evidently will not change if we add to the boundary $z = 0$ a thin additional layer of the thickness ΔZ which has the same optical properties as those of the primary medium.

This means that different supplementary phenomena of scattering and absorption will in this case exactly compensate each other. Writing this condition of compensation, we can obtain an equation for $r(\eta, \xi)$. The decisive point was the understanding of the fact that no quantity immediately related to the internal layers enters into this equation. The equation contains only the unknown function $r(\eta, \xi)$. The same equation which was derived in this way was then obtained also from the integral equation of radiative equilibrium itself.

The demand that $r(\eta, \xi)$ must remain unchanged when a supplementary layer is added to the boundary is called "*invariance principle*."

In the simplest case of monochromatic and isotropic elementary processes we obtain from this principle that $r(\eta, \xi)$ as a function of two variables must have the following structure:

$$r(\eta, \xi) = \frac{\lambda}{4} \eta \, \frac{\varphi(r)\varphi(\xi)}{\eta + \xi} \qquad (1)$$

where λ is the ratio of the scattering coefficient to the extinction (absorption + scattering) coefficient and φ is an auxilary function *of only one variable* which satisfies a very simple functional equation

$$\varphi(\eta) = 1 + \frac{\lambda}{2} \eta \int_1^1 \frac{\varphi(r)\varphi(\xi)}{\eta + \xi} \, d\xi. \qquad (2)$$

Thus, instead of searching for a family of solutions of a complicated linear integral equation for different values of η, we can obtain $r(\eta, \xi)$ solving only one, very simple nonlinear functional equation.

In later work, we showed how one can treat the more complicated cases of anisotropic scattering (the case of nonspherical indicatrix), applying the same invariance principle.

It was shown also that the case of finite optical thickness τ_0 can be treated by some generalization of the same principle. Adding a layer of thickness ΔZ to one of the boundaries we must in this case subtract the layer of the same thickness from the other boundary and require that both the function $r(\eta, \xi)$, which describes the diffuse reflection, and $S(\eta, \xi)$, which describes the capability of diffuse transmission, will remain unchanged. Of course, in this case both these functions will depend also on the value of the finite optical thickness τ_0, which enters as a parameter.

In such a way a powerful tool was developed for the solution of the most sophisticated problems of the transfer of radiation and of neutrons, which often are of much more general nature than the problems of diffuse reflection and transmission.

Before World War II, working on the problems of galactic absorption, I introduced a formalized scheme of the absorbing dust layer in the Galaxy which causes the general absorption in this stellar system. It consists of randomly distributed discrete clouds arranged between two parallel planes. The random distribution of absorbing clouds will cause the fluctuations in the apparent distribution of surface brightness of the Milky Way. Of course, this model was constructed only for the study of brightness fluctuation and is inadequate for application to other problems, for example, to the dynamics of absorbing material.

In a wonderful way the principle of invariance has opened the possibility of reducing the theory of fluctuations of brightness in such a model to a very simple functional equation. The problem was treated further in a series of papers by S. Chandrasenhar and G. Münch in a much more complete form.

During the years after the World War, I succeeded in showing that it is also possible to apply the principle of invariance to some *nonlinear problems* of radiative transfer. However, no result of outstanding significance has been achieved here.

More recently, the Armenian mathematician R. V. Ambartsumian has shown that the invariance principle is widely applicable in the new mathematical field developed by him—the combinatorial integral geometry.

At the same time, it is necessary to admit that, notwithstanding its logical beauty and simplicity, the astrophysical applications of the invariance principle are somewhat restricted owing to some assumptions of geometrical nature (in some cases the plane layers, in others the homogeneity of the medium).

Here I shall confess that I always disapproved of the methods of ad hoc hypotheses and "models" used in astrophysics without any reasonable restriction by many theoreticians. Apparently this distrust is based on the arbitrary character of such approaches and the frequent failures of their application. Of course, the theory of radiative transfer as well as invariance principles, unlike such ad hoc models, are rather useful *mathematical and logical tools* of investigation. However, some narrow assumptions that we sometimes make developing these tools are reminiscent of "model making." This is the reason why I could not restrict my studies solely to this direction of thought. I always tended to find new ways for the direct use of empirical data with the purpose of finding the regularities and laws governing the astrophysical phenomena. In this connection, I was always of the opinion that the approach that is now called "inverse problems" is very promising.

2. *Inverse Problems*

Just after my graduation from the University of Leningrad my attention was attracted to the following question: in what degree does the totality of empirical data of atomic physics (the frequencies of spectral lines, the transition probabilities, etc.) define the system of laws and rules of quantum mechanics or more specifically the form of the Schrödinger equation? Very soon I came to the conclusion that rigorous solution of this problem was beyond my capabilities, and I decided to concentrate on some limited and modest problem of the same kind. I found that the following narrow question is more suitable: *in what degree do the eigenvalues of an ordinary differential operator determine the functions and parameters entering into that operator?* Even the solution of this inverse problem is connected with many difficulties. Therefore, I restricted myself to publishing in 1929 in *Zeitschrift für Physik* a paper which contained the statement of the general problem and the proof of the theorem that among all strings the homogeneous string is uniquely defined by the set of its oscillation-frequencies. Apparently during the fifteen subsequent years nobody has taken notice of that paper (when an *astronomer* is publishing a *mathematical* paper in a *physical* journal, he cannot expect to attract too many readers). However, beginning in 1944, that work was continued in papers by a number of outstanding mathematicians who have succeeded in obtaining many interesting results related to the "inverse Sturm-Lionville problem."

As regards myself I tried persistently during many years to find other cases where we can directly derive from observational data regularities and laws of nature. One cannot forget that the greatest astronomical discovery in history—the establishment of Kepler's laws of planetary motions—was in essence obtained as the solution to the following inverse problem: two planets are moving around the Sun in closed orbits of which one is completely situated inside of other (for simplicity we suppose that both orbits are in the same plane). The motions are periodical but the periods are not commensurable. The observer on the internal planet is continuously measuring the longitudes of the outer planet and of the Sun. It is necessary to determine from these observations the form and relative sizes of two orbits as well as the velocities in the different points of the orbits. As the result of the solution of this problem, Kepler's laws were found. It is true that after the trajectory of Mars was found by Kepler, the method of "trial and error" was applied in order to present this trajectory as one of known geometrical figures, but it is clear that the main results have been achieved by means of analysis and solution of *the inverse problem.*

There were also other interesting examples of solution of outstanding inverse problems in classical astronomy. However, there had been few such cases in astrophysics. A well-known example is the derivation of the space distribution of stars in a spherical stellar cluster from the observed distribution in the projection on the sky. The problem was reduced to the integral equation of Abel, which has a simple solution.

In one of his popular papers Eddington put forward the following question: is it possible to find the distribution function $\varphi(\xi, \eta, \delta)$ of the components of stellar space velocities in the solar neighborhood *from radial velocities alone* without making any special assumption on the form of φ. This problem was solved in a paper that I wrote in 1935 and was presented by A. S. Eddington for publication in *Monthly Notices of The Royal Astronomical Society.*

It was shown in that paper that mathematically the problem is reducible to the problem of finding the values of a function of three coordinates in the velocity space when the values of the integrals of this function over any plane in that space are given as a function of three parameters defining the plane. The problem is soluble in a finite form and the very first trials have shown the applicability of this method to the existing data on radial velocities. I think that now, when we have much richer catalogues of radial velocities, it is worth while to try again to apply the solution.

Quite recently the inverse problem approach has found wide application to the statistics of flare stars in open clusters and associations. Let us consider here one of the simplest problems related to these flare stars. There are serious reasons to believe that the sequence of flares of any flare star is of the type of the Poisson stationary process with some mean frequency of occurrence v. Then it is possible to show that between the mathematical expectations n_k of the numbers of stars that have flared k times during the total duration τ of observations, we have the relation

$$n_0 = \frac{n_1^2}{2n_2}. \qquad (3)$$

According to the definition, n_0 is the expectation of the number of flare stars that have not flared during the whole time of observation. In other words, it is the number of flare stars that are not yet discovered. Therefore, adding this n_0 to the sum $n_1 + n_2 + \cdots$ of all stars observed in flares (this means to the number of *known* flare stars) we can obtain the total number N of flare stars in the given stellar aggregate. Of course, in practice instead of mathematical expectations of n_1 and n_2 we use the observed numbers of stars that have flared once and twice respectively and consider the resulting value of n_0 as approximate. For the validity of (3) it is necessary to assume that all flare stars have the same mean frequency of flares. We have strong indications that this assumption is definitely wrong. It is easy to show, however, that in this case we have instead of (3) the inequality

$$n_0 \geq \frac{n_1^2}{2n_2} \tag{4}$$

which gives the possibility of estimating the lower boundary of the total number of flare stars. In this way the result was obtained that the total number of flare stars in Pleiades must be larger than one thousand. Initially this estimate was considered excessive, since the total mass of Pleiades from the virial theorem is only about 400 solar masses. But now there is no doubt about such a high number of flare stars in Pleiades.

Later, the much more complicated and subtle problem of determination of the distribution function $f(v)$ of mean frequencies of stellar flares among the stars of the stellar aggregate under consideration was treated.

It is interesting that in this case we come to an inverse problem where the distribution of the first observed flares of different stars during the whole period of monitoring of the aggregate plays the role of "known function." Since the "first flare" is at the same time the moment of the discovery of a flare star this means that the knowledge of distribution of discoveries is essential for finding $f(v)$. Thus *the chronology of discoveries* of flare stars contains important information about $f(v)$. It was shown also that the distribution of "second flares" expressed as a function of the elapsed monitoring time contains again very important information about $f(v)$. But the moment of the second flare means the moment of the confirmation of discovery. Therefore, the *chronology of confirmations* is of importance for the problem under consideration.

The significance of inverse problems for astrophysics was discussed in detail by us in a special paper presented recently at the International Symposium on the Fundamental Problems of Theoretical and Mathematical Physics in Dubna (August, 1979) which will appear in the proceedings of that symposium. There we have given other examples.

3. *The Empirical Approach to the Evolutionary Processes in the Universe*

From the very beginning of my work in astrophysics I have been interested in the problems of the origin and evolution of stars and galaxies. It was clear to me that the old approach by means of global cosmogonic hypotheses or speculative models could hardly bring serious results. It was clear that one must proceed from empirical data.

The evolutionary processes in the Universe are of exceedingly complicated and diverse nature. Therefore, there is no chance of understanding them using a small number of speculative models or hypotheses. Instead of making more or less arbitrary assumptions, we must analyze patiently the empirical data so far obtained and try to deduce from them all possible conclusions on different links of many evolutionary chains existing in reality. It was necessary to find some general idea about the ways of doing this. In the mid-1930s I decided to apply this approach in my work on these problems.

The idea was to find the cases where it is relatively easy to deduce from the present state of the astronomical body or the system the direction of changes in the state of the body (or system), in other words to find the cases where we can conclude from simple considerations the evolutionary trend at the given phase without the knowledge of all other phases. Thus, roughly speaking, we must try to gather the information on the first derivative of the state we observe. Such an approach can give us in many cases the possibility of connecting the different observed states of some objects into evolutionary chains without artificial assumptions. In some cases, such chains (or rather pieces of chains) may be themselves very short but persistent work will in due time bring success in solving more and more complicated problems. Of course, I don't consider this approach as my invention. To me it was important that I decided to follow this approach as strictly as possible.

Thus, when studying the problems of *planetary nebulae*, I found that they are not in an equilibrium state. But the decisive significance of the observed structure of the emission lines of these objects for the understanding of the direction of evolution of planetaries was formulated by Zanstra. He concluded that the only explanation of the unusual appearance of those lines is the expansion of nebulae. Thus, it soon became clear that the planetaries are the results of ejection of the outer layer of their central stars.

When I analyzed the effect on interactions of members of a *stellar cluster* from close mutual passages during their motion, the conclusion was inevitable that the clusters are subject to the process of evaporation. In the case of open clusters, this process must be relatively rapid, having the time scale of the order of 10^8–10^9 years. This is a short time compared to the time scale of the Galaxy.

Thus, it was shown that the open clusters that now exist in the Galaxy are relatively young and rapidly changing systems and that the general stellar field of the Galaxy is steadily growing in the number of stars at the cost of disintegration of clusters. At the same time, the formation of clusters from individual field stars is practically impossible.

After World War II, I found that the much more extended groups of stars and of diffuse nebulae that have received the name of *stellar associations* are much younger than the ordinary open clusters. They contain often hot giants (O and B stars) and always a large percentage of variable dwarfs (T-Tauri variables and flare stars). The age of many associations is between 10^6 and 10^7 years. Their very existence is the proof of two fundamental facts concerning the birth of stars in the Galaxy: 1. the formation of stars is a process continuing through the present epoch of the evolution of our Galaxy, and 2. the formation of stars proceeds in relatively large groups (associations and clusters).

The subsequent discovery of the fact that the stellar associations contain the multiple stars of a special type—the so-called Trapezium-type systems—has shown that, in the associations, subgroups that are younger than the associations as a whole (the age between 10^5 and 10^6 years) exist.

On the other hand, already in the 1930s I tried to study the statistics of the elements of orbits of *double stars* in the Galaxy to obtain some indications about the direction of their dynamical evolution. The final conclusion was that the wide pairs are rapidly disintegrating. Therefore, the very existence of some very wide pairs puts an upper limit on the age of the Galaxy at least in its present state. This limit is quite independent of any cosmological consideration and is of the order of 10^{10} years.

For achieving further progress in the understanding of evolutionary processes it was necessary not only to avoid the use of speculative models but also to rid ourselves of some *superstitions*, which at the first glance appear like quite natural assumptions, remaining from classical cosmogonies. First is the idea that in the first phase of any process of formation of astronomical bodies or systems their state is always the state of nebular matter. Even now this opinion prevails among many theoreticians. However, it is difficult to find direct evidence for such an assumption in observational data.

The second superstition closely connected with the first is the complete disregard of the problem of the origin of nebulae. However, considering in some detail the situation in our Galaxy as well as in the external galaxies, one can see that all kinds of nebulae (and not only planetary or cometary nebulae) are in the state of rapid change. Their lifetime must be orders of magnitude shorter than the lifetime of the majority of stars or planets. It is, therefore, quite natural to try to begin the analysis of evolutionary processes by the study of changes in nebulae. In the case of planetaries one can see almost directly their origin in the ejection of outer layers of a star and their final destiny in complete dissipation in the surrounding space. The radio-nebulae, of which the best example is the Crab nebula, are the result of supernovae explosions and dissipate in a similar manner. There are many evidences of expansion of some massive diffuse nebulae. The same is true for the so-called *compact H II regions*. Therefore, the fact is that almost everywhere we observe directly or indirectly the formation of nebulae by way of ejections from the stars and their groups. But the evidences in favor of the opposite processes (collapses of nebulae, accretion of nebular material) are infrequent and at times very dubious. Of course, it cannot be excluded that much more definite evidences of such type can be found in the future, but the present-day picture of the Universe is dominated by processes of explosions, ejections from massive bodies, and subsequent formation of such short-living objects as nebulae. It seems that if the problem of the origin and evolution of nebulae had been formulated earlier, the solution of many more general problems connected with evolutionary processes in the Universe could have been reached much more rapidly.

One of the most intriguing questions about stellar associations is that some of them are *expanding* or contain expanding groups of stars. In our first papers on stellar associations (1947–1951) the prediction was made that the expansion is a general phenomenon among associations. From the observed proper motions in the association Perseus II, Professor A. Blaauw concluded that it is indeed in the state of expansion. Later he found the expansion phenomenon in a part of the Scorpio-Centaurus association. At the same time, in many other associations no appreciable expansion has been found. Of course, these negative conclusions are definitive only for a number of nearby associations. Therefore, there are only one or two cases where we certainly have no simple expansion phenomenon. At the same time, the existence of at least some expanding groups is the evidence of some kind of explosion processes connected with the birth or with the early stage of evolution of young stellar groups. Here again, the empirical data are against the theories of condensation of diffuse matter into the stars.

In the years 1955–1965 I turned my attention to the phenomena in and around the nuclei of galaxies. In the past, astronomers and particularly theoreticians showed little interest in the properties of the nuclei of the galaxies. In a report delivered to the Solvay Conference of 1958 I showed that these nuclei are often centers of large scale *activity* which proceeds in different forms. It was shown that the radio galaxies are not the products of collisions of galaxies, as it was accepted at that time, but are the systems in which ejections of tremendous scale from the nucleus have taken place. As a consequence of such ejections, clouds of high energy particles are being formed.

The subsequent discovery of *quasars* added one more form of nuclear activity by which a considerable part of liberated energy is emitted as the nonstellar optical radiation of the nucleus. In such cases, the luminosity of the nucleus often exceeds 10^{11} or 10^{12} times (sometimes even more) the luminosity of our Sun.

In another important development, the astronomers B. Markarjan, E. Khatchikjan, and others who work with me at the Byurakan Observatory initiated a more systematic observational study of the optical manifestations of activity in galaxies such as the ultraviolet excess and strong emission lines. The tenfold increase of the number of known Seyfert galaxies, which was one of the results of this work, has opened new possibilities of understanding active nuclear processes.

At the symposia organised in 1966 at Byurakan Observatory and in 1970 at the Vatican Academy of Sciences, the different forms of the activity of nuclei including the phenomena in QSO's and in Seyfert galaxies were thoroughly discussed. Since then, a huge volume of observational work has been carried out at different observatories in the world in order to reach better understanding of the processes involved. However, the theoretical interpretation has made little progress as yet.

While the observed forms of the activity of nuclei speak directly in favor of the fundamental nature of explosion and expansion processes taking place in central parts of galaxies, many theoreticians are still constructing models of nuclear phenomena in which the ejection processes are preceded by some form of collapse of great amounts of diffuse matter. According to such models, the ejections are only the secondary consequences of more fundamental processes of collapse. It is hardly necessary to say that I am very skeptical about such a speculative mode of thinking. There is no evidence even for the possibility of such a course of events. It seems that such an approach is the remnant of the old notion that the evolutionary processes in the Universe are always going in the direction of contraction and condensation.

4. *Concluding Remarks*

In conclusion I would like to give my evaluation of the degree of success in each of three directions of study discussed above.

1. I am very much satisfied that during the thirty-seven years that elapsed after the publication of my first paper on the application of the invariance principle, many new important results have been achieved. The subject has been extensively developed in the brilliant work of Professor V. Sobolev and his group. Professor R. Bellman has introduced "invariant imbedding." Professor S. Chandrasekhar's participation in the development of the field has greatly inspired me as well as the young people who worked directly with me. I look forward to an even wider application of the

invariance principle to many problems of mathematical physics and to other branches of exact science.

2. It is evident that the success in the field of inverse problems of astrophysics is very modest. It happened that from my youth I was more enthusiastic with this direction of thinking than with the two others. This shows that in all cases success depends not so much on the personal wishes or abilities of the investigator but in large degree on the general state of affairs in the scientific field under consideration and, of course, on the difficulty of the task itself. But I have no doubt that this direction is a very promising one for astrophysics. I take this opportunity to express my conviction that this direction will have in the future great significance for cosmology.

3. The observational approach to the problems of evolution in astrophysics, which remains the essence of the third direction of my studies, has in the last decades rapidly spread to the whole of astrophysics. It has penetrated deeply every domain of our science. Perhaps it is not an exaggeration to say that, partly owing to this approach and owing to the persistent work of a whole generation of astrophysicists, the science of astrophysics is now transformed into an evolutionary discipline.

Now everybody will agree that the problem of the origin and evolution of celestial bodies cannot be solved by one or by a small number of speculative models. The scope of the field will increase and widen with new discoveries in astrophysics.

My skeptical attitude to the existing formal theoretical models proposed so far is confirmed by the fact that almost all the new interesting discoveries, which were extremely abundant during the last three decades, proved to be great surprises for such models. The attempts to accommodate these models to new observational discoveries usually does not help very much. Let us consider two cases of the complete failure of the speculative approach.

(*a*) Many theoreticians are convinced that they have now a more or less comprehensive theory of stellar evolution. Hundreds of models were calculated, particularly for the early phases of evolution. However, the theory has completely failed to predict such an important phenomenon as flare stars. There is no doubt now that the majority of stars after the period of their formation (T-Tauri stage) go through this phase of evolution. Therefore, one of the first tasks of every evolutionary theory must be the explanation of peculiarities of the flare processes. However, the majority of models are even now neglecting this requirement.

(*b*) The situation is even worse with the *problem of fuors* (this term is used in USSR for the stars of FU Orionis type). The fact that the stage of a fuor plays an important role in the life of at least some categories of stars is rather fatal for many speculative theories. But the state of affairs is more serious than it seems from first glance. It appears now that there is a whole sequence of different stars, which in their photometric behavior are more or less similar to fuors. The P Cygni star, which brightened almost four centuries ago, is an example. It is well known that in every spiral or irregular galaxy we have many supergiants of P Cygni type. Therefore, processes of the brightening of pre-P Cygni type stars are very important for the understanding of the evolution of supergiants.

There is no doubt that observational studies of such stars and processes will bring interesting empirical material for the picture of the stellar evolution.

In order to be understood correctly I would like to add some words to my criticism of speculative thinking and of models. Of course, nobody can deny the part which the model approach, as well as the speculative thinking, plays in science. The first of the directions to which I devoted my work (the invariance principle) is much nearer to the construction of models than to empirical studies. My point is, however, that in the concrete cases discussed above the building of models without taking properly into account the vast empirical material now in our possession can hardly give good results.

Nature keeps still many of its secrets. Our aim is to disclose them. It is natural to try this by *observing* the places where they are hidden. We can hardly reach our aim by only theorizing.

THE MASSES OF CEPHEIDS[1] ✵2161

Arthur N. Cox

Theoretical Division, Los Alamos Scientific Laboratory,
University of California, Los Alamos, New Mexico 87545

I INTRODUCTION

About ten years ago it became apparent that the masses of Cepheids predicted from the theory of stellar evolution were larger than those predicted by pulsation theory. This mass anomaly for the classical Cepheids was displayed by Christy (1968) and Stobie (1969a,b,c) using nonlinear hydrodynamic calculations and by Cogan (1970) using linear theory. Rodgers (1970) has also discussed the several mass anomalies in some detail. These mass anomalies, and some others to be discussed, have not yet been completely resolved, but many of the discrepancies have been alleviated mostly by an increase in the Cepheid luminosities and a decrease in their surface temperatures.

Masses of stars are almost always derived by analysis of binary star motions. Perhaps 27% of the Cepheids are binaries with secondary masses similar to the Cepheid (Madore 1977), but only one seems to have a short enough period and is bright enough for us to observe an accurate orbit and obtain a useful mass. This is SU Cyg, which has a pulsation period of 3.8 days. No definitive mass is yet available, but it is known (B. F. Madore and N. R. Evans, private communication, Fernie 1979b) that its mass is very likely closer to the one obtained from evolution theory than any anomalously low value as often obtained from some aspects of pulsation theory.

The Christy (1968, 1974) and Stobie (1969a,b,c) work showed that the phase of the light and velocity curve bumps, which varies with period according to the Hertzsprung (1926) progression, could not be matched by their hydrodynamic calculations unless the mass (the so-called bump mass) used in the calculations was 30–40% less than that given by evolution theory for any given luminosity. This progression of light and velocity curve bumps is such that at a period like 5 or 6 days there is a bump very low on

---

[1] The US Government has the right to retain a nonexclusive, royalty-free license in and to any copyright covering this paper.

descending light, at 10 days the bump is at the peak of the light curve, and at 10–15 days the bump is very low on rising light. Simon & Schmidt (1976) have shown that this bump occurs when the ratio of the second overtone radial pulsation mode period to the fundamental mode period is near 0.5. With this correlation, if always rigorously true, the light curve bump phase and bump masses can now be studied by linear theory alone.

Cogan (1970) studied the 13 period-luminosity-color calibrating Cepheids of Sandage & Tammann (1969) to find that there was another separate mass anomaly. In this case, the calibrating Cepheids have reasonably well-known mean luminosities (L) and mean effective surface temperatures (T_{eff}) which can be used to calculate mean radii (R). Use of the period–mean density relation (M and R in solar units)

$$Q_i = \Pi_i \sqrt{M/R^3} \qquad (1)$$

with the known period Π (called P by observers) for the mode i ($=0$ for fundamental mode) and the Q_i, obtained from linear pulsation theory, gave pulsation masses that were anomalously low by perhaps 30–40%. However, the pulsation mass anomaly was less than for the bump masses.

A further mass anomaly was pointed out by Petersen (1973) in his considerations of the double-mode or beat Cepheids. Takeuti (1973) and King et al. (1975) also discussed this problem together with the pulsation mass anomaly. These double-mode variables with periods ranging from 2.14 days (TU Cas) to 6.29 days (V367 Sct) seem to be definitely pulsing in the radial fundamental and first overtone modes. The observed period ratios which range from 0.697 (V367 Sct) to 0.710 (U Tr A) (Stobie 1977) indicate beat masses typically 1/3 those expected from evolution theory.

There is one other mass determination that gives a slight mass anomaly, and that is the one based on radii measured by the Baade (1926)–Wesselink (1946) method. This mean radius is obtained by considering the relative luminosity of the intrinsic variable star at two different phases (1 and 2) when the color and presumably the surface brightness is the same. Thus the luminosity difference is due only to the differing surface area and not the T_{eff} or color-dependent surface brightness. The gravity dependence of the surface brightness considered by Wooley & Carter (1973) is very small. The varying microturbulence effects may not be so small, however (Evans 1979). Integration of a simultaneously obtained velocity curve between these two phases gives $R_2 - R_1$ which can be plotted against R_2/R_1 from the photometry for a number of equal color phase pairs. The intercept of this R/R_0 versus $R - R_0$ curve at $R - R_0 = 0$ and $R/R_0 = 1$ gives a mean radius (R_0) which, through the period–mean density relation gives a Wesselink mass. These masses, discussed by Cox (1979) using Balona (1977), Evans (1977), and Scarfe (1977) radii, seem to be normal (evolutionary) except perhaps for periods longer than 15 days where they seem 30% lower than evolution theory values. This last anomaly may be a real one due to mass loss in the B star stage for those stars with initial mass of greater than about 10 $M_\odot$ (Sreenivasen & Wilson 1978), though pulsation masses for longer period Cepheids do not entirely support this mass loss.

A brief mention should be made of a method for getting masses that is moderately successful at least for the lower luminosity δ Scuti variables. These variables are in the same pulsation instability strip on the Hertzsprung-Russell diagram. Cepheid multicolor photometry, calibrated by detailed models of stellar atmospheres and their emergent spectrum, allows a measurement of the effective temperature, $T_{\rm eff}$, and the stellar gravity, g. Pel (1978) has tabulated these $T_{\rm eff}$ and $\log g$ data for many Cepheids. If

$$g = GM/R^2 \tag{2}$$

and use is made of either the Cox, King & Stellingwerf (1972) or Faulkner (1977) forms for Q_0 (M, R, $T_{\rm eff}$), one gets two equations for the two unknowns, M and R. Unfortunately, the theoretical period–mean density relation for a typical mass (7 $M_\odot$) and a typical $T_{\rm eff}$ (6000 K) is

$$M \sim R^{2.44}/\Pi_0^{1.38} \tag{3}$$

with a similar relation using the evolution masses and radii (see Fernie 1968) giving the relations

$$R \sim g^{2.3} \quad \text{and} \quad M \sim g^{5.6}. \tag{4}$$

This is not useful when the error in g is often a factor of two. Only for the lower δ Scuti variables, where g can be more accurately measured and Q is constant, can the relevant relations

$$R \sim g \quad \text{and} \quad M \sim g^3 \tag{5}$$

give useful radii and masses.

Equation (3) can be written as $\Pi_0 = R^{1.77}/M^{0.72}$ which can be compared to the observational P-R-M relation $P_0 = R^{1.68}/M^{0.72}$ for the 16 calibrating Cepheids (1.95–45.1 days) studied by Cox (1979), if a least squares fit is made for A and B ($=0.18$) in the equation:

$$\log P - 1.5 \log R + 0.72 \log M = A + B \log R.$$

Two of the four anomalies discussed above are essentially reconciled. They are the pulsation masses of Cogan (1970) and Rodgers (1970) and the Wesselink radius masses, both to be discussed in detail later. The masses that depend on the period ratio, Π_2/Π_0 for bump Cepheids and Π_1/Π_0 for the beat Cepheids, remain anomalous relative to evolution theory masses. Several ideas are currently being pursued to explain these remaining discrepancies.

II MODELS TO RECONCILE MASS ANOMALIES

Mass loss in the red giant stage prior to the blue loops in the H-R diagram that produce Cepheids is a possible but not likely explanation for a bump or beat Cepheid mass anomaly. This problem has been reviewed by Iben (1974a,b). Stellar evolution theory has consistently predicted that the blue looping depends critically on the hydrogen content profile in the outer regions of the stellar core (Durand, Eoll & Schlesinger 1976), and any more than 20% mass loss (Siquig & Sonneborn 1976) will prevent any blue looping. Smaller mass losses that prevent looping were derived by Lauterborn, Refsdal & Weigert (1971). Losses of 80% or more give blue helium stars (Forbes 1968), but they certainly are not appropriate here. Mass loss is really not desired anyway because pulsation masses now agree with evolution masses even for those Cepheids which happen to have light curve bumps.

Perhaps a different composition would give evolution tracks that give a different mass-luminosity law in the blue loops. This is not likely because the works of Carson & Stothers (1976; using several opacity tables) and Becker, Iben & Tuggle (1977) give only a moderate dependence of the M-L relation even for large changes in the helium content, Y, or the heavier element content Z. Again the agreement of evolution and pulsation masses indicates that the evolution tracks are substantially correct.

Possibly the period ratios Π_2/Π_0 or Π_1/Π_0 might differ between the full amplitude real Cepheids and the infinitesimal amplitude linear theory used to discuss the bump and beat Cepheid period ratios and masses. Stellingwerf (1975) for an apparently permanent double-mode case and Cox, Hodson & King (1978) for a mode-switching case looked into the Π_1/Π_0 ratio in hydrodynamic models and found that the ratio does not change from a typical linear or nonlinear theory value of 0.73–0.74 to the observed values of 0.70–0.71.

Cogan (1977) proposed that deep and very efficient convection might change the envelope structure to one with a less steep density gradient. This would increase Π_0 more than Π_1 and reduce the ratio Π_1/Π_0. Deupree (1977) and earlier Cox et al. (1977) showed that this very strong convection is not possible even at very cool effective temperatures that are beyond the red edge of the pulsation instability strip.

Carson & Stothers (1976) and Deupree (1978) discussed the possible effect of rotation on levitation of the outer envelope layers to change Π_1/Π_0. Only for very large rotation velocities, which could easily be observed but are not, can there be any rotation effect.

In a series of papers (Cox et al. 1977, Cox, Michaud & Hodson 1978, Cox & Hodson 1978, Cox, King & Hodson 1978, and Cox, Hodson & King 1979) a helium-enhanced layer has been proposed that can lower the envelope density gradient and lower the conventional linear theory ratios Π_1/Π_0 and Π_2/Π_0. The enhancement needs to be $Y = 0.48$ to depths of 10^{-3} of the mass for the triple-mode Cepheid AC Andromedae ($3\,M_\odot$) at 0.7 day (Cox, King & Hodson 1978) and $Y = 0.65$ to almost that mass depth for the double-mode Cepheids ($5\,M_\odot$) (Cox, Hodson & King 1979). For the bump Cepheids ($7\,M_\odot$) (Cox, Michaud & Hodson 1978) the enriched layer has a Y value of 0.75 but it involves less than 10^{-4} of the mass. This helium-enriched layer is presumed to be due to a helium-deficient wind that blows away more hydrogen than helium similar to the case that seems to exist for the sun (Feldman et al 1977). While there are questions as to how long the enriched layer can be stable against downward mixing, Sonneborn, Kuzma & Collins (1979) have shown that it is not easily observable in the emergent spectrum.

Stothers (1979) proposes a tangled magnetic field to change the envelope structure and give lower Π_1/Π_0 and Π_2/Π_0. The internal pressure is then partly from magnetic forces and not all from matter, thus giving a lower density and a lower density gradient. The surface magnetic fields required are not too different from those actually observed in some Cepheids (Babcock 1958, Kraft 1967, Weiss & Wood 1975, and Wood, Weiss & Jenkner 1977), and the fields required are of the order of 10^4 gauss at the bottom of the pulsing layers.

Another recent suggestion by J. P. Cox (1980) is that there is a nonradial mode contamination of the two radial modes seen in the bump and beat Cepheids—especially in the latter case where Π_1 and Π_0 appear separately and cause beats. The bump Cepheid case is clearly a resonance of radial modes that locks-in Π_2 even though it does not have an exact period commensurability and even though Π_2 is frequently not even pulsationally unstable in the linear theory. Both periods in beat Cepheids give no indication of any nonradial component though it is conceivable that one mode is the fundamental radial mode and the other is, for example, the p mode with the latitudinal quantum number $l = 0$, 1, 2, or 3. These latter modes are not known to be pulsationally unstable, however, according to Osaki (1977); only high l values are unstable in the Dziembowski (1977) calculations. The period ratios need to be reduced for both bump and beat Cepheids to reconcile their masses, and a mechanism that reduces the period ratio for the clearly radial bump Cepheids will probably also apply to the beat Cepheids.

We describe below six ways of deriving Cepheid masses. Only anomalies involving the period ratios Π_1/Π_0 and Π_2/Π_0 seem to remain now with the others largely resolved. The hypotheses of surface helium enhancement, tangled surface magnetic fields, and possibly nonradial effects seem to be the only viable ones to resolve mass anomalies for the bump and beat Cepheids.

III EVOLUTION AND THEORETICAL MASSES

Evolution calculations of intermediate mass stars that can predict Cepheids date from 1962 (Hayashi, Hoshi & Sugimoto 1962). They are reviewed by Iben (1967 and 1974a). The most recent results are by Becker, Iben & Tuggle (1977) using the Los Alamos opacities (Cox & Tabor 1976) and by Carson & Stothers (1976) using unpublished Carson opacities. These two sets of results are similar. The Becker, Iben & Tuggle (1977) mass-luminosity relation for the blue loops (4–13 $M_\odot$) is

$$\log L = j + k \log M \tag{6}$$

with

$$j = 0.46 - 41(Z-0.02) + 6.6(Y-0.28) - 5(Y-0.28)^2$$
$$k = 3.68 + 21(Z-0.02) - 4.5(Y-0.28) + 11(Y-0.28)^2.$$

For the first crossing of the pulsation instability strip for all these masses (between about 6000 K and 7000 K for 3 $M_\odot$), the luminosities are about a factor of 3 lower and the duration typically about 1/10 of that for the loops.

Thus, if the luminosity of a Cepheid can be established due to membership in a galactic cluster or association or some other way, the mass follows immediately from Equation (6). Sixteen cases (Cox 1979) plus 9 more given by Cogan (1978a) are known with reasonably accurate luminosities, but maybe even some of these should not be included. A typical error of 0.25 magnitude in M_{bol} leads to an error of 26% in L and much less than 10% in M_{ev}.

If the Cepheid is on its first crossing of the pulsational instability strip in the H-R diagram, there is the possibility that Equation (6) underestimates the mass. For 7 $M_\odot$ with $Y = 0.28$ and $Z = 0.02$ the luminosity in the loops should be about 1.45×10^{37} erg/s (3700 $L_\odot$). This is equivalent to a mass of 8.2 $M_\odot$ if the Cepheid is only on its first crossing. The crossing durations would indicate that a few of the calibrating Cepheids might be of higher mass than the standard M_{ev}. VY Car and T Mon might be such cases with the latest M_{ev} values, respectively, of 9.4 and 10.8 $M_\odot$ but the pulsation mass M_{pul} (to be discussed in detail in the next section) at about 13 and 16 $M_\odot$ (Cox 1979). SV Vul ($M_{ev} = 12\ M_\odot$, $M_{pul} = 15\ M_\odot$) may be too massive to even have a blue loop, but Fernie (1979a) finds that a decreasing period indicates an evolution to the blue. In any event, for this high mass there is the complicating B star stage mass-loss problem that upsets the accurate use of Equation (6).

A new theoretical mass for Cepheids has been introduced by Cox et al. (1977). Four equations are solved simultaneously for the unknowns M, R, L, and Q_i. These are Equations (1), (6), the Faulkner (1977) expression for Q_0 (M, R, L, T_{eff}), and the T_{eff} definition relation

$$L = 4\pi R^2 \sigma T_{\text{eff}}^4. \tag{7}$$

We assume that both Π_i and T_{eff} are known; an error even as large as 10% in T_{eff} results in less than 15% error in M_{th}. This theoretical mass can be derived for any Cepheid without knowledge of the luminosity, but it does use the evolution-theory mass-luminosity relation. As discussed by Cox (1979), the evolution and theoretical masses agree to better than fifteen percent for 15 of the 16 Cepheids (TW Nor and 9 others given by Cogan 1978a excluded) with reasonably known luminosities.

The value of this theoretical mass is that it enables comparison between theory and observation for any Cepheid with a known period and at least an approximate T_{eff} value. Many bump, beat, or Wesselink masses can then be compared to the theoretical mass which is predominantly controlled by the evolution track data.

IV PULSATION MASSES

Recent improvements in the distance scale for galactic clusters (Hanson 1977) and in the corrections for interstellar reddening (Pel 1978) have made the Cepheids appear to have larger intrinsic luminosities, larger radii, and consequently larger masses than given by Schmidt (1976). The 0.26 magnitude increase in the distance modulus together with Sandage & Tammann (1969) and van den Bergh (1977) magnitude and color data has been used by Cox (1979) to derive new luminosities which produce new photometric radii and finally new masses through the period–mean density relation (Equation 1).

Additionally, the multicolor work of Dean, Warren & Cousins (1978) has shown that the reddenings based on spectral types (often the G-band strength) and the observed colors (Kraft 1963) are usually overestimated. Thus, the Cepheids are cooler than believed hitherto and, at any fixed luminosity, larger in radius. The increase in distance scale and decrease in reddening corrections have produced radii and masses very close to evolution (and theoretical) masses.

Extensive theoretical studies have attempted to reconcile the evolution and the earlier pulsation masses. Fricke, Stobie & Strittmatter (1971) and

Iben & Tuggle (1972a) came to the conclusion that the distance modulus to the calibrating Cepheid clusters should be increased by 0.3 mag, or alternatively the log T_{eff} values were too large by 0.025 (350 K). A possibly larger Y by 0.14 was dismissed as making the blue edge too blue leaving a large gap between it and the bluest Cepheids. Further discussion of the Sandage & Tammann (1968, 1969) period-luminosity-color relation by Iben & Tuggle (1972b) showed that these data even seemed inconsistent with pulsation theory. King et al. (1975) discussed the $(B-V)_0$–log T_{eff} relation to see how it might be changed to give agreement between evolution and pulsation masses. They concluded that the short period Cepheids should be cooler by 300 K and those with longest periods by about 650 K for the anomaly to disappear. Again, cooling the Cepheids without increasing their luminosity leaves a large gap in T_{eff} between the blue edge and the bluest Cepheids.

The new larger distance modulus is almost as much as desired, even if that was the only correction. The resulting changes in the colors and temperatures range up to 0.25 in $\Delta(B-V)_0$ or $\Delta \log T_{eff} = 0.044$ (~ 550 K) cooler giving the largest correction to pulsation masses for the longest periods as desired by King et al. (1975).

Iben & Tuggle (1975) have discussed in detail the Cepheid problems with a preliminary increase in the cluster distance modulus given by van Altena (1974). Many definitive theoretical data are also given which are used to discuss the mass-luminosity relation for our galaxy, the Magellanic clouds, and M31.

It is important to note that the proposals to increase the surface helium content or to get surface-layer pressure from a tangled magnetic field do not destroy the recently acquired agreement between the evolution and pulsation masses. These envelope structures (as do the Carson and Stothers homogeneous Carson opacity model structures) increase fundamental periods by typically 5% and the masses by 10%. Detailed data are given by Cox (1979).

Tables 1 and 2 give the evolution, theoretical, and pulsation masses for the 16 best-known calibrating Cepheids using, respectively, homogeneous and inhomogeneous models. Note that these masses are algebraically related and therefore not completely independent. The surface helium enrichment is not proposed for masses greater than 8 $M_\odot$, and therefore for the last six Cepheids, the Table 1 rather than Table 2 values are preferred.

V BUMP CEPHEID MASSES

Nonlinear results for many Cepheid models, calculated by Christy (1966, 1968) and Stobie (1969b,c), as well as by Adams, Castor & Davis (1980), and

Table 1 Evolution, theoretical, and pulsation masses; new distance scale and temperatures

Cepheid	Period (d)	T_{eff}(K)	$L_{obs}/L_\odot$	$M_{ev}/M_\odot$	Q_{th}(d)	$M_{th}/M_\odot$	$R_{th}/R_\odot$	$L_{th}/L_\odot$	$R_{obs}/R_\odot$	Q_{pul}(d)	$M_{pul}/M_\odot$
SU Cas	1.950	6339	955	4.84	0.0364	4.61	23.6	798	25.8	0.0362	5.95
EV Sct	3.090	6187	1148	5.09	0.0374	5.41	33.3	1434	29.8	0.0381	4.01
CF Cas	4.870	5895	1820	5.77	0.0387	6.11	45.9	2249	52.1	0.0395	4.62
UY Per	5.360	5848	2512	6.29	0.0390	6.28	49.1	2497	49.3	0.0390	6.33
CV Mon	5.380	6064	2344	6.18	0.0389	6.61	50.2	3012	44.3	0.0398	4.75
VY Per	5.530	5943	2818	6.49	0.0391	6.51	50.7	2837	50.5	0.0391	6.44
U Sgr	6.740	5734	3890	7.09	0.0398	6.72	57.8	3191	63.8	0.0392	8.79
DL Cas	8.000	5848	3715	7.00	0.0403	7.40	66.3	4553	59.9	0.0411	5.68
S Nor	9.750	5502	4467	7.36	0.0412	7.37	74.4	4494	74.2	0.0412	7.31
TW Nor	10.790	5617	2884	6.53	0.0415	7.90	81.1	5797	57.2	0.0457	3.35
VX Per	10.890	5685	5888	7.93	0.0415	8.06	82.2	6245	79.8	0.0417	7.47
SZ Cas	13.620	5919	8511	8.77	0.0422	9.32	99.0	10644	88.5	0.0432	6.99
VY Car	18.930	5140	11220	9.45	0.0443	8.75	116.8	8429	134.8	0.0432	12.75
T Mon	27.020	5099	18621	10.85	0.0462	9.95	150.4	13537	176.4	0.0448	15.13
RS Pup	41.390	5309	22387	11.40	0.0484	12.41	208.6	30587	178.5	0.0506	8.49
SV Vul	45.100	5038	30200	12.37	0.0493	11.93	215.3	26432	230.2	0.0485	14.13

Table 2 Evolution, theoretical, and pulsation masses; new distance scale and temperatures; inhomogeneous model Q_0 values

Cepheid	Period(d)	T_{eff}(K)	$L_{obs}/L_\odot$	$M_{ev}/M_\odot$	Q_{th}(d)	$M_{th}/M_\odot$	$R_{th}/R_\odot$	$L_{th}/L_\odot$	$R_{obs}/R_\odot$	Q_{pul}(d)	$M_{pul}/M_\odot$
SU Cas	1.950	6339	955	4.84	0.0375	4.55	23.1	760	25.8	0.0367	6.12
EV Sct	3.090	6187	1148	5.09	0.0390	5.31	32.2	1342	29.8	0.0398	4.36
CF Cas	4.870	5895	1820	5.77	0.0409	5.97	43.9	2062	52.1	0.0415	5.11
UY Per	5.360	5848	2512	6.29	0.0413	6.13	46.9	2280	49.3	0.0408	6.92
CV Mon	5.380	6064	2344	6.18	0.0410	6.46	48.1	2763	44.3	0.0420	5.27
VY Per	5.530	5943	2818	6.49	0.0413	6.35	48.5	2595	50.5	0.0408	7.04
U Sgr	6.740	5734	3890	7.09	0.0423	6.54	55.0	2889	63.8	0.0408	9.52
DL Cas	8.000	5848	3715	7.00	0.0429	7.20	63.1	4114	59.9	0.0435	6.35
S Nor	9.750	5502	4467	7.36	0.0442	7.15	70.3	4010	74.2	0.0436	8.18
TW Nor	10.790	5617	2884	6.53	0.0445	7.66	76.7	5178	57.2	0.0484	3.78
VX Per	10.890	5685	5888	7.93	0.0444	7.82	77.7	5584	79.8	0.0441	8.35
SZ Cas	13.620	5919	8511	8.77	0.0452	9.04	93.6	9514	88.5	0.0459	7.89
VY Car	18.930	5140	11220	9.45	0.0481	8.44	109.3	7384	134.8	0.0459	14.40
T Mon	27.020	5099	18621	10.85	0.0502	9.59	140.7	11836	176.4	0.0475	17.01
RS Pup	41.390	5309	22387	11.40	0.0525	11.98	195.4	26833	178.5	0.0539	9.64
SV Vul	45.100	5038	30200	12.37	0.0535	11.51	201.5	23134	230.2	0.0517	16.01

Y. A. Fadeyev in Moscow (private communication), have unmistakably shown that the phases of bumps in the light and velocity curves indicate a mass anomaly of about 40%. Fricke, Stobie & Strittmatter (1971, 1972) have pursued this problem only to conclude that there is something wrong with pulsation theory. Extensive calculations by many workers have shown that, while light curve bumps are not easily calculated or understood, bumps on the surface radial velocity curve occur more or less at the same phase from one computer program to another, all indicating a mass of about $4\,M_\odot$ (i.e., the 10-day-period Goddard model) for a luminosity appropriate to a $6\,M_\odot$ evolution track.

Numerous other calculations involving velocity curve bumps have been given by King et al. (1973), Karp (1975a,b,c), Vemury & Stothers (1978), Davis & Davison (1978), and Hodson & Cox (1980). The Goddard model results are discussed by Fischel & Sparks (1975) and Deupree (1976).

The Karp (1975a) $5\,M_\odot$ model has a bump at too late a phase for its period of 12 days. These nonlinear results and discussion seem to permit as mechanisms for bumps both Christy echos and atmospheric oscillations for periods of 10–15 days. Gough, Ostriker & Stobie (1965) previously considered that the light and velocity curve bumps could be only an atmospheric phenomenon.

The unpublished Carson opacities have been used in the Vemury and Stothers models. The main difference between the Carson and Los Alamos opacities in this case is the variation of the opacity with temperature for $T \cong 40{,}000\,\text{K}$ where helium is undergoing the bulk of its second ionization. The Carson opacities have a more pronounced variation with T giving an actual increase in opacity with T instead of the less steep decline in the Los Alamos opacities with increasing T. A stronger change in opacity gives more pulsation-driving as discussed by J. P. Cox (1974). This helium ionization drives the amplitude of the Vemury and Stothers $7\,M_\odot$ 8.7-day model to a very large value (71 km/s corrected for limb darkening), much larger than for bump Cepheids which have lower amplitudes (<50 km/s) than those with both shorter and longer periods (Sandage & Tammann 1971, Cogan 1980). It also appears that their bumps are caused by surface oscillations of shock waves and not by the deep-seated echos of Christy (1974) or by the resonances of Simon & Schmidt (1976) that are seen in the results of Hodson & Cox (1980) for a surface helium enhanced model.

A frequent problem, namely, that of obtaining smooth light curves from hydrodynamic calculations, has been studied by Davis & Davison (1978) and Adams, Castor & Davis (1980) with a very fine computing mesh of mass zones which moves to stay with the high temperature-gradient hydrogen ionization zone. Only models with homogeneous composition were calculated. They confirmed the concept that bumps are echos of surface layer compression from the center a little over a period later. The phases of the light and velocity curve bumps were found to be not always exactly the same, as Fricke, Stobie & Strittmatter (1972) had found earlier.

Perhaps a better concept of the cause of the bumps is that the bumps are a resonance of the second overtone mode with the fundamental. Comparison of the extensive Stobie (1969b,c) data on bump phases with linear theory period ratios shows that, between 0.53 and 0.50 for the ratio Π_2/Π_0, the bumps are on the falling part of the light curve (5–10 days), and between 0.50 and 0.46 (10–15 days) the bumps are during rising light. The stimulated second overtone, which is frequently not unstable in linear theory, is locked into the fundamental period and gives the bump. Comparisons of the King et al. (1973) and Hodson & Cox (1980) linear and nonlinear periods show that the bump or the locked-in second overtone does not change the fundamental mode period more than a few percent, which is the usual agreement between these two theories.

The echo idea is not always borne out, whereas a resonance always produces bumps even for inhomogeneous Cepheid models and high luminosity RR Lyrae variables (BL Her stars). Vemury and Stothers question this, however, because their bumps occur over a slightly wider range of Π_2/Π_0, and are not centered on $\Pi_2/\Pi_0 = 0.5$. It is easily possible to even get bumps in velocity curves due to surface disturbances which confuse the interpretations. Hodson & Cox (1980) show that, while the light and velocity curve bumps occur throughout the star as Christy has demonstrated, it is not always possible to trace an echo inward from the surface and back out. A full understanding of the bump phenomenon and their occasional absence is obscured by details of the equation of state and opacity variations during the calculations. In this reviewer's opinion it seems that the resonance of Π_0 and Π_2 is the proper concept, and if this is true, then the bump Cepheids are really just a special case of double-mode Cepheids.

Fricke, Stobie & Strittmatter (1972), in order to get bump masses, made a plot of the period (P) times the bump phases (ϕ) (after zero velocity at minimum radius plus unity) versus the mean Cepheid radius ($R/R_\odot$)

$$P\phi = 0.25 R/R_\odot \tag{8}$$

similar to a relation given by Christy (1970). In this relation ϕ refers to the bump in the velocity curve which runs about 0.2 of the period later than in the light curve. Further, the phase used for the longer periods, when the bump is on rising light, is not for the actual bump but for the real velocity maximum! Use of this relation for 13 Cepheids gave bump radii and masses from the period–mean density relation that were just a bit over 0.7 the evolution masses.

Surface helium enhancement to $Y = 0.75$ in 10^{-5}–10^{-4} of the stellar mass (the two surface convection zones) reduces the ratio Π_2/Π_0 by about 0.05 and gives the resonances for periods between about 5 and 15 days, as observed. Details have been given by Cox, Michaud & Hodson (1978) and Cox & Hodson (1978) who have investigated the optimum amount and depth of this high helium layer. Some important problems with this enriched layer, which apparently cannot be easily detected in the emergent spectrum, are that the inverted mean molecular weight gradient may cause too rapid mixing to re-establish a homogeneous star, and the helium-deficient Cepheid wind, which presumably causes this enrichment, may act too slowly relative to the evolution lifetime. Further, the very blue edges predicted by Cox & Hodson (1978) may conflict with observations and with the instability strip width studied by Deupree (1980). In any case, the enhanced helium does not change the theoretical periods enough to upset the previously discussed agreement between evolution, theoretical, and pulsation masses.

Period ratios can also be reduced with the assumption of a tangled magnetic field in the convection zones. Stothers (1979) has given details of how magnetic pressure can keep the density gradient smaller, as enhanced helium does, and make the fundamental period increase more than the second overtone. Here again the Π_2/Π_0 resonance can indicate evolution theory masses for bump Cepheids.

VI DOUBLE-MODE AND TRIPLE-MODE CEPHEID MASSES

Petersen (1973) pointed out that the period ratio for the double-mode or beat Cepheids indicates very low masses (1.2–2.2 $M_\odot$) when homogeneous models are assumed. Two periods (almost surely the fundamental and first overtone modes) in the two period–mean density and the $Q(M, R)$ relations with known Π_i and Q_i allow determination of both the M and R. This very large mass anomaly of typically a factor of three was further investigated by King et al. (1975) and later Stobie (1976, 1977). Cox & Cox (1976) felt that there was no change in the period ratio between linear and nonlinear theory even in the multimode case (Stellingwerf 1975) and that the low masses seemed to be correct. Nevertheless, strong mode coupling was stressed by Faulkner (1977). Since the period ratio was indeed unchanged by the presence of the other mode at all stages in a mode-switching case (Cox, Hodson & King 1978) or in an apparently permanent mixture of modes (Stellingwerf 1975), the conclusion seemed to be that these double-mode Cepheids were very different from the normal blue looping stars of mass between 4 and 13 solar masses.

Petersen (1974, 1978) investigated homogeneous models to see how important possible errors in the equation of state and opacity would be. It appears that the models were giving reliable periods and, therefore, reliable masses.

This best Cepheid discrepancy has grown recently because detailed investigations of 8 double-mode Cepheids by Balona & Stobie (1979) and of TU Cas by Niva & Schmidt (1979) have produced reliable Wesselink radii which indicate evolutionary masses. Stobie & Balona (1979) get evolutionary radii and masses from the ratio of the radial velocity and light amplitudes for the two modes. D. H. McNamara (private communication) finds that multicolor photometry gives normal colors as for single-mode Cepheids. Even before these data were obtained, the double-mode Cepheid V367 Sct, located most likely in NGC 6649 (Madore & van den Bergh 1975, Flower 1978, Barrell et al 1980), was found to have an evolution and theoretical mass of about $7 M_\odot$, a pulsation mass of $5 M_\odot$ with the new distance scale, and a double-mode mass of $2 M_\odot$ using homogeneous model pulsation constants. Thus evolution, theoretical, and pulsation masses here conflict with the masses determined from the period ratio Π_1/Π_0.

In order to rectify this very large anomaly it was suggested by Cox et al. (1977) that the problem lies in the structure of the envelope. The large helium content postulated ranged from 0.55–0.75 (Cox, Michaud & Hodson 1978 and Cox & Hodson 1978) and is best set for double-mode Cepheids at $Y = 0.65$ down to a level of 250,000 K in about 10^{-3} of the stellar mass (Cox, Hodson & King 1979). A typical period ratio for the evolution value of $5 M_\odot$, Π_1/Π_0 is reduced from 0.74 to about 0.71 matching the observed period ratios and their constancy with period as discussed by Cogan (1978b).

Takeuti (1979) shows that, with helium enhancement, the convection should be much stronger than thought previously and might produce the Hα emission that Barrell (1979) has found for some double-mode Cepheids.

Simon (1979) has suggested that the double-mode phenomenon is a resonance between the two angular frequencies $\omega_0 + \omega_1$ and ω_3 (the third overtone) for Cepheids and $\omega_0 + \omega_1$ and ω_4 for the double-mode δ Scuti variables. Nonlinear calculations by Simon, Cox & Hodson (1980) show, however, that this resonance seems to impede double-mode behavior rather than promote it. Only near the ratio $\Pi_2/\Pi_0 = 0.5$ is there even a close approach to the case where the fundamental and first overtone modes want to switch to each other as Stellingwerf (1975, 1976) has discussed. Actually, no reproducible nonlinear calculation is available that demonstrates double-mode behavior, and this in itself is a very important problem.

The Stothers (1979) tangled magnetic field reduces the period ratio

Π_1/Π_0 just as the enhanced helium does. Therefore, it is also possible to explain the period ratios with evolution masses.

It is interesting to speculate that a combination of enhanced helium content and a magnetic field could exist simultaneously. Both can make the period ratio less and more consistent with evolution and other aspects of pulsation theory. Further, the magnetic viscosity may slow the downward mixing and destruction of the enhanced helium layer.

The mass of the double-mode Cepheids is anomalous with conventional envelope structures, but so also is the actual phenomenon of multiple-mode pulsation. Stellingwerf (1975) suggested that his strictly periodic solutions for $1.6\,M_\odot$ were unstable to perturbations from the other mode (first overtone mode in fundamental limit cycle or fundamental mode in first overtone limit cycle) for cool temperatures. Indeed, these temperatures were very likely cooler than the red edge. Hodson & Cox (1976) were not able to confirm this result using more conventional models with better mass zoning and a deeper lower fixed boundary. Not even higher (but still anomalous) mass ($3.3\,M_\odot$) models by Simon, Cox & Hodson (1980) could be made to pulsate simultaneously in two modes. Only for definite cases of mode switching can multiple-mode behavior be seen in hydrodynamic calculations of either homogeneous or inhomogeneous models.

Clues as to the T_{eff} values of the double-mode Cepheids may help identify their structure, but Cogan (1978a) finds them all across the instability strip. The strip is narrow, however, covering 0.03 in $\Delta \log T_{\text{eff}}$, and it is conceivable that these short period Cepheids are all near a transition line between fundamental and first overtone pulsation modes.

Barrell et al. (1980) have reviewed the data on amplitude and period changes for TU Cas and U Tr A, the two shortest period double-mode Cepheids. Hodson, Stellingwerf & Cox (1979) and Niva (1979) found that TU Cas had a decaying light and radial velocity amplitude for the overtone component. Faulkner & Shobbrook (1979) found a growing first overtone in U Tr A. Barrell et al. (1980) then found period changes consistent with redward evolution of a $5\,M_\odot$ star for TU Cas and blueward evolution of the same mass star for U Tr A. This is further evidence for evolution masses of these two double-mode Cepheids, but it leaves the question that, if the mode switching is so fast, indeed as fast as Stellingwerf (1975) predicted, then why do so many short period Cepheids (approximately 25%; Stobie 1977) show the two modes?

The mass of the only triple-mode Cepheid AC Andromedae has most recently been discussed by Fitch & Szeidl (1976) and Cox, King & Hodson (1978). It now appears that the mass is $3\,M_\odot$ on a first crossing of the instability strip. In order to get the observed periods, however, a deep

Table 3 Theoretical and Wesselink radius masses

Cepheid	Period (d)	T_{eff}(K)	Q_{th}(d)	$M_{th}/M_\odot$	$R_{th}/R_\odot$	$L_{th}/L_\odot$	$R_{Wes}/R_\odot$	Q_{Wes}(d)	$M_{Wes}/M_\odot$	$L_{Wes}/L_\odot$
U Aql	7.024	5708	0.0399	6.79	59.4	3318	64.6	0.0394	8.50	3919
FF Aql	4.471	5986	0.0385	6.02	43.3	2134	43.2	0.0385	5.97	2120
V496 Aql	6.807	5502	—	—	—	—	45.5	0.0382	6.88	2352
η Aql	7.177	5612	0.0400	6.37	57.0	2630	36.4	0.0452	2.13	1074
			0.0401	6.69	59.9	3146	54.2	0.0409	5.17	2578
V Car	6.696	5644	—	—	—	—	69.0	0.0393	9.86	4178
UX Car	3.682	6146	0.0398	6.56	57.0	2918	49.2	0.0411	4.49	2173
ER Car	7.719	5520	0.0379	5.76	37.9	1813	36.8	0.0380	5.32	1709
V Cen	5.494	5958	0.0404	6.74	62.7	3227	46.7	0.0435	3.24	1791
XX Cen	10.954	5644	0.0390	6.51	50.5	2845	48.4	0.0393	5.81	2611
RY CMa	4.678	5881	0.0415	8.00	82.2	6073	58.4	0.0456	3.45	3062
R Cru	5.826	5881	0.0386	5.99	44.5	2091	37.2	0.0400	3.77	1464
S Cru	4.690	5936	0.0393	6.55	52.5	2912	47.1	0.0401	4.94	2348
T Cru	6.733	5526	0.0386	6.07	44.7	2199	43.6	0.0388	5.66	2088
AG Cru	3.837	6253	0.0399	6.38	56.6	2645	55.4	0.0401	6.03	2532
β Dor	9.842	5532	0.0379	6.00	39.4	2104	33.2	0.0391	3.80	1491
W Gem	7.920	5760	0.0412	7.46	75.2	4684	64.6	0.0427	5.08	3458
			0.0403	7.22	65.3	4156	60.0	0.0410	5.78	3506
R Mus	7.510	5806	—	—	—	—	69.0	0.0400	8.36	4637
			0.0401	7.14	63.0	3995	52.0	0.0418	4.36	2718
S Mus	9.660	6000	0.0408	8.27	77.3	6859	59.9	0.0434	4.33	4114
S Nor	9.754	5502	—	—	—	—	61.0	0.0431	4.53	4267
			0.0412	7.37	74.5	4496	53.4	0.0451	3.26	2312
BF Oph	4.068	5860	0.0383	5.63	39.9	1662	50.4	0.0376	10.96	2650
AP Pup	5.084	5840	0.0389	6.14	47.2	2289	48.8	0.0387	6.73	2451
RV Sco	6.061	5888	0.0394	6.67	54.1	3109	44.4	0.0410	4.01	2096
V636 Sco	6.797	5538	0.0399	6.43	57.1	2712	47.2	0.0417	3.96	1854
S Sge	8.382	5594	0.0406	7.09	67.1	3903	56.7	0.0422	4.62	2785
			—	—	—	—	60.0	0.0416	5.32	3119
U Sgr	6.745	5734	0.0398	6.72	57.8	3195	55.3	0.0401	5.98	2925
W Sgr	7.595	5760	0.0402	7.10	63.3	3904	57.3	0.0410	5.47	3197
X Sgr	7.012	6176	0.0397	7.55	61.8	4916	63.1	0.0396	7.99	5125
Y Sgr	5.773	5663	0.0394	6.20	51.1	2375	56.0	0.0388	7.95	2853

MASSES OF CEPHEIDS

AP Sgr	5.058	5760	0.0389	6.01	46.7	2120	36.8	0.0410	3.27	1319
BB Sgr	6.637	5557	0.0399	6.39	56.2	2663	55.2	0.0400	6.11	2571
R TrA	3.389	6029	0.0377	5.42	35.2	1451	35.8	0.0376	5.66	1498
S TrA	6.323	5682	0.0396	6.46	54.6	2770	54.6	0.0394	6.40	2749
T Vel	4.640	5780	0.0387	5.83	43.8	1893	39.7	0.0394	4.50	1556
V Vel	4.371	6253	0.0383	6.33	43.5	2564	35.8	0.0397	3.78	1733
AH Vel	4.227	6065	0.0383	5.99	41.8	2092	44.6	0.0380	7.16	2381
—	—	—	—	—	—	—	47.0	0.0378	8.30	2644
BG Vel	6.924	5676	0.0399	6.70	58.6	3157	53.3	0.0407	5.23	2609
ζ Gem	10.154	5594	0.0413	7.67	77.4	5190	67.6	0.0426	5.43	3959
—	—	—	—	—	—	—	72.4	0.0419	6.45	4541
RT Aur	3.728	6187	0.0379	5.84	38.4	1910	24.7	0.0419	1.91	791
—	—	—	—	—	—	—	28.2	0.0404	2.63	1031
—	—	—	—	—	—	—	28.0	0.0405	2.58	1016
U Car	38.756	5352	0.0479	12.24	200.0	29028	140.1	0.0538	5.30	14249
l Car	35.535	4937	0.0480	10.58	179.8	16985	146.9	0.0511	6.55	11343
SU Cas	1.949	6339	0.0364	4.61	23.6	797	35.2	0.0372	15.89	1770
δ Cep	5.366	6113	0.0389	6.68	50.3	3123	44.6	0.0397	4.87	2457
—	—	—	—	—	—	—	57.0	0.0384	9.48	4014
X Cyg	16.387	5287	0.0435	8.59	106.8	7885	91.0	0.0454	5.78	5725
—	—	—	—	—	—	—	99.0	0.0443	7.10	6776
CD Cyg	17.071	5417	0.0436	9.03	111.5	9469	85.7	0.0469	4.75	5595
T Mon	27.020	5099	0.0462	9.95	150.4	13537	111.1	0.0508	4.85	7383
—	—	—	—	—	—	—	120.0	0.0494	5.78	8613
—	—	—	—	—	—	—	144.0	0.0467	8.92	12403
Y Oph	17.122	5754	0.0433	9.82	115.4	12908	102.6	0.0445	7.29	10210
—	—	—	—	—	—	—	79.0	0.0483	3.92	6053
AW Per	6.463	5991	0.0395	7.01	57.2	3735	49.3	0.0407	4.75	2770
—	—	—	—	—	—	—	70.0	0.0389	12.41	5585
RS Pup	41.388	5309	0.0484	12.41	208.6	30585	183.7	0.0501	9.09	23721
RZ Vel	20.398	5572	0.0443	10.07	128.8	14158	104.1	0.0469	5.95	9243
T Vul	4.436	6113	0.0384	6.18	43.5	2342	40.9	0.0388	5.22	2066
—	—	—	—	—	—	—	52.0	0.0378	10.23	3341
SV Vul	45.100	5038	0.0493	11.93	215.3	26432	146.2	0.0564	4.89	12184
RX Aur	11.610	5801	0.0417	8.50	87.1	7601	59.0	0.0464	3.28	3488
TT Aql	13.740	5439	0.0426	8.33	95.3	7034	72.0	0.0460	4.19	4014

Table 4 Theoretical and Wesselink radius masses, inhomogeneous model Q_0 values

Cepheid	Period (d)	T_{eff}(K)	Q_{th}(d)	$M_{th}/M_\odot$	$R_{th}/R_\odot$	$L_{th}/L_\odot$	$R_{Wes}/R_\odot$	Q_{Wes}(d)	$M_{Wes}/M_\odot$	$L_{Wes}/L_\odot$
U Aql	7.024	5708	0.0425	6.60	56.5	2999	64.6	0.0412	9.26	3919
FF Aql	4.471	5986	0.0404	5.89	41.6	1965	43.2	0.0401	6.48	2120
—	—	—	—	—	—	—	45.5	0.0396	7.39	2352
V496 Aql	6.807	5502	0.0427	6.19	54.0	2362	36.4	0.0451	2.12	1074
η Aql	7.177	5612	0.0427	6.50	56.8	2834	54.2	0.0432	5.78	2578
—	—	—	—	—	—	—	69.0	0.0409	10.67	4178
V Car	6.696	5644	0.0424	6.38	54.2	2635	49.2	0.0434	5.01	2173
UX Car	3.682	6146	0.0396	5.65	36.5	1686	36.8	0.0395	5.75	1709
ER Car	7.719	5520	0.0432	6.54	59.4	2894	46.7	0.0455	3.54	1791
V Cen	5.494	5958	0.0412	6.36	48.3	2603	48.4	0.0412	6.38	2611
XX Cen	10.954	5644	0.0445	7.76	77.7	5426	58.4	0.0485	3.90	3062
RY CMa	4.678	5881	0.0407	5.85	42.6	1918	37.2	0.0421	4.18	1464
R Cru	5.826	5881	0.0415	6.39	50.1	2655	47.1	0.0422	5.49	2348
S Cru	4.690	5936	0.0407	5.93	42.9	2021	43.6	0.0405	6.18	2088
T Cru	6.733	5526	0.0426	6.20	53.7	2378	55.4	0.0423	6.72	2532
AG Cru	3.837	6253	0.0397	5.88	38.0	1958	33.2	0.0411	4.20	1491
β Dor	9.842	5532	0.0442	7.23	71.0	4182	64.6	0.0453	5.71	3458
W Gem	7.920	5760	0.0429	7.02	62.0	3750	60.0	0.0433	6.46	3506
—	—	—	—	—	—	—	69.0	0.0418	9.15	4637
R Mus	7.510	5806	0.0426	6.95	59.9	3613	52.0	0.0444	4.91	2718
S Mus	9.660	6000	0.0435	8.04	73.4	6185	59.9	0.0462	4.92	4114
—	—	—	—	—	—	—	61.0	0.0460	5.14	4267
S Nor	9.754	5502	0.0442	7.15	70.3	4012	53.4	0.0474	3.59	2312
BF Oph	4.068	5860	0.0403	5.50	38.3	1529	50.4	0.0382	11.30	2650
AP Pup	5.084	5840	0.0411	5.99	45.1	2093	48.8	0.0403	7.31	2451
RV Sco	6.061	5888	0.0417	6.50	51.6	2832	44.4	0.0435	4.50	2096
V636 Sco	6.797	5538	0.0426	6.24	54.1	2439	47.2	0.0439	4.38	1854
S Sge	8.382	5594	0.0434	6.89	63.6	3502	56.7	0.0447	5.19	2785
—	—	—	—	—	—	—	60.0	0.0441	5.96	3119
U Sgr	6.745	5734	0.0423	6.54	55.0	2892	55.3	0.0422	6.63	2925
W Sgr	7.595	5760	0.0428	6.90	60.2	3526	57.3	0.0433	6.12	3197
X Sgr	7.012	6176	0.0420	7.37	59.0	4485	63.1	0.0413	8.71	5125
Y Sgr	5.773	5663	0.0418	6.04	48.6	2153	56.0	0.0405	8.65	2853

MASSES OF CEPHEIDS

AP Sgr	5.058	5760	0.0412	5.86	44.6	1934	36.8	0.0431	3.62	1319
BB Sgr	6.637	5557	0.0425	6.22	53.3	2398	55.2	0.0422	6.79	2571
R TrA	3.389	6029	0.0395	5.32	34.0	1349	35.8	0.0390	6.07	1498
S TrA	6.323	5682	0.0421	6.29	52.1	2507	54.6	0.0417	7.06	2749
T Vel	4.640	5780	0.0408	5.69	41.9	1732	39.7	0.0413	4.97	1556
V Vel	4.371	6253	0.0401	6.20	41.9	2376	35.8	0.0419	4.21	1733
AH Vel	4.227	6065	0.0402	5.86	40.2	1934	44.6	0.0392	7.63	2381
					—	—	47.0	0.0388	8.75	2644
BG Vel	6.924	5676	0.0425	6.51	55.7	2851	53.3	0.0430	5.83	2609
ζ Gem	10.154	5594	0.0442	7.44	73.2	4639	67.6	0.0452	6.12	3959
					—	—	72.4	0.0444	7.24	4541
RT Aur	3.728	6187	0.0396	5.73	37.0	1776	24.7	0.0444	2.14	791
					—	—	28.2	0.0427	2.94	1031
					—	—	28.0	0.0428	2.89	1016
U Car	38.756	5352	0.0520	11.81	187.3	25473	140.1	0.0570	5.95	14249
l Car	35.535	4937	0.0522	10.19	167.9	14817	146.9	0.0538	7.27	11343
SU Cas	1.949	6339	0.0375	4.55	23.1	760	35.2	0.0359	14.78	1770
δ Cep	5.366	6113	0.0410	6.53	48.2	2868	44.6	0.0418	5.39	2457
					—	—	57.0	0.0395	10.04	4014
X Cyg	16.387	5287	0.0470	8.30	100.2	6944	91.0	0.0482	6.53	5725
					—	—	99.0	0.0472	8.05	6776
CD Cyg	17.071	5417	0.0470	8.73	104.7	8359	85.7	0.0499	5.37	5595
T Mon	27.020	5099	0.0502	9.59	140.7	11836	111.1	0.0533	5.33	7383
					—	—	120.0	0.0523	6.47	8613
					—	—	144.0	0.0498	10.16	12403
Y Oph	17.122	5754	0.0466	9.51	108.7	11459	102.6	0.0474	8.26	10210
					—	—	79.0	0.0516	4.47	6053
AW Per	6.463	5991	0.0418	6.84	54.6	3404	49.3	0.0430	5.31	2770
					—	—	70.0	0.0398	12.98	5585
RS Pup	41.388	5309	0.0525	11.98	195.4	26832	183.7	0.0534	10.33	23721
RZ Vel	20.398	5572	0.0478	9.73	121.1	12503	104.1	0.0500	6.78	9243
T Vul	4.436	6113	0.0403	6.04	41.8	2163	40.9	0.0405	5.71	2066
					—	—	52.0	0.0385	10.62	3341
SV Vul	45.100	5038	0.0535	11.51	201.5	23134	146.2	0.0587	5.29	12184
RX Aur	11.610	5801	0.0446	8.25	82.4	6802	59.0	0.0495	3.74	3488
TT Aql	13.740	5439	0.0459	8.06	89.7	6231	72.0	0.0488	4.71	4014

surface helium enhancement to $Y = 0.48$ seems necessary. This deep layer in 5×10^{-4} of the total mass might be consistent with a weak helium-enriching wind and a slow downward mixing due to the inverted μ gradient during the slow first crossing to the red in the instability strip.

VII WESSELINK RADIUS MASSES

Balona (1977) and Evans (1977) have used new and older data and compiled lists of radii obtained from many applications of the Wesselink method. Similar data are available from Cogan (1978a). The recent analysis of these radii by Cox (1979) shows that the radius errors are probably much larger than stated. For 69 cases (53 Cepheids) using homogeneous model Q_0 values the ratio of the Wesselink to theoretical mass is 0.93 if the period is less than 10 days and 0.60 if the period is greater. The inhomogeneous composition models with larger Q_0 values increase the masses by the square of the Q_0. Thus the Wesselink to theoretical mass ratio becomes 1.02 below 10 days and 0.70 above. With both homogeneous and inhomogeneous models, a mass anomaly is indicated above 10 days, and that is tentatively suggested to be due to maybe 25 or more percent mass loss in the main sequence or post–main sequence B star stage discussed by many investigators, including Sreenivasen & Wilson (1978).

Tables 3 and 4 give the theoretical and Wesselink radius masses of Cox (1979). Figures 1 and 2 show how the ratio of these masses vary with period.

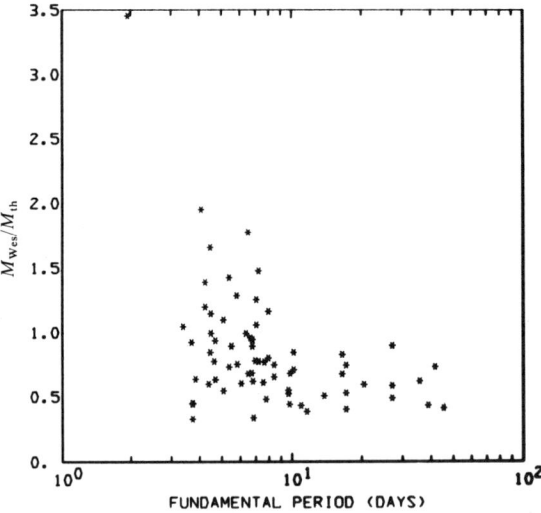

Figure 1 The ratio of the Wesselink radius mass, $M_{\rm Wes}$, to theoretical mass, $M_{\rm th}$, is plotted versus period for 69 cases with measured fundamental pulsation mode periods, $T_{\rm eff}$ values, and Wesselink method radii. Homogeneous $Y = 0.28$ and $Z = 0.02$ composition is assumed.

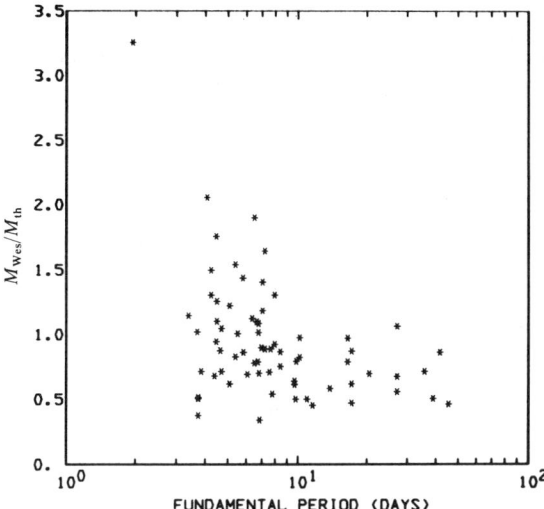

Figure 2 The ratio of the Wesselink radius mass M_{Wes}, to theoretical mass, M_{th}, is plotted versus period for 69 cases with measured fundamental pulsation mode periods, T_{eff} values, and Wesselink method radii. Inhomogeneous composition models with $Y = 0.75$ at the surface and $Y = 0.28$ deeper than 80,000 K are used for the pulsation constants.

VIII DISCUSSION

It now appears that observations on the one hand and evolution and linear pulsation theories on the other agree about as well as they can for the masses of Cepheids. The remaining problems concern those Cepheids with bumps in their light and velocity curves and those displaying two or three modes simultaneously.

Cox (1979) discusses 6 cases where masses are known by several methods. The two best cases are the cluster Cepheids U Sgr (in M25) and S Nor (in NGC6087) whose data are given in Table 5. V367 Sct is discussed below. Note that, as expected, the evolution (M_{ev}) and theoretical (M_{th}) masses for U Sgr and S Nor agree well regardless of whether the models are homogeneous (H) or inhomogeneous (I). The pulsation masses (M_{pul}) are a

Table 5 Cepheid masses

Cepheid	Π_0(days)	M_{ev}	M_{th}(H)	M_{th}(I)	M_{pul}(H)	M_{pul}(I)	M_ϕ(H)	M_ϕ(I)	M_{Wes}(H)	M_{Wes}(I)
U Sgr	6.74	7.1	6.7	6.5	8.8	9.5	4.0	5.3	6.0	6.6
S Nor	9.75	7.4	7.4	7.2	7.3	8.2	3.8	—	3.3	3.6

[a] The new distance scale, reddenings, and the Kraft (1961) temperature scale are used for the evolution, theoretical, and pulsation masses. Bump phase masses are from Fricke, Stobie & Strittmatter (1972) and Cox et al. (1977). Wesselink-radii masses are from Balona (1977) and Cox (1979).

bit larger, and actually favor the homogeneous models. Older bump masses are anomalous but the inhomogeneous models of Cox et al. (1977) raise these closer to the evolution mass. Inhomogeneous models with surface $Y_s = 0.75$ raise the Wesselink radius mass by about 10%. The conclusions to be reached are that the inhomogeneous models help, but do not solve the mass anomalies perfectly. Low bump masses are not likely to be correct in the light of the evolution, theoretical, pulsation, and Wesselink radii masses (at least for U Sgr).

It is harder to make a case for evolution masses for the double-mode Cepheids. The only beat Cepheid that is in a cluster is V367 Sct (6.29 days) with $M_{ev} = 6.9$, $M_{th}(H) = 7.4$, $M_{pul}(H) = 5.0$, and the beat masses $M_B(H) = 2.3$, $M_B(I) = 5.2$, using the data by Cox (1979). It does appear that the inhomogeneous models do reconcile the mass expected from evolution theory for such a period and luminosity.

V367 Sct may not be a member of the cluster NGC 6649 according to Flower (1978). The luminosity of the Cepheid is not large enough, or perhaps all the stars in the cluster were not formed at the same time. On the other hand, Barrell et al. (1980) have shown that the radial velocity of V367 Sct is consistent with cluster membership. In any case, a mass of $7\,M_\odot$ is reasonable for V367 Sct with its 6.29-day period.

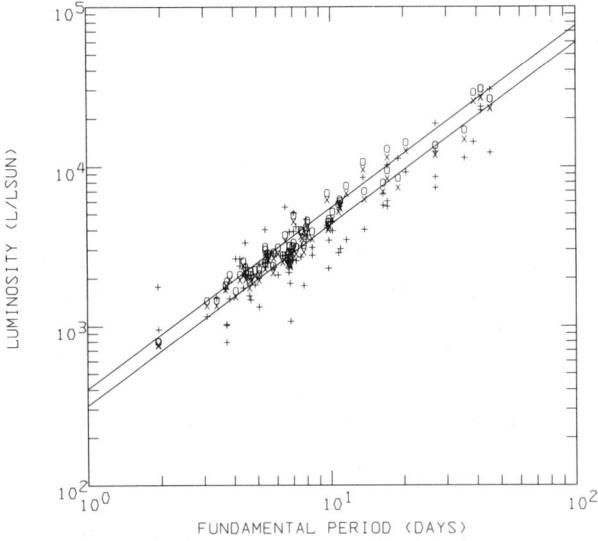

Figure 3 The luminosity is plotted versus fundamental period for 16 calibrating Cepheids and 69 cases (53 Cepheids) with Wesselink radius data. The original Sandage & Tammann (1969) and newly corrected higher luminosity ridge line relations are plotted. Symbols: plus signs, observation; open circles, homogeneous model theoretical luminosities; and crosses, inhomogeneous model theoretical luminosities.

A question remains as to whether the two periods of double-mode Cepheids are correctly identified. The bump Cepheids clearly need Π_2/Π_0 near 0.5 whether their mass is normal or anomalously low. Yet their masses calculated by other methods are normal. Thus one would look for a way to change Π_2/Π_0 without changing the mass. For these bump Cepheids this involves change of envelope structure with a very large Y_s or a large tangled field ($\sim 10^4$ gauss at the bottom of the convection zone). Since this change of envelope structure is needed for 7 $M_\odot$ bump Cepheids, it is reasonable that it also is needed for 5 $M_\odot$ beat Cepheids (Cox, Hodson & King 1979) making it unnecessary to appeal to abnormal pulsation modes.

IX PERIOD-LUMINOSITY-COLOR AND PERIOD-RADIUS RELATIONS

The data of Cox (1979) can be displayed in a way not previously discussed. The 16 calibrating Cepheids have both luminosities and radii derived from observation. Also, the Wesselink radii (69 cases, 53 Cepheids) together with the T_{eff} values from Dean, Warren & Cousins (1978) give the luminosities. All of these luminosities are plotted as + signs in the period-luminosity diagram of Figure 3. The radii are plotted in the period radius diagram of Figure 4. On these two diagrams are also plotted the theoretical lumino-

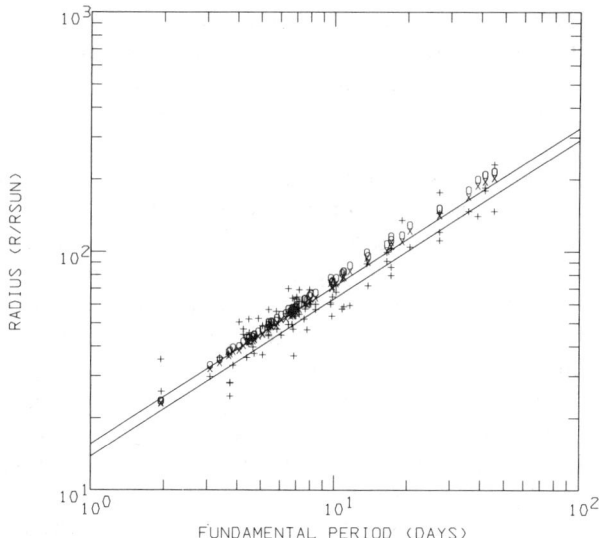

Figure 4 The radius is plotted versus fundamental period for 16 calibrating Cepheids and 69 cases (53 Cepheids) with Wesselink radii. The two lines are observational results discussed in the text. The symbols mean the same as for Figure 1.

sities and radii consistent with the theoretical masses. The radii and luminosities marked with open circles are from homogeneous models and the X points are from the inhomogeneous envelope models that reconcile bump and beat masses with evolution theory masses.

The Sandage & Tammann (1969) period-luminosity relation along their ridge line is also drawn on Figure 3. Above that is given the relation corrected for the 0.26 magnitude distance-modulus increase in the distance scale. The scatter in the observational and theoretical points is due mostly to the color effect, equivalent to an instability strip width of about 1200 K. Theoretically, this width of 20% in $T_{\rm eff}$ at a given luminosity would translate into a width of 40% in R and a width of 60% in period. Therefore, almost half of the observational scatter in Figure 3 is due to the color effect, which is seen clearly in the scatter of the theoretical homogeneous and inhomogeneous model (O and X) data.

To obtain these two lines in Figure 3, it was necessary to use the basic P-L-C relation and the period-color relation as observed to be along the ridge line in the H-R diagram. The width of the P-C plot is due to the fact that a line of constant period in the H-R diagram (just slightly steeper than a line of constant radius due to a larger mass and lower Q_0 at the higher-luminosity blue edge) and ranges from the red to the blue edges covering a range of approximately 1200 K, This temperature range corresponds to a $(B-V)$ color range if about 0.49 magnitude at constant L and 0.40 magnitude at constant P (Sandage & Tammann 1969). The fact that the Sandage & Tammann $(B-V)_0$ may be wrongly corrected for too much reddening, as discussed above, probably does not affect the lines in Figure 3 because the $(B-V)_0$ is merely a parameter that does not enter into the ridge line in the P-L relation.

Use of the theoretical (Kraft 1961) bolometric corrections as a function of color when the color is about 0.1 magnitude ($\Delta \log T_{\rm eff} = -0.02$) larger does not cause a decrease of more than a few hundredths of a magnitude in $M_{\rm bol}$. This has been ignored compared to the 0.26 magnitude distance modulus increase. For these P-L lines, $M_{\rm bol}$ for the sun was taken to be 4.75.

The agreement between the corrected observed period-luminosity ridge line relation and the theoretical luminosities (O and X) in Figure 3 is very comforting. It shows that the mass anomaly is no longer very important, and any individual Cepheid may, if necessary, have its theoretical mass changed slightly by using a slightly different Y for its evolutionary track. A much larger Y (by 0.1 say) implies a $\Delta D \log L = 3.68 \, \Delta \log M$ of (0.61–0.34 log M) according to Becker, Iben & Tuggle (1977). With a fixed L, this Y increase corresponds to a mass decrease of 16–29% for, respectively, 13 and 4 $M_\odot$ which are the limits for blue loops. An increase of 0.01 in Z increases the M for a fixed L by lesser amounts.

The higher luminosity Cepheids seem to have a period that is longer than both the Sandage & Tammann lines and the theoretical results. This effect is due to the mass loss hypothesized for masses greater than $10\,M_\odot$ and periods greater than about 15 days. Longer periods result when the mass for a given luminosity is reduced from its no-mass-loss evolution value since $\Pi \sim M^{-1/2}$.

Further combination of the P-L-C and the P-C ridge line relations with the theoretical $(B-V)_0 - T_{\text{eff}}$ and bolometric correction relations and the definition relation $L \sim R^2 T_{\text{eff}}^4$ allows elimination of the L, C, and T_{eff} dependences. Thus a P-R relation discussed by Balona (1977) can be obtained. In our case, for both the P-L and P-R plots, we have used the Sandage & Tammann period-color ridge line for only galactic Cepheids rather than for all Cepheids as Balona did. The relation plotted (after the $+0.26$ distance modulus and corresponding $+0.052 \log R$ changes) is $\log R = 0.661 \log P + 1.192$ (upper line).

The fact that the T_{eff} values should be 2–3% smaller and the $(B-V)_0$ should be perhaps up to 0.1 magnitude larger than used by Sandage & Tammann causes the decrease in M_{bol} by the above-mentioned few hundredths of a magnitude due to the bolometric correction. Now this is coupled with a color increase of about 0.1 magnitude ($\Delta \log T_{\text{eff}} = -0.02$) due to the fact that the Cepheids are now thought to be cooler. Both effects make the $\Delta \log R = -2\Delta \log T_{\text{eff}} - (DM_{\text{bol}}/5)$ about 0.05 larger with the T_{eff} change being the most important. If this increase in R were to be applied to the observational ridge line in Figure 4, the agreement between theory and observation would be even better.

The lines of constant mass in Figure 4 have a slightly smaller slope than the lines shown. The theoretical and observational relations are given by Equation (3). The radius at the blue edge is about 5% larger, and at the red edge about 5% smaller, than the P-R lines indicate. This color effect is $\Delta \log R / \Delta (B-V)_0 \cong -0.04/0.40 = -0.1$ at a constant period, and it can be compared to the Sandage & Tammann value of -0.09.

For further information and many references about Cepheid masses and many other current theoretical problems, the reader should refer to an excellent review by J. P. Cox (1980). My thanks to D. S. King, B. C. Cogan, and J. P. Cox for many fruitful discussions.

Literature Cited

Adams, T. F., Castor, J. I., Davis, C. G. 1980. *Proc. Conf. Current Problems in Stellar Pulsation Instabilities*, ed. W. M. Sparks. NASA, GSFC

Baade, W. 1926. *Astron. Nachr.* 228:359

Babcock, H. W. 1958. *Ap. J. Suppl.* 3:141

Balona, L. A. 1977. *MNRAS* 178:231

Balona, L. A., Stobie, R. S. 1979. *MNRAS* 189:659

Barrell, S. L. 1979. *Ap. J. Lett.* 226:L141

Barrell, S. L., Cogan, B. C., Faulkner, D. J., Shobbrook, R. R. 1980. *Highlights of Astronomy. Joint Discussion VI. Stellar Instabilities*. Dordrecht: Reidel

Becker, S. A., Iben, I., Tuggle, R. S. 1977. *Ap. J.* 218:633
Carson, T. R., Stothers, R. 1976. *Ap. J.* 204:461
Christy, R. F. 1966. *Ap. J.* 145:340
Christy, R. F. 1968. *Q. J. R. Astron. Soc.* 9:13
Christy, R. F. 1970. *J. R. Astron. Soc. Can.* 64:8
Christy, R. F. 1974. *Mem. Soc. R. Sci. Liege* 8:173
Cogan, B. C. 1970. *Ap. J.* 162:139
Cogan, B. C. 1977. *Ap. J.* 211:890
Cogan, B. C. 1978a. *Ap. J.* 221:635
Cogan, B. C. 1978b. *Ap. J. Lett.* 225:L39
Cogan, B. C. 1980. *Ap. J.* In press
Cox, A. N. 1979. *Ap. J.* 229:212
Cox, A. N., Cox, J. P. 1976. In *Multiple Periodic Variable Stars*, ed. W. S. Fitch, pp. 115–31. Dordrecht: Reidel
Cox, A. N., Deupree, R. G., King, D. S., Hodson, S. W. 1977. *Ap. J. Lett.* 214:L127
Cox, A. N., Hodson, S. W. 1978. *IAU Symp. No. 80. The H-R Diagram*, ed. A. G. D. Philip, D. S. Hayes, p. 237. Dordrecht: Reidel
Cox, A. N., Hodson, S. W., King, D. S. 1978. *Ap. J.* 220:996
Cox, A. N., Hodson, S. W., King, D. S. 1979. *Ap. J. Lett.* 230:L109
Cox, A. N., King, D. S., Hodson, S. W. 1978. *Ap. J.* 224:607
Cox, A. N., Michaud, G., Hodson, S. W. 1978. *Ap. J.* 222:621
Cox, A. N., Tabor, J. E. 1976. *Ap. J. Suppl.* 31:271
Cox, J. P. 1974. *Rep. Prog. Phys.* 37:563
Cox, J. P. 1980. *Proc. Conf. Current Problems in Stellar Pulsation Instabilities*, ed. W. M. Sparks. NASA, GSFC
Cox, J. P., King, D. S., Stellingwerf, R. S. 1972. *Ap. J.* 171:93
Davis, C. G., Davison, D. K. 1978. *Ap. J.* 221:929
Dean, J. F., Warren, P. R., Cousins, A. W. J. 1978. *MNRAS* 183:569
Deupree, R. G. 1976. *Proc. Solar and Stellar Pulsation Conf., LA-6544-C*, ed. A. N. Cox, R. G. Deupree, p. 229
Deupree, R. G. 1977. *Ap. J.* 215:232
Deupree, R. G. 1978. *Ap. J.* 223:982
Deupree, R. G. 1980. *Ap. J.* 236:225
Durand, R. A., Eoll, J. G., Schlesinger, B. M. 1976. *MNRAS* 174:641
Dziembowski, W. 1977. *Acta Astron.* 27:95
Evans, N. R. 1977. *Ap. J.* 209:135
Evans, N. R. 1979. *Proc. Conf. Current Problems in Stellar Pulsation Instabilities*, ed. W. M. Sparks. NASA, GSFC
Faulkner, D. J. 1977. *Ap. J.* 218:209
Faulkner, D. J., Shobbrook, R. R. 1979, *Ap. J.* 232:197
Feldman, W. C., Asbridge, J. R., Bame, S. J., Gosling, J. T. 1977. *The Solar Output and Its Variations*, ed. O. R. White, p. 351. Boulder, Colo: Assoc. Univ. Press
Fernie, J. D. 1968. *Ap. J.* 151:197
Fernie, J. D. 1979a. *Ap. J.* 231:841
Fernie, J. D. 1979b. *Publ. Astron. Soc. Pac.* 91:67
Fischel, D., Sparks, W. M. 1975. *Cepheid Modeling*. NASA SP-383
Fitch, W. S., Szeidl, B. 1976. *Ap. J.* 203:616
Flower, P. J. 1978. *Ap. J.* 224:948
Forbes, J. E. 1968. *Ap. J.* 53:495
Fricke, K., Stobie, R. S., Strittmatter, P. A. 1971. *MNRAS* 154:23
Fricke, K., Stobie, R. S., Strittmatter, P. A. 1972. *Ap. J.* 171:593
Gough, D. O., Ostriker, J. R., Stobie, R. S. 1965. *Ap. J.* 142:1649
Hanson, R. B. 1977. *IAU Symp. No. 80. The H-R Diagram*, ed. A. G. D. Philip, D. S. Hayes, p. 154. Dordrecht: Reidel
Hayashi, C., Hoshi, R., Sugimoto, D. 1962. *Prog. Theor. Phys., Suppl.* 22
Hertzsprung, E. 1926. *Bull. Astron. Inst. Neth.* 3:115
Hodson, S. W., Cox, A. N. 1976. *Proc. Solar and Stellar Pulsation Conf. LA-6544-C*, ed. A. N. Cox, R. G. Deupree, p. 202
Hodson, S. W., Cox, A. N. 1980. *Proc. Nonradial and Nonlinear Stellar Pulsation Workshop*, ed. H. A. Hill, W. Dziembowski. New York: Springer
Hodson, S. W., Stellingwerf, R. F., Cox, A. N. 1979. *Ap. J.* 229:642
Iben, I. 1967. *Ann. Rev. Astron. Astrophys.* 5:571
Iben, I. 1974a. *Ann. Rev. Astron. Astrophys.* 12:215
Iben, I. 1974b. *IAU Symp. No. 59. Stellar Instability and Evolution*, ed. P. Ledoux, A. Noels, A. W. Rodgers, p. 3. Dordrecht: Reidel
Iben, I., Tuggle, R. S. 1972a. *Ap. J.* 173:135
Iben, I., Tuggle, R. S. 1972b. *Ap. J.* 178:441
Iben, I., Tuggle, R. S. 1975. *Ap. J.* 197:39
Karp, A. H. 1975a. *Ap. J.* 199:1448
Karp, A. H. 1975b. *Ap. J.* 200:354
Karp, A. H. 1975c. *Ap. J.* 201:641
King, D. S., Cox, J. P., Eilers, D. D., Davey, W. R. 1973. *Ap. J.* 182:859
King, D. S., Hansen, C. J., Ross, R. R., Cox, J. P. 1975. *Ap. J.* 195:467
Kraft, R. P. 1961. *Ap. J.* 134:616
Kraft, R. P. 1963. *Stars and Stellar Systems 3*, ed. K. Aa. Strand, p. 421. Univ. Chicago Press
Kraft, R. P. 1967. *Aerodynamical Phenomena in Stellar Atmospheres*, ed. R. N. Thomas, p. 207. London: Academic
Lauterborn, D., Refsdal, S., Weigert, A. 1971. *Astron. Astrophys.* 10:97
Madore, B. F. 1977. *MNRAS* 178:505
Madore, B. F., van den Bergh, S. 1975. *Ap. J.* 197:55

Niva, G. D. 1979. *Ap. J. Lett.* 232:L43
Niva, G. D., Schmidt, E. G. 1979. *Ap. J.* 234:245
Osaki, Y. 1977. *Publ. Astron. Soc. Jpn.* 29:235
Pel, J. W. 1978. *Astron. Astrophys.* 62:75
Petersen, J. O. 1973. *Astron. Astrophys.* 27:89
Petersen, J. O. 1974. *Astron. Astrophys.* 34:309
Petersen, J. O. 1978. *Astron. Astrophys.* 62:205
Rodgers, A. W. 1970. *MNRAS* 151:133
Sandage, A., Tammann, G. A. 1968. *Ap. J.* 151:531
Sandage, A., Tammann, G. A. 1969. *Ap. J.* 157:683
Sandage, A., Tammann, G. A. 1971. *Ap. J.* 167:293
Scarfe, C. D. 1977. *Ap. J.* 209:141
Schmidt, E. G. 1976. *Ap. J.* 203:466
Simon, N. R. 1979. *Astron. Astrophys.* 75:140
Simon, N. R., Cox, A. N., Hodson, S. W. 1980. *Ap. J.* In press
Simon, N. R., Schmidt, E. G. 1976. *Ap. J.* 205:162
Siquig, R. A., Sonneborn, G. 1976. *Bull. Am. Astron. Soc.* 8:320
Sonneborn, G., Kuzma, T. J., Collins, G. W. 1979. *Ap. J.* 232:807
Sreenivasan, S. R., Wilson, W. J. F. 1978. *Astrophys. Space Sci.* 53:193
Stellingwerf, R. F. 1975. *Ap. J.* 199:705
Stellingwerf, R. F. 1976. *Multiple Periodic Variable Stars*, ed. W. S. Fitch, p. 153. Dordrecht: Reidel
Stobie, R. S. 1969a. *MNRAS* 144:461
Stobie, R. S. 1969b. *MNRAS* 144:485
Stobie, R. S. 1969c. *MNRAS* 144:511
Stobie, R. S. 1976. *IAU Symp. No. 29. Multiple Periodic Variable Stars*, ed. W. S. Fitch, p. 87. Dordrecht: Reidel
Stobie, R. S. 1977. *MNRAS* 180:631
Stobie, R. S., Balona, L. A. 1979. *MNRAS* 189:627
Stothers, R. 1979. *Ap. J.* 234:257
Takeuti, M. 1973. *Publ. Astron. Soc. Jpn.* 25:567
Takeuti, M. 1979. Preprint
van Altena, W. F. 1974. *Publ. Astron. Soc. Pac.* 86:217
van den Bergh, S. 1977. *IAU Symp. No. 37. Decalages vers le Rouge et Expansion de l'Univers*
Vemury, S. K., Stothers, R. 1978. *Ap. J.* 225:939
Weiss, W. W., Wood, H. J. 1975. *Astron. Astrophys.* 41:165
Wesselink, A. J. 1946. *Bull. Astron. Inst. Neth.* 10:91
Wood, H. J., Weiss, W. W., Jenkner, H. 1977. *Astron. Astrophys.* 61:181
Wooley, R., Carter, B. 1973. *MNRAS* 162:379

INFRARED SPECTROSCOPIC OBSERVATIONS OF THE OUTER PLANETS, THEIR SATELLITES, AND THE ASTEROIDS

×2162

Harold P. Larson

Department of Planetary Sciences, University of Arizona, Tucson, Arizona 85721

I INTRODUCTION

The exploration of objects in the outer solar system by infrared spectroscopic methods has been an exciting and productive area of planetary astronomy in the 1970s. Relatively few groups specialize in planetary IR astronomy, partly because of the frequent need to employ unique instrumentation at specialized observing facilities. Although this activity has not received the coordinated publicity of NASA's program of active exploration of the solar system by spacecraft, many of these earth-based efforts are equally pioneering in their scope. The field of planetary IR astronomy is summarized in several reviews that provide useful background and historical perspective for this chapter. Fink & Larson (1979a) prepared a general review of solar system astronomy that includes instrumental considerations and examples of observational data. The proceedings of the Ottawa symposium on planetary atmospheres (Jones 1977) contains useful summaries of current research with emphasis on the outer planets. The most extensively studied outer planet at IR wavelengths is Jupiter; a comprehensive review of its IR spectrum was prepared by Ridgway, Larson & Fink (1976). The only high resolution observations beyond 1 μm of Uranus, Neptune, and Titan were published by Fink & Larson (1979b). The growing body of literature on IR spectral reflectance measurements of surfaces includes several valuable summaries. Pollack et al. (1978) describe their recent results for the Galilean satellites, and they review previous work using ground-based and airborne observatories. The surface reflectances of the satellites of Saturn are described by Cruikshank (1979). Finally, remote mineralogical studies of asteroid surfaces by IR methods are summarized in Larson & Fink (1977) and Larson & Veeder (1979).

The primary goal of this review is to delineate the spectroscopic data base for planetary studies. Details of the many diverse interpretations that follow from these observations must be left to the original papers. This review includes atmospheric studies, which have a long observational history, and studies of surface mineralogies, which are relatively recent applications of IR methods. The atmospheric section includes the planets Jupiter, Saturn, Uranus, and Neptune and those satellites with atmospheres (Titan and Triton). Infrared studies of surfaces include the satellites of Jupiter and Saturn, the asteroids, and Pluto.

Some restrictive assumptions are required to keep the chapter to a manageable length. Infrared will imply the wavelength region between 1–100 μm, and spectroscopy will designate any technique employing resolving powers $\lambda/\Delta\lambda \geq 50$. In view of the space required to present spectra, only selected examples illustrating very recent or unique results will be reproduced here.

The immediate goals of most spectroscopic studies of planetary atmospheres can be classified into several broad groups. The detection of atmospheric constituents is a major effort, especially when a spectral region is being explored for the first time. Closely related to this activity is the determination of upper limits to the presence of certain key molecules. These data are fundamental to cataloguing the constituents of planetary atmospheres and to identifying the chemical processes that explain their presence. A second major observational goal is to establish the column abundances of known atmospheric constituents. These numbers are equally important to building chemical models of planetary atmospheres. Elemental abundance ratios ($^{12}C/^{13}C$, D/H, C/H, etc.) derived from these measurements permit comparisons of a particular planet with other chemical systems and with terrestrial, solar, and cosmic values. Differences in these ratios can be related to evolutionary processes and to chemical fractionation mechanisms. A third major use of IR spectral observations is to establish thermal properties of planetary atmospheres. Various types of temperature measurements are possible, such as the brightness temperature and the effective temperature. Spectroscopic measurements also contribute to establishing the vertical thermal structure of a planetary atmosphere (P-T profile, for example).

These various results that can be derived from spectroscopic measurements directly contribute to comprehensive models of the physical and chemical structures of planetary atmospheres. Specific areas include stratospheric phenomena (photochemistry), cloud formation and composition, and thermochemical descriptions of the unobservable interior.

These generalized goals are similar to those pursued in IR stellar spectroscopy, of course. When applied to the outer solar system the major distinguishing characteristic is temperature; the warmest temperature encountered is about 300 K for Jupiter at 5 μm. Thus, spectral analyses of planetary atmospheres primarily involve vibration-rotation absorption bands of molecular species that are thermochemically stable in low temperature regimes.

For IR spectral studies of surfaces, the primary interpretive goal is to identify surface minerals through their spectroscopically active components (Fe^{2+}, $CO_3^=$, etc.). Diagnostic capabilities may often be restricted to the identification of classes of minerals (carbonates, for example) rather than of specific minerals ($FeCO_3$, $MgCO_3$, etc.). Abundances derived from relative intensity measurements are quite uncertain since their determination requires knowledge of particle sizes and the degree to which one mineral phase is dispersed in another, none of which can be established by remote observations.

In spite of these limitations, however, spectral observations of the surfaces of satellites and asteroids are contributing fundamental insights into the origin and early evolution of objects in the outer solar system. Specific areas include the major role of volatiles (ices) in the bulk composition of satellites, the existence of magmatic differentiation in asteroid-sized bodies by an intense, early thermal heating episode, and the origin of carbonaceous matter. Although these objects may be considered insignificant based upon their size, their surfaces have uniquely preserved evidence of nebular composition and early solar system evolution that has been eradicated on the planets. Moreover, some aspects of the topics associated with surface studies (short-lived radionuclides, organic synthesis, etc.) are closely related to fundamental problems in astrophysics (nucleosynthesis, chemistry of the interstellar medium).

Many developments in astronomical instrumentation, particularly of spectroscopic systems, have been driven by the needs of planetary astronomy. Since there are only a relatively few solar system objects available to observe, compared to the number of celestial sources, there is a need to study them as intensively and completely as possible. This requires instruments that provide high spectral resolution, extensive wavelength coverage, high efficiency, and temporal and spatial resolution. An excellent review of the instruments employed in IR astronomy is given by Soifer & Pipher (1978). They emphasized that evaluation of overall system performance at IR wavelengths requires inclusion of the detector system and the observing site as integral components of the instrument. These considerations apply equally to planetary and stellar observations, of course. The rapid growth of IR planetary spectroscopy in the past decade has been a

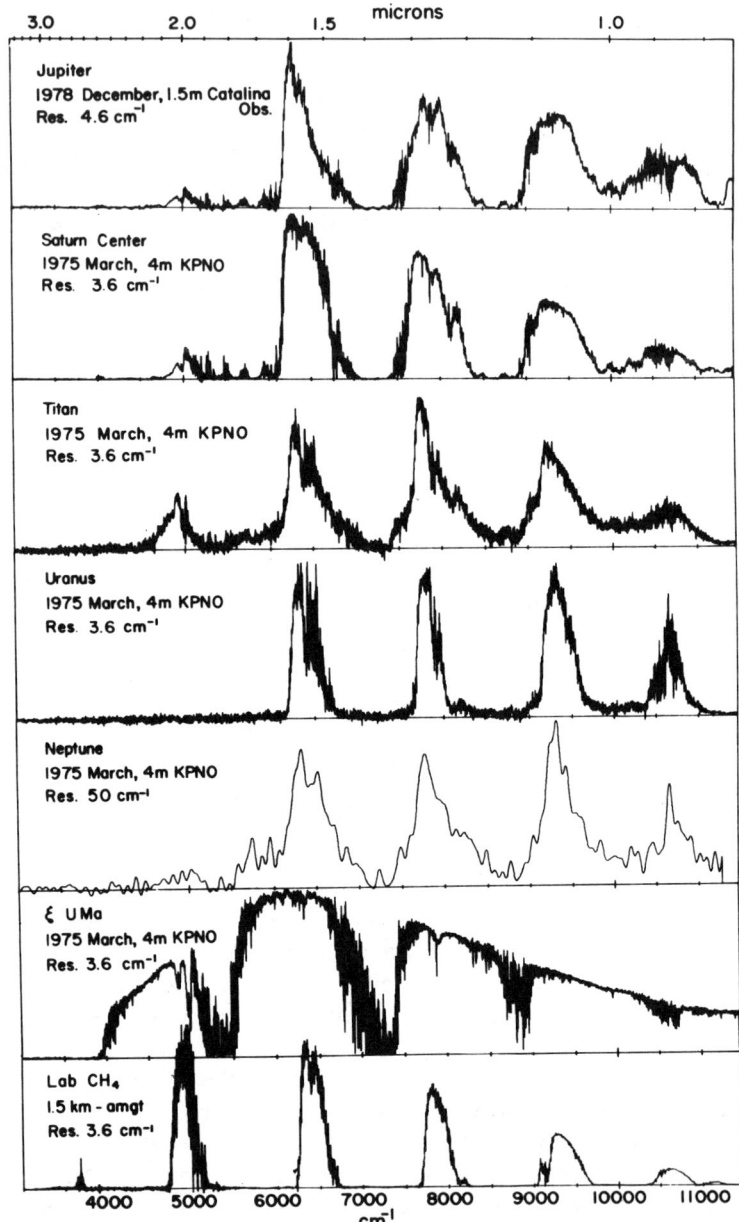

Figure 1 Comparison of the near-IR spectra of the outer planets and Titan. All spectra were produced by a Fourier spectrometer. The noise level in each spectrum is seen near 3 μm where there is no signal. The spectrum of the solar-type comparison star indicates the instrumental response (0.8–2.5 μm) and the locations of the terrestrial atmospheric transmission windows. The laboratory spectrum of CH_4 reveals the narrow, isolated regions of transmission of this gas at the high abundances characterizing these planetary atmospheres. Figure reproduced from Fink & Larson (1979b).

direct consequence of this careful attention to IR system engineering. Developments of special significance include multiplex spectrometers, special sites such as the Kuiper Airborne Observatory (KAO), and InSb IR detector systems. Detailed discussions of modern IR spectroscopic techniques and references to individual programs are available in the Soifer & Pipher review.

II ATMOSPHERES

Figure 1 illustrates a problem common to all ground-based IR observations: objects can be studied only in certain wavelength intervals, or atmospheric transmission windows, which are defined by molecular absorptions in the earth's atmosphere and by constituents in the planet's atmosphere. The solar-type stellar comparison spectrum (ξ U Ma) in Figure 1 defines the regions of transparency in our atmosphere in the near IR (0.8–2.5 μm). Strong absorptions by H_2O at 1.1, 1.4, 1.9, and 2.7 μm restrict astronomical observations to three major near-IR windows, centered approximately at 1.25, 1.6, and 2.2 μm. These positions correspond to the centers of the J, H, and K photometric bands, respectively.

The IR spectra of all of the solar system objects in Figure 1 exhibit even narrower transmission characteristics than would be expected on the basis of the telluric spectrum alone. This is explained with the help of the bottom curve in Figure 1, a high abundance laboratory spectrum of CH_4. This spectrum closely simulates many characteristics of the outer planet data. Thus, CH_4 is the major, spectroscopically active constituent in the atmospheres of the outer planets, and a knowledge of its spectrum is essential to interpreting the planetary observations. Other planetary absorbers include H_2, whose 1–0 pressure-induced dipole spectrum near 2.5 μm is especially prominent in spectra of the giant planets. Gaseous NH_3 is also strongly absorbing in the IR, but only Jupiter exhibits strong NH_3 bands since it is frozen out at the lower atmospheric temperatures of the more distant planets.

In view of these restrictions on the spectral distribution of flux, it is frequently convenient to discuss the IR spectra of the outer planets in terms of their isolated transmission windows. Table 1 lists several groupings of these windows that form a convenient basis for the following discussion. For wavelengths $\lambda < 3$ μm the planetary spectral irradiance is reflected solar radiation. Spectral analyses in this region have emphasized abundance determinations of known atmospheric constituents such as H_2, CH_4, and NH_3. These measurements also contribute to establishing the scattering properties of the atmosphere. For wavelengths $\lambda > 3$ μm the flux received is thermal emission from the planet. The 3–6 μm region has been particularly productive for detecting trace atmospheric constituents with

Table 1 Atmospheric transmission windows of the outer planets

Spectral region	Position of planetary window (μm)	Spectroscopic analyses	Telescope facility required
Near-IR	1.2, 1.6, 2.0	H_2, CH_4, NH_3 abundances, scattering models, overtone and combination bands	Ground-based, high-altitude (2.0 μm)
Transition	2.7	Fundamental bands, trace constituents, solid NH_3	High-altitude
Middle-IR	5.0, 10.0	Fundamental bands, trace constituents, thermal profiles of atmosphere	Ground-based, high-altitude
Far-IR	20–100	Pure rotation bands, thermal emission	High-altitude

abundances (with respect to H_2) in the 10^{-5}–10^{-9} range. At wavelengths of 10 μm and longer, spectral observations are particularly important to establishing thermal properties of planetary atmospheres. Between these two regimes there is a transition region containing a poorly studied planetary window centered at 2.7 μm. The frequent requirement in Table 1 for an airborne telescope facility for spectroscopic studies of the outer planets is a consequence of partial or total obscuration of these planetary windows by terrestrial absorptions.

Jupiter

Infrared spectroscopic observations of Jupiter have been available for more than three decades (Kuiper 1947). In subsequent years as numerous instrumental techniques were being developed for observational astronomy, Jupiter was frequently observed, but in 1969 the knowledge of the composition of its atmosphere was still limited to H_2, CH_4, and NH_3, none of which depended upon IR measurements. At about this time, however, two very significant events occurred that, in retrospect, strongly influenced future research objectives for Jupiter. The first of these was the development of high resolution methods of Fourier spectroscopy and their spectacular results for Venus and Mars, which eclipsed their equally unmatched results on Jupiter and Saturn (Connes et al. 1969). The second event was the publication of Jupiter's middle infrared spectrum (3–14 μm) by Gillett, Low & Stein (1969) which showed Jupiter's unexpectedly high 5 μm flux. The observational opportunities on Jupiter that were revealed by this low resolution work strongly influenced the development of higher resolution Fourier methods which now dominate IR spectroscopic studies of planetary atmospheres. Special emphasis was given applications of FTS devices on aircraft and in wavelength regions where the multiplex advantage could not be realized. Additional historical perspective on IR spectral studies of Jupiter is included in articles by de Bergh & Maillard (1977) and Ridgway et al. (1976a).

Table 2 lists recent observational work that is influencing interpretive studies of Jupiter's atmosphere. The large number of entries illustrates the attention planetary astronomers are giving this unique planet. Note the dependence of these endeavors upon high-level technology, including

Table 2 Infrared spectroscopic observations of Jupiter

Spectral range (cm^{-1})	Resolution (cm^{-1})	Instrument/Telescope[a]	Reference
6000–11,800	0.09	FTS/Palomar 5 m	Lecacheux et al. 1976
4000–6000	0.5	FTS/Hawaii 2.2 m	Martin et al. 1976
2800–6200	0.25	FTS/KAO 0.9 m	H. P. Larson & U. Fink, unpublished observations
1700–2300	0.5	FTS/KAO 0.9 m	Larson et al. 1977
1700–2300	0.2	FTS/McDonald 2.7 m	Beer 1975
1250–2000	~67	CVF/KAO 0.9 m	Russell & Soifer 1977
710–3300	~20	CVF/UAO 0.7, 1.5 m	Gillett et al. 1969
770–1250	~7	Grating/Tenerife 1.5 m	Aitken & Jones 1972
750–875	1.3	FTS/KPNO 1.5 m	Ridgway 1974
725–900	1.3	FTS/Hawaii 2.2 m	Combes et al. 1974
850–1000	~4	Grating/Lick 3 m	Lacy et al. 1975
755–850	0.5	FTS/SAO 1.5 m	Tokunaga et al. 1976
700–1600	10	FTS/KAO 0.9 m	Encrenaz et al. 1976
420–840	~4	FTS/KAO 0.9 m	Aumann & Orton 1976
750–1400	3	FTS/KPNO 1.5 m	Ridgway et al. 1976b
820–1010	1.6	FTS/ESO 1.5 m	Combes et al. 1976
720–1010	1.5	FTS/Hawaii 2.2 m	Encrenaz et al. 1978
1100–1200	2	FTS/Hawaii 2.2 m	Encrenaz et al. 1979
740–980	0.05, 0.3	FTS/KPNO 4 m	Tokunaga et al. 1979
60–220	4	FTS/Balloon-borne 0.4 m	Furniss et al. 1976
100–470	5	FTS/KAO 0.9 m	Erickson et al. 1978
400–600	0.5	FTS/SAO 1.5 m	Tokunaga et al. 1977
250–625	~8	Grating/Learjet 0.3 m	Houck et al. 1975
100–300	1.6	FTS/KAO 0.9 m	Aumann & Orton 1978
105–140	0.02	FTS/KAO 0.9 m	Baluteau et al. 1978
180–2500	4.3	FTS/Voyager	Hanel et al. 1979

[a] CVF: Circular Variable Filter
ESO: European Southern Observatory
FTS: Fourier Transform Spectrometer
KAO: Kuiper Airborne Observatory
KPNO: Kitt Peak National Observatory
SAO: Smithsonian Astrophysical Observatory
UAO: University of Arizona Observatories.

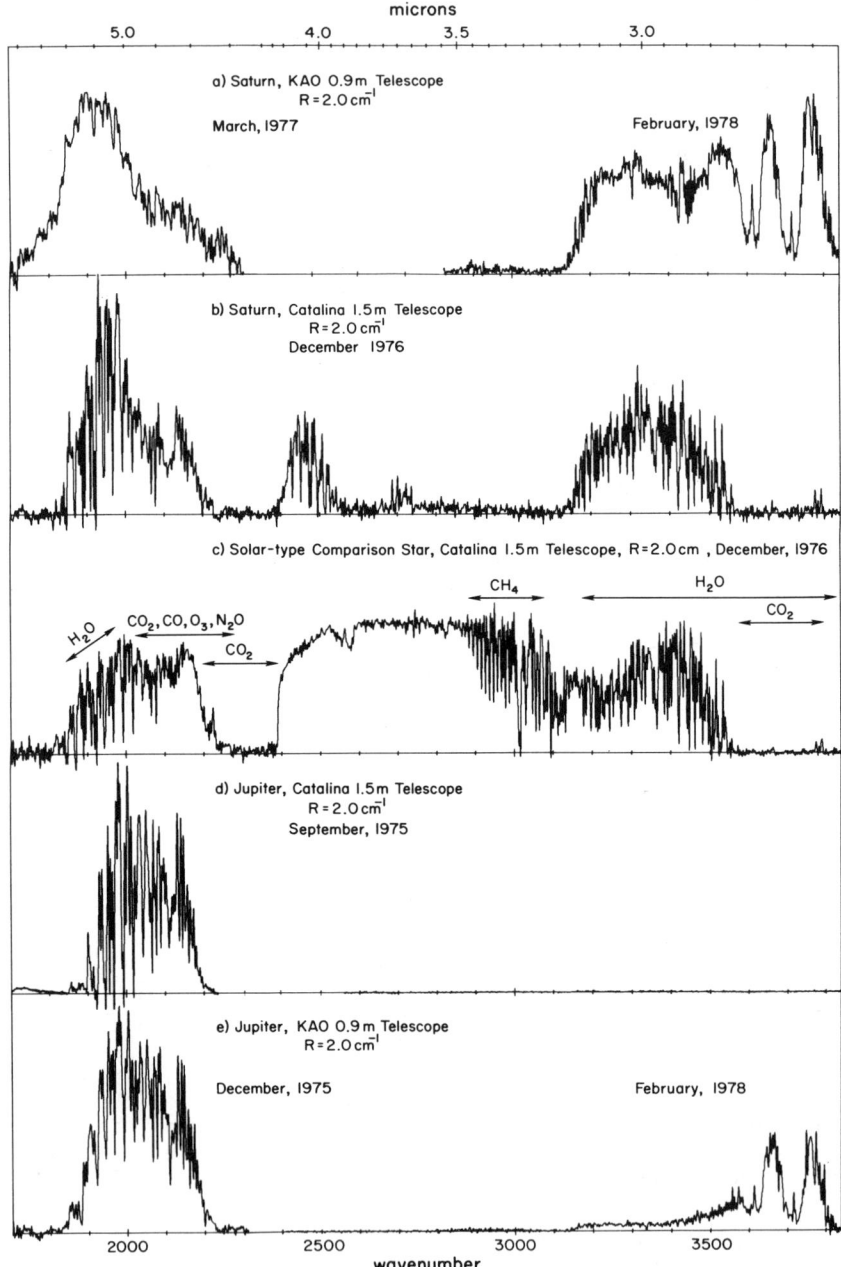

complex multiplex methods (21 entries), and aircraft, balloon, and spacecraft observing sites (11 entries).

As indicated in Figure 1, all of the outer planets can be observed in the near IR from conventional telescopes since the planetary windows are well placed with respect to the terrestrial windows. Table 2 does not include all references to near-IR observations of Jupiter; the single entry (Lecacheux et al. 1976) is the highest resolution spectrum currently available. More than a dozen additional near-IR spectra are listed in de Bergh & Maillard (1977).

By contrast, only a single set of observations defines completely the Jovian windows centered at 2.0 and 2.7 μm (H. P. Larson and U. Fink, unpublished observations, 1979). This is due to complete obscuration of these regions by terrestrial H_2O, thus requiring a high altitude observing site. Figure 2 provides an overview of both Jupiter's and Saturn's spectrum in the 2.7 μm region. The spectrum of the stellar comparison object in the middle of this composite defines the telluric spectrum. Obscuration by H_2O is total from 2.5–2.85 μm, and no ground-based observations are possible. Throughout the 3 μm region, strong H_2O and CH_4 lines dominate the terrestrial spectrum, and astronomical observations are difficult at best. Ground-based observations of Jupiter and Saturn are plotted in Figure 2 immediately above and below the stellar comparison spectrum. Jupiter provides no flux throughout the terrestrial 3 μm window, but Saturn does have a spectrum in this region.

The top and bottom plots in Figure 2 are airborne observations of Jupiter and Saturn. The greatly reduced terrestrial H_2O abundance (10 microns of precipitable H_2O) at aircraft altitude (12.4 km) permits almost complete spectroscopic definition of the 2.7 μm planetary windows. On Jupiter this window is a very narrow region between 2.6 and 2.9 μm, that, unfortunately, is coincident with strong telluric CO_2 bands. Saturn's 2.7 μm window is much broader, extending from 2.6–3.2 μm. The terrestrial CO_2 bands also partially block Saturn's high altitude spectrum near 2.8 μm, but the obvious structure in Saturn's airborne spectrum around 3 μm is not terrestrial. In spite of the narrow spectral bandwidth of these 2.7 μm planetary windows, numerous spectral analyses are possible. Some of these are discussed in the section for Saturn.

No other spectra exist for the outer planets at 2.0 and 2.7 μm, but mention

---

← *Figure 2* Comparison of the middle-IR spectra of Jupiter and Saturn from a conventional telescope and from the Kuiper Airborne Observatory. The spectrum of the solar-type comparison star illustrates the complex terrestrial absorption spectrum that plagues interpretation of high resolution astronomical observations in this wavelength region. Only high altitude observations (top, bottom spectra) reveal fully the two major planetary transmission windows in the 2.5–5.6 μm region. These planetary windows are defined primarily by CH_4, PH_3, and, on Jupiter, NH_3 absorptions. Figure reproduced from Larson et al. (1980).

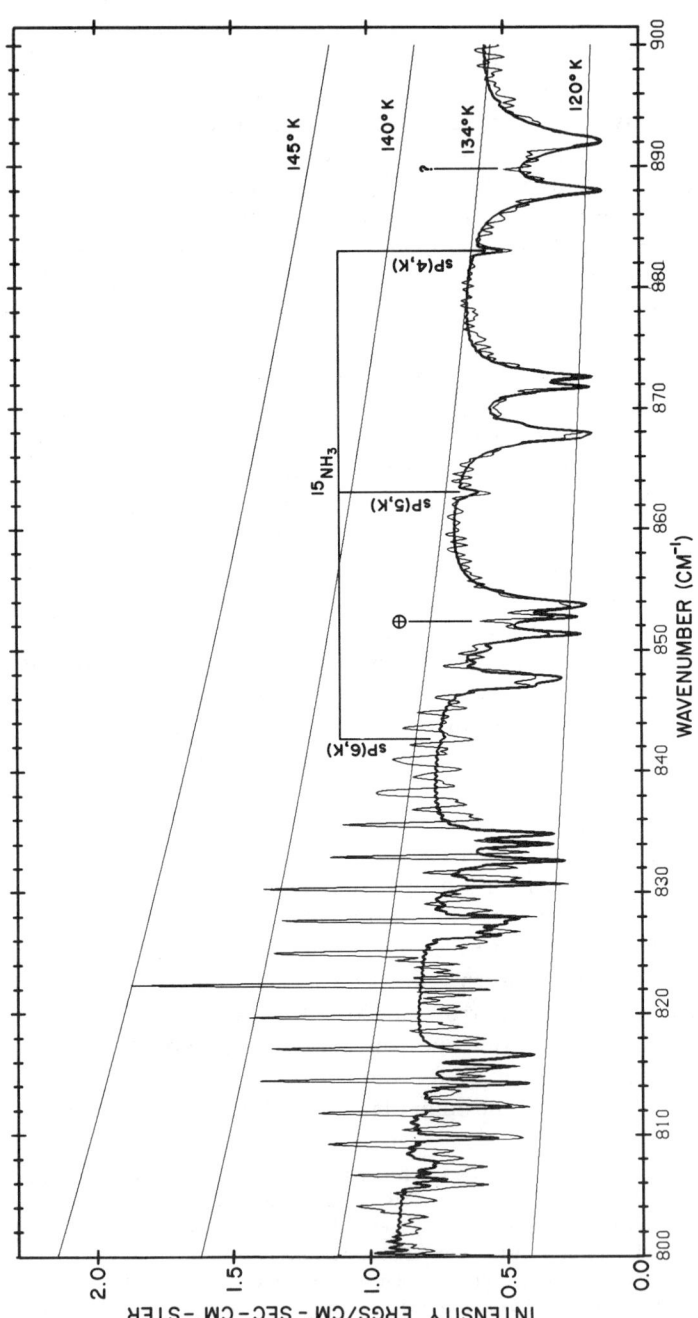

Figure 3 An example of a high resolution spectrum of Jupiter in the 10 μm region. The strong absorption lines are $^{14}NH_3$, and weak lines of $^{15}NH_3$ are indicated. Ethane (C_2H_6) in emission is very obvious in the 820 cm^{-1} region. A synthetic spectrum (solid line) and various blackbody curves accompany the observations as part of their interpretation by thermal modeling techniques. Figure reproduced from Tokunaga et al. (1979).

must be made here of the Stratoscope balloon photometry of Jupiter (Danielson 1966) in this spectral region. These data first revealed the existence of these two planetary windows, and their interpretation emphasized the role of the pressure-induced dipole spectrum of molecular hydrogen in the near-IR spectra of the outer planets.

Knowledge of Jupiter's 5 μm spectrum is also based upon just a few observations. These data have been a prolific source of new trace atmospheric constituents, in part because Jupiter's 5 μm flux emerges from breaks in its cloud cover, thereby permitting observations to deep, hot ($\sim$ 300 K) atmospheric levels.

Table 2 includes many studies of Jupiter in the 10 μm region. Spectroscopic analyses of these data have concentrated on trace constituent detections and thermal models. A useful discussion of these results is included in Tokunaga et al. (1979). Figure 3, from their paper, illustrates the appearance of high resolution, high signal-to-noise spectra of Jupiter in the 10 μm region. This figure indicates how 10 μm observations permit trace constituent analyses (C_2H_6 in emission, $^{15}NH_3$ in absorption) and thermal modeling by spectrum synthesis.

The recent observations of Jupiter from Voyager (Hanel et al. 1979) include all of the 10 μm region. These data provide an unprecedented combination of spectral resolution, spatial resolution, and spectral bandwidth for identifying local compositional and thermal variations in Jupiter's atmosphere.

Beyond 10 μm, earth-based observations of Jupiter depend entirely upon high altitude sites due to terrestrial H_2O interference. The spectrum of Jupiter recorded by Erickson et al. (1978) is reproduced in Figure 4 to illustrate recent advances in far-IR observational capabilities. About 75% of Jupiter's thermal flux is emitted in this spectral region (20–100 μm). Analyses of Jupiter's thermal emission spectrum include the NH_3 rotation spectrum and the collision-induced H_2 continuum, as illustrated in Figure 4.

The following discussion of the chemical and thermal properties of Jupiter's atmosphere is based largely upon the spectral data in Table 2. It has long been assumed that the gross composition of Jupiter's atmosphere is explained by chemical thermodynamic equilibrium applied to a mixture of H_2, CH_4, NH_3, and other trace constituents. An assumption of solar elemental abundances is also part of this model. These expectations have led to a general interpretive framework for predicting other properties of a planetary atmosphere, such as cloud formation and composition, and the presence of trace constituents (Lewis 1969). Spectroscopic determination of atmospheric composition is the principal test for such models. Prinn & Owen (1976) prepared an excellent review of the interrelated roles of atmospheric chemistry and planetary spectroscopy for Jupiter, but the principles are generally applicable to all the planets.

Table 3 lists the known constituents of Jupiter's atmosphere relative to hydrogen, the most abundant molecule. Infrared spectra acquired during the 1970s have provided the evidence for 9 of the 11 newly detected species. Knowledge of the presence of each of these molecules has contributed some useful insight into Jupiter's atmospheric chemistry. Many of the entries in Table 3 were not expected on the basis of the simple equilibrium model. The addition of nonequilibrium mechanisms to the basic model are required to account for the observed composition. The detection of the simple hydrocarbons C_2H_2 and C_2H_6, for example, identified photochemical destruction of CH_4 in the stratosphere as a disequilibrating process (Strobel 1974). The gaseous hydrides PH_3 and GeH_4 detected on Jupiter are chemically compatible with a reducing atmosphere, but not under the conditions of pressure, temperature, and water vapor abundance in the observable levels of Jupiter's atmosphere. One disequilibrating mechanism that can account for their presence without seriously violating the equilibrium model is rapid vertical convection (Prinn & Lewis 1975). This cycle brings unexpected species from hot, interior layers where they are

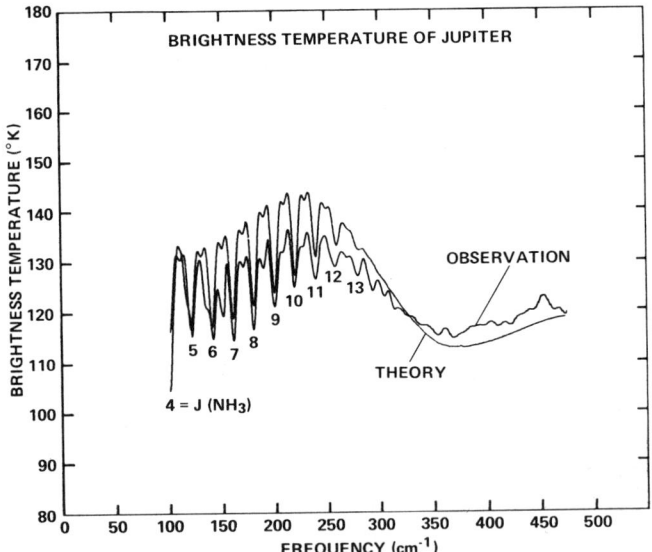

Figure 4 An example of far-IR spectroscopic observations of Jupiter. About 75% of Jupiter's thermal emission is contained in the above spectral passband. The expected blackbody curve ($T_{eff} \sim 125$ K) is modified by absorptions due to the rotational spectrum of NH_3 and the collision-induced spectrum of H_2. A synthetic spectrum accompanies the observations to illustrate the interpretation of these data. Figure reproduced from Erickson et al. (1978).

chemically stable to the cooler top of the atmosphere where they are observed. The time scale of their transfer must be short enough to avoid their depletion by chemical reactions, many of which involve H_2O. A severe test of this scenario is that it must apply consistently to all molecules contained in these rising columns. Among the equilibrium species predicted at the bottom of this cycle are PH_3, GeH_4, H_2O, CO, H_2S, and HF (Barshay & Lewis 1978), all of which have strong IR bands in one of Jupiter's atmospheric windows. The first four molecules have already been detected, and recent observational work includes searches for the remaining two. Thus, the equilibrium model has withstood the onslaught of the many molecules recently detected on Jupiter, although speculation still accompanies some aspects of its revisions at upper atmospheric levels. The present concensus is that, based upon its success with Jupiter, the global atmospheric chemistry of each of the outer planets can be modeled from thermochemical equilibrium processes applied to a solar distribution of the elements.

It is not easy to proceed from the detection of a molecule to an absolute measure of its abundance. The Jovian CH_4 abundance alone has been the subject of extensive research (see summary in Ridgway et al. 1976a). In spite of great improvements in the quality of the spectral data, the derived temperatures, pressures, and abundances of Jovian CH_4 still show large uncertainties and scatter. Table 4 lists representative measurements of the

Table 3 Composition of the jovian atmosphere

Constituent	Abundance relative to H_2	Region where detected (μm)	Reference
H_2	1	0.8	Kiess et al. 1960
HD	1×10^{-4}	0.7	Trauger et al. 1977
He	0.12	–	Carlson & Judge 1977
CH_4	2×10^{-3}	0.7	Wildt 1932
CH_3D	1.5×10^{-7}	5.0	Beer & Taylor 1973
$^{13}CH_4$	$^{12}C/^{13}C \sim 110 \pm 35$	1.1	Fox et al. 1972
NH_3	2×10^{-4}	0.7	Wildt 1932
$^{15}NH_3$	$^{15}N/^{14}N \sim 3 \times 10^{-3}$	10.0	Encrenaz et al. 1978
H_2O	1×10^{-6}	5.0	Larson et al. 1975
PH_3	$\sim 10^{-6}$	5.0, 10.0	Ridgway et al. 1976b (10μ) Larson et al. 1977 (5μ)
GeH_4	6×10^{-10}	5.0	Fink et al. 1978
CO	$\sim 10^{-9}$	5.0	Beer 1975, Larson et al. 1978
C_2H_2	5×10^{-8}	10.0	Ridgway 1974
C_2H_6	3×10^{-5}	10.0	Ridgway 1974

H_2, CH_4, and NH_3 abundances on Jupiter. This short compilation is sufficient to reveal the large differences between independent measurements of a given atmospheric constituent. One generalization does apply to these determinations, however. Abundances derived from absorption bands at visible wavelengths are consistently higher than values extracted from bands at near-IR wavelengths. These differences are explained by two serious interpretive problems: inadequacy of laboratory measurements for such spectral analyses, and uncertainty concerning the mechanism of spectral line formation in the planetary atmosphere. Laboratory data cannot simulate empirically all conditions (temperature, pressure, broadening agents, scattering mechanisms, etc.) characterizing a real planetary atmosphere. Consequently, direct comparison of laboratory and planetary spectra is inappropriate even when simplifying approximations are used, such as adopting a reflecting layer model rather than a more complex multiple scattering regime. To further complicate matters, the spectrum of CH_4 itself is poorly understood. Only a few bands, such as $3v_3$ at 1.1 μm, have been analyzed, but recent long path laboratory studies of CH_4 by Fink & Larson (1979b) indicate that even this widely used band may be unreliable for quantitative measurements.

The formation of spectral lines in planetary atmospheres most certainly involves scattering phenomena. In only a vaguely understood way, scattering in Jupiter's atmosphere is more effective at short wavelengths. The result is that bands "see" longer effective absorbing paths than at longer wavelengths, where the appropriate model may be closer to simple transmission to some opaque layer, such as a cloud top, then reflection back to the top of the atmosphere. There is not available for quantitative use, however, a Jovian atmospheric model that consistently accounts for observed abundances using bands of different strengths formed at different wavelengths.

In view of these problems, recent interpretative work is directed to

Table 4 Abundance measurements on Jupiter

Constituent	Abundance	Spectral region (μm)	Reference
H_2	56^{+12}_{-6} km-amagat	1.11	Lecacheux et al. 1976
	17^{+6}_{-3}	1.24	de Bergh et al. 1977
	56	0.82, 0.64	Margolis & Hunt 1973
CH_4	52 ± 6 m-amagat	1.10	de Bergh et al. 1976
	150	0.50	Lutz et al. 1976
NH_3	3 ± 1	1.56	Owen et al. 1976
	13 ± 3	0.64, 1.08	Mason 1970
	13 ± 3	0.64	Encrenaz et al. 1974

Table 5 Elemental abundance ratios on Jupiter

Element ratio	Jovian ratio	Terrestrial or solar ratio
C/H	6.2×10^{-4}	4.6×10^{-4}
D/H	$<2.3 \times 10^{-5}$	1.5×10^{-4}
N/H	$<7 \times 10^{-5}$	$9.8^{+2.5}_{-2.0} \times 10^{-4}$
$^{12}C/^{13}C$	89^{+12}_{-10}	90 ± 15
$^{14}N/^{15}N$	270	270
C/N	19^{+13}_{-7}	$4.8^{+2.8}_{-1.8}$

improving the quality of laboratory measurements, to refining models of spectral line formation, and most important, to seeking ways to measure abundances that are model-independent. The last consideration is especially important when determining elemental abundance ratios from observed gaseous constituents. Combes & Encrenaz (1979) presented rational criteria for selecting spectral lines for use in abundance ratio calculations. Their approach assumes that a judicious choice of lines will lead to abundance measurements that will be independent of the scattering mechanisms actually present in spectral line formation. Table 5 lists their elemental abundance ratios for Jupiter. Internal consistency should be high since their criteria were uniformly imposed, but some significant differences still remain between their measurements and others. Combes & Encrenaz review these various conflicting results and they suggest possible remaining interpretive problems. It is their opinion that Jupiter's observed elemental and isotopic composition, as summarized in Table 5, is the same as that of the primordial solar nebula.

Saturn

Saturn's H_2 and CH_4 constituents produce a series of isolated transmission windows throughout the IR that are similar to Jupiter's (see Figures 1 and 2). Absorption by gaseous NH_3 on Saturn is much less obvious, compared to Jupiter, due to vapor pressure saturation at Saturn's lower atmospheric temperatures. The production of moderate and high resolution spectra of Saturn is more difficult, and much of this planet's IR spectrum is only now being explored spectroscopically. Reflected solar flux levels at near-IR wavelengths are lower due to distance considerations, and thermal emission is less due to lower atmospheric temperatures. In observations of Saturn most existing spectroscopic instruments are working close to some fundamental performance limit, such as detector noise in the near IR, or background thermal noise at longer wavelengths. These problems are compounded by the generally poor placement of Saturn's expected

Table 6 Infrared spectroscopic observations of Saturn

Spectral region (cm^{-1})	Resolution (cm^{-1})	Instrument/Telescope[a]	Reference
6000–11,800	0.2	FTS/Palomar 5 m	Lecacheux et al. 1976
4000–6000	0.5	FTS/Hawaii 2.2 m	Martin et al. 1976
2800–6200	0.5	FTS/KAO 0.9 m	Larson et al. 1980
1700–6200	0.25	FTS/UAO 1.5 m	Larson et al. 1980
740–1300	~15	CVF/UCSD-Minn. 1.5 m	Gillett & Forrest 1974
910–1000	~4	Grating/Lick 3 m	Bregman et al. 1975
900–1000	2.5	FTS/ESO 1.5 m	Encrenaz et al. 1975
780–860	1	FTS/SAO 1.5 m	Tokunaga et al. 1975
400–590	22	FTS/SAO 1.5 m	Tokunaga et al. 1977
100–450	5	FTS/KAO 0.9 m	Erickson et al. 1978

[a] See Table 2 for key to abbreviations.

windows with respect to terrestrial windows. While high altitude sites such as aircraft can minimize terrestrial obscuration, the limited observing time available and adverse environmental factors such as vibrations severely limit the spectral resolution and signal-to-noise ratios that can be achieved on the fainter objects in the outer solar system.

Table 6 lists the IR spectroscopic observations of Saturn. As for Jupiter, Saturn has been extensively observed in the near IR ($\lambda < 2\ \mu$m), but to emphasize recent work at longer wavelengths, Table 6 identifies only the highest resolution observations in this spectral region. Compared to the entries in Table 2 for Jupiter, knowledge of Saturn's IR spectrum for $\lambda > 2\ \mu$m is based upon fewer measurements, most of which are at lower spectral resolution. Each of Saturn's major atmospheric transmission windows has been observed at least once, however. The 3 μm region of Saturn's spectrum promises to be especially interesting. As previously noted in the discussion of Figure 2, this region cannot be studied on Jupiter because of NH_3 absorptions. As a consequence, the broader spectral bandwidth of Saturn's 2.7 μm window provides unique opportunities for spectral analyses of the outer planets. The initial results reported by Larson et al. (1980) are summarized in Figure 5. A portion of Saturn's high altitude

→

Figure 5 Comparison of high resolution observations of Saturn in the 3 μm region from a conventional telescope and from the Kuiper Airborne Observatory. These are extracts from the spectral composite in Figure 2. The comparison objects in curves (*a*) and (*f*) illustrate the vast differences in the telluric spectrum at the two observing sites. The ground-based spectrum of Saturn (curve *e*) is indistinguishable from the telluric spectrum, whereas the high altitude planetary observations (curve *c*) reveal numerous absorptions originating in its atmosphere. Phosphine (PH_3, curve *d*) accounts for many features of Saturn's spectrum, but NH_3, which is a known constituent of Saturn's atmosphere, produces no absorption at the position of the very strong v_1 fundamental band (curve *b*). Figure reproduced from Larson et al. (1980).

INFRARED PLANETARY ASTRONOMY 59

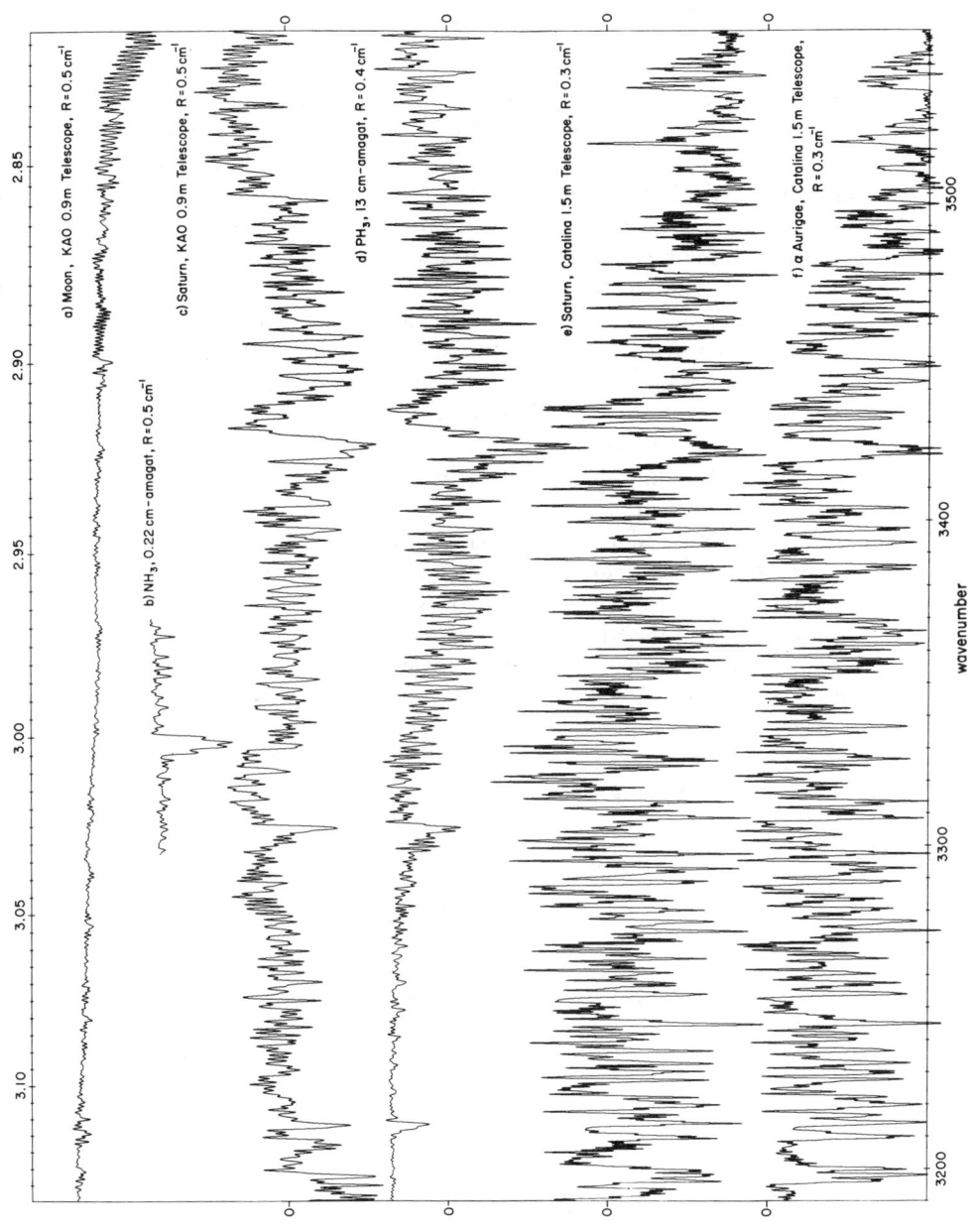

spectrum near 3 μm is reproduced at its full 0.5 cm^{-1} resolution (spectrum c). The lunar comparison spectrum (curve a) at the top of Figure 5 demonstrates that no significant terrestrial absorptions affect interpretation of this region of Saturn's spectrum; therefore, the complex bands observed in Saturn's high altitude spectrum must originate in its atmosphere. A laboratory spectrum of PH_3 (curve d) accounts for many of Saturn's spectral features. Since this is a relatively weak overtone or combination band of PH_3, its presence on Saturn implies a large PH_3 abundance. Larson et al. estimate the P/H ratio on Saturn to be about equal to the solar value, in agreement with analyses at longer wavelengths (Gillett & Forrest 1974, Bregman et al. 1975).

The short segment of Saturn's high altitude spectrum in Figure 5 is informative in other ways. For example, the absorption band in the 3.07 μm region of Saturn's spectrum has been tentatively identified by G. Bjoraker et al. (1980, in preparation) as C_2H_6. On the other hand, no evidence is seen in Figure 5 for absorption by either gaseous or solid NH_3. Curve (b) in Figure 5 is a low abundance laboratory spectrum of NH_3 that Larson et al. used to establish an upper limit to NH_3 of 0.11 cm-amagat. This is compared to the detected amount of NH_3 on Saturn of 200 cm-amagat at 0.65 μm (Encrenaz et al. 1974). Results such as these significantly challenge existing physical and chemical models of Saturn's atmosphere. Figure 5 also contains ground-based observations of Saturn (curve e). The stellar comparison spectrum (curve f) reveals, however, that terrestrial H_2O so dominates Saturn's spectrum that no interpretation is possible.

In a more general context, the interpretations of IR observations of Saturn are proceeding along similar lines as for Jupiter. Much progress is being made in cataloguing Saturn's atmospheric constituents; these are

Table 7 Composition of Saturn's atmosphere

Constituent	Abundance (cm-amag)	Region where detected (μm)	Reference
H_2	5.7×10^6	0.8	Münch & Spinrad 1962
HD	D/H $\sim 5 \times 10^{-5}$	0.7	Trauger et al. 1977
CH_4	1.5×10^4	1.0	Wildt 1932
CH_3D	2.6	5	Fink & Larson 1978
$^{13}CH_4$	$^{12}C/^{13}C \sim 89$	1.1	Combes et al. 1977
NH_3	200	0.65	Encrenaz et al. 1974
NH_3	<0.11	3	Larson et al. 1980
PH_3	6–34	3, 5, 10	Larson et al. 1980, Bregman et al. 1975
C_2H_6	~ 3	3, 10	Bjoraker et al. 1980, in preparation Tokunaga et al. 1975

listed in Table 7. All molecules detected on Saturn have also been found on Jupiter. Predictions of Saturn's atmospheric composition are less detailed than for Jupiter, but a thermochemical equilibrium model based upon a solar distribution of the elements seems a reasonable initial hypothesis. Thus, an atmosphere containing H_2, He, CH_4, and NH_3, and their isotopic species, accounts for most of the entries in Table 7. The wide disparity in the observed NH_3 abundances on Saturn constitutes an interesting modeling problem, however. As still is the case with Jupiter, a major interpretive goal is to explain these differences with a comprehensive model of spectral line formation over an extended spectral region that includes saturation and scattering phenomena.

The presence of C_2H_6 on Saturn seems well established with two independent detections. The most likely production mechanism of C_2H_6 is CH_4 photolysis, as on Jupiter.

Similarly, the detection of abundant PH_3 on Saturn, now confirmed in three atmospheric windows, parallels the situation for Jupiter. It is less obvious, however, that the mechanism responsible for the Jovian PH_3, rapid vertical convection, is appropriate for Saturn. Since all measurements of Saturn's PH_3 abundance indicate at least a solar mixing ratio relative to H_2, Saturn's upper atmosphere may be a chemical sink for phosphorous, a rather different picture than Jupiter's dynamic meteorology. Due to the very recent availability of the observations themselves, however, none of these possibilities has yet been carefully studied.

Uranus, Neptune, Titan, and Triton

These are the remaining bodies in the outer solar system known to possess atmospheres. Few IR spectroscopic observations are available; all represent unique, pioneering efforts to provide initial characterizations of these atmospheres.

A CH_4 atmosphere was recently reported on Triton (Cruikshank & Silvaggio 1979). It is described as very tenuous, with a CH_4 partial pressure (1×10^{-4} bar) controlled by solid CH_4 deposits on Triton's cold (~ 58 K) surface. This contrasts with the more complex atmospheres on Uranus, Neptune, and Titan where clouds, aerosols, and larger CH_4 abundances are present.

The spectral composite in Figure 1, from Fink & Larson (1979b), constitutes the only near-IR ($\lambda < 2.5\,\mu\text{m}$) data suitable for comparative spectroscopic analyses of Uranus, Neptune, and Titan. With the inclusion of spectra of Saturn, Jupiter, and CH_4, Figure 1 provides by inspection alone numerous insights into the characteristics of these atmospheres. Each spectrum is unique, but all are dominated by strong CH_4 absorptions, as seen upon comparing the planetary spectra with the long path laboratory spectrum of CH_4. This composite is especially effective in revealing the different scattering properties of these atmospheres. Features diagnostic of scattering effects in the formation of the planetary spectral lines include the more gradual slopes to the sides of the CH_4 windows, as in the spectrum of Neptune in Figure 1, and residual intensity at the bottoms of the strong CH_4 bands. The spectrum of Uranus in Figure 1 most closely resembles the laboratory CH_4 spectrum. This indicates a deep, clear planetary atmosphere, further supported by the absence on Uranus of any transmission in

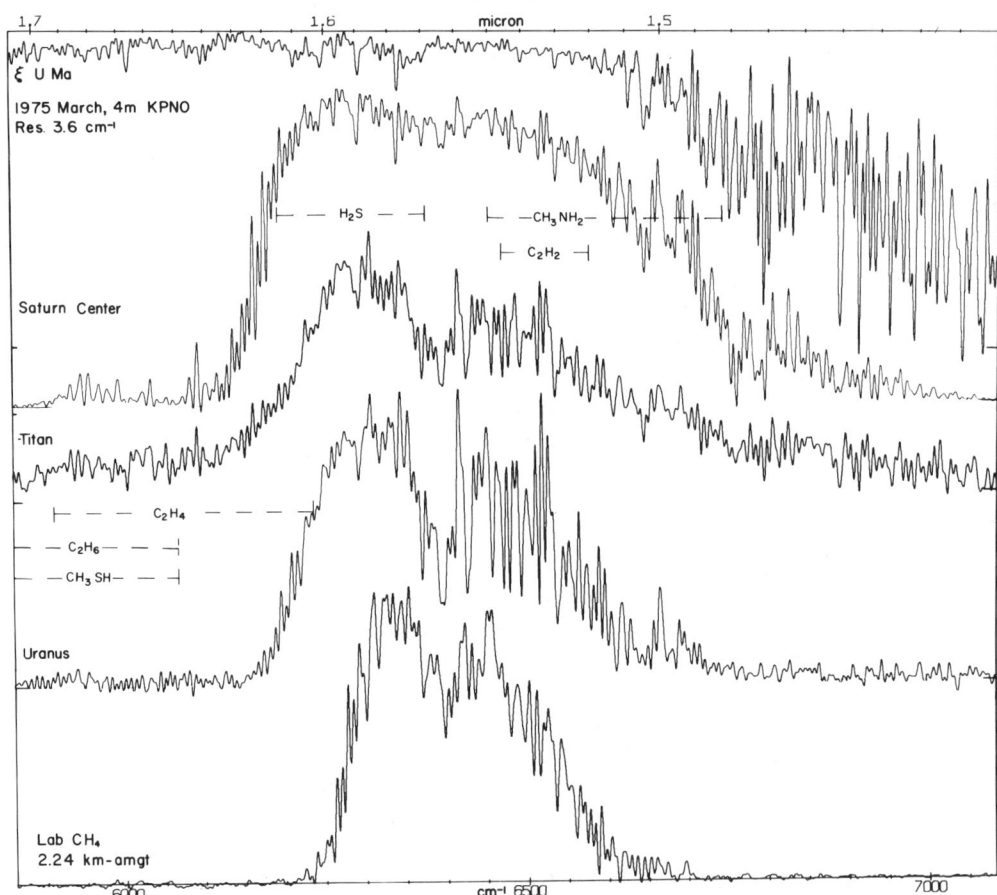

Figure 6 Illustration of the high resolution spectral detail in the 6500 cm^{-1} CH_4 window in the atmospheres of Saturn, Uranus, and Titan (see Figure 1 for overview). The telluric absorptions are identified in the spectrum of the solar-type star, and the planetary absorptions are all associated with the spectrum of CH_4 (bottom curve). The locations of several trace constituents are indicated, none of which have been detected in this window. Figure reproduced from Fink & Larson (1979b).

the 2 μm region, a consequence of very strong absorption by the 1–0 pressure-induced dipole spectrum of H_2.

In addition to providing this overview of the near-IR spectra of outer solar system objects, the data in Figure 1 can be presented in other ways to emphasize the unique attributes of each atmosphere. For example, Figure 1 does not reveal the full spectral resolution of these observations. The 1.6 μm transmission window is expanded in Figure 6 to show all spectral detail. These high resolution spectra permit searches for trace constituents, and they reveal more fully the CH_4 features, which are developed more strongly on Uranus and Titan than on Saturn or in the long path laboratory spectrum. The locations of overtone and combination bands of several molecules compatible with H_2-rich atmospheres are indicated in Figure 6. In their spectral analyses, however, Fink & Larson found no evidence for atmospheric constituents on these objects other than CH_4 and, except for Titan, H_2.

Another way of presenting the observations in Figure 1 is to convert them to absolute reflectivities, or albedos. The outer planet albedo curves from Fink & Larson are summarized in Figure 7. The distinctly different appearances of these curves emphasizes the individuality of these objects. Upon combining these various spectral plots with other lines of reasoning, Fink & Larson describe each of these atmospheres as follows.

Saturn: Its albedo peaks at 0.7 μm, decreasing both to the UV and IR. Methane absorption lowers the IR albedo, while aerosols reduce the UV albedo. The degree of scattering and aerosol absorption on Saturn is the lowest among the outer planets. A distinct cloud deck limits penetration of visible and near-IR radiation; therefore, a reflecting layer model may be an appropriate choice for near-IR spectral analyses.

Titan: The shape of Titan's albedo curve resembles Saturn's, implying similar mechanisms. Since the albedo is only about half that of Saturn, however, the aerosol absorption and scattering mechanism are more effective. No evidence for H_2 is seen.

Uranus: Its high visible albedo drops off rapidly at IR wavelengths due to strong CH_4 absorption. The most consistent atmospheric model is Rayleigh scattering in a deep, clear atmosphere depleted of CH_4 in its upper levels. More H_2 is seen on Uranus than on any other planet.

Neptune: Its albedo is similar to that of Uranus, but particulate scattering is more consistent with the character of the CH_4 windows. A high cloud layer (0.7 atm level) is indicated, possibly variable in height.

The interpretation of these near-IR spectra of the outer planets has required new laboratory studies of CH_4 at very high abundances. These laboratory data reveal spectral behavior that, according to Fink & Larson's

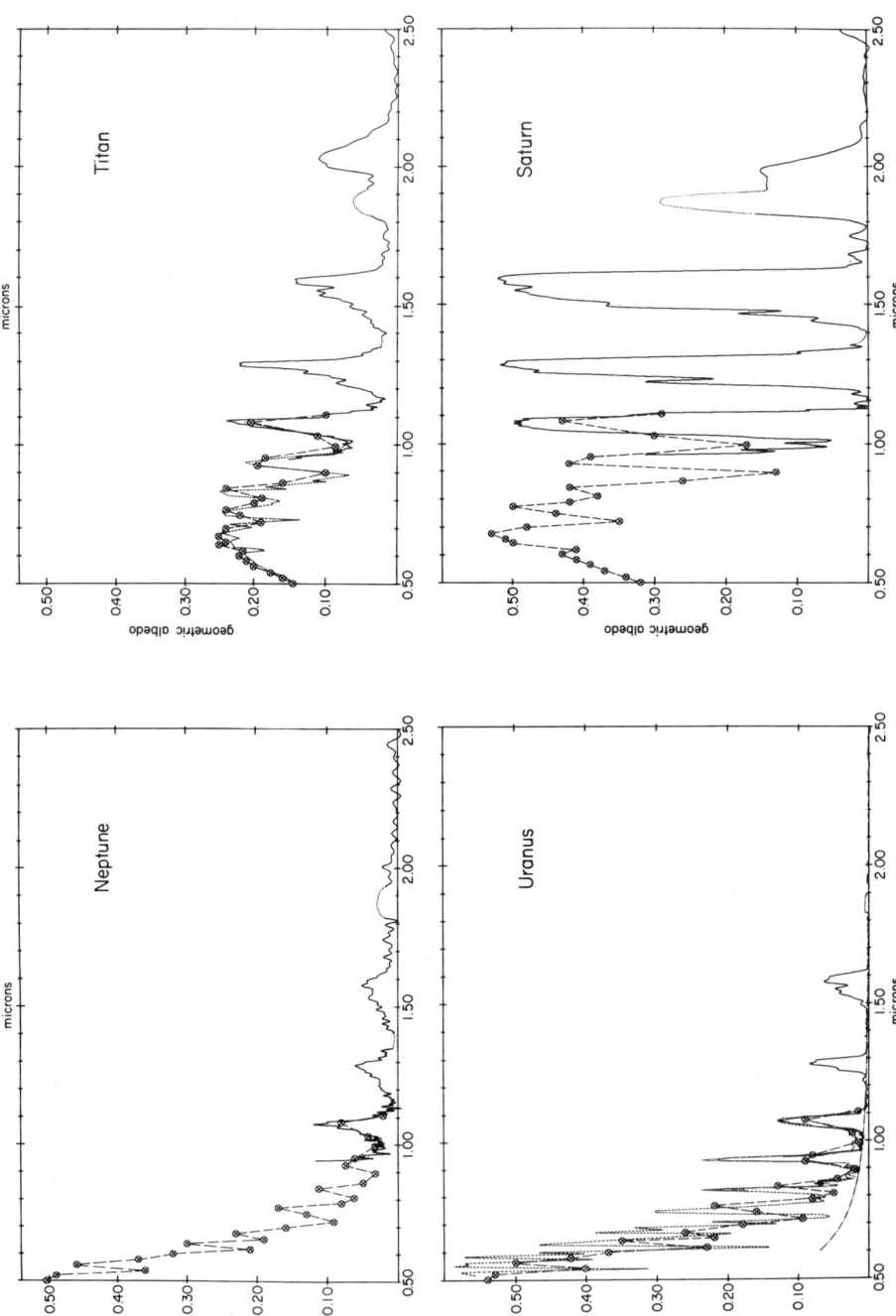

Figure 7 The geometric albedos of Uranus, Neptune, Saturn, and Titan. The IR sections of these curves were derived from the spectra in Figure 1, and the visible portions were drawn from other published measurements. The different characteristics of these curves reveal clearly the different scattering properties of each of these atmospheres. Figure reproduced from Fink & Larson (1979b).

Table 8 Abundances of H_2 and CH_4 in the outer solar system

Abundance	Jupiter	Saturn	Titan	Uranus	Neptune
CH_4 (IR bands, km-am)	0.12	0.16	1.06	1.6	0.7
CH_4 (visible bands, km-am)	0.17	0.14	0.14	7.4	8.8
H_2 (km-am)	70	70	—	400	400
C/H_2	1.6×10^{-3}	2.3×10^{-3}	—	4×10^{-3}	1.8×10^{-3}

analysis, precludes use of those CH_4 bands ($3\nu_3$, for example) that have previously been used in analyses of planetary spectra. They propose, alternatively, the use of two other near-IR CH_4 bands (6400, 7900 cm^{-1}) for abundance determinations. Table 8 summarizes the CH_4 abundances that Fink & Larson derived from the spectra in Figure 1 and their long path CH_4 comparison spectra. The internal consistency of these measurements should be high, which strengthens the authors' conclusions that these atmospheres all have approximately the same C/H value (2×10^{-3}), which is enhanced over the solar value (0.75×10^{-3}). Recall, however, that Combes & Encrenaz (1979) also determined abundance ratios in the outer planets on an internally consistent basis, and that they did not find an enhancement in the C/H, or in any other, ratio.

The entries in Table 8 provide other interesting comparisons with previous work. They do not show for Jupiter and Saturn the large differences between abundance measurements at visible and near-IR wavelengths. As previously noted in the discussion of the extensive interpretive history of Jupiter and Saturn, and summarized in Table 4, these discrepancies have affected the credibility of all analyses. Thus, the agreement for these particular entries in Table 8 is an encouraging sign that a more realistic interpretive framework is being generated for these two planets. Abundance measurements on Neptune and Titan, however, display large differences between visible and near-IR wavelengths. Scattering effects are presumably responsible for these differences. Certainly, scattering is evident, and different, for each of these objects (see Figure 1). Fink & Larson believe that their near-IR measurements are less sensitive to these processes, and that their abundances are closer to one-way transmission values.

It is likely that only atmospheric entry probes will establish the true basis for the formation of spectral lines in these atmospheres. The current availability of new observational data, however, has forced a stimulating reassessment of the analytic techniques used for modeling planetary atmospheres with remote spectroscopic observations.

Table 9 Infrared spectroscopic studies of surfaces in the outer solar system

Object	Spectral range (μm)	Spectral resolution (cm^{-1})	Instrument/Telescope[a]	Reference
Galilean satellites	1.25–4.0	120	FTS/KPNO 1.5 m	Pilcher et al. 1972
Galilean satellites	0.8–4.0	25	FTS/UAO 1.5 m	Fink et al. 1973
Galilean satellites	0.7–5.5	~90	CVF/KAO 0.9 m	Pollack et al. 1978
Galilean satellites	0.65–2.5	~90	CVF/Hawaii 2.2 m	Clark & McCord 1979
Io	0.9–2.7	3.4	FTS/KPNO 4 m, UAO 2.3 m	Fink et al. 1976a
Io	2.8–4.2	~33	CVF/Hawaii 2.2 m	Cruikshank et al. 1978
Io	1.25–5.4	~60	CVF/UAO 1.5 m	Witteborn et al. 1979
Saturn's satellites (S3, S4, S5, S8)	0.8–2.5	50	FTS/UAO 2.3 m	Fink et al. 1976b
Asteroids (Nos. 1, 2, 4, 5, 8, 12, 18, 29, 39, 324, 349, 433)	0.8–2.5	25	FTS/UAO 1.5, 2.3 m	Larson & Veeder 1979

[a] See Table 2 for key to abbreviations.

Beyond 2.5 μm, IR spectroscopic observations exist only for Titan. In the 8–13 μm region the definitive spectrum remains that of Gillett (1975). Very recent observations from the KAO extend the wavelength coverage to 16–30 μm (McCarthy et al. 1979). These results have established a thermal inversion model of Titan's atmosphere that contains CH_4 and simple gaseous hydrocarbons (C_2H_6, possibly C_2H_4, C_2H_2) produced by CH_4 photolysis.

III SURFACES OF SATELLITES AND ASTEROIDS

Infrared spectroscopic studies of surfaces in the outer solar system have developed more slowly than atmospheric studies. A major consideration has been the slowly emerging awareness that these objects have their own unique story to tell about solar system origin and evolution. The stunning images of the Galilean satellites by Voyager have certainly removed all doubt about the individuality of such bodies, but a decade ago when remote IR spectroscopic observations of surfaces were just becoming technically feasible there were legitimate questions about the relevance of studying such small bodies, especially since the faintness of these objects required long integration times on large aperture telescopes to achieve even low spectral resolution. The few groups that persevered in the development of remote mineralogical analyses did establish the legitimacy of such endeavors before Voyager. Not all of this effort is classified as IR spectroscopy, as defined at the beginning of this chapter. The techniques of visible spectrophotometry and broad- and narrow-band photometry have been very influential in creating initially an interpretive framework for all remote spectral analyses; the review papers cited in the introduction should be consulted for details. The specific contributions of IR spectroscopic studies of surfaces are summarized in Table 9. The following material illustrates the nature of some of the observational results and the conclusions that can be drawn from them.

The most abundant materials expected on the surfaces of objects in the outer solar system are divided into two groups: condensed volatiles, or ices, and various silicate minerals. The spectral signatures of these two groups are distinctly different at IR wavelengths. The composite of reflection spectra of ices in Figure 8 illustrates the diagnostic features of this group of solids. Water ice has been found on numerous surfaces (Europa, Ganymede, Tethys, Dione, Rhea, and Iapetus) and on Saturn's rings. Solid CO_2 has been established as a component of the Martian polar caps, and solid CH_4 has been detected on Pluto.

The IR reflection spectra of three common rock-forming silicate minerals are illustrated in Figure 9. Compared to the ice spectra in Figure 8, these

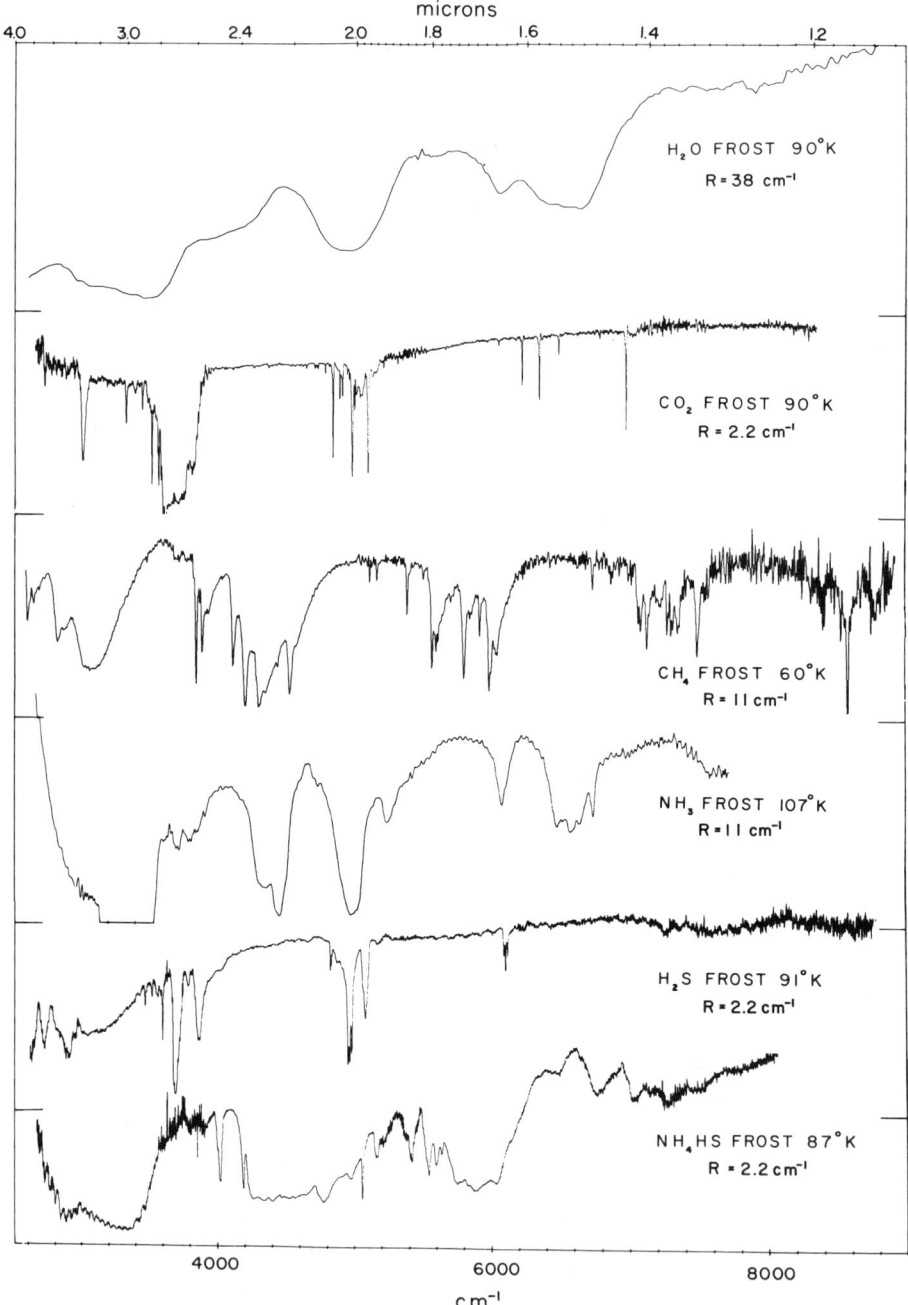

Figure 8 Representative spectra of frosts in the near-IR. All of these solids are cosmochemically acceptable candidates for surface constituents on outer solar system bodies. Figure reproduced from Larson & Fink (1977).

mineral features are broader and more subdued, although they are still uniquely different. Each of the three minerals in Figure 9 has been identified on asteroid surfaces. Additional mineral groups are also cosmochemically acceptable candidates for surface constituents on outer solar system objects. These include hydrated silicates which show strong, diagnostic bands of H_2O, and spectrally featureless minerals such as FeS, Ni-Fe, and carbon.

The appearance of these various mineral spectra on extraterrestrial bodies is illustrated with the two data sets in Figures 10 and 11. The IR spectral reflectivities of four of Saturn's satellites are displayed in Figure 10. All show evidence of H_2O ice as a major surface mineral. This identification is readily apparent upon comparison of the satellite spectra with reflection spectra of Saturn's rings and laboratory H_2O frosts.

The spectral reflectivities of six asteroids are presented in Figure 11. Compared to the icy satellites, these surfaces exhibit more variety in their spectral features, indicating very different surface mineralogies. The surfaces of Ceres and Pallas, two of the largest asteroids, are characterized by low albedo, spectrally featureless IR reflectances that are considered mineralogically similar to primitive carbonaceous chondrite mixtures (hydrated silicates, opaques). The middle pair of spectra in Figure 11 (Victoria and Eros) represent a broad compositional class loosely described as stony-irons, meaning mixtures of elemental Ni-Fe and high temperature

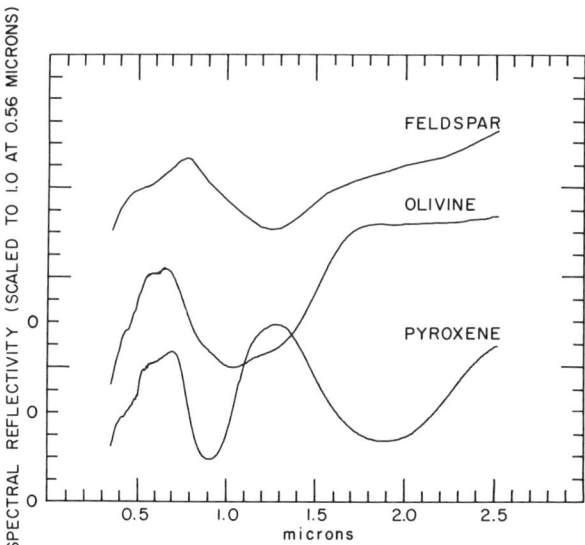

Figure 9 Representative near-IR spectra of three rock-forming mineral groups. Each of these minerals has been identified on various asteroid surfaces by means of remote spectroscopic observations. Figure reproduced from Larson & Fink (1977).

silicates (pyroxenes, olivines) as are found in some types of meteorites. The two spectra at the top of Figure 11 are spectrally unique among available asteroid data. The pyroxene group of minerals is especially obvious with its pair of absorption bands at 1 and 2 μm (compare with the laboratory spectra of silicates in Figure 9). These two objects are similar spectrally to achondritic meteorite assemblages, implying an igneous differentiation episode in their evolution.

These two examples of observational data emphasize the need for broad spectral bandwidth in surface studies. The broad, subdued diagnostic spectral features of most minerals cannot be convincingly revealed by low information (low resolution, restricted spectral range) methods. Due to the

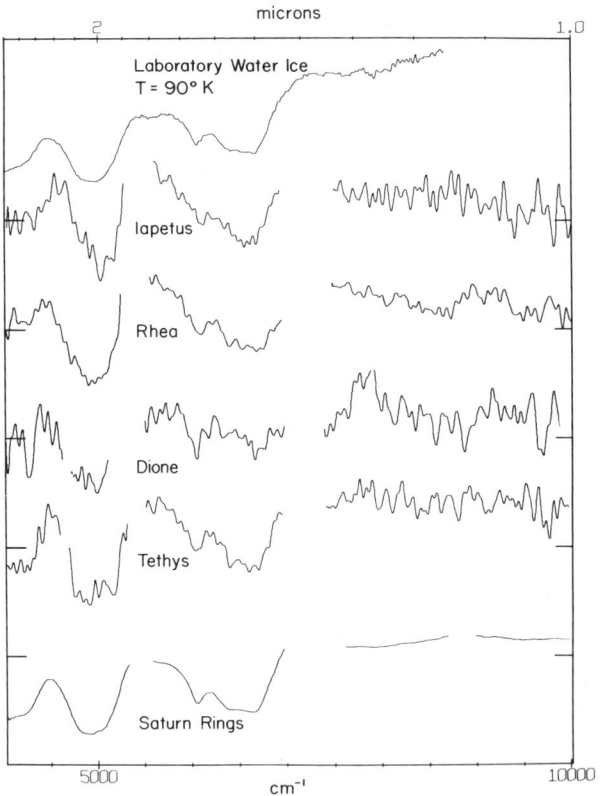

Figure 10 Infrared spectral reflectances of the surfaces of four of Saturn's satellites. The presence of H_2O frost on each of these bodies is apparent upon comparison of their spectra with spectra of Saturn's rings and laboratory H_2O frosts. The spectral composites in Figures 8 and 9 help establish that no other mineral can account for the broad spectral features observed on the satellites. Figure reproduced from Fink et al. (1976b).

INFRARED PLANETARY ASTRONOMY 71

faintness of most of the asteroids and satellites, however, high resolution, broad-band spectroscopic observations are not always feasible. For certain special situations narrow-band photometric observations can provide mineralogically significant data. If the passbands of narrow filters are appropriately located with respect to the characteristic absorption feature

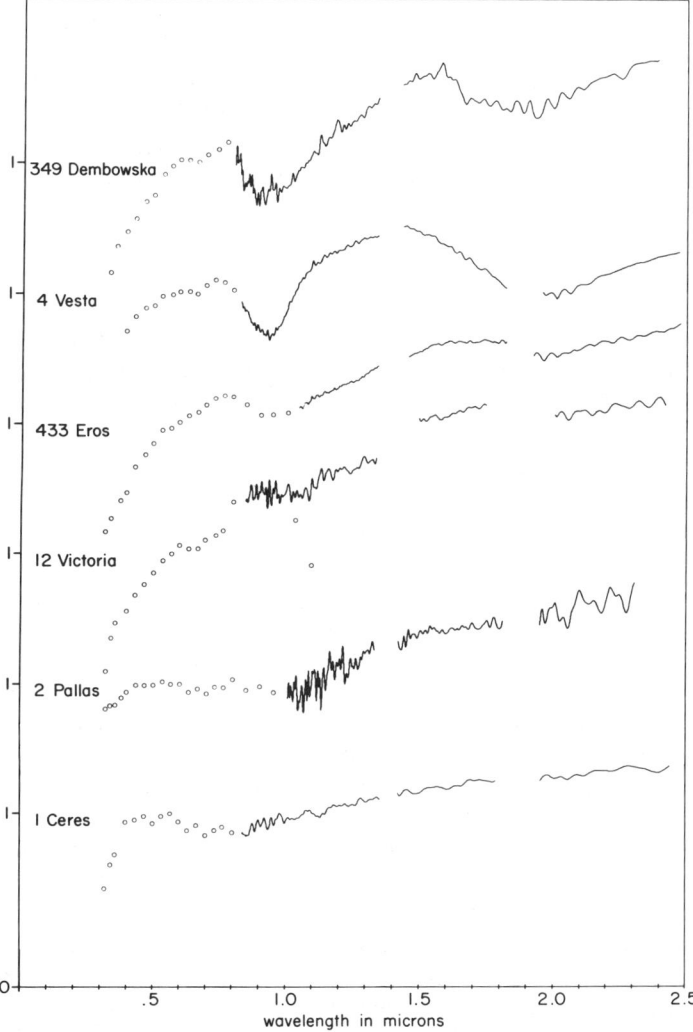

Figure 11 Infrared spectral reflectances of the surfaces of six mineralogically distinct asteroids. Measurements at visible wavelengths have been appended to the IR data to increase their interpretive value. Some of the minerals in Figure 9, particularly the pyroxenes, are especially evident in the asteroidal spectra. Figure reproduced from Larson & Veeder (1979).

of a particular mineral, relative flux measurements may support detection of that material. This procedure has been effectively applied to asteroids (hydrated silicates, Lebofsky 1978) and Pluto (solid CH_4, Cruikshank et al. 1976).

The above discussion emphasizes that the primary use of IR reflectance measurements of surfaces is to provide compositional information. As with planetary atmospheres, however, the next step is to compare the detected constituents to predictions based upon cosmochemical modeling of the primordial solar nebula. The solar nebula condensate beyond the asteroid belt is expected to be rich in ices (see review by Lewis 1973). The densities and compositions of the satellites of the outer planets thus provide important observational constraints to the low temperature condensation-accretion process. The spectroscopic detection of H_2O on numerous satellites verifies that this mineral group contributed substantially to the formation of planetesimals in the outer solar system. The subsequent evolution of rocky-ice satellites has been extensively modeled (see Consolmagno & Lewis 1978, for an overview). Melting and differentiation of the interior is predicted from heat released by radionuclides in the silicate components. Thus, the satellites as a class have experienced geochemical activity that could lead to substantial individualism, as has been confirmed by near-IR reflectance measurements of the Galilean satellites.

Compositional analyses of asteroid surfaces are potentially very informative. They apply to the transitional region between the terrestrial planets with their highly differentiated structures, and the relatively unevolved giant planets with their hydrogen-rich atmospheres. Moreover, the large population of asteroids contains fragments of collisionally destroyed larger bodies, thus permitting remote spectral observations of interior mineral assemblages. Not surprisingly, however, attempts to combine spectral reflectance measurements into a consistent model of asteroid origin and evolution have not been entirely convincing. While the task is aided by having in hand specimens (meteorites) of extraterrestrial mineral assemblages to study, the interrelated origins of asteroids and meteorites is to be demonstrated, not assumed, in the interpretation of spectral reflectance data.

The following conclusions are currently supported by IR observations. The surfaces of some asteroids are mineralogically compatible with the relatively unevolved low temperature remnant of nebular condensation (carbonaceous chondrite–type assemblage). An overview of this conclusion that emphasizes the role of recent IR reflectance measurements is found in Larson et al. (1979). A few asteroids show surface mineralogies composed of high temperature silicates that are spectrally indistinguishable from products of magmatic differentiation. The IR reflectance measurements that support this interpretation are summarized in Feierberg et al. (1980). Many asteroids, however, fall into some intermediate class that has no well-established meteoritic analogue. This diversity of spectral types indicates that the asteroids experienced considerable evolution, probably driven by intense but short-lived thermal heating episodes early in their evolution. A convincing explanation of why some asteroids appear primitive, while others are highly evolved, is still lacking. The asteroids as a class, however, are among the most promising places to look for samples of the low temperature nebular condensate and for evidence of events that occurred early in the evolution of the solar system.

It is appropriate to conclude this section with a brief discussion of Io. As is evident in Table 9, planetary astronomers have given this innermost satellite of Jupiter special attention. Its near-IR reflectance is not unusual, but other types of observations of this satellite have revealed strange behavior. The observed phenomena include post-eclipse brightening at visible wavelengths, anomalous thermal flux levels at mid-IR wavelengths, emission from atomic sodium, and narrow absorption features near 4.0 μm. No consistent explanation of all of these effects existed before Voyager. The Voyager experiments revealed, however, volcanic activity and a sulfur-based surface chemistry that completely changes the perspective for interpreting remote observations of this satellite. It may still not be possible to reconcile all of the observed phenomena, and a review at this time would certainly be premature. A useful summary of the present status of speculations concerning Io is available in the special *Io* issue of *Nature* (Vol. 250, No. 5725, 1979). The startling awareness that volcanism is an active geochemical force on a small satellite is a very important reminder that the outer solar system may contain still other exotic surprises, and that both the very small and the very large objects must be examined by all imaginative means to search for them.

IV SUMMARY

Imagery and spectroscopy are the two complementary, high information content observational techniques that will continue to provide important first-time looks at outer solar system bodies. For earth-based IR spectroscopic methods, this means continued efforts to achieve higher spectral resolution on Jupiter and Saturn at mid-IR wavelengths where the potential for new discoveries is still very high. In the near-IR, however, higher spectral resolution as an observational goal may be less productive than increased efforts to interpret what data already exist, particularly with respect to understanding the CH_4 spectrum. The trends for surface studies include increasing the number of objects observed and extending measurements to wavelengths beyond 2.5 µm. Many of these efforts already make extensive use of special high altitude observing sites such as the Kuiper Airborne Observatory, and this dependence will increase as earth-based opportunities are more thoroughly exploited. A natural extension of these studies will be to use an earth-orbiting, cryogenically-cooled ($T < 20$ K) telescope such as the SIRTF. The unprecedented sensitivity of such a telescope for $3 < \lambda < 1000$ µm will permit first-time spectroscopic observations of many solar system bodies.

ACKNOWLEDGMENTS

It is a pleasure to thank the many individuals who submitted material for inclusion in this review. This chapter was prepared with the support of NASA Grant NGR 03-002-332.

Literature Cited

Aitken, D. K., Jones, B. 1972. *Nature* 240:230
Aumann, H. H., Orton, G. S. 1976. *Science* 194:107
Aumann, H. H., Orton, G. S. 1978. *Bull. Am. Astron. Soc.* 10:564
Baluteau, J. P., Marten, A., Bussoletti, E., Anderegg, M., Moorwood, A., Beckman, J. E., Coron, N. 1978. *Astron. Astrophys.* 64:61
Barshay, S. T., Lewis, J. S. 1978. *Icarus* 33:593
Beer, R. 1975. *Ap. J. Lett.* 200:L167
Beer, R., Taylor, F. W. 1973. *Ap. J.* 179:309
Bregman, J. D., Lester, D. F., Rank, D. M. 1975. *Ap. J. Lett.* 202:L55
Carlson, R. W., Judge, D. L. 1977. In *Jupiter*, ed. T. Gehrels, p. 418. Tucson: Univ. Arizona Press
Clark, R. N., McCord, T. B. 1979. *Icarus*. In press
Combes, M., Encrenaz, T. 1979. *Icarus* 39:1
Combes, M., Encrenaz, T., Vapillon, L., Zéau, Y. 1974. *Astron. Astrophys.* 34:33
Combes, M., Maillard, J. P., de Bergh, C. 1977. *Astron. Astrophys.* 61:531
Combes, M., Encrenaz, T., Berezne, J., Vapillon, L., Zéau, Y. 1976. *Astron. Astrophys.* 50:287
Connes, J., Connes, P., Maillard, J. P. 1969. *Atlas of Near Infrared Spectra of Venus, Mars, Jupiter and Saturn*. Paris: CNRS
Consolmagno, G. J., Lewis, J. S. 1978. *Icarus* 34:280
Cruikshank, D. P. 1979. *Rev. Geophys. Space Phys.* 17:165
Cruikshank, D. P., Silvaggio, P. M. 1979. *Ap. J.* 233:1016
Cruikshank, D. P., Pilcher, C. B., Morrison, D. 1976. *Science* 194:835
Cruikshank, D. P., Jones, T. J., Pilcher, C. B. 1978. *Ap. J. Lett.* 225:L89
Danielson, R. E. 1966. *Ap. J.* 143:949
de Bergh, C., Maillard, J. P. 1977. See Jones 1977, p. 9
de Bergh, C., Maillard, J. P., Lecacheux, J., Combes, M. 1976. *Astron. Astrophys.* 29:307
de Bergh, C., Lecacheux, J., Maillard, J. P. 1977. *Astron. Astrophys.* 56:227
Encrenaz, T., Owen, T., Woodman, J. H. 1974. *Astron. Astrophys.* 37:49
Encrenaz, T., Combes, M., Zéau, Y., Vapillon, L., Berezne, J. 1975. *Astron. Astrophys.* 42:355
Encrenaz, T., Gautier, D., Michel, G., Zéau, Y., Lecacheux, J., Vapillon, L., Combes, M. 1976. *Icarus* 29:311
Encrenaz, T., Combes, M., Zéau, Y. 1978. *Astron. Astrophys.* 70:29
Encrenaz, T., Combes, M., Zéau, Y. 1979. *Astron. Astrophys.* In press
Erickson, E. F., Goorvitch, D., Simpson, J. P., Strecker, D. W. 1978. *Icarus* 35:61
Feierberg, M. A., Larson, H. P., Fink, U., Smith, H. A. 1980. *Geochim. Cosmochim. Acta.* 44:513
Fink, U., Larson, H. P. 1978. *Science* 201:342
Fink, U., Larson, H. P. 1979a. In *Fourier Transform Infrared Spectroscopy: Applications of Chemical Systems, Vol. 11*, ed. J. R. Ferraro, L. J. Basile, p. 243. New York: Academic
Fink, U., Larson, H. P. 1979b. *Ap. J.* 233:1021

Fink, U., Dekkers, N. H., Larson, H. P. 1973. *Ap. J. Lett.* 179:L155
Fink, U., Larson, H. P., Gautier, T. N. 1976a. *Icarus* 27:439
Fink, U., Larson, H. P., Gautier, T. N., Treffers, R. R. 1976b. *Ap. J. Lett.* 207:L63
Fink, U., Larson, H. P., Treffers, R. R. 1978. *Icarus* 34:344
Fox, K., Owen, T., Mantz, A. W., Rao, K. R. 1972. *Ap. J. Lett.* 176:L81
Furniss, I., Jennings, R., King, K. J. 1976. In *Far Infrared Astronomy*, ed. M. Rowan-Robinson, p. 71. Oxford: Pergamon
Gillett, F. C. 1975. *Ap. J. Lett.* 201:L41
Gillett, F. C., Forrest, W. J. 1974. *Ap. J. Lett.* 187:L37
Gillett, F. C., Low, F. J., Stein, W. A. 1969. *Ap. J.* 157:925
Hanel, R., Conrath, B., Flasar, M., Kunde, V., Lowman, P., Maguire, W., Pearl, J., Pirraglia, J., Samuelson, R., Gautier, D., Gierasch, P., Kumar, S., Ponnamperuma, C. 1979. *Science* 204:972
Houck, J. R., Pollack, J., Schaack, D., Reed, R. A., Summers, A. 1975. *Science* 189:720
Jones, A. V., ed. 1977. *Proc. Symp. Planetary Atmospheres*. Ottawa, Canada
Kiess, C. C., Corliss, C. H., Kiess, H. K. 1960. *Ap. J.* 132:221
Kuiper, G. P. 1947. *Ap. J.* 106:251
Lacy, J. H., Larrabee, A. I., Wollman, E. R., Geballe, T. R., Townes, C. H., Bregman, J. D., Rank, D. M. 1975. *Ap. J. Lett.* 198:L145
Larson, H. P., Fink, U. 1977. *Appl. Spectrosc.* 31:386
Larson, H. P., Veeder, G. J. 1979. In *Asteroids*, ed. T. Gehrels, p. 724. Tucson: Univ. Arizona Press
Larson, H. P., Fink, U., Treffers, R. R., Gautier, T. N. 1975. *Ap. J. Lett.* 197:L137
Larson, H. P., Treffers, R. R., Fink, U. 1977. *Ap. J.* 211:972
Larson, H. P., Fink, U., Treffers, R. R. 1978. *Ap. J.* 219:1684
Larson, H. P., Feierberg, M. A., Fink, U., Smith, H. A. 1979. *Icarus* 39:257
Larson, H. P., Fink, U., Smith, H., David, D. S. 1980. *Ap. J.* Submitted
Lebofsky, L. A. 1978. *MNRAS* 182:17
Lecacheux, J., de Bergh, C., Combes, M., Maillard, J. P. 1976. *Astron. Astrophys.* 53:29

Lewis, J. S. 1969. *Icarus* 10:393
Lewis, J. S. 1973. *Space Sci. Rev.* 14:401
Lutz, B. L., Owen, T., Cess, R. D. 1976. *Ap. J.* 203:541
Margolis, J. S., Hunt, G. E. 1973. *Icarus* 18:593
Martin, T. Z., Cruikshank, D. P., Pilcher, C. B., Sinton, W. M. 1976. *Icarus* 27:391
Mason, H. P. 1970. *Astrophys. Space Sci.* 7:424
McCarthy, J. R., Pollack, J. B., Houck, J. R., Forrest, W. J. 1979. Preprint
Münch, G., Spinrad, H. 1962. *Mem. Soc. R. Sci. Liège* 7:541
Owen, T., McKellar, R., Lecacheux, J., deBergh, C., Encrenaz, T., Maillard, J. P. 1976. *Astron. Astrophys.* 54:291
Pilcher, C. B., Ridgway, S. T., McCord, T. B. 1972. *Science* 178:1087
Pollack, J. B., Witteborn, F. C., Erickson, E. F., Strecker, D. W., Baldwin, B. J., Bunch, T. E. 1978. *Icarus* 36:271
Prinn, R. G., Lewis, J. S. 1975. *Science* 190:274
Prinn, R. G., Owen, T. 1976. In *Jupiter*, ed. T. Gehrels, p. 319. Tucson: Univ. Ariz. Press
Ridgway, S. T. 1974. *Ap. J. Lett.* 187:L41
Ridgway, S. T., Larson, H. P., Fink, U. 1976a. In *Jupiter*, ed. T. Gehrels, p. 348. Tucson: Univ. Ariz. Press
Ridgway, S. T., Wallace, L., Smith, G. R. 1976b. *Ap. J.* 207:1002
Russell, R. W., Soifer, B. T. 1977. *Icarus* 30:282
Soifer, B. T., Pipher, J. L. 1978. *Ann. Rev. Astron. Astrophys.* 16:335
Strobel, D. F. 1974. *Ap. J. Lett.* 192:L47
Tokunaga, A., Knacke, R. F., Owen, T. 1975. *Ap. J. Lett.* 197:L77
Tokunaga, A., Knacke, R. F., Owen, T. 1976. *Ap. J.* 209:294
Tokunaga, A. T., Knacke, R. F., Owen, T. 1977. *Ap. J.* 213:569
Tokunaga, A. T., Knacke, R. F., Ridgway, S. T., Wallace, L. 1979. *Ap. J.* 232:603
Trauger, J. T., Roesler, F. L., Mickelson, M. E. 1977. *Bull. Am. Astron. Soc.* 9:516
Wildt, R. 1932. *Nachr. Ges. Akad. Wiss. Goettingen Math. Phys. Kl. Fachgruppe 2* 1:87
Witteborn, F. C., Bregman, J. D., Pollack, J. B. 1979. *Science* 203:643

Ann. Rev. Astron. Astrophys. 1980. 18 : 77–113

FORMATION OF THE TERRESTRIAL PLANETS

×2163

George W. Wetherill
Department of Terrestrial Magnetism, Carnegie Institution of Washington, Washington, DC 20015

1 INTRODUCTION

In attempting to describe the formation of the terrestrial planets, one immediately faces the problem of where to start. These bodies comprise less than 0.5% of the mass of the planetary system, which in turn is only ~0.1% of the mass of the Sun. Because the formation of the Sun is the principal event involved, it may seem logical to first discuss the formation of the Sun before considering such small details as terrestrial planets. On the other hand, at present the formation of stars of one solar mass ($1\ M_{\odot}$) is poorly understood (Larson 1978). For this reason perhaps it would be preferable to start with the wealth of detailed information available from observation of the Earth, Moon, and other planets. However, the problem of even approximately uniquely inverting these observations to obtain the conditions of planet formation is far beyond our present ability. Until the "forward problems" of planetary growth are much better in hand, requests for the luxury of uniqueness are premature.

For these reasons, a starting point intermediate between these alternatives has been chosen. Regardless of how stars like the Sun form, at some time matter in the form of grains or small bodies must have aggregated to form the terrestrial planets. The possible mechanisms by which this may be accomplished are

1. Gravitational collapse. Matter is aggregated into larger objects under the influence of a collective gravitational field following the development of a gravitational instability.
2. Accumulation. Bodies grow by the gradual addition of matter, following two-body collisions and coherence.

Processes 1 and 2 may occur either with or without the presence of a dynamically significant gas phase.

0066-4146/80/0915-0077$01.00

The assemblage of these smaller bodies and gas in the form of a flattened rotating disk is generally referred to as the solar nebula. If the term is defined this broadly, all current theories involve a solar nebula of some kind.

In this review emphasis will be placed on the present state of our understanding of the ways in which processes 1 and 2 operate, with special attention given to their relevance to the formation of the terrestrial planets. It should not be assumed that these processes are mutually exclusive. It is just as likely that both were important at different stages of planet formation. Choice of the most probable chain of events whereby the planets were actually formed will not be possible before we understand these processes well enough to permit prediction of the properties of the resulting planets.

This approach permits starting at some intermediate stage with a collection of matter that shows some promise of being capable of evolving into the present terrestrial planets, and following the evolution of that matter by the mechanisms described above. It is necessary but not sufficient to show that this intermediate stage can indeed lead to the observed planets. It is also necessary to show that the intermediate stage arises naturally as a by-product of the formation of the Sun. Moreover, the same theory of solar formation must provide appropriate starting points for the formation of the giant planets, asteroids, comets, and satellite systems. This requirement will become more demanding as our knowledge of star formation improves.

1.1 *General Observational and Theoretical Background*

Astronomical observation (Strom, Strom & Grasdalen 1975) shows that star formation is occurring in the Galaxy in regions, such as the Orion nebula, in which hydrogen is in the molecular form, and in which the density of gas and dust is 10^3–10^6 times greater than the average interstellar density. Most observational data on newly formed stars has been obtained from the more massive O and B stars in these regions. However, pre–main-sequence stars of 1–2 $M_\odot$ exhibiting an infrared excess have been observed in the same regions. Therefore there is some observational support for the existence of "stellar nebulae" in the vicinity of young stars similar in mass to the Sun. However, few details concerning the quantity and distribution of the circumstellar dust are available. Infrared data show that it is likely that other low-mass stars are surrounded by dense, opaque dust clouds. This heavy obscuration precludes learning much about these objects by observation.

From the theoretical side, the circumstances under which the material of a molecular cloud collapses to form a single star like the Sun are also quite obscure. When rotation is included, hydrodynamic numerical simulations of such collapse appear to show that two or more gravitationally bound stars of comparable mass are usually formed, rather than a single star like the Sun (Bodenheimer & Black 1978). It is possible that single stars are formed in the more infrequent circumstance of clouds with very low angular momenta, or result from ejection of single stars from multiple star systems, or from aggregation of binary systems in an accretion disk. It is not at all clear if it is appropriate to associate the observed objects that appear to be surrounded by opaque clouds with these single stars.

In summary, although it is plausible to associate the formation of the planets with the aggregation of dust and gas in a primitive "solar nebula," neither theory nor observation of star formation provide quantitative constraints on the properties of this nebula. Since the oldest known stars of the galactic disk population appear to be about the same age as the solar system, ~ 5 billion years (Demarque & McClure 1977), it is even possible that the Sun and planets were formed during a unique epoch in galactic evolution, and that presently observed processes of star formation may not be directly applicable to the formation of the solar system.

In contrast to the rather obscure information provided by stellar astronomy, there is a fairly large body of observational data relevant to terrestrial planet formation which is obtainable from these bodies themselves, as well as from meteoritic data. Numerous examples can be given:

It is known that the relative proportions of the refractory chemical elements in the Earth, Moon, and meteorite parent bodies are very similar to those in the Sun and in average solar system matter. On the other hand, the elements that for the most part fail to form nonvolatile compounds are severely depleted in these bodies (Ganapathy & Anders 1974).

The nonzero eccentricities and inclinations of the planets and the magnitude and direction of their axial rotation vectors carry information relevant to conditions at the time of formation even though this relationship is poorly understood at present (Safronov 1969, Harris 1977).

Interpretation of radiogenic isotopic data permits inferences regarding the time scale of planet formation. For example, lunar and meteoritic data show that heat sources sufficient to produce internal igneous differentiation were present very near to the time of formation (Allègre et al. 1975, Papanastassiou & Wasserburg 1975; reviewed by Wetherill 1975; Kirsten 1978).

The observed cratering record constrains the mass and size distribution of the bodies remaining following the principal stage of planet formation (Wetherill 1977).

Even though much of these abundant data are of considerable precision, at present they fail to constrain our ideas concerning planet formation sufficiently well to allow us to make firm inferences regarding physical conditions in the solar nebula prior to planet formation. However, by an iterative process, it may be hoped that continued comparison of theory and observation will lead to a picture of terrestrial planet formation that is not only generally accepted, but has a good chance of being more or less true.

1.2 Scope of This Review

It is probably unnecessary to say that we are still far from this goal. Nevertheless, the volume of work both published and in progress directed toward this goal is large and increasing. The purpose of this review is to summarize recent theoretical investigations directed toward terrestrial planet formation per se, rather than to discuss general theories of star and planet formation. The views of various authors concerning the more general problems, as well as references to earlier work, can be found in two recent collections of papers (Dermott 1978, Gehrels 1978), in several books (Safronov 1969, Alfvén & Arrhenius 1976), as well as in a number of articles (Woodward 1978, Cameron 1978b, Prentice 1978, Horedt 1979). Although the evolution of the spin and orbital angular momenta of the terrestrial planets is an important aspect of the subject of this review, the present author lacks the inclination to undertake this, and defers to the forthcoming review by Harris (1981). An extensive discussion of the origin of the Earth and the Moon, with emphasis on geochemistry, has been given by Ringwood (1979).

The two general mechanisms of terrestrial planet formation, gravitational collapse and accumulation, will be discussed for the cases of both a gaseous solar nebula and a gas-free nebula. Emphasis will be placed on concepts, which will be elucidated by simplistic theoretical expressions. Readers will find at least as much complexity as they wish in the references cited.

2 GRAVITATIONAL INSTABILITY

2.1 Formation of Giant Gaseous Protoplanets by Gravitational Instability

Gravitational instability occurs when the density of an assemblage of dust and gas is so large that a positive perturbation in density will grow with time. Under these circumstances the mutual gravitational attraction of the material overcomes the effects of internal kinetic energy (temperature), rotation, and magnetic fields, all of which tend to inhibit gravitational instability. In the absence of rotation and magnetic fields, the critical wavelength (λ_c) of a perturbation is given by

$$\lambda_c^2 = (\pi c^2/G\rho) \tag{1}$$

where c is the sound velocity $(\gamma RT/\mu)$, ρ is the density, γ is the ratio of specific heats, T is the temperature (K), μ is the molecular weight, and G is the gravitational constant. Equation (1) is known as the Jeans criterion (Jeans 1929). Instability will result for perturbations of wavelength $\lambda > \lambda_c$, and the smallest perturbation that can grow will be given by $\lambda = \lambda_c$.

Although this criterion must be modified in the presence of rotation and magnetic fields, both of these effects oppose the development of instability, requiring even larger values of λ. Therefore Equation (1) will provide information regarding the *minimum* dimension and mass of a terrestrial planet that can form as a result of gravitational instability in a nebula of given density.

For a body of mass M, the density of a spherical body with a diameter equal to λ_c will be

$$\rho = (6M/\pi\lambda_c^3). \tag{2}$$

Substituting (2) into (1) gives

$$\lambda_c = (6MG/\pi^2 c^2) = (6MG\mu/\pi^2 \gamma RT). \tag{3}$$

For the case of formation of a gravitational instability in a circumsolar nebula, it is also necessary that the density be high enough to prevent tidal disruption, i.e. greater than the Roche density ρ_R:

$$\rho_R = (3M_\odot/2\pi a^3) \tag{4}$$

where $M_\odot$ is the solar mass, and a is the semimajor axis. This implies that

$$(\lambda_c/2) \leqq a(M/2M_\odot)^{1/3} \cong 1.15L \tag{5}$$

where L is the radius of the collinear Lagrangian point in the restricted 3-body problem (e.g. Blanco & McCuskey 1961, Szebehely 1967).

In the presence of nebular gas, the appropriate mass of the Earth will be the present Earth's mass (6×10^{27} g) augmented by the complement of volatile gas (mostly H_2 and He) which in material of average solar system composition is associated with the nonvolatile elements that form the Earth. Addition of this material increases the mass by a factor of ~ 375, and $M = 2.25 \cdot 10^{30}$ g. Use of Equation (3), with $T = 300$ K, $\mu = 2$, $\gamma = 1.4$ (corresponding to H_2) gives a critical wavelength

$$\lambda_c = 5.2 \times 10^{12} \text{ cm} = 0.35 \text{ AU}.$$

The corresponding density given by Equation (2) will be

$$\rho = 3.1 \times 10^{-8} \text{ g cm}^{-3}.$$

At the distance of the Earth from the Sun with the present value of the Sun's mass, the critical Roche density, below which the matter will be unstable with respect to tidal disruption, is $2.9 \times 10^{-7}\,\mathrm{g\,cm^{-3}}$. Therefore an Earth "protoplanet," satisfying the Jeans criterion, will not be stable with respect to tidal disruption. Assumption of a lower value of the temperature in (3) will lead to even lower densities, and greater tidal instability.

If the plausible assumption is made that the density in the solar nebular becomes higher as the Sun is approached, the density found above for the gravitational instability leads to a total mass of the nebula within 1 AU of

$$M(<1\,\mathrm{AU}) > 1.31 \times 10^{32}\,\mathrm{g},$$

assuming a mean thickness of the nebula of 0.1AU. This is about 30 times the mass required to form the terrestrial planets. If a higher temperature, rotation, or magnetic field are invoked in order to achieve tidal stability, the quantity of excess mass becomes enormous, and no way to remove a mass so large is known.

For these reasons, formation of the terrestrial planets by gravitational instability near their present positions *after* the formation of the Sun doesn't appear to be possible. Furthermore, a fully developed planet formed in this way would resemble Jupiter, not Earth or Venus. Therefore, theories proposing that the terrestrial planets formed by gravitational instability must postulate the formation of these planets at greater heliocentric distances, the protoplanets moving sunward along with the gas that is to form the Sun. When a protoplanet is sufficiently close to the protosun, the gaseous envelope of the collapsing protoplanet can then be removed by tidal disruption leaving behind a nonvolatile core, which becomes the terrestrial planet (Donnison & Williams 1975, Cameron 1978a).

The physics of this chain of events is very complex. No secure calculations of the expected number, position, and size distribution of the gaseous protoplanets have been presented. The detailed nature of the redistribution of mass and angular momentum during the collapse of the solar nebula has not been worked out. Calculations have been presented regarding the evolution and differentiation of the protoplanets (DeCampli & Cameron 1979, Cameron 1979). When one considers the complexity of this problem, it is hard to be at all certain that all the required events will occur on schedule. For example, the timing of solar and protoplanetary growth must be so well coordinated that the observed number of terrestrial planets will enter the region of tidal instability after the formation of a nonvolatile core but before the collapse of a significant fraction of their gaseous envelopes. In order to produce the gravitational instabilities necessary to form the terrestrial planets, the solar nebula must contain an excess mass of about $1\,M_\odot$ which will have to be removed by rather uncertain mechanisms.

Because of these quantitative uncertainties, much work remains to be done before this can be regarded as a probable mechanism for the formation of terrestrial planets. Even if it is granted that Jupiter and Saturn formed in this manner, it seems just as likely that the inner solar system would be populated by a large mass of solid debris with smaller dimensions: a mixture of solid material which aggregated in the absence of gravitational instability, protoplanets which were tidally disrupted at an early stage of their evolution, and fragments of protoplanetary cores disrupted by close approaches to one another or by collisions. If this is the case, the formation of the terrestrial planets will consist of the accumulation of this debris, along the lines of the accumulation theories to be discussed later.

If the terrestrial planets are indeed the fully collapsed cores of giant gaseous protoplanets, the time scale of planetary formation is rather well fixed at $\sim 10^5$ years, because of the required simultaneity of the growth of the Sun and planets. The time scale is determined by the time required for the Sun to complete its growth following the stage in which the nebular density becomes sufficiently large to produce gravitational instabilities. This time scale is shorter than the predicted time scale for terrestrial planet formation by accumulation by a factor of 100 to 1000.

2.2 Gravitational Instability in a Central Dust Layer

Although it is likely that initially a protosolar cloud would be highly turbulent, energetic considerations preclude maintenance of much turbulence beyond the stage at which the Sun approached its final mass. Unless terrestrial planets formed by the growth and stripping of giant gaseous protoplanets, as discussed in Section 2.1, the planets were formed from a nonturbulent nebula. It is also energetically impossible for a flat solar nebula of dimensions comparable to the solar system to remain at temperatures above the condensation temperature of silicates for times longer than about 10^3 years (Ter Haar 1950).

Under these conditions solid grains will spiral down through the gas toward the central plane of the nebula, and form a flat central dust layer. The time scale for the formation of the dust layer will be short: $\sim 10^5$ yr for 1 μm particles, 10 yr for particles that have grown to 1 cm diameter (Safronov & Ruskol 1957, McCrea 1960, Weidenschilling 1977a, 1979, Kusaka, Nakano & Hayashi 1970). As Edgeworth (1949) and Gurevich & Lebedinskii (1950) pointed out, if the dust layer is sufficiently thin, gravitational instabilities will develop within the dust layer itself, even though the nebula as a whole is gravitationally stable. The formation and evolution of these dust layer instabilities have been discussed by several authors (Safronov 1960, 1969, 1975, Lyttleton 1961, 1972, Goldreich & Ward 1973, Polyachenko & Fridman 1972). Following the discussion by Goldreich & Lynden-Bell (1965) the Jeans criterion for the gravitational stability of a thin rotating layer is given by a dispersion relationship for the propagation of a sound wave

$$p = p_0 \exp\{2\pi i(x-vt)/\lambda\} \tag{6}$$

where p is the perturbation in pressure, λ the wavelength of the perturbation, and v the velocity of propagation of a perturbation of wavelength λ. The dispersion of this velocity is given by

$$v^2 = c^2 - G\sigma\lambda + (\omega^2\lambda^2/4\pi^2) \tag{7}$$

where σ is the density per unit area of the layer, ω is the angular velocity, and c is the sound speed, which for the dust layer is taken to be comparable to the velocity associated with the random kinetic energy of the dust particles. When v^2 is negative, a disturbance will grow exponentially. v^2 will not be negative for any value of λ unless

$$c < (\pi G\sigma/\omega). \tag{8}$$

This is a necessary condition for an instability to develop.

For a "minimum solar nebula," i.e. one containing no more dust than that required to form the observed terrestrial planets, $\sigma \sim 10 \text{ g cm}^{-2}$. Use of Equation (8) then implies a random velocity of the dust particles $c \lesssim 10 \text{ cm s}^{-1}$. This velocity is quite low, but in the absence of turbulence or significant gravitational perturbations, there is no compelling reason that velocities could not fall below this value. The approximate thickness of this dust layer will be given by the condition that c must be less than that required for a particle at the midplane to rise to the surface against the gravitational field of the layer, given by

$$d = (c^2/2\pi G\sigma) \cong 240 \text{ km}. \tag{9}$$

At the Earth's distance the corresponding density is $\sim 4 \cdot 10^{-7} \text{ g cm}^{-2}$, slightly above the Roche limit for tidal disruption by the Sun.

The size range of the dust layer instabilities will correspond to those values of λ for $v^2 < 0$ in Equation (7). At the critical value of the velocity c given by Equation (8)

$$\lambda_c = (2\pi c/\omega) \sim 3 \cdot 10^8 \text{ cm} \tag{10}$$

which will be associated with a mass

$$m_c = \tfrac{4}{3}\pi(\sigma/d)(\lambda_c/2)^3 = 2.4 \times 10^{18} \text{ g}. \tag{11}$$

About 5×10^9 dust condensations of this kind are required to form the terrestrial planets.

At first these dust condensations will be very tenuous assemblages. The subsequent evolution of these low density condensations into solid bodies is complex, and has been discussed by Safronov (1969) and Goldreich & Ward (1973). Following their formation, the dust condensations will contract until an equilibrium state is reached determined by the threshold for rotational instability. This equilibrium will correspond to solid densities of ~ 3 g cm^{-3} only for small initial condensations $m \sim 10^{-4}\, m_c$. Goldreich & Ward consider the evolution of a cluster of these smaller bodies derived from a single condensation of mass m_c. Their relative velocity is reduced by gas drag and they coalesce to form solid bodies of mass m_c ($\sim 2 \times 10^{18}$ g) on a time scale of ~ 100 yr. Safronov does not consider gas drag, but rather evaluates the averaging of rotational spin vectors which occur during the growth of bodies of mass m_c by dissipative collisions, and concludes that this effect will decrease the rotational velocity enough to permit the dust condensations to contract to solid densities in $\sim 10^4$ yr at the Earth's distance from the sun, by which time the bodies will have grown to masses of $\simeq 10^{20}$ g. Although these arguments are quite different, it seems likely that at least one of these processes will operate and that the initial diffuse dust condensations will evolve into solid bodies of $\lesssim 10$ km diameter on time scales of $\lesssim 10^3$ yr. Following the formation of these solid bodies it is unlikely that further gravitational instabilities can occur in the terrestrial planet region, and the subsequent growth of planets must occur by a series of accretional collisions whereby smaller bodies are swept up by larger ones, either in the presence or absence of a gaseous resisting medium.

The possibility of building ~ 10 km bodies on a short time scale eliminates the often-mentioned problem of understanding how bodies could stick to one another and grow, prior to being sufficiently large for gravitational binding to become important. This seems pleasing because our ignorance of the physical and chemical nature of these primitive bodies causes hypothetical sticking mechanisms to appear ad hoc and be, therefore, suspect. However, as Weidenschilling (1977a) has pointed out, quite low sticking efficiencies suffice to produce ~ 1 km bodies prior to their settling into an extremely thin dust layer. In this case the effect of differential gas drag, collisions, and the mutual gravitational perturbations of the bodies would result in velocities sufficiently high to preclude the onset of gravitational instability. Therefore it would be a mistake to conclude that the solar system *must* have developed dust-layer instabilities simply because this does not require specification of sticking processes that are poorly understood, but that quite possibly may have occurred anyway.

3 ACCUMULATION OF PLANETESIMALS BY GRADUAL SWEEPING UP OF MATTER

The collective gravitational forces of the solar nebula may have been insufficient to cause its material to aggregate into larger bodies. In this case growth can still occur by the collision and cohesion of individual bodies. Cohesion can result from gravitational attraction or by chemical or other physical attractive forces such as magnetism. The effects of nongravitational cohesion ("sticking") are not well understood because of lack of detailed knowledge of the physical conditions that prevailed in the solar nebula. However, it is plausible that for small objects (e.g. $\lesssim 1$ cm diameter) cohesion will primarily depend on nongravitational forces, whereas for bodies $\gtrsim 1$ km diameter, gravitational attraction will dominate. In the intermediate size range, both of these forces will be relatively weak, and it is difficult to say which will be the more important.

The minimum collision velocity of two bodies in heliocentric orbit is their gravitational escape velocity:

$$v_e = [2G(m_1+m_2)/(r_1+r_2)]^{1/2}. \tag{12}$$

If two bodies impact at this velocity they will adhere to one another because collisions will never be perfectly elastic, and if any energy is lost in the collision the bodies will be gravitationally bound. At higher collision velocities it is energetically possible for the bodies to bounce apart without cohering, and growth can occur only if the collision is sufficiently dissipative. At impact velocities above ~ 5 m s^{-1} (~ 10 miles h^{-1}) even fairly strong objects, for example automobile fenders, will deform and break, and collisions will tend to be fairly dissipative. This velocity corresponds to the escape velocity of a solid body about 7 km in diameter, i.e. about the size of the bodies that may result from gravitational instabilities in a central dust layer. Under these circumstances impacts at 1.5 times the escape velocity or even higher may still permit cohesion.

For bodies with relative velocity v prior to a close encounter, the impact velocity v_i will be

$$v_i = (v^2+v_e^2)^{1/2}. \tag{13}$$

Safronov (1969) makes considerable use of a useful dimensionless parameter θ (which I have termed the "Safronov number"),

$$\theta = (v_e^2/2v^2). \tag{14}$$

Safronov (1969, Equation 7.12) defines θ by an equation of the form of Equation (14), but defines v to be the relative velocity of the body with respect to a body moving in a circular orbit with semimajor axis corresponding to the instantaneous heliocentric distance of the encounter. When the accumulation of the terrestrial planets is calculated (Safronov's Equation 9.12) the "planetary embryos" are assumed to be moving in circular orbits, and his definition is then equivalent to the relative velocity between the two bodies, given by (14). Kaula (1979a) apparently defines the reference orbits in a still different way. In this article, θ will be used in a general way to represent equations of the form of (14). When appropriate, the exact way in which the velocity is defined will be indicated by subscripts on θ. It should be noted these distinctions may be significant (cf Section 3.1.2).

For $v_i = 1.5v_e$, the value of θ calculated from (14) is $\theta = 0.4$. Lower values of v correspond to higher values of θ. Although the actual value of θ for which cohesion is possible obviously depends on the unknown physical properties of the colliding bodies, it is likely that values of $\theta \gtrsim 1$ will result in cohesion, whereas values of $\theta \lesssim 0.2$ will result in fragmentation without cohesion. Experimental data relevant to this matter have been provided by Hartmann (1978).

For smaller bodies in the meter size range the situation is more complex. Gravitational forces are weak and quite low velocities are sufficient to exceed the escape velocity. On the other hand high impact velocities (very small values of θ) may facilitate sticking. For example, rifle bullets may remain imbedded in a target at velocities $\sim 1 \text{ km s}^{-1}$, whereas if they are simply thrown at the target at $\sim 5 \text{ m s}^{-1}$ they will merely bounce off. In such circumstances there will be a range of θ extending from ~ 1 to some very low value, depending on the nature of the bodies, for which coherence does not take place. For values of θ above or below this region coherence can occur.

Whether or not growth of terrestrial planets will occur by continued two-body collisions will therefore depend on whether or not

1. their relative velocity remains sufficiently low ($\theta \gtrsim 1$),
2. a sufficient number of bodies remain in intersecting orbits.

Accumulation of solid bodies may take place in the presence of a gas phase, or under conditions in which gas drag forces are negligible. Gas drag will lead to lower relative velocities, which in itself will promote growth. However, gas drag will lead to circularization of orbits, which can reduce the number of bodies in intersecting orbits, thus inhibiting growth.

If it is assumed that the terrestrial planets accumulated by a gradual accumulation process, it is not certain whether this accumulation took place prior to the loss of residual gas from the interplanetary medium, or afterward. It is sometimes written that accumulation of planets requires the presence of a gas phase to slow down the colliding bodies, and that this can be used as a constraint on the timing of the growth of the Earth and the dissipation of nebular gas. As discussed above, this is not true, provided $\theta \gtrsim 1$ for the largest bodies throughout the course of their growth. Subsequent discussion will show that the opposite problem, too low velocities even in the absence of gas, may prove a more serious obstacle. Because of their hydrogen-rich composition, it is clear that Jupiter and Saturn must have formed in the presence of a gas phase, regardless of whether they were formed by a massive gravitational instability or by an accumulation process. As will be discussed later in this section, the time scale for the formation of Jupiter and Saturn by an accumulative process will be longer than that for the Earth and Venus. Therefore if these gas-rich planets formed by *accumulation* in a gaseous nebula, it is likely that the terrestrial planets also formed in the presence of gas inasmuch as they would be expected to form sooner.

Although they are not compelling, there are some good reasons why this conclusion may be unattractive. If the Earth grew in the presence of nebular gas it might be expected that it would capture much of the gas into its primitive atmosphere. Hydrogen and helium could subsequently be lost by thermal escape or by photoreactions with solar ultraviolet radiation. This is not obviously possible for heavier gases such as the other inert gases, CO_2 and N_2, and polymerized hydrocarbons. However, the Earth is strongly depleted (by factors of 10^4–10^{11}) in these gases, relative to their expected abundance in a solar nebula. Sekiya, Nakazawa & Hayashi (1979) have presented arguments to support the view that these heavier gases may be swept along by hydrogen as it escapes. Their theory requires assumptions regarding the mechanism for hydrogen loss that are not necessarily valid. The loss of these gases is sometimes attributed to a T-Tauri solar "hurricane" before the Sun reached the main sequence (Horedt 1978), although it is not known if such a process could really remove a massive terrestrial atmosphere. It would take place $\sim 10^6$ yr after the formation of the Sun, and the accumulation calculations to be discussed later imply time scales of $\sim 10^8$ yr for the formation of the solid core of Jupiter, during which time the hydrogen must still be present.

Another reason for believing Jupiter formed before the terrestrial planets is that this provides an explanation for the gross mass depletion of the asteroid region and the small mass of Mars. During the final stages of Jupiter's growth residual planetesimals from the Jupiter region would be perturbed by Jupiter into the region between 4.8 and 1.0 AU and the planetesimals may have destroyed or ejected most of the material originally present. This event may also be responsible, in some rather vague way, for the present high relative velocities of the asteroids (Safronov 1969, Wetherill 1972, Kaula & Bigeleisen 1975, Weidenschilling 1975, 1977b).

For these reasons it is worthwhile to retain in one's mind the possibility that the growth mechanisms for Jupiter and Saturn were quite different from those responsible for the growth of the terrestrial planets. For example, in the theory of Cameron (1978b) Jupiter may have been a giant gaseous protoplanet, whereas stable objects of this kind may have failed to develop in the vicinity of the Earth. Although the formation of the giant planets is beyond the scope of this review, it should be pointed out that the complex history of growth and stripping required for the formation of the terrestrial planets by massive gravitational instabilities is less complex for Jupiter and Saturn.

3.1 *Accumulation of Planetesimals from a Gas-Free Circumsolar Swarm of Bodies*

During the earliest stage of solar system history, large amounts of gas must have been present, and therefore in some sense the beginning of planetary growth must have involved a gas phase. For example, if the original ~ 5 km diameter solid bodies resulted from dust-layer gravitational instabilities (see Section 2.2), these bodies were formed in a gaseous nebula. The first stages in their growth must involve the effects of gas drag, as discussed by Goldreich & Ward (1973). However, it is quite possible that this gas was removed by solar UV or corpuscular radiation on a time scale (i.e. 10^4–10^6 yr) which is less than that required for these bodies to grow by mutual collisions into terrestrial planets. If this is so, then the significant body of theory that has been developed for gas-free accumulation can be used to describe the growth of the terrestrial planets. The most complete quantitative discussion of this theory is given in the book by Safronov (1969).

The basic idea of gas-free planetary accumulation is quite simple. As a body moves through the solar nebula it will collide with other bodies. As discussed earlier in this section, if the relative velocities of the bodies are sufficiently low ($\theta \gtrsim 1$), the bodies will cohere to one another and grow. This process will continue until all the material has accumulated into bodies moving in orbits that are sufficiently isolated from one another to preclude further collision and growth.

The theoretical problem that must be addressed is whether or not some plausible initial assemblage of bodies will spontaneously collide and grow into large objects with masses and orbits similar to those of the present terrestrial planets. If such growth is possible, an adequate theory should also quantitatively describe the time scale required for accumulation, as well as the size, velocity, and orbital distribution during the course of accumulation. More detailed work has been carried out concerning this mode of planetary accumulation than for any of the alternatives, but there are still a number of questions that must be resolved more clearly before we can say the theory is reasonably complete, or if indeed it is even possible that gas-free accumulation could have played a major role in the formation of the terrestrial planets.

The initial state usually assumed in the gas-free accumulation problem is a swarm of small (e.g. 1 km diameter) solid bodies moving about the Sun in Keplerian orbits that are nearly coplanar and circular. The total mass and semimajor axis distribution of these bodies is taken to be similar to that of the terrestrial planets into which they are to evolve.

3.1.1 EARLY STAGES OF ACCUMULATION During the early stages of the accumulation process the bodies will be "closely packed" and close encounters can occur even though the eccentricities are very low. For purposes of illustration, consider that all the mass of the terrestrial planets (1.2×10^{28} g) were spread out in a plane disk of width 1 AU consisting of a large number of bodies of mass m' initially in circular orbits. If $m' \lesssim 2 \times 10^{14}$ g the bodies will physically touch when overtaking one another because of their Keplerian motion, even if their gravitational attraction is neglected. As long as $m' \lesssim 4 \cdot 10^{24}$ gm bodies in circular orbits will pass within their spheres of influence as defined in the restricted three-body problem (Tisserand 1896). Very low eccentricities ($\sim 10^{-3}$) are sufficient to permit these ~ 1000 km diameter bodies to collide with one another simply on geometrical grounds.

Numerical simulations of this early stage of planetary growth have been presented by Greenberg et al. (1978). Because of the close packing referred to above, their assumption that the separation of bodies, associated with differences in their semimajor axis and heliocentric distance, can be ignored is reasonable. The dynamical evolution was treated as if the bodies were simply particles confined to a three-dimensional space or "box," as is assumed in the kinetic theory of gases.

In this work it was assumed that the initial state consisted of 10^{12} bodies one kilometer in diameter, moving in low inclination orbits with $e = 0.0001$. As a result of low velocity collisions these bodies grew, at higher velocities fragmentation was assumed to occur, based on the experimental work of Hartmann (1978). Close encounters, which are "near misses" rather than collisions, cause perturbations in the eccentricity of the bodies as a consequence of the mutual gravitational attraction of the bodies. On the average, the effect of these perturbations will be to increase the eccentricity of the bodies, whereas that of collisions will be to circularize the orbits. The evolution of the swarm was followed in successive time steps. The bodies were sorted into size "bins" following each fragmentation or accumulation event and the average eccentricity and inclination for each size bin recalculated. By thus treating all bodies in the same size range as having the same eccentricity and inclination, it was possible to handle this very large number of bodies.

A typical result of Greenberg et al. (1978) is shown in Figure 1. At first the size distribution is sharply peaked near the assumed initial size. Somewhat later, smaller bodies are produced by fragmentation and some larger bodies by accumulation. As time goes on, the region < 1 km begins to fill in and in the 1–10 km range a size distribution develops that is very roughly similar to the power law

$$(dN/dr) \propto r^{-3.4}; \quad (dN/dm) \propto m^{-0.8} \tag{15}$$

predicted by theoretical discussions (Zvyagina, Pechernikova & Safronov 1973, Pechernikova 1974, Pechernikova, Safronov & Zvyagina 1976). During this time the mean velocity remains $\sim 1 \, \mathrm{m \, s^{-1}}$, about the escape velocity of the bodies in which most of the mass is concentrated. However, even at this early stage of accumulation, a few bodies in the size range ~ 400 km are produced. In the calculations it was assumed that the

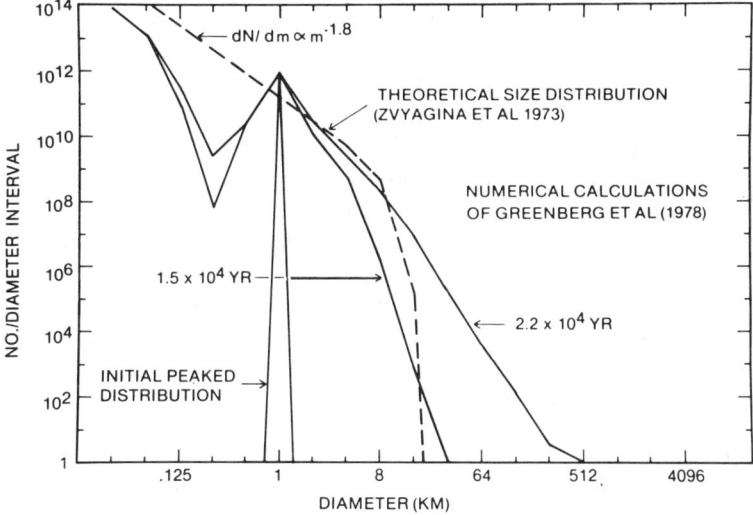

Figure 1 Numerical accumulation calculation by Greenberg et al. (1978). After $\sim 10^4$ years the initial distribution of 10^{12} 1 km diameter bodies has evolved into a distribution containing objects as large as 512 km. The abundance of large bodies relative to smaller objects is in disagreement with theoretical models based on coagulation theory, and arise as a consequence of the gravitational cross section (Equation 16) causing the kernel in the coagulation equation to be nonlinear.

probability of collision was proportional to the "gravitational cross section" of the body. This is larger than the geometrical section by a factor

$$f_g = 1 + (v_e^2/v^2) = 1 + 2\theta. \tag{16}$$

The enhanced cross section is a consequence of "gravitational focusing," whereby the distance of closest approach of the two bodies is less than the distance between their initial trajectories. This is equivalent to the difference between the asymptotic separation and the perigee distance in the hyperbolic two-body problem (e.g. Wood 1961). For the largest bodies of Figure 1, $v_e \sim 1 \text{ km s}^{-1}$ and $f_g \sim 10^6$. Thus the larger bodies are able to sweep up smaller objects much more efficiently, and this effect is responsible for the early appearance of the ~ 400 km bodies.

It is not at all certain that the proposed enhancement in the rate of growth of the large bodies is physically realistic. The use of Equation (16) and the treatment of the encounter between two bodies in Keplerian orbit about a third body (the Sun) are valid approximations when the approach velocity is high enough (Öpik 1951, Cox, Lewis & Lecar 1978). However, when the velocity is so low that the encounter time is a significant fraction of the orbital periods of the encountering bodies, this is not the case (Dole 1962, Giuli 1968, Cox 1978). Under such circumstances the approximations introduced by breaking the actual three-body problem into separate two-body problems fail. If the effect of this failure is to reduce f_g significantly below that given by (16) it is likely that the high mass "tail" of the size distributions shown in Figure 1 will be steeper, and approximate establishment of the power law size distribution (15) will extend to higher masses. If (16) is more or less valid, it is likely that runaway growth of the largest bodies will take place, and the end result of this stage of growth will be the formation of about $1000 \sim 10^{25}$ g bodies in nearly circular, low inclination, noncrossing orbits. It is conceivable that longer range perturbations by protoplanets in the outer solar system, or resonant perturbations by terrestrial planetesimals, could destabilize the situation, produce crossing orbits, and permit further evolution toward the present system of terrestrial planets. But this has not been demonstrated as yet.

3.1.2 THE SAFRONOV STEADY STATE VELOCITY In the event that runaway growth and isolation do not occur and size distribution with most of the mass in the largest bodies, e.g. (15), is established, the analytical results principally developed by Safronov and his co-workers will become applicable. The most important result of these studies is that the competition between mutual collisional damping and gravitational acceleration by the members of the swarm results in a steady state velocity distribution, with the mean velocity being comparable with the escape velocity of the largest body, i.e. $\theta \approx 1$. Relative velocities will then remain low enough to permit the largest bodies to grow, and will "keep step" with the growth of these bodies, increasing with the radius of the largest bodies and never deviating far from their escape velocity. Under such conditions, approximate analytical expressions based on coagulation theory show that a power law of the form (15) represents a quasi-steady state for the size distribution for all but the largest members of the swarm.

The fundamental reason the velocity of the swarm keeps step with the growth of the bodies is that the change in relative velocity accompanying a close gravitational encounter between two bodies of the same mass is given by

$$(\Delta v)_g \sim (2GM/Dv) \tag{17}$$

where M is the mass of the bodies and D is the distance of closest approach (Öpik 1951). The change in kinetic energy (E_g) following a single encounter will be

$$\Delta E_g = \tfrac{1}{2}M(v+\Delta v)^2 - \tfrac{1}{2}Mv^2$$
$$= \tfrac{1}{2}M[2\Delta v - (\Delta v)^2]. \tag{18}$$

After a large number (n) of gravitation encounters the first term on the right of Equation (18) will average to zero and

$$n\overline{\Delta E_g} = \Sigma \Delta E_g = \Sigma \tfrac{1}{2}M(\Delta v)^2. \tag{19}$$

The contribution to $\Sigma \Delta E_g$ in the time interval Δt from those bodies passing at a distance between D and $D + \Delta D$ will be

$$(\Delta E_g)_D = \pi \rho_s v D \Delta D (\Delta v)^2 \Delta t = \Delta t (4\pi \rho_s G^2 M^2 / vD) \Delta D \tag{20}$$

where ρ_s is the mass density of the swarm. Integrating the encounter distance out to the edge of the sphere of influence

$$D = R_i = 2^{1/5}(M/M_\odot)^{2/5} a \tag{21}$$

we obtain

$$\Sigma \Delta E_g = 4\pi \rho_s G^2 M^2 v^{-1} [\ln(R_1/2R)] \Delta t. \tag{22}$$

This increase in kinetic energy will be offset by a loss of energy caused by inelastic collisions

$$\Sigma \Delta E_c = \pi (2R)^2 \varepsilon \rho_s v^3 \Delta t, \tag{23}$$

where ε is a dimensionless factor representing the fractional loss of energy in collisions. In the steady state, the gain and loss will be equal. Equating (22) to (23) gives

$$v^2 = (GM/R)[\varepsilon^{-1} \ln(R_i/2R)]^{1/2} = 0.5 v_e^2 [\varepsilon^{-1} \ln(R_i/2R)]^{1/2}. \tag{24}$$

From Equation (21) it may be seen that $(R_i/R) \propto R^{0.2}$, and thus the ratio $\ln(R_i/2R)$ varies only by a factor of ~ 3 as R grows from 10 km to 5000 km. For $\varepsilon = 1$ and $R = 1000$ km, $v^2 = 0.98 v_e^2$, $\theta = 0.5$ ($\sim$ constant), and $v \approx v_e$. Because v_e is proportionate to R, v will be so also, i.e. v will increase linearly with the radius. This result is not sensitive to the use of the geometrical cross section instead of the gravitational cross section (Equation 16) because gravitational focusing increases both the collision rate and the close encounter rate.

If the size distribution follows a power law, rather than all the masses being equal, the situation is more complex. For power laws of the form

$$(dN/dm) \propto m^{-q} \tag{25}$$

and with $q < 2$, most of the mass will be in the largest bodies. These largest bodies will be responsible for most of the perturbations, and therefore for the increase in energy in the appropriate expression analogous to Equation (22). However, if $q > 1.67$ as well, most of the integrated area will be in the small bodies, and mutual collisions between small bodies will be the dominant mode for energy loss. Moreover, in this case the velocity distribution will be size-dependent. The largest bodies will tend to be in orbits of low eccentricity and inclination because only rare close encounters with other large bodies will increase their velocity. For the smallest bodies, the large ratio of cross section to mass will also result in relatively circular orbits, and maximum velocity will occur at some intermediate value of the mass. Both the numerical investigations and analytical theories referred to earlier suggest that $1.67 < q < 2.0$. Although much work remains to be done on this problem of the coupled mass and velocity distribution, it is quite likely that, except for a few of the largest mass bodies, most of the objects fall in this range where the mass is concentrated in the largest bodies, and the area in the smaller ones.

Detailed analytical theories of the velocity of an accumulating planetesimal swarm with unequal masses have been given by Safronov (1969) and Kaula (1979a). All this work is in turn based on the stellar dynamical methods of Chandrasekhar (1942). In these theories the heliocentric motion of the swarm is not explicitly introduced. However, the velocity gradient resulting from the differential Keplerian velocity is an intrinsic aspect of the problem. Furthermore, the perturbations of the semimajor axes of the bodies are as important as changes in eccentricity in their effect on the velocity distribution.

The results of these analytical theories are in fair agreement (i.e. a factor of ~2) with one another, and with the simplified discussion given above, in predicting values of θ in the range from 1–5. They differ primarily as a consequence of different approaches to the calculation of the mutual gravitational acceleration, and to some extent to differences in the way the circular reference orbits are defined. As mentioned above, there are conceptual problems in treating this problem, which involves Keplerian heliocentric motion in an essential way, by the Chandrasekhar approach, which ignores heliocentric motion.

Numerical calculations for the case of a nonaccumulating swarm of bodies of equal mass have been presented, in which the heliocentric motion is explicitly introduced (Wetherill 1979). These calculations show (Figure 2) that the average velocity approaches an approximate equilibrium value regardless of whether the initial velocities are chosen to be above or below the steady state value. The steady state velocities increase linearly with radius (Figure 3) as expected, and the numerical values of θ are in the range predicted by analytical theories. The numerical calculations also verify that the velocity distribution is nearly Maxwellian, as assumed in the analytical theory. It is also found that mutual perturbations cause a marked diffusion of semimajor axis (Figure 4). This result has no analytical counterpart because differences in semimajor axes are not included in these theories.

It appears that, for a given set of assumptions regarding size distribution and degree of inelasticity of the bodies, the theoretical results can be

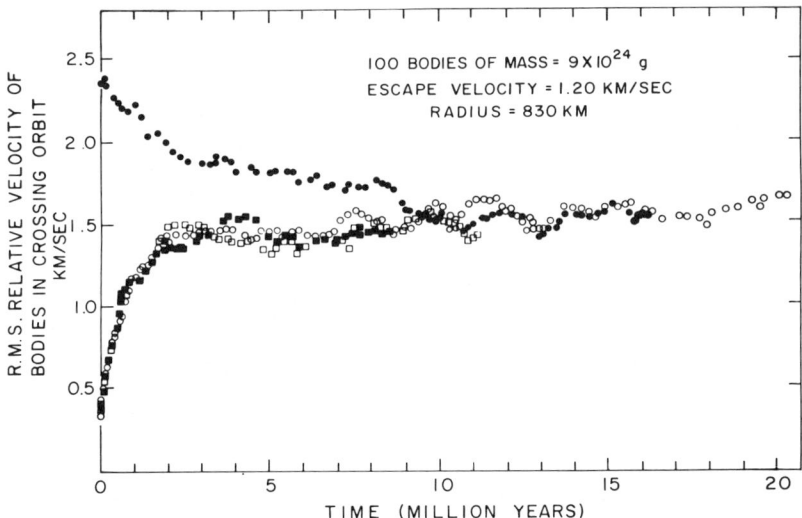

Figure 2 Numerical calculations of the approach to a steady state mean velocity of a nonaccreting swarm of bodies of equal mass. The equilibrium velocity is nearly independent of the initial velocity distribution (Wetherill 1979).

calculated to within a factor of ~2. Further precision, and detailed understanding of the variation of the velocity and size distributions with time and with one another, will require extension and improvement of both the analytical and numerical theories.

3.1.3 TIME SCALE FOR ACCUMULATION   If it is assumed that the velocity distribution given by theory is correct, the time scale for gas-free accumulation of the terrestrial planets is constrained to be in the range 10^7–10^8 yr. This can be seen approximately as follows.

In the "kinetic theory of gas" approach used in the analytical theories the rate of sweep up mass by a planet of mass M from a swarm of density ρ_s with velocity v will be

$$(dM/dt) = \pi R^2 \rho_s v (1 + 2\theta) \tag{26}$$

where the last factor on the right represents the gravitational enhancement of the radius (16). In terms of the radius (R) this is equivalent to

$$(dR/dt) = (\rho_s v / 4\rho_p)(1 + 2\theta) \tag{27}$$

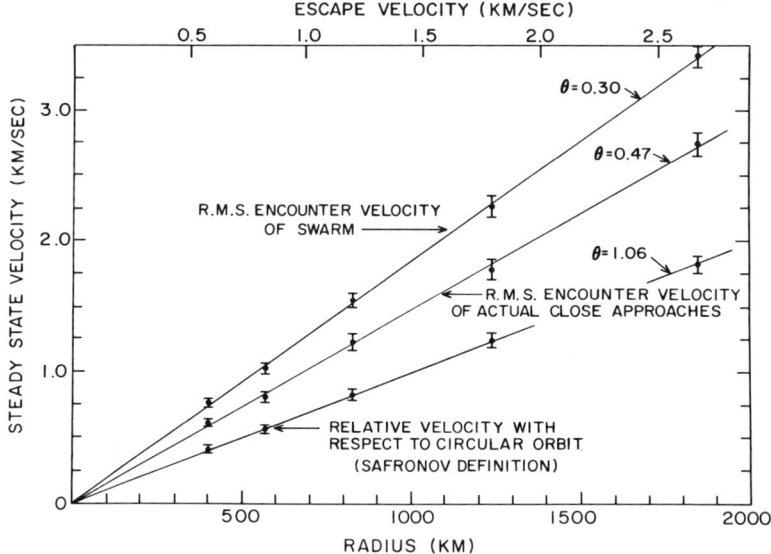

Figure 3 Steady state velocities found by numerical calculations of nonaccreting swarms of bodies of the same size, as a function of radius. The uppermost straight line is the root mean square relative velocity of all the bodies that can encounter one another; the middle line represents the relative velocity weighted according to the probability of encounter; the lowermost line is the mean relative velocity with respect to a circular orbit with semimajor axis equal to the heliocentric distance of the encounter. All of these velocities are linear functions of radius and escape velocity (Wetherill 1979).

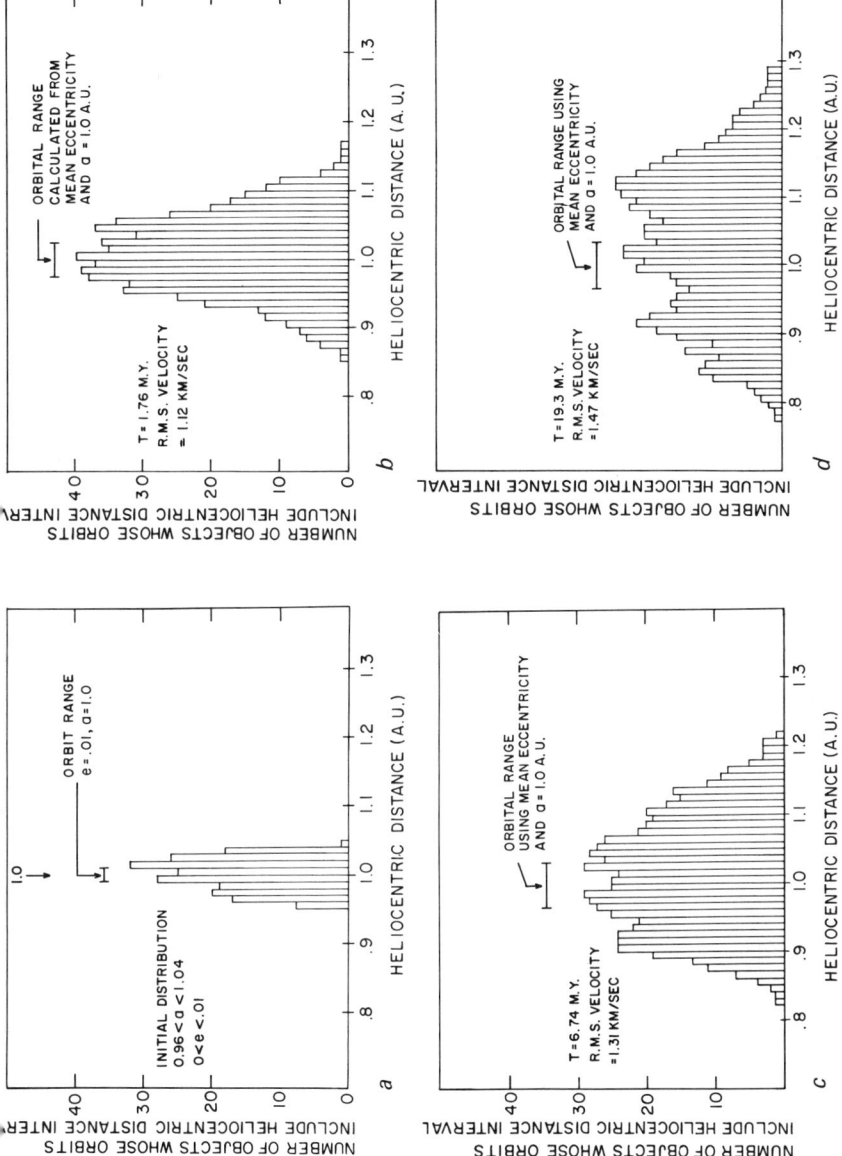

Figure 4 Radial distribution of a swarm of nonaccreting bodies of equal size (9×10^{24} g) as a function of time. The bracket labeled "orbit range" is the range of distance that would be swept out if the radial distribution resulted from all bodies having the same semimajor axis and the mean value of their eccentricity (Wetherill 1979). (*a*) Initial state. (*b*) After 1.76 m.y. Mutual gravitational encounters have caused significant "radial diffusion" even though the velocity has not yet reached the steady state value of 1.45 km/s. (*c*) After 6.74 m.y. The velocity is approaching the steady state value. In a real swarm significant accretion would have taken place. (*d*) After 19.3 m.y. Radial diffusion continues to spread the swarm, as a consequence of conservation of angular momentum in a swarm losing energy by collisions. This case is not physically significant because 19 m.y. is comparable to the accumulation time.

where ρ_p is the density of the solid bodies. The density ρ_s can be estimated by assuming the total mass that is to become a terrestrial planet of mass M_p is distributed over a volume

$$\Omega = 4\pi a \Delta a \Delta z \tag{28}$$

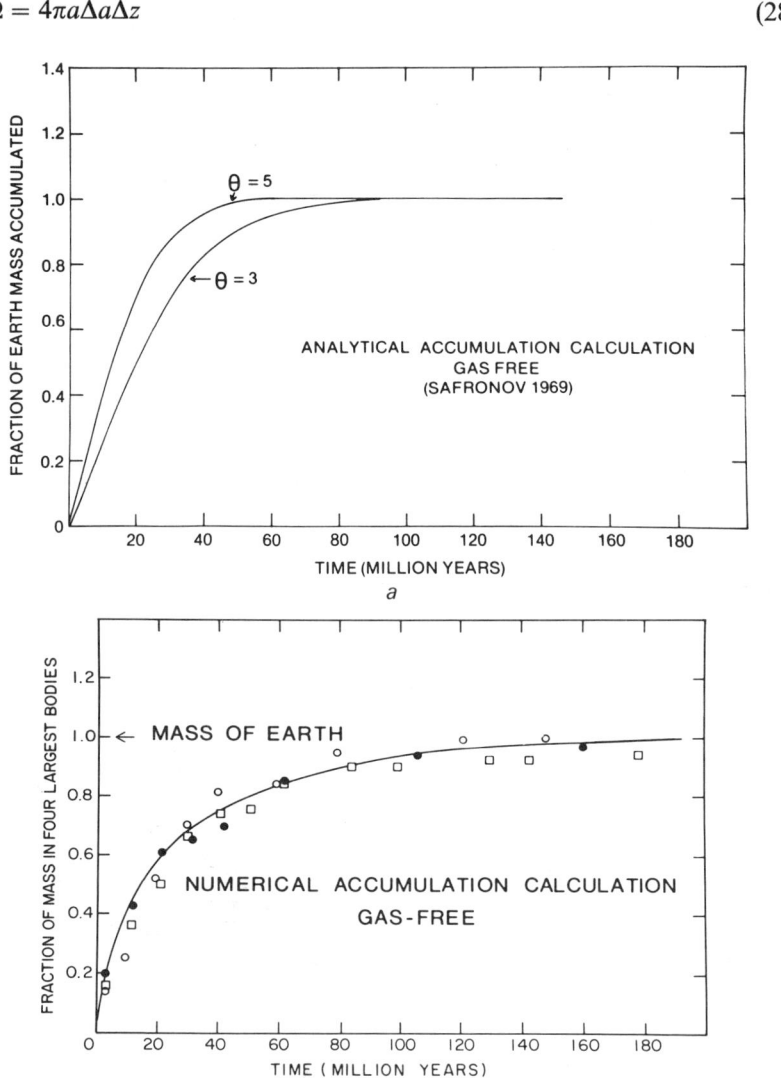

Figure 5a,b Analytical and numerical calculations of accumulation of the Earth as a function of time. (*a*) Analytical accumulation for two values of the Safronov number θ (Equation 14; Safronov 1969). (*b*) Three-dimensional numerical calculations of gas-free accumulation (Wetherill 1979).

where Δz is the thickness of the disc, given by

$$\Delta z \approx a \sin i \approx (av/\sqrt{3}v_c) \tag{29}$$

where v_c is the circular Keplerian velocity $(GM_\odot/a)^{1/2}$ and equipartition is assumed between the vertical, tangential, and radial components of the velocity v. Use of (28) and (29) in (27) gives

$$(dR/dt) = [\sqrt{3}v_c(1+2\theta)M_p]/(16\pi a^2 \Delta a \rho_p) \sim \text{constant}. \tag{30}$$

Substitution of $M_p = 6 \times 10^{27}$ g (Earth mass), $a = 1$ AU, $\Delta a = 0.5$ AU, $v_c = 30$ km s^{-1}, $\rho_p = 5.5$ g cm^{-3}, and $\theta = 3$ gives a radial growth rate

$$(dR/dt) \cong 15 \text{ cm yr}^{-1}. \tag{31}$$

At this rate the growth of the Earth will require 40 million years. The actual time will be longer because the density of the swarm will decrease as the mass is swept up, e.g. when the Earth has grown to 80% of its final radius, ρ_s and (dR/dt) will be a factor of 2 smaller. More detailed analytical calculations of growth time taking this and other factors into consideration have been given by Safronov (1969) under various assumptions and the result is always $\sim 10^8$ yr (Figure 5). The same result is found by numerical calculations (Wetherill 1979; see Figure 5b). One hundred million years seems to be the time scale required for this mode of planet formation to go to completion, under a wide range of circumstances. Equation (30) shows that gas-free growth of a 10 Earth mass Jovian core would require ~ 5 times longer.

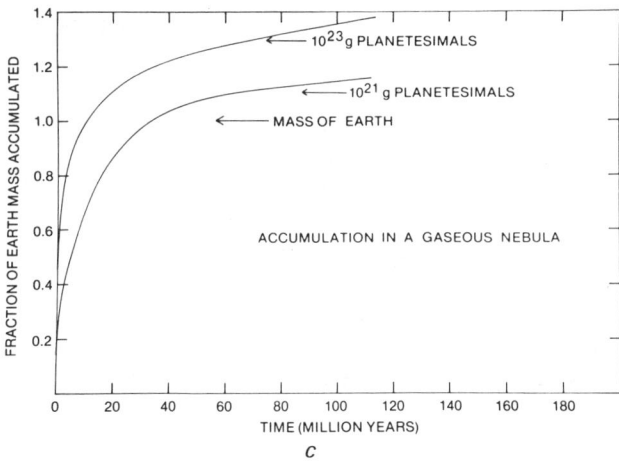

Figure 5c Analytical and numerical calculations of accumulation of the Earth as a function of time. (c) Accumulation in a gaseous nebula, assuming two different sizes for the planetesimals (Hayashi et al. 1977).

Some authors have formally used (26) to infer much more rapid growth rates for the Earth by choosing a value of the velocity that is arbitrarily low, leading to arbitrarily large values of θ. In addition to the problem that v is not a free parameter, but is determined by the mass distribution and physical properties of the swarm, this formal use of (26) is invalid for another reason. When v is too low the orbit of the growing planetary embryo will not intersect all the nearly circular orbits of the bodies in the swarm, and its growth cannot proceed to completion. Weidenschilling (1974) has also shown that for a distribution of semimajor axes the integrated contribution of individual bodies with low values of v converges, and that the singularity as $v \rightarrow 0$ in Equation (16) has no physical significance.

3.1.4 SIMULTANEOUS GAS-FREE ACCUMULATION OF SEVERAL TERRESTRIAL PLANETS In the real solar system the accumulation of the terrestrial planets proceeded more or less simultaneously. Present analytical theories describe only the formation of a single planet because multiplanet growth involves differences in heliocentric distance, which are not introduced in these theories. Older numerical simulations of multiplanet growth are of little relevance because they neglect other essential factors, as discussed by Wetherill (1978). For example, the calculations of Dole (1970) and Isaacman & Sagan (1977) ignore mutual gravitational perturbations, and simply assume velocities that are impossibly high in the gaseous interplanetary medium used in their model. In contrast, Hills (1970) includes gravitational perturbations but ignores collisions.

To my knowledge Cox (1978) is the first work that includes both of these essential phenomena. In his work an assemblage of 100 bodies of equal mass, with total mass equal to that of the present terrestrial planets, is initially confined to the region of the terrestrial planets. The eccentricities are chosen as randomly distributed between zero and a maximum value e_{max}. The inclinations are zero, both initially and throughout the calculation, i.e. the problem is treated two dimensionally. The dynamical evolution of this system is followed. Mutual gravitational perturbations are calculated by a method that has been shown to be an accurate approximation to straightforward numerical integration (Cox, Lewis & Lecar 1978). Coherence and accumulation occur when two bodies collide at sufficiently low energy, and the kinetic energy of their relative motion is dissipated in the collision.

Cox's calculations illustrate clearly a matter of central importance. When the initial eccentricity is chosen to be $\lesssim 0.10$, the final result of the calculation is an assemblage of planets that is too numerous, too small, and too closely spaced (Figure 6a). This is a consequence of averaging of orbits of cohering bodies (Ziglina 1976, Ziglina & Safronov 1976), which causes a number of bodies to evolve into circular orbits as they grow, rather than remain in more eccentric orbits that cross one another. In order for the observed system of terrestrial planets to form, it is necessary that this "isolation of dominant planetary embryos" be delayed until only two large embryos (corresponding to Earth and Venus) and two small embryos (corresponding to Mercury and Mars) remain. The tendency toward isolation must be offset by some process, e.g. higher initial eccentricity (Figure 6b) which opposes the tendency toward circularization.

Extension of Cox's calculations to three dimensions (Wetherill 1979) shows that in the actual problem studied by Cox, the premature embryos were an artifact of the two-dimensional calculations, which underestimate the effect of perturbations caused by encounters, relative to collisions (Figure 6c). There are real effects, such as more collisional damping, which lead to the same problem. In this two-dimensional calculation, collisional damping was probably significantly less than in a real accumulating system. In a real system the 100 large bodies assumed in the calculations would be accompanied by a swarm of small bodies which would cause the entire system to become more dissipative. In the three-dimensional case, it is found that increasing the dissipation by a factor of about three again leads to premature embryos (Wetherill 1979). It is also found that if the four required embryos are placed in nearly circular orbits too early, then unwanted extra embryos will emerge between them (Wetherill 1978 and unpublished work). This effect is related to that of Horedt (1979) who obtained too many small planets with an analytical theory, assuming the feeding zone to be determined solely by the eccentricity of the bodies. Since radial diffusion will occur (Figure 4) Horedt's constraint errs in the direction of being too severe.

These numerical studies of multiplanet accumulation can be thought about in a more general way. Theories of multiplanet accumulation have as their initial state a large number of relatively small bodies in orbits of low eccentricity and inclination, and with a distribution of semimajor axes. They have as their final state an assemblage consisting of a smaller number of larger bodies, also in more or less circular and coplanar orbits. For a theory to be successful the latter assemblage must have at least a high statistical probability of corresponding to something like the observed terrestrial planets in number, size, and semimajor axes. It is significant that the real solar system stopped short of growing the minimum number of terrestrial planets. The numerical studies cited show that for a given initial state there are evolutionary trajectories leading to final states with a range of sizes and number of final planets. The principal parameter defining these trajectories is the ratio of velocity increments caused by gravitational

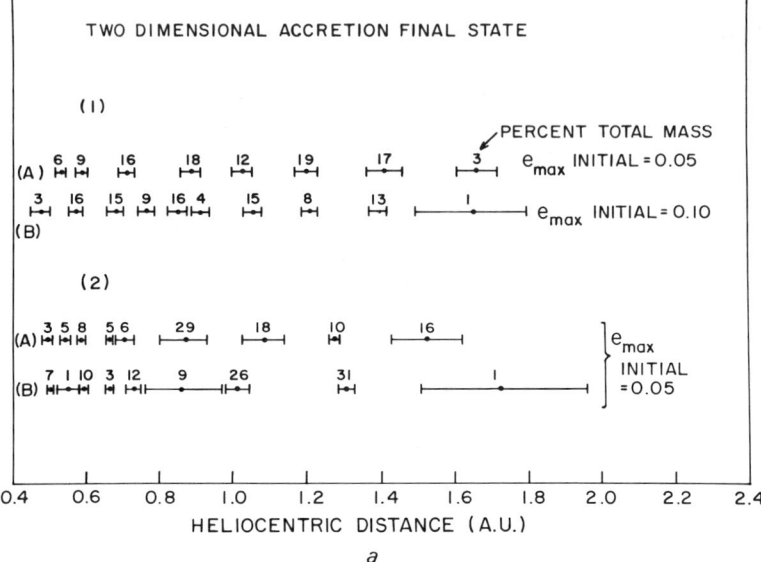

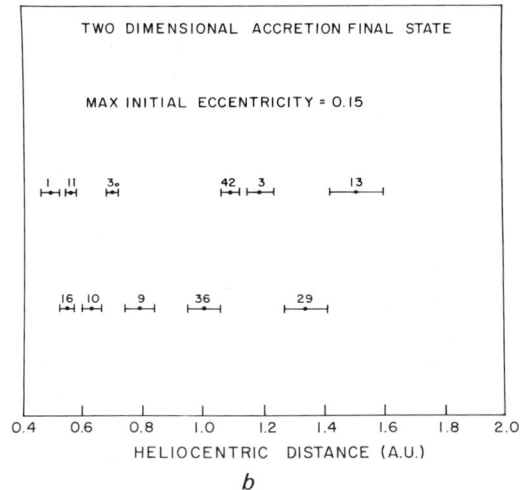

Figure 6a, b Numerical calculations of multiplanet accumulation final states. (*a*) Two-dimensional calculations: (*1*), Cox (1978) and (*2*), Wetherill (1979). For the initial eccentricities chosen, an excessive number of terrestrial planets result. (*b*) Two-dimensional calculations by the same authors showing the smaller number of planets formed if quite large values of initial eccentricity are used.

perturbations to the velocity decrements caused by energy dissipation in collisions, i.e. the value of θ. If this ratio is too low, an excessive number of planets is produced; if the ratio is too high, too few are produced. The actual value of the ratio is determined by the physical properties of the bodies, and to some extent by choice of the initial state. The size distribution of the bodies will be an important, but not an independent, variable, because it will be determined by the physical properties and the initial state.

We can therefore expect that anything that tends to increase this ratio, e.g. more elastic collisions, size distributions weighted toward large bodies, and long range perturbations, will lead to a small number of final planets. Anything that tends to decrease this ratio, e.g. more dissipative collisions or concentration of the mass in small bodies, will lead to a large and possibly excessive number of final planets. The role of gas drag is more complex. Although gas drag tends to damp velocity increments, it also can lead to changes in semimajor axis, which can promote growth by bringing nearly circular orbits into intersection. Gas drag will be discussed further in Section 3.2.

Several authors have recently discussed a model in which the gravitational perturbations are primarily dominated by a single embryo (Levin 1978a,b, Greenberg 1979, Safronov 1979). Safronov (1969) states that the effect will be to "pass control" of the gravitationally induced velocity increments over to the second largest body of the swarm. This is a

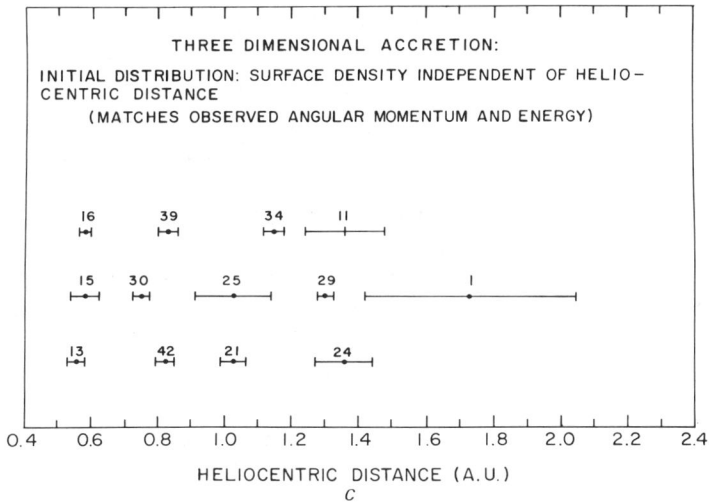

Figure 6c Numerical calculations of multiplanet accumulation final states. (*c*) Three-dimensional calculations with low initial eccentricity ($e_{max} = 0.05$), leading to a small number of planets (Wetherill 1979). In a real solar system greater energy dissipation is probable, which may invalidate this result.

consequence of the Jacobi parameter of the restricted three-body problem being equivalent to the relative velocity of the swarm body and the embryo. Both these quantities tend to be conserved when perturbations from only one embryo become dominant, and the mean velocity increments will be smaller. The effect will be to increase the number of final planets. Because the production of as few as four planets already appears marginal, the conclusion of Levin and Greenberg that low velocities imply shorter time scales for planetary growth is probably incorrect. More likely, the time scales become infinite because needed material is stranded in noncrossing orbits. It is probable that some minimum velocity needs to be maintained during most of the growth following the stage ($\sim 10^{25}$ g bodies) at which the close packing of bodies breaks down. Whether or not this velocity need be as high as that found by merger of the accumulation zones of Earth and Venus (Wetherill 1976) is not known.

If the foregoing discussion is correct, it has the somewhat ironic consequence that the *low* velocities and runaway accretion found by Greenberg et al. (1978) could be necessary during the early stages of accumulation. Such a low velocity may be required in order to remove small bodies from the swarm, and thereby permit *high* velocities to be generated by the relatively unhindered mutual gravitation perturbations of the remaining 10^{25}–10^{26} g bodies during the later stages of growth.

The concept of "jet streams" introduced by Alfvén (1969) may be relevant here. The term envisions an assemblage of colliding bodies gravitationally evolving into a few larger bodies. In some sense those multiplanet accumulation trajectories that lead to a few large planets could be thought of as these jet streams. However, jet streams are said to result when gravitational interactions are negligible, and evolution results from collisional dissipation. This corresponds to an extreme case of zero for the velocity increment/decrement ratio. Numerical studies by Ip (1977), confirmed by unpublished work of the present author, show that insofar as these calculations are relevant to the question, the necessary focusing of semimajor axes does not occur in the absence of mutual perturbations. However, these calculations make the assumption, introduced by Öpik (1951), that the longitudes of perihelion and node are uncorrelated, whereas for the "jet streams" discussed by Alfvén (1969) this correlation is an essential aspect of the concept. The only existing calculations including gravitational perturbations that could bear on the question of whether jet streams can develop are those of Cox (1978), who doesn't discuss the question, and whose results are limited by being two-dimensional. For this reason it is hard to rule out the possibility that in some sense failure of these longitudes to be randomized might be significant. The matter should be investigated further, even though some of the arguments advanced by Alfvén (1971) have been shown to be erroneous (Henon 1978).

During the later stages of growth the kinetic energy accompanying high velocity impact of large bodies with the embryos would be an important source of initial heat (Barrell 1918). If the planets actually formed by gas-free accumulation along the lines discussed in this section, the initial temperatures of the Earth and Venus would be high even though $\sim 10^8$ yr was required for accumulation (Safronov 1969, 1978, Kaula 1979c). Many conventional thermal history discussions (e.g. Hanks & Anderson 1969, Mizutani, Matsui & Takeuchi 1972, Wetherill 1972) incorrectly imply that rapid accumulation ($\sim 10^4$ yr) is required if the initial temperature of the Earth is to be high. This is not the case if terrestrial planets formed in the manner described in this section or, for that matter, in any other way (Section 3.2). It is not clear whether or not smaller bodies (Moon, Mars, Mercury) could be extensively melted by the impacts predicted by this theory (Kaula 1979b,c, Wetherill 1976). If the melting of the Moon was limited to a relatively shallow "magma ocean" (Solomon & Chaiken 1976) the actual melting of the Moon would have been only marginal. Therefore, a correct theory would only predict marginal melting, and it will be difficult to define the physical parameters required in the theory sufficiently to know if the predicted melting is too much or too little.

After the growth of the terrestrial planets is nearly complete, the theory predicts a residual population of large bodies. Some of these will evolve into orbits with lifetimes of 10^8–10^9 yr and will contribute to, perhaps dominate, the cratering history of the Moon and planets for the first $\sim 10^9$ yr of their history. The question of whether or not the observed cratering of the Moon is consistent with gas-free accumulation theory has been discussed (Wetherill 1977), as well as the role this residual swarm may have played in the formation of the asteroidal parent bodies of the differentiated meteorites (Wetherill & Williams 1979, Wasson & Wetherill 1979). Progress on this question will require much more understanding of the importance of tidal disruption of the residual planetesimals by the planets during the first 10^9 yr of solar system history. Existing treatments of tidal disruption consider only the static stability of the body while it is within the Roche limit of the planet (Aggarwal & Oberbeck 1974, Ziglina 1978). However, the ≤ 1 h duration of the passage within the Roche limit may be insufficient to actually fragment and disperse the body.

3.2 *Accumulation in a Gaseous Medium*

If anything like the usual picture of solar system formation from a solar nebula is correct, it must be supposed that at least the earliest stages of growth took place in a gas-rich interplanetary medium, if only to explain the composition of the Sun, Jupiter, and Saturn. It is very uncertain when the interplanetary medium first became essentially gas-free. Presumably this took place after the formation of Jupiter and Saturn, but the time scale for the formation of those planets depends on the way in which they were formed, and the time of gas loss could have been anywhere in the range $\sim 10^5 - \gtrsim 10^8$ yr. Most authors (with the exception of Alfvén & Arrhenius 1976) have assumed that the early ~ 1 km size planetesimals were formed in the presence of at least as much gas as that required to complement the observed nonvolatile mass of the planets ($\sim 0.01\ M_\odot$) and usually enough more (0.03–0.1 $M_\odot$) to permit a nonsolar composition for Jupiter and Saturn and some weakly bound material in the Uranus-Neptune region.

In a solar nebula containing gas and solid particles, the gas phase will be partially supported by a gas pressure gradient effectively changing the gravitational attraction of the Sun. This will cause the velocity v_g of the gas to differ slightly from the Kepler velocity v_k:

$$v_g^2 = v_k^2 + \frac{r}{\rho_g}\frac{dP_g}{dr} \tag{32}$$

where P_g and ρ_g are the pressure and density of the gas, as discussed by Whipple (1973) and Kusaka, Nakana & Hayashi (1970). The gas density and pressure probably decreased with heliocentric distance, and for plausible values of the physical parameters, v_g will be less than v_k by $\sim 0.1\%$. In the absence of gas, solid bodies will move at velocity v_k. Because $v_g < v_k$ they will experience a gas drag force. This drag will cause very small particles to move essentially at the gas velocity. The drag force will have

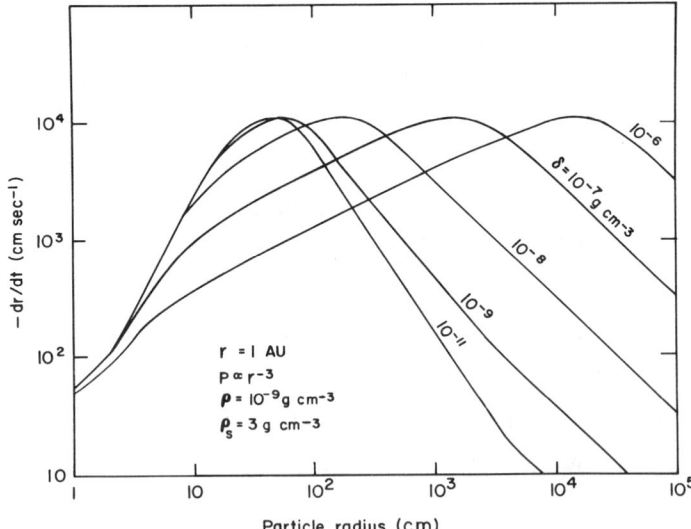

Figure 7 Radial velocity of small bodies, subject to the effects of gas drag and particle-particle collisions (Weidenschilling 1977a,c). δ is the space density of solid particles.

little effect on the motion of very large bodies. Intermediate size bodies will tend to spiral sunward. The resulting radial velocity can be as high as ~ 0.1 km s^{-1} for ~ 1 m size bodies using plausible values of the gas and dust density (Figure 7; Weidenschilling 1977a). In the early stages of planetary growth this size-dependent radial motion could have caused objects to spiral into the Sun, collide with one another, and could have precluded the low dust velocity required for dust layer gravitational instability (Section 2.2). It is possible that such effects may have caused radial separation of iron and silicate particles, and influenced the composition of Mercury (Weidenschilling 1977a).

Assuming that $\sim 10^{18}$ g bodies are formed either by dust layer instability or by sticking processes, the gravitational accumulation of these objects into $\sim 10^{25}$ g objects has been described by Nakagawa (1978). For objects this large, spiraling sunward by gas drag is unimportant. In contrast to the analytical theories discussed in the gas-free case, this analytical theory introduces the radial distribution of the bodies by use of a Fokker-Planck equation (e.g. Richards 1959) for the distribution of the bodies in space and time. Similar use of the Fokker-Planck equation for the gas-free case would be of interest in considering analytically the radial dispersion found numerically (Figure 4). Because this is essentially a diffusion approximation it permits calculation of a characteristic diffusion time for the bodies in phase space and demonstration that the mean square values of the random components of velocity represent a balance between the effect of gravitation encounters and gas drag. The resulting distribution of velocity (rms velocity ~ 16 m s^{-1}) is then used to calculate a growth rate as a function of mass by use of the gravitational cross section (Equation 16) together with Equation (26). It is found that $\sim 10^{22}$ g bodies are formed on a time scale of $\sim 10^4$ yr. As discussed in Section 3.1.1 the bodies are "closely packed" during this stage of growth and the assumption that the accumulation process does not greatly diminish the number of bodies in intersecting orbits is probably a reasonable one.

In many ways the problem calculated analytically by Nakagawa resembles the gas-free numerical calculation of Greenberg et al. (1978), except that in Nakagawa's case gas drag limits the growth of mean velocity, rather than collisional dissipation. Nakagawa calculates a velocity distribution such that for a swarm of equal mass bodies, the root mean square velocity increases with $m^{\sqrt{2}/3}$, rather than with $m^{1/3}$ as found by Safronov (1969) for the equivalent collisional case. (See discussion in Section 3.1.2.) However, Greenberg et al. (1978) find that the velocities of the largest bodies in the accumulating swarm remain very small, as long as most of the mass continues to reside in the smallest bodies. Nakagawa's assumption that the velocity of the larger bodies will be given by the result he found for large bodies of equal mass is not likely to be correct if the result of Greenberg et al. is correct. As long as the assumption of "close packing" remains valid, the effect of the resulting lower velocities of the large bodies will be to cause them to grow even faster than found by Greenberg et al.

Hayashi, Nakazawa & Adachi (1977) consider a later stage of accumulation. It is assumed that, as a result of the process described by Nakagawa, a few $\sim 10^{25}$ g embryos have been formed near the position of the present planets and that the remaining mass is in the form of 10^{21} g bodies. As noted above, because of the difficulty of simultaneously treating by analytical methods the coupled mass and velocity distribution, this distribution may not actually be achieved. It could be that most of the mass will reside in $\sim 10^{25}$ g bodies in nonintersecting orbits, in which case the evolution to be described seems unlikely to occur.

In the presence of gas, capture of the $\sim 10^{21}$ g "planetesimals" by the $\sim 10^{25}$ g embryos will occur if as a result of gas drag and mutual perturbations the body enters the "Hill sphere" of the protoplanet. The Hill sphere is a volume surrounding the embryo bounded radially by the inner and outer collinear Lagrangian points of the restricted 3-body problem (see, for example, Blanco & McCuskey 1961, Szebehely 1967). Its dimensions are of order

$$d \sim \left(\frac{m}{M_0}\right)^{1/3} a \tag{33}$$

and are comparable in size to the "sphere of influence" defined by Equation (21). The probability is high that a body with slightly positive total energy will lose sufficient energy within the Hill sphere to be captured into circum-embryo orbit. This orbit will decay rapidly and the planetesimal will be accumulated by the embryo (Pollack, Burns & Tauber 1979).

As Hayashi, Nakazawa & Adachi point out, the rate of growth of the embryo will be limited by the rate at which planetesimals are supplied to its Hill sphere. For 10^{21} g planetesimals the rate of supply will result from the eccentricity and inclinations being "pumped up" to steady state values $\sim 10^{-3}$ representing a balance between mutual encounters and gas drag. The resulting velocity difference between the planetesimal and the gas will cause it to spiral inward on a time scale of $\sim 10^7$ years at the Earth's distance and $\sim 10^8$ yr at the distance of Jupiter. The increase in Earth mass as a function of time is shown in Figure 5c. Comparison with the gas-free results (Figure 5a,b) indicates that somewhat more rapid accumulation is predicted. However, the result is dependent on the initial size of the planetesimals. If, instead of 10^{21} g objects, the earlier stage of growth considered by Nakagawa resulted in larger bodies in isolated orbits, then the necessary pumping up of eccentricity and inclination would not occur. When the effect of the resulting lower relative velocity between planetesimal and gas is combined with the lower radial velocity characteristic of large bodies because of their small area-to-mass ratio, it is possible that the embryo may not grow much at all. Intermediate conditions would correspond to arbitrarily long accumulation times of $\gtrsim 10^7$ yr. Hayashi, Nakagawa & Adachi calculate more *rapid* accumulation times for large planetesimals. However, this depends on their assumption that these large bodies will have high enough eccentricities for their orbits to remain crossing as they spiral inwards. As discussed in Section 3.1.4, avoidance of premature isolation of embryos is a difficult enough problem in the gas-free case, and the tendency for orbits of very large bodies to become circular and noncrossing will be greater when gas drag is present. As before, long range perturbations might conceivably provide some way out of this problem, but again this may require the prior formation of large planets.

If the Earth formed by accumulation of solid bodies from a gaseous nebula, then a large quantity of gas would also have accumulated. Hayashi, Nakazawa & Mizuno (1979) calculate this to be $\sim 3 \cdot 10^{26}$ g, i.e. 5% of the present Earth mass and 6×10^4 times more massive than the present atmosphere. These workers calculate that this optically opaque atmosphere could lead to surface temperatures as high as 4000 K, and that early melting would occur even though the time scale for accumulation is $\sim 10^7$ yr. Thus all the models of terrestrial planet formation considered, giant gaseous protoplanets, gas-free accumulation, and accumulation in a gaseous nebula, lead to an initial Earth in which very extensive melting and differentiation are predicted. The time scale for accumulation doesn't make any difference in this regard. The situation for the Moon is less clear. The theory of Hayashi, Nakazawa & Mizuno would lead to low lunar initial temperatures because the primordial atmosphere of such a small planet would be very small.

The massive primordial atmosphere would consist principally of H_2 and He and could escape (Hunten & Donahue 1976). It is not clear what would be expected for the other volatile species expected in the solar nebula, e.g. H_2O, CO_2, Ne, Xe. Mizuno, Nakazawa & Hayashi (1979) consider the question of the dissolution of primordial rare gases in the melted Earth, and conclude that quantities of neon 2–400 times the present atmospheric inventory should have dissolved and been trapped in the Earth's interior. A much greater quantity should have been present in the primordial atmosphere. Sekiya, Nakazawa & Hayashi (1979) argue that Kr and Xe may be carried along with H_2 and He if the rate of outflow is sufficiently large. When this problem is fully understood, it may be that the present chemical and isotopic composition of the Earth's atmosphere and interior may provide evidence for or against accumulation in a gaseous solar nebula.

To the author's knowledge, no results of multiplanet accumulation in the presence of a gas phase have been presented as yet. It is hard to guess how this would affect the discussion in Section 3.1.4. During the later stages of accumulation the gas drag forces may permit the removal of smaller bodies from the swarm and allow the velocities of the large bodies to be "pumped up" more readily. On the other hand, the increased dissipation resulting from gas drag would work in the opposite direction and it isn't clear which effect would be more important. Another problem requiring investigation is the effect of small quantities of gas, such as might be released by collisions of planetesimals containing volatile material in concentrations similar to those of carbonaceous meteorites.

4 CONCLUDING REMARKS

Initial states intermediate in the evolution of the solar nebula have been described which are at least consistent with our very imperfect knowledge of star formation. The possible role of the two growth mechanisms, gravitational instability and gradual accumulation, in transforming these initial states into the observed planets has been discussed. In the course of this discussion it should have been obvious that there are many uncertain "leaps of faith" that must be taken. The elimination of some of these uncertainties is straightforward and simply requires some hard work, e.g. the clarification of radial diffusion in multiplanet accumulation. Other questions, such as the size distribution of fragments from tidally disrupted 10^{25} g bodies, will probably be very intractable.

Also it should not be imagined that the initial states discussed are the only ones possible. It does not take very much imagination to think of others and thereby get your name on a theory. However, until we understand how to use the basic mechanisms to go from initial states to prediction of their observable consequences, these speculations are of limited usefulness.

In view of this situation, it seems premature to become very vehement in the advocacy of any particular one of these "theories." This is not a time for "great debates" regarding solar system origin, but rather for the patient solving of difficult theoretical problems, gathering of data, and searching for ways to confront theoretical predictions with observable data.

Not long ago alternative theories of solar system origin were not much more than a source of entertainment to most working scientists. To some extent this view is still prevalent. However, today implicit and explicit ideas concerning solar system formation influence the context and significance of a large fraction of current planetological and geological research. It may be expected that many more able workers will soon be carrying out serious work on problems of solar system origin. It is important that this does not lead to the Tower of Babel, but rather that they try to understand one another and to explain their own work to others as well as they are able. If this happens, the increased interest in this field could lead to remarkably rapid progress on this ancient and fundamental scientific question.

ACKNOWLEDGMENTS

I wish to thank A. G. W. Cameron, L. P. Cox, R. Greenberg, A. W. Harris, W. M. Kaula, G. V. Pechernikova, V. S. Safronov, A. V. Vityazev, S. Weidenschilling, and I. N. Ziglina for valuable discussions and C. Hayashi for illuminating correspondence. A grant from the NASA Planetary Program Office helped make these discussions and the calculations reported herein possible. I also want to express appreciation to L. Finger for maintaining and improving the computing facilities at the Geophysical Laboratory of the Carnegie Institution of Washington and to Mary Coder for help in the preparation of the manuscript.

Literature Cited

Aggarwal, H. R., Oberbeck, V. R. 1974. *Ap. J.* 191:577–88
Alfvén, H. 1969. *Astrophys. Space Sci.* 4:84–102
Alfvén, H. 1971. *Science* 173:522–28
Alfvén, H., Arrhenius, G. 1976. Evolution of the solar system. *NASA SP-345*
Allègre, C. J., Birck, J. L., Fourcade, S., Semet, M. P. 1975. *Science* 187:436–38
Barrell, J. 1918. In *The Evolution of the Earth and Its Inhabitants*, pp. 1–44. New Haven, Conn: Yale Univ. Press
Blanco, V. M., McCuskey, S. W. 1961. *Basic Physics of the Solar System.* Reading, Mass: Addison-Wesley. 307 pp.
Bodenheimer, P., Black, D. C. 1978. In *Protostars and Planets*, ed. T. Gehrels, pp. 288–322. Tucson: Univ. Ariz. Press. 756 pp.
Cameron, A. G. W. 1978a. In *The Origin of the Solar System*, ed. S. F. Dermott, pp. 49–74. New York: Wiley, 668 pp.
Cameron, A. G. W. 1978b. *The Moon and the Planets* 18:5–40
Cameron, A. G. W. 1979. *The Moon and the Planets* 21:173–83
Chandresekhar, S. 1942. *Principles of Stellar Dynamics.* Univ. Chicago Press. 251 pp.
Cox, L. P. 1978. PhD thesis. M.I.T., Cambridge, Mass.

Cox, L. P., Lewis, J. S., Lecar, M. 1978. *Icarus* 34:415–27
DeCampli W. M., Cameron, A. G. W. 1979. *Icarus* 38:367–91
Demarque, P., McClure, R. D. 1977. In *The Evolution of the Galaxies and Stellar Populations*, ed. B. M. Tinsley, R. B. Larson, pp. 199–217. New Haven, Conn: Yale Univ. Press. 449 pp.
Dermott, S. F., ed. 1978. *The Origin of the Solar System.* New York: Wiley. 668 pp.
Dole, S. H. 1962. *Planet. Space Sci.* 9:541–53
Dole, S. H. 1970. *Icarus* 13:494–508
Donnison, J. R., Williams, I. P. 1975. *MNRAS* 172:257–69
Edgeworth, K. E. 1949. *MNRAS* 109:600–9
Ganapathy, R., Anders, E. 1974. *Proc. Lunar Sci. Conf. 5th*, pp. 1181–1206
Gehrels, T., ed. 1978. *Protostars and Planets.* Tucson: Univ. Ariz. Press. 756 pp.
Giuli, R. T. 1968. *Icarus* 8:301–23
Goldreich, P., Lynden-Bell, D. 1965. *MNRAS* 130:97–124
Goldreich, P., Ward, W. R. 1973. *Ap. J.* 183:1051–61
Greenberg, R. 1979. *Icarus* 39:140–51
Greenberg, R., Wacker, J. F., Hartmann, W. K., Chapman, C. R. 1978. *Icarus* 35:1–26
Gurevich, L. E., Lebedinskii, A. I. 1950. *Izv. Akad. Nauk SSSR, Ser. Fiz.* 14:765–99

Hanks, T. C., Anderson, D. L. 1969. *Phys. Earth Planet. Inter.* 2:19–29
Harris, A. W. 1977. *Icarus* 31:168–74
Harris, A. W. 1981. To be published in *Ann. Rev. Earth Planet. Sci.* 9
Hartmann, W. K. 1978. *Icarus* 33:50–61
Hayashi, C., Nakazawa, K., Adachi, I., 1977. *Publ. Astron. Soc. Jpn.* 29:163–96
Hayashi, C., Nakazawa, K., Mizuno, H. 1979. *Earth Planet. Sci. Lett.* 43:22–28
Henon, M. 1978. *Science* 199:692–93
Hills, J. G. 1970. *Nature* 225:840–42
Horedt, G. P. 1978. *Astron. Astrophys.* 64:173–78
Horedt, G. P. 1979. *The Moon and the Planets* 21:63–121
Hunten, D. M., Donahue, T. M. 1976. *Ann. Rev. Earth Planet. Sci.* 4:265–92
Ip, W.-H. 1977. *Proc. Lunar Sci. Conf. 8th*, pp. 67–77
Isaacman, R., Sagan, C. 1977. *Icarus* 31:510–33
Jeans, J. H. 1929. *Astronomy and Cosmogony.* Cambridge, England: Cambridge Univ. Press. 420 pp.
Kaula, W. M. 1979a. *Icarus.* 40:262–75
Kaula, W. M. 1979b. *Geol. Assoc. Can. Spec. Pap.* In press
Kaula, W. M. 1979c. *J. Geophys. Res.* 84:999–1008
Kaula, W. M., Bigeleisen, P. E. 1975. *Icarus* 25:18–33
Kirsten, T. 1978. In *The Origin of the Solar System*, ed. S. F. Dermott, pp. 267–346. New York: Wiley. 668 pp.
Kusaka, T., Nakano, T., Hayashi, C. 1970. *Prog. Theor. Phys.* 44:1580–95
Larson, R. B. 1978. In *Protostars and Planets*, ed. T. Gehrels, pp. 43–57. Tucson: Univ. Ariz. Press. 756 pp.
Levin, B. J. 1978a. *Letters to Astron. Zh.* 4:102–7
Levin, B. J. 1978b. *The Moon and the Planets* 19:289–96
Lyttleton, R. A. 1961. *MNRAS* 122:399–407
Lyttleton, R. A. 1972. *MNRAS* 158:463–83
McCrea, W. H. 1960. *Proc. R. Soc. London Ser. A* 256A:245–66
Mizuno, H., Nakazawa, K., Hayashi, C. 1979. Preprint.
Mizutani, H., Matsui, T., Takeuchi, J. 1972. *The Moon* 4:476–89
Nakagawa, Y. 1978. *Prog. Theor. Phys.* 59:1834–51
Öpik, E. J. 1951. *Proc. R. Irish Acad.* 54A:165–99
Papanastassiou, D. A., Wasserburg, G. J. 1975. *Proc. Lunar Sci. Conf. 6th*, pp. 1467–89
Pechernikova, G. V. 1974. *Astron. Zh.* 51:1305–14. Transl. in *Sov. Astron.-AJ* 18:1305–18
Pechernikova, G. V., Safronov, V. S., Zvygina, E. V. 1976. *Astron. Zh.* 53:612–19. Transl. *Sov. Astron.-AJ* 20:346–50
Polyachenko, V. L., Fridman, A. M. 1972. *Astron. Zh.* 49:157–64. Transl. in *Sov. Astron.-AJ* 16:123–31
Pollack, J. B., Burns, J. A., Tauber, M. E. 1979. *Icarus* 37:587–611
Prentice, A. J. R. 1978. In *The Origin of the Solar System*, ed. S. F. Dermott, pp. 111–61. New York: Wiley. 668 pp.
Richards, P. I. 1959. *Manual of Mathematical Physics.* London: Pergamon. 486 pp.
Ringwood, A. E. 1979. *Origin of the Earth and Moon.* New York: Springer. 295 pp.
Safronov, V. S. 1960. *Vopr. Kosmog.* 7:121
Safronov, V. S. 1969. *Evolution of the Protoplanetary Cloud and Formation of the Earth and Planets.* Moscow: Nauka. Transl. Israel Program for Scientific Translations, 1972. *NASA TTF-677*
Safronov, V. S. 1975. In *Proc. Soviet-American Conf. Cosmochem. of the Moon and Planets*, pp. 624–29. Moscow: Nauka. Also in *NASA SP-370*, pp. 797–803
Safronov, V. S. 1978. *Icarus* 33:3–12
Safronov, V. S. 1979. *Icarus.* In press
Safronov, V. S., Ruskol, E. L. 1957. *Vopr. Kosmog.* 5:22
Sekiya, M., Nakazawa, K., Hayashi, C. 1979. Preprint of manuscript submitted to *Earth Planet. Sci. Lett.*
Solomon, S. C., Chaiken, J. 1976. *Proc. Lunar Sci. Conf. 7th*, pp. 3229–43
Strom, S. E., Strom, K. M., Grasdalen, G. L. 1975. *Ann. Rev. Astron. Astrophys.* 13:187–216
Szebehely, V. 1967. *Theory of Orbits.* New York: Academic. 668 pp.
Ter Haar, D. 1950. *Ap. J.* 111:179–90
Tisserand, F. 1896. *Traité de Mécanique Céleste* V4:200. Paris: Gauthier-Villars
Wasson, J. T., Wetherill, G. W. 1979. *Asteroids*, ed. T. Gehrels, pp. 926–75. Tucson: Univ. Ariz. Press
Weidenschilling, S. J. 1974. *Icarus* 22:426–35
Weidenschilling, S. J. 1975. *Icarus* 26:361–66
Weidenschilling, S. J. 1977a. *MNRAS* 180:57–70
Weidenschilling, S. J. 1977b. *Astrophys. Space Sci.* 51:153–58
Weidenschilling, S. J. 1977c. *Carnegie Inst. Washington Yearb.* 76:755–60
Weidenschilling, S. J. 1979. *Lunar Science X*, pp. 1327–28 Houston: Lunar & Planet. Inst.
Wetherill, G. W. 1972. *Tectonophysics* 13:31–45
Wetherill, G. W. 1975. *Ann. Rev. Nucl. Sci.* 25:283–328
Wetherill, G. W. 1976. *Proc. Lunar Sci. Conf. 7th*, pp. 3245–57

Wetherill, G. W. 1977. *Proc. Lunar Sci. Conf. 8th*, pp. 1–16

Wetherill, G. W. 1978. In *Protostars and Planets*, ed. T. Gehrels, pp. 565–98. Tucson: Univ. Ariz. Press. 756 pp.

Wetherill, G. W. 1979. *Geol. Assoc. Can. Spec. Pap.* In press

Wetherill, G. W., Williams, J. G. 1979. In *Origin and Distribution of the Elements, Second Symposium*, ed. L. H. Ahrens, pp. 19–31. Oxford: Pergamon

Whipple, F. L. 1973. *Evolutionary and Physical Properties of Meteoroids*, ed. C. L. Hemenway, P. M. Millman, A. F. Cook, pp. 355–61. *NASA SP-319*

Wood, J. A. 1961. *MNRAS* 122:79–88

Woodward, P. R. 1978. *Ann. Rev. Astron. Astrophys.* 16:555–84

Ziglina, I. N. 1976. *Astron. Zh.* 53:1288–94. Transl. in *Sov. Astron-AJ* 20:730–733

Ziglina, I. N. 1978. *Izv. Akad. Nauk SSSR Fiz. Zemli* 7:3–10

Ziglina, I. N., Safronov, V. S. 1976. *Astron. Zh.* 53:429–35. Transl. in *Sov. Astron-AJ* 20:224–48

Zvyagina, Y. V., Pechernikova, G. V., Safronov, V. S. 1973. *Astron. Zh.* 50:1261–1273. Transl. in *Sov. Astron-AJ* 17:793–800

STELLAR MASSES

✗2164

Daniel M. Popper

Department of Astronomy, University of California, Los Angeles, California 90024

INTRODUCTION

In an earlier review (Popper 1967), the problems of determining masses from data for eclipsing binaries were discussed and some of the available results tabulated. The present review may be considered a sequel, although the most definitive visual binaries are also included. The same viewpoint is adopted, namely that only individual masses of considerable accuracy, determined directly from the observational data, are treated. In the first section of the earlier review, methods of mass determination based on various hypotheses (moving groups, photometric parallaxes, assumptions about the mass of one component of a binary, etc.) were discussed briefly and will not be referred to. The principal justification for this review lies in the very considerable increase in available data. Two classes of binary systems are not treated, contact binaries of the W UMa class and close binaries with a degenerate component (X-ray binaries, binary pulsars, dwarf novae, novae). The observational status of W UMa systems has been summarized by Binnendijk (1970), while Bahcall (1978) has discussed the masses of collapsed objects in X-ray binaries.

A list of the properties of the components of eclipsing binaries was prepared some years ago by Wood (1963). My commentary on this list (Popper 1966) was directed primarily to the question of the suitability of the observational data for providing reliable properties. More recent tables of the properties of binary star components, in addition to the brief list in my review (Popper 1967), have been published by Cester (1965a, eclipsing binaries; 1965b, visual binaries), Popov (1969a,b), Kříž (1969), McCluskey & Kondo (1972), Giannone & Giannuzzi (1974), Heintz (1978), and Lacy (1977b, 1979). Svechnikov (1969) assumes mass ratios for many systems. Popov, in particular, attempted to assess the uncertainties in the resulting parameters. McCluskey & Kondo, Heintz, and Lacy, as well as Cester, included visual binaries in their discussions. The data on eclipsing binaries included in these lists were taken from a large variety of published sources.

Many of the results are based on spectrographic material obtained with prismatic spectrographs of relatively low dispersion, on photographic or visual light curves, or on studies in which details of the observational material are lacking. Lacy's lists are the most selective.

For most of the eclipsing binaries listed in the following sections, the analyses are based on spectrographic observations obtained with grating spectrographs giving 20 Å mm^{-1} or higher dispersion. In each case I have either examined spectrograms or have satisfied myself from published or unpublished material that the spectrographic material is of adequate quality and free of appreciable systematic errors. Justification for this approach to spectrographic material is contained in a series of papers, "Rediscussion of Eclipsing Binaries," appearing in the *Astrophysical Journal* and is also discussed in the earlier review (Popper 1967). Thus, the present review has more personal and perhaps more consistent standards than discussions in which published results are accepted at face value.

The quality of photometric observations can usually be judged from available light curves. Only photoelectric data are accepted. The form of a light curve is much less clearly dependent on the values of the system parameters than in the case of a velocity curve. Hence, the reliability of a photometric solution is more difficult to judge than that of a spectrographic solution. These matters are discussed in more detail in subsequent sections, as are the criteria for including visual binaries. The comments of Andersen, Clausen & Nordström (1979) on criteria for absolute dimensions should be read in the present context.

Emphasis is placed on the determination of mass because it is the quantity upon which the other properties, such as luminosity and rate of evolution, depend most sensitively. Masses are of little interest, however, without knowledge of other fundamental properties of the stars, which are usually determinable if the masses are. For visual and resolved spectroscopic binaries the masses and absolute visual magnitudes are obtained directly from the observations of a particular system. Evaluation of bolometric luminosities, effective temperatures, and radii, on the other hand, requires the use of scales of temperature, bolometric correction, and absolute radiative flux, calibrated in terms of color index or some other correlate of radiative flux. For eclipsing binaries it is the masses and radii that are obtained directly for individual stars, while effective temperatures, bolometric luminosities, and absolute visual magnitudes require calibrated scales. The scales adopted for these radiative quantities are based primarily on the scales of effective temperature and bolometric correction given by Hayes (1978), which is essentially that of Code et al. (1976) for the hotter stars, and on the scale of absolute visual fluxes compiled by Barnes, Evans & Moffett (1978). While Barnes, Evans & Moffett (BEM) have preferred to employ the color index $V-R$ as the observable parameter correlated with surface flux, their study shows that, except for the coolest stars, the color index $B-V$ serves as well. For a given binary, the color index that is known with greater accuracy, usually $B-V$, is used. The intermediate-band index, $b-y$, correlates closely with $B-V$, so that $b-y$ observations are also employed. In some cases it is helpful to make use of the relationship between spectral type and color index. The list of Hayes (1978), which is in essential agreement with the compilation by FitzGerald (1970), is adopted.

Several considerations have led me to modify slightly the relations between visual absolute flux and color indices published by BEM. First, BEM have adopted linear relations between $F_V \propto$ log of flux and color index over some ranges of temperature. Examination of their data, particularly for the B and A stars, where the angular diameter results for main-sequence stars are most numerous, shows that systematic departures from linear relations are significant. Second, the scatter in the relation between flux and color index is reduced somewhat by considering $b-y$ as well as $B-V$ or $V-R$ for the calibration stars with measured angular diameters. Third, the adopted scales of T_e, F_V, and bolometric correction (B.C.) must be consistent with each other:

$$F'_V(B-V) = \log T_e(B-V) + 0.1 \, \text{B.C.} \, (B-V). \tag{1}$$

The symbol F'_V is used here rather than the F_V of BEM since the flux scale is established in a manner slightly different from that in BEM. The slight difference of a few hundredths of a magnitude in the two flux scales is caused by the use of stellar absolute fluxes rather than the sun's alone and by the adoption of Hayes' (1978) scale of bolometric corrections, which gives -0.14 mag for stars of solar color rather than -0.07. The reason for abandoning the time-honored value -0.07 for the sun as the zero point for bolometric corrections is, in addition to consistency with Hayes' scale, Hayes' (1979) argument that the colors and spectral energy distribution are not so well known now for the sun as they are for spectrophotometric standard stars. In employing Equation (1), Hayes' (1978) scale of bolometric corrections is adopted.

The form of the $F'_V(B-V)$ relation is best defined from angular diameters of main-sequence stars in the range $-0.15 < (B-V) < +0.06$. The zero point of the F'_V scale is then adjusted according to Equation (1) by employing Hayes' T_e's over this color range. Then the form of the T_e-$(B-V)$ relation is modified slightly by requiring that (1) be satisfied at each value of $B-V$. Over the range of temperatures less than about 8000 K, the only main-sequence angular diameters available for temperature and flux calibration are of α CMi and the sun and, for M dwarfs, of YY Gem and CM Dra.

Table 1 Radiative parameters

Spectrum	B–V	b–y	V–R	B.C.	$F'_V{}^a$	$\log T_e{}^a$
Main sequence						
O7	−0.31	—	—	−3.6	4.225	4.585
O8	−0.305	—	—	−3.4	4.211	4.551
O9	−0.30	—	—	−3.2	4.201	4.521
O9.5	−0.295	—	—	−3.1	4.187	4.497
B0	−0.285	—	—	−2.96	4.179	4.475
B0.5	−0.28	—	—	−2.83	4.172	4.455
B1	−0.26	—	—	−2.59	4.159	4.418
B2	−0.24	−0.115	—	−2.36	4.128	4.364
B3	−0.20	−0.090	—	−1.94	4.086	4.280
B5	−0.16	−0.072	—	−1.44	4.046	4.190
B6	−0.14	−0.063	—	−1.17	4.032	4.149
B7	−0.12	−0.052	—	−0.94	4.018	4.112
B8	−0.09	−0.035	—	−0.61	4.002	4.063
B9	−0.06	−0.024	—	−0.31	3.984	4.015
A0	0.00	+0.002	—	−0.15	3.959	3.974
A2	+0.06	+0.031	—	−0.08	3.935	3.943
A5	+0.14	+0.075	—	−0.02	3.909	3.911
A7	+0.19	+0.105	—	−0.01	3.889	3.890
F0	+0.31	+0.195	—	−0.01	3.843	3.844
F2	+0.36	+0.238	—	−0.02	3.824	3.826
F5	+0.43	+0.282	—	−0.03	3.807	3.810
F8	+0.54	+0.350	+0.47	−0.08	3.777	3.785
G0	+0.59	+0.382	+0.50	−0.10	3.762	3.772
G2	+0.63	+0.400	+0.53	−0.13	3.755	3.768
G5	+0.66	+0.417	+0.54	−0.14	3.748	3.762
G8	+0.74	+0.455	+0.58	−0.18	3.720	3.738
K0	+0.82	+0.485	+0.64	−0.24	3.691	3.715
K2	+0.92	+0.525	+0.74	−0.35	3.655	3.690
K5	+1.15	—	+0.99	−0.66	3.567	3.633
K7	+1.30	—	+1.15	−0.93	3.511	3.604
M0	+1.41	—	+1.28	−1.21	3.468	3.589
M1	+1.48	—	+1.40	−1.49	3.422	3.571
M2	—	—	+1.50	−1.75	3.381	3.556
M3	—	—	+1.60	−1.96	3.346	3.542
M4	—	—	+1.70	−2.28	3.300	3.528
M5	—	—	+1.80	−2.59	3.254	3.513
M6	—	—	+1.93	−2.93	3.204	3.497
M7	—	—	+2.20	−3.46	3.113	3.459
M8	—	—	+2.50	−4.0	3.018	3.418
Red giants						
G0	+0.64	—	+0.53	−0.13	3.750	3.763
G5	+0.90	+0.555	+0.685	−0.34	3.642	3.676
G8	+0.95	+0.584	+0.72	−0.38	3.624	3.662

Table 1 (*continued*)

Spectrum	B–V	b–y	V–R	B.C.	$F_V'^a$	$\log T_e^a$
K0	+1.01	+0.617	+0.77	−0.42	3.594	3.636
K1	+1.09	+0.660	+0.81	−0.48	3.581	3.629
K2	+1.16	+0.710	+0.84	−0.53	3.571	3.624
K3	+1.26	—	+0.96	−0.60	3.533	3.593
K4	+1.43	—	+1.06	−0.90	3.501	3.591
K5	+1.51	—	+1.20	−1.19	3.456	3.575
M0	+1.57	—	+1.23	−1.28	3.446	3.574
M1	—	—	+1.28	−1.36	3.430	3.566
M2	—	—	+1.34	−1.52	3.411	3.563

[a] Primary determinants of F_V' and T_e
O7–B1: spectral type
B2–K7, main sequence: $B–V$
M0–M8, main sequence: $V–R$
G0–K2, red giants: $B–V$
K3–M2, red giants: $V–R$

Furthermore, α CMi appears somewhat anomalous in its color-color relationships. The F_V' and T_e scales have been adjusted over the range $+0.2 < B-V < +0.8$ by fitting F_V' to the solar values of T_e and B.C. at $B-V = +0.66$ (Barry, Cromwell & Schoolman 1978, Hayes 1979). Although Hayes' discussion is very comprehensive, a recent determination by Clements & Neff (1979), $(B-V)_\odot = +0.63$, reminds us that the determination of this fundamental quantity is still not completely satisfactory.

The scales for cooler main-sequence stars are even less well established. Hayes' (1978) scales of T_e and B.C. are adopted since they fit the flux for YY Gem better than the BEM flux scale, which is based primarily on angular diameters of giants over this range. For giant stars, the BEM flux scale is adopted since it is based on a considerably greater number of angular diameters than is Hayes'. Throughout the whole range of spectral types, Hayes' list of bolometric corrections has been employed in conjunction with Equation (1). The scales used in the following sections of this review are given in Table 1. For types B1 and earlier, spectral type is taken as the observable quantity with which F_V', T_e, and B.C. are correlated rather than $B-V$, while for stars with $B-V > 1.2$, $V-R$ is a much better discriminant, although Kunkel & Rydgren (1979) note difficulties with $V-R$ for M dwarfs, as does Lacy (1979) for the redder stars.

The principal uses to be made of the values in Table 1 are, in addition to supplying effective temperatures, to determine the absolute visual magnitudes of eclipsing binaries:

$$M_V = -5 \log R - 10 F_V' + C_1, \tag{2}$$

where R is the stellar radius in solar units and C_1 is a constant to be adjusted. The tabular values are also required to evaluate the radii of visual binaries:

$$\log R = -0.2 M_V - 2 F'_V + 0.2 C_1. \tag{3}$$

The value of C_1 employed, 42.255, is 0.11 mag less than obtained from the BEM flux scale. The difference arises from my use of the best-determined angular diameters of 11 stars on the main sequence later than B5 in addition to the sun, which is the sole basis for the BEM value. Higher weight is given to the sun and the four other stars with largest diameters (α CMa, α CMi, α Lyr, and α Aql). Hayes (1979) has stressed the uncertainties introduced by relying too heavily on the sun alone. In computing the luminosities in solar units, solar values must be employed, for which $T_e = 5780$ K, $M_V = +4.83$, and B.C. $= -0.14$ are adopted in connection with the temperatures in Table 1 (eclipsing binaries) and the bolometric corrections (visual binaries).

There remain the effects on the radiative scales of differing surface gravity and chemical composition. There is a considerable range of metallicity, at least among A-type eclipsing binaries (Popper 1971b, Kitamura & Kondo 1978). Because of effects of differing wavelength-dependent line absorption, the relations between color index, absolute visual flux, and effective temperature are likely to be gravity and composition dependent. The data for the calibrating stars with angular diameters are quite inadequate to establish these differences, and recourse must be had to atmospheric models. Models available for this purpose are those of Bell (1971) for temperatures in the range 5000 K to 7400 K and metal abundances equal to solar, 1/4, and 4 times solar, with $\log g = 4.4$, 4.0, and 3.6; of Bell & Gustaffson (1978) for temperatures 5500 K and 6000 K, metal abundances equal to 1/2 and 1/10 solar, $\log g = 3.0$ and smaller; and of Kurucz (1979a) for temperatures 5500 K to 50,000 K, abundances equal to solar and 1/10 solar, and $\log g = 4.5$ and smaller. Results from the more complete grid, including heavy element abundances greater than solar (Kurucz 1979b), are not as yet available. These models, although not completely consistent with each other and stressing lower heavy element abundances rather than the higher abundances encountered in A-type eclipsing binaries, imply that, for a given value of the index $B-V$, stars with greater heavy element abundance have higher values of T_e and larger radiative fluxes in the V band. For most of the stars included in this review, good measures of heavy element abundance are not available. Higher surface gravities lead to small effects in the same direction as higher metal abundance. For the range of abundances likely to be encountered in population I and disk stars and for the range of surface gravities of main-sequence and subgiant stars, the effects appear to correspond to differences up to about 0.1 mag in luminosity for a given value of $B-V$. Uncertainties in the calibrations of the temperature and flux scales are of the same order. Since the emphasis of this review is on stellar masses rather than on radiative properties, further discussion of the rather complex details involved is not warranted. As Lacy (1979) has surmised, the models show that the predicted flux versus color index relations are considerably less sensitive to composition when $V-R$ is used as an index rather than $B-V$. But at the 0.1 mag level, the difference is relatively unimportant in view of uncertainties in the flux and temperature calibrations and in the radii, parallaxes, and, in some cases, color indices of the components of the binary stars. The temperature and flux scales of Table 1 are adopted without adjustment for the small effects discussed here.

For each star with mass and other properties contained in most of the tables of this review, an effort is made to estimate realistic values of the uncertainties of the parameters. Considerable personal judgment is involved in these estimates. In some cases, it is permissible to accept the formal mean errors of published values, for example, for the masses of eclipsing binaries with orbital inclinations, i, close to 90° and with well-separated unblended lines of both components measured on spectrograms of adequate dispersion (greater than that corresponding to 20 Å mm^{-1} in most cases). But for most parameters the formal mean errors cannot be accepted literally. Examples are parallaxes of visual binaries and relative radii of eclipsing binaries, for both of which differences between values from independent investigations are often many times larger than the internal mean errors. In such cases only estimated uncertainties, rather than formal mean errors, can be assigned. I have tried to be conservative in the sense that the adopted uncertainty should not be smaller than a "true" mean error based on all sources of inaccuracy, internal and external. Such estimates should be taken as indications of the ranges within which, in the sense of a mean error, the parameters are thought to lie.

For the sake of convenience, the binary systems considered in this review are grouped into five categories: 1. detached main-sequence eclipsing binaries of types B6 to M; 2. OB eclipsing systems; 3. detached evolved eclipsing systems; 4. resolved spectroscopic binaries; 5. visual binaries; and 6. semidetached eclipsing systems.

DETACHED MAIN-SEQUENCE ECLIPSING SYSTEMS, B6 to M

General Considerations

Among all classes of binary systems, detached main-sequence eclipsing systems, with the spectrum lines of the two components well separated relative to their widths and with light curves showing two well-defined minima, provide the greatest number of reliable masses and radii. With few exceptions, systems earlier than B6 do not have such favorable properties. Problems of adequate dispersion and the selection of lines to measure in order to avoid systematic effects in the velocities were discussed in some detail in the earlier review, as well as elsewhere (Popper 1971a). With good homogeneous spectrographic material, the details of the method of orbit computation are relatively unimportant. In these cases the statistical mean errors of the minimum masses and orbital radii should be realistic approximations to their actual uncertainties.

In systems with these properties, effects of tidal distortion and of reflection on the light curves are usually quite small, as are irregularities associated with gas streams or intrinsic variability, encountered in other classes of close binaries. Then, if both minima are well observed with homogeneous photoelectric observations obtained under good conditions (scatter of points less than 0.01 mag), the values of the relative radii, r_h and r_c (h = hotter component, c = cooler) and of the light ratio L_h/L_c should be derivable with uncertainties of a few percent or less and the inclination i to within a degree or less. Here also the method of solution should be relatively unimportant, provided that all parts of the light curve are represented by the adopted elements without systematic trends in the residuals.

Many papers have been published in recent years describing computer programs and techniques for the analysis of eclipsing binary light curves, and it is not appropriate to review them here. The program of Wood (1972 and subsequent status reports) appears to have had the widest application for relatively undistorted systems. Etzel (1975) has shown that the method of Nelson & Davis (1972), modified by him to provide simultaneous, rather than sequential, adjustment of the elements by the method of least squares, gives results identical to those of Wood's program for spherical stars and is much more efficient of computer time. Numerous applications of Wood's and other computing programs to available observations have appeared in the literature. Cester and his co-workers (Cester et al. 1978c and references therein), in particular, have applied Wood's program to quite a few systems. Judicious use of the Russell model (Russell & Merrill 1952) may provide adequate representation of simple systems.

As mentioned in the introduction, the uncertainties of the parameters derived from a light curve (r_h, r_c, i, L_c/L_h) are much more difficult to evaluate than in the case of a radial velocity curve. Formal mean errors of the parameters in a least-squares adjustment based on several hundred photoelectric observations may be small fractions of a percent. On the other hand, representation of the observations with parameters differing from the "best" set by many times their formal mean errors may be indistinguishable from representation by the "best" set. Moreover, differences between the values of the parameters based on independent observations are usually much larger than the mean errors of either set. The probable reason for this situation is that the values of some of the parameters are very sensitive to small changes in the observations in critical parts of the minima. Even with several hundred observations (the usual number), each section of each minimum will usually be covered by observations on a very small number of nights. It is well known that small unexplained differences are often present between observations on different nights with identical equipment. The existence of such effects invalidates somewhat the use of the method of least squares, which requires that observational errors, particularly in critical parts of a light curve, be randomly distributed. Hence, the fitting of a particular set of observations by least-squares adjustment, while giving the "best fit" to that set in a formal sense, does not necessarily give a realistic evaluation of the uncertainties of the parameters. A better estimate of the real uncertainties of the parameters in a photometric solution may be obtained from the differences in the elements derived for different wavelength bands or, even better, from differences between the results from completely independent sets of equally good observations.

In the solution of a light curve observed in two or more wavelength regions, it is customary to evaluate the color indices of the two components from their light ratios in the two regions, obtained in the photometric solutions. But these light ratios contain uncertainties that are related to uncertainties in the geometric parameters. In the case of an eclipse of appreciable depth, the color of a star being eclipsed is best derived directly from the color index of the light lost during eclipse, without reference to the geometric elements or the corresponding light ratios. The difference in limb darkening at the two wavelengths should be taken into account. Failure to use this procedure may introduce another source of inaccuracy in the fluxes if the two components differ appreciably in color index, and hence an inaccuracy in the luminosities.

An additional source of uncertainty in the photometric parameters is the real indeterminacy in the ratio of the radii of the components in the case of partial eclipses of nearly equal depth. Clausen & Grønbech (1977 and references therein) show that even with excellent observations and use of a computer program this indeterminacy may remain. The ability of a computer to "converge" on a particular solution has, on occasion, led investigators to ignore this basic indeterminacy. Determinations of light ratios from spectrograms (e.g. Petrie 1950) have been employed to remove the indeterminacy, a procedure of limited validity in most cases. Qualitative results on the light ratio from spectrograms may be useful to detect incorrect photometric solutions (e.g. Popper 1976b). But the assumption that the light ratio is given by the ratio of line intensities, when the components are not identical, may not be valid because of the usually unknown dependence of line strength on temperature and gravity, which must be different for the two components if the stars are not identical. Hence, in systems with approximately but not exactly equal partially eclipsed components, neither photometric nor spectrographic analysis may lead to satisfactory light ratios or ratios of the radii. In such systems, differences in mass and in surface brightness (depths of minima) may be well determined, but not differences in radius or luminosity.

It is clear from this discussion that evaluation of meaningful uncertainties of the parameters in a photometric solution is subject to considerable difficulty, particularly when all details of the observations and solution are not provided. Thus, for favorable detached systems, the uncertainties in the masses will be well established, those in the radii less so. The procedures employed for evaluating surface temperatures and luminosities from observed quantities correlated with surface flux is discussed in the introduction. If the photometric observations, precise as they may be, are not accurately tied to one of the standard systems, good estimates or surface fluxes will not be available. The inability to evaluate surface fluxes well is an important cause of uncertainties in the stellar luminosities for a number of systems.

The Available Data

Table 2 contains masses, radii, and color indices for those detached systems with spectral types later than B5 for which the masses and radii are, in my judgment, known to an accuracy of 15% or better. The values in Table 2 are based on my own assessment of published and unpublished material. Table 2 contains a reference list showing where the material may be found from which the values of mass, radius, and color are derived. The list of references is not intended to be complete, but rather gives the sources considered most useful. The catalogues of Koch, Plavec & Wood (1970) and of Batten, Fletcher & Mann (1978) are cited wherever they contain references to material that is used. The Notes to the latter catalogues have proven exceptionally helpful. My apologies are extended to those whose work has not been adequately cited.

In Table 2 the systems are listed in order of decreasing mass of the more massive component of the system. The spectral types are listed for convenience. Except for the primaries in systems with appreciable color excess, the types are not actual evaluations, but are derived from the tabulated values of $(B-V)_0$ by use of the listing in Table 1. For AS Cam, χ^2 Hya, V451 Oph, RX Her, SZ Cen, MY Cyg, and DM Vir, the color excess is derived from multicolor indices. For V805 Aql and CM Lac, the excess is obtained from the observed colors and estimated spectral types of the primary components.

The estimated uncertainties in the masses listed are, for the most part, derived directly from the mean errors of the spectroscopic elements, K_1 and K_2, since these systems are considered to be without important sources of systematic error in the velocities. The estimated uncertainties in the radii are based on an assessment of several factors, and represent a judgment rather than an evaluation. In nearly every system, problems with the photometric solution dominate. For reasons given above, the published mean errors of the photometric elements obtained from a least-squares fit are not adopted. While the sum of the relative radii, $r_1 + r_2$, may be known to an accuracy of 1% or better in the more favorable cases, the individual radii usually are not. I have adopted uncertainties in the range 1–10% for the individual radii, depending on the quality of the observations, the depths of minima, the agreement between different investigations, and the quality of the evidence from spectrograms. In some cases, the components differ clearly in surface brightness, as shown by the different depths of the minima (and indicated by the difference in the listed values of $[B-V]_0$). These components probably differ in radius as well, but the observations are often inadequate to determine the ratio of the radii, a well-known indeterminacy in light-curve analysis, and a ratio of unity has been adopted. These systems are indicated by the footnote symbol c in Table 2. If the reader is interested in the accuracy of the *mean* radius of the components of a system, an uncertainty of about one-half to one-third that given in Table 2 for the individual components may be adopted.

For most of the systems, two or more independent measures of color index are available. Indices on the Strömgren system have been converted to $(B-V)_0$. The tabulation by Crawford & Barnes (1970) establishes a very tight relationship between $(b-y)_0$ and $(B-V)_0$. The dependence on metallicity in this relationship amounts to less than 0.01 mag in $B-V$ and is not clearly established. For reasons discussed in the introduction, the adopted values of $B-V$ have not been "adjusted" for ultraviolet excess or metallicity. Only rarely is the difference between the color indices of the two components well established from the photometry. On the other hand, the ratio of the surface fluxes *is* well determined directly from the depths of the minima for systems with circular orbits and spherical stars, as most of the systems in Table 2 are. In numerous recently published analyses of light curves, this ratio is, unfortunately, not given. The individual color indices listed in Table 2 are, in most cases, derived from the combined color index and from the relation between difference in color index and flux ratio, obtained from Table 1. For some systems (Popper 1968, 1970, 1971b), the differences in color indices derived from the depth ratios are not in good agreement with those given by the colors of the individual components, obtained from the differences in depths in the two colors. Such disagreements are reflected in the uncertainties listed for $(B-V)_0$.

Table 2 Detached main-sequence systems, B6 to M

Binary	Sp[b]	$(B-V)_0$	E_{B-V}	m	R	Vel	l.c.	CI Sp	$\log T_e$	$\log L$	M_V
ζ Phe	B6	−0.145	0.00	3.92	2.85	3		4	4.160	2.50	−0.37
HD6882	—	±0.02	±0.01	±0.05	±0.05		4		±0.03	±0.12	±0.15
$P = 1.^{d}67$	B9	−0.05	—	2.54	1.85			5	4.005	1.51	1.13
	—	0.02	—	0.03	0.04				0.015	0.08	0.10
χ² Hya	B9	−0.09	0.02	3.61	4.39	1		6	4.063	2.49	−0.98
HD96314	—	0.02	0.01	0.08	0.04		6		0.035	0.14	0.10
$P = 2.27$	A0	−0.05	—	2.64	2.16				4.005	1.64	0.79
	—	0.02	—	0.05	0.04				0.02	0.08	0.10
AS Cam	B8	−0.105	0.09	3.31	2.70	1	1	7	4.088	2.17	0.00
HD35311	—	0.02	0.02	0.07	0.20		7	8	0.03	0.14	0.20
$P = 3.43$	B9	−0.045	—	2.51	2.00				4.001	1.56	0.98
	—	0.02	—	0.05	0.15				0.015	0.09	0.20
V451 Oph	B9	−0.06	0.05	2.78	2.65	1	2	9	4.015	1.86	0.30
HD170470	—	0.02	0.02	0.06	0.20			10	0.02	0.10	0.20
$P = 2.20$	A0	−0.02	—	2.36	2.12			11	3.985	1.54	0.95
	—	0.02	—	0.05	0.20				0.012	0.09	0.20
RX Her	B9	−0.06	0.07	2.75	2.44	1	2	9	4.015	1.79	0.48
HD170757	—	0.03	0.03	0.06	0.10			10	0.03	0.12	0.20
$P = 1.78$	A0	−0.02	—	2.33	1.96			12	3.985	1.48	1.12
	—	0.03	—	0.05	0.10				0.02	0.09	0.15
AR Aur[c]	B8	−0.08	0.00	2.48	1.83	1	2	2	4.048	1.67	0.99
HD34364	—	0.02	0.02	0.10	0.05		13	9	0.035	0.14	0.15
$P = 4.13$	B9	−0.04	—	2.29	1.83			10	3.997	1.46	1.19
	—	0.02	—	0.10	0.05			12	0.015	0.06	0.10

STELLAR MASSES 127

Name	Sp											
β Aur^c	A1	+0.035	0.00	2.35	2.49		1			3.955	1.56	0.82
HD40183	—	0.03	0.00	0.03	0.15				9	0.02	0.10	0.20
P = 3.96	A1	+0.035	—	2.27	2.49				11	3.955	1.56	0.82
	—	0.03	—	0.03	0.15				12	0.02	0.10	0.20
SZ Cen	A7	+0.20	0.08	2.28	3.62				15	3.885	1.60	0.61
HD120359	—	0.04	0.03	0.02	0.10		1	16	16	0.05	0.06	0.20
P = 4.11	A7	+0.22	—	2.32	4.55					3.878	1.79	0.17
	—	0.04	—	0.03	0.10					0.015	0.06	0.20
EE Peg^d	A3	+0.10	0.00	2.08	2.05		1		8	3.927	1.28	1.48
HD206155	—	0.02	0.02	0.16	0.10		11	17	9	0.01	0.06	0.15
P = 2.63	F5	+0.45	—	1.32	1.29				10	3.806	0.40	3.68
	—	0.04	—	0.04	0.06					0.01	0.04	0.15
V624 Her^d	Am	+0.19	0.00	2.1	3.0		1		10	3.890	1.47	0.98
HD161321A	—	0.03	0.02	0.3	0.3			1	11	0.012	0.10	0.25
P = 3.90	Am	+0.21	—	1.8	2.2				18	3.882	1.16	1.73
	—	0.03	—	0.2	0.25					0.012	0.10	0.25
V805 Aql^d	A5	+0.13	0.13	2.06	2.10		1	2	11	3.915	1.26	1.52
HD177708	—	0.04	0.03	0.07	0.12		11	11		0.015	0.08	0.20
P = 2.41	F0	+0.28	—	1.60	1.75			13		3.855	0.86	2.50
	—	0.05	—	0.04	0.12					0.02	0.10	0.25
RR Lyn	Am	+0.22	0.00	2.00	2.50		1		10	3.880	1.26	1.49
HD44691	—	0.02	0.01	0.05	0.15		19	20	12	0.01	0.06	0.15
P = 9.95	F0	+0.29	—	1.55	1.93				21	3.850	0.93	2.32
	—	0.02	—	0.03	0.15					0.01	0.08	0.20
WW Aur^c	Am	+0.14	0.00	1.98	1.89		1	1	1	3.910	1.15	1.78
HD46052	—	0.04	0.02	0.05	0.04			22	10	0.015	0.06	0.15
P = 2.52	Am	+0.19	—	1.82	1.89				12	3.890	1.065	1.98
	—	0.04	—	0.05	0.04				22	0.015	0.06	0.15
CM Lac	A2	+0.08	0.03	1.88	1.59		1	2	1	3.935	1.095	1.97
HD209147	—	0.04	0.02	0.09	0.06				9	0.015	0.07	0.15
P = 1.60	F0	+0.28	—	1.47	1.42				14	3.855	0.68	2.95
	—	0.06	—	0.04	0.06					0.02	0.09	0.25

Table 2 (*continued*)

Binary	Sp[b]	$(B-V)_0$	E_{B-V}	m	R	Vel	l.c.	CI Sp	$\log T_e$	$\log L$	M_V
RS Cha[c]	A8	+0.20	0.00	1.86	2.28	1	1	11	3.885	1.21	1.61
HD75747	—	0.03	0.01	0.02	0.15			23	0.01	0.07	0.20
$P = 1.67$	A8	+0.26	—	1.82	2.28				3.863	1.12	1.85
	—	0.03	—	0.02	0.15				0.01	0.07	0.20
MY Cyg	Am	+0.30	0.04	1.81	2.20	1	11	11	3.848	1.03	2.07
HD193637	—	0.03	0.03	0.04	0.04		25	12	0.01	0.04	0.13
$P = 4.00$	Am	+0.31	—	1.78	2.20		26		3.845	1.02	2.11
	—	0.03	—	0.04	0.04				0.01	0.04	0.13
V477 Cyg	A2	+0.07	0.00	1.78	1.52	1	1	9	3.940	1.08	2.03
HD190786	—	0.04	0.02	0.12	0.06		2	10	0.015	0.07	0.15
$P = 2.35$	F2	+0.38	—	1.34	1.20		24	11	3.820	0.41	3.66
	—	0.06	—	0.07	0.05				0.020	0.09	0.20
EI Cep	F0	+0.32	0.00	1.68	2.54	1	13	8	3.840	1.12	1.84
HD205234	—	0.03	0.02	0.03	0.25		27	27	0.01	0.09	0.25
$P = 8.44$	F2	+0.36	—	1.78	2.80			28	3.827	1.15	1.77
	—	0.03	—	0.03	0.25				0.01	0.09	0.20
XY Cet[e]	Am	+0.23	0.00	1.76	1.88	1	27	8	3.875	1.00	2.15
HD18597	—	0.03	0.02	0.02	0.07		29	10	0.01	0.05	0.15
$P = 2.78$	Am	+0.27	—	1.63	1.88			29	3.860	0.94	2.30
	—	0.03	—	0.02	0.07				0.01	0.05	0.15
ZZ Boo[d]	F2	+0.37	0.00	1.72	2.22	1	1	8	3.824	0.94	2.29
HD121648	—	0.02	0.01	0.08	0.10		13	10	0.007	0.05	0.10
$P = 4.99$	F2	+0.37	—	1.72	2.22			30	3.824	0.94	2.29
	—	0.02	—	0.08	0.10				0.007	0.05	0.10

STELLAR MASSES

Star	Sp											
TX Her	A8	+0.26	0.00	1.62	1.58			1		3.863	0.80	2.64
HD156965	—	0.03	0.02	0.04	0.05					0.01	0.05	0.15
$P = 2.06$	F2	+0.36	—	1.45	1.48		13	10		3.827	0.60	3.15
	—	0.03	—	0.03	0.05			12		0.01	0.05	0.13
CW Eri	F2	+0.35	0.00	1.52	2.11	11				3.830	0.92	2.35
HD19115	—	0.03	0.01	0.015	0.15		31	11		0.01	0.07	0.20
$P = 2.73$	F2	+0.39	—	1.28	1.48		32			3.820	0.57	3.23
	—	0.03	—	0.01	0.10					0.01	0.07	0.20
RZ Cha	F5	+0.47	0.00	1.51	2.26	1		11		3.800	0.86	2.51
HD93486	—	0.02	0.00	0.04	0.06		33	33		0.005	0.03	0.08
$P = 2.83$	F5	+0.47	—	1.51	2.26			34		3.800	0.86	2.51
	—	0.02	—	0.04	0.06					0.005	0.03	0.08
BK Peg^d	F8	+0.54	0.00	1.27	1.48	11	11	8		3.785	0.43	3.63
+25°5003	—	0.02	0.01	0.01	0.15					0.005	0.09	0.20
$P = 5.49$	F8	+0.56	—	1.43	2.03					3.780	0.69	3.01
	—	0.02	—	0.02	0.20					0.005	0.09	0.20
DM Vir	F7	+0.47	0.02	1.46	1.76	1		3		3.800	0.64	3.06
HD123423	—	0.02	0.02	0.03	0.08			11		0.005	0.04	0.10
$P = 4.67$	F7	+0.47	—	1.46	1.76			34		3.800	0.64	3.06
	—	0.02	—	0.03	0.08					0.005	0.04	0.10
CD Tau	F7	+0.48	0.00	1.40	1.74	1	1	8		3.798	0.625	3.11
HD34335	—	0.02	0.01	0.05	0.07		35	14		0.004	0.04	0.10
$P = 3.44$	F7	+0.48	—	1.31	1.61			36		3.798	0.56	3.28
	—	0.02	—	0.04	0.07			37		0.004	0.04	0.10
TV Cet	F2	+0.39	0.00	1.39	1.50	1	11	8		3.820	0.58	3.20
HD20173	—	0.03	0.01	0.05	0.05		38	10		0.01	0.05	0.10
$P = 9.10$	F5	+0.46	—	1.27	1.26			38		3.803	0.365	3.75
	—	0.03	—	0.04	0.05					0.007	0.04	0.10
BS Dra	F5	+0.44	0.00	1.37	1.44	1	11	8		3.808	0.50	3.41
HD190020	—	0.01	0.01	0.03	0.05		39	10		0.002	0.03	0.12
$P = 3.36$	F5	+0.44	—	1.37	1.44					3.808	0.50	3.41
	—	0.01	—	0.03	0.05					0.002	0.03	0.12

Table 2. (continued)

Binary	Sp[b]	$(B-V)_0$	E_{B-V}	m	R	Vel	l.c.	CI Sp	$\log T_e$	$\log L$	M_v
HS Hya	F5	+0.43	0.00	1.34	1.36	1	40	8	3.810	0.46	3.52
HD90242	—	0.02	0.01	0.05	0.14			40	0.005	0.09	0.25
P = 1.57	F5	+0.46	—	1.28	1.22				3.803	0.34	3.82
	—	0.02	—	0.05	0.14				0.005	0.10	0.25
V1143 Cyg[c]	F5	+0.45	0.00	1.33	1.31	1	2	1	3.806	0.41	3.64
HD185912	—	0.02	0.01	0.03	0.10			12	0.005	0.07	0.20
P = 7.64	F5	+0.45	—	1.29	1.31				3.806	0.41	3.64
	—	0.02	—	0.03	0.10				0.005	0.07	0.20
VZ Hya	F5	+0.42	0.00	1.23	1.35	1	13	11	3.812	0.46	3.51
HD72257	—	0.03	0.01	0.03	0.04		41	41	0.007	0.04	0.10
P = 2.90	F6	+0.48	—	1.12	1.12		42	43	3.798	0.24	4.07
	—	0.03	—	0.03	0.03		43		0.007	0.04	0.10
UX Men[c]	F8	+0.54	0.00	1.17	1.28	1	44	44	3.785	0.31	3.95
HD37513	—	0.02	0.01	0.05	0.05				0.005	0.04	0.10
P = 4.18	F8	+0.55	—	1.11	1.28				3.782	0.29	3.98
	—	0.02	—	0.05	0.05				0.005	0.04	0.10
WZ Oph	F8	+0.54	0.00	1.12	1.34	1	2	1	3.785	0.35	3.85
HD154676	—	0.02	0.01	0.04	0.03		13	10	0.005	0.03	0.08
P = 4.18	F8	+0.54	—	1.12	1.34				3.785	0.35	3.85
	—	0.02	—	0.04	0.03				0.005	0.03	0.08
UV Leo[c]	G2	+0.62	0.00	0.99	1.08	1	2	1	3.768	0.09	4.52
HD92109	—	0.02	0.00	0.04	0.04		13	10	0.005	0.04	0.10
P = 0.60	G2	+0.62	—	0.92	1.08				3.768	0.09	4.52
	—	0.02	—	0.04	0.04				0.005	0.04	0.10

YY Gem[e]	M1	+1.37	0.00	0.59	0.62	1		46	3.576	−1.13	8.90
+32°1582	—	0.02	0.00	0.015	0.01			47	0.003	0.10	0.10
P = 0.81	M1	+1.37	—	0.59	0.62	1	45		3.576	−1.13	8.90
	—	0.02	—	0.015	0.01				0.003	0.10	0.10
CM Dra[e]	M4	+1.83	0.00	0.24	0.25	48	48		3.500	−2.16	12.77
GL630.1	—	0.02	0.00	0.015	0.01				0.01	0.10	0.10
P = 1.27	M4	+1.83	—	0.21	0.235	48			3.500	−2.16	12.92
	—	0.02	—	0.015	0.01				0.01	0.10	0.10

[a] Uncertainties of scale-dependent quantities do not include effects of uncertainties in T_e and flux scales.
[b] Spectral types are derived primarily from color indices.
[c] Photometry is inadequate to establish ratio of the radii, which are assumed equal, even though colors and/or masses differ significantly.
[d] Unpublished spectrographic material may lead to improved masses and radii.
[e] Absolute magnitudes are derived from parallaxes, and hence are not scale dependent. Colors are V–R rather than B–V.

[f] References

1. Cited in BFM tables and notes
2. Cited in KPW notes
3. J. Andersen, private communication
4. Clausen, Gyldenkerne & Grønbech 1976
5. Dachs 1971
6. Clausen & Nordström 1978
7. Padalia & Srivastava 1975b
8. Popper & Dumont 1977
9. Olson 1975
10. Hildtich & Hill 1975
11. D. M. Popper, unpublished; D. M. Popper and P. B. Etzel, unpublished
12. Lindemann & Hauck 1973
13. Cester et al. 1978b
14. McNamara 1966
15. Johnson et al. 1966
16. Grønbech, Gyldenkerne & Jørgensen 1977
17. Linnell 1973
18. Zissell 1972
19. Kondo 1976
20. Budding 1974
21. Popper 1971b
22. P. B. Etzel, private communication
23. Cousins & Stoy 1963
24. Scarfe, Barlow & Niehaus 1976
25. Williamon 1975
26. Tremko, Papoušek & Vetešnik 1976
27. Kitamura & Kondo 1978
28. van Rijsbergen 1973
29. Okazaki 1978
30. McNamara, Hanson & Wilcken 1971
31. Mauder & Ammann 1976
32. Chen 1975
33. Jørgensen & Gyldenkerne 1975
34. O. J. Eggen, unpublished
35. Srivastava 1976
36. Eggen 1963
37. Wood 1976
38. Jørgensen 1979
39. Güdür et al. 1979
40. Gyldenkerne, Jørgensen & Carstensen 1975
41. Walker 1970
42. Wood 1971
43. Padalia & Srivastava 1975a
44. Clausen & Grønbech 1976
45. Leung & Schneider 1978
46. Lacy 1979
47. Veeder 1974
48. Lacy 1977a

For comparisons of observed properties of the components of close binaries with those of single stars or with those predicted by theory, it is desirable to give values of effective temperatures, luminosities, and absolute magnitudes. The values of these quantities listed in Table 2 are obtained from the observed values of the color indices and radii by using the scales of effective temperatures and radiative fluxes given in Table 1. Since there is not complete agreement among astronomers on these scales, and since they are subject to alteration as more information becomes available, the derived temperatures and luminosities are listed in Table 2 as "scale-dependent" quantities. The uncertainties given for them depend only on those adopted for the color indices and radii, and not on uncertainties in the scales of Table 1 themselves.

To illustrate some of the points discussed here, it may be instructive to treat one system in some detail. Consider RX Her (HD170757), for example. The masses given in Table 2 are adopted directly from the spectrographic analysis (Popper 1959). Table 3 gives solutions by F. B. Wood (1948) and Magalashvili (1953) of their unfiltered photometric observations, obtained with Russell's method. A reanalysis of F. B. Wood's observations by Cester et al. (1978c) using D. B. Wood's (1972) computer program is also listed. In the table, k is the ratio of the radii, r_h and r_c the relative radii of the hotter and cooler component, respectively, i the orbital inclination, L the light of each component, J the surface flux, and $1-l$ the depth of minimum. We may adopt $r_h = 0.230 \pm 0.005$, $r_c = 0.185 \pm 0.005$, $i = 86° \pm 0.4$. The flux ratio should be better determined from the light curve than is the light ratio or the ratio of the radii. The listed values of J_h/J_c are derived from L_h/L_c and k. They should equal the ratio of the observed depths, given in the last column of the table. That they do not is a measure of the inconsistency in the solutions. Petrie's (1950) rather uncertain value of the spectroscopic light ratio is $L_h/L_c = 1.44$. The large disagreement between this value and the value 2.44 given by Cester et al. (1978c), as well as the significant difference between the flux ratios obtained in the Cester et al. solution and that obtained from the depth of minima, are examples of the

Table 3 Photometric solutions of RX Her

Author	k	r_h	r_c	i	L_h/L_c	J_h/J_c	$\dfrac{(1-l)_{pri}}{(1-l)_{sec}}$
Wood	0.81	0.228	0.186	85°6	1.75	1.15	1.22
Magalashvili	0.85	0.223	0.190	86.2	1.60	1.16	1.14
Cester et al.	0.76	0.231	0.174	86.3	2.44	1.38	1.22
	±0.02	±0.002	±0.005	±0.3			

misleading results that can be obtained by allowing a computer program to "converge" to a solution. A more nearly definitive light curve of RX Her is required to establish the luminosity and flux ratios with precision. We may adopt a flux ratio in the V band of 1.15 ± 0.05, a small correction having been made to correct for the shorter effective wavelength of the unfiltered observations. A flux ratio of 1.15 ± 0.05 corresponds to a color difference between the components, $\Delta(B-V) = 0.04 \pm 0.015$ (Table 1).

Strömgren photometry of RX Her by Cameron (1966), Hilditch & Hill (1975), and Olson (1975) leads to values of $(B-V)_0 = -0.035$ and color excess $E_{B-V} = 0.06$, whereas Olson's (1975) photometry of lines in the spectrum leads to $(B-V)_0 = -0.05$. My UBV photometry (Popper 1959) leads to $(B-V)_0 = -0.025$ and $E_{B-V} = 0.07$. $(B-V)_0 = -0.04$ and $E_{B-V} = 0.07$ are adopted as are individual color indices of -0.06 and -0.02, with estimated uncertainties of 0.03 mags. The corresponding spectral types are B9 and A0. The estimate by Hill et al. (1975) outside eclipse is B9.5V. The values of log T_e, log L, and M_V follow. The difference in the values of M_V between the two components, 0.65, corresponds to a light ratio $L_h/L_c = 1.8$, a value not seriously violating the results in Table 3, particularly in view of all the intervening steps and their uncertainties. The values of M_V in Table 2 for RX Her are approximately 0.3 mag less positive than the values given by Lacy (1979). The most important reason for the difference is that Lacy's value of $(V-R)_0$, based on only two observations, corresponds more nearly to $(B-V)_0 = 0.00$ than to our adopted value, -0.04, leading to lower surface fluxes. A similar analysis was undertaken for each of the systems of Table 2.

Double-lined detached systems for which my spectrographic material is not yet complete are PV Cas (B9), EG Ser (A0), WX Cep (A2), GZ CMa (Am), AI Hya (F2), VZ CVn (F2), EW Ori (F8), and HS Aur (K0). In addition, new material is on hand for V624 Her (Am) and ZZ Boo (F2). Of these systems, only for WX Cep, GZ CMa, ZZ Boo, and AI Hya is the photometry reasonably satisfactory, while it is almost completely lacking for EG Ser, EW Ori, and HS Aur. HS Aur could be the important long-sought main-sequence eclipsing binary lying in its properties between the sun and YY Gem. The secondary of UV Psc, a binary with H and K emission (Table 6), may have similar properties (Sadik 1979).

OB ECLIPSING SYSTEMS

General Considerations

As seen in the preceding section, a considerable number of detached main-sequence systems in the spectral range B6–G0 is available for reasonably definitive spectroscopic and photometric analysis. The most important considerations for systems with approximately equal components are 1. that the fractional radii of the stars, $r = R/a$, be sufficiently small so that (*a*) the spectral lines of the two components are well resolved, and (*b*) the proximity effects on the light curve are small, and 2. that lines are present in the spectrum that are strong enough for reliable measurement and yet do not have damping wings causing the profiles of the component lines to overlap significantly. Of the 72 stars listed in Table 2, only nine have values of $r > 0.2$, and, with the possible exception of UV Leo, lines are available in all of them that are free of overlapping.

Nature has been less kind in designing its supply of hotter binaries. Only three systems are known to me with reasonably good spectrographic and photometric coverage having $r < 0.2$ for both components, namely, QX Car, CV Vel, and DI Her. When $r > 0.2$, the rotational velocities are so large that the central intensities of all but the strongest lines are low. But the stronger lines, primarily those of H and He, are just those with broad damping wings (e.g. Leckrone 1971), which tend to overlap in the two components. In a very careful study, Andersen (1975) has shown that even in the case of CV Vel ($r = 0.13$), in which the lines are separated by several times their rotational widths, only lines without appreciable damping wings give velocities not subject to systematic effects. In particular, the strong diffuse triplet lines of He I, which in a considerable fraction of the systems may otherwise be the best lines available to measure, give velocity separations that are systematically too small. The profiles of the hydrogen lines in all these systems are, of course, hopelessly blended. An additional irony is that MgII 4481, which in CV Vel and other systems is the most favorable line, without appreciable wings, may have its shortward displaced component blend with the wing of the longward displaced component of HeI 4471 in systems with large orbital velocities.

Comparison of Andersen's (1975) profile of HeI 4471 in CV Vel with my profile of HeI 4026 in the spectrum of U Oph (Popper 1978), $r \approx 0.25$, shows that systematic effects in the velocities, and hence in the radii and masses, can be expected to be appreciable even in such a relatively favorable system as U Oph. My stated belief (Popper 1974) that, for the best B-type systems, "there are no important sources of systematic error" in the masses and radii, now appears too optimistic. The magnitude of the systematic effects in published results depends not only on the nature of the spectrum and quality of the spectrograms, but also on the method of measurement. Under some circumstances, visual measurement of spectrograms has been shown to result fortuitously in smaller underestimates of line separations than oscilloscopic measurement (Petrie, Andrews & Scarfe 1967, Batten & Fletcher 1971). Andersen (1975 and unpublished) has shown how it is possible to obtain velocities free of the systematic effects of overlapping wings by a careful selection of lines. Modern recording devices or plates with high signal-to-noise ratio, such as the IIIaJ emulsion, and dispersions corresponding to 20 Å mm^{-1} or higher may be required. This latter technique has been applied successfully by Andersen (unpublished) to the additional systems QX Car ($r = 0.14$) and V539 Ara ($r = 0.20$). The orbit of DI Her ($r = 0.06$) is so eccentric that the lines of the components were not well resolved on older material at one nodal passage (Petrie, Andrews & Scarfe 1967). New material with higher dispersion on IIIaJ plates (Popper, unpublished) appears to remove the difficulty. Improvements in the masses and radii of other relatively favorable OB systems listed in Table 4 will require careful re-examination of existing spectrographic material or the acquisition of improved material. In some cases, spectrophotometric modeling of the line profiles may be required. Confidence in results obtained in this way cannot be as great as in results obtained from clearly separated lines.

We are led to the conclusion that only for V539 Ara, QX Car, CV Vel, and probably DI Her, can we be confident that masses of stars earlier than about B6 are currently available with an accuracy (better than 15%) comparable to that achieved for many cooler systems. The published masses of the other systems in Table 4 are likely to be smaller than their true values by unknown amounts up to possibly 20%. The fractional systematic effects in the radii from the same cause are one-third those in the masses. An additional uncertainty in the radii is introduced in a number of systems because of proximity effects on the light curves for $r \gtrsim 0.2$. Sophisticated computer models (e.g. Hill & Hutchings 1970, Wilson & Devinney 1971, Leung & Wilson 1977) have been applied to the light curves of a number of OB systems. The resulting photometric parameters have greater validity than those obtained with less satisfactory models, but inherent uncertainties remain in the solutions, the degree of uncertainty depending on the nature of the light curve and quality of the observations. An added advantage of the computer models is that they provide improved discrimination between detached, semidetached, and contact systems.

The principal criteria employed for including an OB system in Tables 4 and 5 are that the lines of both components be clearly visible in the spectrum and appear at least as well resolved as those of V382 Cyg or U Oph (Popper 1978), and that a photoelectric light curve showing two well-defined minima be available. These criteria will not necessarily exclude contact and semidetached systems.

The radiative properties (fluxes, T_e's, B.C.'s) of B stars may also be less well determined than for cooler stars because of interstellar reddening and the relative insensitivity of color index to radiative flux. For stars of type B1 and earlier, spectral type is taken as the primary determinant of radiative

Table 4 Detached OB systems

Binary	Sp[b]	$(B-V)_0$	E_{B-V}	m	R	Vel	l.c.	Sp	$\log T_e$	$\log L$	M_V
Y Cyg	O9.8	−0.30	0.23	16.7	6.0	1	1	3	4.485	4.45	−3.45
HD198846	±0.5	±0.02	±0.03	±0.5	±0.3				±0.025	±0.11	±0.15
$P = 3{.}^d00$	O9.8	−0.30	—	16.7	6.0		2		4.485	4.45	−3.45
	0.5	0.02	—	0.5	0.3				0.025	0.11	0.15
V478 Cyg[c]	O9.8	−0.29	0.84	15.6	7.3	1	4	5	4.485	4.62	−3.90
HD193611	0.3	0.02	0.03	0.7	0.2			6	0.015	0.07	0.10
$P = 2.88$	O9.8	−0.29	—	15.6	7.3				4.485	4.62	−3.90
	0.3	0.02	—	0.7	0.2				0.015	0.07	0.10
V453 Cyg[c]	B0.4	−0.27	0.46	14.5	8.6	1	1	1	4.47	4.69	−4.15
HD227696	0.3	0.02	0.03	1.2	0.3	7	8	6	0.015	0.07	0.10
$P = 2.88$	B0.7	−0.26	—	11.3	5.4			9	4.445	4.20	−3.10
	0.5	0.02	—	1.0	0.2				0.025	0.10	0.15
CW Cep	B0.4	−0.28	0.67	11.8	5.4	1	1	1	4.47	4.28	−3.15
HD218066	0.5	0.02	0.03	0.2	0.3		8		0.025	0.10	0.15
$P = 2.73$	B0.7	−0.28	—	11.1	5.0		10		4.445	4.13	−2.90
	0.5	0.02	—	0.2	0.3				0.025	0.10	0.15
α Vir[d]	B1.5	−0.25	0.00	10.8	8.1	1		12	4.39	4.33	−3.5
HD116658	—	0.02	0.01	1.3	0.5	11		13	0.04	0.10	0.1
$P = 4.01$	B4	−0.18	—	6.8	4.4				4.23	3.16	−1.5
	—	0.03	—	0.9	0.7				0.06	0.17	0.2
QX Car	B2	−0.23	0.04	9.2	4.3	14		14	4.34	3.59	−2.10
HD86118	—	0.01	0.01	0.15	0.1		14		0.02	0.08	0.10
$P = 4.48$	B2	−0.225	—	8.5	4.0				4.33	3.48	−1.90
	—	0.01	—	0.15	0.1				0.02	0.08	0.10
V539 Ara	B4	−0.19	0.07	6.13	4.4	14		15	4.255	3.26	−1.70
HD161783	—	0.015	0.02	0.07	0.1		15	16	0.03	0.12	0.15
$P = 3.17$	B4	−0.18	—	5.25	3.7				4.23	3.02	−1.25
	—	0.015	—	0.06	0.2				0.03	0.12	0.15

STELLAR MASSES

CV Vel	B3	−0.19	0.03	6.10	4.05	1	17	4.255	3.19	−1.55
HD77464		0.02	0.02	0.04	0.1		17	0.04	0.16	0.15
$P = 6.89$	B3	−0.19		6.00	4.05		18	4.255	3.19	−1.55
		0.02		0.04	0.1		19	0.04	0.16	0.15
BM Ori	B3	−0.21	0.30	5.90	2.90	1	1	4.30	3.08	−1.05
HD37021		0.02	0.03	0.08	0.4			0.04	0.20	0.20
DI Her[c]	B5	−0.175	0.19	5.2	2.55		20	6	4.22	2.65	−0.40
HD175227		0.02	0.02	0.3	0.2		21	22	0.04	0.17	0.25
$P = 10.55$	B5	−0.16		4.6	2.55	7		23	4.19	2.52	−0.25
		0.02		0.3	0.2				0.04	0.17	0.25
U Oph	B5	−0.175	0.22	5.16	3.43	1	1	4	4.22	2.90	−1.00
HD156247		0.01	0.02	0.10	0.07		2	12	0.02	0.08	0.10
$P = 1.68$	B5	−0.165		4.60	3.11		8	22	4.20	2.74	−0.70
		0.01		0.06	0.06		24	23	0.02	0.08	0.10
V760 Sco[e]	B5	−0.165	0.34	5.00	3.00	7	25	25	4.20	2.71	−0.65
HD147683		0.01	0.02	0.12	0.15	25			0.02	0.09	0.15
$P = 1.73$	B5	−0.155		4.65	2.70				4.18	2.53	−0.30
		0.01		0.09	0.15				0.02	0.09	0.15
AG Per	B4	−0.185	0.19	4.53	3.00	1	1	6	4.245	2.89	−0.85
HD25833		0.015	0.02	0.07	0.15		2	22	0.03	0.13	0.15
$P = 2.03$	B5	−0.17		4.12	2.60		26	23	4.21	2.62	−0.35
		0.015		0.06	0.15				0.03	0.13	0.20

[a] Uncertainties of scale-dependent quantities do not include effects of uncertainties in the T_e and flux scales.
[b] Spectral types later than B1 are derived from color indices as well as from appearance of the spectra.
[c] Additional radial velocities are becoming available. Masses and radii provisional.
[d] M_V's are independent of flux scale. Radius of secondary is scale dependent.
[e] References

1. Cited in BFM tables and notes
2. Cited in KPW notes
3. Olson 1968
4. D. M. Popper and P. B. Etzel, unpublished
5. Roman 1956
6. Popper & Dumont 1977
7. D. M. Popper, unpublished
8. Cester et al. 1978c
9. Morgan, Whitford & Code 1953
10. Söderhjelm 1976
11. Dukes 1974
12. Olson 1975
13. Johnson & Morgan 1953
14. Andersen & Clausen 1980
15. Clausen 1979
16. Crawford, Barnes & Golson 1971a
17. Clausen & Grønbech 1977
18. Cousins & Stoy 1963
19. Feast 1954
20. Martynov & Khaliullin 1978
21. S. Catalano and M. Rodonó, private communication and P. B. Etzel, unpublished
22. Hilditch & Hill 1975
23. McNamara 1966
24. Koch & Koegler 1977
25. J. Andersen, J. V. Clausen and B. Nordström, private communication
26. Güdür 1978

properties (Table 1), while for later types the color index, $B-V$, corrected for reddening, is utilized. The difference in spectral type or color index between the components of a system is obtained in most cases from the ratio of the surface fluxes (see Table 1) derived from the light curve. The correction of Strömgren intermediate-band indices for reddening and their conversion to $(B-V)_0$ follow the precepts given by Crawford (1978).

The Available Data

In Table 4 are listed results for the eleven detached systems in which systematic effects on the masses from line blending are considered to be no worse than in the case of U Oph (Popper 1978). Only for QX Car, V539 Ara, CV Vel, and DI Her are these effects probably absent. The systems are listed in order of decreasing mass. The components of α Vir are listed from Table 7, while the hotter component of BM Ori (Popper & Plavec 1976) is also listed. The results for α Vir probably suffer from the effects of line blending similar to those in other systems (Struve et al. 1958). The absolute magnitudes of the components of α Vir are not "scale dependent," since they are obtained directly from interferometric data (Herbison-Evans et al. 1971). The radius of the smaller component is obtained from its absolute magnitude and adopted color index. The B star of known mass in ζ Aur (Popper 1961) is not listed because its radius and absolute magnitude are poorly known.

Three more massive, hotter main-sequence systems with masses as well determined as some of those in Table 4 are listed in Table 5. These systems are all probably contact systems. In contrast to the systems in Table 4, these O-type binaries have mass ratios differing much more from unity than expected from the small differences in their luminosities. Two apparently similar systems are LZ Cep (HD209481; O9; minimum masses 6.2 and 2.9; noneclipsing) and NY Cep (HD217312; O9; minimum masses 27 and 13; possibly eclipsing). See BFM for references. Although NY Cep has been classified as B0, the HeII lines are pronounced on my plates. It is possible that this rather large group of O-type systems contains high-mass counterparts of the W UMa systems.

There is good evidence for stars with higher masses than those discussed here (e.g. Conti & Walborn 1976, Hutchings & Cowley 1976). Hutchings' (1975) discussion contains material on masses of OB stars considerably less well established than is appropriate in this review.

B-type systems for which analyses have been published but with lines even less well resolved than those of V382 Cyg and U Oph (Popper 1978) include SX Aur, TT Aur, BF Aur, and IU Aur. It is not clear whether any of these systems are detached (e.g. Chambliss & Leung 1979, Mammano, Margoni & Stagni 1974). The lines in the spectra of EO Aur (Popper 1978)

Table 5 O-type systems[a]

Binary	Sp	$(B-V)_0$	E_{B-V}	m	R	References[c] Vel	l.c.	$B-V$ Sp	Scale dependent[b] $\log T_e$	$\log L$	M_v
V382 Cyg	O7.3	−0.30	1.00	26.9	9.2	3	2	3	4.575	5.17	−4.75
	±0.5	±0.02	±0.06	±2.7	±0.7		4	6	±0.015	±0.09	±0.20
HD228854	O7.7	−0.30	—	19.0	7.5		5		4.565	4.95	−4.25
$P = 1\overset{d}{.}8.9$	0.5	0.02	—	1.9	0.5				0.015	0.08	0.15
TU Mus	O7.8	−0.30	0.37	23.8	7.3	1	1	1	4.555	4.90	−4.20
HD100213	0.5	0.02	0.04	2.5	0.4		7	6	0.015	0.08	0.15
$P = 1.39$	O8.2	−0.30	—	15.8	6.2		8		4.545	4.72	−3.80
	0.5	0.02	—	1.5	0.3				0.015	0.07	0.15
LY Aur	O9.2	−0.30	0.50	21.7	13.0	1	1	6	4.512	5.23	−5.30
HD35921	0.5	0.02	0.03	1.7	0.8	9	10	11	0.025	0.11	0.20
$P = 4.00$	O9.4	−0.30	—	8.4	10.0			12	4.502	4.96	−4.65
	0.5	0.02	—	0.9	0.8				0.025	0.12	0.20

[a] These are all probably contact systems.
[b] Uncertainties of scale-dependent quantities do not include effects of uncertainties in the T_e and flux scales.
[c] References
1. Cited in BFM tables and notes
2. Cited in KPW notes
3. Popper 1978
4. Bloomer, Burke & Millis 1979
5. Devinney & Twigg 1974
6. Popper & Dumont 1977
7. Eggen 1978
8. Wilson & Rafert 1980
9. D. M. Popper, unpublished
10. Cester et al. 1978a
11. Morgan, Whitford & Code 1953
12. Hill et al. 1975

and VV Ori (Popper, unpublished) are completely unresolved, while V380 Cyg is a system in which I have been unable to find on high-contrast plates the lines of the secondary reported by Batten (1962). The better-studied semidetached systems are listed in a later section.

The references given in Tables 4 and 5 contain sources of the data employed in evaluating the results in the preceding columns. As in Table 2, works that are referred to in the Notes in the *Seventh Catalogue of Spectroscopic Binary Orbits* (BFM) are cited by reference to that catalogue (reference 1 in the tables) rather than to the original papers. The determination of the "scale dependent" parameters for the stars follows the same procedures discussed earlier for detached cooler systems (Table 2). The difference in the derived values of M_V of the components of a system may differ by as much as 0.1 mag from the difference obtained in the photometric solution because of uncertainties in the intervening steps. The tabulated uncertainties in the individual values of M_V are generally less than 2.5 times the uncertainties in log L, since log T_e is more sensitive to color or spectral type than is the surface flux because of the rapidly varying bolometric correction.

DETACHED BINARIES WITH COOL EVOLVED COMPONENTS

Systems Containing Subgiants

Most G-type eclipsing binaries with periods greater than one day, as well as a few F-type systems, contain detached subgiant components. The cooler components, at least, have emission at the H and K lines of Ca II strong enough to be seen outside eclipse on spectrograms of moderate resolution. These are the classical "RS CVn" systems, although the term is currently employed in a much broader sense. Most of the material on which the masses of the components in systems of this type, listed in Table 6, are based is as yet provisional. Even though the systems are totally eclipsing in many cases, the relative radii are generally quite uncertain, both because of limited photometric observations and because of instabilities in most of the light curves. For these reasons estimated uncertainties are not given in Table 6, where the systems are listed in order of decreasing total mass. The radii followed by colons (:) are rough determinations based on the durations of partial and total phases of primary eclipses with the assumption that the orbital inclinations i are 90°. This assumption minimizes the radius of the larger star and maximizes that of the smaller. Unpublished as well as published radial velocities are employed for the masses and radii except for Z Her, AR Lac, and TY Pyx, which are listed in

Table 6 Detached subgiant eclipsing systems

Binary	P	Sp:	$B-V$	$m\sin^3 i$	R	$\log L$	M_V
CQ Aur +31°1179	10^d62	G2 K0	+0.87	1.6 2.0	— —	— —	— —
RZ Eri HD30050	39.28	Am K0	+0.43 +0.91	1.7 1.7	— —	— —	— —
RU Cnc +24°1959	10.18	F5 K1	+0.46 +1.01	1.5 1.5	1.9: 5.0:	0.7 1.0	2.9 2.6
PW Her —	2.88	G2 K0	+0.91	1.4 1.6	— —	— —	— —
SZ Psc HD219113	3.97	F8 K1	+0.83	1.33 1.65	1.6: 4.0:	0.5 0.8	3.5 3.0
RW UMa +52°1579	7.33	F8 K1	+0.50 +1.08	1.5 1.45	2.0: 3.8:	0.7 0.7	2.9 3.4
VV Mon −5°1935	6.05	G2 K0	+0.81	1.45 1.35	1.8 6.0:	0.5 1.3	3.4 1.8
RS CVn HD114519	4.80	F5 K0	+0.42 +0.91	1.35 1.40	1.7: 4.0:	0.7 0.9	3.0 2.7
WW Dra HD150708	4.63	G2 K0	+0.60 +0.97	1.4 1.4	2.3: 3.9:	0.8 0.8	2.8 2.9
AW Her +18°3678	8.81	G0 K1	— —	1.4 1.4	— —	— —	— —
RT CrB HD139588	5.12	G0 G8	+0.70	1.3 1.3	— —	— —	— —
AR Lac HD210334	1.98	G2 K0	— +0.93	1.30 1.30	1.8 3.1	0.5 0.7	3.5 3.3
LX Per +47°781	8.04	G0 K0	+0.55 +0.95	1.23 1.32	1.6: 2.8:	0.5 0.6	3.5 3.6
GK Hya +2°1993	3.59	F8 G8	+0.52 +0.81	1.2: 1.3:	1.4: 3.0:	0.4 0.8	3.7 2.9
MM Her +22°3245	7.96	G2 G8	+0.85	1.2 1.25	1.5: 3:	0.4 0.8	3.9 3.0
TY Pyx HD77137	3.20	G5 G5	+0.69 +0.69	1.21 1.21	1.6 1.7	0.4 0.4	3.9 3.8
Z Her HD163930	3.99	F5 K0	+0.47 +0.91	1.22 1.10	1.6 2.6	0.6 0.6	3.3 3.6
UV Psc HD7700	0.86	G2 K0	+0.65 +0.91	1.2 0.9	1.2: 0.9:	0.2 −0.4	4.4 5.9
UX Com +29°2355	3.64	G2 K1	+0.61 +1.05	0.95 1.12	— —	— —	— —
SS Boo +39°2849	7.61	G0 K1	+0.59 +1.01	1.00 1.00	— —	— —	— —

BFM. Changes and additions to the list published earlier (Popper & Ulrich 1977) are based on additional material. The color indices given are based on published sources listed in BFM and KPW, as well as on my own material (Popper & Dumont 1977). They are generally less well established than for the detached systems of Table 2 because photometric analyses are made uncertain by the nature of the light curves. The spectral types listed are generally only rough estimates based on color indices as well as the appearance of the spectrograms. Thus, uncertainties in the absolute magnitudes are relatively large and are not estimated. For the best-studied and most stable systems, such as TY Pyx and Z Her, they may be as small as 0.1 mag, but more typically they are considerably larger. Masses given to one decimal place and to two decimal places may be considered uncertain by about 0.15 $m_\odot$ and 0.05 $m_\odot$, respectively. UV Psc appears to be the only system in Table 6 in which the cooler component also has the smaller radius (Sadik 1979). The radii are from Sadik's solution. The systems RV Lib and RT Lac, resembling other members of the group spectroscopically but with mass ratios differing greatly from unity, are probably not detached (see Milone 1977).

Many of the properties of this and related groups of binaries have been discussed by Hall (1976). Their evolutionary status has been reviewed by Popper & Ulrich (1977) and by Morgan & Eggleton (1979). The conjecture of the latter authors, that mass ratios in the systems listed in Table 6 differing substantially from 0.96 do so because of observational uncertainties, is without foundation.

The subgiant component of the visual binary ζ Her (Table 8) has properties similar to those of some of the hotter components of the systems in Table 6, although its radius is larger than most.

Systems With Giant or Supergiant Components

The paucity of data on the masses of giants and supergiants is understandable in terms of selection effects. The nearest giants are too distant for good parallaxes. If a potential giant or supergiant is in a small enough binary system to have a good chance of becoming eclipsing as it expands, the star is likely to fill its Roche lobe before reaching the giant stage. Alpha Aur is still the only system containing giants of luminosity class III with the masses obtained directly (G0III, $m = 2.6$; G5III, $m = 2.7$; see Table 7). After α Aur, ϕ Cyg (G8III–IV, minimum masses 2.3 for each component) appears to be the most favorable double-lined giant spectroscopic binary for optical resolution. But according to H. A. McAlister (private communication) ϕ Cyg is not well resolved by the speckle technique. TW Cnc and AL Vel are totally eclipsing systems containing apparently normal red giants near K0. But in neither system has a velocity curve of the hotter (B or A) component been successfully obtained. Uncertain estimates of about 3 $m_\odot$ have been made for each of the giants. DL Vir and probably V635 Mon are eclipsing systems containing red giants, but they are triple systems in which the giant is distant from the eclipsing pair. The minimum mass of the K0II–III star in the triple system β Cap is 3.6 $m_\odot$ (Evans & Fekel 1979). Fekel's (1979) value of 2.9 $m_\odot$ for the minimum mass of the G5III component in the visual-spectroscopic triple system ψ Sgr requires a definitive visual orbit to combine with the spectrographic observations. McLaughlin has given values of 3.6 and 4.7 as minimum masses for the G-type giants in HD 15798/9 and γ Per, respectively, but the individual observations have not been published. References are in BFM. Of the five additional double-lined binaries in BFM containing cool giants, only one has minimum masses less than one solar mass. The weight of all this evidence is that, despite selection effects favoring larger masses, the masses of cool giants in binaries are appreciably greater than the value 1 $m_\odot$ obtained from indirect arguments by Scalo, Dominy & Pumphrey (1978). HD 20301, a double-lined eclipsing system probably consisting of early G giants, may lead to definitive masses (Andersen & Nordström 1977, Olsen 1977).

To the masses of the supergiants in the systems ζ Aur (K4Ib, $m = 8\ m_\odot$, $R_{\min} = 130\ R_\odot$) and 31 Cyg (K4Ib, $m = 10\ m_\odot$, $R_{\min} = 135\ R_\odot$) (Wright 1970), listed in the earlier review (Popper 1967), we may now add the M supergiant in VV Cep (M1Ia, $m = 18\ m_\odot$, $R = 1600\ R_\odot$) (Wright 1977). The velocity curve of the hot companion is poorly determined from Hα emission, and the published mean error of only 4% in its amplitude is probably too optimistic.

RESOLVED SPECTROSCOPIC BINARIES

In the case of a double-lined spectroscopic binary with good orbits of both components, only the value of the inclination, i, is needed in addition for mass determination. Values of i are known for a large number of visual binaries, and in principle all binaries except those with $i = 0$ are spectroscopic binaries. But only in particularly favorable circumstances can one resolve optically the components of binaries so close together as to have orbital velocities large enough for our purposes. Resolution may be attempted by a variety of techniques: (*a*) visual measurement; (*b*) visual Michelson interferometry; (*c*) intensity interferometry; (*d*) speckle interferometry; (*e*) amplitude (or electronic Michelson) interferometry; (*f*) lunar occultation. A discussion of interferometric techniques was the subject of IAU Colloquium No. 50 (In press). Method (*f*) is quite limited in its application to the binary problem (see Evans & Fekel 1979 for its application to β Cap), and results from various approaches to method (*e*) are yet to appear. Programs of speckle interferometry of spectroscopic binaries are being undertaken by McAlister (1976) and by Morgan et al. (1978). Method (a) has led to satisfactory results for δ Equ and ADS10598, while each of methods (b), (c), and (d) has borne fruit for one system, namely α Aur, α Vir, and 12 Per, respectively. Comments on these systems follow. Results are listed in Table 7.

ADS10598 Although the lines of the two components of this 5th mag G-type binary are marginally resolved on the small number of available spectrograms (Batten, Fletcher & West 1971), the masses resulting from combining the velocity differences with the visual orbit are of acceptable accuracy. The uncertainties in the masses (Table 7) are increased over those given in Batten, Fletcher & West to allow for the effects of uncertainties in the visual elements, particularly in the phasing of the spectroscopic observations. Photometry is from Johnson et al. (1966). The properties of the components are nearly enough identical for us to neglect their

Table 7 Resolved spectroscopic binaries

Binary	Sp	$B-V$	m	M_V	Scale dependent[a]		
					$\log T_e$	$\log L$	$\log R$
ADS10598	G8	+0.72	1.11	4.95	3.745	−0.03	0.01
HD158614	—	±0.02	±0.10	±0.07	±0.005	±0.03	±0.02
$P = 46\overset{d}{.}1$	G8	+0.72	1.11	4.95	3.745	−0.03	0.01
		0.02	0.10	0.07	0.005	0.03	0.02
δ Equ	F7	+0.50	1.19	3.95	3.794	0.32	0.085
HD202275	—	0.01	0.02	0.05	0.003	0.02	0.01
$P = 2083\overset{d}{.}5$	F7	+0.50	1.19	3.95	3.794	0.32	0.085
	—	0.01	0.02	0.05	0.003	0.02	0.01
α Aur	G0III	+0.66	2.55	0.25	3.765	1.84	0.91
HD33959	—	0.03	0.2	0.15	0.01	0.06	0.045
$P = 104.0$	G5III	+0.94	2.65	0.10	3.662	1.98	1.17
	—	0.04	0.5	0.15	0.015	0.06	0.04
α Vir	B1.5	−0.25	10.8	−3.5	4.390	4.33	0.91[b]
HD116658	—	0.02	1.3	0.1	0.04	0.10	0.025
$P = 4.0$	B4	−0.18	6.8	−1.5	4.235	3.16	0.62
	—	0.03	0.9	0.2	0.06	0.17	0.07
12 Per	F9	+0.55	1.19	3.8	3.782	0.38	0.14
HD16739	—	0.04	0.20	0.2	0.010	0.08	0.05
$P = 331.0$	F9	+0.62	1.04	4.1	3.770	0.23	0.09
	—	0.04	0.15	0.2	0.008	0.08	0.05

[a] Uncertainties of scale-dependent quantities do not include effects of uncertainties in the T_e and flux scales.
[b] Radius of primary not scale dependent.

differences. The parallax obtained, 0″.060 ± 0″.002, may be compared to the trigonometric value 0″.051.

δ Equ The properties of the nearly identical components of δ Equ appear firmly established (Popper & Dworetsky 1978, Hans et al. 1979). The parallax derived by combining spectroscopic and visual observations, 0″.055, agrees with the trigonometric value, 0″.053.

α Aur Uncertainties in the masses are caused by the difficulty of measuring the lines of the hotter component and by the sensitivity to a small uncertainty in the inclination of 43°. The data are discussed in BFM. The colors of the composite system correspond more closely to types G8III and G0III and a magnitude difference, $\Delta V = 0.15$, than to Wright's (1954) types of G5III and G0III with $\Delta V = 0.25$. These uncertainties are reflected in those of the luminosities and radii. The angular diameters of the components given by Blazit et al. (1977) are of insufficient precision to aid in this problem.

α Vir The data for α Vir, resolved by intensity interferometry, are taken from Herbison-Evans et al. (1971) and Shobbrok, Lomb & Herbison-Evans (1972). The radius of the primary component is obtained from its angular diameter, rather than from the magnitude and color index, which are used for the other stars listed in Table 7. The spectral type of the secondary is uncertain (Struve et al. 1958), leading to the large uncertainty in its radius. A value of $B-V = -0.18$ is adopted for this component, corresponding to type B4. Difficulties with the radial velocities remain, particularly for the secondary component.

12 Per McAlister's (1978) analysis of his speckle observations of 12 Per give $i = 123° \pm 2°$. Reanalysis of Colacevich's (1941) spectroscopic observations gives $P = 331^{d}.2 \pm 0^{d}.15$, $K_1 = 21.8 \pm 0.8$, $K_2 = 24.8 \pm 1.4$, $e = 0.69 \pm 0.02$. The masses in Table 7 result from combining McAlister's inclination with these spectroscopic results. The parallax used in obtaining M_V, 0″.047 ± 0″.003, is obtained from McAlister's values of i and a'' and the spectroscopic value of $a \sin i$. The trigonometric parallax is 0″.040. The magnitude difference between the components is poorly known. The color difference, which is unknown, was adopted to correspond to the difference in masses.

The bolometric corrections and surface fluxes, used in deriving the luminosities and radii of the stars listed in Table 7, are from Table 1, as are the effective temperatures.

χ Dra (Breakiron & Gatewood 1974), ADS 11060 (Batten et al. 1979), and ψ Sgr (Fekel 1979) are examples of resolved spectroscopic binaries for which the masses are not as yet obtained with sufficient accuracy for our purposes. A more general discussion of combining available radial velocity and astrometric data is given by Halliwell (1980).

VISUAL BINARIES

General Considerations

Critical lists of the masses and other properties of visual binaries have been given by van de Kamp (1958) and by Lacy (1977b). Less restricted lists are those of Harris, Strand & Worley (1963) and Heintz (1978). Veeder (1974) has discussed M dwarfs. It is rather sad, in view of the very great amount of difficult observing for more than 150 years, that the number of visual binaries for which masses are known to an accuracy of about 20% is not more than a dozen or so. Lacy's list contains only three systems not listed by van de Kamp nearly 20 years earlier, and one of these is of doubtful quality. There are, of course, orbits of many more visual binaries that are adequate for determining the sum of the masses of the components. The compilation by Finsen & Worley (1970) contains nearly 200 systems considered by the authors to have definitive or reliable orbits. The parallax of a system must be known to an accuracy of 7% for 20% accuracy in the masses. It is true that individual parallax determinations, as well as the mean of the determinations from two or more observatories, may have formal mean errors considerably less than 0".01. But random differences between the well-observed values for a given star from two observatories, as well as systematic differences between observatories (see, for example, Vasilevskis 1966, Gliese 1972, Upgren 1977) and somewhat uncertain corrections from relative to absolute parallaxes, make it appear unwise to assume at this time that the true uncertainty of a parallax is less than 0".01.

The major axes of the better-determined orbits of eclipsing binaries have uncertainties considerably less than 10%. If results from visual binaries are to be of comparable accuracy, the ratio of angular major axis to parallax should have an uncertainty of less than 10%, as well. Hence, in the present context, parallaxes less than 0".1 are of little use in mass determination. Examination of existing material on parallaxes shows that, for reliable values, in addition to the limiting size, there should be at least three reasonably accordant independent determinations. Van de Kamp & Worth (1971) discuss further this matter of the accuracy in parallaxes required for stellar masses in connection with η Cas. For purposes other than mass determination or for studies in which masses of high reliability are not required, less rigid standards would be appropriate.

Programs designed to obtain parallaxes of higher accuracy have been undertaken at the Lick (Vasilevskis 1969), Yerkes (van Altena 1971), and US Naval (Worley 1966) Observatories. But binary stars with good orbits do not appear to have priority in these programs. Greatly improved accuracy is predicted for parallaxes from space stations and with other improved techniques (Lacroute 1977, Høg & Fogh Olson 1977, Connes 1979, Høg 1979). In principle, means are at hand or may become available for deriving masses of many more of the visual binaries with good orbits than are presently available. It it not clear, however, that improved results for visual binary systems will be forthcoming soon.

The accuracy with which the masses are known in a visual binary system depends also on the reliability of the double star measures and on the uncertainty in the orbital solutions. The catalogue of Finsen & Worley (1970) classifies orbital solutions according to quality. Only systems with orbits corresponding to their classes 1 ("definitive") or 2 ("reliable") are considered as candidates. Even among these systems there are a few cases in which independent solutions lead to values of $a^3 P^{-2}$ differing by 20%, where a is the angular semimajor axis and P is the orbital period. One reason for such differences lies in the choice and weighting of observations. Eggen (1956) has emphasized this point. In six of the 29 systems listed by Eggen with orbits of quality 1 or 2, his value of $a^3 P^{-2}$ differs from that obtained from the data listed in Finsen & Worley by more than 15%. The average difference is 13%. The most extreme case is the bright system, ψ Vel, with nearly equal components (van den Bos 1945), for which values of a from observations of different experienced observers range from 0$.''$76 to 1$.''$11. The matter of systematic errors in visual double star observations was discussed in IAU Colloquium No. 10 (Heintz 1971), and is clearly a matter of personal opinion. The use of photoelectric scanning techniques (Rakos 1965, Franz 1970) does not appear to have led to published results as yet. The use of speckle interferometry was referred to in the previous section.

For orbits of high quality, uncertainties introduced by the method of orbital solution are of less importance than uncertainties in the parallaxes. In recent years, most orbital solutions have been carried out with modern digital computing systems. Their use for this purpose has been discussed by Wielen (1962), Eggen (1967), Heintz (1967), and Morbey (1975). Eichhorn & Clary (1974) point out that for very close systems the usual first-order least-squares treatment may not be adequate. Among the more active workers applying machine methods have been Heintz (e.g. 1976) and Starikova (e.g. 1978).

A final consideration concerning masses is the mass ratio, obtained photographically except for the wider pairs. The most recent compilation of mass ratios I am aware of is that of Harris, Strand & Worley (1963). The Sproul Observatory observers appear to have been the most active in this important enterprise in recent years, as they have been in obtaining additional parallaxes of visual binaries. The reliability of the mass ratio for a close system is difficult to estimate, being dependent on the effects of blending of images under different seeing conditions. Tests carried out by

Feierman (1971) imply that there are greater uncertainties in the correction from photocenter to barycenter in the blended stellar images than had been assumed.

In order to obtain luminosities and radii to correlate with the masses, we need, in addition to parallaxes, apparent magnitudes and color indices or spectral types on standardized systems. For binaries with separations closer than a few arcseconds, data for the individual components must often be based on estimates rather than on measures. For the nearby stars under consideration here, the relevant data have been compiled by Gliese (1969). Worley (1969) has shown that the better estimates of magnitude differences are usually reliable to within 0.1 mag. For the fainter components of close systems, the bolometric corrections and surfaces fluxes (and consequently their radii) may be quite uncertain.

The Available Data

According to the principles adopted in this review, we consider only visual binary systems with parallaxes greater than $0''.09$, with at least three independent determinations of the parallax within 10% of their mean value, and with orbits of quality 1 or 2. There are 14 such systems, if we include L726-8 (Worley & Behall 1973) and o^2 Eri BC (Heintz 1974), for which discussions subsequent to the Finsen-Worley catalogue may provide sufficient improvement in the visual orbit to put them marginally into quality category 2. The properties of these 14 systems are listed in Table 8. Fourteen is really a pitifully small number in view of the great effort by many observers. The only systems in Table 8 not included in the list carefully selected 20 years ago by van de Kamp (1958) are L726-8, Wolf 630, and HR6426. The systems in the table not included in Lacy's (1977b) list are o^2 Eri BC, ξ Boo, HR6426, and ζ Her.

To illustrate the problems encountered in attempting to add visual binary systems to the list of objects with well-defined properties, we may consider the data for HR6426 and ζ Her.

HR6426 is GL 667 (Gliese 1969), and is often referred to as M1b 4. Finsen & Worley (1970) list two orbits, *both* of quality 1. The two values of $a^3 P^{-2}$ differ by 14%. The reasonably accordant parallaxes lead to $\pi = 0''.14 \pm 0''.01$. (The ± 0.01 is adopted from the discussion above.) In this case, the mass ratio, $m_B(m_A + m_B)^{-1} = 0.41 \pm 0.03$, is obtained from micrometer measures relative to nearby stars (Hirst 1947) rather than from the usual long-focus photographic astrometry, so the problems of blended images are avoided. The resulting masses (Table 8) have nearly acceptable uncertainties in the context of this review, but the spectral types and magnitude difference in HR6426 are poorly known. Composite types from K2 to K5 may be found in the literature and visual magnitude differences of 0.9 and 1.5 mag. Thus, the luminosities and radii are subject to more uncertainty than is desirable for a first-rate system. UBV photometry of the combined light (Johnson et al. 1966) shows $V = 5.91$, $B-V = +1.04$, $U-B = +0.82$. While these colors can be matched by combining those of main-sequence stars of types K2-3 and K5 with a V magnitude difference of 1.0 mag, the radii and luminosities obtained from these data can hardly be considered definitive determinations. That the radius of the less luminous component is found to be greater than that of the more luminous one is undoubtedly a consequence of incorrect differences in magnitude and color.

Zeta Her illustrates a somewhat different set of problems. There are four parallax determinations within a range of $0\rlap{.}''016$. In this case, $0\rlap{.}''01$ uncertainty in the parallax may be somewhat of an overestimate, and the uncertainty of the masses from this cause may be taken as 25%. The value of a given by Eggen (1956) is 3.5% less than the value adopted by Finsen & Worley (1970), and the values of the mass ratio found in the literature range from 0.34 to 0.42. Thus, the uncertainty in the masses depends principally, though not entirely, on that in the parallax. The visual magnitude difference between the components of ζ Her is well determined as 2.65 ± 0.1 (Worley 1969). But the deconvolution of the individual color indices requires an assumption as to the "normality" of the colors, as in the case of HR 6426.

The apparent magnitudes and parallaxes in Table 8 are taken from Gliese (1969) and Veeder (1974) except for the parallaxes of L726-8 (Worley & Behall 1973) and 70 Oph (Worth & Heintz 1974). An uncertainty of $\pm 0\rlap{.}''01$ is adopted for each value of π except for L726-8 ($\pm 0\rlap{.}''02$) and ζ Her ($\pm 0\rlap{.}''008$). An uncertainty of $\pm 0\rlap{.}''01$ is assumed in a for orbits of quality 1, $\pm 0\rlap{.}''02$ for quality 2, and $\pm 0\rlap{.}''04$ for quality 2-3. The periods are assumed known without sensible error. The values of a and P are taken from Finsen & Worley (1970), except for L726-8 (Worley & Behall 1973) and o^2 Eri BC (Heintz 1974). The sources for the mass ratios are given in the table. An uncertainty of ± 0.02 is adopted for the ratios except for the wider pairs and systems with large values of the magnitude difference, for which ± 0.01 is adopted. In cases where the ratio is *assumed* to be unity, larger values are adopted (± 0.03 for ζ Her and ± 0.05 for Wolf 630 and Fu 46). In most cases, the uncertainty in the masses is dominated by that in the parallax. The masses in Table 8 differ but little from those given by Lacy (1977b), although his mean errors are usually smaller than those obtained here, since he has adopted published mean errors of the parallaxes. There is good evidence (e.g. Lippincott & Hershey 1972, Probst 1977, Hershey 1978) for stellar masses less than 0.1 $m_\odot$, but they are as yet not known with sufficient accuracy for inclusion here.

Table 8 Visual binaries

Binary	Sp	V	$(B-V)$ or $(V-R)$[b]	π''	a''	$m_B/\Sigma m$	m	M_V	Ref.[d] for $m_B/\Sigma m$	Scale dependent[a]		
										$\log T_e$	$\log L$	$\log R$
α CMa	A1V	−1.46	0.00	0.377	7.50	0.30	2.20	1.42	1	3.975	1.37	0.225[c]
HD48915	—	±0.02	±0.01	+0.010	±0.01	±0.02	±0.2	±0.06	2	±0.005	±0.025	±0.015
$P = 50\overset{y}{.}1$	wA	8.3	−0.12:	—	—	—	0.94	11.2		—	—	—
		0.1					0.1	0.1				
α CMi	F5IV-V	0.37	+0.42	0.287	4.55	0.27	1.77	2.66	1	3.816	0.82	0.314[c]
HD61421	—	0.02	0.01	0.010	0.01	0.01	0.2	0.08	2	0.002	0.03	0.020
$P = 40.6$	w	10.7:	—	—	—	—	0.65	13.0:		—	—	—
							0.1					
ζ Her	G0IV	2.89	+0.64	0.104	1.37	0.35	1.25	3.00	2	3.765	0.74	0.35
HD150680	—	0.02	0.01	0.008	0.01	0.03	0.3	0.17	3	0.003	0.07	0.035
$P = 34.4$	K0V	5.49	+0.75	—	—	—	0.70	5.60	4	3.735	−0.28	−0.10
	—	0.03	0.05	—	—	—	0.2	0.17		0.015	0.07	0.05
α Cen	G2V	−0.01	+0.68	0.743	17.56	0.45	1.14	4.34	1	3.755	0.12	0.105
HD128620/1	—	0.02	0.03	0.01	0.01	0.01	0.05	0.03	2	0.010	0.01	0.02
$P = 79.9$	K0V	1.33	+0.88	—	—	—	0.93	5.68		3.700	−0.28	−0.025
	—	0.03	0.03	—	—	—	0.04	0.04		0.010	0.015	0.02
γ Vir	F0V	3.49	+0.36	0.094	3.75	0.50	1.08	3.36	1	3.826	0.52	0.13
HD110379/80	—	0.03	0.01	0.01	0.01	0.01	0.35	0.23	2	0.002	0.09	0.05
$P = 171.4$	F0V	3.49	+0.36	—	—	—	1.08	3.36		3.826	0.52	0.13
	—	0.03	0.01	—	—	—	0.35	0.23		0.002	0.09	0.05
η Cas	G0V	3.45	+0.57	0.172	11.99	0.38	0.91	4.63	1	3.777	0.06	−0.01
HD4614	—	0.02	0.02	0.01	0.02	0.01	0.15	0.15	5	0.005	0.06	0.03
$P = 480$	M0V	7.51	+1.40	—	—	—	0.56	8.69	6	3.590	−1.13	−0.23
	—	0.05	0.15	—	—	—	0.10	0.15		0.02	0.06	0.12
ξ Boo	G8V	4.68	+0.71	0.148	4.92	0.45	0.90	5.53	2	3.745	−0.27	−0.115
HD131156	—	0.03	0.02	0.01	0.01	0.02	0.20	0.15	7	0.005	0.06	0.03
$P = 152$	K4V	6.84	+1.10	—	—	—	0.72	7.69		3.645	−0.96	−0.26
	—	0.05	0.08	—	—	—	0.15	0.15		0.02	0.06	0.07

STELLAR MASSES   151

HD165341	—	0.02		0.01	0.02	0.15	5.78	1	3.721	−0.53	−0.10
$P = 88.1$	K5V	5.94					0.10	2	0.01	0.04	0.03
	—	0.03	+1.16			0.61	7.48	8	3.631	−0.84	−0.17
	—	0.06	0.10			0.10	0.12		0.03	0.05	0.08
HR6426	K3V	6.3	+0.94	0.137	1.82	0.78	7.0	9	3.685	−0.77	−0.24
HD156384	—	0.1	0.06	0.01	0.01	0.2	0.2		0.015	0.085	0.06
$P = 42.1$	K5V	7.2	+1.20			0.54	7.9		3.622	−0.98	−0.22
	—	0.4	0.12			0.1	0.4		0.025	0.2	0.12
Wolf 630	M4.5V	9.76	+1.63[b]	0.161	0.22	0.50	10.80	4	3.538	−1.62	−0.37
HD152751	—	0.05	0.1	0.01	0.01	0.05	0.15		0.015	0.15	0.10
$P = 144.5$	M4.5V	9.76	+1.63[b]			0.42	10.80		3.538	−1.62	−0.37
	—	0.05	0.1			0.10	0.15		0.015	0.15	0.10
Fu 46	M3V	9.96	+1.58[b]	0.153	0.71	0.50	10.90	1	3.546	−1.71	−0.43
HD155876	—	0.03	0.03	0.01	0.01	0.05	0.15	2	0.005	0.07	0.04
$P = 13.0$	M3.5V	10.39	+1.70[b]			0.30	11.30	4	3.528	−1.73	−0.41
	—	0.10	0.06			0.07	0.17		0.01	0.10	0.06
Kr 60	M3V	9.85	+1.76[b]	0.250	2.38	0.36	11.85	1	3.518	−1.86	−0.46
+56°2783	—	0.10	0.03	0.01	0.01	0.03	0.15	2	0.005	0.07	0.04
$P = 44.4$	M4.5V	11.30	+1.89[b]			0.16	13.30	10	3.503	−2.30	−0.64
	—	0.30	0.09			0.02	0.30	11	0.015	0.15	0.09
o^2Eri BC	wA	9.52	+0.03	0.207	6.94	0.43	11.1	2			
HD26976 BC	—	0.02	0.02	0.01	0.04	0.07	0.1	12			
$P = 252$	M4.5V	11.2	+1.75[b]			0.16	12.8		3.520	−2.25	−0.66
	—	0.15	0.15			0.03	0.15		0.025	0.20	0.14
L726-8	M5.5V	12.50	+2.32[b]	0.385	2.06	0.49	15.45	13	3.443	−2.83	−0.79
Star B = UVCet	—	0.1	0.05	0.02	0.04	0.02	0.15		0.007	0.07	0.04
$P = 26.5$	M5.5V	13.10	+2.45			0.11	16.00		3.425	−2.98	−0.82
	—	0.2	0.10			0.02	0.25		0.015	0.15	0.08

[a] Uncertainties in scale-dependent quantities do not include effects of uncertainties in T_e and flux scales.
[b] Color indices are $V–R$.
[c] These radii are not scale dependent.
[d] References for mass ratios

1. van de Kamp 1954
2. Strand & Hall 1954
3. Wyller 1955
4. Harris, Strand & Worley 1963
5. Strand 1969
6. van de Kamp & Worth 1971
7. Hershey 1977
8. Worth & Heintz 1974
9. Hirst 1947
10. Lippincott 1953
11. Wanner 1967
12. Heintz 1974
13. Worley & Behall 1973

The values of V and the color indices $B-V$ or $V-R$ are from Gliese (1969) and Veeder (1974) when available. The uncertainties of the color indices, upon which the fluxes required for the luminosities and radii depend, are quite large for most of the cooler stars, as shown in Table 8. The discussion above of HR6426 and ζ Her illustrates some of the problems encountered in separating the colors of the components. In some cases, the observations themselves are of low weight, and when values of both $V-R$ and $B-V$ are available, they may not be compatible. Kunkel & Rydgren (1979) noted difficulties in transforming observations to $V-R$ indices for M dwarfs. For the fainter members of the closer binaries, the apparent magnitudes may

Table 9 The nearest visual binaries with orbits of quality 1–4

Binary	A or GL[a]	Spectra[b]	Orbit[c] quality	Period	Par.	n[d]	Needed[e]
α Cen	GL559	G2, K0	1	80^y	0″.75	4	—
L726-8	GL65	M5.5	2–3	27	0.38	3	(o), (p)
α CMa	A5423	A1, w	1	50	0.37	6	—
α CMi	A6251	F5, w	1	41	0.29	6	—
61 Cyg	A14636	K5, K7	4	650	0.29	6	o
+59°1915	A11632	M4, M5	4	450	0.28	6	o
Ross 614	GL234	M7, ?	3	16	0.25	6	r
Kr 60	A15972	M3, M4.5	1	44	0.25	4	—
Wolf 424	GL473	M5.5	4	16	0.23	5	r
o^2Eri BC	A3093	w, M4.5	2–3	250	0.20	5	(o)
70 Oph	A11046	K0, K5	1	88	0.20	4	—
η Cas	A671	G0, K5	2	480	0.17	4	(o)
Wolf 630	GL644	M4.5	1	1.7	0.16	5	—
Fu 46	GL661	M3	1	13	0.15	3	(p)
ξ Boo	A9413	G8, K4	1	150	0.15	3	(p)
HR486-7	GL66	K2, K5	4	480	0.15	2	o, p
HR6426	GL667	K3, K5	1	42	0.14	3	(o), (p)
ξ UMa	A8119	G0	1	60	0.13	1	p
−21°1051	GL185	M1, ?	3	48	0.13	1	o, p
μ Her BC	A10786	M4	1	43	0.12	2	p
HR6416	GL666	G8, M0	4	700	0.12	2	o, p
HD115953	A8862	M2, ?	2	49	0.11	1	(o), p
ζ Her	A10157	G0IV, K0	1	34	0.10	4	—
γ Vir	A8630	F0	1	170	0.09	3	(p)
HD99279	GL428	K7, M0	3	421	0.09	3	o, p

[a] A = Aitken Double Star Catalogue; GL = Gliese Catalogue of Nearby Stars, 1969 edition.
[b] All are main sequence unless otherwise indicated.
[c] As given in Finsen & Worley (1970).
[d] Number of independent parallax determinations within 10% of their mean [Jenkins (1963) and later determinations].
[e] o: orbit; p: parallax; r: resolution.

also not be well established. All these sources of uncertainty, in addition to those in the parallaxes, enter into the calculations of log L and log R. The values of T_e and F'_V are obtained from Table 1.

It should be clear from this brief discussion that determination of the radii of visual binaries, as well as their masses, is beset with more serious and a greater variety of difficulties than in the cases of the better eclipsing binaries. While most of the values of log R in Table 8 are in reasonable agreement with those of Lacy (1977b) for the stars in common, there are a few significant differences, and my m.e.'s are generally larger than Lacy's because of the considerations discussed above. The radii listed in Table 8 for α CMa and α CMi are taken directly from their angular diameters (Hanbury Brown, Davis & Allen 1974).

In Table 9 are listed, in order of increasing distance, 24 visual binaries with $\pi \geq 0''.09$ and with orbits of quality 4 or better. For the systems with inadequate data, notes specify the inadequacies that would have to be removed for the systems to be added to Table 8. Additional nearby systems of particular interest are the astrometric binaries $+20°2465 = $ GL388 (M4, $\pi = 0''.21$, $P = 26^y$) and μ Cas (sdG5, $\pi = 0''.13$, $P = 18^y$), for which resolution of the components would provide masses of importance, and AC $+59°24610/1 = $ GL169 (M5+w, $\pi = 0''.19$, $P = 350^y$) for which an improved parallax and orbit could give us an additional direct determination of the mass of a white dwarf.

The systems listed in Table 10 are those with parallaxes between $0''.05$ and $0''.09$ and orbits of quality 1 or 2. A new standard of accuracy for parallaxes, applied to some of these systems, would appreciably increase our knowledge of stellar masses, particularly among the G and K main-sequence stars, where the eclipsing binaries provide little information. A number of these systems of smaller parallax are included in the compilation of masses by Harris, Strand & Worley (1963).

As an alternative to improved parallax determination, a great increase in the precision of radial velocities (e.g. Campbell & Walker 1979), to the order of ± 50 m s^{-1}, would place many visual binaries in the category of "resolved spectroscopic binaries" discussed above. The velocity system would have to be stable and the velocities measured over time intervals comparable to the orbital periods. Hence, this improvement in technique, which may be forthcoming, would rescue good visual orbits less rapidly than could a program of more reliable parallaxes.

In addition to the systems listed in Tables 9 and 10, there are nearby binaries of relatively short period with orbits and parallaxes as yet inadequately known (e.g. Heintz 1979), as well as undiscovered binaries, that can be expected to provide significant mass determinations.

Table 10 Visual binaries with parallaxes between 0".05 and 0".09 and orbits of quality 1 or 2

Binary	A or GL[a]	Spectrum[b]	Orbit[c] quality	Period	Par.[d]
	A490	F8	2	6.9	0".063
	A520	G5	1	25	0.074
	GL60	K3	1	4.6	0.057
	A1865	M0	2	25	0.064
ε Cet	GL105.4	F8	2	2.7	0.069
	A6420	G1	1	23	0.069
	A6664	M0	2	59	0.057
	A7114	M1	2	40	0.066
10 UMa	GL332	F5	2	22	0.074
	A7284	K3	1	34	0.058
ψ Vel	GL351	F2	1	34	0.065
α Com	A8804	F5	1	26	0.053
	A9031	K2	2	155	0.086
η CrB	A9617	G2	1	42	0.060
	A9689	K5	2	54	0.09[e]
	A9716	K0	1	56	0.053
	A10075	K2	2	236	0.058
η Oph	A10374	A2	2	88	0.051
	A10585	M0	2	60	0.055
	A10598	G8IV-V	1	46	0.052
	A10660	G1	2	76	0.067
99 Her	A11077	F7	2	55	0.061
	A11871	G0	2	61	0.054
γ CrA	GL743.1	F8	1	120	0.054
τ Cyg	A14787	F0	1	50	0.050
85 Peg	A17175	G3	1	26	0.084

[a] A = Aitken Double Star Catalogue; GL = Gliese Catalogue of Nearby Stars, 1969 edition.
[b] All are main sequence except A10598.
[c] Finsen & Worley (1970).
[d] Jenkins (1963).
[e] Spectroscopic parallax.

SEMIDETACHED SYSTEMS

General Considerations

Semidetached systems are close binaries in which the less massive component, which is usually the cooler and larger one, fills the zero-velocity surface passing through the inner Lagrangian point. The more complex physical nature of these systems leads in most cases to difficulties in the analyses of observational data, so that the fundamental properties of the components cannot usually be determined as reliably as for many detached systems (Table 2). These difficulties include relative faintness of the cooler component so that its spectral lines are difficult to measure; ellipticity and reflection in the light curves, making analyses less definitive; gas streams giving rise to nonorbital displacements of spectrum lines associated with the hotter component as well as to nongeometrical effects in the light curves. Even the decision that a system is semidetached may not always be based on clear evidence, since both the mass ratio and radii of the components must be known in order to make the decision.

Systems to be included in this section are those in which the lines of both components are measured throughout the orbits. It is convenient to distinguish somewhat arbitrarily three groups of semidetached eclipsing systems. First are the more massive systems in which the hotter component is an early B-type star and the cooler of type B or early A (V Pup, V356 Sgr, u Her, Z Vul). The system μ_1 Sco is omitted because its low orbital inclination, near 60°, makes its properties even less well defined than in the other systems. Second are the more "typical" Algol systems of lower mass, in which the more massive component lies in the range from middle B to early F, and the other is of type F or later. Third are the later subgiant (RV Lib, RT Lac) and giant (RZ Cnc, AR Mon) semidetached systems.

While the first and third groups, with the two components not differing greatly in spectral type, have limited membership among observed eclipsing binaries, there are large numbers of typical Algol systems. In most of them the lines of the cooler components are difficult to measure, if indeed they are visible in spectrograms of the resolution available. Most success in obtaining velocity curves of the cooler components has resulted from use of the D lines, although these lines have their own difficulties including, in addition to their weakness in most cases, blending with interstellar or circumstellar lines, blending with water-vapor lines, and blending of the lines of the two components. Although lines of the hotter components in most typical Algol systems are strong and readily measurable, it has been realized ever since Struve's early work on U Cep that the measured velocities may not represent orbital motion because of the effects of material not moving with the photosphere of the star. Even when the lines appear reasonably sharp and symmetrical (e.g. RY Gem, McKellar 1949, or TT Hya, D. M. Popper and M. Plavec, unpublished), they may show considerable departures from orbital motion, whereas the D lines of the cooler components appear to represent true orbital motion. Because of the small amplitudes of velocity variation for the hotter components in a number of systems, these effects can make determination of the masses of the cooler components very uncertain.

Table 11 Hot semidetached systems

Binary	Sp	$(B-V)_0$	E_{B-V}	m	R	Vel	l.c.	$B-V$ Sp	$\log T_e$	$\log L$	M_V
V Pup	B1	−0.26	0.03	17	6.3	1	2	5	4.42	4.20	−3.35
HD65818		±0.02	±0.02	±1	±0.3		3	6	±0.04	±0.16	±0.25
$P = 1\overset{d}{.}45$	B2	−0.24	—	9	5.3		4	7	4.36	3.85	−2.65
		0.03	—	1	0.3				0.06	0.25	0.20
V356 Sgr	B4	−0.18	0.20	12.1	6.0	1	8	1	4.23	3.45	−2.30
HD173787		0.02	0.03	1.1	0.7				0.04	0.16	0.30
$\dot{P} = 8.90$	A2	+0.08	—	4.7	14.0				3.935	3.00	−2.75
		0.03	—	0.6	1.5				0.012	0.10	0.25
u Her	B2.5	−0.22	0.03	7.3	4.8	1	9	11	4.32	3.60	−2.20
HD156633		0.02	0.02	1.0	0.4		10	12	0.04	0.16	0.25
$P = 2.05$	B7	−0.12	—	2.7	4.3			13	4.11	2.65	−1.10
		0.03	—	0.3	0.2			14	0.04	0.16	0.20
Z Vul	B3	−0.19	0.23	5.4	4.7	1	1	1	4.255	3.30	−1.85
HD181987		0.02	0.03	0.3	0.2		15	16	0.04	0.16	0.20
$P = 2.455$	A2	+0.04	—	2.3	4.5			17	3.955	2.07	−0.45
		0.03	—	0.1	0.2				0.015	0.06	0.15

[a] Uncertainties of scale-dependent quantities do not include effects of uncertainties in the T_e and flux scales.
[b] References
1. Cited in BFM tables and notes
2. Cited in KPW notes
3. Schneider, Darland & Leung 1979
4. Eaton 1978a
5. Johnson et al. 1966
6. Crawford, Barnes & Golson 1970
7. Hiltner, Garrison & Schild 1969
8. Wilson & Caldwell 1978
9. Eaton 1978b
10. Söderhjelm 1978
11. Kopylov 1958
12. Olson 1968
13. Crawford, Barnes & Golson 1971b
14. McNamara 1966
15. Cester et al. 1977
16. Olson & Weiss 1974
17. Olson 1975

The Available Data

Observational results for the three groups of semidetached eclipsing binaries are presented in Tables 11, 12, and 13. For reasons discussed above, these results are generally less well established than for the detached eclipsing systems of Tables 2 and 4. For the more massive systems, listed in Table 11, an attempt is made to estimate the uncertainties of the results, although these estimates are less well grounded than for the detached system. For V Pup and u Her, in particular, the velocity curves are quite uncertain. The Roche-lobe–filling less-massive component in V Pup is the smaller one, in contrast to nearly all other semidetached systems.

Table 12 Algol systems

Binary	P	Sp[a]	m	R	log L	M_V
RY Per[b]	6^{d}86	B4	5:	3.4:	3.0	−1.1
HD17034	—	F5	0.8::	7	1.9	0.0
RS Vul	4.48	B5	4.5	4.1	2.9	−1.3
HD180939	—	G0:	1.4	5.6	1.5	0.9
U Sge	3.38	B8	5.7	4.1	2.6	−1.0
HD181182	—	G5	1.9	5.3	1.3	1.5
Algol	2.87	B8	3.7	3.1	2.2	−0.2
HD19356	—	G8	0.8	3.2	0.8	2.8
S Cnc	9.48	A0	2.4	2.1	1.5	1.1
HD74307	—	K2:	0.2::	5	1.0	2.6
RY Gem[b]	9.30	A2	2.6	3.3	1.8	0.2
HD58713	—	K2:	0.6:	6	1.1	2.3
TT Hya[b]	6.95	A2	2.6:	2:	1.3	1.4
HD97528	—	K1:	0.7::	5:	1.1	2.1
XY Pup[b]	13.78	A2	2.3:	—	—	—
HD67862	—	K	0.3::	—	—	—
AS Eri	2.66	A3	1.9	1.8	1.2	1.7
HD21985	—	K0:	0.2	2.2	0.4	3.9
TW Dra	2.81	A3	1.7	2.4	1.4	1.1
HD139319	—	K0:	0.8	3.4	0.8	2.9
AW Peg[b]	10.62	A5:	2.0:	—	—	—
HD207956	—	K1	0.3::	—	—	—
RY Aqr	1.97	A5	1.3	1.5:	0.9	2.3
HD203069	—	K1:	0.3:	1.7:	0.2	4.4
TW And	4.12	A8	1.8	2.2	1.1	1.9
+32°4756	—	K2:	0.4	3.4	0.7	3.4

[a] Types of cooler components based on colors and depths of minima of light curves as well as on appearance of the spectra.

[b] Strong double Hα emission.

Table 12 lists results for those "typical" Algol systems for which velocity curves of the cooler stars have been measured. Except for AS Eri (Popper 1973), AW Peg (Hilton & McNamara 1961), Algol (Tomkin & Lambert 1978), and U Sge (Tomkin 1979), the masses are based on unpublished velocities for the cooler components and published (BFM) and unpublished velocities for the hotter ones. Work is continuing on these systems. Photometric analyses, required for the radii, are obtained from KPW. These analyses are often uncertain because of shallow secondary minima, large reflection effects, effects of gas streams, unrealistic photometric models, etc. With a few exceptions, both the photometric observations of Algol systems and their interpretation require thorough investigation. For systems with partial eclipses, the spectral types of the cooler components are estimated from the photometry (depths of minima, color indices), in which cases corrections for reflected light may be large and uncertain. On the basis of available information, the cooler components have colors and types near K1 IV except for AW Peg, RY Per, RS Vul, and U Sge, for which types of F5:, F5:, G0:, and G5 are indicated. Instead of attempting to estimate uncertainties in the masses and radii, I employ single (:) or double (::) colons to indicate that the digit immediately preceding the colon is poorly or very poorly known. With few exceptions, the luminosities listed are quite uncertain because of poorly known radii and temperatures.

Table 13 Cool semidetached systems

Binary	Sp	$B-V$	m	R	$\log T_e$	$\log L$	M_V
					Scale dependent[a]		
RZ Cnc	K1	+1.08	3.20	10.2	3.630	1.50	1.35
HD73343	—	±0.03	±0.15	±0.5	±0.006	±0.05	±0.15
$P = 21\overset{d}{.}6$	K4	+1.40	0.54	12.2	3.585	1.45	1.90
	—	0.06	0.05	0.7	0.006	0.05	0.20
AR Mon	K0	+0.95	2.70	10.8	3.660	1.65	0.85
HD37364	—	0.04	0.10	1.0	0.012	0.10	0.25
$P = 21.2$	K3	+1.30	0.80	14.2	3.590	1.62	1.25
	—	0.06	0.05	1.5	0.010	0.10	0.30
RV Lib	G5:		2.2:	—	—	—	—
HD128171	—	+1.04	—	—	—	—	—
$P = 10.7$	K2:		0.4:	—	—	—	—
RT Lac	G8:	—	0.6	4.6	3.72:	1.15:	2.10:
HD209318	—	—	—	—	—	—	—
$P = 5.1$	K1:	—	1.5	4.3	3.66:	0.85:	2.90:

[a] Uncertainties given for scale-dependent quantities do not include effects introduced by uncertainties in the flux and T_e scales.

Results for the small group of cooler semidetached systems are in Table 13. The available material and analyses of the giants RZ Cnc and AR Mon are from Popper (1976a). No photometric solution is available for RV Lib, the semidetached nature of which is inferred from the considerable inequality of the masses. According to Milone's (1977) analysis, the hotter component of RT Lac fills its Roche lobe, although the nature of the components of this system is still in question. The velocity curves of both systems are based on unpublished material. See also Popper & Ulrich (1977). No evaluation of realistic uncertainties of the parameters of these two systems is possible at present.

CONCLUDING REMARKS

The attempt has been made in this review to gather together our current knowledge of the masses of the components of detached binary systems when they can be determined with reasonable accuracy. An effort to provide realistic estimates of uncertainties of the tabulated parameters has also been made. A great deal of work is in progress on various aspects of binary star observation that will add significantly to the store of fundamental properties. The present compilation constitutes a progress report, subject to continuous revision. I am aware that the selection of data and their treatment reflect a personal point of view, and that other reviewers would make different selections.

In the tabulations in this review, those parameters of the stars derived directly from observations are distinguished from those that depend, in addition, on adopted scales of radiative parameters. In order to evaluate radii of visual binaries and luminosities of all the stars, it has been necessary to adopt scales of surface flux, bolometric correction, and effective temperature. Since the emphasis here is on masses, an exhaustive study of these scales has not been attempted. Many readers will have their own views of the appropriateness of the entries in Table 1. Values of the "scale dependent" parameters in the other tables can be adjusted accordingly.

Difficulties restricting the reliability with which the masses and other properties can be determined are discussed for each group of systems. It is to be expected that improved techniques in observation and analysis will lessen the effects of some of these difficulties, enabling knowledge of fundamental stellar properties to be placed on an improved basis. Comparison of the number of detached main-sequence eclipsing systems and the uncertainties of their properties (Table 2) with the corresponding quantities for visual binaries (Table 8), which furnish most of our data for main-sequence stars cooler than G0, shows that lower main-sequence masses are much less well known than masses of hotter main-sequence stars. As discussed in the section on visual binaries, major improvements in the techniques of parallax determination, in particular, are required before stars of the later types are on equal footing with the late B, A, and F stars insofar as masses are concerned.

Except for the white dwarfs, none of the stars in detached systems listed in Tables 2, 4, 7, and 8, including the giants in α Aur, departs from a single smooth mass-luminosity relation by an amount significantly greater than the uncertainties in the parameters. As in earlier listings, apparent outstanding exceptions to this statement in BFM, such as EO Aur, WX Cep, and VV Ori, have been shown (Popper 1978 and unpublished; Andersen 1976) to be based on misinterpretation of spectroscopic data. An additional apparent exception, HD208095, is under investigation.

The results listed in Tables 2, 4–9, and 11–13 can be displayed in a variety of two-dimensional diagrams with a variety of scales, and compared with predictions of a variety of theories. Examples of such comparisons are in Popper et al. (1970; interior composition), Lacy (1977b; interior models), and Popper & Ulrich (1977; evolutionary status). An approach that has been little used is to require, for a system with components of appreciably different mass, that the models predict, with the same composition and age, the radii and luminosities of *both* components to within the accuracy of the results derived from observation.

While models for single stars, appropriate for comparisons with observed results for components of detached binaries, may be in a reasonably satisfactory state, both theory and observation are much less satisfactory for semidetached systems (Tables 11–13), as emphasized in several contributions to IAU Symposium No. 88 on close binary stars: observation and interpretation, held in Toronto, Ontario, Canada in August 1979 (in press).

ACKNOWLEDGMENTS

My greatest debt is to the large number of astronomers who have contributed useful data on the properties of binary stars. To the extent that I have overlooked, misrepresented, or failed to acknowledge some of this work, I offer regrets. It is not possible to guarantee that no errors are present in any of the tabulations. I hope they are minimal in number and insignificant in magnitude. If there are some aspects of this work that appear to have ancestry in the pioneering studies of Wyse (1934) and Kuiper (1938), the connection is not entirely coincidental. My interest in binary stars was stimulated by contact with them during their most active years in binary star studies. Work on close binaries, including preparation of this review, is supported at the University of California, Los Angeles, by continuing grants from the National Science Foundation.

Literature Cited

Andersen, J. 1975. *Astron. Astrophys.* 44:355
Andersen, J. 1976. *Astron. Astrophys.* 47:467
Andersen, J., Clausen, J. V. 1980. In preparation
Andersen, J., Clausen, J. V., Nordström, B. 1979. In *Close Binary Stars: Observation and Interpretation, IAU Symposium No. 88*, ed. M. Plabec, R. K. Ulrich, D. M. Popper. Dordrecht: Reidel. In press
Andersen, J., Nordström, B. 1977. *Astron. Astrophys. Suppl.* 29:309
Bahcall, J. N. 1978. *Ann. Rev. Astron. Astrophys.* 16:241
Barnes, T. G., Evans, D. S., Moffett, T. J. 1978. *MNRAS* 183:285
Barry, D. C., Cromwell, R. H., Schoolman, S. A. 1978. *Ap. J.* 222:1032
Batten, A. H. 1962. *Publ. Dom. Astrophys. Obs., Victoria, BC* 12:91
Batten, A. H., Fletcher, J. M. 1971. *Astrophys. Space Sci.* 11:102
Batten, A. H., Fletcher, J. M., Mann, P. J. 1978. *Publ. Dom. Astrophys. Obs., Victoria, BC* 15:121 (cited as BFM)
Batten, A. H., Fletcher, J. M., West, F. R. 1971. *Publ. Astron. Soc. Pac.* 83:149
Batten, A. H., Morbey, C. L., Fekel, F. C., Tomkin, J. 1979. *Publ. Astron. Soc. Pac.* 91:304
Bell, R. A. 1971. *MNRAS* 154:343
Bell, R. A., Gustaffson, B. 1978. *Astron. Astrophys. Suppl.* 34:229
BFM: see Batten, Fletcher & Mann 1978
Binnendijk, L. 1970. *Vistas Astron.* 12:217
Blazit, A., Bonneau, D., Josse, M., Koechlin, L., Labeyrie, A., Onéto, J. L. 1977. *Ap. J. Lett.* 217:L55
Bloomer, R. H., Burke, E. M., Millis, R. L. 1979. *Bull. Am. Astron. Soc.* 11:439
Breakiron, L. A., Gatewood, G. 1974. *Publ. Astron. Soc. Pac.* 86:448
Budding, E. 1974. *Astrophys. Space Sci.* 30:433
Cameron, R. C. 1966. *Georgetown Obs. Monogr. No. 21*
Campbell, B., Walker, G. A. H. 1979. *Publ. Astron. Soc. Pac.* 91:540
Cester, B. 1965a. *Z. Ap.* 62:191
Cester, B. 1965b. *Mem. Soc. Astron. Ital.* 36:215
Cester, B., Fedel, B., Giuricin, G. Mardirossian, F., Pucillo, M. 1977. *Astron. Astrophys.* 61:469
Cester, B., Fedel, B., Giuricin, G., Mardirossian, F. 1978a. *Astron. Astrophys.* 62:291
Cester, B., Fedel, B., Giuricin, G., Mardirossian, F., Mezzetti, M. 1978b. *Astron. Astrophys. Suppl.* 32:351
Cester, B., Fedel, B., Giuricin, G., Mardirossian, F., Mezzetti, M. 1978c. *Astron. Astrophys. Suppl.* 33:91

Chambliss, C. R., Leung, K.-C. 1979. *Ap. J.* 228:828
Chen, K.-Y. 1975. *Acta Astron.* 25:89
Clausen, J. V. 1979. *Astron. Astrophys. Suppl.* 36:45
Clausen, J. V., Grønbech, B. 1976. *Astron. Astrophys.* 48:49
Clausen, J. V., Grønbech, B. 1977. *Astron. Astrophys.* 58:131
Clausen, J. V., Gyldenkerne, K., Grønbech, B. 1976. *Astron. Astrophys.* 46:205
Clausen, J. V., Nordström, B. 1978. *Astron. Astrophys.* 67:15
Clements, G. L., Neff, J. S. 1979. *Astron. Astrophys.* 75:193
Code, A. D., Davis, J., Bless, R. C., Hanbury Brown, R. 1976. *Ap. J.* 203:417
Colacevich, A. 1941. *Oss. Mem. Arcetri* 59:16
Connes, P. 1979. *Astron. Astrophys.* 76:L11
Conti, P. S., Walborn, N. R. 1976. *Ap. J.* 207:502
Cousins, A. W. J., Stoy, R. H. 1963. *R. Obs. Bull. No. 64*
Crawford, D. L. 1978. *Astron. J.* 83:48
Crawford, D. L., Barnes, J. V. 1970. *Astron. J.* 75:978
Crawford, D. L., Barnes, J. V., Golson, J. C. 1970. *Astron. J.* 75:624
Crawford, D. L., Barnes, J. V., Golson, J. C. 1971a. *Astron. J.* 76:621
Crawford, D. L., Barnes, J. V., Golson, J. C. 1971b. *Astron. J.* 76:1058
Dachs, J. 1971. *Astron. Astrophys.* 12:286
Devinney, E. J., Twigg, L. W. 1974. *Bull. Am. Astron. Soc.* 6:335
Dukes, R. J. 1974. *Ap. J.* 192:81
Eaton, J. E. 1978a. *Acta Astron.* 28:63
Eaton, J. E. 1978b. *Acta Astron.* 28:601
Eggen, O. J. 1956. *Astron. J.* 61:361
Eggen, O. J. 1963. *Astron. J.* 68:697
Eggen, O. J. 1967. *Ann. Rev. Astron. Astrophys.* 5:105
Eggen, O. J. 1978. *Astron. J.* 83:288
Eichhorn, H., Clary, W. G. 1974. *MNRAS* 166:433
Etzel, P. B. 1975. Masters thesis. San Diego State Univ., San Diego, Calif.
Evans, D. S., Fekel, F. C. 1979. *Ap. J.* 228:497
Feast, M. W. 1954. *MNRAS* 114:246
Feierman, B. H. 1971. *Astron. J.* 76:73
Fekel, F. C. 1979. PhD dissertation. Univ. Texas, Austin, Texas
Finsen, W. S., Worley, C. E. 1970. *Republic Obs. Circ.* 7:203
FitzGerald, M. P. 1970. *Astron. Astrophys.* 4:234
Franz, O. G. 1970. *Lowell Obs. Bull.* 7:191
Giannone, P., Giannuzzi, M. A. 1974. *Astrophys. Space Sci.* 26:289
Gliese, W. 1969. *Veroeff. Astron. Rechen-Inst. Heidelberg No. 22*

Gliese, W. 1972. *Q. J. R. Astron. Soc.* 13:138
Grønbech, B., Gyldenkerne, K., Jørgensen, H. E. 1977. *Astron. Astrophys.* 55:401
Güdür, N. 1978. *Astrophys. Space Sci.* 57:17
Güdür, N., Gülmen, O., Ibanoğlu, C., Bozkurt, S. 1979. *Astron. Astrophys. Suppl.* 36:65
Gyldenkerne, K., Jørgensen, H. E., Carstensen, E. 1975. *Astron. Astrophys.* 42:303
Hall, D. S. 1976. In *Multiple Periodic Variable Stars. IAU Colloq. No. 29*, ed. W. S. Fitch, p. 287. Dordrecht: Reidel
Halliwell, M. J. 1980. In preparation
Hanbury Brown, R., Davis, J., Allen, L. R. 1974. *MNRAS* 167:121
Hans, E. M., Scarfe, C. D., Fletcher, J. M., Morbey, C. L. 1979. *Ap. J.* 229:1001
Harris, D. L., Strand, K. Aa., Worley, C. E. 1963. In *Basic Astronomical Data*, ed. K. Aa. Strand, Chap. 15. Chicago: Univ. Chicago Press
Hayes, D. S. 1978. In *The HR Diagram, IAU Symposium No. 80*, ed. A. G. D. Philip, D. S. Hayes, p. 65. Dordrecht: Reidel
Hayes, D. S. 1979. *Dudley Obs. Rep. 14*, p. 297
Heintz, W. D. 1967. *Acta Astron.* 17:311
Heintz, W. D. 1971. *Astrophys. Space Sci.* 11:133
Heintz, W. D. 1974. *Astron. J.* 79:819
Heintz, W. D. 1976. *Ap. J.* 208:474
Heintz, W. D. 1978. *Double Stars*. Dordrecht: Reidel
Heintz, W. D. 1979. *Publ. Astron. Soc. Pac.* 91:490
Herbison-Evans, D., Hanbury Brown, R., Davis, J., Allen, L. R. 1971. *MNRAS* 151:161
Hershey, J. L. 1977. *Astron. J.* 82:179
Hershey, J. L. 1978. *Astron. J.* 83:308
Hilditch, R. W., Hill, G. 1975. *Mem. R. Astron. Soc.* 79:101
Hill, G., Hilditch, R. W., Younger, F., Fisher, W. A. 1975. *Mem. R. Astron. Soc.* 79:131
Hill, G., Hutchings, J. B. 1970. *Ap. J.* 162:265
Hiltner, W. A., Garrison, R. F., Schild, R. E. 1969. *Ap. J.* 157:313
Hilton, W. B., McNamara, D. H. 1961. *Ap. J.* 134:839
Hirst, W. P. 1947. *Union Obs. Circ.* 105:172
Høg, E. 1979. *Astron. Astrophys.* 75:L4
Høg, E., Fogh Olson, H. J. 1977. *IAU Highlights in Astron.* 4(I):362
Hutchings, J. B. 1975. *Publ. Astron. Soc. Pac.* 87:529
Hutchings, J. B., Cowley, A. P. 1976. *Ap. J.* 206:490
Jenkins, L. F. 1963. *General Catalogue of Trigonometric Stellar Parallaxes*. New Haven: Yale Univ. Obs.
Johnson, H. L., Mitchell, R. I., Iriarte, B., Wiśniewski, Z. 1966. *Comm. Lunar Planet. Lab.* 4:99
Johnson, H. L., Morgan, W. W. 1953. *Ap. J.* 117:313
Jørgensen, H. E. 1979. *Astron. Astrophys.* 72:356
Jørgensen, H. E., Gyldenkerne, K. 1975. *Astron. Astrophys.* 44:343
Kitamura, M., Kondo, M. 1978. *Astrophys. Space Sci.* 56:341
Koch, R. H., Koegler, C. A. 1977. *Ap. J.* 214:423
Koch, R. H., Plavec, M., Wood, F. B. 1970. *Publ. Univ. Penn. Astron. Ser.* Vol. XI (cited as KPW)
Kondo, H. 1976. *Tokyo Ann.* 16:1
Kopylov, I. M. 1958. *Publ. Crimean Astrophys. Obs.* 20:156
KPW: see Koch, Plavec & Wood 1970
Kříž, S. 1969. *Bull. Astron. Inst. Czech.* 20:202
Kuiper, G. P. 1938. *Ap. J.* 88:472
Kunkel, W. E., Rydgren, A. E. 1979. *Astron. J.* 84:633
Kurucz, R. L. 1979a. *Cent. Astrophys. Preprint No. 1050*
Kurucz, R. L. 1979b. *Cent. Astrophys. Preprint No. 1111*
Lacroute, P. 1977. *IAU Highlights in Astron.* 4(I):353
Lacy, C. H. 1977a. *Ap. J.* 218:444
Lacy, C. H. 1977b. *Ap. J. Suppl.* 34:479
Lacy, C. H. 1979. *Ap. J.* 228:817
Leckrone, D. S. 1971. *Astron. Astrophys.* 11:387
Leung, K.-C., Schneider, D. P. 1978. *Astron. J.* 83:618
Leung, K.-C., Wilson, R. E. 1977. *Ap. J.* 211:853
Lindemann, E., Hauck, R. 1973. *Astron. Astrophys. Suppl.* 11:119
Linnell, A. P. 1973. *Astrophys. Space Sci.* 22:13
Lippincott, S. L. 1953. *Astron. J.* 58:135
Lippincott, S. L., Hershey, J. L. 1972. *Astron. J.* 77:679
Magalashvili, N. I. 1953. *Abastumani Bull.* 15:3
Mammano, A., Margoni, R., Stagni, R. 1974. *Astrophys.* 35:143
Martynov, D. Ya., Khaliullin, Kh. F. 1978. *Astron. Tsirk. No. 1016*
Mauder, H., Ammann, M. 1976. *Mitt. Astron. Ges.* 38:231
McAlister, H. A. 1976. *Publ. Astron. Soc. Pac.* 88:317
McAlister, H. A. 1978. *Ap. J.* 223:526
McCluskey, C. E., Kondo, Y. 1972. *Astrophys. Space Sci.* 17:134
McKellar, A. 1949. *Publ. Dom. Astrophys. Obs. Victoria, B.C.* 8:235
McNamara, D. H. 1966. In *Spectral Classification and Multicolor Photometry, IAU Symposium No. 24*, ed. K. Lodén, L. O. Lodén, V. Sinnerstad, p. 190.

Dordrecht: Reidel
McNamara, D. H., Hanson, H. K., Wilcken, C. K. 1971. *Publ. Astron. Soc. Pac.* 83:192
Milone, E. F. 1977. *Astron. J.* 82:998
Morbey, C. L. 1975. *Publ. Astron. Soc. Pac.* 87:689
Morgan, B. L., Beddoes, D. R., Scanlan, R. J., Dainty, J. C. 1978. *MNRAS* 183:701
Morgan, J. G., Eggleton, P. P. 1979. *MNRAS* 187:661
Morgan, W. W., Whitford, A. E., Code, A. D. 1953. *Ap. J.* 118:318
Nelson, B., Davis, W. D. 1972. *Ap. J.* 174:617
Okazaki, A. 1978. *Astrophys. Space Sci.* 56:293
Olsen, E. H. 1977. *IAU Inf. Bull. Variable Stars No. 1317*
Olson, E. C. 1968. *Ap. J.* 153:187
Olson, E. C. 1975. *Ap. J. Suppl.* 29:43
Olson, E. C., Weiss, E. W. 1974. *Astron. J.* 79:642
Padalia, T. D., Srivastava, R. K. 1975a. *Astrophys. Space Sci.* 35:249
Padalia, T. D., Srivastava, R. K. 1975b. *Astrophys. Space Sci.* 38:87
Petrie, R. M. 1950. *Publ. Dom. Astrophys. Obs., Victoria, BC* 8:319
Petrie, R. M., Andrews, D. H., Scarfe, C. D. 1967. In *Determination of Radial Velocities and Their Application. IAU Symposium No. 30*, ed. A. H. Batten, J. F. Heard, p. 221. Dordrecht: Reidel
Popov, M. V. 1969a. *Sov. Astron.—AJ* 12:640
Popov, M. V. 1969b. *Sov. Astron.—AJ* 12:1033
Popper, D. M. 1959. *Ap. J.* 129:659
Popper, D. M. 1961. *Ap. J.* 134:828
Popper, D. M. 1966. *Trans. IAU* XIIB:485
Popper, D. M. 1967. *Ann. Rev. Astron. Astrophys.* 5:85
Popper, D. M. 1968. *Ap. J.* 154:191
Popper, D. M. 1970. *Ap. J.* 162:925
Popper, D. M. 1971a. *Ap. J.* 166:361
Popper, D. M. 1971b. *Ap. J.* 169:549
Popper, D. M. 1973. *Ap. J.* 185:265
Popper, D. M. 1974. *Ap. J.* 188:559
Popper, D. M. 1976a. *Ap. J.* 208:142
Popper, D. M. 1976b. *Astrophys. Space Sci.* 45:391
Popper, D. M. 1978. *Astrophys. J. Lett.* 220:L11
Popper, D. M., Dumont, P. J. 1977. *Astron. J.* 82:216
Popper, D. M., Dworetsky, M. M. 1978. *Publ. Astron. Soc. Pac.* 90:71
Popper, D. M., Jørgensen, H. E., Morton, D. C., Leckrone, D. S. 1970. *Ap. J. Lett.* 161:L57
Popper, D. M., Plavec, M. 1976. *Ap. J.* 205:462
Popper, D. M., Ulrich, R. K. 1977. *Ap. J. Lett.* 212:L131

Probst, R. G. 1977. *Astron. J.* 82:656
Rakos, K. D. 1965. *Appl. Opt.* 4:1453
Roman, N. G. 1956. *Ap. J.* 123:246
Russell, H. N., Merrill, J. E. 1952. *Contrib. Princeton Univ. Obs. No. 26*
Sadik, A. R. 1979. *Astrophys. Space Sci.* 63:351
Scalo, J. M., Dominy, J. F., Pumphrey, W. A. 1978. *Ap. J.* 221:616
Scarfe, C. D., Barlow, D. J., Niehaus, R. J. 1976. *Astrophys. Space Sci.* 39:129
Schneider, D. P., Darland, J. J., Leung, K.-C. 1979. *Astron. J.* 84:236
Shobbrok, R. R., Lomb, N. R., Herbison-Evans, D. 1972. *MNRAS* 156:165
Söderhjelm, S. 1976. *Astron. Astrophys. Suppl.* 25:151
Söderhjelm, S. 1978. *Astron. Astrophys.* 66:161
Srivastava, R. K. 1976. *Astrophys. Space Sci.* 40:15
Starikova, G. A. 1978. *Sov. Astron. Lett.* 4:52
Strand, K. Aa. 1969. *Astron. J.* 74:760
Strand, K. Aa., Hall, R. G. 1954. *Ap. J.* 120:322
Struve, O., Sahade, J., Huang, S.-S., Zebergs, V. 1958. *Ap. J.* 128:310
Svechnikov, M. N. 1969. *Uch. Zap. Ural. Gos. Univ. No. 88*
Tomkin, J. 1979. *Ap. J.* 231:495
Tomkin, J., Lambert, D. L. 1978. *Ap. J. Lett.* 222:L119
Tremko, J., Papoušek, J., Veteŝnik, M. 1976. *Bull. Astron. Inst. Czech.* 27:125
Upgren, A. R. 1977. *Vistas Astron.* 21:241
van Altena, W. F. 1971. *Astron. J.* 76:932
van de Kamp, P. 1954. *Astron. J.* 59:447
van de Kamp, P. 1958. *Encyclopedia of Physics*, ed. S. Flügge. Vol. L, p. 187. Berlin: Springer
van de Kamp, P., Worth, M. D. 1971. *Astron. J.* 76:1129
van den Bos, W. H. 1945. *Astron J.* 51:198
van Rijsbergen, R. 1973. *Mitt. Astron. Ges.* 32:278
Vasilevskis, S. 1966. *Ann. Rev. Astron. Astrophys.* 4:57
Vasilevskis, S. 1969. *Bull. Am. Astron. Soc.* 1:209
Veeder, G. J. 1974. *Astron. J.* 79:1056
Walker, R. L. 1970. *Astron. J.* 75:720
Wanner, J. F. 1967. *Sky Telesc.* 33:16
Wielen, R. 1962. *Astron. J.* 67:599
Williamon, R. M. 1975. *Astron. J.* 80:976
Wilson, R. E., Caldwell, C. N. 1978. *Ap. J.* 221:917
Wilson, R. E., Devinney, E. J. 1971. *Ap. J.* 166:605
Wilson, R. E., Rafert, J. B. 1980. In preparation
Wood, D. B. 1971. *Astron. J.* 76:701
Wood, D. B. 1972. *A Computer Program for Modeling Non-spherical Eclipsing Binary*

Star Systems. Greenbelt, Md: Goddard Space Flight Cent.
Wood, D. B. 1976. *Astron. J.* 81:855
Wood, F. B. 1948. *Ap. J.* 110:465
Wood, F. B. 1963. In *Basic Astronomical Data*, ed. K. Aa. Strand, Ch. 19. Chicago: Univ. Chicago Press
Worley, C. E. 1966. *Vistas Astron.* 8:33
Worley, C. E. 1969. *Astron. J.* 74:764
Worley, C. E., Behall, A. L. 1973. *Astron. J.* 78:650
Worth, M. D., Heintz, W. D. 1974. *Ap. J.* 193:647
Wright, K. O. 1954. *Publ. Dom. Astrophys. Obs., Victoria, BC* 10:1
Wright, K. O. 1970. *Vistas Astron.* 12:147
Wright, K. O. 1977. *J. R. Astron. Soc. Can.* 71:152
Wyller, A. A. 1955. *Astron. J.* 60:39
Wyse, A. B. 1934. *Lick Obs. Bull.* 17:37
Zissell, R. 1972. *Astron. J.* 77:610

THE STRUCTURE OF EXTENDED EXTRAGALACTIC RADIO SOURCES

✳2165

George Miley
Sterrewacht Leiden, P.O. Box 9513, 2300 RA Leiden, The Netherlands

1 INTRODUCTION

> The very story of the development of the social enterprises of science, technology, economy, and politics throughout the ages does in itself indicate something of the nature of the connections between them.
>
> J. D. Bernal, *Science in History*

The largest, the furthest, the most powerful, and to some of us the most fascinating objects known in the Universe are to be found among the radio sources associated with some elliptical galaxies and QSOs. In recent years it has become apparent that they are also objects of considerable beauty.

This article deals with the appearances of strong extended extragalactic radio sources. Included as "extended" are sources whose sizes are comparable with or larger than a typical galactic diameter, and as "strong," those with intrinsic luminosities at 408 MHz greater than $\sim 10^{23}$ W Hz^{-1}.[1] Compact radio sources have been reviewed recently by Kellermann (1978). Although there are several similarities between the radio properties of the nuclei of spirals and ellipticals (de Bruyn 1978), extended radio emission from spiral galaxies (van der Kruit & Allen 1976) will not be discussed here.

Previous reviews on the structure of extended extragalactic sources have been given by Moffet (1966), Miley (1976), and Willis (1978). Moffet (1975) and De Young (1976) have given more general reviews, the latter

---

[1] We adopt a Hubble constant of $H = 75$ km s^{-1} Mpc^{-1} throughout and have applied the relevant scaling to all parameters quoted.

emphasizing the theoretical problems involved. Here we concentrate on the observations, stressing the results and literature of the past five years.

Three important types of inference can be attempted from radio source structure. First, the structures, combined with a modicum of imagination, provide clues to the processes that produce the radio sources. Second, with the addition of a large dollop of assumptions, the source structures give information about the physical parameters that characterize the environment of the sources. Third, stretching our credulity and fantasy to the full, source structure can be used to probe the geometry of the Universe. In this article we discuss only the first two. For an account of structure measurements applied to cosmology, see Miley (1976) and Ekers & Miley (1977).

Before considering the present state of our knowledge we review the changes that have taken place in our understanding of the structure of extended radio sources over the last three decades. The main factor in this evolution has been a steady progress in instrumentation and observational techniques. Therefore, although a detailed discussion of instrumental methods (e.g. Cohen 1969) is beyond the scope of this article, a brief summary of the main techniques used for measuring source structure is appropriate. The general properties of commonly observed source morphologies are considered in Section 2. Before dealing with the detailed characteristics of radio components (Section 4), we give an account in Section 3 of how some relevant physical parameters can be derived from the structures. Finally, the relationship of radio-source morphologies to the properties of the parent galaxies and clusters is reviewed in Section 5.

1.1 Techniques for Measuring Source Structure

Because of their limited resolution, single dishes are useful only for studying the largest extragalactic sources at short wavelengths. Until recently, owing to receiver instabilities, dynamic ranges greater than 10 to 1 were difficult to achieve. With new observing techniques incorporating multiple beams (Emerson et al. 1979) and cleaning methods (Reich et al. 1978) much larger dynamic ranges have been obtained with the 100-m Effelsberg telescope in Germany.

Several tools have been developed to achieve the higher resolutions needed for studying the structure of extragalactic radio sources. In the technique of interplanetary scintillations, measurements of the degree and time scale of scintillation (flickering) of a source and their variation with distance from the Sun are used to provide useful information on small-scale structure ($<1''$; Hewish et al. 1964, Cohen et al. 1967, Little & Hewish 1968). The scintillation method has the advantage of providing relatively high resolution at low frequencies. However, all but the crudest inferences about structure depend on the model assumed for the solar wind. Moreover, only sources within a few tens of degrees from the Sun are observed to scintillate and can be studied extensively.

Another way to achieve high resolution is to observe a source as it is occulted by the moon. The variation of flux density then gives its one-dimensional brightness distribution convolved with the diffraction pattern of the moon (Getmansev & Ginzburg 1950, Scheuer 1962, von Hoerner 1964). In contrast to other techniques, the resolution that can be obtained with lunar occultations is not strongly frequency dependent and is mainly governed by signal to noise. A resolution of a few tenths of an arcsecond is readily achievable. There are however three disadvantages of this method. First, it can only be applied to sources that lie on the moon's path. Second, it is difficult to reconstruct the two-dimensional structure without observing several occultations covering a large range of orientations. Third, dynamic ranges of more than 10 to 1 cannot easily be obtained. Despite these limitations the occultation technique has scored two big successes. The occultation of 3C 273 (Hazard et al. 1963) was one of the most crucial observations leading to the discovery of quasars. Also, it was by means of occultation measurements that the relation between the angular sizes and flux densities of radio sources was discovered (Swarup 1975, Kapahi 1975).

By far the most important tool for investigating the structure of radio sources has been interferometry (see Fomalont & Wright 1974 for an excellent account). This uses two or more telescope elements extending over a baseline B to give an angular resolution of λ/B radians. A Michelson two-element interferometer was first applied to astronomy in 1946 by Ryle & Vonberg (1948). Since then the method has undergone many sophisticated developments along two broad fronts.

First, the technique of earth rotational aperture synthesis (Ryle & Hewish 1960) made it possible to produce two-dimensional maps of source structures by Fourier transforming the complex fringe amplitudes using digital computers. The power of aperture synthesis was subsequently increased by the development of restoration methods such as "clean" which removed the distortion in maps due to incomplete sampling of the aperture (Högbom 1974, Schwarz 1978). Several arrays of telescopes linked by cables were built primarily to map the brightness and polarization (Conway & Kronberg 1969, Weiler 1973) distributions of radio sources. Particularly important contributions in delineating radio-source structure on the scale of seconds to minutes of arc have been made by the "One Mile" and "5 km" telescopes at Cambridge, England, the Westerbork telescope in the Netherlands, and NRAO interferometer system at Green Bank, West Virginia. The VLA (Very Large Array) at present being built by the NRAO can attain higher resolution than any of these. Even in its unfinished state the VLA is now routinely producing maps with resolution of better than 1″ and sensitivity better than 1 mJy.[2]

This resolution is not high enough to probe the range of angular scales within individual sources, which sometimes exceeds 10^6. However, from the beginning of interferometry a second distinct path evolved in which detailed mapping was subordinated to a quest for higher and higher resolution. At Jodrell Bank, England, radio links were used to connect the telescope elements (Brown et al. 1955, Elgaroy et al. 1962). In this way, by 1966 baselines of more than 100 km and resolutions of $\sim 1/20″$ were achieved, at the cost of sacrificing fringe phase information (Palmer et al. 1967). Next, accurate atomic clocks and video tape recorders, which became available during the sixties, were used to develop linkless interferometers. The first examples were completed almost simultaneously in Canada (Broten et al. 1967) and the US (Bare et al. 1967). Very long baseline interferometry (VLBI) with angular resolutions of a fraction of a milliarcsecond could then be carried out. Since the absolute phase of the interferometer fringes cannot yet be generally measured using VLBI, unambiguous Fourier reconstruction of maps is impossible. However, considerable information about fine-scale structure in sources can be extracted by means of the many sophisticated model-fitting schemes now available and from these models a pseudo-map is often produced. The so-called "closure phases," which can be measured with interferometer arrays (Jennison 1958, Readhead & Wilkinson 1978), are particularly valuable in constructing VLBI pseudo-maps.

Aperture-synthesis and VLBI arrays now in operation are listed by Fomalont (1979) and Moffet (1979) respectively.

1.2 Strategies of Measuring Source Structure

There are two complementary strategies for obtaining basic information about radio sources from measurements of structure. One can extrapolate conclusions drawn from detailed measurements of individual source morphologies to the whole population of radio sources, or one can try to infer properties of radio sources from statistical analyses of well-defined samples. Highly detailed measurements can only be made for a typical source, e.g. the closest (Cygnus A, Virgo A, Centaurus A) or the largest (3C 236, DA 240, NGC 315, NGC 6251). On the other hand, the statistical approach limits the range of properties one can investigate. And even when samples are chosen with the greatest of care, selection effects that are difficult to take into account often remain.

---

[2] $1 \text{ Jansky} = 10^{-26} \text{ W m}^{-2} \text{ Hz}^{-1} = 10^3 \text{ mJy}$.

1.3 Source Structure—The Story So Far

Almost thirty years ago Jennison & Das Gupta (1953) made the fundamental and exciting discovery that Cygnus A, the strongest observed extragalactic radio source, consists of two components which symmetrically straddle the associated optical galaxy and exceed it in size by an order of magnitude. It was gradually established during the next several years that double structure is a very common property of radio sources (Allen et al. 1963, Maltby & Moffet 1963) and that there are often bright regions of emission or "hot spots" within each of the two radio lobes (Allen et al. 1963).

These surprisingly accurate deductions about radio-source structures were made from an analysis of the interferometer fringe amplitudes. Since the mid-sixties earth-rotation aperture synthesis, including fringe-phase data, has been used to map the brightness distributions of several hundred radio sources. The maps confirmed the general features of radio sources that had been extracted from the fringe-amplitude data. Bright hot spots were found frequently at the edges of the intrinsically strongest double sources (e.g. Macdonald et al. 1968, Miley & Wade 1971, Fanaroff & Riley 1974).

Simple models were proposed to explain these observed features. One picture (Burbidge 1967) viewed the extended components as being composed of many thousands of condensed objects each having a mass of $\sim 10^3 \, M_\odot$. An alternative model explained the structure of the radio components in terms of evolving plasmons produced in discrete outbursts from the parent nucleus, whose morphologies were governed by interaction with an intergalactic medium (e.g. van der Laan 1963, De Young & Axford 1967, Christiansen 1969, Mills & Sturrock 1970).

An intergalactic medium was also invoked to interpret the more complicated morphologies that were often revealed by detailed radio maps. In particular the tadpole-like "head-tail" radio sources associated with some galaxies in clusters (Ryle & Windram 1968, Hill & Longair 1971) were explained as double sources deformed by the motion of their parent galaxies through an intracluster medium (Miley et al. 1972).

Most of these measurements were made at frequencies of 1400 MHz and below. During the sixties advances in microwave electronics permitted interferometers to be operated at higher frequencies, resulting in the discovery that compact radio cores with relatively flat spectra are frequently embedded in the nuclei of all types of extended radio galaxies and quasars (Dent & Haddock 1965, Barber et al. 1966, Palmer et al. 1967, Mitton 1970). It became clear that the nuclei of these systems frequently remained active for a considerable fraction of the radio-source lifetimes.

These discoveries stimulated the development of models of radio sources in which the lobes are continuously powered by energy channeled or "beamed" quasi-continuously from their nuclei to hot spots in the lobes (Rees 1971, Scheuer 1974, Blandford & Rees 1974, Lovelace 1976, Blandford 1976, Benford 1979, Wiita 1978a,b).

During the past five years considerable progress has been made on the observational front. First there was the discovery that (at least for some nearby radio galaxies) narrow *jets* of radio emission emanate from the nuclear cores and extend far out into the radio lobes (Section 4.4). These jets constituted beautiful circumstantial evidence in favor of beam-type models and probably provide a new way of studying energy transport in radio sources. Second, the detection of *optical emission* has been reported both from some of these radio jets and from radio hot spots in the lobes (Section 4.2). Third, various *alignments* have been demonstrated both within the radio sources themselves and between radio and optical structures (Sections 4.3.3 and 5.2). These imply that the nuclear axis that defines the radio-source phenomenon often stays fixed to within a few degrees during the source lifetime and is fundamentally related to the axis of symmetry of the associated stellar system. Fourth, *multifrequency comparisons* of the total intensity and polarization distributions for several sources have yielded considerable information on the magnetic field distributions and on the physical conditions within the sources and in their environment (Sections 3 and 4).

2 BASIC RADIO SOURCE STRUCTURE

> Not chaos-like, together crushed and bruised,
> But as the world harmoniously confused;
> Where order in variety we see,
> And where all things differ, all agree
>
> Alexander Pope, "Windsor Forest"

2.1 *Commonly Observed Morphologies*

We shall first describe the classes of radio structure that are commonly seen and then show how these morphologies can be ordered in various sequences.

2.1.1 NARROW EDGE-BRIGHTENED DOUBLE SOURCES These sources, which are variously referred to as "classic doubles," "symmetric doubles," "Cygnus A-type doubles," "Type I doubles," and "triples," form the major constituent ($\sim 70\%$) of low frequency radio surveys at high flux levels (3C, 4C, Parkes). They have length-to-width ratios $\gtrsim 4$, and lobes that are brightened at their edges and symmetrically straddle an elliptical galaxy or

QSO. Most radio galaxies with 178-MHz luminosities $\gtrsim 10^{25}$ W Hz^{-1} and most extended radio sources associated with quasars are of this type.

When studied at high resolution ($\gtrsim 5$ pixels across the source) the edge brightening is usually seen to be due to intense compact regions of emission or "hot spots" at the outer edges of the radio lobes. The hot spots have sizes that are typically ~ 1 kpc (Kapahi 1978), i.e. two orders of magnitude smaller than the source as a whole (see Section 2.3). Extended tails or bridges often reach from the hot spots towards the associated galaxy or QSO.

A flat-spectrum radio core is usually associated with the nucleus of the parent galaxy or QSO (see Section 4.3). For this reason these sources are sometimes called "triple." The example shown in Figure 1 is 3C 452 (Högbom 1979).

The overall structure of edge-brightened doubles has several important morphological properties.

(*a*) *Collinearity*   As can be seen from Figure 1 the hot spots line up accurately with the nuclear radio core. This is a common property of edge-brightened doubles (e.g. Hargrave & McEllin 1975). In Cygnus A the hot spots and the core are collinear to better than 1° (Miley & Wade 1971, Hargrave & Ryle 1976).

(*b*) *Overall symmetry*   The distributions of flux ratios and positional asymmetries between opposite lobes of edge-brightened double sources have been studied at 1.4 GHz (Fomalont 1969, Mackay 1971) and at 5 GHz (Riley & Jenkins 1977, Katgert-Merkelijn et al. 1980, Longair & Riley

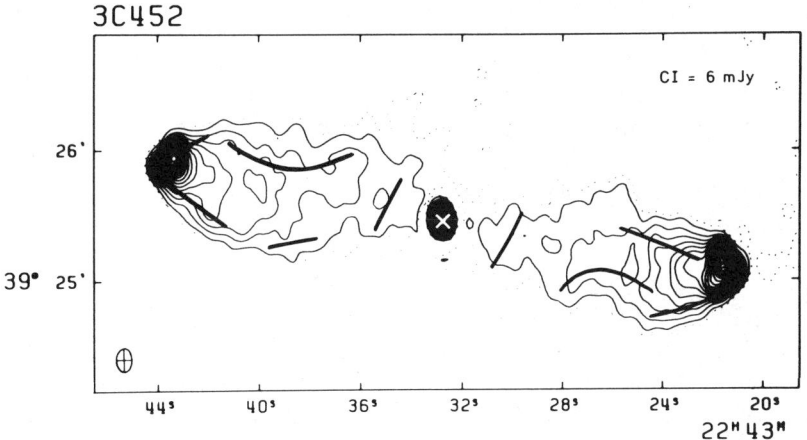

Figure 1   3C 452, a typical narrow edge-brightened double source. Westerbork 5 GHz intensity contour map with magnetic field directions superimposed. The cross indicates the position of the nucleus of the associated galaxy (from Högbom 1979 and kindly supplied by the author).

1979). The reader should be cautious in reading too much into the detailed statistics because the likelihood of an optical identification is biased in favor of symmetric sources. It is clear, however, that components of limb-brightened doubles are quite similar. The strongest component is observed to be more than twice as strong as the weaker one in only about 30% of the sources. The components are also usually symmetrically distributed with respect to their parent nuclei; the distances to opposite extremities rarely differ by more than 50%. Components of sources associated with quasars tend to have less similar flux ratios and to be more asymmetric than those associated with galaxies. For the radio galaxies the components closest to the nuclei tend to be the brightest, but this trend is not apparent in the case of the quasars.

In some sources (e.g. 3C 273; Conway & Stannard 1975), extended structure is seen only on one side of the nucleus, the flux ratio between opposite lobes usually exceeding 10:1. These highly asymmetric sources are sometimes called "D2 doubles" (Miley 1971, Conway & Stannard 1975, Davis et al. 1977, 1978) or "C-sources" (Readhead et al. 1978a). It is not clear whether these sources are members of a different species or whether they represent the tail of a continuous distribution of relative brightness. Three of them have recently been mapped with the VLA (Perley & Johnston 1979). Such sources make up $\sim 10\%$ of all extended quasars (Stannard & Neal 1977, Miley & Hartsuijker 1978). One-sided sources identified with galaxies are rarer but are still occasionally seen (e.g. Argue et al. 1978). Apparent differences between radio galaxies and quasars may merely reflect a dependence on luminosity, since the more luminous sources are almost all quasars; see Section 5.2.

(c) *Alignment* The agreement between the orientation of the cores and that of the overall structure of symmetric edge-brightened doubles is discussed in Section 4.3.3. Alignment on an intermediate scale is seen in the bridges that extend from the hot spots to the cores and in which enhancements are sometimes present (e.g. Goss et al. 1977).

2.1.2 NARROW EDGE-DARKENED DOUBLE SOURCES Several narrow double sources have brightness distributions that gradually die away at their extremities. The closest radio galaxy Centaurus A (Cooper et al. 1965) falls into this class. Figure 2 shows 3C 449, another good example (Perley et al. 1979, Bystedt & Högbom 1979). 3C 31 (Burch 1977b) and the two radio galaxies that make up 3C 402 (Riley & Pooley 1975) are also of this type.

These radio sources usually have radio cores (Section 4.3). When studied with high enough resolution, narrow collimated radio jets are often seen to emanate from their nuclei (Section 4.4). They are sometimes called "3C 31-types" (Simon 1978).

EXTENDED RADIO SOURCES 173

2.1.3 WIDE DOUBLE SOURCES   A few relatively relaxed doubles have fatter lobes with length-to-width ratios of 2 or 3. Figure 3 shows 3C 310 (van Breugel 1980a). Other examples are Virgo A (Andernach et al. 1979), Fornax A (Cameron 1971), Centaurus A (Cooper et al. 1965), and 3C 314.1 (Miley & van der Laan 1973).

2.1.4 NARROW TAILED SOURCES   These have a tadpole-like asymmetric structure, with a high brightness head that coincides with the nucleus of an elliptical galaxy and a diffuse tail that often extends for several hundred kpc. Alternative names by which this species is known are "head tail" or "narrow angle tail" (NAT) source. In some cases their morphology is observed to be two-pronged with a double tail and/or a double head. The ability to distinguish between "twin tailed" and "single tail" sources sometimes depends on the resolution and dynamic range of the observations.

Shown in Figure 4 is 3C 83.1B/NGC 1265 (twin) and in Figure 6 IC 310 (single), both radio galaxies in the Perseus Cluster. Other examples of narrow twin-tailed sources are 3C 129 (Miley 1973), B2 1108 + 41 (Rudnick & Owen 1977, Valentijn 1979a), B2 1502 + 28 (Valentijn 1979a), and 4C 39.49 (Miley & Harris 1977).

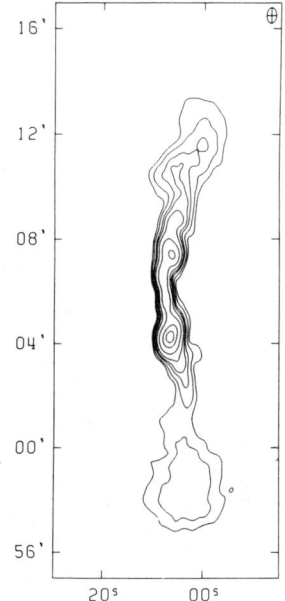

Figure 2   3C 449, a narrow edge-darkened double source. Westerbork 1.4 GHz intensity contour map (from J. Bystedt and J. A. Högbom, in preparation, and kindly supplied by the authors).

174   MILEY

Radio cores are usually found in tailed sources (see Section 4.3). When studied with sufficiently high resolution the heads are often resolved into narrow jets of radio emission (see Section 4.4).

2.1.5 WIDE-TAILED SOURCES   The structure of these sources is intermediate between that of the narrow edge-darkened doubles and the narrow twin-tailed sources. A common alias by which they are known is "WAT" (wide-angle tail). Shown here, in Figure 5, is 3C 465 (van Breugel 1980c). Other examples are IC 708 (Vallée & Wilson 1976, Wilson & Vallée 1977, Vallée et

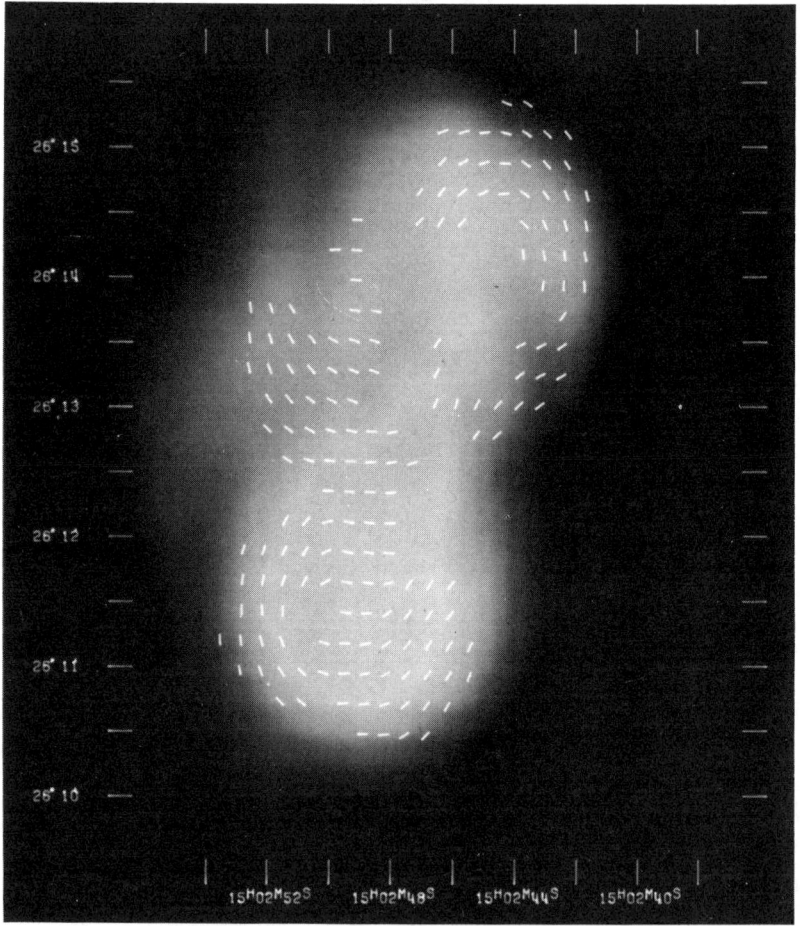

Figure 3   3C 310, a wide double source. Radiophoto of the total intensity distribution (Westerbork 5 GHz) with magnetic field directions superimposed (from van Breugel 1980a and kindly supplied by the author).

al. 1979), B2 0658+33B, B2 0836+29, and B2 0838+32A (Valentijn 1979a), NGC 6034 (Valentijn & Perola 1978), and 4C 48.21 (Owen & Rudnick 1976).

2.1.6 CLUSTER HALOS Several clusters of galaxies have associated large radio components with steep nonthermal spectra ($\alpha \lesssim -1.2$) which often

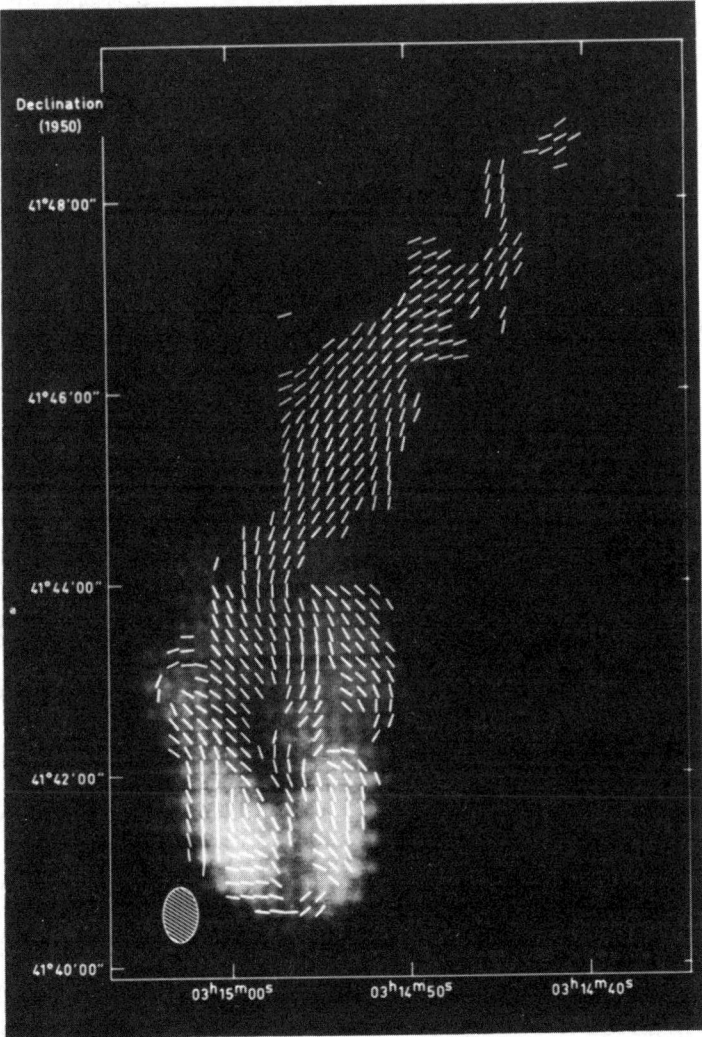

Figure 4 3C 83.1B/NGC 1265, a twin-tailed source in the Perseus Cluster. Radiophoto of the total intensity distribution (Westerbork, 1.4 GHz) with magnetic field directions superimposed (from Miley et al. 1975).

dominate the cluster radio emission at frequencies below 100 MHz (e.g. Baldwin & Scott 1973, Slingo 1974). Instruments operating at this low frequency have insufficient resolution to map them and they are extremely difficult to detect at high frequencies owing to their faintness and confusion from other sources in the cluster. Therefore little is known about them. The sources of this class to have been studied in most detail are Coma C, a smooth diffuse halo in the Coma Cluster with a size of $\sim 40'$ (1.2 Mpc; Jaffe et al. 1976, Jaffe 1977, Valentijn 1978, Hanisch et al. 1979) and Abell 2256 where the situation is complicated (Bridle & Fomalont 1976, Masson & Mayer 1978, Bridle et al. 1979b). Other possible examples are in Abell 1367 (Gavazzi 1978) and Abell 2139 (Harris & Miley 1978), but the cluster halo 3C 84 B reported to be in the Perseus Cluster (Ryle & Windram 1968) has been shown to be nonexistent (Gisler & Miley 1979, Birkinshaw 1980, Jaffe & Rudnick 1979).

It has been suggested that cluster halos are the remnants of tailed radio

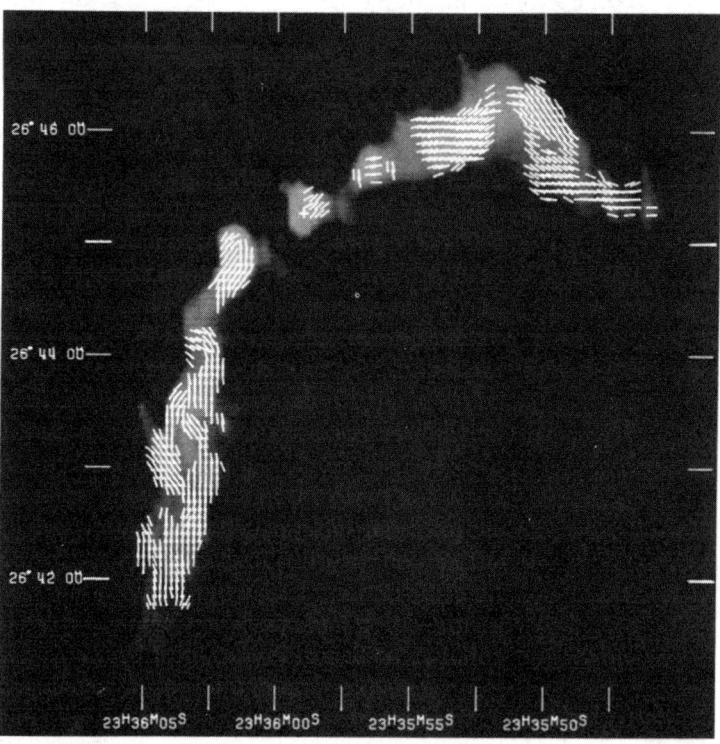

Figure 5 3C 465/NGC 7720, a wide-angle tailed source. Radiophoto of the total intensity distribution with magnetic field directions superimposed (from van Breugel 1980c and kindly supplied by the author).

galaxies whose spectra are steepened by synchrotron losses (e.g. Harris & Miley 1978) or interfaces between two subclusters undergoing coalescence (Harris et al. 1980).

2.2 Taxonomy of Extended Sources

The taxonomic approach to radio morphology attempts to recognize recurrent patterns that could give useful phenomenological information about the mechanisms involved. Pattern recognition depends on the amount of detail delineated by a given observation, i.e. both on the number of pixels (beam elements) across the source and on the dynamic range in brightness that can be studied.

Bearing these remarks in mind we now consider some similarities and differences between the various types of radio sources described in Section 2.1. The duplicity of the large-scale structure, the widespread presence of radio cores, and the nuclear collimation in wide sources (see Section 4.3) all suggest that the mechanisms for producing the radio sources are similar and occur in the parent galaxy nuclei.

However, as we have seen, there are also several differences. Three gross structural properties have emerged as being particularly important in studying these morphological differences. In order of the increasing amount of relative detail needed to discern them they are 1. the overall bending at the outer extremities, 2. the edge-brightening, 3. rotational symmetry. These properties define three sequences along which the large-scale structure of almost all sources can be classified, allowing for projection effects.

2.2.1 BENDING SEQUENCE—RADIO TRAILS The bending sequence is illustrated in Figure 6. The simplest measure of source bending is the angle subtended at the radio core or nucleus by lines drawn to the extremities of opposite lobes. An analysis of the distribution of this "opening angle," χ, for a sample of 44 tailed and double sources has been given by Valentijn (1979b). Opening angles between 0° (narrow-tailed source) and 180° (double source) are seen, with a possible small zone of avoidance $45° < \chi < 90°$. The bending sequence extends fairly continuously from double sources to narrow tailed sources, so any discussion of bending that includes only double sources (Harris 1974, Ingham & Morrison 1975) seems unnecessarily restrictive. The opening angle seems to be correlated with the absolute magnitude of the parent galaxy; narrow tails are associated with fainter galaxies than the wider tails (Rudnick & Owen 1977, McHardy 1979, Simon 1978, Valentijn 1979b). The bending is most simply explained as distortion of an initially collimated double morphology by translational motion with respect to a surrounding intergalactic medium (Miley et al. 1972).

Within this picture, therefore, tailed sources are viewed as trails or fossil records deposited by active galaxies. This interpretation is supported by (a) similarities between the fine-scale structure of tailed and double sources (Section 4.4; Miley 1974) and (b) the preferential occurrence of the most bent sources in dynamically active clusters containing hot gas (Section 5.4). A number of radio-trail models have been proposed to account for the detailed brightness and polarization distributions (e.g. Jaffe & Perola 1973, Pacholczyk & Scott 1976, Cowie & McKee 1975, Jones & Owen 1979).

2.2.2 EDGE-BRIGHTENING SEQUENCE—MORPHOLOGY AND LUMINOSITY The edge-brightening sequence is illustrated in Figure 7. There is a striking correlation between the degree of edge brightness of a source and its luminosity. This was first noted by Fanaroff & Riley (1974) who examined the structure of 3C sources as a function of luminosity at 178 MHz (P_{178}). For $P_{178} \lesssim 10^{26}$ W Hz^{-1} ("Class I") almost all sources have a relatively relaxed morphology, while for $P_{178} \gtrsim 10^{26}$ W Hz^{-1} ("Class II") almost all are edge-brightened doubles.

This critical luminosity is tantalizingly close to the break in the radio luminosity function for elliptical galaxies, which, scaling from Auriemma et al. (1977), occurs at $\sim 1.9 \times 10^{25}$ W Hz^{-1}. The break in the luminosity function separates sources that show strong population evolution (the more luminous sources) from those that do not.

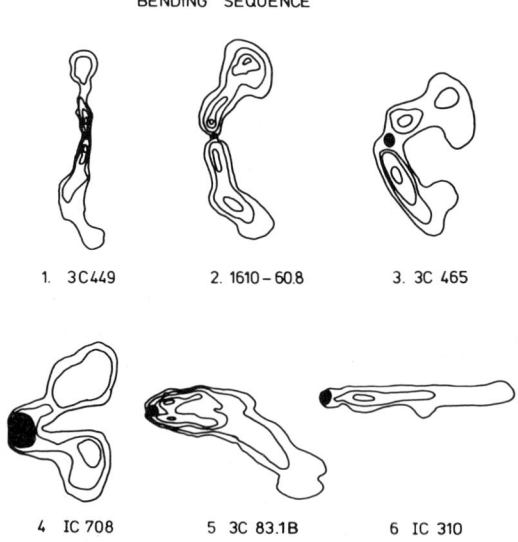

Figure 6 A sequence of radio-source bending illustrated with schematic drawings of actual sources.

Further work (Jenkins & McEllin 1977, Speed & Warwick 1978) suggested that for the edge-brightened doubles themselves the fractional flux in the hot spots increases with luminosity. However, Kapahi has pointed out that this might be caused by selection. Owing to the limited resolution on the maps of the strong sources, linear size evolution of the hot spots might cause the more distant (stronger) ones to appear more intense.

The lower luminosity (Class I) sources are a mixed bag. They include wide doubles, edge-darkened doubles, tailed, and complex sources. There are indications that the degree of bending (the opening angle in the bending sequence) decreases with luminosity, the wide-angle tails being more luminous than the narrow-angle tails (Owen & Rudnick 1976, Simon 1978, Valentijn 1979b). This can be qualitatively understood: One would expect outbursts in the more powerful sources to involve more kinetic energy, and the distorting pressure of an intergalactic medium on their evolution to be consequently smaller. Alternatively, the lower luminosity sources are presumably associated with less massive galaxies which may on average be moving faster.

2.2.3 ROTATIONAL SYMMETRY   Also shown in Figure 7 are some examples of two-dimensional rotational symmetry (S or Z shape) which is seen in

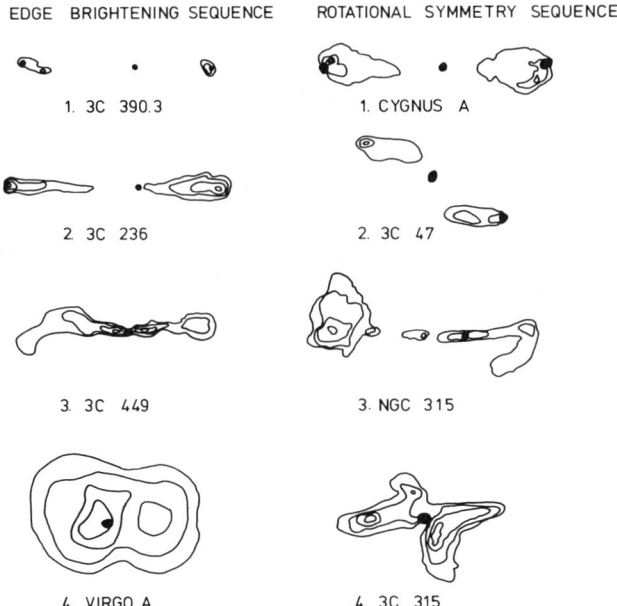

Figure 7 Sequences of edge-brightness and rotational symmetry illustrated with schematic drawings of actual sources.

many radio sources (Harris 1974, Miley 1976). This symmetry was first observed in Centaurus A (Cooper et al. 1965) and is particularly evident in 3C 47 (Pooley & Henbest 1974, Miley & Hartsuijker 1978), 3C 272.1 (Jenkins et al. 1977), NGC 315 (Bridle et al. 1976, Bridle et al. 1979a), and the outer components of Cygnus A (Hargrave & Ryle 1974). A possible cause is distortion of the structure by rotation or shear in the intergalactic medium. But to explain the Z shapes in giant sources such as NGC 315 such motions must take place over scales of more than a megaparsec. An alternative hypothesis is a swinging of the fundamental nuclear ejection axis during the lifetime of the source (Miley 1976). The symmetries in 3C 315 (Northover 1976, Högbom 1979) and B2 0055+26/4C 26.03/NGC 326/(Ekers et al. 1978a) are indeed suggestive of precessing nuclear beams. A scenario in which such precession could occur was sketched by Rees (1978b). Misalignment between the rotational axis of the nuclear machine (a black hole) and the angular momentum axis of the galaxy is supposed to take place as a result of galaxy merging. It is noteworthy that both 3C 315 and B2 0055+26 are associated with close pairs of galaxies. Precession could also occur as the central black hole was accreting material from a cloud or disk with a different angular momentum direction from that of the galaxy as a whole. See Section 5.3.

2.3 *Overall Linear Sizes of Sources*

The linear sizes of extended extragalactic sources span on enormous range, from sources like 3C 346 (Pooley & Henbest 1974) which are no larger than their parent galaxies to giant sources like 3C 236 with a total size in its largest direction of 4 Mpc (Willis et al. 1974, Strom & Willis 1979). The size distribution and its dependence on source structure and luminosity have been considered by Gavazzi & Perola (1978) using well-defined 3C and B2 samples. To minimize cosmological effects and ensure completeness they restricted their study to (104) elliptical galaxies with $z < 0.1$. Unfortunately, this inevitably excluded the highest luminosity sources, which tend to be the most distant. The luminosities considered ranged from $P_{178} \sim 10^{23.4}$ to $\sim 10^{27}$ W Hz^{-1}.

The sizes of edge-brightened doubles are well defined. The linear size function for these sources indicates a slight increase of size with luminosity ($\propto P^{1.4}$), with a median value of ~ 170 kpc at 10^{25} W Hz^{-1}. Most have sizes between 150 and 300 kpc with a few ($\lesssim 5\%$) that exceed a megaparsec. Only 10% of the two-sided edge-brightened sources are smaller than 130 kpc, suggesting that either the outward motion of source components decelerates rapidly after reaching this size or that the smaller sources are effectively invisible. The one-sided sources (Section 2.1.1) appear to be smaller than the symmetric ones, with typical sizes of a few tens of kpc (Readhead et al. 1978a).

The situation for the more diffuse (Class I) sources is unclear. The median size (~ 150 kpc) of the relaxed doubles and the sizes of their largest members are comparable with those of the edge-brightened doubles. In the data of Gavazzi & Perola there is no significant dependence of the total linear size on luminosity for the diffuse sources, but the uncertainties are large. Because of their ill-defined extremities, estimates of the total sizes of the Class I sources are affected by the sensitivity of measurement. However, sometimes the diffuse sources have regions of enhanced emission *within* their lobes and the separation of these hot spots provides a less subjective yardstick than the total linear extent of the source. The linear distance of those hot spots from their parent nuclei seems to increase with source luminosity (Birkinshaw et al. 1978). In view of the similar results for sources of diverse morphologies the increase in separation of the hot spots with the total source luminosity is quite well established. Such an effect would be intuitively expected, for example, if the more powerful sources are associated with more powerful nuclear ejecta or beams which can penetrate further into the surrounding medium.

It should be stressed that great caution should be exercised in interpreting linear size functions since the results often depend on whether the function is plotted linearly or logarithmically (Ekers & Miley 1977).

3 PHYSICAL PARAMETERS

In the early treatizes on this subject, the mean value assigned to π will be found to be 40.000000. Later writers suspected that the decimal point had been accidentally shifted, and that the proper value was 400.00000.

Lewis Carroll, "The new method of evaluation as applied to π," from *Notes by an Oxford Chiel*

Total intensity and polarization data can be used to extract information about several physical parameters intrinsic to radio sources and thereby help place constraints on some of the mechanisms involved. Since the advent of computers and more recently of pocket calculators, this "interpretation of the data" has become such an automatic ritual that the many assumptions inherent in each step are sometimes forgotten. We shall review here some of the arguments most commonly used in deriving physical parameters from intensity and polarized brightness distributions. In order to make the uncertainties more explicit, the formulas will be expressed as much as possible in terms of observed quantities. For block diagrams of the various steps and assumptions that are made the reader is referred to Miley (1976).

3.1 Total Intensity Distributions

3.1.1 ENERGETICS The energetics of radio sources are important not only because of their relation to the source-production mechanism but because they also play an important role in all considerations of how radio sources are held together or confined. Total intensity distributions provide information about the minimum energies involved. A good discussion of source energetics is given by Moffet (1975) and in more detail by Pacholcyzk (1970). The minimum energy condition corresponds almost (but not quite) to equipartition of energy between relativistic particles and magnetic field.

For a region in a synchrotron radio source delineated by an ellipse of angular diameters θ_x and θ_y in orthogonal directions, we can write the minimum energy density as

$$u_{me} = (7/3)(B_{me}^2/8\pi) = 0.0928 \, B_{me}^2 \text{ erg cm}^{-3} \qquad (1)$$

where the corresponding magnetic field is

$$B_{me} = 5.69 \times 10^{-5} \left[\frac{(1+k)}{\eta}(1+z)^{3-\alpha} \frac{1}{\theta_x \theta_y s \sin^{3/2}\phi} \right.$$
$$\left. \times \frac{F_0}{v_0^\alpha} \frac{v_2^{\alpha+1/2} - v_1^{\alpha+1/2}}{\alpha + \frac{1}{2}} \right]^{2/7} \text{ gauss.} \qquad (2)$$

Here k is the ratio of energy in the heavy particles to that in the electrons, η is the filling factor of the emitting regions, z is the redshift, θ_x and θ_y (arcsec) correspond either to the source/component sizes or to the equivalent beam widths, s (kiloparsec) is the path length through the source in the line of sight, ϕ is the angle between the uniform magnetic field and the line of sight, F_0 (Jy or Jy per beam) is the flux density or brightness of the region at frequency v_0 (GHz), v_1 and v_2 (GHz) are the upper and lower cut off frequencies presumed for the radio spectrum, and α is the spectral index $[F(v) \propto v^\alpha, v_1 < v < v_2]$.

Apart from the basic assumptions that the radiation is synchrotron emission and that the radio and optical emission are redshifted by the same amount there are several more mundane uncertainties inherent in these formulas.

First, k is unknown and could have a value between 1 and 2000 (Pacholczyk 1970). Indirect arguments have usually led to canonical values of $k = 1$ or $k = 100$ (eg. Moffet 1975). $k = 1$ is clearly more consistent with the use of the minimum energy condition. Note that a difference of 100 in the assumed k results in an order-of-magnitude difference in the minimum energy densities derived. Second, to obtain s, the path through the source along the line of sight, one must make some assumptions about the symmetry and distance of the source. Frequently cylindrical symmetry is assumed, with s equal to the width of the source in the plane of the sky. Third, the formulas depend on the form of the source spectrum, but this dependence is weak. For extended sources $\alpha \lesssim -0.6$, and v_1 dominates over v_2. Usually $v_1 = 0.01$ GHz is assumed. Fourth, the $\sin \phi$ term is unknown in individual cases. It arises because the emission measure depends on the perpendicular component of the magnetic field and the visible radiation is beamed from electrons moving towards us. Fifth, and perhaps most irritating, is the strong dependence on the filling factor. It is possible that on a scale much less than an arcsecond or a kiloparsec the radiation is clumpy or filamentary. This would result in much greater local energy densities and less total energy.

Taking $k = 1, \eta = 1, \sin \phi = 1, v_1 = 0.01$ GHz, $v_2 = 100$ GHz, $\alpha = -0.8$ we obtain approximate expressions for the minimum energy condition,

$$B_{me} \simeq 1.4 \times 10^{-4}(1+z)^{1.1}v_0^{0.22}\left[\frac{F_0}{\theta_x\theta_y s}\right]^{2/7} \text{ gauss,} \quad (3)$$

$$u_{me} \simeq 1.9 \times 10^{-9}(1+z)^{2.2}v_0^{0.44}\left[\frac{F_0}{\theta_x\theta_y s}\right]^{4/7} \text{ erg cm}^{-3}. \quad (4)$$

The minimum energy condition is particularly sensitive to angular size. Because of the limited range of brightnesses that can be studied with a particular instrument, values of B_{me} derived from measurements made with the same telescope usually do not differ by much more than a factor of four or five. Typical values obtained for B_{me} are $\sim 10^{-6.5}$ for cluster halos, $10^{-5.5}$ G for the diffuse lobes, $\sim 10^{-5}$ G for the hot spots, and $\sim 10^{-3}$ G for the tiny flat spectrum cores. Resultant total source energies can reach 10^{60} ergs. However, it must be stressed that there is little evidence that the minimum energy condition is actually obeyed.

3.1.2 SOURCE CONFINEMENT   In order to account for the nonspherical shapes of radio sources the particles and field must be prevented from dispersing at the relativistic internal sound speed of $c/\sqrt{3}$. The pressure exerted by the relativistic gas in the radio source, $u/3$, must therefore be balanced by some pressure, P, or by inertial drag. Several mechanisms have been considered for confinement (see, for example, Longair et al. 1973, Pacholczyk 1977) and in each case the pressure balance condition can be used to calculate limits for the various parameters that characterize the resisting forces.

Table 1 Some possible confinement mechanisms

	Mechanism	Approximate Restriction
Internal:	Gravitational: Point mass M at center of spherical source, diam. θ (arcsec), distance D (Mpc)	$M \gtrsim 7 \times 10^7 (D\theta)^2 B_{me} M_\odot$
	Inertial: Blob, translational velocity, v_t Source separation/component size, R/r	$\rho_{int} \gtrsim 3 \times 10^{-22} (R/r)^2 B_{me}^2 (v_t/c)^{-2}$ gm cm^{-3}
External:	Static thermal: Temperature T	$\rho_{ext} T \gtrsim 2.5 \times 10^{-10} B_{me}^2$ gm cm^{-3} deg
	Dynamic: Ram pressure, translational velocity v_t	$\rho_{ext} \gtrsim 7 \times 10^{-23} B_{me}^2 (v_t/c)^{-2}$ gm cm^{-3}
	Magnetic	$B_{ext} \gtrsim B_{me}$

Some possible confining mechanisms are listed in Table 1 together with the restrictions implied by pressure balance. Their applicability to the various regions of radio sources will be considered in Section 4. In real life several of these processes may be occurring simultaneously, and the detailed hydrodynamic interactions will probably result in a considerably more complicated situation than can be treated by simple static-pressure balance arguments.

3.1.3 MAGNETIC FIELD FROM COMPARISON WITH X RAYS   An independent method of obtaining information about the distribution of magnetic field strengths in extended sources is based on the fact that the same relativistic electrons that produce the radio synchrotron emission will scatter photons of the microwave background to X rays. The ratio between radio and X-ray surface brightness depends on the magnetic field strength (Perola & Reinhardt 1972, Bridle & Feldman 1972, Costain et al. 1972, Harris & Romanishin 1974, Harris & Grindlay 1979). The process is reviewed by Gursky & Schwartz (1977).

The expression for the resultant magnetic field strength given by Harris & Grindlay (1979) can be approximated to within about $\pm 10\%$ between $-0.6 > \alpha > -1.4$ and rewritten as

$$B = \{6.6 \times 10^{-40}(4800)^{\alpha}(1+z)^{3-\alpha}F_R F_x^{-1} v_R^{-\alpha} E_x^{\alpha}\}^{1/1-\alpha} \text{ gauss} \tag{5}$$

where F_R is the radio flux density (in Jy) at frequency v_R (GHz), F_x is the X-ray flux (erg cm^{-2} Hz^{-1}) at energy E_x (keV), α is the spectral index ($F_R \propto v_R^{\alpha}$), and z is the redshift.

The main problems in using this method to determine the magnetic field strengths in extended radio sources arise from the required X-ray surface brightness sensitivity and the difficulty of separating possible inverse Compton X rays from the thermal X-ray distribution. However, one can at least obtain an upper limit for the inverse Compton contribution and hence a lower limit on the magnetic field strengths. Until now this has been done using only a few relatively low resolution X-ray measurements. For Centaurus A Cooke et al. (1978) obtained $B \gtrsim 7 \times 10^{-7}$ G and for Cygnus A Fabbiano et al. (1979) obtained $B \gtrsim 1.6 \times 10^{-6}$ G indicating that the extended lobes are not too far from equipartition. With high sensitivity X-ray observations and resolution comparable to the radio maps, this method promises to provide more information about magnetic fields in extended radio sources. A beginning is now being made with the Einstein Observatory, which, although providing an enormous improvement in X-ray sensitivity and resolution, can detect extended inverse Compton emission only from a few of the strongest sources.

3.1.4 AGES—IN SITU ACCELERATION   A problem frequently considered in discussions of source morphology concerns the place where the observed synchrotron electrons are accelerated. Does this occur entirely in the nucleus, mainly in the hot spots, or is there an accelerating process at work throughout the entire source?

For the acceleration to have taken place at an angular distance θ (arcsec) from where the radiation is observed, the electron must have traveled to its present position in time,

$$t_d = 4.7 \times 10^6 D_\theta \theta v^{-1} (\sin \psi)^{-1} \text{ yr}, \tag{6}$$

where D_θ (Mpc) is the "angular size" distance (McVittie 1965), v (km s^{-1}) is the velocity at which electrons travel from the nucleus to the position under study, ψ is the angle between this direction and the line of sight. For an isotropic distribution of pitch angles the average age of a radiating synchrotron electron, t_r, is (van der Laan & Perola 1969, De Young 1976)

$$t_r = 0.82 B^{1/2} (B^2 + B_R^2)^{-1} (1+z)^{-1/2} v_*^{-1/2} \text{ yr} \tag{7}$$

where B (G) is the magnetic field strength in the source, $B_R = 4 \times 10^{-6} (1+z)^2$ G is the equivalent magnetic field strength of the microwave background. v_* (GHz) is the frequency above which an exponential drop in flux will occur, and usually exceeds v_0, the observing frequency. A necessary condition for acceleration to have occurred at a distance θ is $t_d = t_r$, i.e.

$$5.7 \times 10^6 D_\theta \theta v^{-1} (\sin \psi)^{-1} B^{-1/2} (B^2 + B_R^2)(1+z)^{1/2} v_*^{1/2} < 1. \tag{8}$$

To apply this condition we need to have some information about v and B. Of course we can always take $v < c$, but in most cases there is evidence that the velocities are considerably smaller. If we accept the radio-trail hypothesis (Section 2.2.1), for tailed sources v is the velocity of the parent galaxy. A reasonable upper limit to v of a few thousand km s^{-1} is then given by the distribution of radial velocities of galaxies in the cluster. Likewise, for the lobes of double sources, the observed asymmetry has been used as a statistical argument that the outward electron velocities are smaller than $0.2\,c$ (Sections 4.2 and 4.4). An additional possible upper limit to the velocities is the generalized sound speed in the source $(u/\rho)^{1/2}$ which in the equipartition case reduces to the Alfvén velocity $B/(4\pi\rho)^{1/2}$. However, this is less certain since resonant damping by thermal protons in the source might permit considerably higher velocities (Holman et al. 1979).

There are two ways of specifying B. The first is to assume that the minimum energy ($\sim$ equipartition) condition holds and to calculate B from (2) or (3). The weaknesses of this approach are the lack of evidence for equipartition and the uncertainties in the formulas. (Note that minimum energy does not imply minimum B.) Also, the assumption that the pitch angles θ are isotropically distributed may be invalid. In that case one must replace B by $B_\perp = B \sin \phi$, so electrons flowing along the magnetic field lines will suffer smaller energy losses and have prolonged lives (Spangler 1979).

One can avoid most of these limitations by choosing $B = B_R/\sqrt{3}$, the value that gives a maximum radiative lifetime (van der Laan & Perola 1969). The necessary condition for remote acceleration then becomes

$$8.0 \times 10^3 (1+z)^{3.5} v_*^{1/2} D_\theta \theta v^{-1} < 1. \tag{9}$$

In Section 4 evidence will be presented that localized particle acceleration in radio sources may be widespread and almost certainly occurs in jets where optical emission has been seen. An imaginative assortment of complicated mechanisms has been invoked to produce this in situ acceleration (see, for example, Pacholczyk & Scott 1976, Lacombe 1977, Blandford & Ostriker 1978, Eichler 1979, Eilek 1979, Ferrari et al. 1979, Smith & Norman 1979), but in view of the many assumptions and free parameters inherent in these models, they will not be discussed here.

3.2 Polarization Distributions

In the absence of a thermal plasma, synchrotron radiation is polarized orthogonally to the magnetic field direction by a percentage $100[(3-3\alpha)/(5-3\alpha)][(B_u^2/(B_u^2+B_r^2)]$ where α is the spectral index ($S \propto v^\alpha$), and B_u and B_r are the respective field strengths of the uniform and random components of the magnetic field (Gardner & Whiteoak 1966, Moffet 1975). Hence simple inspection of the polarization distributions should apparently give information about the direction and turbulence of the magnetic field in a source. However, at all but the shortest wavelengths the interpretation of polarization distributions is made both more complicated and more informative by Faraday rotation. In a magnetoionic medium the plane of polarization of linearly polarized radiation is rotated through an angle $\Delta\chi$ proportional to the square of the wavelength: $\Delta\chi = 5.73 \times 10^{-3} R\lambda^2$ deg, where $R = 812\int n_t(s) B_\parallel(s) \, ds \sim 8.1 \times 10^8 n_t B_\parallel s$ rad m^{-2} is the "rotation measure", λ(cm) is the wavelength, s (kpc) is the path length through the medium, $n_t(s)$ (cm^{-3}) is the density of thermal electrons, and $B_\parallel(s)$ is the component of magnetic field parallel to the path, i.e. in the line of sight.

Integrated rotation measures have been published for several hundred sources (Vallée & Kronberg 1975, Haves 1975). Although a small proportion of sources such as 3C 123 and 3C 427.1 have intrinsic rotation measures that probably exceed several hundred rad m^{-2} (Kronberg & Strom 1977, Riley & Pooley 1978), more than three quarters have absolute values smaller than 50 rad m^{-2}. Hence, the directions indicated by the measured polarization distributions for $\lambda \lesssim 6$ cm will usually be within $10°$ of the unrotated angles (i.e. perpendicular to the uniform magnetic field projected on the plane of the sky). Maps of projected magnetic field directions $B_\perp$ are therefore relatively easy to produce from polarization distributions measured at short wavelengths.

In principle, a comparison of polarization distributions at several wavelengths gives the distribution of R across a source and hence information about variations in $n_t B_\parallel$. Unfortunately, matters are complicated by (a) ambiguities in R due to the limited number of observing frequencies, (b) difficulties in separating the in-source contribution to R from foreground rotation, (c) uncertainties in the path length s through the source, (d) effects of nonuniformity of the magnetic field within the source, including field reversals, (e) smearing due to the presence of several independent "cells" within the same observing beam.

Until now most multifrequency polarization mapping has been carried out using two (or, in a few special cases, three) observing wavelengths λ_1, λ_2. There is consequently an ambiguity of $n\pi$ in χ and a corresponding ambiguity in R of $3.14 \times 10^4/(\lambda_2^2 - \lambda_1^2)$ rad m^{-2}.

Separation of the rotation that occurs within the source from that in the foreground has been attempted in two ways. First, the foreground rotation is obtained as the average of integrated rotation measures of several nearby sources. Second, one ignores the component of rotation that is constant across the source and examines only point-to-point variations in R, on the assumption that all these relatively small-scale changes in R are produced within the source.

One can guess at the path length s using symmetry arguments and assumptions about the distance, as in the preceding discussion of the derivation of the minimum-energy conditions.

A proper treatment of the effects of a tangled magnetic field and of beam smearing must take into account the distribution of the percentage polarization as a function of wavelength. The complex polarization as a function of wavelength can be evaluated in terms of source parameters (Burn 1966, Gardner & Whiteoak 1966). To obtain meaningful information from available data, an unduly large number of assumptions must be made about the statistical properties of the inhomogeneities and the source geometry. Such analyses have been carried out for 3C 465 (van Breugel 1980c) and Virgo A (Forster 1980). A slightly different (Monte Carlo) model fitting procedure has been used by Burch (1979b) in a study of 3C 47, 3C 79, 3C 219, 3C 234, 3C 300, 3C 382, and 3C 430. The results (electron densities of from 10^{-5} to 10^{-3} cm^{-3} corresponding to total masses of $\sim 10^{11}\ M_\odot$) are all quite similar to those given by back of the envelope calculation using the simple formulas in this section. However, the more sophisticated analyses illustrate that multifrequency polarization comparisons made, using many more frequencies than now available, could furnish useful information about the intrinsic distribution of n_t and B *along the line of sight.*

4 THE BUILDING BLOCKS OF EXTENDED RADIO SOURCES

> The fundamental cause of the development of a thing is not external but internal; it lies in the degree of contradiction within the thing.
>
> Mao tse Tung, "On Contradiction"

4.1 *Diffuse Emission*

Most radio sources are associated with some diffuse emission. The low-brightness extremities of tailed sources and narrow relaxed doubles have similar diffuse appearances, and the wide doubles have relaxed amorphous lobes. In edge-brightened doubles the diffuse emission takes the form of low-brightness bridges reaching from the outer hot spots towards the nuclei, but the distinction between bridges and radio jets (Section 4.4) is not always clear.

4.1.1 PROPERTIES The diffuse emission is characterized by a relatively steep spectrum ($-0.7 \gtrsim \alpha \gtrsim -1.2$). For tails the spectrum is sometimes observed to steepen with distance from the parent galaxy, indicating that radiative losses of the type treated in Section 3.1.4 may be occurring (e.g. Miley 1974, Schilizzi & Ekers 1975, Valentijn & Perola 1978). Similar behavior has been observed in the edge-darkened double 3C 31. However, the form of the spectral steepening in this source, one of the few that has been mapped at more than two frequencies, is not that expected from simple synchrotron losses (Burch 1977b). In contrast to the spectral variations observed in the low luminosity sources, spectral steepening has been reported in an inward direction, away from the outer edges of some edge-brightened doubles (Burch 1977a,b, 1979a,b, Dreher 1979, Högbom 1979) but not others (Jenkins & Scheuer 1976, Gopal-Krishna 1977, Gopal-Krishna & Swarup 1977).

Despite the spectral steepening that is sometimes observed, application of the arguments in Section 3.1.4, with the equipartition proviso, suggests that localized particle acceleration occurs frequently in tails (e.g. Wilson & Vallée 1977, Hintzen et al. 1977, Ekers et al. 1978b, Baggio et al. 1978, Simon 1979, Downes 1980) and also throughout the lobes of some double sources (Burch 1977b, Willis & Strom 1978, Strom & Willis 1979, Burch 1979b, van Breugel 1980a).

Fine structure is often seen in the bridges of edge-brightened doubles. These enhancements may correspond to periods of increased activity in the nuclear history, or pinpoint regions where localized particle acceleration is occurring. High sensitivity measurements of the wide double 3C 310 have also revealed weak fine structure on a scale of a few kiloparsecs within its diffuse lobes (van Breugel 1980a).

The diffuse emission is usually highly polarized with polarizations often approaching 60% at frequencies above 1 GHz indicating the presence of a well-ordered magnetic field (Section 3.2). The field direction is generally observed to be circumferential (e.g. Fomalont 1972, Miley & van der Laan 1973, Strom et al. 1978, Willis & Strom 1978, Andernach et al. 1979, Burch 1979b, De Young et al. 1979, van Breugel 1980a, Strom & Willis 1979). See also Figures 1, 3, and 5. For the diffuse bridges and tails the known polarization directions therefore indicate *average* fields directed along the components (e.g. Miley 1976, Miley et al. 1975, Högbom 1979, Burch 1979b), although locally the magnetic field directions are often more complex, particularly near regions of enhancement in total intensity.

4.1.2 CONFINEMENT AND SHAPING   How are the diffuse emitters held together? Evidence that the confining agent is thermal pressure of a hot gas comes from X-ray measurements (e.g. Gursky & Schwartz 1977, Mushotzky et al. 1978), which show that, at least in some of the rich clusters that house radio sources with diffuse lobes, there is an intergalactic medium with $\rho \sim 10^{-27.5}$ g cm^{-3} and $T \sim 10^{7.5}$ K. By the arguments of Section 3.1.2 the pressure of such a gas would be sufficient to confine the lobes provided their internal energy is not much greater than the minimum permitted values ($B_{me} \sim 10^{-5.5}$ G). If thermal pressure does indeed confine tails, the discovery of two tailed galaxies in poor clusters (Ekers et al. 1978b) suggests either that an intergalactic density of $\sim 10^{-27.5}$ gm cm^{-3} may be widespread in the Universe or, as is perhaps more likely, that the sources are confined by giant circumgalactic gaseous halos with dimensions comparable with or larger than the extended radio emission (Norman & Silk 1979).

One of the properties of the diffuse emission that has received considerable attention is the curvature frequently occurring in the tails of tailed radio galaxies. Significantly curved tails have been observed in 3C 83.1B, 3C 129 (Miley et al. 1972, Miley 1973), IC 711 (Vallée & Wilson 1976, Wilson & Vallée 1977), 5C 4.81 (Jaffe et al. 1976), 1200+519/4C 51.29, and 1700+397/4C 39.49 (Miley & Harris 1977). Several explanations have been proposed for this curvature—all within the context of the radio-trail model (Section 2.2.1). It has been suggested that the tail traces a curved orbit of the parent galaxy in the gravitational field of the cluster (Miley et al. 1972) or of a neighbouring galaxy (Byrd & Valtonen 1978). Alternatively, the tail may be distorted by large-scale shear in the velocity field of the intergalactic gas (Jaffe & Perola 1973) or by buoyancy forces in the gravitational field of the cluster (Cowie & McKee 1975). Probably more than one of these mechanisms contributes to bending radio tails. Interpretation of the observations is complicated by effects of projection.

Buoyancy is a particularly interesting possibility. It would cause tails lighter than their surroundings to rise away from the cluster centers. According to Cowie & McKee (1975) denser "detached plasmoids" may be left behind in this process. This may be occurring in the case of 3C 83.1B/NGC 1265 in the Perseus Cluster whose low brightness structure is shown in Figure 8. It has been suggested (Gisler & Miley 1979) that the weak extension to the northeast is such a "heavy tail," which traces the actual orbit of NGC 1265, while the main body of the tail has been bent

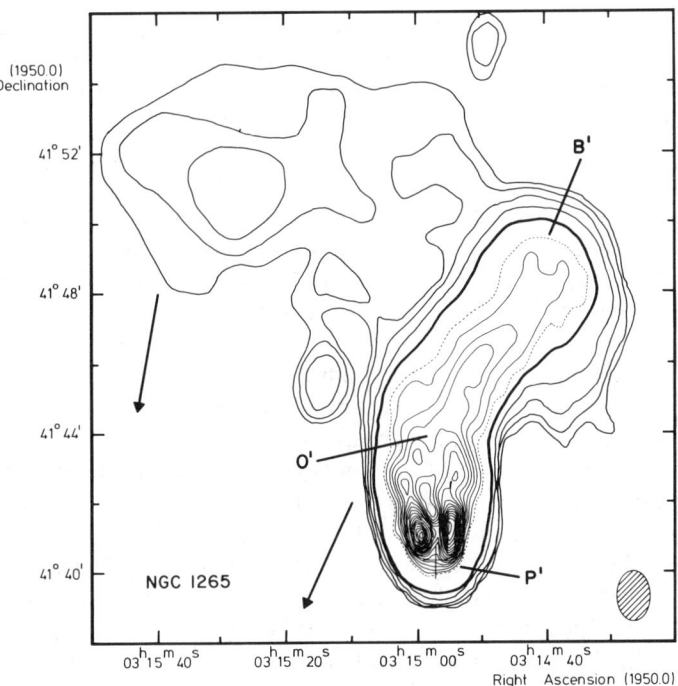

Figure 8 3C 83.1B/NGC 1265, the outer contours are from Westerbork 0.6-GHz intensity measurements and the inner contours were taken from higher resolution 1.4-GHz data (from Gisler & Miley 1979). The cross shows the position of the nucleus of NGC 1265 and the arrows indicate the direction to the center of the Perseus Cluster. The weak northeastern extension may represent the trajectory of the galaxy and the main tailed structure a buoyant tail floating away from the center of the Perseus Cluster.

away from the cluster center by buoyancy forces. On this basis, making several additional assumptions about the geometry, one obtains a lower limit to the mass of the Perseus Cluster of $\sim 10^{15}\ M_\odot$, similar to that obtained independently from the virial theorem. There are also indications that buoyancy plays a part in shaping the structures of diffuse emission in double radio galaxies, e.g. 3C 264, 3C 315 (Northover 1976), and 3C 449 (Bystedt & Högbom 1979).

4.2 Hot Spots

As we have seen in Section 2.2.1, hot spots occur at the outer edges of the most luminous sources and are sometimes seen within the diffuse lobes of the weaker ones.

4.2.1 PROPERTIES Hot spots have sizes that are typically a few kpc (see, for example, Readhead & Hewish 1976, Kapahi 1978). Hot spots in opposite lobes are usually collinear with the central radio core (Section 2.1.1). There are suggestions from the scintillation data that hot spots occur less frequently in sources whose overall sizes are larger than ~ 200 kpc (Readhead & Hewish 1976), or alternatively that the hot spots in the larger sources are bigger.

Hot spots have spectra similar to or slightly flatter than the surrounding diffuse emission (e.g. Hargrave & Ryle 1976, Jenkins & Scheuer 1976, Gopal-Krishna 1977, Gopal-Krishna & Swarup 1977, Burch 1977a, Spangler & Meyers 1978, Högbom 1979, Burch 1979b). Their integrated polarizations at frequencies above 1 GHz are typically 20–30% (see Hargrave & Ryle 1974, Strom & Willis 1979). Circumferential magnetic fields are indicated (see, for example, Dreher 1979) with the average direction of the magnetic field usually oriented roughly perpendicular to the overall extent of the radio component.

Although the presence of hot spots is clearly connected with the total radio-source luminosity, studies of the variation of hot-spot properties with luminosity are difficult to carry out. Such studies are largely confined to the most luminous and distant sources for which cosmological effects are important. The best-established correlation is undoubtedly the increase with luminosity of the separation of the hot spots from their parent nuclei (Section 2.2.3). Less credence should be given to the report that the average hot spot size increases with luminosity (Readhead & Hewish 1976). The initial interpretation of the (scintillation) data on which this report was based has been contradicted by high resolution interferometry (Kapahi & Schilizzi 1979).

During the last few years there have been several searches for optical emission from hot spots. Evidence has been found for such emission in 3C

285 (Tyson et al. 1977), in 3C 265 and 3C 390.3 (Saslaw et al. 1978), in 3C 33 (Simkin 1978), and in NGC 7385 (Simkin & Ekers 1979). The observed radio polarizations of the hot spots would argue against a thermal origin for the optical emission (Saslaw et al. 1978), although in the case of 3C 33 Simkin has claimed that there may be weak emission lines present. The most plausible mechanisms for producing the optical emission would be direct synchrotron radiation or inverse Compton scattering of the microwave background by the electrons that produce the radio emission (Saslaw et al. 1978).

4.2.2 ORIGIN   The collinearity of the hot spots with the nucleus in some double radio sources (Section 2.1.1) indicates that hot spots play a fundamental role in the radio-source phenomenon. It has been suggested that they are (*a*) regions where acceleration of electrons to relativistic energies occurs through energy pumped in from the nucleus (e.g. Rees 1971), (*b*) places where the bulk kinetic energy of the relativistic electrons created in the nucleus is thermalized to produce visible radiation (e.g. Blandford & Rees 1974), (*c*) the accumulation of plasmons, multiply ejected from the nucleus and decelerated by ram pressure (Christiansen et al. 1977), or (*d*) the location of compact supermassive objects ejected from the nucleus which produce and accelerate the relativistic electrons through interaction with circumgalactic gas (Saslaw et al. 1974).

In the first two cases the hot spots represent the end of a beam and they move outwards from the nucleus as the beam rams through the ambient medium. Although the hot spots need not be strictly "confined" in such a situation their motion will be governed primarily by ram pressure forces (e.g. Blandford & Rees 1974), and from the arguments of Section 3.1.2 an external medium with a density of $\rho_{ext} \gtrsim 10^{-28}$ gm cm^{-3} is needed. This lower limit was calculated using an upper limit of 0.1 c for the outward velocities of the hot spots, since for much greater outward speeds relativistic effects would result in a higher proportion of very unequal doubles than is observed (Mackay 1971, Katgert-Merkelijn et al. 1980, Longair & Riley 1979). Similar confinement arguments apply to the multiple-plasmon models.

To hold the hot spots together by gravitation (Burbidge 1967, Callahan 1976, Flasar & Morrison 1976) requires compact objects of mass $\gtrsim 10^9$ $M_\odot$. This is unlikely. First, there is no evidence for structure in the hot spots on a scale much smaller than a kiloparsec. Second, in order to explain the observed alignment of the extended emission with the cores (Section 4.3.3), and with the optical galaxies and quasar polarizations (Sections 5.3), a polar ejection mechanism for the multiple ejection of compact objects from the nucleus is required. The slingshot process (Saslaw et al. 1974), the only mechanism to have been proposed for such ejection, is equatorial and therefore does not preserve memory of direction. Moreover it has difficulty in explaining the detailed brightness distributions of head-tail galaxies (see, for example, Downes 1980). Attempts to reconcile this theory with observations (e.g. Valtonen 1977) appear contrived.

4.3 Cores

4.3.1 TWO TYPES OF CORES Core emission in extended radio sources is of two types. First, there are the ultracompact flat-spectrum ($\alpha < -0.4$) components with sizes $\ll 1$ pc buried deep in the nuclei of the parent galaxies or QSOs. These components have similar properties to the (isolated) compact variable sources (Kellermann 1978). During the last decade compact cores have been shown to be present in extended radio sources of every morphological type. Flat-spectrum radio cores also frequently occur in the nuclei of elliptical and S0 galaxies that are not extended radio sources (Ekers 1978, Crane 1979), and at a weaker level in some spirals (de Bruyn 1978). Even our own galaxy has a tiny flat-spectrum component near its center (Oort 1977 and references therein).

Second, about a quarter of the cores found in radio galaxies have steep spectra with indices smaller than -0.4 (Bridle & Fomalont 1978). These have typical sizes of a few kpc and are clearly different from the ultracompact cores. The best-studied examples are in galaxies—Virgo A/M 87 (Turland 1975a, Forster et al. 1978), Fornax A (Geldzahler & Fomalont 1978), and the giant edge-brightened double 3C 236 (Fomalont & Miley 1975, Fomalont et al. 1979, Schilizzi et al. 1979). A steep-spectrum core has also been found in the quasar 3C 207 (Joshi & Gopal-Krishna 1977). The steep-spectrum cores in Virgo A and 3C 236 have been mapped in detail and in both cases they have basically double structure with compact flat-spectrum cores imbedded within them. The limited information available suggests that the spectral index distribution across steep-spectrum cores is remarkably constant (Berlin et al. 1975, Fomalont et al. 1979). Steep-spectrum cores having the steepest spectra tend to be the most luminous (Bridle & Fomalont 1978); a similar behavior is observed for the integrated luminosities and spectra of extended sources (e.g. Blumenthal & Miley 1979). Isolated steep-spectrum sources with sizes of a few kiloparsecs but no detectable extended emission frequently occur in spiral and Seyfert galaxies (Ekers 1978, de Bruyn & Wilson 1978). Despite the similarities it is not clear whether these isolated sources have any connection with the steep-spectrum kiloparsec cores in extended radio sources.

Since radio-core emission is almost certainly an indication of relatively recent nuclear activity, we may glean information about radio-source evolution by investigating possible relationships between the cores and the associated extended emission.

4.3.2 CORE LUMINOSITIES First we shall examine the relative strengths of cores in different types of extended radio sources. For many sources, measurements are of insufficient resolution to separate the flat- and steep-spectrum cores. Therefore, most studies have been made using combined core fluxes. Neither Fanti & Perola (1977) nor Bridle & Fomalont (1978) find evidence that the core luminosity, P_{core}, increases with that of the extended emission, P_{ext}, for radio galaxies, although Fanti & Perola cannot exclude a weak dependence.

Extended radio sources associated with QSOs tend to have relatively more powerful radio cores than those associated with galaxies (Ekers & Miley 1977, Riley & Jenkins 1977, Miley & Hartsuijker 1978, Owen et al. 1978b). Further data (G. K. Miley, T. Heckman, and R. Fanti, in preparation) indicates that extended quasars have cores that are on average about a factor of twenty stronger than radio galaxies having the same total luminosity. In addition, there appears to be a weak dependence of the core luminosity on the extended emission ($P_{core} \propto P_{ext}^{1/2}$) which would be consistent with the earlier work of Fanti & Perola (1977).

4.3.3 CORE STRUCTURES In several cases the core structures have been determined, primarily with VLBI techniques, and the morphologies of the cores have been related to those of the extended emission. This has been done for the flat-spectrum cores in Cygnus A (Kellermann et al. 1975), 3C 111 (Pauliny-Toth et al. 1976), 3C 273 and 3C 345 (Readhead et al. 1979), 3C 390.3 (E. Preuss et al., in preparation), NGC 6251 (Readhead et al. 1978b, Cohen & Readhead 1979), Virgo A/M 87 (Kellermann et al. 1977 and references therein), and 3C 84/NGC 1275 (Pauliny-Toth et al. 1976), and for the steep-spectrum cores in 3C 236 (Wilkinson 1972, Fomalont & Miley 1975, Fomalont et al. 1979, Schilizzii et al. 1979), Virgo A/M 87 (Turland 1975a, Forster et al. 1978), Centaurus A/NGC 5128 (Christiansen et al. 1977), and 3C 84/NGC 1275 (Miley & Perola 1975).

For the narrow sources whose extended emission is linear and symmetrically distributed (e.g. Cyg A, 3C 111, 3C 236, 3C 390.3, NGC 6251) the cores are observed to be *aligned* to within a few degrees of the outer lobes, even though the scales involved differ by as much as 5×10^6. Assuming the core emission to have been produced by relatively recent nuclear activity allows one to conclude that the orientation of the nuclear powerhouses must have remained fixed throughout the lifetimes of these sources. For the giant sources 3C 236 and NGC 6251 the light travel time from their nuclei to their outer edges is $\sim 10^7$ yr implying a memory for direction in excess of this time. The only relevant direction that could feasibly remain fixed for such a long time is the angular momentum axis of a rotating compact object buried in the nucleus. Despite the overall agreement between the position angles of the core and extended structures in those sources there are fine-scale misalignments of a few degrees apparent both in the steep-spectrum core of 3C 236 (Fomalont & Miley 1975, Schilizzi et al. 1979) and in the flat-spectrum core of NGC 6251 (Cohen & Readhead 1979). These may reflect bending in the cores similar to the wiggles observed in some radio jets (Section 4.4.2).

In contrast to the aligned cores in symmetric narrow sources, the four known cases of sources with one-sided extended emission (S or D2 sources in Section 2.1.1) have compact cores that are considerably bent (Readhead et al. 1978a). Two explanations have been proposed for this misalignment of cores in asymmetric sources. First, the cores may be bent by a pressure gradient in the nucleus. Second, a slight bending in the core might be apparently magnified if the radiating particles were moving at relativistic

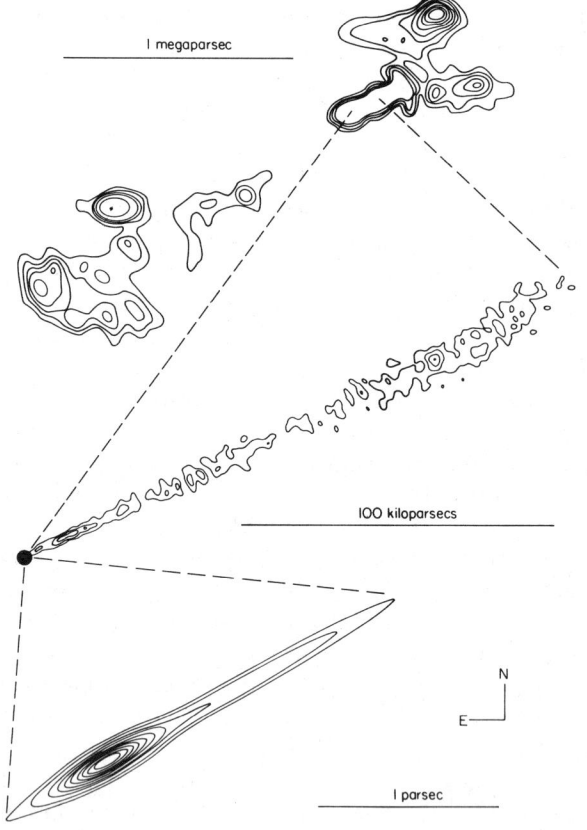

Figure 9 The radio source associated with NGC 6251 showing the alignment of the core and jet relative to the overall source structure (from Readhead et al. 1978a and kindly supplied by Marshall Cohen).

speeds along a narrow beam pointing approximately in our direction (Scheuer & Readhead 1979).

So far we have only considered the structure of cores in narrow sources. For the two wide doubles for which there is information, the cores and the extended structures are grossly misaligned, in Virgo A/M 87 by $\sim 70°$ and in Centaurus A/NGC 5128 by $\sim 45°$. It is possible that the nuclear axes of the wide sources change appreciably during the source lifetime (Miley 1976). Alternatively there may be a shearing pressure gradient in the medium surrounding these galaxies which disrupts the beam and distorts the morphology of the outer extended radio emission. Such a process may well be occurring in 3C 84/NGC 1275 which has a radio core that is aligned on scales of a parsec to 2 kpc but whose extended emission is completely amorphous. 3C 84 is located at the center of the Perseus Cluster, which X-ray observations indicate contains some of the densest and hottest gas of any known cluster. Virgo A and Centaurus A also have strong X-ray halos.

4.4 Jets

Long before celestial radio sources were dreamed of, an optical jet was seen in the nebula M 87 (Curtis 1918). More than fifty years later a radio counterpart of this jet was found in the associated radio source Virgo A (Hogg et al. 1969, Wilkinson 1974, F. N. Owen and P. Hardee, in preparation). See Figure 10. The radio and optical spectra of this jet and its high optical polarization (Baade 1956, Hiltner 1959, Schmidt et al. 1978) are

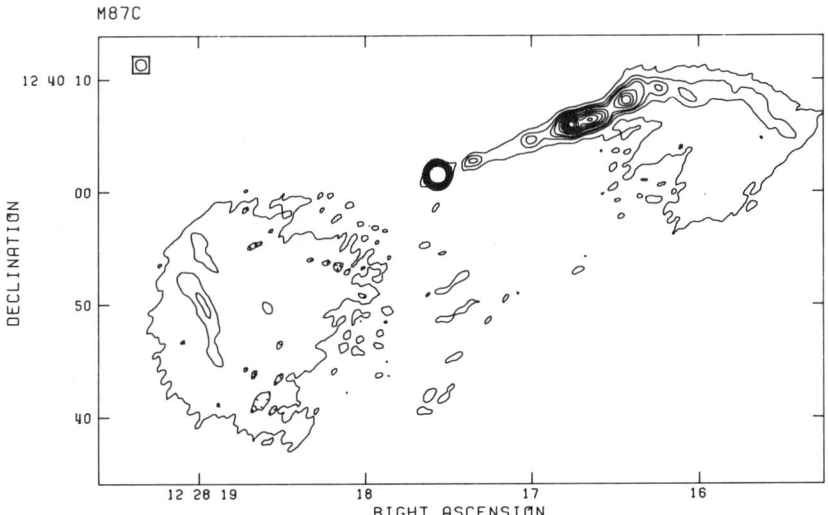

Figure 10 The jet and steep spectrum core of Virgo A/M 87 (VLA, 5 GHz) (from F. N. Owen and P. Hardee, in preparation and kindly supplied by the authors).

all indicative of nonthermal emission. On the identification of 3C 273, the brightest known quasar, an optical jet was also seen in the direction of the extended radio emission (Hazard et al. 1963). For several years no further examples of jets were recognized; M 87 and 3C 273 were regarded as exceptional, but in fact their most exceptional property is probably their proximity.

The situation changed in the mid-seventies. Long narrow radio structures were found to emanate from the nuclei of the radio galaxies 3C 219 (Turland 1975b) and B 0844+31 (van Breugel & Miley 1977) and it was realized that analogous jet-like structures were also present at the heads of several tailed galaxies (van Breugel & Miley 1977). Several additional examples of jets were then recognized and at the time of writing about twenty radio jets are known. Some of the most prominent examples are in the double sources NGC 315 (Bridle et al. 1976, Bridle et al. 1979a), 3C 31 (Burch 1977b, 1979c), NGC 6251 (Waggett et al. 1977), 3C 449 (Perley et al. 1979, Bystedt & Högbom 1979) B 0844+31 (van Breugel 1980b), and HB 13 (Masson 1979) and the tailed sources 3C 83.1B/NGC 1265 (Miley et al. 1975, Owen et al. 1978a), 3C 129 (van Breugel & Miley 1977, Owen et al. 1979, Downes 1980), and 3C 465 (van Breugel 1980c). Some of these are illustrated in Figures 11, 9, 12, 14, 13 and 3.

To date, most radio jets have been found in nearby relatively low

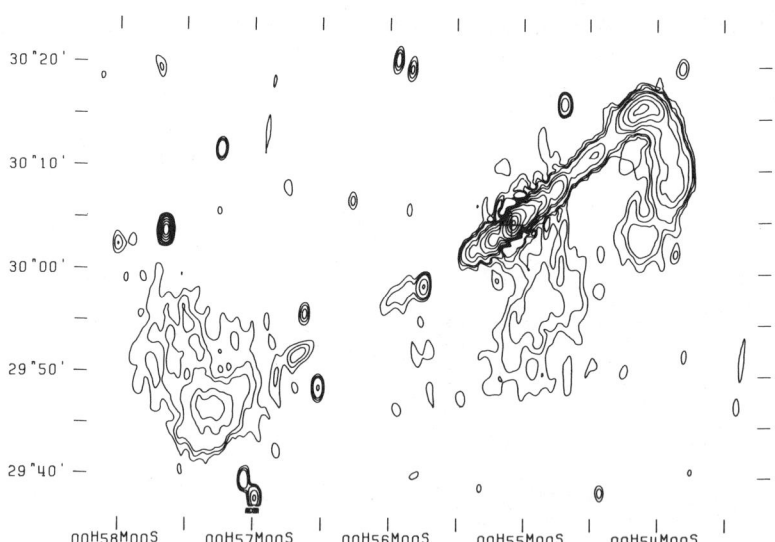

Figure 11 The 1-Mpc radio source associated with NGC 315 (Westerbork 0.6 GHz) showing the jet and its overall rotational symmetry. The cross indicates the position of the associated galaxy (from Willis 1978 and kindly supplied by the author).

EXTENDED RADIO SOURCES 199

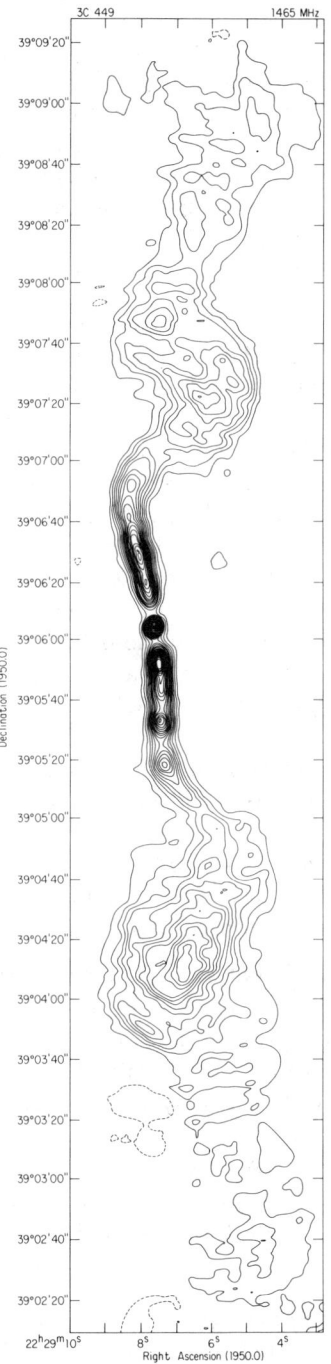

Figure 12 The inner region of 3C 449 (see Figure 2) showing the jet (VLA, 1.4 GHz) (from Perley et al. 1979 and kindly supplied by Ed. Fomalont).

luminosity sources. This may be partly because nearby sources can be studied in greater detail. However, in the more luminous edge-brightened doubles the distinction between the features previously called "bridges" and "jets" is often unclear. Several quasars, e.g. 1004+13, 1047+09, 1058+11 (Miley & Hartsuijker 1978), and 4C 32.69 (Potash & Wardle 1979) have large-scale structures reminiscent of jets. Higher resolution observations with the VLA have recently confirmed the presence of jets in 4C 32.69 (Potash & Wardle 1980) and 1004+13 (E. Fomalont and G. K. Miley, in preparation).

4.4.1 ORIGIN Before the widespread existence of jets was suspected, a considerable body of evidence had accumulated that extended radio sources are powered quasi-continuously from the nuclei of their parent galaxies (Section 1.4). Although the existence of visible jets was not explicitly predicted by models of the energy transport, it seemed logical to associate these narrow structures that emanate from the radio cores and penetrate for several hundred kiloparsecs into the radio lobes with the umbilical cords that carry the energy required to nurture the extended emission. There are several byproducts of the energy-pumping process that could produce the observed radiation. Possibilities are collisions between freshly ejected plasmons and older "relics" (Christansen et al. 1977), or shocks within a relativistic beam (Rees 1978a, Blandford & Königl 1979). Dissipation at the edges of a beam has been suggested as a possible alternative (Blandford & Rees 1978), but since there is no indication that the sides of jets are enhanced (Perley et al. 1979), this is unlikely to be the dominant process for jets in low luminosity sources.

Within the context of these various mechanisms one can list three properties of the energy transport process that may exert a dominating influence on the observed luminosities and other properties of the jets.

1. The angle that the jet is inclined to the line of sight. If the jets represent outflow at relativistic velocities one might observe a dependence of intensity on aspect (cf Rees 1978a, Blandford & Königl 1979).
2. The amount of energy being conducted to the lobes. The highly energetic processes required to generate radiation in the jets may vary sporadically as a result of unsteady nuclear activity (Rees 1978a).
3. The efficiency of the energy transport. Inefficiency may be related to the degree to which energy transport is relativistic (Rees 1978a) or to the amount of interstellar matter entrained in the beam (Blandford & Königl 1979).

There is as yet an insufficiently large sample to investigate these questions properly, but simple interpretation of some of the individual morphological

EXTENDED RADIO SOURCES   201

properties of jets provides several clues as to their nature and to the processes involved. Since, at the time of writing, sources containing radio jets are being studied intensively by the VLA, the following discussion is inevitably very preliminary.

4.4.2 SOME PROPERTIES   The few radio jets that have been studied in detail show considerable diversity. For example, the lengths of the jets range from ~ 2 kpc in Virgo A to ~ 260 kpc in NGC 315 and their spectral indices from ~ -0.5 to ~ -0.8. Three properties of jets deserve special mention.

First, in several cases (e.g. 3C 449, 3C 83.1B/NGC 1265, NGC 315, 3C

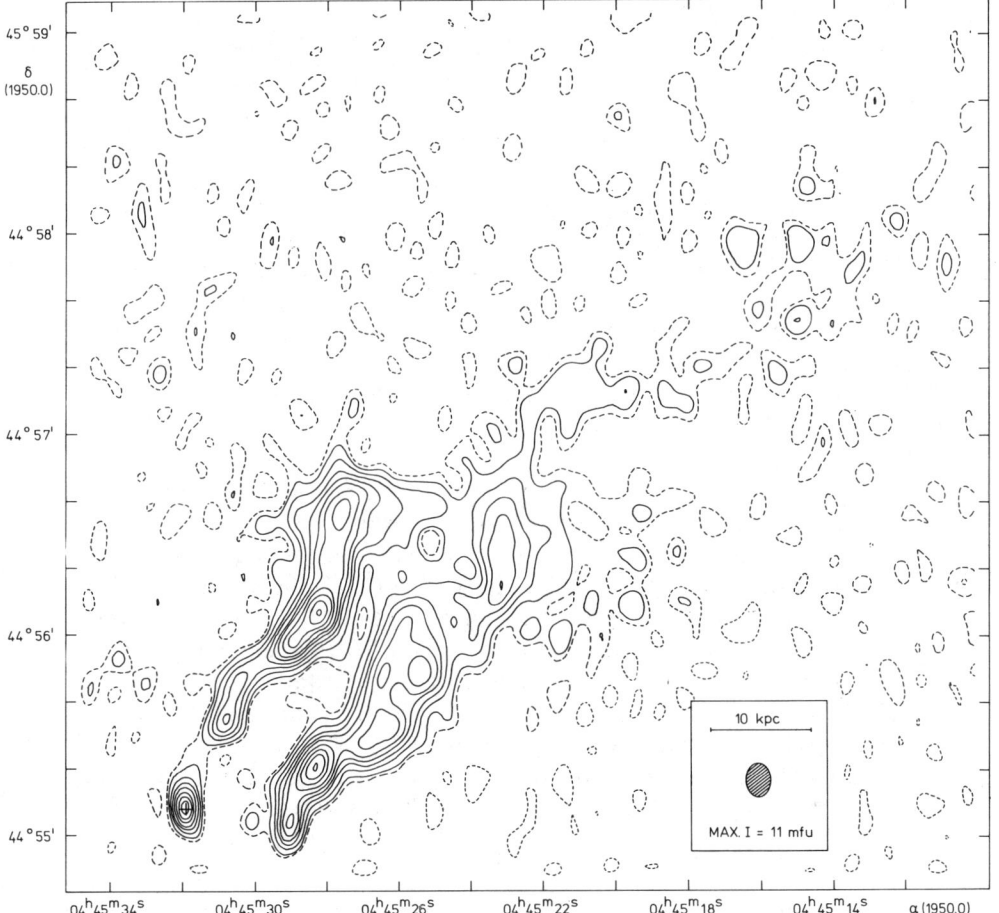

Figure 13   The head of the tailed source 3C 129 showing the symmetric wiggles (from van Breugel & Miley 1977).

219) there appear to be "gaps" of a few kpc between the nuclear core and the location where the jets are seen to begin (Perley et al. 1979). This may be caused by large variations in the power of the nuclear machine, or to more efficient and hence less visible energy transport near the nucleus due to a more highly collimated and less randomized beam of relativistic electrons.

Second, several jets are observed to have wiggles on a scale of about 30 kpc. Examples are B 0844+31 and 3C 129 (van Breugel & Miley 1977, Owen et al. 1979), 3C 31 (Burch 1977a,b, 1979a,c, Fomalont et al. 1979), 3C 449 (Perley et al. 1979, Bystedt & Högbom 1979), HB 13 (Masson 1979), and 3C 83.1B/NGC 1265 (Owen et al. 1978a). Figure 13 shows the symmetric wiggles clearly apparent in the head of the tailed source 3C 129. The wiggles have been attributed either to precessional motion of the central nuclear machine during the source lifetime (van Breugel & Miley 1977) or orbital motion of the parent galaxy (Blandford & Icke 1978, Bystedt & Högbom 1979).

Third, an intriguing aspect of jets that have been surveyed is their asymmetry. Except for 3C 449 (Perley et al. 1979) and 3C 83.1B/NGC 1265 (Owen et al. 1978a), the flux ratio between opposing jets is much larger than the flux ratio between the corresponding extended lobes (e.g. Willis et al. 1978, Cohen & Readhead 1979). Several possible explanations have been proposed to explain how asymmetric jets can power symmetric radio sources. Because of the existence of some two-sided jets the possibility that jets are accretion wakes formed by the motion of their parent galaxies through an intergalactic medium (Yabushita 1979) can be rejected. Another possibility is that the brightness of the jets (reflecting the efficiency of energy transport) may be strongly influenced by a nonuniform galactic environment, but it is not clear why the properties of the environment should differ on opposite sides of the nucleus. Third, opposing jets may have similar intrinsic intensities with the receding one apparently weakened by the Doppler effect (Rees 1978a, Blandford & Königl 1979). Typical flow velocities of $\sim 0.3\ c$ would then be implied. A powerful argument against this view is the observation that some one-sided jets bend without undergoing significant intensity changes (E. van Groningen, C. Norman, and G. K. Miley, in preparation). Fourth, the nuclear engine may flip, supplying energy to the low lobes alternately (Willis et al. 1978). Again it is not clear how this could occur. A fifth possibility is that the beam shines through a clumpy nucleus and that these clumps occasionally block the beam.

4.4.3 OPTICAL EMISSION Optical emission from jets can provide unique information about the physics involved. Since synchrotron decay times for electrons emitting in the optical are typically ~ 10 yr, optical synchrotron features precisely locate regions where the electrons are accelerated (cf Section 3.1.4). An analogy with the optical/radio jet in M 87 prompted a deep search for optical counterparts of four other radio jets (Butcher et al. 1980). In two cases (3C 66 and 3C 31) optical jets were found with similar ratios of optical to radio emission as in M 87. These measurements suggest that a continuous nonthermal spectrum extending from $\sim 10^9$ Hz to $\sim 10^{15}$ Hz is a fairly common property of jets in radio galaxies and that acceleration of the relativistic particles occurs along the jet. The recent detection of X-ray emission from the M 87 jet (P. Gorenstein, E. Schreier, and E. Feigelson, private communication) shows that some jets emit over eight decades of frequency.

M 87 is the only nonthermal jet whose optical properties have been studied in some detail (e.g. de Vaucouleurs et al. 1968). Optical polarization measurements of the M 87 jet (Schmidt et al. 1978) show a fairly ordered magnetic field that changes direction from knot to knot. The changes in the optical polarization angle along the jet are mimicked closely by the 6-cm radio polarization angle. This suggests that the electrons that produce the optical and radio emission originate in roughly the same regions of the magnetic field and that there is little Faraday rotation within the knots. However, the optical and 6-cm radio polarization angles are offset from each other by $\sim 75°$ indicating considerable foreground rotation. The foreground rotation may be produced within the 2-kpc steep-spectrum core of Virgo A (Turland 1975a, Forster 1980) or perhaps within one of the 40-kpc extended radio lobes ($B_{me} \sim 4 \times 10^{-6}$ G; Andernach et al. 1979) which coexists with the X-ray halo ($n_t \sim 10^{-2}$ cm^{-3}; Catura et al. 1972).

Optical jets with significant thermal emission have been reported in the giant radio galaxies Centaurus A (Blanco et al. 1975, Dufour & van den Bergh 1978) and DA 240 (Burbidge et al. 1975, 1978). Although an X-ray jet has been seen in Centaurus A (Schreier et al. 1980), a radio jet has been seen in neither of these galaxies. The absence of an *observed* radio jet in Centaurus A may merely reflect the fact that this southern source is inaccessible to the most sensitive radio telescopes with sufficient resolution. Other extended radio sources in which jets may have been seen are the quasar PKS 0837-12, which has a visible extension along its radio axis (Wehinger & Wyckoff 1978), and some of the 18 candidates noted by Ghigo (1978).

4.4.4 CONE ANGLES AND MAGNETIC FIELDS/FREEDOM FOR JETS? Are the jets in pressure equilibrium with their surroundings? Model builders prefer jets to be free rather than confined. In a free jet the thermal electron density exceeds that of the surroundings, and Kelvin-Helmholtz instabilities, which may destroy a confined jet (Blandford & Pringle 1976, Turland & Scheuer 1976), would probably be unimportant.

For a free jet the cone angle subtended at the nucleus should be constant along its length. In the giant source NGC 6251 (Readhead et al. 1978a) this angle is indeed observed to be constant over a range of more than 10^5 in length.

Also, simple arguments (Blandford & Rees 1978) predict that as a freely expanding jet widens (radius r increases) the parallel component of magnetic field should drop as r^{-2}, whereas the perpendicular component should vary only as r^{-1}. This is consistent with the magnetic field configuration observed in both 3C 31 and NGC 315 (Fomalont et al. 1980) where parallel fields close to the nucleus change to predominantly perpendicular fields after about 2 kpc. Moreover, in 3C 449 (Perley et al. 1979) and NGC 6251 (Readhead et al. 1978b) the magnetic field calculated from the minimum energy condition (Section 3.3.1) B_{me} scales roughly as r^{-1}, a further pointer that these jets are free, as well as weak evidence that the conditions within the jet are close to equipartition. On the assumption that the jet in 3C 449 is free and in equipartition together with several of the arguments of Section 3.1.2, Perley et al. (1979) find the velocity of electron flow along the jet to be ~ 1200 km s^{-1}.

However, there are a number of indications that this picture of a free jet may be oversimplified and not universally applicable.

First, the cone angles of some jets are not constant. For both NGC 315 (Bridle et al. 1979a) and 3C 449 (Perley et al. 1979), the jets shown in Figures 11 and 12, the cone angle initially decreases with distance, most of this "collimation" occurring within 30 kpc of the nucleus. This suggests that the beam can be influenced, perhaps even focused, by a circumgalactic medium. A further indication that jets may be affected by the galaxies through which they pass comes from 3C 83.1B/NGC 1265, the best-studied tailed radio galaxy (Miley et al. 1975, Owen et al. 1978a). Its jet, seen in Figure 14, has been observed to flare out about 10 kpc from the nucleus and this abrupt widening has been cited as evidence for the presence of a hot ($\sim 10^7$ K) dense ($\sim 10^{-2}$ cm^{-3}) interstellar medium in the central part of the elliptical galaxy (Owen et al. 1978a, Jones & Owen 1979).

Second, there are differences in the magnetic field configurations observed in various jets. Although in most jets in low luminosity radio galaxies the field is observed to be predominantly perpendicular to the jets, this may not be the case for high luminosity sources. In 3C 219, the only high luminosity source whose jet has been studied, the polarization distribution indicates a magnetic field aligned *along* the jet (Burch 1979b). Similarly, the bridges in most high luminosity sources have predominantly parallel magnetic fields (e.g. Haves & Conway 1975, Miley 1976). It is unclear whether there is a distinction between a "bridge" in an edge-brightened double source and a "jet." The question of the magnetic field structure in jets is therefore complex. The interpretation is especially difficult because polarization position angles give merely the *projected* magnetic field directions. Fomalont et al. (1980) suggest that the three-dimensional field configuration in 3C 31 and NGC 315 may be helical.

Third, the surface brightness of the jet in 3C 31 decreases roughly as r^{-1}, compared with r^{-3} predicted by simple synchrotron theory (Valtonen 1979, Fomalont et al. 1979, Burch 1979c). Thus, to compensate for adiabatic losses there is likely to be an additional source of energy available to accelerate more particles or to amplify the field. As we have seen, even more stringent evidence for localized acceleration in jets is provided by the detection of optical emission.

Future searches for correlations between the radio and optical properties

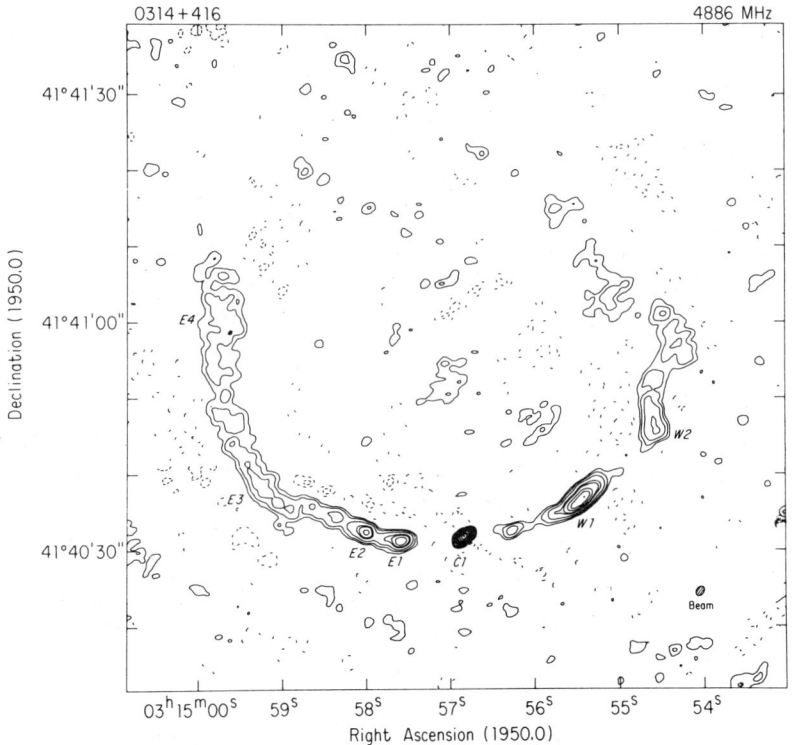

Figure 14 The head of the tailed source 3C 83.1B/NGC 1265 (VLA, 5 GHz) (from Owen et al. 1978a). See also Figure 8.

of jets and other radio-source parameters may help to elucidate the energy transport mechanism further.

The breakdown of a radio source into its elementary component parts has been useful in discussing the various processes involved, but it has inevitably been oversimplified. We have already remarked on the often arbitrary distinction between a "jet" and a "bridge." It is also not always possible to clearly distinguish between a "core" and a "jet." When the cores are observed with sufficient resolution jet-like structures are observed within them, e.g. the steep-spectrum core of Virgo A/M 87 (Figure 10) or the flat-spectrum core of NGC 6251 (Figure 9). The flat-spectrum cores, steep-spectrum cores, and jets are almost certainly all manifestations of different regimes in the energy transport process.

5 RADIO-SOURCE STRUCTURE AND THE ENVIRONMENT

> But it is an error to suppose, because we resolve to confine our attention to a certain number of the properties of an object that we therefore conceive, or have an idea of, the object denuded of its other properties.
>
> J. S. Mill, *Logics and Mathematics Based on Observation*

5.1 *Characteristic Scales*

Some typical scales observed in extended radio sources and their environment are compared in Table 2. Note the similarities between the characteristic radio and optical scales. These suggest that the radio-source phenomenon may be intimately connected with properties of the source environment. Here we shall consider possible relationships.

5.2 *Nuclear Activity*

Do the structures of extended radio sources identified with QSOs differ from those associated with galaxies? Although from the radio maps alone one cannot state with certainty whether an individual radio source is a

Table 2 Some characteristic scales observed in extended radio sources

Length (pc)	Radio	Optical
$\sim 10^{-3}$		Nuclear nonthermal continuum
$\sim 10^{-1}$	Flat-spectrum core	Broad (permitted) line region
$\sim 10^{3}$	Steep-spectrum core	Narrow (forbidden) line region
$\sim 10^{4.5}$		Galaxy
$\sim 10^{5.2}$	Extended radio source	Galactic halo

quasar or a radio galaxy there are several *statistical* differences between the two classes of objects.

As we saw in Section 2.1.1 quasars tend to have more pronounced hot spots and their structures are more asymmetric than sources associated with galaxies. However, because the radio luminosities of quasars are systematically larger than those of radio galaxies, it is impossible to state whether these effects are causally related to the nature of the parent optical objects or to the luminosity of the associated extended sources.

Nevertheless, a real difference is indicated by the fact that the radio cores of quasars are stronger than those of radio galaxies for sources that have extended emission of similar power. It seems reasonable to ascribe this difference to the existence of a correlation between the radio and optical core luminosities in extended radio sources. This is not a one-to-one relationship, however. Perhaps the optical core luminosity measures the instantaneous nuclear activity whereas the radio core represents the activity integrated over a longer period. A very likely possibility is that extended quasars are indeed radio galaxies undergoing transient active periods. The weak optical nonthermal emission seen in the nuclei of some radio galaxies (Yee & Oke 1978) could well be associated with the simmering remnant of a QSO. The more pronounced asymmetry between opposite lobes (Section 2.1.1) would then be a relativistic effect of orientation produced by a more energetic pumping of energy into the hot spots.

The most violent examples of ("present") nuclear activity are probably BL Lac objects. Although differences between their properties and those of quasars appear to be increasingly blurred, BL Lac objects are nominally characterized by high optical polarization, rapid variability at optical infrared and radio wavelengths, and weak or in most cases no observed emission lines. Until recently BL Lac objects were thought to possess little or no extended radio emission, but a survey by Weiler & Johnson (1980) has shown that 22 out of 42 objects surveyed have radio emission with sizes $\gtrsim 1''$. There is therefore no evidence that the properties of this emission are qualitatively different from those of extended radio emission in quasars and radio galaxies.

Emission lines in the parent nuclei of radio sources are strongly connected with their radio structure. Almost all extended quasars exhibit strong broad emission lines. The fraction is about half for the most luminous (edge-brightened-double) 3CR radio galaxies and only 10% for radio galaxies with $P_{178} \lesssim 10^{26}$ W Hz^{-1} (Hine & Longair 1979). For the limb-brightened double radio galaxies the ratio of core to extended flux increases with the width of the emission lines in a fairly continuous sequence from the extended radio galaxies to the extended quasars. This is also consistent with models of radio sources in which a more active nucleus pumps more energy into the extended radio lobes.

Emission lines can also help differentiate quasars that possess extended radio structure from those that do not. A large proportion of all quasars have little or no associated extended structure; for a few of these, upper limits of $\sim 1\%$ have been set on the ratio of extended to compact structure at 610 MHz (Miley & Hartsuijker 1978). The QSOs with the broadest and most irregular emission lines seem to be associated with the extended radio sources, whereas those with relatively narrow emission lines have compact radio structures (Miley & Miller 1979). There is also a tendency for FeII emission lines to be found preferentially in quasars without extended radio structures (Setti & Woltjer 1977, Miley & Miller 1979). These relations, while not properly understood, indicate that the broad line emission regions are closely connected with the production of radio sources.

5.3 Angles and Things

All relations between the structure of radio sources and the properties of their associated galaxies and quasars give basic information concerning the formation of sources. During the last few years several statistical relationships have been discovered between gross optical features and radio structures.

One of the most important relations concerns the orientation of double radio sources relative to that of their parent galaxies. Recently, careful studies have been carried out to investigate this, with samples of 17 (Guthrie 1979) and 78 (Palimaka et al. 1979) objects respectively. In each case rigorous selection criteria were applied and only objects with accurately known radio and optical orientation were included. Both studies convincingly show (in contradiction to some earlier work, e.g. Gibson 1975, Sullivan & Sinn 1975) that the radio structures are preferentially but not exclusively aligned along the (projected) minor axes of the elliptical galaxies. This effect appears strongest for the largest radio sources, implying that alignment of the radio-source production axis with the parent galaxy minor axis may favor the development of giant radio sources.

If most radio galaxies rotate as nearly oblate systems, such an alignment would indicate that the radio-source axes are predominantly oriented along the angular momentum axes of the galaxies. The dynamics of five narrow edge-brightened double radio galaxies, 3C 33, 98, 184.1, 218, and Cygnus A have been studied (Simkin 1977, 1979) and in all cases the rotation axes of the gas and/or stars do seem indeed to be directed to within a few degrees of the radio structures.

Yet more evidence that the radio axis is related to the rotation axes of galaxies is provided by the dust lanes that are often seen to cross the optical images of radio galaxies. A study of eight radio galaxies with dust lanes (Kotanyi & Ekers 1979, Kotanyi 1979, Ekers et al. 1978c) shows that the seven unambiguous cases have dust lanes perpendicular to the axes of the associated radio sources. But again this alignment is not exclusive (Butcher et al. 1980).

Alignment has also been observed between the extended radio structure and the optical polarization of QSOs. Most QSOs are weakly polarized at the 1% level or less. A comparison of the radio orientation and the optical polarization angles for 24 quasars shows that for 20 of them there is agreement to 30° (Stockman et al. 1979, Angel & Stockman 1980). Although the reason for this alignment is unknown, one can draw the elementary conclusion that the long term memory of direction established for edge-brightened double radio galaxies is also a feature of edge-brightened double quasars.

The apparently simple dynamical behavior of galaxies associated with edge-brightened double sources contrasts with the disturbed optical appearance and complex spectroscopic properties of the few wide extended radio galaxies whose distributions have been studied spectroscopically — e.g. NGC 5128/Centaurus A (Blanco et al. 1975, Osmer 1978, Dufour et al. 1979, Graham 1979), NGC 1316/Fornax A (Schweizer 1979), M87/Virgo A (Young et al. 1978, Sargent et al. 1978), and NGC 1275/3C 84 (Burbidge & Burbidge 1965, Rubin et al. 1977). In both Fornax A and Centaurus A, disks of gas and dust are misaligned with the major stellar axes and it has been suggested that the optical properties of these giant elliptical galaxies have been affected by merging. A connection between wide radio sources and galaxy merging would be consistent with Rees's (1978b) suggestion (Section 2.2.3) that merging can cause misalignment between the angular momentum axis of a galaxy and of the nuclear machine responsible for radio emission. The radio ejection axis would then precess, causing a wide radio source to be produced with a wide rotationally symmetric morphology. It may be significant that in Centaurus A the dust lane (Graham 1979) is warped in the same sense as the extended radio emission (Cooper et al. 1965).

Finally, we mention another possible connection between radio and optical morphology which could, if confirmed, be profoundly important. Some years ago it was discovered that two of the brightest tailed radio galaxies, NGC 1265/3C 83.1B and NGC 7720/3C 465, have asymmetric optical brightness distribution, both having the steepest gradients in the direction of their radio heads (Bertola & Perola 1973). On the radio-trail picture this might imply the existence of a relation between the optical asymmetry and the direction of motion of a galaxy in a rich cluster. Since this effect was reported, the sample of tailed radio galaxies has been enlarged considerably and work is at present in progress to investigate whether there is indeed a statistically significant relation between optical asymmetry and radio morphology (E. A. Valentijn et al., in preparation).

5.4 Clustering of Galaxies

In recent years there have been several studies on the effect of cluster membership on the structure of radio galaxies. There is general agreement that sources inside clusters have more complex structures than those outside (Lari & Perola 1977, Rudnick & Owen 1977, Burns & Owen 1977, Simon 1978, McHardy 1979). The more bent sources have a larger density of galaxies surrounding them (Stocke 1979) and of the 26 narrow-tailed sources catalogued by Valentijn (1979b) all except two occur in rich clusters. Three narrow-tailed sources occur in the Perseus Cluster (Gisler & Miley 1979) and five others in Abell 2256 (Bridle & Fomalont 1976, Bridle

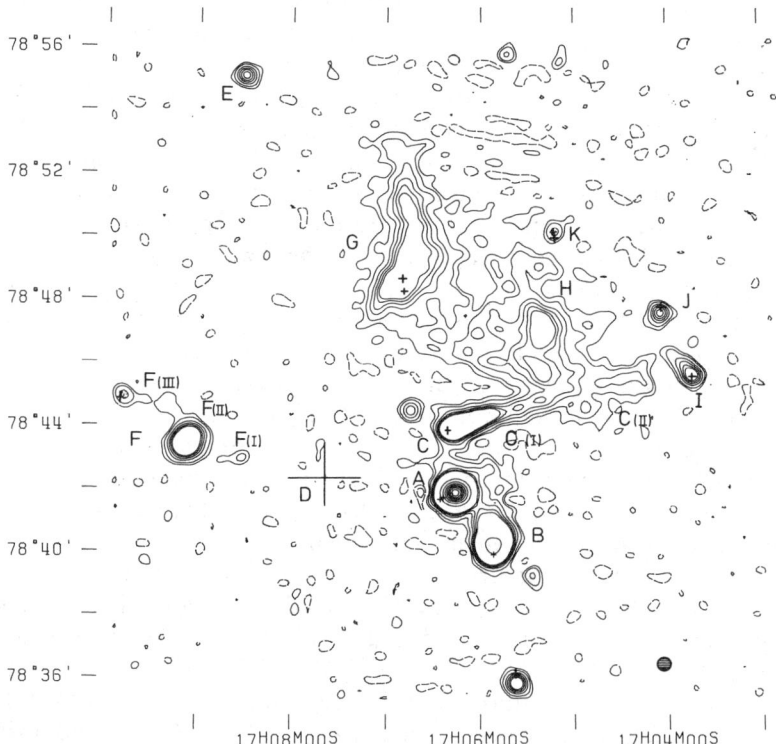

Figure 15 Intensity contours of 1.4-GHz emission from the cluster Abell 2256 (Westerbork). The crosses mark the positions of associated galaxies. The cluster contains at least 4 and possibly 8 tailed radio galaxies (from Bridle et al. 1979b).

et al. 1979b). Both these clusters have large intrinsic X-ray powers and their galaxies have high radial velocity dispersions. Dynamically active clusters containing dense hot gas are therefore good spawning grounds for tailed sources, as would be expected from the radio-trail hypothesis (Section 2.2.1).

Because of the preferential occurrence of distorted radio galaxy morphologies in rich clusters it has been suggested that quasars with bent structures may be suitable indicators of X-ray clusters at intermediate and high redshifts (Hintzen & Scott 1978).

In Section 2.2.1 we saw that more highly bent radio sources tend to have optically fainter parent galaxies. Another form in which this correlation manifests itself is the fact that radio sources associated with dominant cluster galaxies tend to be only slightly bent, whereas the narrow-tailed sources tend to be associated with nondominant galaxies (Rudnick & Owen 1977, McHardy 1979, Simon 1978, Valentijn 1979b).

If an intergalactic medium plays a role in confining radio sources, one might well expect the physical sizes of sources to be smaller inside clusters, where the medium is probably denser and hotter. Whether cluster membership influences radio-source sizes is still controversial. McHardy (1979) finds that 4C sources in Abell clusters are not significantly larger than field sources. On the other hand Stocke (1979) and Guindon (1979) concluded that source sizes are systematically larger in regions of low galactic density. In both the latter studies the density of galaxies was determined in the vicinity of strong radio galaxies.

6 CONCLUSIONS

Life is the art of drawing conclusions from insufficient premises.

Samuel Butler, *Notebooks*

The enormous amount of structural data gathered in recent years suggests the following description of the radio-source phenomenon. A "machine" imbedded in the nucleus of a galaxy ejects a collimated flux of energy in two opposite directions along its angular momentum axis (Section 4.3.3) in the form of a recurrent spluttering of plasmons or a quasi-continuous "beam" of relativistic particles. As a fairly direct manifestation of this nuclear activity we see radio and optical ("quasar") continuum cores and broad line emission regions (Sections 4.3 and 5.2). Provided the nuclear activity is strong enough, prolonged enough, and/or stable enough in direction, an extended radio source is built up. The overall morphology of the extended lobes is determined by the degree of activity, as well as the amount of translational and/or rotational motion of the nuclear machine and/or of the surrounding medium (Sections 2.2, 3 and 4).

What is the nuclear machine? There are several pointers that it may be a rotating black hole that derives its energy from accretion (Lynden-Bell 1969). First, one can cite the elephantine memory for direction discussed in Sections 4.3 and 5.2. Second, photometric and spectroscopic studies of M 87 provide some direct evidence for the presence of a black hole in this, one of the closest radio galaxies (Sargent et al. 1978, Young et al. 1978). Third, collimated extragalactic radio sources have several properties in common with the nonthermal radio emission associated with some galactic X-ray stars (e.g. Braes & Miley 1973). For example the triple source associated with Sco X-1 is uncannily like a miniature edge-brightened extragalactic radio source. SS 433 also shows extended radio emission symmetrically distributed with respect to an X-ray star (Spencer 1979; W. Gilmore and E. R. Seaquist, in preparation; R. Hjellming, K. J. Johnson, and G. K. Miley, in preparation). The only energy sources that have been put forward as possible power houses for both X-ray stars and galactic nuclei are accreting black holes. Fourth, the energy budget is consistent with the appetite of a not too hungry black hole. The most energetic radio sources require total energies in excess of 10^{60} ergs supplied over a period probably longer than 10^8 yr (an age derived by combining the most plausible outflow speeds in Section 4.4 with the typical source sizes for Section 2.3). The resultant energy needs of $\gtrsim 10^{-2} M_\odot$ per year could well be supplied by an accreting black hole buried in a dense galactic nucleus.

Although the above scenario is certainly very oversimplified and possibly conceptually wrong, it is, in my view, the most plausible unified picture consistent with currently available data.

In the near future we shall certainly see further progress on the observational side. The completed VLA will map the detailed morphologies of the jets and hot spots, while more sophisticated optical detectors will provide visual information. The resultant statistical studies should enlarge our knowledge of the energy transport and particle acceleration processes. Second, and perhaps more important, there is the possibility of tackling one of the most intriguing remaining problems—the nature of the nuclear energy conversion and collimation processes. There must be a fundamental mechanism which works over a vast range of physical conditions which can produce collimated radio sources with sizes spanning more than a factor of 10^7 and energies more than a factor of 10^{20}. During the next few years we may expect to see an attack on this problem from several separate wavelength regions. More sensitive VLBI measurements of radio cores of both galactic and extragalactic sources will become available together with the Einstein Observatory's X-ray data. In addition high resolution optical observations from the space telescope should provide considerable information about the dynamics of the inner regions of active galaxies. A third fruitful avenue of research promises to be a continuation of the work described in Section 5.3, carrying out more detailed comparisons between the morphologies of radio sources and the structures and dynamics of their parent galaxies.

Nearly three decades ago an historic encounter took place in which Baade wagered Minkowski a bottle of premium whisky that the visible object identified with Cygnus A was a colliding pair of galaxies. Baade apparently lost the bet. More recently it has been suggested that the hypothetical nuclear black holes that power extended extragalactic radio sources are fueled by infall of material to the nuclei (see, for example, Rees 1978b). The rate of infall may well be influenced by galaxy merging, a process now thought to play an important role in the evolution of galaxies (e.g. Silk & Norman 1979). It is ironic that the wheel of theory shows signs of moving almost full circle. Surely Baade posthumously deserves at least half a bottle of medium quality whisky?

ACKNOWLEDGMENTS

Review articles inevitably reflect the personal preferences and prejudices of the author. I am grateful to the many friends and colleagues who over the years were instrumental in forming my present conditioned outlook. Without risk of incrimination I also wish to thank Wil van Breugel, Ed Fomalont, Dan Harris, Harry van der Laan, Colin Norman, Richard Strom, Tony Willis, and Lorenzo Zaninetti for critically reading the manuscript and Lenore for patiently typing it. I acknowledge the receipt of a NATO research grant (no. 1828).

Literature Cited

Allen, L. R., Hanbury Brown, R., Palmer, H. P. 1963. *MNRAS* 125:57
Andernach, H., Baker, J. R., von Kap-herr, A., Wielebinski, R. 1979. *Astron. Astrophys.* 74:93
Angel, J. R. P., Stockman, H. S. 1980. *Ann. Rev. Astron. Astrophys.* 18:321
Argue, A. N., Riley, J. M., Pooley, G. G. 1978. *Observatory* 98:132
Auriemma, C., Perola, G. C., Ekers, R., Fanti, R., Lari, C., Jaffe, W. J., Ulrich, M. H. 1977. *Astron. Astrophys.* 57:41
Baade, W. 1956. *Ap. J.* 123:550
Baggio, R., Perola, G. C. Tarenghi, M. 1978. *Astron. Astrophys.* 70:303
Baldwin, J. E., Scott, P. F. 1973. *MNRAS* 165:259
Barber, D., Donaldson, W., Miley, G. K., Smith, H. 1966. *Nature* 209:753
Bare, C., Clark, B. G., Kellermann, K. I., Cohen, M. H., Jauncey, D. L. 1967. *Science* 157:189
Benford, G. 1979. *MNRAS* 183:29
Berlin, A. B., Gol'nev, V. Ya., Esepkina, N. A., Zverev, Yu. K., Ipatov, A. V., Kaidanovskii, N. L., Korol'kov, D. V., Lavrov, A. P., Pariiskii, Yu. N., Soboleva, N. S., Stotskii, A. A., Timofeeva, G. M., Shivris, O. N. 1975. *Sov. Astron. Lett.* 1:234, *Pis'ma Astron. Zh.* 1:3

Bertola, F., Perola, G. C. 1973. *Astrophys. Lett.* 14:7
Birkinshaw, M. 1980. *MNRAS* 190:793
Birkinshaw, M., Laing, R., Scheuer, P., Simon, A. 1978. *MNRAS* 185:39P
Blanco, V. M., Graham, J. A., Lasker, B. M., Osmer, P. S. 1975. *Ap. J. Lett.* 198:L63
Blandford, R. D. 1976. *MNRAS* 176:465
Blandford, R. D., Icke, V. 1978. *MNRAS* 185:527
Blandford, R. D., Königl, A. 1979. *Astrophys. Lett.* 20:15
Blandford, R. D., Ostriker, J. P. 1978. *Ap. J. Lett.* 221:L29
Blandford, R. D., Pringle, J. E. 1976. *MNRAS* 176:443
Blandford, R. D., Rees, M. J. 1974. *MNRAS* 169:395
Blandford, R. D., Rees, M. J. 1978. *Physica Scripta* 17:265
Blumenthal, G., Miley, G. K. 1979. *Astron. Astrophys.* 80:13
Braes, L. L. E., Miley, G. K. 1973. *Proc. IAU Symp. 55, X-Ray and Gamma-Ray Astronomy*, p. 86
Bridle, A. H., Feldman, P. A. 1972. *Nature Phys. Sci.* 235:168
Bridle, A. H., Fomalont, E. B. 1976. *Astron. Astrophys.* 52:107
Bridle, A. H., Fomalont, E. B. 1978. *Astron. J.* 83:704
Bridle, A. H., Davis, M. M., Meloy, D. A., Fomalont, E. B., Strom, R. G., Willis, A. G. 1976. *Nature* 262:179
Bridle, A. H., Davis, M. M., Fomalont, E. B., Willis, A. G., Strom, R. G. 1979a. *Ap. J. Lett.* 228:L9
Bridle, A. H., Fomalont, E. B., Miley, G. K., Valentijn, E. A. 1979b. *Astron. Astrophys.* 80:201
Broten, N. W., Clarke, R. W., Legg, T. H., Locke, J. L., McLeish, C. W., Richards, R. S., Yen, J. L., Chisholm, R. M., Galt, J. A. 1967. *Nature* 216:44
Brown, R. H., Palmer, H. P., Thompson, A. R. 1955. *Philos. Mag.* 46:857
Burbidge, E. M., Burbidge, G. R. 1965. *Ap. J.* 142:1351
Burbidge, E. M., Smith, H. E., Burbidge, G. R. 1975. *Ap. J. Lett.* 199:L137
Burbidge, E. M., Smith, H. E., Burbidge, G. R. 1978. *Ap. J.* 219:400
Burbidge, G. R. 1967. *Nature* 216:1287
Burch, S. F. 1977a. *MNRAS* 180:623
Burch, S. F. 1977b. *MNRAS* 181:599
Burch, S. F. 1979a. *MNRAS* 186:293
Burch, S. F. 1979b. *MNRAS* 186:519
Burch, S. F. 1979c. *MNRAS* 187:187
Burn, B. J. 1966. *MNRAS* 133:67
Burns, J. O., Owen, F. N. 1977. *Ap. J.* 217:34
Butcher, H., van Breugel, W. J. M., Miley, G. K. 1980. *Ap. J.* 235:749
Byrd, G. G., Valtonen, M. J. 1978. *Ap. J.* 221:481
Bystedt, J., Högbom, J. A. 1979. Preprint submitted to *Nature*
Callahan, P. S. 1976. *MNRAS* 174:587
Cameron, M. J. 1971. *MNRAS* 152:439
Catura, R. C., Fischer, P. C., Johnson, H. M., Meyerott, A. J. 1972. *Ap. J. Lett.* 177:L1
Christiansen, W. A. 1969. *MNRAS* 145:327
Christiansen, W. A., Pacholczyk, A. G., Scott, J. S. 1977. *Nature* 266:593
Christiansen, W. N., Frater, R. H., Watkinson, A., O'Sullivan, J. D., Lockhart, I. A., Goss, W. M. 1977. *MNRAS* 181:183
Cohen, M. H. 1969. *Ann. Rev. Astron. Astrophys.* 7:619
Cohen, M. H., Readhead, A. C. S. 1979. *Ap. J. Lett.* 233:L101
Cohen, M. H., Gundermann, E. J., Harris, D. E. 1967. *Ap. J.* 150:767
Conway, R. G., Kronberg, P. P. 1969. *MNRAS* 142:11
Conway, R. G., Stannard, D. 1975. *Nature* 255:310
Cooke, B. A., Lawrence, A., Perola, G. C. 1978. *MNRAS* 182:661
Cooper, B. F. C., Price, R. M., Cole, D. J. 1965. *Aust. J. Phys.* 18:589
Costain, C. H., Bridle, A. H., Feldman, P. A. 1972. *Ap. J. Lett.* 175:L15
Cowie, L. L., McKee, C. F. 1975. *Astron. Astrophys.* 43:337
Crane, P. C. 1979. *Astron. J.* 84:281
Curtis, H. D. 1918. *Lick Obs. Publ.* 13:11
Davis, R. J., Stannard, D., Conway, R. G. 1977. *Nature* 267:596
Davis, R. J., Stannard, D., Conway, R. G. 1978. *MNRAS* 185:453
de Bruyn, A. G. 1978. *Proc. IAU Symp. 77, The Structure and Properties of Nearby Galaxies*, p. 205
de Bruyn, A. G., Wilson, A. S. 1978. *Astron. Astrophys.* 64:433
Dent, W. A., Haddock, F. T. 1965. *Nature* 205:487
de Vaucouleurs, G., Angione, R., Fraser, C. W. 1968. *Astrophys. Lett.* 2:141
De Young, D. S. 1976. *Ann. Rev. Astron. Astrophys.* 14:447
De Young, D. S., Axford, W. I. 1967. *Nature* 216:129
De Young, D. S., Hogg, D. E., Wilkes, C. T. 1979. *Ap. J.* 228:43
Downes, A. 1980. *MNRAS* 190:261
Dreher, J. W. 1979. *Ap. J.* 230:687
Dufour, R. J., van den Bergh, S. 1978. *Ap. J. Lett.* 226:L73
Dufour, R. J., van den Bergh, S., Harvel, C. A., Martins, D. H., Schiffer, F. H. III, Talbot, R. J. Jr., Talent, D. L., Wells, D. C. 1979. *Astron. J.* 84:284
Eichler, D. 1979. *Ap. J.* 229:419

Eilek, J. A. 1979. *Ap. J.* 230:373
Ekers, R. D. 1978. *Proc. IAU Symp. 77, The Structure and Properties of Nearby Galaxies*, p. 221
Ekers, R. D., Miley, G. K. 1977. *Proc. IAU Symp. 74*, p. 109
Ekers, R. D., Fanti, R., Lari, C., Parma, P. 1978a. *Nature* 276:588
Ekers, R. D., Fanti, R., Lari, C., Ulrich, M.-H. 1978b. *Astron. Astrophys.* 69:253
Ekers, R. D., Goss, W. M., Kotanyi, C. G., Skellern, D. J. 1978c. *Astron. Astrophys.* 69:L21
Elgaroy, O., Morris, D., Rowson, B. 1962. *MNRAS* 124:395
Emerson, D. T., Klein, U., Haslam, C. G. T. 1979. *Astron. Astrophys.* 76:92
Fabbiano, G., Doxsey, R. E., Johnston, M., Schwartz, D. A., Schwarz, J. 1979. *Ap. J. Lett.* 230:L67
Fanaroff, B. L., Riley, J. M. 1974. *MNRAS* 167:31p
Fanti, R., Perola, G. C. 1977. *Proc. IAU Symp. 74*, p. 171
Ferrari, A., Trussoni, E., Zaninetti, L. 1979. *Astron. Astrophys.* 79:190
Flasar, F. M., Morrison, P. 1976. *Ap. J.* 204:352
Fomalont, E. B. 1969. *Ap. J.* 157:1027
Fomalont, E. B. 1972. *Astrophys. Lett.* 12:187
Fomalont, E. B. 1979. *IAU Comm. 40 Rep.*, ed. H. van der Laan. *IAU Trans.* 17A(3):149
Fomalont, E. B., Miley, G. K. 1975. *Nature* 257:99
Fomalont, E. B., Wright, M. C. H. 1974. *Galactic and Extragalactic Radio Astronomy*, chap. 10, ed. G. L. Verschuur, K. I. Kellermann. Berlin: Springer
Fomalont, E. B., Miley, G. K., Bridle, A. H. 1979. *Astron. Astrophys.* 76:106
Fomalont, E. B., Bridle, A. H., Willis, A. G., Perley, R. A. 1980. Preprint
Forster, J. R. 1980. *Ap. J.* In press
Forster, J. R., Dreher, J., Wright, M. C. H., Welch, W. J. 1978. *Ap. J. Lett.* 221:L3
Gardner, F. F., Whiteoak, J. B. 1966. *Ann. Rev. Astron. Astrophys.* 4:245
Gavazzi, G. 1978. *Astron. Astrophys.* 69:355
Gavazzi, G., Perola, G. C. 1978. *Astron. Astrophys.* 66:407
Geldzahler, B. J., Fomalont, E. B. 1978. *Astron. J.* 83:1047
Getmansev, G. G., Ginzburg, V. L. 1950. *Zh. Eksp. Teor. Fiz.* 20:347
Ghigo, F. D. 1978. *Astron. J.* 83:1363
Gibson, D. M. 1975. *Astron. Astrophys.* 39:377
Gisler, G. R., Miley, G. K. 1979. *Astron. Astrophys.* 76:109
Gopal-Krishna. 1977. *MNRAS* 181:247
Gopal-Krishna, Swarup, G. 1977. *MNRAS* 178:265
Goss, W. M., Wellington, K. J., Christiansen, W. N., Lockhart, I. A., Watkinson, A., Frater, R. H., Little, A. G. 1977. *MNRAS* 178:525
Graham, J. A. 1979. *Ap. J.* 232:60
Guindon, B. 1979. *MNRAS* 186:117
Gursky, H., Schwartz, D. A. 1977. *Ann. Rev. Astron. Astrophys.* 15:541
Guthrie, B. N. G. 1979. *MNRAS* 187:581
Hanisch, R. J., Matthews, T. A., Davis, M. M. 1979. *Astron. J.* 84:946
Hargrave, P. J., McEllin, M. 1975. *MNRAS* 173:37
Hargrave, P. J., Ryle, M. 1974. *MNRAS* 166:305
Hargrave, P. J., Ryle, M. 1976. *MNRAS* 175:481
Harris, A. 1974. *MNRAS* 166:449
Harris, D. E., Grindlay, J. E. 1979. *MNRAS* 188:25
Harris, D. E., Miley, G. K. 1978. *Astron. Astrophys. Suppl.* 34:117
Harris, D. E., Romanishin, W. 1974. *Ap. J.* 188:209
Harris, D. E., Kapahi, V. K., Ekers, R. D. 1980. *Astron. Astrophys. Suppl.* 39:215
Haves, P. 1975. *MNRAS* 173:553
Haves, P., Conway, R. G. 1975. *MNRAS* 173:53p
Hazard, C., Mackey, M. B., Shimmins, A. J. 1963. *Nature* 197:1037
Hewish, A., Scott, P. F., Wills, D. 1964. *Nature* 203:1214
Hill, J. M., Longair, M. S. 1971. *MNRAS* 154:125
Hiltner, W. A. 1959. *Ap. J.* 130:340
Hine, R. G., Longair, M. S. 1979. *MNRAS* 188:111
Hintzen, P., Scott, J. S. 1978. *Ap. J. Lett.* 224:L47
Hintzen, P., Scott, J. S., Tarenghi, M. 1977. *Ap. J.* 212:8
Högbom, J. A. 1974. *Astron. Astrophys. Suppl.* 15:417
Högbom, J. A. 1979. *Astron. Astrophys. Suppl.* 36:173
Hogg, D. E., Macdonald, G. H., Conway, R. G., Wade, C. M. 1969. *Astron. J.* 74:1206
Holman, G. D., Ionson, J. A., Scott, J. S. 1979. *Ap. J.* 228:576
Ingham, W., Morrison, P. 1975. *MNRAS* 173:569
Jaffe, W. J. 1977. *Ap. J.* 212:1
Jaffe, W. J., Perola, G. C. 1973. *Astron. Astrophys.* 26:423
Jaffe, W. J., Rudnick, L. 1979. *Ap. J.* 233:453
Jaffe, W. J., Perola, G. C., Valentijn, E. A. 1976. *Astron. Astrophys.* 49:179

Jenkins, C. J., McEllin, M. 1977. *MNRAS* 180:219
Jenkins, C. J., Scheuer, P. A. G. 1976. *MNRAS* 174:327
Jenkins, C. J., Pooley, G. G., Riley, J. M. 1977. *Mem. RAS* 84:61
Jennison, R. C. 1958. *MNRAS* 118:276
Jennison, R. C., Das Gupta, M. K. 1953. *Nature* 172:996
Jones, T. W., Owen, F. N. 1979. *Ap. J.* 234:818
Joshi, M. N., Gopal-Krishna. 1977. *MNRAS* 178:717
Kapahi, V. K. 1975. *MNRAS* 172:513
Kapahi, V. K. 1978. *Astron. Astrophys.* 67:157
Kapahi, V. K., Schilizzi, R. T. 1979. *Nature* 277:610
Katgert-Merkelijn, J., Lari, C., Padrielli, L. 1980. *Astron. Astrophys. Suppl.* 40:91
Kellermann, K. I. 1978. *Physica Scripta* 17:257
Kellermann, K. I., Clark, B. G., Niell, A. E., Shaffer, D. B. 1975. *Ap. J.* 197:L113
Kellermann, K. I., Shaffer, D. B., Purcell, G. H., Pauliny-Toth, I. I. K., Preuss, E., Witzel, A., Graham, D., Schilizzi, R. T., Cohen, M. H., Moffet, A. T., Romney, J. D., Niell, A. E. 1977. *Ap. J.* 211:658
Kotanyi, C. G. 1979. *Astron. Astrophys.* 74:156
Kotanyi, C. G., Ekers, R. D. 1979. *Astron. Astrophys.* 73:L1
Kronberg, P. P., Strom, R. G. 1977. *Ap. J.* 215:438
Lacombe, C. 1977. *Astron. Astrophys.* 54:1
Lari, C., Perola, G. C. 1977. *Proc. IAU Symp. 79, The Large Scale Structure of the Universe*, p. 137
Little, L. T., Hewish, A. 1968. *MNRAS* 138:393
Longair, M. S., Riley, J. M. 1979. *MNRAS* 188:625
Longair, M. S., Ryle, M., Scheuer, P. A. G. 1973. *MNRAS* 164:243
Lovelace, R. V. E. 1976. *Nature* 262:649
Lynden-Bell, D. 1969. *Nature* 223:690
Macdonald, G. H., Kenderdine, S., Neville, A. C. 1968. *MNRAS* 138:259
Mackay, C. D. 1971. *MNRAS* 154:209
Maltby, P., Moffet, A. T. 1963. *Ap. J. Suppl.* 7:141
Masson, C. R. 1979. *MNRAS* 187:253
Masson, C. R., Mayer, C. J. 1978. *MNRAS* 185:607
McHardy, I. M. 1979. *MNRAS* 188:495
McVittie, G. C. 1965. *General Relativity and Cosmology*. London: Chapman & Hall. 2nd ed.
Miley, G. K. 1971. *MNRAS* 152:477
Miley, G. K. 1973. *Astron. Astrophys.* 26:413
Miley, G. K. 1974. *Proc. IAU Symp. 58, Formation and Dynamics of Galaxies*, p. 109
Miley, G. K. 1976, *Proc. NATO Summer School, Physics of Non-Thermal Radio Sources*, p. 1, Dordrecht: Reidel
Miley, G. K., Harris, D. E. 1977. *Astron. Astrophys.* 61:L23
Miley, G. K., Hartsuijker, A. P. 1978. *Astron. Astrophys. Suppl.* 34:129
Miley, G. K., Miller, J. M. 1979. *Ap. J. Lett.* 228:L55
Miley, G. K., Perola, G. C. 1975. *Astron. Astrophys.* 45:223
Miley, G. K., van der Laan, H. 1973. *Astron. Astrophys.* 28:359
Miley, G. K., Wade, C. M. 1971. *Astrophys. Lett.* 8:11
Miley, G. K., Perola, G. C., van der Kruit, P. C., van der Laan, H. 1972. *Nature* 237:269
Miley, G. K., Wellington, K. J., van der Laan, H. 1975. *Astron. Astrophys.* 38:381
Miley, G. K., Wellington, K. J., van der Laan, H. 1975. *Astron. Astrophys.* 38:381
Mills, D. M., Sturrock, P. A. 1970. *Astrophys. Lett.* 5:105
Mitton, S. 1970. *Astrophys. Lett.* 6:161
Moffet, A. T. 1966. *Ann. Rev. Astron. Astrophys.* 4:145
Moffet, A. T. 1975. *Stars and Stellar Systems*, Vol. IX, p. 211. Univ. Chicago Press
Moffet, A. T. 1979. *IAU Comm. 40 Rep.*, ed. H. van der Laan. *IAU Trans.* 17A(3):155
Mushotzky, R. F., Serlemittos, P. J., Smith, B. W., Boldt, E. A., Holt, S. S. 1978. *Ap. J.* 255:21
Norman, C., Silk, J. 1979. *Ap. J. Lett.* 233:L1
Northover, K. J. E. 1976. *MNRAS* 177:307
Oort, J. H. 1977. *Ann. Rev. Astron. Astrophys.* 15:295
Osmer, P. S. 1978. *Ap. J. Lett.* 226:L79
Owen, F. N., Rudnick, L. 1976. *Ap. J. Lett.* 205:L1
Owen, F. N., Burns, J. O., Rudnick, L. 1978a. *Ap. J. Lett.* 226:L119
Owen, F. N., Porcas, R. W., Neff, S. G. 1978b. *Astron. J.* 83:1009
Owen, F. N., Burns, J. O., Rudnick, L., Greisen, E. W. 1979. *Ap. J. Lett.* 229:L59
Pacholczyk, A. G. 1970. *Radio Astrophysics*, Chap. 7. San Francisco: Freeman
Pacholczyk, A. G. 1977. *Radio Galaxies*, Chap. 5. Oxford: Pergamon
Pacholczyk, A. G., Scott, J. S. 1976. *Ap. J.* 203:313
Palimaka, J. J., Bridle, A. H., Fomalont, E. B., Brandie, G. W. 1979. *Ap. J. Lett.* 231:L7
Palmer, H. P., Rowson, B., Anderson, B., Donaldson, W., Miley, G. K., Gent, H., Adgie, R. L., Slee, O. B., Crowther, J. H. 1967. *Nature* 213:789
Pauliny-Toth, I. I. K., Preuss, E., Witzel, A.,

Kellermann, K. I., Shaffer, D. B. 1976. *Astron. Astrophys.* 52:471
Perley, R. A., Johnston, K. J. 1979. *Astron. J.* 84:1247
Perley, R. A., Willis, A. G., Scott, J. S. 1979. *Nature* 281:437
Perola, G. C., Reinhardt, M. 1972. *Astron. Astrophys.* 17:432
Pooley, G. G., Henbest, S. N. 1974. *MNRAS* 169:477
Potash, R. I., Wardle, J. F. C. 1979. *Astron. J.* 84:707
Potash, R. I., Wardle, J. F. C. 1980. Preprint
Readhead, A. C. S., Hewish, A. 1976. *MNRAS* 1976:571
Readhead, A. C. S., Wilkinson, P. N. 1978. *Ap. J.* 223:25
Readhead, A. C. S., Cohen, M. H., Pearson, T. J., Wilkinson, P. N. 1978a. *Nature* 276:768
Readhead, A. C. S., Cohen, M. H., Blandford, R. D. 1978b. *Nature* 272:131
Readhead, A. C. S., Pearson, T. J., Cohen, M. H., Ewing, M. S., Moffet, A. T. 1979. *Ap. J.* 231:299
Rees, M. J. 1971. *Nature* 229:312, 510
Rees, M. J. 1978a. *MNRAS* 184:61p
Rees, M. J. 1978b. *Nature* 275:516
Reich, W., Kalberla, P., Reif, K., Neidhöfer, J. 1978. *Astron. Astrophys.* 69:165
Riley, J. M., Jenkins, C. J. 1977. *Proc. IAU Symp. 74, Radio Astronomy and Cosmology*, p. 237
Riley, J. M., Pooley, G. G. 1975. *Mem. RAS* 80:93
Riley, J. M., Pooley, G. G. 1978. *MNRAS* 183:245
Rubin, V. C., Ford, W. K., Peterson, C. J., Oort, J. H. 1977. *Ap. J.* 211:693
Rudnick, L., Owen, F. N. 1977. *Astron. J.* 82:1
Ryle, M., Hewish, A. 1960. *MNRAS* 120:220
Ryle, M., Vonberg, D. D. 1948. *Proc. R. Soc. London Ser. A.* 193:98
Ryle, M., Windram, M. D. 1968. *MNRAS* 138:1
Sargent, W. L. W., Young, P. J., Boksenberg, A., Shortridge, K., Lynds, C. R., Hartwick, F. D. A. 1978. *Ap. J.* 221:731
Saslaw, W. C., Valtonen, M. J., Aarseth, S. J. 1974. *Ap. J.* 190:253
Saslaw, W. C., Tyson, J., Crane, P. 1978. *Ap. J.* 222:435
Scheuer, P. A. G. 1962. *Aust. J. Phys.* 15:333
Scheuer, P. A. G. 1974. *MNRAS* 166:513
Scheuer, P. A. G., Readhead, A. C. S. 1979. *Nature* 277:182
Schilizzi, R. T., Ekers, R. D. 1975. *Astron. Astrophys.* 40:221
Schilizzi, R. T., Miley, G. K., van Ardenne, A., Baud, B., Baath, L., Rönnäng, B. O., Pauliny-Toth, I. I. K. 1979. *Astron. Astrophys.* 77:1
Schmidt, G. D., Peterson, B. M., Beaver, E. A. 1978. *Ap. J. Lett.* 220:L31
Schreier, E. J., Feigelson, E., Delvaille, J., Giacconi, R., Grindlay, J., Schwartz, D. A., Fabian, A. C. 1980. *Ap. J. Lett.* 234:L39
Schwartz, U. J. 1978. *Astron. Astrophys.* 65:345
Schweizer, F. 1979. Preprint
Setti, G., Woltjer, L. 1977. *Ap. J. Lett.* 218:L33
Silk, J., Norman, C. A. 1979. *Ap. J.* 234:86
Simkin, S. M. 1977. *Ap. J.* 217:45
Simkin, S. M. 1978. *Ap. J. Lett.* 222:L55
Simkin, S. M. 1979. *Ap. J.* 234:56
Simkin, S. M., Ekers, R. D. 1979. *Astron. J.* 84:56
Simon, A. J. B. 1978. *MNRAS* 184:537
Simon, A. J. B. 1979. *MNRAS* 188:637
Slingo, A. 1974. *MNRAS* 168:307
Smith, M. D., Norman, C. A. 1979. *Astron. Astrophys.* 81:282
Spangler, S. R. 1979. *Ap. J. Lett.* 232:L7
Spangler, S. R., Meyers, K. A. 1978. *Astron. J.* 83:547
Speed, B., Warwick, R. S. 1978. *MNRAS* 182:761
Spencer, R. E. 1979. *Nature* 282:482
Stannard, D., Neal, D. S. 1977. *MNRAS* 179:719
Stocke, J. 1979. *Ap. J.* 230:40
Stockman, H. S., Angel, J. R. P., Miley, G. K. 1979. *Ap. J. Lett.* 227:L55
Strom, R. G., Willis, A. G. 1979. *Astron. Astrophys.* In press
Strom, R. G., Willis, A. G., Wilson, A. S. 1978. *Astron. Astrophys.* 68:367
Sullivan, W. T. III., Sinn, L. A. 1975. *Astrophys. Lett.* 16:173
Swarup, G. 1975. *MNRAS* 172:501
Turland, B. D. 1975a. *MNRAS* 170:281
Turland, B. D. 1975b. *MNRAS* 172:181
Turland, B. D., Scheuer, P. A. G. 1976. *MNRAS* 176:421
Tyson, J. A., Crane, P., Saslaw, W. C. 1977. *Astron. Astrophys.* 59:L15
Valentijn, E. A. 1978. *Astron. Astrophys.* 68:449
Valentijn, E. A. 1979a. *Astron. Astrophys. Suppl.* 38:319
Valentijn, E. A. 1979b. *Astron. Astrophys.* 78:367
Valentijn, E. A., Perola, G. C. 1978. *Astron. Astrophys.* 63:29
Vallée, J. P., Kronberg, P. P. 1975. *Astron. Astrophys.* 43:233
Vallée, J. P., Wilson, A. S. 1976. *Nature* 259:451
Vallée, J. P., Wilson, A. S., van der Laan, H. 1979. *Astron. Astrophys.* 77:183
Valtonen, M. J. 1977. *Ap. J.* 213:356
Valtonen, M. J. 1979. *Ap. J. Lett.* 227:L79

van Breugel, W. J. M. 1980a. *Astron. Astrophys.* 81:265
van Breugel, W. J. M. 1980b. *Astron. Astrophys.* 81:275
van Breugel, W. J. M. 1980c. *Astron. Astrophys.* In press
van Breugel, W. J. M., Miley, G. K. 1977. *Nature* 265:315
van der Kruit, P. C., Allen, R. J. 1976. *Ann. Rev. Astron. Astrophys.* 14:417
van der Laan, H. 1963. *MNRAS* 126:535
van der Laan, H., Perola, G. C. 1969. *Astron. Astrophys.* 3:468
von Hoerner, S. 1964. *Ap. J.* 140:65
Waggett, P. C., Warner, P. J., Baldwin, J. E. 1977. *MNRAS* 181:465
Wehinger, P. A., Wyckoff, S. 1978. *MNRAS* 184:335
Weiler, K. W. 1973. *Astron. Astrophys.* 26:403
Weiler, K. W., Johnson, K. J. 1980. *MNRAS* 190:269
Wiita, P. J. 1978a. *Ap. J.* 221:41
Wiita, P. J. 1978b. *Ap. J.* 221:436
Wilkinson, P. N. 1972. *MNRAS* 160:305
Wilkinson, P. N. 1974. *Nature* 252:661
Willis, A. G. 1978. *Physica Scripta* 17:243
Willis, A. G., Strom, R. G. 1978. *Astron. Astrophys.* 62:375
Willis, A. G., Strom, R. G., Wilson, A. S. 1974. *Nature* 150:625
Willis, A. G., Wilson, A. S., Strom, R. G. 1978. *Astron. Astrophys.* 66:L1
Wilson, A. S., Vallée, J. P. 1977. *Astron. Astrophys.* 58:79
Yabushita, S. 1979. *MNRAS* 188:59
Yee, H. K. C., Oke, J. B. 1978. *Ap. J.* 226:753
Young, P. J., Westphal, J. A., Kristian, J., Wilson, C. P., Landauer, F. P. 1978. *Ap. J.* 221:721

INTERSTELLAR SHOCK WAVES

✳2166

Christopher F. McKee
Departments of Physics and Astronomy, University of California, Berkeley, California 94720

David J. Hollenbach
NASA—Ames Research Center, Moffett Field, California 94035

1 INTRODUCTION

Shock waves occur in compressible media when pressure gradients are large enough to generate supersonic compressive motions. Because the shock propagates faster than the characteristic signal velocity, the medium ahead of the shock cannot dynamically respond to the shock until the shock strikes. The shock then compresses, heats, and accelerates the medium. For a nonradiative shock, the jump in density is limited to a factor of four (for a specific heat ratio $\gamma = 5/3$); the change in pressure across the shock is comparable to the dynamic pressure $\rho_0 v_s^2$, where ρ is the density, v_s the shock velocity, and the subscript 0 denotes quantities ahead of the shock.

A shock produces an irreversible change in the medium through which it passes. The increase in entropy across the shock front is produced by either collisions between particles in the shocked and unshocked fluids or by the generation and dissipation of plasma turbulence. The thickness of the shock front is generally small compared to other hydrodynamic length scales so shock fronts can usually be approximated as discontinuities.

Most shocks in the interstellar medium (ISM) are thought to be produced by stars (see Table 1). When an early-type star first turns on, it rapidly ionizes the gas in its vicinity, raises its pressure, and drives a shock into the ambient medium at $v_s \leqslant 10$ km s^{-1} (Spitzer 1978). Early-type stars are observed to have strong winds. If most of the wind energy is radiated away (the momentum-conserving case, Steigman et al. 1975), then the wind usually does not have a strong effect on the expansion of the HII region. However, if a significant fraction of the wind energy is not radiated away

Table 1 Interstellar shock parameters

	HII regions		SNRs	
Parameters	Ionization-driven[a]	Wind-driven[b]	Adiabatic[c]	Radiative[c]
$t(10^4 \text{ yr})$	$\dfrac{170}{v_{s6}^{7/3}}\left(\dfrac{L_{48}^*}{n_0^2}\right)^{1/3}$	$\dfrac{290}{v_{s6}^{5/2}}\left(\dfrac{L_{w36}}{n_0}\right)^{1/2}$	$\dfrac{14}{v_{s7}^{5/3}}\left(\dfrac{E_{51}}{n_0}\right)^{1/3}$	$\dfrac{8.3\,E_{51}^{0.32}}{v_{s7}^{1.45}\,n_0^{0.36}}$
$R(\text{pc})$	$\dfrac{31}{v_{s6}^{4/3}}\left(\dfrac{L_{48}^*}{n_0^2}\right)^{1/3}$	$\dfrac{50}{v_{s6}^{5/2}}\left(\dfrac{L_{w36}}{n_0}\right)^{1/2}$	$\dfrac{36}{v_{s7}^{2/3}}\left(\dfrac{E_{51}}{n_0}\right)^{1/3}$	$\dfrac{27\,E_{51}^{0.32}}{v_{s7}^{0.45}\,n_0^{0.36}}$
$N_{\text{tot}}(10^{19}\,\text{cm}^{-2})$	$\dfrac{3.2(L_{48}^* n_0)^{1/3}}{v_{s6}^{4/3}}$	$\dfrac{5.2(L_{w36} n_0)^{1/3}}{v_{s6}^{3/2}}$	$\dfrac{3.7(E_{51} n_0^2)^{1/3}}{v_{s7}^{2/3}}$	$\dfrac{2.7\,E_{51}^{0.32}\,n_0^{0.64}}{v_{s7}^{0.45}}$

[a] L_{48}^* is the ionizing photon luminosity in units of 10^{48} photons s^{-1}. v_{s6} = shock velocity in units of 10^6 cm s^{-1}. n_0 is the ambient hydrogen density. Temperature in HII region = 8000 K. Ref.: Spitzer (1978).
[b] L_{w36} is the wind luminosity in units of 10^{36} erg s^{-1}. Energy-conserving case assumed. Ref.: Weaver et al. (1977).
[c] E_{51} is the explosion energy in units of 10^{51} erg. Note that v_s is in units of 10^7 cm s^{-1}. Ref.: Chevalier (1974).

(the energy-conserving case, Castor et al. 1975), then the shock produced by the expanding HII region may be significantly strengthened by the wind. The strongest shocks in the ISM are produced by supernovae. In the adiabatic phase of their evolution, supernova remnants (SNRs) drive shocks into the ISM with velocities of up to about 10^4 km s^{-1} and produce copious X-ray emission. Radiative losses become dynamically important when the shock velocity drops to about 200 km s^{-1}; the subsequent evolution of the SNR described in Table 1 is based on the numerical results of Chevalier (1974). SNR evolution is more complicated if the ambient medium is not homogeneous (McKee & Ostriker 1977). Interstellar shocks can also be produced by the collision of two interstellar clouds (Stone 1970). The origin and propagation of shocks in the ISM are more extensively discussed in the recent reviews by Chevalier (1977a) and McCray & Snow (1979). On a larger scale, spiral density waves in galaxies appear to be accompanied by interstellar shocks many kiloparsecs long (Roberts 1969).

Interstellar shocks thus cover a wide range of parameters: velocities of $1-10^4$ km s^{-1}, pre-shock densities of $\sim 10^{-2}-10^7$ cm^{-3}, and post-shock temperatures $\sim 10^2-10^9$ K. The strength of a shock is indicated by the Mach number $\mathcal{M} = v_s/c_0$, where c_0 is the signal velocity ahead of the shock; $\mathcal{M}$ can range up to $\sim 10^3$, far larger than in laboratory shocks. The interstellar medium is magnetized, so interstellar shocks are hydromagnetic. The magnetic field B affects the Mach number, the structure of the shock front, and the maximum amount of post-shock compression. In contrast to stellar shocks and most laboratory shocks, neither the pre-shock nor the post-shock medium is in thermodynamic equilibrium, and radiation emitted by the shock can generally escape freely.

In this review we focus on the structure of interstellar shocks, not on the propagation. As indicated above, shocks are an ubiquitous feature in the ISM, and a discussion of the propagation of interstellar shocks would essentially cover the entire dynamics of the ISM. On the other hand, the structure of a shock determines the spectrum of the emitted radiation and the chemistry occurring in the shock, both of which are directly observable.

An interstellar shock can be approximately divided into three regions (Figure 1): (a) the radiative precursor which may heat, ionize, and dissociate the upstream gas; (b) a relatively thin region (which we shall term the "shock front") in which elastic collisions or plasma instabilities excite the translational degrees of freedom in the gas and produce the increase in the temperature, density, and entropy of the gas; and (c) the post-shock relaxation layer in which inelastic collisional processes (ionization, dissociation, collisional excitation, recombination, and molecule formation) cause the gas to cool and relax to its final composition. In laboratory or atmospheric shocks there is often a fourth layer in which the shocked gas

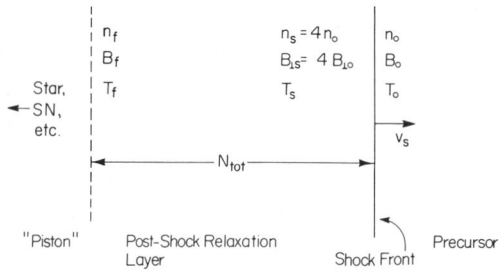

Figure 1 Schematic structure of interstellar shocks showing the radiative precursor, the shock front itself, and the post-shock relaxation layer. The "piston" driving the shock might be supernova ejecta or a stellar wind; once the shock becomes radiative and a dense shell forms, the piston might be the hot gas in the interior of a SNR or wind-driven HII region. The density n, temperature T, and component of magnetic field B perpendicular to the shock velocity are indicated in the pre-shock gas (n_0, etc.), just behind the shock (n_s, etc.), and far behind the shock (n_f, etc.). N_{tot} is the column density of swept up gas.

gets into LTE and radiates like a blackbody. In interstellar shocks such a layer would exist only if the dust associated with the interstellar gas were optically thick to its own thermal radiation, which occurs only for very large post-shock column densities, $N \gtrsim 10^{27}(n_0 v_{s7}^3)^{-1/2}$ cm^{-2}, where $v_{s7} = v_s/(10^7$ cm s$^{-1})$ (Hollenbach & McKee 1979). The precursor and the post-shock relaxation layer have nearly Maxwellian particle distributions and are described by the usual hydrodynamic equations. The dominant feature of the shock front itself is the viscous layer in which the relative motion of the pre-shock and post-shock gas is thermalized, and the particle distributions are far from Maxwellian. The structure of collision-dominated shock fronts is reasonably well understood, but the same is not true for collisionless shocks (see Section 3). Processes in a collisionless shock front determine the electron-ion temperature ratio T_e/T_i, which is not necessarily unity as it is for collisional shocks. Furthermore, there is the possibility of particle acceleration (see Section 7.4) and nonthermal emission from a collisionless shock front. Determining the structure of collisionless shock fronts is one of the most important unsolved problems in plasma astrophysics.

More detailed discussions of shocks can be found in the monographs by Zel'dovich & Raizer (1966) and Tidman & Krall (1971) and in the review by Chu & Gross (1969).

2 JUMP CONDITIONS

Since the thickness of the shock front is generally very small compared to hydrodynamic length scales, the hydrodynamic equations in the vicinity of the shock front reduce to their one-dimensional, steady-state form in a coordinate system moving with the shock. For example, the equation of continuity becomes $d(\rho v_\parallel)/dz = 0$, where z is the distance normal to the plane of the shock and $v_\parallel \equiv v_z$ is the fluid velocity parallel to $\mathbf{v}_s$ measured in the shock frame. Integration of the hydrodynamic equations across the shock front gives the Rankine-Hugoniot relations, or jump conditions; for a hydromagnetic shock, Maxwell's equations are needed as well. The jump conditions can be written in a form sufficiently general that they apply inside the shock front as well as across it. They simplify considerably, however, if they are restricted to apply outside the shock front, where viscous effects, relative drifts between different particle species, and (usually) heat conduction are negligible; furthermore, outside the shock front the field is usually frozen to the plasma so that $\mathbf{E} + (\mathbf{v}/c) \times \mathbf{B} = 0$. The jump conditions for an arbitrary hydromagnetic shock are then (Tidman & Krall 1971, Hollenbach & McKee 1979)

$$[\rho v_\parallel] = 0 \tag{1}$$

$$[\rho v_\parallel^2 + p + B_\perp^2/8\pi] = 0 \tag{2a}$$

$$[\rho v_\parallel v_\perp - B_\parallel B_\perp/4\pi] = 0 \tag{2b}$$

$$[v_\parallel(\tfrac{1}{2}\rho v^2 + u + p) + (v_\parallel B_\perp^2 - v_\perp B_\perp B_\parallel)/4\pi + F] = 0 \tag{3}$$

$$[B_\parallel] = 0 \qquad [v_\parallel B_\perp - v_\perp B_\parallel] = 0 \tag{4}$$

where the square brackets indicate the quantity inside is to be evaluated on either side of the shock front and the difference taken, and where the coordinate system is such that the shock velocity is normal to the plane of the shock (i.e., $v_\perp = 0$ ahead of the shock so that the shock is not oblique). Equation (1) is mass conservation, Equation (2), momentum flux conservation, and Equation (3), energy flux conservation; F is the flux of electromagnetic wave energy, and u is the internal energy density. If the effects of thermal conduction are to be included, then the heat flux q must be added to the expression in brackets in Equation (3).

The jump conditions simplify considerably if $B = 0$ and if the internal energy is related to the pressure by $u = p/(\gamma - 1)$:

$$[\rho v] = 0 \tag{5}$$

$$[p + \rho v^2] = 0 \tag{6}$$

$$\left[\rho v\left(\tfrac{1}{2}v^2 + \frac{\gamma}{\gamma - 1}p\right) + F\right] = 0. \tag{7}$$

The equations of mass and momentum conservation (5 and 6) apply to any plane hydrodynamic jump independent of the amount of energy emitted or absorbed at the front. In particular they apply to ionization fronts, which are classified as R-type if they are supersonic relative to the gas ahead of the front and D-type otherwise, and as strong or weak depending on the magnitude of the density jump (see Spitzer 1978). In this terminology, shocks correspond to strong R fronts.

2.1 The Shock Front

In most cases the shock front is sufficiently thin that inelastic processes are negligible inside it. Consequently the quantities F and $n_j v_\parallel$ do not change from just ahead of the shock front to just behind it, where n_j is the density of species j. For a gas dynamic shock front ($B = 0$), solution of the jump conditions (5–7) gives the following general results (Chu & Gross 1969): (a) shocks are *supersonic* relative to the gas ahead of the shock but *subsonic* relative to the gas behind; (b) shocks are compressive (p_s/p_0, ρ_s/ρ_0, T_s/T_0 all > 1, where p_s, etc., are evaluated just behind the shock front); (c) the compression increases monotonically with the Mach number $\mathcal{M}$; and (d) strong shocks ($\mathcal{M} \to \infty$) have $p_s = 2\rho_0 v_s^2/(\gamma+1)$, $\rho_s/\rho_0 = (\gamma+1)/(\gamma-1)$, and $kT_s = 2(\gamma-1)(\gamma+1)^{-2}\mu_s v_s^2$, where μ_s is the mean mass per particle just behind the shock front. Note that p_s/p_0 and T_s/T_0 are unbounded as $\mathcal{M} \to \infty$, whereas ρ_s/ρ_0 is bounded and approaches 4 for $\gamma = 5/3$.

The above expression for the post-shock temperature follows directly from energy conservation in the frame of the shocked fluid: the upstream kinetic energy per particle $1/2\mu_s(v_s-v_1)^2$ must equal the downstream thermal energy per particle $3/2 kT_s$, where v_1 is the post-shock velocity in the shock frame. In the usual case in which $\gamma = 5/3$, the post-shock temperature is

$$T_s = 3\mu_s v_s^2/16k \qquad (8)$$

or $1.38 \times 10^5 \, v_{s7}^2$ K for a fully ionized gas, $2.9 \times 10^5 v_{s7}^2$ K for a neutral atomic gas, and $5300 v_{s6}^2$ K for a molecular gas, where $v_{s6} = v_s/(10^6 \text{ cm s}^{-1})$, etc., and cosmic abundances with $n(\text{He})/n(\text{H}) = 0.1$ have been assumed. Bear in mind that T_s is the mean post-shock temperature; in a collisionless shock the electron and ion temperatures need not be equal. Under very unusual circumstances ($d^2p/dV^2 < 0$, where V is the specific volume), item (b) is violated and rarefaction shocks can occur (Zel'dovich & Raizer 1966, Bezzerides et al. 1978).

In hydromagnetic shocks the flux-freezing condition $\mathbf{E} + (\mathbf{v}/c) \times \mathbf{B} = 0$ is applied only *outside* the shock front, but it ensures that flux is conserved *across* the shock irrespective of the processes that occur in the front itself. Unless $B_\perp = 0$ there will be an electric field $\mathbf{E} = -(\mathbf{v}/c) \times \mathbf{B} \neq 0$ in the shock. However, unless $B_\perp/B_\parallel > (c/v)$, one can transform to a frame moving in the plane of the shock in which $\mathbf{E} = 0$ so that $\mathbf{v}$ and $\mathbf{B}$ are parallel on both sides of the shock front, and the magnetic field lines are stationary. For the simple case in which $B_{\parallel} = 0$, Equations (1) and (4) give $B_{\perp}/\rho$ = constant. Since the jump in ρ across the shock is bounded, so is the jump in B. Hence, for very strong shocks the magnetic pressure $B^2/8\pi$ will be negligible, and the hydromagnetic jump conditions reduce to the gas dynamic ones. For example, Hollenbach & McKee (1979) found that Equation (8) for T_s is accurate to within 30% for Alfvén Mach numbers $\mathcal{M}_A \equiv v_s/v_{A0} \equiv v_s/(B_0^2/4\pi\rho_0)^{1/2} > 6$, provided the pre-shock gas pressure is negligible. In general the presence of a magnetic field reduces T_s for a given shock velocity.

Strong hydromagnetic shocks such as these are termed fast, or super-Alfvénic. They occur only if $v_s > v_{A0}$ and can result from the steepening of the fast, or magnetosonic, wave. Slow, or sub-Alfvénic ($v_s < v_{A0}$) shocks also occur; they are based on the slow mode.

In some situations the downstream field can be time-dependent so that the rms field exceeds the mean field. In this case the momentum and energy fluxes of the hydromagnetic waves must be included in the jump conditions. In fact, oscillatory magnetic fields that overshoot the steady-state value given by the usual jump conditions have been observed behind the Earth's bow shock under conditions when the thermal pressure exceeds the magnetic pressure upstream (Russell & Greenstadt 1979). The oscillations tend to die out sufficiently far behind the shock. Chevalier (1977b) has suggested that if such enhanced fields occurred in interstellar shocks they would increase the synchrotron emission from relativistic electrons in SNRs. At present it is not clear what conditions are necessary for producing the time-dependent fields, nor if they extend far enough behind the shock for Chevalier's effect to be observable.

In some cases the effects of thermal conduction are not confined to the thin shock front, and the heat flux must be included in the energy jump condition. Solinger et al. (1975) studied the case of an isothermal blast wave in which thermal conduction inside a nonradiative SNR is so efficient that it eliminates any internal temperature gradients. Because energy is added to the post-shock gas from the interior of the SNR, the shock compression is only 2.4 instead of 4 (they assumed $T_e = T_i$ everywhere, which is not very realistic—Lerche & Vasyliunas 1976). C. F. McKee and J. M. Shull (in preparation) have studied conduction behind shocks in which a high metal abundance leads to very rapid cooling; in this case, energy is drained from the electrons just behind the shock front, and the compression is increased somewhat. However, the post-shock electron temperature T_{es} can drop by only a factor of a few before the heat flux saturates (Cowie & McKee 1977).

The jump conditions for interstellar shocks are also affected by cosmic rays, which pervade most of the ISM and often have a pressure comparable to that of the magnetic field and the thermal gas (Wentzel 1971). For perpendicular shocks ($B \simeq B_\perp$), the cosmic rays are tightly coupled to the gas by the field. For parallel shocks ($B \simeq B_\parallel$), Wentzel argues that the cosmic rays are coupled to the gas by hydromagnetic waves provided the shock velocity exceeds the critical velocity $v_{\text{crit}} \sim 1.5 v_A + (10^8 n_{\text{HI}})$ cm s^{-1} for exciting the waves. (This assumes that the cosmic rays have an energy density close to that in the solar neighborhood and that v_{crit} is determined by protons with energies ~ 1 GeV). Wentzel argues that the interaction between the cosmic rays and gas in a parallel shock with $v_s > v_{\text{crit}}$ heats the gas and cools the cosmic rays, but this ignores the particle acceleration processes discussed in Section 7. In any case, the cosmic ray pressure and enthalpy must be included in the momentum and energy jump conditions, respectively, whenever the cosmic rays are coupled to the gas.

2.2 The Post-Shock Relaxation Layer

In contrast to the shock front, the post-shock relaxation layer is dominated by inelastic collisions and by the emission and absorption of radiation. In many cases—especially for very fast shocks—the cooling time behind the shock exceeds the age of the shock; since the radiation emitted from the shocked gas does not affect the dynamics, such shocks are said to be nonradiative (see Section 4). In nonradiative shocks there is no clear distinction between the relaxation layer and the overall flow behind the shock. On the other hand, when the cooling time is less than the age, the shock is said to be radiative, and the relaxation layer is well-defined (see Sections 5 and 6). The criterion separating radiative and nonradiative shocks can be simply expressed in terms of the cooling column density $N_{\text{cool}} = n_0 v_s t_{\text{cool}}$, where t_{cool} is the cooling time behind the shock front. For example, for $10 \gtrsim v_{s7} \gtrsim 0.6$, Hollenbach & McKee (1979) estimated the cooling column density to 10^4 K to be

$$N_{\text{cool}}(T_s \to 10^4 \text{ K}) \simeq 2 \times 10^{17} v_{s7}^{4.2} \text{ cm}^{-2}, \tag{9}$$

independent of the ambient density n_0. A radiative shock is one in which the column density of the shocked gas N_{tot} is greater than N_{cool}.

In radiative shocks, the momentum jump condition (Equation 2a) determines the increase in density with the drop in temperature behind the shock front. The final compression is limited by gas pressure if the temperature stabilizes at a value T_f that is sufficiently high, by magnetic pressure if B becomes sufficiently large, or possibly by cosmic ray pressure if particle acceleration is sufficiently efficient (see Section 7.4). In a strong shock the maximum post-shock density n_m allowed by the compressed magnetic field is given by $(n_m/n_0)^2 B_{0\perp}^2/8\pi = \rho_0 v_s^2$ or

$$n_m = 77(n_0^{3/2} v_{s7}/B_{0\perp-6}) \text{ cm}^{-3} \tag{10}$$

where $B_{0\perp-6} = B_{0\perp}/10^{-6}$ G. In the ISM, the strength of the magnetic field is generally correlated with the density (Mouschovias 1976); for $B_{0\perp} \sim 10^{-6} n_0^{1/2}$ G the maximum compression is $n_m/n_0 = 77 v_{s7}$. Corresponding to the density n_m, one can define a characteristic temperature T_m by $n_m k T_m = n_0 \mu_s v_s^2$; for a fully ionized gas this gives $T_m = 10^4 (B_{0\perp-6} v_{s7} n_0^{-1/2})$ K. In terms of T_m, the perpendicular Alfvén Mach number is $\mathcal{M}_A = 3.8 T_s/T_m$. The condition that the cooled post-shock gas be magnetically supported rather than thermally supported is $T_f \lesssim T_m$. For $T > T_m$ the cooling behind the shock front is isobaric (provided the cosmic ray pressure gradient is negligible); at $T = T_m$ the density is within a factor 2 of its final value, so that for $T < T_m$ the cooling is approximately isochoric.

3 STRUCTURE OF SHOCK FRONTS

3.1 Collisional Shocks

The structure of collisional shock fronts (see Zel'dovich & Raizer 1966) is much better understood than that of collisionless shock fronts. In a perfect fluid, a shock front is a discontinuity. Thermal conduction in the absence of viscosity can provide a smooth transition through a shock front only for weak shocks; viscosity is essential to provide the required dissipation in strong shocks. A reasonably good approximate description of strong collisional shock fronts is provided by the Mott-Smith (1951) model: Two interpenetrating Maxwellian distributions with fixed temperatures, representing the shocked and unshocked fluids, interact via collisions which cause particles to go from the unshocked to the shocked distribution. The shock thickness Δ_s is of order $(n_s \sigma)^{-1}$, the mean free path in the shocked fluid, where σ is the effective hard sphere cross section, and n_s is the density of the shocked fluid.

In an unmagnetized, fully ionized gas the viscous ion shock is embedded in an electron thermal precursor driven by conduction which has a thickness $(m_i/m_e)^{1/2}$ times the ion shock thickness, where m_i is the ion mass; a significant amount of ion heating occurs in this precursor due to electron-ion collisions (Jaffrin & Probstein 1964). An electrical potential jump $e\phi \sim kT_s$ occurs across the shock front in order to prevent the shocked electrons from escaping upstream. If the gas is not fully ionized, the large ion-atom cross section usually keeps the ions and atoms at nearly the same velocity (Jaffrin 1965). For example, Dalgarno's (1960) results give $\sigma_{cx} = 5 \times 10^{-15} T_4^{-0.16}$ cm^2 for the H-H$^+$ charge exchange cross section in the range $0.1 \lesssim T_4 \equiv T/10^4$ K $\lesssim 100$.

The structure of collisional shocks in a weakly ionized ($n_e = 5 \times 10^{-4} n$) atomic plasma with a perpendicular magnetic field was investigated by Mullan (1971). He assumed $v_A \sim$ few km s^{-1}, $v_s \sim$ 5–10 km s^{-1}, and either H$^+$ or C$^+$ ions. Radiation and the effects of finite electrical resistivity were ignored. Because of the weak coupling between ions and atoms the increase in B and n_e occurred in front of and over a much greater distance than the increase in the atom density n_a. The thickness Δ_B of this magnetic/ionic precursor can be estimated by balancing the $\Delta B^2/8\pi$ force on the ions against the ion-atom drag (cf. Spitzer 1978): $\Delta_B \sim B_s^2/8\pi n_{is} n_{ao} m_H \langle \sigma v \rangle v_s$, where we set the relative ion-atom drift velocity equal to v_s. For modest $\mathcal{M}_A$ one has $n_a \Delta_B \sim (n_a/n_i) \sigma_{cx}^{-1} \sim 10^{14}(n_a/n_i)$ cm^{-2}. Mullan suggested that shocks provide an effective mechanism for removing magnetic flux from a collapsing protostar; since a shock displaces the ions and their associated magnetic flux a distance of only $(n_i \sigma_{cx})^{-1}$ relative to the atoms, a number of such shocks would be required. However, the effectiveness of this shock-driven ambipolar diffusion would be enhanced in molecular clouds, where both the ionization and the cross section are lower. Elmegreen et al. (1978) have shown that the gas well behind the shock front can be heated by ambipolar diffusion, which is enhanced by the shock compression.

3.2 Collisionless Shocks

Low density plasmas often have collisional length scales far larger than the length scales characterizing the collective behavior of the plasma. For example, the mean free paths for ion-ion and electron-electron collisions equal $8.4 \times 10^{14} v_7^4/n_e$ cm, where v is the particle velocity. By contrast, the Debye length is $6.9(T/n_e)^{1/2}$ cm, the ion Larmor radius is $r_{Li} = 1.0 \times 10^9 v_7/B_{-6}$ cm (where **v** is normal to **B**), and the ion inertial length is $c/\omega_{pi} = (v_A/v) r_{Li} = 2.3 \times 10^7 n_e^{-1/2}$ cm, where ω_{pi} is the ion plasma frequency. On both observational and theoretical grounds, it is expected that shock fronts in low density plasmas will be much thinner than a mean free path, the dissipation being due to turbulent collective electric and magnetic fields rather than collisions. Such shocks are termed collisionless.

The best-studied astrophysical collisionless shock is the Earth's bow shock, which occurs in the solar wind when the wind strikes the Earth's magnetosphere. Typical conditions are $v_{s7} \sim 3-5$, $n_e \sim 2-10$ cm^{-3}, $B_0 \sim 50$ μG, $T_{e0}/T_{i0} \sim 2-5$, $\mathcal{M}_A \sim 3-5$, and $\beta \equiv 8\pi nk(T_e + T_i)/B^2$ ranging from much less than one to greater than 100. Data from the ISEE (International Sun Earth Explorer) satellites (Russell & Greenstadt 1979, Bame et al. 1979) show that the bow shock has a well-defined structure of thickness $\sim$ few $c/\omega_{pi} \lesssim r_{Li}$ when the shock is quasi-perpendicular ($\theta_B \gtrsim 50°$, where θ_B is the angle between **v**$_s$ and **B**$_0$), whereas it is ill-defined and much thicker ($\Delta_s \sim 10^2 c/\omega_{pi}$) when it is quasi-parallel ($\theta_B \lesssim 45°$), at least in the one instance it was observed in this orientation. Profiles of T_e, n_e, and B are similar. Temporary field compressions up to a factor 13 were observed in quasi-perpendicular shocks with $\beta \gtrsim 2$. The electron temperature T_{es} behind the shock was generally much smaller than the mean in

Equation (8). The average jump was $T_{es}/T_{e0} = 2.7$, the value expected for adiabatic compression by a factor 4 (Gronenschild & Mewe 1979); the most probable value was $T_{es}/T_{e0} = 1.7$, and the peak value was 9.5. In principle, thermal conduction away from the shock front could lower T_e, but observations of the heat flux (Ogilvie & Scudder 1979) indicate that this is not the case. Caution should be exercised in applying results on the bow shock to other astrophysical shocks because (a) the transverse extent of the bow shock ($\sim 10^{10}$ cm) is far smaller than shocks associated with SNRs, etc., (b) the bow shock is open, extending along the magnetotail, rather than closed as SNRs are, and (c) the bow shock is subsonic relative to the upstream electrons, which is not always true for other astrophysical shocks.

The best evidence on much stronger collisionless shocks, with $\mathcal{M} > 10^2$, comes from studying the young SNRs, Cas A and Tycho. X-ray observations of Cas A with the Einstein X-ray Observatory (Murray et al. 1979) show a shell of X-ray emission coincident with a plateau of radio emission (Bell 1977) and lying outside a complex bright X-ray region. Murray et al. (1979) interpret this X-ray shell as emission from shocked ISM at a temperature of 4–6 $\times$ 10^7 K. Observations with HEAO-1 up to energies of 25 keV, far higher than the 4 keV limit of the Einstein Observatory, show that some of the emission from Cas A is characterized by a temperature of at least 10^8 K; Tycho has a component at $T \gtrsim 8.4 \times 10^7$ K (Pravdo & Smith 1979).

These high electron temperatures lead to two important conclusions. First, there is a substantial amount of collisionless heating of the electrons. Since the time for the electrons and ions to reach equipartition by collisions behind a strong shock is 2.1 $T_e^{3/2}/n_0$ s (Spitzer 1962), the maximum T_e due to collisions with the shock heated ions with $n_s = 4n_0 = 6$ cm^{-3} (Murray et al. 1979) in the 300 year life of Cas A would be 4×10^6 K, far lower than the 10^7 and 10^8 temperatures observed. (A corollary is that the X-ray emitting electrons need not be Maxwellian.) The maximum plausible value for v_s for Cas A is 6000 km s^{-1} (Kamper & van den Bergh 1976), which corresponds to no deceleration; the HEAO-1 data then imply $T_{es}/T_s \gtrsim 0.2$. For Tycho, Chevalier et al. (1980) estimate $v_s \simeq 2300$ km s^{-1} so that the mean temperature of the hot X-ray emitting electrons is about T_s; since Tycho has suffered substantial deceleration (Kamper & van den Bergh 1978), this mean temperature could exceed T_{es} somewhat.

The second conclusion, which follows from the first, is that collisionless shocks exist even under the extreme conditions present in these young SNRs. In principle, a thin collisional shock based on charge exchange in a magnetized medium could occur, but it could not produce the high T_e that is observed. From the spherical symmetry of the X-ray emission, the shocks appear to surround the SNRs; hence they exist under a variety of conditions and are not localized to particular directions. The X-ray observations also show that the shocked electrons are confined to the interior of the shock; there is no large thermal wave preceding the shock front as suggested by Chevalier (1975a).

The theory of collisionless shocks is very incomplete. Little progress appears to have been made since the reviews by Tidman & Krall (1971) and Biskamp (1973), and we draw heavily upon their discussion below. Two main types of collisionless shock can be distinguished: *laminar* shocks, in which the density, magnetic field, etc. are steady in time though perhaps oscillatory in space, and *turbulent* shocks. Laminar shocks correspond to solitons (nonlinear solitary waves) with dissipation. Their thickness is set by dispersion effects: a nonlinear wave steepens until it reaches a characteristic size set by the dependence of the phase velocity on wavelength (i.e., by the dispersion relation for the wave). Laminar shocks exist only at lower Mach numbers ($\mathcal{M} < \mathcal{M}^*$); above $\mathcal{M}^*$ the wave breaks and becomes turbulent. For example, ion acoustic solitons have $\mathcal{M}^* \simeq 1.6$ and magnetosonic solitons propagating perpendicular to B have $\mathcal{M}^*_{\rm ms} \lesssim 2$, where we define the magnetosonic Mach number as $\mathcal{M}_{\rm ms} \equiv v_s [v_A^2 + k(T_e + T_i)/m_i]^{-1/2}$.

In turbulent shocks, the structure is determined by the nonlinear state of one or more plasma instabilities. First consider the case of no magnetic field, which is only of academic interest for interstellar plasmas. The existence of high Mach number shocks in this case is unproven. The most obvious dissipation mechanism is the two stream instability between the shocked and unshocked ions, but this instability is shorted out by the electrons unless they are extremely hot ($T_e > 1.5\ T_s$ for hydrogen; e.g. McKee 1970). Large amplitude electrostatic waves can reduce the required electron temperature somewhat (Papadopoulos & McBride 1973). An alternative possibility is that the dissipation is provided by the electromagnetic streaming instability (Weibel 1959, Davidson et al. 1972), which involves the formation of current filaments rather than charge bunches. Although at present there is no demonstrated mechanism for forming collisionless shocks at high $\mathcal{M}$ with $B = 0$, it is important to remember that the collisionless length scales are generally so much smaller than the collisional mean free path that it is possible in principle for a very weak instability to govern the shock structure.

In the case of a perpendicular magnetic field, a collisionless shock must form: since the ions are gyrating around the magnetic field, it is impossible to have two ion beams streaming through each other perpendicular to B on a scale larger than the ion gyroradius. One suggestion for the shock structure in this case is the modified two stream instability, in which the presence of B prevents the electrons from shorting out the ion-ion two stream instability (Papadopoulos et al. 1971a,b). The shock thickness is $\Delta_s \sim 3(m_e/m_i)^{1/2} r_{Li}$. The theory breaks down for $\mathcal{M}_{ms} > \sqrt{8}$; in principle this instability can supply part of the shock dissipation at higher Mach numbers since the local value of $\mathcal{M}_{ms}$ decreases through the shock front.

The existence of high Mach number shocks propagating along B is unproven, but very likely in view of the large number of possible instabilities. Parker (1961) and Kennel & Sagdeev (1967) have suggested that the counterstreaming ion beams could drive the firehose instability. Cipolla et al. (1977) have shown that such ion beams drive whistlers unstable.

The problem of determining the amount of electron heating in a collisionless shock front is a thorny one. In a strong shock the dominant interaction is between the ions since it is their energy that must be thermalized. There are a number of instabilities connecting the electrons and ions, such as the ion acoustic instability, which can result from current driven by the compression of the magnetic field. Electrons can also be heated by turbulent low frequency, large amplitude electrostatic waves generated by the ion-ion instability: although the electrons are nearly adiabatic as they cross any one wave, they can in principle interact with enough waves in crossing the shock front that significant heating results. For example, in one simulation of the one-dimensional ion-ion instability McKee (1970) found an electron heating of $\Delta T_e = 0.3\ T_s$ in a case in which $T_{eo} \sim 2\ T_s$. McKee (1974) argued that approximate equipartition between electrons and ions should occur behind very strong collisionless shocks ($T_{es} \sim T_{is} \sim T_s$), although no conclusive evidence was presented. Resonant interaction with high frequency waves can result in more efficient electron heating, superthermal electron generation, and nonthermal radiation at the plasma frequency and its first harmonic. Smith & Krall (1974) argued that such waves do not occur in the shock front, but Lampe & Papadopoulos (1977) have challenged their conclusion.

In summary, the Earth's bow shock is an ideal laboratory for studying moderate Mach number ($\mathcal{M} \lesssim 10$) collisionless shocks, although its small extent ($\lesssim 10^{10}$ cm) compared to other astrophysical shocks may lead to some significant differences. In the quasi-perpendicular case, the thickness of the bow shock is of order the ion inertial length c/ω_{pi}, whereas in the quasi-parallel case it is about 100 times larger. X-ray observations of young SNRs show that collisionless shocks exist at much higher Mach numbers as well and that significant collisionless heating of the electrons occurs ($T_{es} \gtrsim 0.2\ T_s$ and possibly $T_{es} \sim T_s$). A variety of instabilities have been suggested as the source of the dissipation in collisionless shocks, but at present our understanding of the structure of these shocks at high Mach numbers is minimal. The existence of arbitrarily strong collisionless shocks propagating perpendicular to B is definite, parallel to B is almost definite, and in the absence of B is likely. The role of collisionless shocks in accelerating particles to relativistic energies will be discussed in Section 7.4.

4 NONRADIATIVE SHOCKS

Shocks are nonradiative if the shocked column density $N \sim n_0 v_s t$ is less than the characteristic cooling column density N_{cool} (see Section 2), so that the emitted radiation does not affect the dynamics of the gas behind the shock front. Shocks are nonradiative because v_s is large (as in young SNRs), n_0 is low (as in stellar winds in early type stars), t is small (as in young SNRs), or the shocked gas is convected away (as in the case of the shocks associated with the interaction between the solar wind and the ISM). Shocks associated with SNRs, which have been extensively studied recently, are reviewed here; the dynamics of winds in early type stars are discussed by Weaver et al. (1977) and the solar wind shock is discussed by Holzer (1972).

Young SNRs have shock velocities $v_s \gtrsim 10^3$ km s^{-1} and primarily radiative X rays. Recently, however, Chevalier & Raymond (1978) and Chevalier et al. (1980) have demonstrated that a detectable amount of Balmer line emission will result if the shock propagates into a partially neutral gas. The existence of the neutral atoms near the SNR limits the ionizing luminosity of the supernova; radiation from the shock itself cannot pre-ionize the gas if N/N_{cool} is sufficiently small (see Section 5). The neutral atoms are unaffected by the passage of the collisionless shock, and since they have the same velocity as the ambient medium, they are termed "slow" atoms. Behind the collisionless shock a population of "fast" atoms is built up by charge exchange with the hot shocked ions. Both populations of atoms are subject to destruction by collisional ionization by the shocked electrons and ions; the slow atoms are also destroyed by charge exchange since the resulting ions are picked up and accelerated by the magnetic field in the shocked plasma. The ratio of fast to slow atoms is $\langle \sigma_{\text{cx}} v \rangle / \langle \sigma_{\text{ion}} v \rangle$, where σ_{ion} is the ionization cross section. This is a sensitive function of shock velocity: at high v_s slow atoms dominate since σ_{cx} is small, whereas for $v_s \lesssim 2 \times 10^3$ km s^{-1} the fast atoms dominate. Narrow and broad emission line components are produced by collisional excitation of slow and fast atoms, respectively; charge exchange into excited states adds to the intensity of the broad components. Since the intensities depend primarily on the number of excitations per ionization, they are relatively independent of T_e for $T_e \gg 10^5$ K. Forbidden lines are expected to be weak compared to H and He lines since the collision strengths are small at the high electron temperatures behind the shock. Effects of the cool electrons produced by ionization of the neutrals and effects of cascading from higher excited states have not been included in the estimates of forbidden line strengths, however. Their results successfully account for observations of the optical spectrum of Tycho and suggest a revised distance of 2.3 kpc (compared to the canonical 6 kpc) and a shock velocity of 2300 km s^{-1}. Their results can also be applied to older SNRs, such as the Cygnus Loop, where they identify the high velocity clouds observed in Hα by Kirshner & Taylor (1976) as shocks propagating through clouds of partially neutral gas. Because the velocities are low ($\lesssim 300$ km s^{-1}), the spectrum should be dominated by the broad component, in accord with observation. The main prediction of this charge exchange model is that the forbidden lines should be weak, in contrast to the model in which the clouds have been hydrodynamically accelerated by the blast wave and are being photoionized by radiation from the conductive interfaces (McKee et al. 1978). Recently, a faint filament 4'–6' outside the bright filaments in the Cygnus Loop has been observed to have only Balmer emission in the range 4500–6900 Å (Raymond et al. 1980b), supporting the charge exchange model.

The fast atoms are also detectable in absorption (Cowie et al. 1979, Jenkins et al. 1980). For stellar background sources, the shock velocity must be large enough to shift the fast atoms' absorption out of the corresponding stellar absorption features. For Lα and Lβ, the required velocities are about 400 and 200 km s^{-1}, respectively. The expected HI column density is about $v_s x_0(\text{HI})/4\langle \sigma_{\text{ion}} v \rangle \sim 10^{14.5-15.5} x_0(\text{HI})$ cm^{-2} for $100 < v_s < 5000$ km s^{-1}, where the pre-shock HI fraction $x_0(\text{HI})$ depends on both N/N_{cool} and v_s. HI absorption lines are also detectable in the radiatively cooling gas farther behind the shock front, and Cowie et al. (1979) have detected Lδ and Lε lines in Orion at $v \lesssim 100$ km s^{-1}.

X-ray emission from interstellar shocks has primarily been studied for the nonradiative phases of SNR evolution. The X-ray continuum is largely determined by T_{es} (Gorenstein et al. 1974) and the variation of T_e behind the shock front under the competing effects of ion heating (if $T_{es} < T_s$) and adiabatic expansion (Itoh 1977). The X-ray line spectrum is determined primarily by time-dependent ionization (Gorenstein et al. 1974), which has been observationally indicated in Vela X (Moore & Garmire 1976), the Cygnus Loop (Stark & Culhane 1978), and Cas A (Pravdo & Smith 1979). Detailed calculations of the time-dependent ionization and resulting X-ray spectrum by Itoh (1979) and Gronenschild & Mewe (1979) show that the characteristic ionization temperature T_{ionzn} — as determined from those ions with the highest concentrations — can be several times less than the actual electron temperature T_e. This is the reverse of the time-dependent cooling calculations of Shapiro & Moore (1976), for example, in which $T_e < T_{ionzn}$. As a result, the emitted spectrum has a higher line/continuum ratio for $\lambda \gtrsim 20$ Å and is softer than an equilibrium spectrum at the same T_e. Itoh (1979) finds that the temperature inferred from the line-dominated portion of the spectrum is virtually constant at about 2.5×10^6 K for SNRs between 2500 and 15,000 years old, which is up to three times smaller than the actual temperature (cf. Gronenschild & Mewe 1979).

The X-ray emitting plasma in nonradiative shocks can also be studied in the optical region of the spectrum through observations of coronal line emission. The [Fe X]λ6374 line has been observed in the Cygnus Loop (Shcheglov 1968) and Puppis A (Lucke et al. 1979); the generally stronger [Fe XIV]λ5303 line has been observed in the Cygnus Loop (Woodgate et al. 1974, 1977), Vela X (Woodgate et al. 1975), Puppis A (Lucke et al. 1979), and in the LMC (Dopita & Mathewson 1979). In the Cygnus Loop, the [Fe XIV] emission extends about 5' or 1 pc beyond the bright optical filaments (Woodgate et al. 1977), as does the X-ray emission (Tuohy et al. 1979).

The effects of inhomogeneities in the ISM present major difficulties in interpreting both the optical and X-ray spectra from the hot plasma behind SNR shocks, and as yet these effects have not been taken into account in an adequate manner. Not only do regions of different temperatures, etc., contribute to the emission along a given line of sight, but the thermal evaporation of clouds by a hotter ambient medium can alter both the dynamics of the SNR (McKee & Ostriker 1977) and the emissivity of the plasma (Chevalier 1975b, McKee & Cowie 1977, Tsunemi & Inoue 1979). Observations with high spatial resolution from the Einstein X-ray Observatory should help unravel these effects.

Nonradiative shocks can produce an appreciable infrared luminosity from dust grains embedded in the hot, shocked gas (Ostriker & Silk 1973, Silk & Burke 1974). Only a small fraction ($\sim 10\%$) of the energy of the hot gas can be radiated by the dust grains before the grains are destroyed by sputtering. In an extensive analysis of destruction mechanisms for interstellar dust, Draine & Salpeter (1979) showed that refractory grains are destroyed by SNR blast waves with $v_s > 300$ km s^{-1}.

5 RADIATIVE ATOMIC SHOCKS

A shock enters its radiative phase when the downstream shocked gas has sufficient time to cool significantly, i.e., when $N > N_{cool}$. In atomic shocks with pre-shock ionization levels that are consistent with the radiation field produced by the shock itself (see below) and the ambient interstellar UV field, one finds

$$N_{cool} \simeq 10^{19}-10^{20} \text{ cm}^{-2}, \quad 0.01 < v_{s7} < 0.2$$
$$N_{cool} \simeq 6.4 \times 10^{15} v_{s7}^{-4.0} \text{ cm}^{-2}, \quad 0.2 \lesssim v_{s7} \lesssim 0.6$$

(11)

and N_{cool} for $0.6 \leq v_{s7} \leq 10$ is given by Equation (9). The intermediate and low velocity column densities are taken from numerical results of Shull & McKee (1979) and D. J. Hollenbach and C. F. McKee (in preparation). Generally SNRs have not swept up enough material to enter the radiative phase until $v_{s7} \lesssim 2.2\, n_0^{0.12} E_{51}^{0.06}$; winds from early-type stars drive radiative shocks into *neutral* gas when $v_{s7} \lesssim 0.2(n_0 L_{w36}^3/L_{48}^{*2})^{0.2}$ (see Table 1 and Hollenbach et al. 1976). However, radiative shocks with v_{s7} as high as 20 and $N \simeq 10^{24}$ cm^{-3}, resulting from massive cloud-cloud collisions, have been proposed as sources for emission lines in QSOs (Daltabuit & Cox 1972 and Daltabuit et al. 1978).

Inelastic collisional and radiative processes intricately interact in the post-shock relaxation layer which characterizes a radiative shock. Inelastic collisional processes sputter and heat dust grains and act to ionize, excite, cool, and eventually recombine the atoms and ions. The resultant emission from collisionally excited ions radiates both upstream and downstream and transfers energy from the hot post-shock region to these cooler regions. Under suitable conditions, the upstream "precursor" field photoionizes the pre-shock gas and may photodesorb or heat and sublimate the icy mantles from grains. Likewise, the downstream radiative flux can photoionize a denser Stomgren layer, extending the recombination zone at $T \simeq 10^4$ K to $N \simeq 10^{18}$–10^{19} cm^{-2} and can dominate the heating of grains and the ionization of metals to $N \simeq 10^{21}$ cm^{-2}.

In this section we shall focus on atomic line emission behind $v_s \simeq 30$–200 km s^{-1} shocks since their optical emission has been observed. Pikel'ner's (1954) pioneering study of this problem is discussed by Kaplan (1966). Another classic early paper on atomic interstellar shocks by Field et al. (1968) treated $v_s = 4$–16 km s^{-1} shocks incident upon $n_0 = 10$ cm^{-3} atomic clouds and calculated the resultant weak infrared line strengths produced by atomic fine structure lines and trace quantities of H_2. Hill & Hollenbach (1978) treated higher density ($n_0 \simeq 10^{3-4}$ cm^{-3}) atomic shocks in the same low velocity regime. Cox (1972) was the first to use the hydrodynamic flow equations outlined in Section 2 to calculate the optical spectrum from a fast $v_s \simeq 100$ km s^{-1} shock. He found that the optical recombination lines are dominated by the effects of photoionization due to emission from the hot gas just behind the shock front, in accord with Parker's (1967) suggestion. However, his assumption that the initial post-shock ionization was given by collisional equilibrium at T_s overestimated the level of ionization and underestimated the cooling rates. Dopita (1976, 1977) and Raymond (1976, 1979) extended Cox's work by starting from an assumed pre-shock ionization and following the resultant post-shock ionization and recombination of ionic species. Dopita systematically studied the effects of varying the metal abundances, but he ignored the important cooling by the $\lambda 304$ line of He$^+$. Shull & McKee (1979) improved upon these calculations by iterating to obtain self-consistent pre-shock ionization levels created by the precursor field as well as including important charge exchange mechanisms (see also Butler & Raymond 1980). Unless otherwise noted, the following discussion in Sections 5.1–5.4 is from the work of Raymond and Shull & McKee. Most of the relevant collisional and radiative rates can be found conveniently tabulated in Raymond (1976).

5.1 Excitation Processes for Atomic Lines and Post-Shock Cooling

Generally, inelastic electronic collisions are the dominant post-shock cooling mechanism because of their higher cross sections and velocities, but protons have larger rate coefficients for exciting neutral fine structure lines [e.g., OI(63μ)] and hydrogen atoms are the dominant collisional agents in low ionization gas. In addition to collisions, recombinations of ionized hydrogen and helium are a significant source of the line emission from these species.

In high temperature ($10^4 < T < 5 \times 10^5$ K) gas the dominant cooling is provided by the collisional excitation of permitted lines of H and various ionic species of He, C, N, and O. Because of their higher abundances, H and He$^+$ often dominate when they are not completely ionized. The precursor field from a 100 km s^{-1} shock ($T_s \simeq 1.4 \times 10^5$ K) yields fractional abundances $x(\text{H}^0) \simeq 0.3$, $x(\text{He}^0) \simeq 0.7$, and $x(\text{He}^+) \simeq 0.3$ in the pre-shock gas. Clearly, then, nonequilibrium cooling will initially be much larger than equilibrium cooling rates which would assume the hydrogen and helium to be nearly completely ionized at $T_s = 1.4 \times 10^5$ K. In addition, hydrogen and helium can radiate considerable energy before becoming collisionally ionized. At $T = 1.4 \times 10^5$ K, the He$^+$ radiates 290 eV at 304 Å for every ionization to He^{++}. Therefore, the pre-shock ionization is a crucial parameter in determining the post-shock structure. Lyman continuum and HeII 304 photons are the dominant photoionizing agents.

Since the dominant coolants in high temperature gas are permitted or semi-forbidden transitions, collisional de-excitation is unimportant, and the cooling rate is proportional to n^2, the cooling time proportional to n^{-1}, and the cooling column density to n^0 (see Equation 11). The change in the form of the cooling column density at $v_{s7} = 0.6$ is caused by the small rates of excitation and ionization of hydrogen atoms by collisions with other hydrogen atoms compared with the electron excitation rates. At temperatures of $T \simeq 10^4$ K, H atom rates are 10^{-4}–10^{-5} of the electron rates according to Drawin (1969). Therefore, as the peak ionization fraction behind the shock drops below 0.35 for $v_{s7} \lesssim 0.6$, the cooling time and column density increase significantly.

For atomic gas with $T \lesssim 5 \times 10^3$ K, found downstream in shocks with

$v_{s7} \gtrsim 0.2$ and directly behind the shock front in slower shocks, the cooling is mediated by fine structure transitions of the metals. These spin-flip transitions have low spontaneous transition rates, and the excited states can be collisionally de-excited in the post-shock gas. Assuming that the interstellar FUV field ionizes all metals with ionization potentials below 13.6 eV and that abundances are roughly solar, the dominant cooling for post-shock densities $n \lesssim 10^4$ cm^{-3} comes from the CII (156 μ) and the Si II (35 μ) lines. At higher densities the OI (63 μ) line dominates the CII (156 μ) line, which collisionally quenches. Furthermore, as will be discussed below, there are important chemical and photoionization heating sources. In contrast to the situation in the high temperature gas, the cooling column density now becomes a complicated function of n_0 and cannot be readily given a simple form. Roughly, the cooling column density to 100 K lies between 10^{19} cm^{-2} and 10^{20} cm^{-2} for $n_0 \lesssim 10^3$ cm^{-3} (D. J. Hollenbach and C. F. McKee, in preparation).

5.2 *Radiative Transfer Processes*

Shocks with $v_{s7} \gtrsim 0.2$, $T_s \gtrsim 10^4$ K, produce substantial intensities of visible and FUV radiation, and those with $v_{s7} \gtrsim 0.5$ produce significant numbers of Lyman continuum photons. The heating and ionization rates of these photons in both the upstream and downstream gas are determined by the mean intensity of the radiation J. The radiation field is highly anisotropic because the lateral extent of the shock L is much greater than the thickness of the relaxation layer D. The anisotropy can be characterized by the parameter $\mathcal{R}_{max} = 4\pi J_s/F_s$, where F_s is the flux at the shock front $\mathcal{R}_{max}$ depends on the internal absorption in the emitting slab of shocked gas and on L/D. Cox (1972) and Shull & McKee (1979) used $\mathcal{R}_{max} = 3$, whereas Raymond also considered $\mathcal{R}_{max} = 1$ and 0.3, corresponding to $L/D = 5$ and 2, respectively, for negligible internal absorption.

The level of ionization of hydrogen produced by the ionizing precursor ranges from 0.012 at $v_{s7} = 0.6$ to 1 at $v_{s7} = 1.1$ (Shull & McKee 1979). At sufficiently high velocities the precursor ionization will drop below unity because the post-shock column density is inadequate to produce enough ionizing photons; for an adiabatic blast wave we estimate that this occurs for $v_{s7} > 4.1(E_{51}n_0^2)^{0.068}$.

Resonant scattering in the lines H(np–1s), He^0(n^{1}P–1^{1}S), and He$^+$(np–1s) increases the mean intensity of these photons in the region around the emitting slab and decreases their penetrating power. It also makes the intensity of these and other UV resonance lines dependent upon the orientation of the shock (Raymond et al. 1980a). Much of the shock energy ultimately emerges as Lyα and is resonantly scattered until absorbed by grains in a column density $N \simeq 10^{20}$ cm^{-2}. The Lyα is the prime heat source for both pre-shock and post-shock grains if $v_{s7} \gtrsim 0.2$ and will evaporate icy matter in pre-shock grains if $n_0 v_{s7}^3 \gtrsim 4 \times 10^4 \, a_\mu$, where a_μ is the grain radius in microns; photodesorption by Lyα removes icy mantles on pre-shock grains if $v_{s7} > 1.3(5 \times 10^{-3}/Y_{pd})^{1/2} a_\mu^{1/2}$ where Y_{pd} is the photodesorption yield (Hollenbach & McKee 1979).

Photoionization of atoms and ions proceeds downstream from the emitting slab as well as in the pre-shock gas. The effect, ultimately, is to maintain the ionization level above that of collisional equilibrium at the local temperature, the reverse of the situation in the immediate post-shock gas. For example, in a $v_{s7} = 1$ shock the HeII $\lambda 304$ photons keep oxygen doubly ionized down to $T \simeq 30,000$ K, increasing the intensity of the OIII lines, lowering the temperature which they indicate, and weakening the OII lines. Furthermore, while cooling to 10,000 K has been achieved in $N_{cool} \simeq 10^{17}$ cm^{-2}, the photoionization of recombining gas at $T \simeq 5,000$–10,000 K maintains $T \simeq 10^4$ K to $N \simeq 10^{18}$–10^{19} cm^{-2}, the reciprocal of the photoionization cross section of neutral hydrogen for $\lambda \simeq 300$ Å photons.

5.3 Observational Characteristics of Radiative Shocks

Historically, the optical signature of a shocked ($v_{s7} \gtrsim 0.6$) nebula compared with HII regions and planetary nebulae has been (a) the appearance of strong lines from low excitation species like OI, OII, NI, NII, SII, and CaII compared with Hβ and (b) high values of temperature-sensitive ratios like [OIII] $\lambda 4363/\lambda 5007$. The former derives from the fact that the post-shock gas has a small fully ionized zone ($\sim N_{cool} \simeq 10^{17}$ cm^{-2}), and a large recombination zone ($N_{rec} \simeq 10^{19}$ cm^{-2}). This contrasts with HII regions and planetary nebulae where the fully ionized zone, $N \simeq 10^{20}$–10^{21} cm^{-2}, dominates the recombination zone. Hence, in HII regions Hβ is primarily produced in the fully ionized zone, whereas behind shocks Hβ is only produced in the recombination zone along with the low excitation lines. The temperature-sensitive ratio [OIII] $\lambda 4363/\lambda 5007$ is an indication of the high post-shock temperature ($T \simeq 30,000$ K) where lines of high excitation are produced compared with the $T \lesssim 10^4$ K of photoionized gas.

For shocks with speeds $v_{s7} > 0.6$, the hydrogen and helium optical lines are mostly produced by recombination with little contamination from collisional excitation. Therefore their ratios do not differ significantly from HII regions. However, slower shocks with $v_{s7} < 0.6$ do not ionize appreciable hydrogen, the Balmer lines are significantly collisionally excited, and the Hα/Hβ and Lα/Hβ ratios increase. Shull & McKee (1979) give Hα/Hβ = 4.2 for $v_{s7} = 0.4$, in contrast to the recombination value Hα/H$\beta \simeq 3.0$.

Table 2 presents specific examples of optically observed interstellar nebulae including HII regions, planetary nebulae, SNRs, HH objects, and

Table 2 Optical emission line ratios in nebulae[a]

Lines	OI 6300+ 6363	OII 3726+ 3729	OII 7320+ 7330	OIII 4363/ (4959+ 5007)	NI 5198+ 5200	NII 6548+ 6584	SII 4069+ 4076	SII 6717+ 6731	Hα 6563	Ref.[e]
$n_{cr}(T=10^4 \text{ K})$[b]	1(6)	3(3)	6(6)	7(5)	2(3)	8(4)	4(6)	6(3)		
Photoionized Regions										
ηCar (HII)	1	155	3	0.004	<1	59	<1	24	316	1
Ori (HII)	1	204	6	0.008	1	95	2	24	282	2
NGC 7662 (PN)	<1	13	2	0.01	<1	12	1	1	258	3
Possible Shocked Regions										
N49 (SNR)	167	648	—	0.05	8.3	108	36	362	296	4
Cygnus Loop (SNR)	62	1300	52	0.06	—	345	19	284	308	5
HH1[c]	61	186	—	—	4.5	151	67	191	266	6
HH43S	630	—	—	—	160	184	122	1235	720	7
Burnham's Nebula	171	171	—	—	22	189	67	277	340	8
NGC 1502	199	800	40	0.06	20	400	45	201	286	9
Radiative Shock Model										
$v_s = 80$ km s^{-1}	42[d]	196	8.4	0.01	34[d]	123	10[d]	151[d]	346	10
$v_s = 130$ km s^{-1}	242	855	41	0.04	170	395	46	601	298	10

[a] All line intensities scaled to Hβ = 100, except the [OIII] λ 4363/(λ4959 + 5007) ratio.
[b] Collisional suppression of the line important above n_{cr}; 1(6) = 1 × 10^6.
[c] [OII] λ 3726 + 3729 and [SII] λ 6716 + 6731 are somewhat collisionally suppressed.
[d] These ratios have a minimum at $v_s = 80$ km s^{-1} and increase rapidly at lower velocities. $n_0 = 10$ cm^{-3} and $B_0 = 10^{-6}$ G are the pre-shock conditions for the calculations at both shock velocities.

[e] References
1. Peimbert et al. 1978: position 3a
2. Peimbert & Torres-Peimbert 1977: position 3a
3. Kirkpatrick 1972
4. Osterbrock & Dufour 1973
5. Miller 1974: position 1
6. Böhm et al. 1976
7. Dopita 1978b
8. Schwartz 1975
9. Fosbury et al. 1978
10. Shull & McKee 1979

nebulae surrounding T Tauri stars as well as with two theoretical models of shocks with $v_{s7} = 0.8$ and 1.3. Only those observed lines or line ratios that are particularly sensitive to the presence of radiating shock waves are included. Clearly, the supernova remnants, HH1, and Burnham's nebula show characteristic shock features.

One of the ultimate goals in studying theoretical models of shock emission lines is to be able to deduce pre-shock density, shock velocity, and metal abundances from relative line strengths. Dopita (1977) discusses a scheme to obtain these parameters and applies it (Dopita et al. 1977) to calculating metal abundances in supernova remnants seen in SMC, LMC, and the Galaxy. Raymond (1979) cautions about the difficulty in disentangling these parameters from the subtle effects of magnetic field, pre-shock ionization, and lateral extent of the shock. Furthermore, observed regions may include a spectrum of shock velocities and pre-shock densities, rendering the deduction of a single parameter, like abundance, even more difficult.

5.4 Shock Modeling of Observed Nebulae

5.4.1 SUPERNOVA REMNANTS The optical emission line spectra of a number of SNRs have been carefully observed and modeled with shock parameters. The Cygnus Loop filaments were among the first to be extensively observed, and Cox (1972) used Parker's (1967) observations to derive shock parameters $n_0 \simeq 6$ cm^{-3} and $v_s \simeq 100$ km s^{-1}. The observations of Miller (1974) of three positions in the Loop were used by Raymond (1979) to obtain the similar result $n_0 \simeq 10$ cm^{-3} and $v_s \simeq 90$ km s^{-1}. Raymond found that [OIII]/Hβ at Miller's position 3 is significantly larger than predicted by the shock calculations, which he attributed to deviations from steady flow behind the shock front.

Recently, satellite UV observations have permitted the first observation of the many strong UV lines emitted by shocks in SNRs such as the Cygnus Loop. IUE observations of Miller's position 1 by Benvenuti et al. (1979) indicated a shock velocity of 120 km s^{-1}, somewhat higher than Raymond's optical estimate. They inferred that C^{+3} was more depleted than C^{+2}, which in turn may have been more depleted than C^+, and they suggested that this reflected the injection of carbon atoms from sputtering and shattering graphite grains (see, for example, Draine & Salpeter 1979). Subsequently, Benvenuti et al. (1980) found an improved fit between shock models and the data by assuming a range of shock velocities between 90 and 130 km s^{-1}. Observations of the Loop down to 912 Å were made by Shemansky et al. (1979) on Voyager, allowing detection of CIII λ977 and CII λ1037. Benvenuti et al. (1980) suggest that other strong lines in their data are the Lyman series of hydrogen, shifted by about 8 Å due to instrumental problems. Further observations of the Loop were made by Raymond et al. (1980a), who also found a higher shock velocity ($v_s \simeq 130$ km s^{-1} at Miller's position 3) than had been inferred from the optical data. Assuming a single shock velocity (in contrast to Benvenuti et al. 1980), they argue that their observations strengthen the case for nonsteady flow behind the shock. They make the important point that finite optical depth in the resonance lines makes the observed line intensities dependent upon the orientation of the shock relative to the line of sight, and they present observational evidence for this effect for CII λ1335. This effect significantly complicates the interpretation of UV data on shocks, such as in the case of searching for evidence of grain destruction in shocks.

Osterbrock & Dufour (1973) and Osterbrock & Costero (1973) observed N49, a SNR in the LMC, and Vela; applying Cox's (1972) models, they obtained $v_s \simeq 100$ km s^{-1} for N49 and $v_s \simeq 141$ km s^{-1} for the Vela remnant. Dopita (1976, 1977) applies his own model to N49 to obtain $n_0 = 56$, $v_s = 130$ km s^{-1} and slightly depleted abundances of N, O, S, and Ne. However, later improvements in Cox's and Dopita's model by Shull & McKee (1979), Raymond (1979), and Butler & Raymond (1980) make these parameter choices somewhat uncertain.

Dopita et al. (1977) observed a number of SNRs in the Galaxy and the Magellanic Clouds. These observations, along with the previous observations of Cygnus, Vela, and N49 already referenced, were modeled with Dopita's shock program to determine n_0, T_s, and the abundances of O, N, and S. With a total of five remnants in the Galaxy, eight in the LMC, and one in SMC, they find a strong gradient in chemical evolution, increasing from the SMC to the LMC to the Galaxy, consistent with theories of enrichment (Talbot & Arnett 1973). Oxygen and sulfur increase by a factor of 5 and nitrogen by a factor of 25 from the SMC to the Galaxy. However, there is a great deal of scatter in twelve of the fourteen remnants, and much of the observed gradient is due to the high metal abundance of Pup A in the Galaxy and the low metal abundance observed in N19, the lone SMC object.

In several SNRs there is evidence for slow, radiative shocks which are driven into ambient clouds or filaments by a fast, nonradiative blast wave (Bychkov & Pikel'ner 1975, McKee & Cowie 1975; see the review by Chevalier 1977a). In Cas A the shocked clouds appear as "quasi-stationary flocculi" with velocities of -65 to -450 km s^{-1} (R. Minkowski, unpublished; van den Bergh 1971), which have been numerically modeled by Sgro (1975). In the Cygnus Loop, the bright optical filaments ($v \sim 100$ km s^{-1}) are attributed to shocked clouds, while the X-ray emission is attributed to the nonradiative blast wave ($v_b \sim 450$ km s^{-1}). In both cases, pressure balance gives the velocity v_c of the cloud shock as $v_c \simeq v_b(n_0/n_c)^{1/2}$ where n_c

is the pre-shock cloud density. Shocked clouds have also been observed in Vela (Jenkins et al. 1976a,b) and in the LMC SNRs (Dopita 1979, Dopita & Mathewson 1979). However, in the remnant S147, Kirshner & Arnold (1979) have found that, contrary to this picture, the high density knots are not moving slower than the low density knots and that the sharp filaments are apparently rope-like rather than sheets, since the latter would have low velocities.

SNRs can also be observed in absorption in the UV if a bright early-type star is found behind them. Such measurements provide column densities and velocities of various ionization stages of C, N, Si, S, and Fe. One of the main features of these observations is that the abundance of Si, Fe, and Ca (observable in the optical) is higher in intermediate and high velocity ($v \gtrsim 30$ km s^{-1}) gas than in low velocity gas. These results are consistent with models of sputtering and shattering of grains behind $v_s > 30$ km s^{-1} shocks (see reviews of the observational and theoretical evidence for grain destruction behind shocks in McCray & Snow 1979 and Hollenbach & McKee 1979). Jenkins et al. (1976a,b) have constructed a detailed model of the Vela remnant from absorption measurements toward several stars behind the remnant. Raymond (1979) discusses the difficulty in modeling the Vela observations with shock models: the angle between the line of sight and the flow direction is unknown, and the assumption of steady flow used in all shock models may be inappropriate. (The models predict far too much OI and NI, which is also a problem in the Cygnus Loop.) One component seen at -90 km s^{-1} contains OVI, implying a shock faster than 150 km s^{-1} seen at an angle, or significant ionization by X rays or cosmic rays produced by the remnant.

5.4.2 HH NEBULAE AND NEBULAE ASSOCIATED WITH T TAURI STARS A second important class of objects that are likely to be shocked nebulae are certain HH objects and nebulae associated with T Tauri type stars. Schwartz (1975) has shown that Burnham's Nebula, which is associated with T Tauri, has an emission spectrum very similar to HH1 and somewhat similar to the supernova remnant N49 (see Table 2). Of particular note are the strong OI, OII (the $\lambda 3726$ and $\lambda 3729$ lines are probably collisionally suppressed), NI, NII, and SII lines as well as a number of weaker low excitation lines of CaII, MgI, and FeII that are not ordinarily found in appreciable strength in gaseous nebulae. Schwartz concludes that Burnham's Nebula arises from a wind from T Tau which drives a 100 km s^{-1} shock into ambient intracloud material of density $n_0 \simeq 100$ cm^{-3}. Furthermore, objects like HH1 may be younger versions of T Tau, with such high mass loss rates as to obscure the central star, leaving only shocked clouds in the vicinity—the HH nebulosities—to be optically observed. Maran et al. (1979) propose a 200 km s^{-1} shock in T Tau whose radiation may pump the observed $v = 1 \to 0S(1)$ infrared transition of H$_2$ in T Tau (Beckwith et al. 1978b) and may produce the anomalous OI $\lambda 6300$ to Hα line ratio.

Schwartz (1978) has amplified his original discussion with detailed observations of HH1 and HH2 and has presented a global model for the excitation of some, if not all, HH objects. A wind from a nearby star is incident upon cloudlets, accelerating them and setting up standing bow shocks on the star side of the cloudlets. The bow shocks produce the optical radiation. For the knots composing HH1 and HH2, a characteristic stellar wind velocity of about 100 km s^{-1} and a mass loss rate of $\simeq 10^{-5}$–10^{-6} $M_\odot$ yr^{-1} from the obscured star is required. The pre-shock stellar wind density $n_0 \simeq 200$ cm^{-3} places the cloudlets at 10^{17} cm from the star. Böhm (1978) has proposed a shock model with an energy source embedded in each nebulosity; however, this model does not explain the lack of spatial coincidence between the nebular condensations and the IR sources observed near HH objects.

Dopita (1978a,b) has obtained extensive absolute spectrophotometry of HH objects and quantitatively matches the observed line strengths and ratios with radiative shocks. His derived parameters for the objects HH1, 2, 3, 24, 32, 43, 46, and 101 all lie in the range $n_0 = 30$–300 cm^{-3} and $v_s = 40$–150 km s^{-1}. An interesting feature of his observations is the occasional observation of high Hα/Hβ ratios, of order 7, correlated with high OI and NI intensities and low absolute Hβ fluxes. We suggest that this might be characteristic of low velocity ($v_s \simeq 20$–30 km s^{-1}) neutral shocks.

More recently, Schwartz & Dopita (1980) have observed several other HH objects and have further substantiated Schwartz's (1978) model of shocked cloudlets accelerated in strong pre–main-sequence stellar winds.

5.4.3 SHOCKS IN AND NEAR HII REGIONS Lasker (1977) has observed emission nebulae, often ringlike, with SII $\lambda 6716$, 6731 line strengths as strong as Hα in the LMC. The observed flows suggest 20–40 km s^{-1} shocks which may produce sufficient heating to strengthen the SII lines; the ionization source comes from external stars. Brand & Mathis (1978) qualitatively estimate the increase in various line intensity ratios in weak shocks ($v_s < 30$ km s^{-1}) in HII regions and find that the SII 6716, 6731 to Hα ratio does indeed increase, partly because the compression behind the shock drives S^{++} to S^+. Manfroid (1977) notes that compressed shells passing through HII regions shield the outer regions from photoionization and that recombinations there lead to higher S^+ abundances and higher intensities. The origin of the shells in the LMC nebulae remains uncertain; stellar winds or supernovae are the likely sources.

Gull & Sofia (1979) have used optical observations of wind-driven bow shocks in HII regions near ζ Oph and LL Ori to derive mass loss rates from the stars. The position of the head of the bow coupled with known physical parameters yield mass loss rates of $2.2 \times 10^{-8}\, M_\odot\, \text{yr}^{-1}$ and $1.5 \times 10^{-8}\, M_\odot\, \text{yr}^{-1}$.

Hill & Hollenbach (1978) have made global models of expanding compact HII regions (D-type ionization fronts) which drive shocks into the ambient medium ($n_0 = 10^3$–$10^4\, \text{cm}^{-3}$). They show that the FUV (11 eV–13 eV) radiation from the star dissociates essentially all molecules in the shocked shell and in the pre-shocked gas until late in the evolution of the system, when the shock is traveling at $v_s \lesssim 3\, \text{km s}^{-1}$. Therefore, as the shock decelerates from 10 km s^{-1} to 3 km s^{-1}, the shocked gas is atomic even though the densities are high, and the OI (63 μ) line is generally the strongest observable line emission. However, the OI (63 μ) intensity from the nearby ionization front/recombination zone on the star-side of the shocked shell is larger by a factor of 3–10 than the intensity from the shock front, and cannot presently be spatially resolved from the shock contribution.

6 MOLECULES IN SHOCK WAVES

The main observational effects of shock waves on molecules include (a) acceleration, compression, and heating, (b) production of high abundances of certain molecular species such as OH, H_2O, CH^+, OCS, SO, H_2S, and SiO through high temperature chemical reactions in the $T \gtrsim 1000$ K post-shock gas, and (c) excitation of rotational/vibrational infrared lines in the heated post-shock gas. Unfortunately none of these effects is unique to shocks: molecular clouds may be accelerated by other mechanisms including gravity, radiation pressure, and the rocket effect; gas phase chemistry may result in similarly high abundances of target molecular species when other parameters such as the cosmic ray or X-ray ionization rates are changed from the normative values; and high molecular gas temperatures may be achieved by other mechanisms than shock heating. A combination of these observations is required in order to determine the presence of a shock. Radio measurements are most often used for (a) and (b), although, in the case of diffuse clouds with background early-type stars, UV absorption measurements can be made of trace amounts of CH^+, OH, and H_2O. Infrared measurements have been hampered by the effects of the Earth's atmosphere and the lack of sensitive detectors. Recent improvements in detector technology and in high altitude observing sites will make IR spectroscopy extremely favorable for shock identification since the infrared lines will provide measures of velocities, abundances, and levels of excitation in the line-emitting regions and will penetrate through regions of high optical extinction. The visible and UV generally are unsuitable for study of molecules in shock waves because the high extinction associated with molecular regions usually prevents absorption measurements and because the temperatures required for molecular emission at these wavelengths lead to molecular dissociation.

For shocks incident upon relatively dense ($n_0 \gtrsim 10^{2-3}$ cm^{-3}) ambient media, H_2, CO, OH, or H_2O molecules may dominate the post-shock cooling and therefore determine the structure behind the shock front. A shock velocity of about 25 km s^{-1} produces enough thermal energy in the post-shock gas ($\simeq 4.5$ eV per hydrogen molecule) to dissociate all the molecules. Indeed, if H_2 vibrational levels are in LTE ($n \gtrsim 10^{5-6}$ cm^{-3} required at 5000 K), dissociative H_2 cooling dominates other cooling mechanisms for $T \gtrsim 3000$ K, and molecular hydrogen is completely dissociated at $v_s = 25$ km s^{-1} (Kwan 1977). However, Dalgarno & Roberge (1979) show that vibrational/rotational cooling will dominate dissociative cooling by neutral collisions at low densities because the dissociative rates from the ground vibrational state ($v = 0$) are much smaller than the rates from higher v. At low pre-shock densities, therefore, only a fraction of the available thermal energy goes into dissociation, and higher shock velocities are required to dissociate all the molecules. Hollenbach & McKee (1980) show that a combination of dissociation from $v = 0$ by neutral collisions as well as dissociation by collisions with electrons ensures the complete destruction of H_2 at $v_s \simeq 50$–55 km s^{-1} for $n_0 \lesssim 10^3$ cm^{-3}. Furthermore, once H_2 dissociates, the resultant hydrogen atoms chemically destroy the other molecules as shall be discussed in Section 6.1. Therefore, the critical velocity v_d for the destruction of all molecules ranges from $v_d \simeq 25$ km s^{-1} for $n_0 \gtrsim 10^4$ cm^{-3} to $v_d \simeq 50$–55 km s^{-1} for $n_0 \lesssim 10^3$ cm^{-3}. Nondissociative shocks ($v_s \ll 30$ km s^{-1}) mainly heat the ambient molecules and create new molecular abundance ratios by the effects of increased post-shock density and temperature on the formation rates of key molecules. Dissociative shocks ($v_s \gg 30$ km s^{-1}) destroy pre-existing molecules but re-form them in the cooling post-shock gas.

6.1 Chemistry Behind Shock Fronts

As discussed by Aannestad (1973), Iglesias & Silk (1978), Elitzur & de Jong (1978), Elitzur (1979b), Hollenbach & McKee (1979, 1980), and Elitzur & Watson (1980), the gas phase chemistry in warm, $10^4 > T \gtrsim 10^3$ K, post-shock gas is generally dominated by a small number of rapid reactions that require ~ 0.1 eV–1 eV of activation energy. The core of this gas phase network is illustrated in Figure 2 which schematically traces the dominant formation and destruction routes of OH, CO, H_2O, and O_2 given the relative abundance of H and H_2. The rate coefficients for the reactions on

Figure 2 are all roughly equal ($\simeq 10^{-10}$ cm^3 s^{-1}) for $T \gg 1000$ K. The figure illustrates, therefore, that if hydrogen is mostly molecular, then oxygen is found mostly in O_2 and H_2O and carbon in CO. On the other hand, if hydrogen is atomic, the hydrogen atoms chemically dissociate heavy molecules, and the carbon and oxygen will mostly be in atomic form as well.

Furthermore, a comparison of relevant column densities (equivalent to time scales since $N = n_0 v_s t$) illustrates that N_{cool} and the hydrogen molecule formation column density $N_{\text{H2 form}}$ are relatively large compared with the column density N_{chem} for the reaction network in Figure 2 to reach equilibrium. If T is of order several thousand degrees,

$$N_{\text{chem}} \simeq 10^{17} \eta'^{-1} v_{s7} \text{ cm}^{-2},$$

$$N_{\text{H2 form}} \gtrsim 10^{24} \eta'^{-1} v_{s7} \text{ cm}^{-2}, \qquad (12)$$

$$N_{\text{cool}} \simeq 10^{19} \text{ to } 10^{23} \text{ cm}^{-2},$$

where the compression factor η' is the ratio of the post-shock density at several thousand degrees to n_0. When H_2 is the dominant coolant, the comparison of N_{chem} to N_{cool} reflects the smaller cross sections for vibrational excitation of H_2 compared with chemical cross sections. When OH or H_2O is the dominant coolant, the relatively small N_{chem} arises since H or H_2 must only make one effective collision with each OH or H_2O molecule to equilibrate the chemistry, whereas 10^{3-4} hydrogen nuclei must inelastically collide with each OH or H_2O molecule to cool the gas.

Consequently, behind dissociative shocks and behind shocks incident upon atomic gas, the concentration of CO, O_2, OH, and H_2O produced in the warm $T > 10^3$ K, post-shock region depends in large part on the fraction $x(H_2)$ of molecular hydrogen formed in the warm compressed gas, which is about $x(H_2) \simeq N_{\text{cool}}/N_{\text{H2}}$ for $N_{\text{cool}} < N_{\text{H2}}$. The formation rate of molecular hydrogen depends on the gas temperature T, the grain tempera-

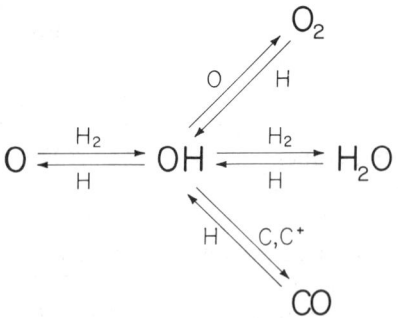

Figure 2 High temperature gas phase chemistry in interstellar shocks.

ture T_{gr} (typically 15 K ≪ T_{gr} < 150 K), and the ionization (see Hollenbach & McKee 1979 for a detailed discussion). Most of the H_2 is formed on grain surfaces provided (a) the refractory grains survive the shock (v_{s7} ≲ 3; Draine & Salpeter 1979), and (b) T_{gr} is low enough (≲ 50–100 K) that H atoms that strike the surface form molecules with atoms lodged in impurity sites, lattice defects, or semichemical valence bonds before evaporating. The grains are heated primarily by radiation from the cooling shocked gas, and condition (b) requires $n_0 v_{s7}^3$ ≲ $3 \times 10^4 \, a_\mu$ cm^{-3} for N ≲ 10^{20} cm^{-2} and $n_0 v_{s7}^3$ ≲ $2 \times 10^6 \, a_\mu$ cm^{-3} otherwise. An additional formation process, which is important behind ionizing shocks (v_{s7} ≳ 0.7) above 2000 K, is the H$^-$ process: H + e → H$^-$ + γ followed by H$^-$ + H → H_2 + e.

Although Figure 2 illustrates the heart of the $T > 10^3$ K chemistry, gas behind dissociative shocks is affected by the FUV radiation from the hot post-shock gas and the abundances of H$^+$ and He$^+$ in the recombining gas. Much of the FUV intensity is in Lyα so that photoreactions involving this line are especially important. As an example, the column density N_γ for photodissociation of H_2O by Lyα behind the shock front is of order $N_\gamma \simeq 10^{16} \, v_{s7}^{-2}$ cm^{-2} which is comparable to N_{chem} (see Equation 12) for v_{s7} ≳ 1. The radiation field not only affects the chemistry through photodissociation and photoionization, but it also heats the grains and therefore determines the rate of H_2 formation. Furthermore, the reactions of O_2, H_2O, and CO with H$^+$ and He$^+$ can deplete the abundances of these molecules. These mechanisms and the relevant rate coefficients and cross sections are treated in detail in Hollenbach & McKee (1979).

Another reaction in the high temperature gas which does not greatly affect the overall chemistry but which produces observable quantities of CH$^+$ is the reaction C$^+$ + H_2 → CH$^+$ + H. Elitzur & Watson (1978, 1980) model post-shock gas to explain the high CH$^+$ abundances seen in diffuse clouds toward early-type stars. For $x(H_2)$ ≲ 0.01 the column density of $N(CH^+)$ depends linearly on $x(H_2)$. Furthermore, $N(CH^+)$ rises monotonically with v_s when $v_s < 12$ km s^{-1}. The observed column densities, e.g. $N(CH^+) \simeq 10^{13}$ cm^{-2} seen in absorption toward Zeta Ophiucus, require $x(H_2)$ ≳ 0.01 and $v_s \simeq 12$ km s^{-1}. In turn, $x(H_2) \simeq 0.01$ yields approximately $N(OH) \simeq 3 \times 10^{13}$ cm^{-2} and $N(H_2O) \simeq 2 \times 10^{11}$ cm^{-2} by the reaction network of Figure 2, with FUV destruction of H_2O by the ambient interstellar field included. Higher abundances of H_2 would result in the production of more $N(H_2O)$ than the current observational upper limit. The OH abundance is close to detection limits, and may have been observed in the pre-shock and post-shock gas by Crutcher (1979).

As the gas temperature drops to T ≲ 10^3 K, the rate coefficients for the reactions in Figure 2 drop because of the activation energies involved, and the relative importance of the "cold chemistry" rates (a recent review of these reactions is found in Prasad & Huntress 1979), the photoreaction rates, and the rates involving H^+ and He^+ increase. Furthermore, as the gas cools, the gas phase column density N_{chem} becomes larger than N_{cool}, and the chemistry is no longer in equilibrium with the local concentration of H and H_2. Consequently, in a nondissociative shock incident upon a molecular cloud, all of the oxygen not locked up in CO can be channeled to H_2O by the high temperature reactions and can remain "frozen" in H_2O for large post-shock column densities before returning to its cold equilibrium abundance (Iglesias & Silk 1978). Elitzur (1979a) uses this mechanism to explain the large H_2O abundances seen in Orion. Lada et al. (1978) invoke a similar mechanism to explain the high abundance and velocity of SiO and OCS in Orion. Formation of SiO proceeds in the high temperature gas where OH may be abundant by the reaction $OH + Si \rightarrow SiO + H$. Formation of OCS is initiated by the high temperature reaction $H_2 + S \rightarrow HS + H$ followed by $HS + CO \rightarrow OCS + H$. Both molecules are assumed to persist far downstream in the cold post-shock gas because of low destruction rates. Similarly, Hartquist et al. (1980) predict high post-shock abundances of SO and H_2S.

Since the infrared lines from a given molecular species are often optically thin behind strong shocks, their intensities are proportional to the hot post-shock column density of that species provided by the chemistry outlined above. Behind dissociative shocks ($v_s \gtrsim 25$–35 km s^{-1}) incident upon relatively low density ambient gas ($n_0 \simeq 10^3$ cm^{-3}), the numerical code of D. J. Hollenbach and C. F. McKee (in preparation) shows that the cooling column density to 100 K, N_{100}, is approximately 10^{19} hydrogen nuclei cm^{-2}. Only a fraction of this column density is molecular, however, 10^{16} H_2 molecules cm^{-2} and 10^{12} CO, H_2O, and OH molecules cm^{-2}. Complete re-formation of the H_2 molecules (and of many of the other molecules) occurs in a post-shock column density $N_{H2\,form} \simeq 10^{21}$ cm^{-2} in the cool, $T < 100$ K, gas. UV absorption or radio measurements would generally detect this cold post-shock gas; the infrared emission from the warm gas is below present sensitivity. On the other hand, when $n_0 \gtrsim 10^4$ cm^{-3}, $N_{H2\,form} \simeq N_{cool} \simeq 10^{21}$–$10^{23}$ cm^{-2}, the greater column density being due to a combination of collisional suppression of atomic and molecular cooling and significant heating by H_2 formation.

Behind slow ($v_s \lesssim 25$ km s^{-1}) nondissociative shocks incident upon $n_0 \lesssim 10^3$ cm^{-3} ambient gas, $N_{100} \simeq 10^{18}$–10^{19} cm^{-2} with dominant coolants H_2, OI (63 μ), and CII (156 μ). As discussed above, such gas may produce $N(CH^+) \simeq 10^{13}$ cm^{-2} in the hot post-shock region. If $10^7 > n_0 \gtrsim 10^4$ cm^{-3}, N_{100} increases to 10^{19}–10^{20} cm^{-2}, assuming solar gas abundances, and is determined primarily by the dominant H_2O cooling. To achieve higher column densities of warm molecular gas such as may be needed in Orion (see below), the H_2O abundance must be suppressed.

6.2 Infrared Cooling and Rotational/Vibrational Cooling by Molecules

The dominant molecular coolants in the post-shock gas are H_2, which is homonuclear and hence has negligible line opacity, and CO, OH, and H_2O, all of which are dipolar and have large line opacities. An accurate treatment of the cooling involves solving for the populations of each of the vibrational/rotational levels at each point in the shocked gas and then solving the radiative transfer equation for each line. For H_2 the problem is simplified because the lines are optically thin for the conditions of interest. Hollenbach & McKee (1979), Shull & Hollenbach (1978), and Elitzur & Watson (1980) have solved for the level populations and determined the H_2 cooling rate due to rotational excitation by $H-H_2$ collisions; the former authors gave analytic approximations to the cooling and discussed H_2-H_2 collisions and H_2 vibrational cooling as well. These H_2 lines are observable in the infrared, $\lambda \sim 2$ μm for lower vibrational levels and $\lambda \sim 5$–30 μm for the lower rotational lines (see Section 6.3 below).

Calculation of the cooling by the polar molecules is complicated by the lack of collisional excitation rates for most of the transitions and by the formidable radiative transfer problem. Detailed calculations of cooling by the lower rotational levels have been given by de Jong (1973) for H_2O, by de Jong et al. (1975) for CO, and by Goldsmith & Langer (1978) for a number of molecules; Hollenbach & McKee (1979) extended these results to high levels by assuming a constant, geometrical de-excitation cross section for all levels. In each case the radiative transfer was approximated by the escape probability formalism, in which the cooling due to the transition $J \to J-1$ is written $n_J A_J \Delta E_J \varepsilon_J$, where n_J is the population of level J, A_J is the spontaneous transition probability, ΔE_J is the photon energy, and ε_J is the probability that the emitted photon actually escapes.

The relative importance of the different molecular coolants may be summarized as follows (Hollenbach & McKee 1979). For densities $n \lesssim 10^6 T_3^{1/2}$ cm^{-3} and column densities $N \lesssim 10^{21}(1+10^5 \text{cm}^{-3} n^{-1})$ cm^{-2}, cooling by ^{12}CO, OH, and H_2O is unaffected by collisional de-excitation or line opacity, and the relative contributions of the different coolants is simply proportional to their concentrations. At higher values of n or N, CO cooling is suppressed and OH and H_2O are the dominant coolants. The most intense lines of CO, OH, and H_2O occur in the far infrared because of the close spacing of the lines, roughly at wavelengths $\lambda_{CO} \simeq 137\, T_3^{-0.5}$ μm, $\lambda_{OH} \simeq 44\, T_3^{-0.5}$ μm, and $\lambda_{H_2O} \simeq 55\, T_3^{-0.5}$ μm, where $T_3 = T/1000$ K. Vibrational cooling by molecules other than H_2 is generally negligible compared to vibrational cooling by H_2 or to the cooling associated with collisional dissociation of H_2. H_2 generally dominates the cooling at lower densities provided the temperature is not too low, e.g. $n \lesssim 10^4$ cm^{-3} at T

~ 1000 K. When $N \gtrsim 10^{22}$ cm^{-2}, dust may play an indirect role in cooling the gas by absorbing photons in optically thick lines and re-radiating their energy in the continuum, thereby increasing their effective escape probability (see also Leung 1978).

6.3 Observations of Shocked Molecules

The observation of vibrationally excited H_2 in the KL-BN region of the Orion molecular cloud is perhaps the most convincing evidence of shocked interstellar molecular gas. Although H_2 can be vibrationally excited by the absorption of UV photons and subsequent cascade or by the process of formation from two hydrogen atoms, the observation of several different vibrational transitions around 2μ indicated that the H_2 was collisionally excited in a gas of $T \simeq 2000$ K and $n \gtrsim 10^5$ cm^{-3} (Gautier et al. 1976). Furthermore, maps of the region in the strong 1-0 S(1) line indicated that the hot H_2 is found in thin sheets (Grasdalen & Joyce 1976, Beckwith et al. 1978a). Kwan & Scoville (1976) noted the coincidence of the hot H_2 with the broad molecular line emission around KL (e.g. CO was observed to emit over a full range of 120 km s^{-1}) and suggested an explosive event to accelerate the molecules and a shock with $v_s \simeq 60$ km s^{-1} and $n_0 \simeq 5 \times 10^3$ cm^{-3} to explain the H_2 emission. However, shocks with $v_s \simeq 10$–25 km s^{-1} and $n_0 \simeq 3 \times 10^5$ cm^{-3} were required to match the observed intensities of the H_2 lines (Hollenbach & Shull 1977, London et al. 1977, Kwan 1977); unfortunately, the neglect of H_2O cooling in these models resulted in an overestimation of the H_2 intensities. Furthermore, Beckwith et al. (1979) and Simon et al. (1979) have recently shown that the emission originated in a region with four magnitudes of extinction at 2μ. Therefore, the intrinsic line intensities are a factor of 50 times higher than the original estimates. In addition, the lines appear to be quite broad with Δv (FWHM) $\simeq 18$–50 km s^{-1} and detectable features at ± 50 km s^{-1} from the line center (Nadeau & Geballe 1979, Ogden et al. 1979, as well as measurements of the 0-0 S(2) line of H_2 at 12.28 μm observed by Beck et al. 1979). Therefore, shocks with higher velocities that dissociate and re-form H_2 must be considered as well as slower shocks passing through rapidly moving clumps of gas.

Another good example of observations of shocked molecules came from a combination of infrared and radio measurements of the supernova remnant IC 443. DeNoyer (1978) first observed a dense ($n > 200$ cm^{-3}) high velocity ($\Delta v \sim 40$ km s^{-1}) HI condensation (1×3 pc) within the remnant. Such high density material was suspected to harbor molecules, and the theory of McKee & Cowie (1975) suggested the presence of shocks. Therefore radio observations were made which uncovered OH in absorption (DeNoyer 1979a) and CO in emission (DeNoyer 1979b). Both molecules had components with high negative velocities of -15 km s^{-1} to

-40 km s^{-1} relative to the pre-shock ambient gas at about -3 km s^{-1}. Furthermore, the post-shock component showed an OH/CO ratio approximately 100 times the pre-shock value, possibly indicative of post-shock chemistry. Treffers (1979) observed the 1-0 S(1) line from vibrationally excited H$_2$ at ~ -33 km s^{-1}, and modeled the observation with a $v_s \simeq 30$ km s^{-1} shock incident upon $n_0 \simeq 500$ cm^{-3} ambient gas.

Other observations give more ambiguous evidence for shocks in molecular gas (Beckwith 1980). A number of radio observations of OH, CO, and H$_2$CO reveal spatial configurations and velocity components suggestive of shock waves (e.g., Crutcher 1979, Dickel 1974, Dickel et al. 1976, Cornett et al. 1977). Occasionally, enhanced CO brightness temperature is observed behind the presumed shock front as well as increased column densities (e.g., Elmegreen & Moran 1979, Lada & Black 1976, Wootten 1977, Kutner et al. 1976, 1979). However, the CO temperature increase could be associated with grain heating from nearby luminous stars rather than shock heating. Finally, the observations of high rotational states ($J = 4, 5,$ and 6) of molecular hydrogen seen in diffuse clouds with UV absorption measurements (Spitzer & Cochran 1973, Spitzer et al. 1974, Spitzer & Morton 1976) have led to the theoretical hypothesis of dense post-shock sheets of high pressure molecular gas near the background early-type star which UV pumps the molecules to high rotational states (Jura 1975a,b, Hollenbach et al. 1976).

7 CURRENT PROBLEMS AND APPLICATIONS

7.1 Initiation of Star Formation by Radiative Shock Waves

Roberts (1969) and Shu et al. (1972) originally suggested that the passage of shock waves associated with galactic spiral density waves would result in the compression of interstellar clouds and thereby trigger star formation. Woodward (1976) has numerically calculated the effect of such shocks on clouds within the context of the two-phase model of the ISM. The ram pressure of the shocked intercloud gas flowing past the cloud produces gravitationally bound subregions from the action of the Rayleigh-Taylor instability on the front cloud surface. The structure of the density wave shock in a two-phase ISM was analyzed by Shu et al. (1972). Such a shock can also occur in the three-phase model of the ISM (McKee & Ostriker 1977) despite the fact that the shock velocity is much less than the sound speed of the hot phase ($T \sim 10^6$ K): interstellar clouds act like atoms in a gas and undergo a collisional shock (Shu 1978, Cowie 1980, Norman & Silk 1980a).

On a smaller scale Elmegreen & Lada (1977), Elmegreen & Elmegreen

(1978), and Lada et al. (1979) have proposed that star formation is triggered in the compressed gas layer behind shock waves driven by expanding HII regions, stellar winds, cloud-cloud collisions, or supernova explosions. They argue that the effect of expanding HII regions around OB associations is the dominant mechanism. Elmegreen & Elmegreen (1978) review supporting observations and suggest that star formation commences in the compressed post-shock layer behind a plane-parallel shock once a critical column density N_x of hydrogen nuclei has been swept up, where $N_x \simeq 1.6 \times 10^{22} \, n_0^{0.5} v_{s7}$ cm^{-2}. This mechanism can lead to a chain reaction of sequential formation of subgroups in OB associations, the shock wave generated by one group leading to the formation of the next, as outlined by Elmegreen & Lada (1977). On a larger scale, Gerola & Seiden (1978) hypothesize that supernova explosions can trigger a sequential chain of star formation in neighboring molecular clouds which ultimately leads to large spiral-like features in galaxies. Norman & Silk (1980b) suggest that the stellar winds from pre–main-sequence T Tauri stars may drive shocks and clump gas in molecular clouds, leading to the subsequent formation of more pre–main-sequence stars.

7.2 Interstellar Masers

The high abundances of OH and H_2O, the high temperatures and densities, and the high velocities relative to the ambient gas produced behind shock waves are quite suggestive of interstellar masering regions. Litvak (1972) noted that post-shock chemistry and radiation fields would be favorable to masering conditions. Elitzur & de Jong (1978) suggest that OH masers, and possibly low velocity H_2O masers, are produced in the compressed gas between the ionization front and the shock front around expanding HII regions. High abundances of H_2O are produced by the reaction chain of Figure 2 behind the shock front, and the H_2O is photodissociated by the FUV from the early-type star to produce the OH. The OH inversion is collisionally pumped, possibly by streams of charged particles that are moving through the neutral gas as the magnetic field undergoes ambipolar diffusion (Elitzur 1979c). The pump mechanism for H_2O is unknown. Black & Hartquist (1979) calculate that observable H_2 rotational line emission may result from the ambipolar heating; however, they ignored OH and H_2O cooling which will suppress the H_2 intensities. Gwinn et al. (1973) propose that weak OH masers can be pumped by collisional dissociation of H_2O in the high temperature post-shock gas around expanding HII regions.

Strel'nitskii & Sunyaev (1973) speculate that the temperatures and densities behind slow ($v_s \lesssim 20$ km s^{-1}) shocks moving through high velocity ($v_s \simeq 100$ km s^{-1}) clouds near early-type stars may result in high velocity H_2O masers. D. J. Hollenbach and C. F. McKee (in preparation) find that high velocity H_2O masers may be produced in the hot post-shock gas if $n_0 \simeq 10^7$ cm^{-3}, $v_s \gtrsim 30$ km s^{-1}, and the effect of the heating by reformation of the H_2 molecules is included. The pumping mechanism in this case is collisional, as discussed by Shmeld, Strel'nitskii & Muzylev (1976). J. C. Tarter and W. J. Welch (in preparation) propose that the time scales for the variation of high velocity H_2O masers can be understood if the masing occurs in the pre-shock gas. In this model, two clouds with $n_0 \simeq 10^9$ cm^{-3} collide at $v_s \simeq 100$ km s^{-1}; intense radiation from the shock fronts heats the pre-shock grains, whose IR radiation rapidly diffuses through the pre-shock gas, radiatively pumping the H_2O inversion. The maser turns off when the gas, heated by collisions with grains and H_2O, is warmed to the local grain temperature, and LTE is achieved.

7.3 Stability of Shocks

Nonradiative shocks are almost always stable against wrinkling the surface for $B = 0$ (Erpenbeck 1962, Swan & Fowles 1975) and for fast shocks with B parallel or perpendicular to $\mathbf{v}_s$ (Gardner & Kruskal 1964). For a polytropic magnetized gas a sufficient condition for stability is $\gamma < 3$ (Gardner & Kruskal 1964). The reason for the stability is simple (Chevalier & Theys 1975): If the shock surface is wrinkled, then the fluid velocity ahead of the shock is no longer parallel to the local $\mathbf{v}_s$—the shock is oblique. Now, the component of the fluid velocity parallel to $\mathbf{v}_s$ is reduced by a factor ρ_0/ρ_s across the shock, whereas the perpendicular component (which lies in the local shock plane) is unchanged; hence the streamlines are deflected away from the shock normal. For a wrinkled shock front this implies that the density—and thus the pressure—drops behind the advanced parts of the shock front and rises behind the retarded parts, so that the shock front straightens out. Hence it is stable.

Several assumptions in these analyses of shock stability limit their applicability to interstellar shocks. First, radiation is not included; this can be allowed for approximately by taking $\gamma \to 1$. (Note that even in a radiative shock, the front is always subsonic relative to the shocked gas.) Second, it was assumed that the energy density depends on only two variables, such as p and ρ, with no dependence on the chemical or ionization state of the gas. Finally, the shock was treated as a discontinuity, so the results only apply to perturbations with wavelengths large compared to the total shock thickness. Nonradiative interstellar shocks, in which the shocked gas is virtually all ionized, all neutral atomic, or all molecular, satisfy these assumptions and are stable. Radiative interstellar shocks are probably stable against long wavelength perturbations, though this is not certain.

Thermal instability can occur in the cooling gas in a radiative shock, but it affects the appearance of the shock more than the dynamics (Avedisova 1974, McCray et al. 1975, Mufson 1975). The instability develops for $d \log \Lambda/d \log T < 2$ (Field 1965), where $n^2\Lambda$ is the cooling rate per unit volume, and it proceeds at approximately constant pressure. Eventually all the shocked gas reaches approximately the same temperature and density; the effect of the instability is that positive density fluctuations will pass through the temperature domain of maximum optical emission sooner than average. The large compression behind the shock front leads to a sheet-like structure for the optical emission, which is broken up into successive sheets by the instability. In the more general case considered by Mufson (1975) in which the wave vector **k** of the instability is not parallel to $\mathbf{v}_s$, the sheets will be wavy, and if the brightness is nonuniform, the wavy sheets may appear filamentary. Avedisova (1974) and Mufson (1975) suggested that filamentary structure parallel to the magnetic field would arise because thermal conduction suppresses the instability along **B**; such filaments would be shorter than the thickness of the radiative cooling layer (McCray et al. 1975). Sheet-like and filamentary structure is commonly observed in SNR shells, and these authors attribute this structure to thermal instability. Note that in this case proper motion velocities would not exactly correspond to fluid velocities.

Chevalier & Theys (1975) studied the effect of a clump with twice the mean density on the propagation of a decelerating radiative shock. The results showed the possible nonlinear evolution of a thermal instability in the shock. The process described above, which stabilized nonradiative shocks against wrinkling of the shock front, in this case funneled mass onto the clump and made it grow. Their treatment of the cooling below 10^4 K was quite approximate, and their results might have been substantially altered by the inclusion of a magnetic field, which limits the compression.

Avedisova (1974) has suggested that a decelerating radiative shock is Rayleigh-Taylor unstable. At first sight this seems plausible since it appears that the low density, hot shocked gas is decelerating the cold dense layer behind. In fact, however, the shocked gas is accelerated from $(3/4)v_s$ to v_s as it cools, so the radiative cooling layer is stable provided it is sufficiently thin. L. L. Cowie and J. H. Krolik (unpublished) have also reached this conclusion based on much more extensive analysis.

The stability of ionization-shock fronts has recently been analyzed by Giuliani (1979) under the assumptions that the gas is isothermal, the wavelength of the wrinkling of the surface is large compared to the thickness of the gas between the ionization and shock fronts, and self-gravity is negligible. He finds the system is unstable in an oscillatory manner. Presumably the increased density behind the retarded parts of the front gives a higher pressure (because the ionized gas is assumed isothermal) which then drives the retarded region ahead of the mean position of the front. A detailed understanding of the energy source which drives the instability is lacking, however.

7.4 Particle Acceleration in Shocks

High energy particles are observed in association with solar flares, interplanetary shocks, and SNRs, and this has led to a number of analyses of particle acceleration by shocks (see Blandford 1979 for a recent review). One mechanism for particle acceleration by shocks relies on the electric field $\mathbf{E} = -(\mathbf{v}_s/c) \times \mathbf{B}$ in the plane of the shock (e.g. Pesses 1979), but this is efficient only if $\mathbf{B}_0$ is nearly perpendicular to $\mathbf{v}_s$. An alternate mechanism of potentially wider applicability is first order Fermi acceleration (Axford et al. 1977, Bell 1978a,b, Blandford & Ostriker 1978, Krymsky 1977; see also Fisk 1971): Suprathermal particles are scattered upstream by turbulence in the shock front; since these particles are then streaming through the unshocked fluid, they generate Alfvén turbulence which scatters them back across the shock. Thus the shocked and unshocked fluids act like converging magnetic mirrors which efficiently accelerate particles. The steady state suprathermal particle distribution is determined by balancing the diffusion upstream against the convection downstream. Of particular importance, the energy spectrum of the accelerated particles for strong shocks depends only on the compression ratio $\eta \equiv n_s/n_0$. In terms of $n_{st}(E)$, the density of suprathermal particles with energy above E, one finds $d\ n_{st}/d(\ln E) \propto \hat{p}^{-3/(\eta-1)}$, or $E^{-3/2(\eta-1)}$ in the nonrelativistic case and $E^{-3/(\eta-1)}$ in the relativistic case, where $\hat{p}$ is the particle momentum. This result is quite general and does not depend on adopting a particular form for the diffusion coefficient of the suprathermal particles in the turbulence. The observed cosmic ray spectrum is recovered for $\eta \sim 3$. Since the relativistic electron spectrum is steeper than the nonrelativistic proton spectrum, the number of electrons above 1 GeV is expected to be substantially less than the number of protons, as observed in the cosmic rays. Blandford & Ostriker (1980) have developed a detailed theory of cosmic ray acceleration in interstellar shocks.

The problem of injection of the suprathermal particles, i.e., determining which particles will be accelerated, is not clearly addressed in the above papers. Eichler (1979) has suggested that the energy density of the suprathermal particles becomes large enough to significantly modify the structure of the shock front and thereby regulate the injection. The high energy particles diffuse farther upstream than the lower energy ones and begin to decelerate the flow of the unshocked fluid (as observed in the shock frame). This reduces the compression and hence the acceleration of the lower energy particles; in a steady state, just enough low energy particles are accelerated that the high energy particles can mediate the shock. An ingenious, but not rigorous, argument shows that exactly half the post-shock pressure is due to the suprathermal particles if the overall compression is $\eta = 4$. However, the details of the shock structure and the efficiency of electron acceleration in this case remain to be worked out.

The central hypothesis of the theory of Fermi acceleration by shocks is that suprathermal particles are efficiently scattered in both the shocked and unshocked fluids. The validity of this assumption does not depend on quasi-linear theory (which is notoriously unreliable in the interplanetary medium—Jokipii 1979), but the analysis of this assumption does. Thus our discussion of the possible limitations on the scattering of the particles in the unshocked gas will not be rigorous. First, Bell (1978a) has pointed out that the density of neutral hydrogen ahead of the shock front must be sufficiently low that the Alfvén waves are not damped by charge exchange collisions. Under the assumption that the particles are resonantly scattered, the waves with the lowest growth rate are those associated with the particles of lowest density and thus of highest energy. Hence, charge exchange damping decreases the acceleration efficiency above some critical particle energy. The condition that this critical energy be at least 1 GeV for protons can be shown to be

$$v_{s7} > 3.5 \, T_3^{0.1} x_e^{1/8} [x(H^0)]^{1/4} n^{1/8}. \tag{13}$$

Since the neutral fraction is $x(H^0) \sim 0.01$ for $v_{s7} > 1.1$ (Shull & McKee 1979) the effective minimum velocity for acceleration of relativistic protons is about 100 km s^{-1}; it can be lower if an external ionization source keeps the gas highly ionized. Note that the characteristic diffusion length of the suprathermal particles is $(1/\pi)(c/\omega_{pi})(n_0/n_{st})$, which is usually large compared to the thickness of the collisionless shock but small compared to collisional lengths.

Two further limitations on the validity of the scattering hypothesis have been suggested by Holman et al. (1979). They show that ion cyclotron damping prevents the suprathermal particles from resonantly scattering through zero velocity in the frame of the unshocked fluid, and they argue that therefore the suprathermal particles will stream ahead of the shock without further acceleration. However, they also argue that the suprathermals will be subject to mirror scattering and this streaming avoided if the amplitude of the magnetic turbulence is large enough; it is easily shown that if the suprathermal energy density is of order $\rho_0 v_s^2$ as argued by Eichler (1979), then the criterion for mirror scattering is well satisfied. Hence, the effects of ion cyclotron damping are not inconsistent with Fermi acceleration in shocks.

Their second, and potentially more serious, objection is based on calculations by Morrison (1979): that if the growth rate of the instability driving the Alfvén turbulence is comparable to the real part of the frequency, then the phase velocity of the waves is increased. In principle, this increase can be so large that the waves move with the suprathermal particles, the scattering becomes inelastic, and the acceleration becomes a second order process, which is much less efficient. A rough estimate gives Re $\omega/\text{Im }\omega \simeq \pi^{-1}(v_A^2/vv_s)(n_e/n_{st})$, where ω is the frequency, so that this potential problem is most severe when n_{st} is large, as in Eichler's (1979) model. It remains to be shown, however, that the actual increase in the phase velocity of the waves is large enough to significantly weaken the acceleration and that this difficulty persists when the waves are nonlinear. Other limitations on the theory of Fermi acceleration in shocks are discussed by Achterberg & Norman (1980), who analyze particle acceleration in solar flares.

Finally, we briefly comment on the implications of efficient particle acceleration by shocks on the overall shock structure under the assumption that about half the post-shock pressure is in suprathermal particles, as suggested by Eichler (1979). The post-shock temperature T_s will then be about half the value given in Equation (8); correspondingly the hot electrons in Cas A and Tycho will be that much closer to equipartition with the protons. If the suprathermals are relativistic, then the effective γ behind the shock front will be less than 5/3, and the compression will exceed 4. Whether the suprathermals can prevent a large compression of the gas in a radiative shock depends on how readily they can diffuse in the post-shock relaxation region: the diffusion coefficient would have to be small enough to maintain a doubling of the suprathermal pressure over a distance $N_{cool}/4n_0$ in order to avoid the compression. Further study of this problem is required.

7.5 Shocks in Galactic Nuclei

Galactic nuclei are known to generate supersonic gas velocities, and therefore radiative shock waves may contribute significantly to the resultant emission. Minkowski & Osterbrock (1959) first suggested that the broad emission lines observed in active elliptical galaxies such as NGC 1052 arise in shocked gas. Although broad emission lines in most active galactic nuclei are now often interpreted as arising in photoionized gas, Koski & Osterbrock (1976) supported the shock model for NGC 1052 with observations of [OIII], which implied $T_e > 30{,}000$ K. Fosbury et al. (1978) fitted detailed shock models to the optical spectrum of NGC 1052 (see Table 2) and found $v_s \approx 130$ km s^{-1}, $n_0 \approx 3$ cm^{-3}. Daltabuit & Cox (1972) and Daltabuit et al. (1978) have suggested that the radiative shocks produced, for example, when two clouds of radius 1 pc and density 10^7 cm^{-3} collide at a relative speed of order 1000 km s^{-1}, can produce QSO-like emissions. Much of the line emission arises in regions photoionized by the hot gas behind the shock front, so the line spectrum closely resembles conventional photoionization spectra. Their attempt to explain the continuum spectrum is unsuccessful, however. Kent & Sargent (1979) consider the possibility that the "low velocity" filament in the Seyfert galaxy NGC 1275 may be excited by a $v_s \simeq 100$ km s^{-1} shock mechanism since the line ratios resemble those of Vela. Ford & Butcher (1979) infer that the filaments in the nucleus of M87 are shock-excited. Thompson et al. (1978) report the detection of the 1-0 S (1) 2 μ vibrational transition in molecular hydrogen from the Seyfert galaxy NGC 1068. The intensity indicates that if thermally excited, 6000 $M_\odot$ of molecular hydrogen exist at $T > 1000$ K. This source is 10^5 times brighter than the Orion source of vibrationally excited H$_2$, and indicates a large amount of turbulence in the molecular medium of this galaxy (Beckwith 1980).

The center of the Galaxy may have also undergone periods of activity, generating shock waves that may drive material observed to be radially expanding from the galactic center. The features include the 3 kpc arm, the molecular ring at 190 pc, and various other radially expanding HI features. The expanding features and various shock models to model them are reviewed in Oort (1977), Kato (1977), and Saito & Saito (1977).

ACKNOWLEDGMENTS

We wish to thank J. Arons, A. Dalgarno, M. Dopita, D. Eichler, C. Norman, M. Shull, and J. Silk for helpful comments. This work was supported in part by NSF grant AST 77-23069.

Literature Cited

Aannestad, P. A. 1973. *Ap. J. Suppl.* 25:223–52
Achterberg, A., Norman, C. A. 1980. *Astron. Astrophys.* In press
Avedisova, V. S. 1974. *Sov. Astron.-AJ.* 18:283–88
Axford, W. I., Leer, E., Skadron, G. 1977. Paper presented at Int. Cosmic Ray Conf., 15th, Plodiv, Bulgaria
Bame, S. J., Asbridge, J. R., Gosling, J. T., Halbig, M., Paschmann, G., Sckopke, N., Rosenbauer, H. 1979. *Space Sci. Rev.* 23:75–92
Beck, S. C., Lacy, J. H., Geballe, T. R. 1979. *Ap. J. Lett.* 234:L213–16
Beckwith, S., Persson, S. E., Neugebauer, G., Becklin, E. E. 1978a. *Ap. J.* 223:464–70
Beckwith, S., Gatley, I., Matthews, K., Neugebauer, G. 1978b. *Ap. J.* 223:L41–45
Beckwith, S., Persson, S. E., Neugebauer, G. 1979. *Ap. J.* 227:436–40
Beckwith, S. 1980. In *IAU Symp. No. 87, Interstellar Molecules*, ed. B. Andrews, Vol. 87. Dordrecht: Reidel. In press
Bell, A. R. 1977. *MNRAS* 179:573–86
Bell, A. R. 1978a. *MNRAS* 182:147–56
Bell, A. R. 1978b. *MNRAS* 182:443–55
Benvenuti, P., D'Odorico, S., Dopita, M. A. 1979. *Nature* 277:99–102
Benvenuti, P., Dopita, M. A., D'Odorico, S. 1980. *Ap. J.* In press
Bezzerides, B., Forslund, D. W., Lindman, E. L. 1978. *Phys. Fluids* 21:2179–85
Biskamp, D. 1973. *Nucl. Fusion* 13:719–40

Black, J. H., Hartquist, T. W. 1979. *Ap. J. Lett.* 232: L179–82
Blandford, R. D., Ostriker, J. P. 1978. *Ap. J. Lett.* 221: L29–32
Blandford, R. D. 1979. In *Workshop on Particle Acceleration Mechanisms in Astrophysics*, ed. J. Arons, C. E. Max, C. F. McKee, pp. 333–50. New York: AIP. 433 pp.
Blandford, R. D., Ostriker, J. P. 1980. *Ap. J.* In press
Böhm, K. H., Siegmund, W. A., Schwartz, R. D. 1976. *Ap. J.* 203: 399–409
Böhm, K. H. 1978. *Astron. Astrophys.* 64: 115–18
Brand, P. W. J. L., Mathis, J. S. 1978. *Ap. J.* 223: 161–67
Butler, S. E., Raymond, J. C. 1980. *Ap. J. Lett.* In press
Bychkov, K. V., Pikel'ner, S. B. 1975. *Pis'ma Astron. Zh.* 1: 29–35. Transl. in *Sov. Astron. Lett.* 1: 14–20
Castor, J., McCray, R., Weaver, R. 1975. *Ap. J. Lett.* 200: L107–10
Chevalier, R. A. 1974. *Ap. J.* 188: 501–16
Chevalier, R. A. 1975a. *Ap. J.* 198: 355–60
Chevalier, R. A. 1975b. *Ap. J.* 200: 698–708
Chevalier, R. A., Theys, J. C. 1975. *Ap. J.* 195: 53–60
Chevalier, R. A. 1977a. *Ann. Rev. Astron. Astrophys.* 15: 175–96
Chevalier, R. A. 1977b. *Nature* 266: 701–2
Chevalier, R. A., Raymond, J. C. 1978. *Ap. J. Lett.* 225: L27–30
Chevalier, R. A., Kirshner, R. P., Raymond, J. C. 1980. *Ap. J.* 235: 186–95
Chu, C. K., Gross, R. A. 1969. In *Advances in Plasma Physics*, ed. A. Simon, W. B. Thompson. 2: 139–202. New York: Wiley. 202 pp.
Cipolla, J. W., Golden, K. I., Silevitch, M. B. 1977. *Phys. Fluids* 20: 282–90
Cornett, R. H., Chin, G. Knapp, G. R. 1977. *Astron. Astrophys.* 54: 889–94
Cowie, L. L., McKee, C. F. 1977. *Ap. J.* 211: 135–46
Cowie, L., Laurent, C., Vidal-Madjar, A., York, D. G. 1979. *Ap. J.* 229: L81–85
Cowie, L. L. 1980. *Ap. J.* 236: 868–79
Cox, D. P. 1972. *Ap. J.* 178: 143–57
Crutcher, R. M. 1979. *Ap. J. Lett.* 231: L151–53
Dalgarno, A. 1960. *Proc. Phys. Soc.* 75: 374–77
Dalgarno, A., Roberge, W. G. 1979. *Ap. J. Lett.* 233: L25–28
Daltabuit, E., Cox, D. 1972. *Ap. J. Lett.* 173: L13–17
Daltabuit, E., MacAlpine, G. M., Cox, D. P. 1978. *Ap. J.* 219: 372–80
Davidson, R. C., Hammer, D. A., Haber, I., Wagner, C. E. 1972. *Phys. Fluids* 15: 317–33
de Jong, T. 1973. *Astron. Astrophys.* 26: 297–313
de Jong, T., Chu, S.-I., Dalgarno, A. 1975. *Ap. J.* 199: 69–78
DeNoyer, L. K. 1978. *MNRAS* 183: 187–93
DeNoyer, L. K. 1979a. *Ap. J. Lett.* 228: L41–43
DeNoyer, L. K. 1979b. *Ap. J. Lett.* 232: L165–68
Dickel, H. R. 1974, *Astron. Astrophys.* 31: 11–16
Dickel, J. R., Dickel, H. R., Crutcher, R. M. 1976. *Publ. Astron. Soc. Pac.* 88: 840–43
Dopita, M. A. 1976. *Ap. J.* 209: 395–401
Dopita, M. A. 1977. *Ap. J. Suppl.* 33: 437–49
Dopita, M. A., Mathewson, D. S., Ford, V. L. 1977. *Ap. J.* 214: 179–88
Dopita, M. A. 1978a. *Astron. Astrophys.* 63: 237–41
Dopita, M. A. 1978b. *Ap. J. Suppl.* 37: 117–44
Dopita, M. A. 1979, *Ap. J. Suppl.* 40: 455–74
Dopita, M. A., Mathewson, D. S. 1979. *Ap. J. Lett.* 231: L147–50
Draine, B. T., Salpeter, E. E. 1979. *Ap. J.* 231: 438–55
Drawin, H. W. 1969. *Z. Phys.* 225: 483–94
Eichler, D. 1979. *Ap. J.* 229: 419–23
Elitzur, M., de Jong, T. 1978. *Astron. Astrophys.* 67: 323–32
Elitzur, M., Watson, W. D. 1978. *Ap. J. Lett.* 222: L141–4 (Erratum 226: L157)
Elitzur, M. 1979a. *Ap. J.* 229: 560–66
Elitzur, M. 1979b. *Astron. Astrophys.* In press
Elitzur, M. 1979c. *Astron. Astrophys.* In press
Elitzur, M., Watson, W. D. 1980. *Ap. J.* In press
Elmegreen, B. G., Lada, C. J. 1977. *Ap. J.* 214: 725–41
Elmegreen, B. G., Elmegreen, D. M. 1978. *Ap. J.* 220: 1051–62
Elmegreen, B. G., Dickenson, D. F., Lada, C. J. 1978. *Ap. J.* 220: 853–63
Elmegreen, B. G., Moran, J. M. 1979. *Ap. J. Lett.* 227: L93–96
Erpenbeck, J. J. 1962. *Phys. Fluids* 5: 1181–87
Field, G. B. 1965. *Ap. J.* 142: 531–67
Field, G. B., Rather, J. D. G., Aannestad, P. A., Orszag, S. A. 1968. *Ap. J.* 151: 953–75
Fisk, L. 1971. *J. Geophys. Res.* 76: 1662–75
Ford, H. C., Butcher, H. 1979. *Ap. J. Suppl.* 41: 147–68
Fosbury, R. A. E., Mebold, U., Goss, W. M., Dopita, M. A. 1978. *MNRAS* 183: 549–68
Gardner, C. S., Kruskal, M. D. 1964. *Phys. Fluids* 7: 700–6
Gautier, T. N. III, Fink, U., Treffers, R. R., Larson, H. P. 1976. *Ap. J. Lett.* 207: L129–34
Gerola, H., Seiden, P. E. 1978. *Ap. J.* 223: 129–39
Giuliani, J. L. 1979. *Ap. J.* 233: 280–93
Goldsmith, P. F., Langer, W. D. 1978. *Ap. J.* 222: 881–95

Gorenstein, P., Harnden, F. R., Tucker, W. H. 1974. *Ap. J.* 192:661–76
Grasdalen, G. L., Joyce, R. R. 1976. *Bull. Am. Astron. Soc.* 8:349
Gronenschild, E. H. B. M., Mewe, R. 1979. *Astron. Astrophys.* In press
Gull, T. R., Sofia, S. 1979. *Ap. J.* 230:782–85
Gwinn, W. D., Blackman, G. L., Turner, B. E., Goss, W. M. 1973. *Ap. J.* 179:789–813
Hartquist, T. W., Oppenheimer, M., Dalgarno, A. 1980. *Ap. J.* 236:182–88
Hill, J. K., Hollenbach, D. J. 1978. *Ap. J.* 225:390–404
Hollenbach, D. J., Chu, S.-I., McCray, R. 1976. *Ap. J.* 208:458–67
Hollenbach, D. J., McKee, C. F. 1979. *Ap. J. Suppl.* 41:555–94
Hollenbach, D. J., McKee, C. F. 1980. *Ap. J. Lett.* In press
Hollenbach, D. J., Shull, J. M. 1977. *Ap. J.* 216:419–26
Holman, G. D., Ionson, J. A., Scott, J. S. 1979. *Ap. J.* 228:576–81
Holzer, T. E., 1972. *J. Geophys. Res.* 77:5407–31
Iglesias, E. R., Silk, J. 1978. *Ap. J.* 226:851–57
Itoh, H. 1977. *Publ. Astron. Soc. Jpn.* 29:813–30
Itoh, H. 1979. *Publ. Astron. Soc. Jpn.* In press
Jaffrin, M. Y., Probstein, R. F. 1964. *Phys. Fluids* 7:1658–74
Jaffrin, M. Y. 1965. *Phys. Fluids* 8:606–25
Jenkins, E. B., Silk, J., Wallerstein, G. 1976a. *Ap. J.* 209:L87–92
Jenkins, E. B., Silk, J., Wallerstein, G. 1976b. *Ap. J. Suppl.* 32:681–714
Jenkins, E. B., Silk, J., Wallerstein, G. 1980. *Ap. J.* In press
Jokipii, J. R. 1979. In *Workshop on Particle Acceleration in Astrophysics*, ed. J. Arons, C. E. Max, C. F. McKee, pp.1–9. New York: AIP. 433 pp.
Jura, M. 1975a. *Ap. J.* 197:575–80
Jura, M. 1975b. *Ap. J.* 197:581–85
Kamper, K., van den Bergh, S. 1976. *Ap. J. Suppl.* 32:351–66
Kamper, K. W., van den Bergh, S. 1978. *Ap. J.* 224:851–56
Kaplan, S. A. 1966. *Interstellar Gas Dynamics*, pp. 40–74. Oxford: Pergamon. 126 pp.
Kato, T. 1977. *Publ. Astron. Soc. Jpn.* 29:369–85
Kennel, C. F., Sagdeev, R. Z. 1967. *J. Geophys. Res.* 72:3303–26
Kent, S. M., Sargent, W. L. W. 1979. *Ap. J.* 230:667–80
Kirkpatrick, R. C. 1972. *Ap. J.* 176:381–93
Kirshner, R. P., Taylor, K. 1976. *Ap. J. Lett.* 208:L83–86
Kirshner, R. P., Arnold, C. N. 1979. *Ap. J.* 229:147–52
Koski, A. T., Osterbrock, D. E. 1976. *Ap. J. Lett.* 203:L49–51
Krymsky, G. F. 1977. *Dokl. Akad. Nauk SSSR* 234:1306–7
Kutner, M. L., Evans, N. J. II, Tucker, K. D. 1976. *Ap. J.* 209:452–61
Kutner, M. L., Guelin, M., Evans, N. J., Tucker, K. D., Miller, S. C. 1979. *Ap. J.* 227:121–25
Kwan, J. 1977. *Ap. J.* 216:713–23
Kwan, J., Scoville, N. 1976. *Ap. J.* 210:L39–43
Lada, C. J., Black, J. H. 1976. *Ap. J. Lett.* 203:L75–79
Lada, C. J., Oppenheimer, M., Hartquist, T. W. 1978. *Ap. J. Lett.* 226:L153–56
Lada, C. J., Elmegreen, B. G., Blitz, L. 1979. In *Protostars and Planets*, ed. T. Gehrels, pp. 341–67. Tucson: Univ. Ariz. Press. 649 pp.
Lampe, M., Papadopoulos, K. 1977. *Ap. J.* 212:886–90
Lasker, B. M. 1977. *Ap. J.* 212:390–96
Lerche, I., Vasyliunas, V. M. 1976. *Ap. J.* 210:85–99
Leung, C. M. 1978. *Ap. J.* 225:427–41
Litvak, M. M. 1972. In *Atoms and Molecules in Astrophysics*, ed. T. R. Carson, M. J. Roberts, pp. 201–76. New York: Academic. 352 pp.
London, R., McCray, R., Chu, S.-I. 1977. *Ap. J.* 217:442–47
Lucke, R. L., Zarnecki, J. C., Woodgate, B. E., Culhane, J. L., Socker, D. G. 1979. *Ap. J.* 228:763–70
Manfroid, J. 1977. *Astron. Astrophys.* 58:437–38
Maran, S. P. Hobbs, R. W., Brown, R. L., Jura, M., Knapp, G. R. 1979. *Astron. J.* 84:1709–12
McCray, R., Stein, R. F., Kafatos, M. 1975. *Ap. J.* 196:565–70
McCray, R., Snow, T. P. 1979. *Ann. Rev. Astron. Astrophys.* 17:213–40
McKee, C. F. 1970. *Phys. Rev. Lett.* 24:990–94
McKee, C. F. 1974. *Ap. J.* 188:335–39
McKee, C. F., Cowie, L. L. 1975. *Ap. J.* 195:715–26
McKee, C. F., Cowie, L. L. 1977. *Ap. J.* 215:213–25
McKee, C. F., Ostriker, J. P. 1977. *Ap. J.* 218:148–169
McKee, C. F., Cowie, L. L., Ostriker, J. P. 1978. *Ap. J. Lett.* 219:L23–26
Miller, J. S. 1974. *Ap. J.* 189:239–48
Minkowski, R., Osterbrock, D. 1959. *Ap. J.* 129:583–92
Moore, W. E., Garmire, G. P. 1976. *Ap. J.* 206:247–53
Morrison, P. Jr. 1979. PhD thesis. Univ. Maryland, College Park
Mott-Smith, H. 1951. *Phys. Rev.* 82:885–92
Mouschovias, T. 1976. *Ap. J.* 207:141–58

Mufson, S. L. 1975. *Ap. J.* 202:372–88
Mullan, D. J. 1971. *MNRAS* 153:145–70
Murray, S. S., Fabbiano, G., Fabian, A., Epstein, A., Giacconi, R. 1979. *Ap. J. Lett.* 234:L69–72
Nadeau, D., Geballe, T. 1979. *Ap. J. Lett.* 230:L169–72
Norman, C., Silk, J. 1980a. *IAU Symp. No. 87, Interstellar Molecules*, ed. B. Andrews. Dordrecht: Reidel. In press
Norman, C., Silk, J. 1980b. *Ap. J.* In press
Ogden, P. M., Roesler, F. L., Larson, H. P., Smith, H. A., Reynolds, R. J., Scherb, F. 1979. *Ap. J. Lett.* 233:L21–24
Ogilvie, K. W., Scudder, J. D. 1979. *Space Sci. Rev.* 23:123–33
Oort, J. H. 1977. *Ann. Rev. Astron. Astrophys.* 15:295–362
Osterbrock, D. E., Costero, R. 1973. *Ap. J. Lett.* 184:L71–74
Osterbrock, D. E., Dufour, R. J. 1973. *Ap. J.* 185:441–51
Ostriker, J. P., Silk, J. 1973. *Ap. J. Lett.* 184:L113–16
Papadopoulos, K., Davidson, R. C., Dawson, J. M., Haber, I., Hammer, D. A., Krall, N. A., Shanny, R. 1971a. *Phys. Fluids* 14:849–57
Papadopoulos, K., Wagner, C. E., Haber, I. 1971b. *Phys. Rev. Lett.* 27:982–86
Papadopoulos, K., McBride, J. B. 1973. *Phys. Fluids* 16:711–13
Parker, E. N. 1961. *J. Nucl. Energy C* 2:146–53
Parker, R. A. R. 1967. *Ap. J.* 149:363–72
Peimbert, M., Torres-Peimbert, S. 1977. *MNRAS* 179:217–34
Peimbert, M., Torres-Peimbert, S., Rago, J. F. 1978. *Ap. J.* 220:516–24
Pesses, M. E. 1979. In *Workshop on Particle Acceleration in Astrophysics*, ed. J. Arons, C. E. Max, C. F. McKee, pp. 107–13. New York: AIP. 433 pp.
Pikel'ner, S. B. 1954. *Izv. Krym. Astrofiz. Obs.* 12:94–110
Prasad, S. S., Huntress, W. T. 1979. *Ap. J. Suppl.* 42: No. 3
Pravdo, S. H., Smith, B. W. 1979. *Ap. J. Lett.* 234:L195–98
Raymond, J. C. 1976. *Theoretical Models of Shock Waves in the Interstellar Medium*, pp. 58–168. PhD thesis. Univ. Wisc., Madison
Raymond, J. C., Black, J. H., Dupree, A. K., Hartmann, L., Wolff, R. S. 1980a. *Ap. J.* In press
Raymond, J. C., Davis, M., Gull, T., Parker, R. A. R. 1980b. In preparation
Raymond, J. C. 1979. *Ap. J. Suppl.* 39:1–27
Roberts, W. W. 1969. *Ap. J.* 158:123–44
Russell, C. T., Greenstadt, E. W. 1979. *Space Sci. Rev.* 23:3–37
Saito, M., Saito, Y. 1977. *Publ. Astron. Soc. Jpn.* 29:387–414
Schwartz, R. D. 1975. *Ap. J.* 195:631–42
Schwartz, R. D. 1978. *Ap. J.* 223:884–900
Schwartz, R. D., Dopita, M. A. 1980. *Ap. J.* 236:543–48
Sgro, A. G. 1975. *Ap. J.* 197:621–34
Shapiro, P. R., Moore, R. T. 1976. *Ap. J.* 207:460–83
Shcheglov, P. V. 1968. *Sov. Astron.-AJ.* 11:567–71
Shemansky, D. E., Sandel, B. R., Broadfoot, A. L. 1979. *Ap. J.* 231:35–47
Shmeld, I. K., Strel'nitskii, V. S., Muzylev, V. V. 1976. *Sov. Astron.-AJ* 20:411–18
Shu, F. H., Milione, V., Gebel, W., Yuan, C., Goldsmith, D. W., Roberts, W. W. 1972. *Ap. J.* 173:557–92
Shu, F. H. 1978. In *Structure and Properties of Nearby Galaxies, IAU Symp. No. 77*, ed. E. M. Berkhuijsen, R. Wielebinski, pp. 139–45. Dordrecht: Reidel. 307 pp.
Shull, J. M., Hollenbach, D. J. 1978. *Ap. J.* 220:525–37
Shull, J. M., McKee, C. F. 1979. *Ap. J.* 227:131–49
Silk, J., Burke, J. R. 1974. *Ap. J.* 190:11–18
Simon, M., Righini-Cohen, G., Joyce, R. R., Simon, T. 1979. *Ap. J. Lett.* 230:L175–78
Smith, D. F., Krall, N. A. 1974. *Ap. J. Lett.* 194:L163–65
Solinger, A., Rappaport, S., Buff, J. 1975. *Ap. J.* 201:381–86
Spitzer, L. 1962. *Physics of Fully Ionized Gases*, p. 135. New York: Wiley. 170 pp. 2nd ed.
Spitzer, L., Cochran, W. D. 1973. *Ap. J. Lett.* 186:L23–27
Spitzer, L., Cochran, W. D., Hirshfeld, A. 1974. *Ap. J. Suppl.* 28:373–95
Spitzer, L., Morton, W. A. 1976. *Ap. J.* 204:731–49
Spitzer, L. 1978. *Physical Processes in the Interstellar Medium*, pp. 246–96. New York: Wiley. 318 p.
Stark, J. P. W., Culhane, J. L. 1978. *MNRAS* 184:509–22
Steigman, G., Strittmatter, P. A., Williams, R. E. 1975. *Ap. J.* 198:575–82
Stone, M. E. 1970. *Ap. J.* 159:277–92
Strel'nitskii, V. S., Sunyaev, R. A. 1973. *Sov. Astron.-AJ.* 16:579–84
Swan, G. W., Fowles, G. R. 1975. *Phys. Fluids* 18:28–35
Talbot, R. J., Jr., Arnett, W. D. 1973. *Ap. J.* 186:51–68
Thompson, R. I., Lebofsky, M. J., Rieke, G. H. 1978. *Ap. J. Lett.* 222:L49–54
Tidman, D. A., Krall, N. A. 1971. *Shock Waves in Collisionless Plasmas*. New York: Wiley. 175 pp.
Treffers, R. R. 1979. *Ap. J. Lett.* 233:L17–20

Tsunemi, H., Inoue, H. 1979. *Publ. Astron. Soc. Jpn.* In press
Tuohy, I. R., Nousek, J. A., Garmire, G. P. 1979. *Ap. J. Lett.* 234: L101–5
van den Bergh, S. 1971. *Ap. J.* 165: 457–69
Weaver, R., McCray, R., Castor, J., Shapiro, P., Moore, R. 1977. *Ap. J.* 218: 377–95
Weibel, E. S. 1959. *Phys. Rev. Lett.* 2: 83–84
Wentzel, D. G. 1971. *Ap. J.* 170: 53–64
Woodgate, B. E., Angel, J. R. P., Kirshner, R. P. 1974. *Ap. J. Lett.* 188: L79–82
Woodgate, B. E., Angel, J. R. P., Kirshner, R. P. 1975. *Ap. J.* 200: 715–18
Woodgate, B. E., Kirshner, R. P., Balon, R. J. 1977. *Ap. J. Lett.* 218: L129–31
Woodward, P. R. 1976. *Ap. J.* 207: 484–501
Wootten, H. A. 1977. *Ap. J.* 216: 440–45
Zel'dovich, Ya. B., Raizer, Yu. P. 1966. *Physics of Shock Waves and High Temperature Hydrodynamic Phenomena*, Vols. 1, 2. New York: Academic. 916 pp.

Ann. Rev. Astron. Astrophys. 1980. 18 : 263–88

ENVELOPES AROUND LATE-TYPE GIANT STARS

⁎2167

B. Zuckerman
Astronomy Program, University of Maryland, College Park, Maryland 20742
and Astronomy Department, University of Texas at Austin, Austin, Texas 78712

1 INTRODUCTION

Giant stars have long been a favorite of optical astronomers. Now the cool circumstellar matter that surrounds most giants can also be studied in the infrared and microwave domains. Because of the very many published research papers that relate to circumstellar material, the present review cannot do justice to all of the early classical work in this field. Emphasis is placed, rather, on the most recent studies in which appropriate references to classical papers may be found (the opposite is not likely to be the case). We consider the following topics: chemical composition, physical properties, masers, mass loss, and late stellar evolution. Because of space limitations, some topics have received less coverage than their importance would otherwise warrant.

2 CHEMICAL COMPOSITION

The chemical composition of circumstellar material is determined by the elemental mixture and the physical conditions in the circumstellar environment. In many cases the elemental composition can best be determined from studies, not of the circumstellar shell itself, but of the underlying photosphere. Except, possibly, for a few unusual objects (e.g. FG Sge), it is safe to assume that the elemental mix in a stellar atmosphere is the same as the mixture was during the preceeding 10^4 years when the material in the circumstellar envelope was being lost from the star. Therefore, elemental and isotopic abundances that are determined for the photosphere from infrared and optical spectroscopy may usually be used as a constraint to

0066-4146/80/0915-0263$01.00

help us understand chemical processes, such as isotopic fractionation and dust grain formation, that may take place in the circumstellar envelope.

Our knowledge of the elemental and isotopic compositions of the photospheres of cool stars has recently been reviewed by Merrill & Ridgway (1979) and by Lambert (1980). In terms of the appearance of the stellar spectrum, the most important abundance ratio is [C]/[O]. Stellar spectra are characterized as oxygen-rich (K and M stars, O/C > 1), carbon-rich (R and N stars, C/O > 1), and S-type (C/O ~ 1). There exist recent catalogues of carbon stars (Stephenson 1973) and S stars (Stephenson 1976, Yorka & Wing 1979). Spectra and spectral classification of S stars are discussed by Scalo & Ross (1976), Wyckoff & Clegg (1978), and Ake (1979). Most S stars probably have $0.95 \lesssim C/O \lesssim 1.00$.

2.1 Gas Phase Composition

Hydrogen is the most abundant element in most stars. We are interested in whether it is in atomic or molecular form in the circumstellar envelopes. The 21-cm line of atomic hydrogen has been searched for, but not detected, toward various evolved stars (Zuckerman et al. 1980). The best limits were obtained for the very cold carbon stars IRC+10216 and CIT6 and oxygen-rich stars IRC+10011 and NML Tau. In all cases no more than 10% by mass of the circumstellar hydrogen is in atomic form. The best limit, $\lesssim 1\%$ H by mass, is for IRC+10216 if the model of Kwan & Hill (1977) is adopted or $\lesssim 0.3\%$ H by mass if, instead, more recent $J = 2 \to 1$ CO data (Knapp et al. 1980) are utilized.

That the preponderance of the hydrogen is molecular is not surprising if chemical equilibrium abundances are achieved in the photosphere and then these same abundances are maintained in the circumstellar shell. However, Balmer emission in Miras indicates that at least some of the hydrogen is atomic at least some of the time (possibly due to shocks in the atmosphere). Also, the ambient interstellar radiation field will photodissociate molecular hydrogen far from the star. Thus, the 21-cm limits place constraints on the importance of these and other mechanisms of dissociation. For α Ori the much higher photospheric temperature suggests that much (most) of the hydrogen will be atomic but, unfortunately, strong 21-cm emission from background hydrogen rendered the Arecibo measurement nearly useless in spite of considerable effort to subtract out this background.

For other elements the [C]/[O] ratio is probably the most important in determining the overall chemical composition in the envelope. From photospheric spectra one can easily deduce if C/O is greater or less than unity. But quantitative determinations of C/O in M giants are still lacking.

For carbon stars IRC+10216 is the prototype and 17 molecules have been identified in its envelope either in the radio (CO, CN, CS, HCN, HNC,

C_2H, SiS, SiO, HC_3N, HC_5N, HC_7N, C_4H, C_3N, CH_3CN) or in the infrared (C_2H_2, CH_4, NH_3) or in both. Except for a small amount of SiO no oxygen-containing molecules (other than CO of course) have been detected. Indeed, the [SiS]/[SiO] ratio is greater than unity as expected only in a carbon-rich environment (Tsuji 1973). Also striking is the absence of the molecular ions HCO^+ and N_2H^+ (Wannier & Linke 1978) which are abundant in interstellar molecular clouds. This suggests that positive-ion neutral-molecule reactions are not important in the envelope around IRC +10216. Indeed McCabe et al. (1979, see also Clegg 1980 and Clegg & Wootten 1980) have attempted to match the measured abundances with a "freeze-out" model. This supposes that chemical equilibrium holds close to the photosphere but that in the circumstellar shell, temperature and pressure drop quickly enough that no significant additional chemistry takes place. The observed (HCN)/(HNC) ratio ($\gtrsim 100$) in IRC+10216 is very different from the interstellar ratio (~ 1) but nicely matches the ratio expected for chemical equilibrium at ~ 1000 K. More generally, the freeze-out model can match the relative abundances of most of the observed molecules but at pressures that correspond to mass loss rates at least 100 times larger than observed. Also, the agreement with observations is much poorer if carbon is allowed to condense into graphite grains—believed to be a likely occurrence in IRC+10216 [but see Lewis & Ney (1979) for the opposite view]. Finally, the model cannot match the observed abundances of NH_3, CN, SiO, and SiS all of which may either react on or be removed by grains. Betz & McLaren (1980) suggest, for example, that the relative abundance of NH_3 is rather more in equilibrium with the dust temperature than frozen at the photospheric value.

The molecular inventory in envelopes around oxygen-rich stars is sparser: OH, H_2O, and SiO masers (Engels 1979); thermal millimeter wavelength emission from SiO and CO (Morris et al. 1979, Lo & Bechis 1977, Zuckerman et al. 1977, 1978); NH_3 and CO absorption in the infrared (Betz & McLaren 1980, Hall 1980, Bernat et al. 1979); and possibly H_2O in emission and absorption in the infrared (Hinkle 1978, Hinkle & Barnes 1979, Tsuji 1978b). SiS emission is not detected (P. Palmer and B. Zuckerman, 1978, private communication), so [SiS]/[SiO] is less than unity as expected in an oxygen-rich environment.

Calculations of Tsuji (1973) and Vardya (1966) suggest that SiO is the dominant gas phase carrier of silicon in these stars. Millimeter wavelength observations of Morris et al. (1979) and Lambert & Vanden Bout (1978) suggest that $\sim 99\%$ of the silicon is in the grains in the outer parts of the circumstellar shell but this is unlikely to be the case in the inner shell where the SiO maser is produced. Thus the bulk of the silicon is probably incorporated into grains between 10^{14} and 10^{16} cm from the central star in agreement with the inner dust boundary determined from 11-μm interferometry for a limited sample of stars (see Section 3.1.2).

At temperatures above 500 K, OH molecules are destroyed within the characteristic expansion time by the exothermic reaction $OH + H_2 \rightarrow H_2O + H$. The OH maser molecules, observed mainly in the outer parts of the circumstellar envelope, may be produced by formation on the surface of dust grains or by dissociation of H_2O molecules either by interstellar ultraviolet photons or by collisions with high speed grains (Goldreich & Scoville 1976, Kwok 1976). An additional source of ultraviolet photons might be a small, relatively hot companion to the M star such as the companion of Mira herself. Searching for such companion stars with IUE appears feasible (A. G. Michalitsianos, 1979, private communication).

2.2 Isotopic Abundances

There exist only a few measurements of isotopic abundance ratios in circumstellar material. Bernat et al. (1979) observed the ^{12}CO and ^{13}CO fundamentals at 4.6 μm in α Ori and found $\zeta \equiv [^{12}C]/[^{13}C] \approx 6$. This agrees with one of two discrepant photospheric values for ζ in α Ori. Morris (1980) derives a similar low value for ζ ($=8\pm3$) in carbon-rich CRL 2688. On the other hand, microwave measurements suggest that in many stars with large mass loss rates ζ is fairly large. Specifically, in the carbon star IRC +10216, $\zeta \gtrsim 30$ (Wannier & Linke 1978, Kwan & Hill 1977), in the carbon star CIT6, $\zeta \gtrsim 20$ (Knapp et al. 1979), and in the (probably carbon-rich) planetary nebula NGC7027, $\zeta \gtrsim 20$ (Mufson et al. 1975). This ratio for IRC +10216 is in some disagreement with a 4.6-μm measurement of $\zeta \approx 20$ (Barnes et al. 1977) for the innermost part of the circumstellar shell. This might be due to an actual radial variation in ζ throughout the envelope but, more likely, based on more recent observations, the infrared ratio requires revision upward to a value between 50 and 100 (D. N. B. Hall, 1979, private communication). Lower limits to the $[^{12}C]/[^{14}C]$ ratio of 10^4 and 10^3 have been measured in the infrared (Barnes et al. 1977) and microwave (Rodriguez-Kuiper et al. 1977) spectra of IRC+10216.

For other elements Wannier & Linke (1978) measured $[^{14}N]/[^{15}N]$ and $[^{16}O]/[^{18}O]$ ratios in IRC+10216 that are significantly larger than terrestrial. $[^{17}O]/[^{18}O]$ also seems to be enhanced relative to its terrestrial value. Qualitatively, but not necessarily quantitatively, similar results on the oxygen isotopes in IRC+10216 have been obtained in the infrared (Barnes et al. 1977, Rank et al. 1974). Finally, the $[^{32}S]/[^{34}S]$ ratio in IRC +10216 agrees with the terrestrial value (Wannier & Linke 1978).

A large $[^{14}N]/[^{15}N]$ ratio is indicative of cold CNO processing. Since $[C]/[O] > 1$ in conjunction with $[C]/[H] \gtrsim$ solar (Kwan & Hill 1977) implies carbon production in the 3α process, we have evidence in IRC +10216 for the products of two different nuclear burning processes.

2.3 Dust Grains

There appears to be some, although by no means universal, agreement that the grains in oxygen-rich stars are mainly "silicates" (e.g. Mg_2SiO_4) and in carbon stars mainly graphite and SiC (see, for example, Gilman 1969). The identifications rest largely on the 9.7-μm "silicate" feature and the 11.2-μm emission feature of SiC (Gilra 1973, Treffers & Cohen 1974). SiC particles may also be responsible for the strong violet opacity in carbon stars (Gilra 1973) although Bregman & Bregman (1978) argue in favor of the C_3 pseudo-continuum originally proposed long ago to explain the violet opacity (see Gilra 1973 for a history).

The otherwise featureless infrared continuum observed in carbon stars and planetary nebulae (Mathis 1978) is often attributed to graphite (but see Czyzak & Santiago 1973, Lewis & Ney 1979). The absence of a clear spectral signature leaves open the possibility of other types of grains. Webster (1978) has suggested that carbyne, a high temperature allotrope of carbon (Whittaker 1978), might form in carbon-star atmospheres before graphite but Clegg (1980) has argued that this seems unlikely in such cool atmospheres. The rare R Cr B stars are much hotter (~ 6000 K) and, conceivably, conditions might be conducive to the formation of some carbyne but, even so, little would be produced on a galactic scale. If expanding supernovae remnants are an important source of refractory interstellar grains (Dwek & Scalo 1980) and part of the remnant is dense and carbon-rich, then carbyne might form. Progenitors of some Type II supernovae may be massive carbon stars and the R Cr B stars may be the progenitors of some Type I supernovae (Wheeler 1978).

Another possibility is that the grains are made of amorphous carbon (Savage & Mathis 1979) possibly with an admixture of hydrogen in some kind of soot (Clegg 1980). However, infrared features near 3.3, 6.0, and 6.8 μm attributable to various stretching and bending vibrations in CH, CO, CH_2, and CH_3 (Willner et al. 1980) are absent or very weak in IRC+10216 and GL 3068 (Merrill & Stein 1976, Jones et al. 1978, S. P. Willner, 1979, private communication). Thus, some form of "pure" carbon dust is probably more abundant than hydrocarbon dust around carbon stars (S. P. Willner, 1979, private communication).

Two more exotic suggestions are organic polymers (Hoyle & Wickramasinghe 1977) and tholins (Sagan & Khare 1979) both of which are sufficiently nonspecific that they possibly could form in both protostellar nebulae and evolved stars. However, the organic polymer idea has been often criticized (see Egan & Hilgeman 1978, Sagan & Khare 1979) and the tholins are really envisaged as the products of protostellar nebulae and, therefore, fortunately, need not concern us here.

How large are circumstellar grains? Theoretical estimates based on considerations of nucleation and sputtering rates are inconclusive (Salpeter 1974, Kwok 1975, Lefèvre 1979, Phillips 1979) and so are the observations. For example, the presumed graphite and SiC grains in IRC+10216 are estimated to have radii $a \sim 1$ µm from a fit to the far-infrared spectrum (Campbell et al. 1976). But impure grains with enhanced far-IR emissivities could be a lot smaller. In the carbon-rich Egg Nebula (CRL 2688) the very large percentage polarization in the scattered light implies $a < 0.1$ µm (Schmidt et al. 1978) whereas if the grains in planetary nebulae (PN's) are graphite then the absence of a $\lambda 2200$ Å feature in their spectra implies $a > 0.04$ µm (Mathis 1978). It might be argued (Kwok 1980) that the grains in PN's form in the ionized gas and therefore may be different from grains that form in the neutral clouds around IRC+10216 and CRL 2688. However, since molecular clumps exist inside PN's (Beckwith et al. 1978) the bulk of the dust emission may be associated with this neutral gas rather than the ionized gas (see also Scalo & Shields 1979, Shields 1980).

Models for oxygen-rich Miras yield silicate grain radii between 10^{-1} and 10^{-2} µm (Cahn & Wyatt 1978b, Michalitsianos & Kafatos 1978) or between 1 and 10^{-1} µm (Phillips 1979). Shawl (1975) and Materne (1976) deduce grain dimensions $\lesssim 0.1$ µm around Miras from polarization data (but the polarization may be partially atmospheric, see Section 3.2, rather than circumstellar). For nonvariable M stars, Kwok (1975) estimated $10^{-2} \lesssim a \lesssim 10^{-1}$ µm and Jura & Jacoby (1976) estimated that $a \lesssim 0.05$ µm for the grains in the circumstellar shell around α Ori and predicted that polarized scattered light from the shell might be detected (McMillan & Tapia 1978).

3 PHYSICAL PROPERTIES

The radial and azimuthal distribution of material in the circumstellar shell yields information on the physical mechanisms responsible for mass loss. For example, an r^{-2} density distribution is suggestive of a steady wind whereas a nonuniform density profile is suggestive of ejection of matter in discrete puffs, perhaps the result of periodic shock waves or even more dramatic abrupt ejection events as in some models of the origin of planetary nebulae.

3.1 *Radial Structure*

We may divide the outer layers of late-type giants into three regions: 1. the photosphere which is gravitationally bound to the star, 2. an overlying chromosphere/corona that exists around some but probably not all late-type giants (mass outflow at $V < V_\infty$ is probably taking place in this region), 3. a cool, expanding circumstellar shell that is not gravitationally bound to the star. Unfortunately, the precise structures of regions 1, 2, and the inner parts of 3 are still not understood, especially for pulsating stars. This situation should improve in the next few years due to advances in infrared (see, for example, Hall 1980) and ultraviolet (IUE) observations. Therefore, our review of photospheric and chromospheric structure will be brief, especially since the photosphere, which has been discussed recently by Merrill & Ridgway (1979), is somewhat beyond the scope of this review (and this reviewer).

3.1.1 PHOTOSPHERE/CHROMOSPHERE Early M giants and supergiants do not appear to be pulsating to a significant degree. Therefore, their atmospheres should be simpler to model than those of long period variables (LPV's). Still there are unresolved problems. For example, in a series of papers, Tsuji (1978c, 1979, and references therein) has attempted to reconcile photometric and interferometric diameters for α Ori. Optical interferometry ("the direct technique") yields a diameter a factor of 1.5–2.0 times larger than does the photometry. Tsuji attributes this discrepancy to scattering by dust inside of $\sim 3R_*$ in the circumstellar envelope. There are two problems with Tsuji's solution. The first is the implied blackbody dust temperatures which are significantly above the ~ 1000 K condensation temperature of magnesium silicates (see, for example, Salpeter 1977). This might not be an insurmountable difficulty in view of the possibility of an inverse greenhouse effect (see Section 3.1.2 and Weymann 1977) that cools the dust. The other problem is that Tsuji used interferometric data of Sutton et al. (1977) which have since been superseded by improved data (Sutton 1979a,b, and 1979, private communication). The flatness of this more recent visibility curve seems to rule out the presence of dust within $3R_*$ of α Ori. If Tsuji's solution is not the correct one then the discrepancy between the photometric and interferometric diameters still remains to be explained. Perhaps the photometric technique is not so accurate for low effective temperatures (Blackwell & Shallis 1977, E. C. Sutton, 1979, private communication). We note in passing a third technique for obtaining photospheric angular diameters, that of model atmospheres (Tsuji 1978a, Scargle & Strecker 1979). For α Ori this technique yields a diameter roughly midway between the other two.

There appears to be a variety of observational evidence (infrared, microwave, and ultraviolet) for a warm (~ 5000 K) chromosphere but probably not a hot ($> 20,000$ K) corona in α Ori (Lambert & Snell 1975, Linsky & Haisch 1979, O'Brien & Lambert 1979, Sutton 1979a, Altenhoff et al. 1979, Bowers & Kundu 1979). α Her may also possess a chromosphere (O'Brien & Lambert 1979). Humphreys (1974) and Gilman (1974) have argued for the existence of chromospheres in some late M supergiants (e.g. S

Per and VX Sgr). However, for the latter two stars, Frawley (1977) suggests an alternate model in which all optical and infrared observations can be explained by a combination of photospheric and thermal dust emission and a chromosphere is not required.

Our primary interest in the chromosphere concerns its relationship to mass loss. Hα and calcium H and K line asymmetries (Stencel 1978, Boesgaard & Hagen 1979 and references therein) suggest that outflow is already occurring within the chromospheres of many nonvariable M stars. One possibility is that dust has somehow condensed in the chromosphere and radiation pressure on the dust drives the mass loss [see Weymann (1977) for a summary of the arguments]. Here we mention only a few recent developments all of which seem to go against a dust-driven chromospheric expansion. Most direct is the 11-μm interferometry of Sutton and collaborators (see Section 3.1.2) which does not indicate dust emission inside of $\sim 5R_*$.

Since the scale height that corresponds to thermal motions in the gas is small an additional mechanism is needed to "support" the atmosphere out to the dust condensation point. A recent study of velocity fields in the shell around α Ori suggests that the observed linewidths can be reproduced with marginally subsonic turbulence which may not provide the necessary support (Boesgaard 1979). In this picture upwelling material inside of $\sim 2R_*$ is driven by photospheric convection (Schwarzschild 1975) or by radiation pressure on molecules such as CO (Maciel 1976, 1977). Part of this material returns in free-fall to the photosphere while other material moves outward into the circumstellar envelope possibly becoming supersonic at the top of the chromosphere and driving mass loss through a hydrodynamic expansion of a hot corona (Mullan 1978). Mullan's model is controversial, however (see, for example, Haisch et al. 1980, Linsky & Haisch 1979). A purely thermally driven wind may be inadequate (given the temperatures implied by the IUE data), and it may be necessary to include an additional source of momentum in the wind (e.g. Alfvén waves). Although this would modify the $\dot{M}$'s estimated by Mullan at his supersonic transition locus, its location in the H-R diagram would not change significantly (D. J. Mullan, 1979, private communication).

Models for the photosphere and inner circumstellar shell of LPV's are also in disarray. One concrete advance has been the discovery of a simple method of measuring the systemic velocity of an LPV provided that SiO or CO thermal microwave emission or 1612 MHz OH maser emission is detectable (Reid & Dickinson 1976). Thus it appears that, typically, optical absorption lines are red-shifted with respect to the systemic velocity in LPV's and, also, in M-type supergiants (Humphreys 1975). With this information in hand (Wallerstein 1975, 1977, 1978a) it is possible to analyze the complex emission/absorption line velocity patterns in LPV's in terms of shock models (Wood 1978, 1979, Willson 1978, Willson & Hill 1979, Hill & Willson 1979, Wallerstein 1978a, Hinkle 1978, Hinkle & Barnes 1979). The problem is too many different shock models all purporting to explain the observations! It is not yet even clear if the pulsation is in the fundamental mode (Hill & Willson 1979) or the first overtone (Wood 1978). Perhaps Mira variables are the fundamental mode counterparts of the smaller amplitude semiregular red variables (Willson 1978), or, if Miras are actually pulsating in the first overtone, the semiregular variables are in still higher modes (Wood 1978). The pulsation mode of LPV's is relevant to mass-radius relations (Willson 1978), orbital periods in binary systems (Zuckerman 1979), and evolutionary models (Wood & Cahn 1977, Cahn & Wyatt 1978a, Tuchman et al. 1979).

Mira variables, on the average, seem to lose mass at a greater rate than other M giants and supergiants (Section 5). It is plausible that this increase is due in part to the pulsation and associated shocks but it still has not been decisively shown that pulsation is the *cause* of the increased $\dot M$. What has been shown in the models of Wood (1979) and Willson & Hill (1979) is that certain kinds of shocks can produce mass loss by themselves or greatly increase mass loss that is occurring because of radiation pressure on grains. In the latter case the increased $\dot M$ is a result of a levitation of the atmosphere by the shocks, which increases the density near the sonic point.

Evidence for chromospheric/coronal regions in LPV's is still sparse (e.g. Lambert & Snell 1975). Future observations in the ultraviolet are awaited and, even if chromospheres fail to appear, perhaps indications of small hot companion stars will (A. G. Michalitsianos, 1979, private communication)!

3.1.2 CIRCUMSTELLAR SHELL   In this section we are concerned with the radial dependence of the density, temperature, and velocity of the gas and the dust.

For uniform outflow at constant velocity the gas density varies as r^{-2}. The shape of most quasithermal CO and SiO microwave emission profiles are consistent with $\rho_{gas} \propto r^{-2}$. However, these shapes depend on many factors and tell us little about the rate of change with time of the mass loss rate (Morris 1975, Kuiper et al. 1976, Morris & Alcock 1977, Morris 1980). To obtain this quantity it is necessary to resolve the circumstellar envelope and, in IRC+10216, Wannier et al. (1979) and Knapp et al. (1980) find that the distribution of CO $2 \to 1$ emission suggests a mass loss rate changing with time over the past $\sim 10{,}000$ years. Combination of future molecular observations in the radio and infrared (Betz & McLaren 1980, Ridgway & Hall 1980, Hall 1980) should yield $\rho_{gas}(r)$ for many LPV's and late-type supergiants. In principle, the intensity of scattered KI resonance line emission (see, for example, Bernat et al. 1978, Lambert & Vanden Bout 1978) may also yield the run of gas density with r but, in practice, it may be very difficult to relate KI to hydrogen densities (A. P. Bernat, 1979, private communication).

Where does the dust form? Since oxygen-rich giants and supergiants typically have photospheric temperatures between 2000 and 3500 K and silicates condense at $T_c \lesssim 1200$ K it is to be expected that most of the dust will form at least a few stellar radii from the center of these stars. Weymann (1977) has summarized the case for an inverse greenhouse effect that might permit clean silicate grains to form much closer to the photosphere, but these arguments seem largely vitiated by the 11-μm measurements discussed below. For cool carbon stars the situation is not, a priori, so obvious since $T_\text{photosphere} \sim 2000$ K and SiC and graphite can condense at $T_c \sim 1700$ K.

The dust density distribution around evolved stars was first measured by means of lunar occultations (Toombs et al. 1972, Zappala et al. 1974) but recently infrared interferometry has emerged as a powerful technique (Sutton et al. 1977, 1978, 1979, Sutton 1979a,b, Low 1979, McCarthy et al. 1977, 1978, 1980, McCarthy 1979). This is especially evident in the 11-μm study of o Ceti (Mira) by Sutton et al. (1978). From observations over the course of one year they found that the relative fractions of the 11-μm flux contributed by the central star and the surrounding dust envelope vary throughout the light cycle. The dust emission is maximum at maximum stellar luminosity due to an increase in grain temperature and, possibly, the spatial extent over which the dust emits. By relating the change in the bolometric luminosity to the change in grain temperature and emission, Sutton et al. show that most of the dust emitting at 11 μm is at 300 K and that the dust forms in a region with $T \lesssim 700$ K.

More generally, but still for a limited sample of stars, Sutton's (1979a) visibility curves suggest that if $\rho_\text{dust} \propto r^{-2}$, then most of the dust condensation begins near a radius at which $T \sim T_c/2$ (i.e. beyond $10R_*$). Also, $\rho_\text{dust} \propto r^{-2}$ appears to fit the fringe visibility curves for IRC+10216 much better than the two-shell model suggested by the earlier lunar occultation data and, indeed, Crabtree & Martin's (1979) model of IRC+10216 with $\rho_\text{dust} \propto r^{-2}$ is consistent with the occultation data.

Alternatively, typical visibility curves can be fitted with dust condensation beginning near the radius at which $T \sim T_c$ provided that $\rho_\text{dust} \propto r^{-1.5}$. What might cause an $r^{-1.5}$ dependence? $\dot{M}$ could be declining as a function of time but this seems unlikely to be true in all cases. The gas could be decelerating but this seems unlikely at large r. Perhaps the most likely explanation for an $r^{-1.5}$ dependence is dust formation between 10^{14} and 10^{16} cm from the center of the star. That the dust forms at $r \sim 10^{15}$ cm is consistent with the infrared energy distribution and 10-μm silicate depth in the spectra of OH/IR stars (Werner et al. 1980). At larger distances from the star we have the scattered light profiles at 4500 Å of McMillan & Tapia (1978) who find ρ_{dust} somewhere in the range $r^{-1.5}$–$r^{-3.0}$ (with r^{-2} preferred) around α Ori. These results apply to r between 4×10^{16} and 2×10^{17} cm.

An additional observation that is consistent with dust formation far above the stellar photosphere is the detection of CO $J = 1 \to 0$ emission toward six rather bright N- and S-type LPV's (χ Cyg, V Cyg, R Scl, W Aql, V Hya, and R And). Excepting perhaps in χ Cyg, the mass loss rates in these stars are probably $\gtrsim 3 \times 10^{-6}$ $\dot{M}_\odot$/year (see Zuckerman 1978 for a summary). Therefore, the hydrogen projected density between the photosphere ($R_* \sim 3 \times 10^{13}$ cm) and infinity is $\gtrsim 3 \times 10^{23}$ cm^{-2}. If the dust-to-gas ratio in these molecular envelopes is comparable to the ratio in the interstellar medium, and if the bulk of the dust forms very near the photosphere, then $A_v \gtrsim 300$ is implied. Since this clearly is not the case (see, for example, Cohen 1979), either (a) the dust-to-gas ratio is much smaller than the interstellar ratio, (b) we are looking through "holes" in a non–spherically symmetric flow; (c) the dust has a very large albedo and scatters mainly in the forward direction; or, most probably, (d) the bulk of the dust forms well away from the photosphere (the dust-projected density toward the star is inversely proportional to the inner dust shell radius). We note that in the model of OH/IR stars adopted by Werner et al. (1980) within $\sim 10^{16}$ cm of the central star, the derived dust-to-gas ratio by mass is comparable to the interstellar value.

As may be inferred from the above discussion, the intensity of infrared emission as a function of distance from the central star is determined by the combined effects of $\rho_{\text{dust}}(r)$ and $T_{\text{dust}}(r)$. The two are not uniquely separable given the present observational state of the art. The dust temperature as a function of r may be estimated by assuming a model for the grains (e.g. "clean" vs "dirty" silicates; Jones & Merrill 1976, Hagen 1978) and an envelope optically thin at wavelengths at which the dust absorbs and scatters radiation from the central star. In this case one may balance the energy absorbed and reemitted by a grain and $T_{\text{dust}}(r) \propto r^{-2/(4+n)}$ when the grain emissivity, ε, is proportional to λ^{-n} (the characteristic wavelength and dust temperature may be related through Wien's law). If the inner portions of the circumstellar shell are optically thick in the infrared then the problem becomes significantly more complex and the radiative transfer must be considered explicitly (Jones & Merrill 1976). Qualitatively, this "trapping" of infrared radiation will raise the temperature of the grains in the inner portion of the shell and lower the grain temperature in the outer portions (assuming that ε decreases with λ), since these grains will be deprived of the chance to absorb the relatively short wavelength stellar radiation.

The run of gas temperature with r in the circumstellar shell is more difficult to obtain. Indeed, at present, there is no really reliable observational or theoretical way to determine $T_{gas}(r)$ (Goldreich & Scoville 1976, Kwan & Hill 1977, Lambert & Vanden Bout 1978). The best bet is to model the thermodynamics of the expanding shell. For both oxygen-rich and carbon-rich stars the major heat input is collisions between gas molecules and dust grains that are driven by radiation pressure at supersonic speeds through the gas. Adiabatic expansion and the emission of line radiation by abundant molecules are the principal sources of cooling. For oxygen-rich stars, H_2O rotational transitions are probably the dominant molecular coolant (Goldreich & Scoville 1976) and, for carbon stars, CO rotational transitions (Kwan & Hill 1977).

Next we consider the run of dust and gas velocity with radial distance from the central star. The situation close to the star is difficult to unravel (Section 3.1.1). As we have seen above, observations suggest that the bulk of the dust forms outside of 10^{14} cm. Inside of this distance, there is already some evidence (although not completely compelling) for outflow at velocities $\lesssim V_\infty/2$. For early M stars, for example, there are Hα and calcium H and K asymmetries (Boesgaard & Hagen 1979); for late M stars there is the spread in H_2O and SiO maser velocities (Spencer et al. 1979, Moran et al. 1979). However, this spread may be due to turbulence rather than outflow (Moran et al. 1979). [See Hall (1980) for a different picture of this inner region.]

Once the dust forms, theoretical estimates (Salpeter 1974, Kwok 1975, Lucy 1976, Phillips 1979) suggest that it rapidly accelerates (drags) the gas up to terminal velocity. Infrared observations of NH_3 (Betz et al. 1979, Betz & McLaren 1980), CH_4 (Hall & Ridgway 1978), CO (Ridgway et al. 1976, Ridgway & Hall 1980), and C_2H_2 and HCN (Ridgway et al. 1978) in IRC +10216 indicate that the gas has already reached terminal velocity (~ 15 km s^{-1}) at $\sim 8R_*$ (10^{15} cm). The coldest NH_3 observed in absorption in VY CMa, VX Sgr, and IRC+10420 also has been accelerated to V_∞ (Betz & McLaren 1980). In all of these stars and α Ori as well (Bernat et al. 1979), the typical infrared linewidths lie between 1 and 4 km s^{-1}. This velocity dispersion is, presumably, a measure of the combined effects of differential expansion, projection effects, and local turbulence along lines of sight to the central star. Of course, each transition samples a different region in the circumstellar shell.

Between 10^{14} and 10^{16} cm the dust formation process continues (the dust is drifting supersonically with respect to the gas and dust formation may still take place beyond the radius at which the gas has reached terminal velocity). At $r \gtrsim 10^{16}$ cm, 1612 MHz OH maser and SiO thermal emission in many oxygen-rich stars (Bowers & Kerr 1977, Nguyen-Q-Rieu et al. 1979, Morris et al. 1979) and CO thermal emission in both oxygen- and carbon-rich stars (Lo & Bechis 1977, Zuckerman et al. 1977, 1978) yield both terminal and systemic velocities. At much greater distances the molecules are photodissociated by the ambient interstellar radiation field and the gas is decelerated by the interstellar medium.

Finally, there is the question of whether the circumstellar material is distributed smoothly (e.g. $\rho_{\text{gas}} \propto r^{-2}$) or in discrete shells. We pointed out above that the 11-μm visibility curves for IRC+10216 can be fitted much better with $\rho_{\text{dust}} \propto r^{-2}$ than with the two-shell model previously suggested but not, in fact, required by the lunar occultation data. For the gas, Ridgway & Hall (1980) detect two kinematic features in the CO absorption spectrum of IRC+10216—one is expanding at V_∞ and the other, due to hotter CO, is expanding 5 km s^{-1} more slowly. Betz & McLaren (1980) find only a single kinematic feature in NH$_3$ absorption against IRC+10216, IRC+10420, VX Sgr, and VY CMa. Two discrete velocity components are evident in high resolution KI (Goldberg et al. 1975) and CO (Bernat et al. 1979) spectra of α Ori and Hall (1980) and A. P. Bernat (1979, private communication) report multiple velocity components in CO absorption toward other stars. Multiple velocity components are also evident in 1612 MHz OH masers. To summarize this point, the distribution of matter appears, more often than not, to be fairly complicated. This may be related, in part, to significant departures from spherical symmetry (see following section).

3.2 Azimuthal Shapes

Nonspherical shapes of circumstellar clouds may be due to rotation, magnetic fields, or nonradial pulsation. Whereas the envelope around α Ori appears to be roughly symmetric (Bernat & Lambert 1976, McMillan & Tapia 1978), in optical photographs some evolved stars display extensive, non–circularly symmetric envelopes, including IRC+10216 (Becklin et al. 1969, McCarthy et al. 1980), VY CMa (Herbig 1972), CRL 2688 (Ney et al. 1975, Schmidt et al. 1978), OH 0739-14 (Cohen & Frogel 1977), CRL 618 (Westbrook et al. 1975), and M1-92 (Herbig 1975). The latter four objects are classified as bi-polar or bow-tie nebulae. All six of these objects show very large percentage polarization ($\sim 30\%$) in scattered light with $\lambda \lesssim 1$ μm (Shawl & Zellner 1970, Herbig 1972, Jura 1975, Schmidt et al. 1978, Kobayashi et al. 1978, Westbrook et al. 1975, Zellner & Serkowski 1972). The polarization position angle of the scattered light is perpendicular to the long axis of the reflection nebulosities. The grain masses in the scattering nebulae in CRL 2688 and M1-92 are $\sim 10^{-4}$ $M_\odot$, comparable to those deduced from the infrared emission in some planetary nebulae (Schmidt et al. 1978). Searches for polarization in a few PN's yielded appreciable polarization only in apparent bi-polars such as M2-9 (G. D. Schmidt, 1979, private communication). Models of nonspherical PN's have been constructed by Phillips & Reay (1977) and by Fabian & Hansen (1979) and of bow-tie nebulae by Morris & Bowers (1980).

The generally much more modest degree of polarization (see, for example, Shawl 1975, Materne 1976) observed in many Mira variables may be photospheric as well as circumstellar in origin (Harrington 1969, Landstreet & Angel 1977, McLean & Clarke 1977, McLean 1979).

Other techniques for measuring departures from spherical symmetry in circumstellar material include infrared interferometry (McCarthy 1979, 1980, Sutton et al. 1979) and mapping of CO mm-λ emission (Wannier et al. 1979). McCarthy et al. (1980) attempt to incorporate all existing data for IRC + 10216 into a consistent picture. To explain the 2–5 μm observations they suggest a flattened disk, radius $\sim 10 R_*$, viewed at low inclination. Perhaps this axial symmetry, if present also in the larger CO emission envelope, can help to explain why spherically symmetric models fail to reproduce the radial variation of the CO intensity (Wannier et al. 1979, Knapp et al. 1980). McCarthy et al. also suggest that the structure of IRC + 10216 may resemble that of the bi-polar nebulae mentioned above.

4 MASERS

Maser emission from SiO, H_2O, and OH molecules has been observed from more than 300 late-type giant stars (Engels 1979). The energies required to populate the maser levels range from 1000's of degrees for SiO to less than one degree for OH. Thus the various masers sample material throughout the entire circumstellar envelope.

Because of the complicated nature of the maser emission and the mandated upper limit to the length of this review, it is not possible to do justice here to all of the fine work that has been carried out on circumstellar masers. Various maser papers are cited throughout the body of the present review in conjunction with other problems. For more comprehensive reviews of circumstellar masers per se the interested reader is referred to Snyder (1980), Goldreich (1980), and Winnberg (1977). In the present section we, rather arbitrarily, confine our remarks to magnetic field strengths and a few peculiar OH maser stars.

Interpretation of VLBI observations of two unusual OH maser stars (U Ori and IRC + 10420) suggests surface magnetic fields of 10–100 gauss (Reid et al. 1979). This might be typical of Mira variables. Stencel & Ionson (1979) have analyzed the peculiar M giant HD 4174 that is purported to have a kilogauss magnetic field but little mass loss. Such large magnetic fields may result in closed magnetic flux loops that inhibit hydrodynamic flow. However, this interpretation is in some doubt since the large magnetic field may not exist (M. H. Slovak, 1979, private communication) and/or the

star may be double and may possess a substantial stellar wind (Smith 1980).
The spectral type (F 8 Ia) of IRC + 10420 is much bluer than other stellar OH masers. VLBI observations by Mutel et al. (1979) and by Benson et al. (1979) suggest that IRC + 10420 is surrounded by a "fossil" dust shell left over from a previous M supergiant phase of evolution (Stothers 1975). Large mass loss from IRC + 10420 would be consistent with its location above Mullan's supersonic transition locus in the HR diagram (Mullan 1978, Stencel 1978).

Perhaps the most fascinating maser star is VY CMa whose evolutionary state is still unknown [see Lada & Reid (1978) for a summary of the arguments in favor of a pre– vs a post–main-sequence status]. Isotopic ratios, such as $[^{12}C]/[^{13}C]$ measured in the infrared, might be a means for deciding this argument. [Geballe et al. (1979) have measured the silicon isotopic ratios to be nearly terrestrial.] Expanding disc or shell models were suggested by VLBI observations of OH and H_2O masers (Reid & Muhleman 1978, Rosen et al. 1978) but more recent OH observations fail to show evidence for either expansion or contraction (Benson & Mutel 1979). McCarthy (1979) has incorporated existing optical, infrared, and microwave data in his Figures 3 and 4.

Another mysterious maser emission source is OH 0739-14 (also known as OH 231.8 + 4.2; Morris & Knapp 1976). In spite of a variety of infrared observations (Gillett & Soifer 1976, Cohen & Frogel 1977, Kleinmann et al. 1978, Kobayashi et al. 1978) the nature of OH 0739-14 is not yet clear. Morris & Bowers (1980) propose that the wide, flat, 1667 MHz OH masers observed toward OH 0739-14, M1-92, and VY CMa can be produced by an expanding, axially symmetric envelope.

There are also a handful of early-type emission line objects with OH maser emission (see, for example, Davis et al. 1979). Among these Vy 2-2 and/or M1-92 may be very young planetary nebulae.

5 MASS LOSS

In the two decades since Deutsch (1956) proved that α Her is losing mass there has been a tremendous amount of interest in the subject of mass loss from stars. The reason is partially observational. The biggest mass losers are very bright stars—O, B, and M giants and supergiants—whose ultraviolet, optical, infrared, and microwave spectra are relatively easy to study. But, in addition, substantial mass loss significantly influences the star's ultimate fate and represents a source of freshly made elements and dust particles for the interstellar medium. Kleinmann & Payne-Gaposchkin (1979) have summarized the properties of the 96 reddest stars in the Two Micron Sky Survey.

In spite of all this interest, for late-type giant stars we still lack definitive answers to many questions: 1. What physical mechanism(s) drive the mass loss? 2. What is the rate of mass loss for a given star? 3. How do circumstellar clouds fare as a source of interstellar grains? 4. What is the upper mass limit for stars that manage to shed sufficient mass during and after their main-sequence evolution to end their lives as white dwarfs?

Various recent reviews relate to these and other problems of mass loss (e.g., Reimers 1977a, Weymann 1977, Salpeter 1977, Wood 1978, Cassinelli 1979). Given the limited length of the present review we can address the above four questions only briefly.

For mass loss driven by radiation pressure the mass loss rate $\dot{M}$ is given by: $\dot{M}V_\infty = (\tau_R L)/c$. Here τ_R is an effective radiation pressure optical depth through the shell (Salpeter 1974, Werner et al. 1980) and L is the stellar luminosity. In this expression one assumes that the shell surrounds the entire star. The momentum in the gas and dust shell is gained from the momentum in the stellar photons—if the shell is optically thin to the stellar radiation ($\tau_R \ll 1$) then $\dot{M}V_\infty$ will be small. In very thick shells in which visible and/or infrared photons are absorbed and/or scattered more than once before escaping, $\tau_R > 1$. The ultimate limit to $\dot{M}V_\infty$ obtainable, in principle, with this mechanism but not likely to ever be approached in actual circumstellar shells is set by energy (rather than momentum) conservation—the photons are absorbed and/or scattered many times before escaping from the shell and give up all their energy to the expansion of the gas.

In Section 3.1.1 we noted that although radiation pressure on dust is probably responsible for acceleration of the gas to V_∞ some other mechanism probably initiates the mass loss. Then grain formation is a consequence of mass loss rather than the contrary (see, for example, Fix & Alexander 1974, Renzini 1977). Hagen (1978) has assembled a variety of arguments that suggest that radiation pressure on grains is not responsible for mass loss in M stars. Since this prospect would surprise and disturb many researchers in this field more compelling evidence must be found. One possibility is observation of CO emission ($J = 1 \to 0, 2 \to 1, 3 \to 2$, etc.) in highly obscured stars. The analysis of Knapp et al. (1980) suggests that $\dot{M}V_\infty$ in some infrared giants exceeds L/c by factors of 3 or more. If so most, or possibly all, of the stars that show CO emission and lie to the left of the diagonal line in Figure 1 of Zuckerman et al. (1977) are also losing mass at a rate large compared to $L/V_\infty c$. Except for NGC7027 and CRL618 these objects are probably all Mira variables and mechanical pulsations are a potential mechanism to generate large $\dot{M}$ (see, for example, Willson & Hill 1979). Pulsational or dynamical instabilities that have been suggested (e.g. Smith & Rose 1972, Härm & Schwarzschild 1975, Tuchman et al. 1979) but not yet shown to actually eject PN's such as NGC7027 and CRL618 might be regarded as extreme cases of this mechanism. Then $\dot{M}V_\infty$ could be $\gg L/c$.

Rotation, in combination with another physical mechanism, may result in mass loss into a disk or torus and, in time, may significantly affect the angular momentum of a star. Rotation has been considered in models of mass loss from M supergiants (Kafatos & Michalitsianos 1979), Miras (Phillips 1979), and the infrared star CIT6 (Mufson & Liszt 1975). However, at least for CIT6, rotationally forced mass ejection is probably not important (Knapp et al. 1979).

Quantitative estimates of $\dot{M}$ in individual stars may be obtained from blue-shifted circumstellar absorption lines, infrared continuum, or microwave emission lines. Unfortunately, at present, none of these techniques are particularly reliable—discrepancies in $\dot{M}$ of one or even two orders of magnitude are not uncommon.

The major uncertainty in estimates of $\dot{M}$ from optical lines is the inner radius R_0 of the circumstellar shell in which the lines are formed (Reimers 1975a,b, Sanner 1976, Bernat 1977, Hagen 1978). $\dot{M}$ deduced from observations is linearly proportional to the value of R_0 chosen for the analysis. The 11-μm interferometry of Sutton (1979a) suggests that the inner radius of the dust shell is many times R_* and, therefore, R_0 may also be large for the gas shell although not necessarily as large as preferred by Bernat (1977).

One way to circumvent the R_0 problem is with visual binary systems such as α Her (Deutsch 1956, Reimers 1977b) and α Sco (Kudritzki & Reimers 1978, van der Hucht et al. 1980). Absorption lines due to the envelope of the M-star primary are observed against the companion star. Unfortunately, even here, $\dot{M}$'s deduced for α Sco differ by an order of magnitude.

At any rate, the various papers referenced above give typical $\dot{M}$'s between 10^{-8} and 10^{-6} $M_\odot$/yr for M giants (including Miras) and 10^{-7} to 10^{-4} $M_\odot$/yr for M supergiants.

Gehrz & Woolf (1971) used infrared continuum observations, assumed mass loss was due to radiation pressure on dust, and, with certain other assumptions, derived $\dot{M}$ for a large number of M giants and supergiants. These $\dot{M}$'s fell in the same range as those deduced from the optical circumstellar absorption lines. The largest $\dot{M}$'s were obtained for the reddest Miras and approached 10^{-5} $M_\odot$/yr. By adopting the mass loss rates of Gehrz & Woolf, Bowers & Kerr (1977) estimated that Miras with OH maser emission contribute a substantial fraction of the total mass lost by stars in the solar neighborhood. Models of OH maser radiation from Mira variables based on an assumed period-density relation and a radiation pressure–driven wind yield $\dot{M}$'s in general agreement with those deduced from optical absorption line and infrared observations (Michalitsiamos & Kafatos 1978).

Much larger mass loss rates are indicated for some infrared stars. $\dot{M}$'s for these objects have been estimated in two different ways. One method is to assume that radiation pressure on dust and $\tau_R \sim 1$ effectively converts all of the photon momentum into particle momentum (e.g., Goldreich & Scoville 1976, Werner et al. 1980). The other is to make a detailed model of the circumstellar shell based either on microwave (Kwan & Hill 1977, Morris 1980) or infrared (Betz et al. 1979) observations. It is still too early to say if the two methods are in agreement (see below). In any event the deduced mass loss rates are very large, between 10^{-5} and 10^{-4} $M_\odot$/yr for the oxygen-rich OH/IR stars and comparable or even larger values for some carbon stars. Results for IRC+10216, the prototype of the latter group, are in some disarray. Originally Kwan & Hill (1977) estimated $\dot{M} \sim$ few $\times 10^{-5}$ $M_\odot$/yr. But more recent infrared and microwave measurements suggest $\dot{M} \gtrsim 10^{-4}$ $M_\odot$/yr (Betz et al. 1979, Knapp et al. 1980). This implies a circumstellar mass of at least a few solar masses if 90″ is the minimum radius of the shell (Wannier et al. 1979).

Estimates of mass loss rates for oxygen-rich infrared stars are also in apparent disagreement. Knapp et al. (1980) find $\tau_R \sim 10$ for IRC+10011. Yet Werner et al. (1980) assume a much smaller value of τ_R ($=2$) for three OH/IR maser sources with 10-μm silicate absorption features that are much deeper than the one in IRC+10011. [Compare Figure 1 in Werner et al. with Figure 4 in Merrill & Stein (1976).]

Total mass loss rates, summed over the entire galaxy, for red giants plus PN's are estimated to be $\sim$ few $\times M_\odot$/yr (e.g., Salpeter 1977, Cahn & Wyatt 1978a). The star formation rate is probably comparable so, if the ejected mass is not much less than the mass going into the stars, the total mass loss rate from protostellar nebulae will be comparable to that from evolved stars. Therefore, at present, it is difficult to choose between a pre- and post–main-sequence origin for the cores of interstellar dust grains. In the models of Dwek & Scalo (1980) grain cores form preferentially in post–main-sequence objects, including the ejecta of supernovae. It is also conceivable that the cores can form in situ in interstellar molecular clouds if radiative association rates are very fast for very large molecules (Smith & Adams 1978). The dust mantles are probably accreted mainly in the interstellar clouds (Salpeter 1977).

Observational evidence suggests that stars with initial masses $\lesssim$ 4–8 $M_\odot$ give rise to planetary nebulae (Tinsley 1975, 1977). Theoretical models lead to estimates that cover this entire range (Scalo 1976, Mengel 1976, Fusi-Pecci & Renzini 1976, Wood & Cahn 1977, Cahn & Wyatt 1978a, Tuchman et al. 1978). Taylor & Manchester (1977) suggest that many of these stars must (eventually) also become pulsars.

6 LATE STELLAR EVOLUTION

6.1 *(Apparently) Single Stars*

Finally we consider the evolutionary state of stars with circumstellar envelopes. As stars evolve up the red giant branch for the second time it is generally agreed that their core masses, luminosities, and mass loss rates steadily increase. For Miras $\sim 10^{-7}$ $M_\odot$/yr is added to the core but, typically, $\gtrsim 10^{-6}$ $M_\odot$/yr is lost via expansion of the circumstellar envelope. So the subsequent state of the red giant star, i.e. planetary nebula or supernova, is determined by its initial mass and the mass loss (see references at end of Section 5). Perhaps the most ambitious model of late stellar evolution is due to Cahn and collaborators (Wood & Cahn 1977, Cahn & Wyatt 1978a,b). With this model and two observables that are relatively easily obtained—the period and the mean spectral class at maximum light—one can estimate the mass and bolometric luminosity of individual Miras. In addition, Cahn & Wyatt (1978b) find a correlation between bolometric luminosity and outflow velocity for Miras. [Dickinson et al. (1975, 1978) had previously shown that outflow velocity and pulsation period are linearly related and Morris et al. (1979) find that the same velocity-period relationship holds for both carbon-rich and oxygen-rich LPV's]. Although it is not yet possible to tell if Cahn's prescription is essentially correct there are already some potential problems. For example, it is assumed that Miras are first overtone pulsators, an assumption that is contested by Hill & Willson (1979). Also, it is not clear that the model can produce a Mira-like IRC+10216 with a very massive envelope ($\sim$ few $M_\odot$, see Section 5).

Alternatively, Willson (1978) using her own model for fundamental mode Mira pulsation derives relationships among mass, radius, and period such that more massive stars have longer periods in agreement with earlier statistical results of Clayton & Feast (1969). On the other (third!) hand, models of OH maser emission from Mira variables by Michalitsianos & Kafatos (1978) show no strong correlation between period and either mass, radius, or luminosity and data summarized by Feast (1979) suggest that there is little difference in the bolometric luminosities of LPV's with periods between 200 and 500 days. So we presently have three different pictures for the relationship between observed and derived quantities for Mira variables!

Within a radius of a few kiloparsecs of the sun we would expect a supernova explosion within the next 1000 years. Therefore, some of the red giant and supergiant stars that we observe are almost certainly pre-supernovae. The question is, which ones? We mentioned in Section 5 that infrared and microwave observations suggest that the circumstellar envelope of IRC+10216 contains a few solar masses. If so, at least some massive stars can shed large amounts of mass very rapidly and may never become supernovae. Another carbon-rich LPV with a substantial molecular envelope is V Hya (Zuckerman et al. 1977) which Wood (1978) suggests may be a pre-supernova if its 18 year secondary period is due to rapid helium shell flashing. Alternatively, if 18 years represents the fundamental mode pulsation period then V Hya may be near the point of dynamical instability (which corresponds to the fundamental period becoming infinitely long) and on the verge of ejecting a planetary nebula. It would be interesting to determine if long secondary periods are a common feature of infrared Mira variables, although this would not be a good thesis project.

R Cr B stars have been proposed as the precursors of supernovae of Type I (Wheeler 1978, but see also Oemler & Tinsley 1979). These stars have not yet been detected by radio astronomers and may have relatively low mass circumstellar envelopes.

Although radio and infrared astronomers may be observing a few pre-supernovae most red giants with substantial circumstellar envelopes will, no doubt, eventually evolve into planetary nebulae. Which of the many objects now under study are actually pre-planetary nebulae (PPN's) is somewhat controversial. One school of thought holds that PPN's are to be found among very red giant stars with CO radio emission (Lo & Bechis 1976, Zuckerman et al. 1976) most of which are suggested to be carbon-rich (Zuckerman et al. 1978, Zuckerman 1978). It was mentioned in Section 2 that [C]/[O] ratios in M stars are still lacking. Conversion of observed CO intensities to mass loss rates requires knowledge of this ratio. Zuckerman et al. assumed that [C]/[O] in late M stars is comparable to or larger than the solar ratio. If, instead, carbon is significantly depleted (e.g. through conversion to nitrogen in the CNO cycle) then their argument that the majority of red-giant stars with the largest mass loss rates are carbon-rich or S-type is weakened.

An alternative view of PPN's is that they are characterized by eruptive emission line objects such as V1016 Cyg and HM Sge (Kwok et al. 1978, Kwok & Purton 1979). Our view is that if these objects are PPN's then they will evolve into only low mass PN's but not the bright $\sim 0.2 \, M_\odot$ PN's whose pictures appear in elementary astronomy texts. Both V1016 Cyg and HM Sge contain only a small amount of ionized gas $\sim 10^{-4} \, M_\odot$ (Davidson et al. 1978, Kwok et al. 1978, Kwok & Purton 1979). There is no evidence for a shell of neutral gas around HM Sge (Davidson et al. 1978) and neither 2 μm H_2 nor CO $J = 1 \rightarrow 0$ emission has been detected from V1016 Cyg

(Beckwith et al. 1978, Zuckerman et al. 1977). Admittedly, given the likely distances to these objects, ~ 3 kpc, CO would be difficult to detect. OH maser radiation might be a better bet but definitive OH observations have, apparently, still not been carried out (Lepine & Rieu 1974, Davis et al. 1979).

A more fundamental criticism of this model of PPN's is due to Kenyon & Cahn (1980) who argue that V1016 Cyg is a binary consisting of a white dwarf and an M-type Mira (see, also, Section 6.2). In general, the PPN candidates preferred by Kwok & Purton can perhaps be described as very young, compact PN's whereas, excepting CRL618 and NGC7027, the suggested PPN's in Table 1 of Zuckerman (1978) are in an earlier evolutionary state.

The situation can be clarified if [C]/[O] ratios can be determined in planetary nebulae. For the gas the best tool seems to be the ultraviolet lines of CIII and CIV. For IC418 and NGC7662, C/O ~ 1 is indicated although some interpretive problems still remain (Harrington et al. 1980, Torres-Peimbert et al. 1980). Even if these problems are resolved, the meaning of C/O ratios near unity may remain ambiguous. If during the PPN phase all of the less abundant of the two elements goes into CO, which is later photodissociated by the central star of the PN, and the bulk of the left over C or O is incorporated into the dust, then the gas may always show C/O ~ 1 largely independent of its origin in C-rich or O-rich material (J. P. Harrington, 1979, private communication). If so, it will be necessary to determine the composition of the dust as well as the gas. This may be difficult since Kwok (1980) argues that the 10-μm silicate feature should not be detectable from ordinary (evolved) planetary nebulae even if the dust condenses in an oxygen-rich environment. To date, the handful of planetary nebulae ($\lesssim 10$) with apparent 10-μm silicate or 11-μm SiC features are roughly divided between the two (Aitken et al. 1979, Willner et al. 1979, Russell et al. 1978, Puetter et al. 1978). Consistent with Kwok's remarks, the objects that show SiC are mainly bona-fide (evolved) planetaries whereas the other objects all appear to be very young or protoplanetary nebulae.

Models for the origin of planetary nebulae can be divided into two classes (e.g. Wood 1978, Scalo & Shields 1979): rapid ejection due to a dynamical instability or by a stellar wind alone. A satisfactory model should account for the multiple shells and giant halos observed in many planetaries (Terzian 1980).

6.2 *Multiple and Symbiotic Systems*

Little is currently known about multiplicities in late-type giant stars. Statistics of main-sequence stars and the nuclei of planetary nebulae (Lutz 1978, Terzian 1980) imply that many (most?) M-type giants are members of multiple systems. During the next few decades it should be possible to measure cyclical changes, due to orbital motion, in photospheric velocities and positions (D. G. Currie, 1979, private communication). In addition, SiO masers may, in principle, be used to track the positions and radial velocities of the circumstellar shells (W. J. Welch, 1979, private communication, Zuckerman 1979) although additional secular variations in the maser emission that are known to exist are still not well understood and may, in practice, introduce problems. Linear diameters of the maser emission regions are $\sim$ few AU (Moran et al. 1979) and positions may now be measured to $0''.1$ with conventional interferometers and may ultimately be measurable to $0''.001$ with "very long baseline" interferometers (W. J. Welch, 1979, private communication).

Once it is established exactly which M giants belong to multiple systems we may better understand the relationship among planetary nebulae, symbiotic stars, peculiar emission line objects, etc. In spite of their suggestive spectra very few symbiotic stars have been convincingly shown to be members of binary systems (AG Peg, AR Pav). A good case in point is R Aqr for which a suggested orbital period of 27 years (Merrill 1950) was not confirmed (Jacobson & Wallerstein 1975). Very recent IUE spectra indicate the presence of a small ionized nebula at R Aqr (Michalitsianos et al. 1980). Emission lines from CIII, CIV, and SIII are not typical of the M7 Mira in R Aqr and are, therefore, suggestive of a hot white dwarf (ex-nova) companion. But, still, the case is not yet completely compelling. Other unusual systems (e.g. V1016 Cyg, HM Sge) produce substantial radio continuum flux (as does R Aqr) and/or a rich ultraviolet spectrum yet the single vs double star picture for many of these objects is still controversial.

The eruptive variables V1016 Cyg and HM Sge have attracted the most attention. Most groups interpret these objects as protoplanetary nebulae rather than as classical symbiotic stars (but see Kenyon & Cahn 1980). This is in agreement with the observation (Wallerstein 1978b) of very large (>1000 km s^{-1}) velocities in HM Sge. Bona fide symbiotic stars do not show velocities >100 km s^{-1} (M. H. Slovak, 1979, private communication). The controversy is over single vs double star models. Double star models are preferred by Ciatti et al. (1978), Davidson et al. (1978), and Puetter et al. (1978) based on infrared and optical data. The infrared variability (Harvey 1974, Slovak 1978) is suggestive, at least in the case of V1016 Cyg, of a Mira with a period $\approx 450^d$. The absence of similar variability at visual wavelengths suggests a second, hotter star. In these models the gas is supplied by the cool star and ionized by the hot star. In the single star models favored by Ahern et al. (1977), Kwok et al. (1978), and Kwok & Purton (1979), the small variability at visual wavelengths must be explained.

ACKNOWLEDGMENT

I am grateful for helpful comments and/or correspondence from members of the astronomy faculties at the University of Texas at Austin and at the University of Maryland and, also, Drs. E. C. Sutton, S. P. Willner, M. Morris, J. H. Cahn, M. Jura, M. W. Werner, and D. J. Mullan and Mr. M. H. Slovak. The preparation of this review was aided by National Science Foundation grants AST 77-28475 and AST 76-17600 to the Universities of Texas and Maryland, respectively.

Literature Cited

Ahern, F. J., Fitzgerald, M. P., Marsh, K. A., Purton, C. R. 1977. *Astron. Astrophys.* 58:35
Aitken, D. K., Roche, P. F., Spenser, P. M., Jones, B. 1979. *Ap. J.* 233:925
Ake, T. B. 1979. *Ap. J.* 234:538
Altenhoff, W. J., Oster, L., Wendker, H. J. 1979. *Astron. Astrophys.* 73:L21
Barnes, T. G., Beer, R., Hinkle, K. H., Lambert, D. L. 1977. *Ap. J.* 213:71
Becklin, E. E., Frogel, J. A., Hyland, A. R., Kristian, J., Neugebauer, G. 1969. *Ap. J. Lett.* 158:L133
Beckwith, S., Persson, S. E., Gatley, I. 1978. *Ap. J. Lett.* 219:L33
Benson, J. M., Mutel, R. L. 1979. *Ap. J.* 233:119
Benson, J. M., Mutel, R. L., Fix, J. D., Claussen, M. J. 1979. *Ap. J. Lett.* 229:L87
Bernat, A. P. 1977. *Ap. J.* 213:756
Bernat, A. P., Hall, D. N. B., Hinkle, K. H., Ridgway, S. T. 1979. *Ap. J. Lett.* 233:L135
Bernat, A. P., Honeycutt, R. K., Kephart, J. E., Gow, C. E., Sanford, M. T., Lambert D. L. 1978. *Ap. J.* 219:532
Bernat, A. P., Lambert, D. L. 1976. *Ap. J.* 210:395
Betz, A. L., McLaren, R. A., Spears, D. L. 1979. *Ap. J. Lett.* 229:L97
Betz, A. L., McLaren, R. A. 1980. *Interstellar Molecules, IAU Symp. No. 87,* ed. B. H. Andrew. In press
Blackwell, D. E., Shallis, M. J. 1977. *MNRAS* 180:177
Boesgaard, A. M. 1979. *Ap. J.* 232:485
Boesgaard, A. M., Hagen, W. 1979. *Ap. J.* 231:128
Bowers, P. F., Kerr, F. J. 1977. *Astron. Astrophys.* 57:115
Bowers, P. F., Kundu, M. R. 1979. *Astron. J.* 84:791
Bregman, J. D., Bregman, J. N. 1978. *Ap. J. Lett.* 222:L41
Cahn, J. H., Wyatt, S. P. 1978a. *Ap. J.* 221:163

Cahn, J. H., Wyatt, S. P. 1978b. *Ap. J. Lett.* 224:L79
Campbell, M. F., Elias, J. H., Gezari, D. Y., Harvey, P. M., Hoffman, W. F., Hudson, H. S., Neugebauer, G., Soifer, B. T., Werner, M. W., Westbrook, W. E. 1976. *Ap. J.* 208:396
Cassinelli, J. P. 1979. *Ann. Rev. Astron. Astrophys.* 17:275
Ciatti, F., Mammano, A., Vittone, A. 1978. *Astron. Astrophys.* 68:251
Clayton, M. L., Feast, M. W. 1969. *MNRAS* 146:411
Clegg, R. E. S. 1980. *MNRAS.* In press
Clegg, R. E. S., Wootten, H. A. 1980. *Ap. J.* In press
Cohen, J. G., Frogel, J. A. 1977. *Ap. J.* 211:178
Cohen, M. 1979. *MNRAS* 186:837
Crabtree, D. R., Martin, P. G. 1979. *Ap. J.* 227:900
Czyzak, S. J., Santiago, J. J. 1973. *Astrophys. Space Sci.* 23:443
Davidson, K., Humphreys, R. M., Merrill, K. M. 1978, *Ap. J.* 220:239
Davis, L. E., Seaquist, E. R., Purton, C. R. 1979. *Ap. J.* 230:434
Deutsch, A. J. 1956. *Ap. J.* 123:210
Dickinson, D. F., Kollberg, E., Yngvesson, S. 1975. *Ap. J.* 199:131
Dickinson, D. F., Reid, M. J., Morris, M., Redman, R. 1978. *Ap. J. Lett.* 220:L113
Dwek, E., Scalo, J. M. 1980. *Ap. J.* In press
Egan, W. G., Hilgeman, T. 1978. *Nature* 273:369
Engels, D. 1979. *Astron. Astrophys. Suppl. Ser.* 36:337
Fabian, A. C., Hansen, C. J. 1979. *MNRAS* 187:283
Feast, M. W. 1979. Paper presented at the XVIIth General Assembly of the IAU
Fix, J. D., Alexander, D. R. 1974. *Ap. J. Lett.* 188:L91
Frawley, W. M. 1977. *Ap. J.* 218:181

Fusi-Pecci, F., Renzini, A. 1976. *Astron. Astrophys.* 46:447
Geballe, T. R., Lacy, J. H., Beck, S. C. 1979. *Ap. J. Lett.* 230:L47
Gehrz, R. D., Woolf, N. J. 1971. *Ap. J.* 165:285
Gillett, F. C., Soifer, B. T. 1976. *Ap. J.* 207:780
Gilman, R. C. 1969. *Ap. J. Lett.* 155:L185
Gilman, R. C. 1974. *Ap. J.* 188:87
Gilra, D. P. 1973. *Interstellar Dust and Related Topics, IAU Symp. No. 52*, ed. J. M. Greenberg, H. C. van de Hulst, p. 517. Dordrecht: Reidel
Goldberg, L., Ramsey, L., Testerman, L., Carbon, D. 1975. *Ap. J.* 199:427
Goldreich, P. 1980. See Betz & McLaren 1980
Goldreich, P., Scoville, N. 1976. *Ap. J.* 205:144
Hagen, W. 1978. *Ap. J. Suppl.* 38:1
Haisch, B. M., Linsky, J. L., Basri, G. S. 1980. *Ap. J.* 235:519
Hall, D. N. B. 1980. See Betz & McLaren 1980
Hall, D. N. B., Ridgway, S. T. 1978. *Nature* 273:281
Härm, R., Schwarzschild, M. 1975. *Ap. J.* 200:324
Harrington, J. P. 1969. *Astrophys. Lett.* 3:165
Harrington, J. P., Lutz, J. H., Seaton, M. H., Strickland, D. J. 1980. *MNRAS.* 191:13
Harvey, P. M. 1974. *Ap. J.* 188:95
Herbig, G. H. 1972. *Ap. J.* 172:375
Herbig, G. H. 1975. *Ap. J.* 200:1
Hill, S. J., Willson, L. A. 1979. *Ap. J.* 229:1029
Hinkle, K. H. 1978. *Ap. J.* 220:210
Hinkle, K. H., Barnes, T. G. 1979. *Ap. J.* 227:923
Hoyle, F., Wickramasinghe, N. C. 1977. *Nature* 268:610
Humphreys, R. M. 1974. *Ap. J.* 188:75
Humphreys, R. M. 1975. *Publ. Astron. Soc. Pac.* 87:433
Jacobson, T. S., Wallerstein, G. 1975. *Publ. Astron. Soc. Pac.* 87:269
Jones, B., Merrill, K. M., Puetter, R. C., Willner, S. P. 1978. *Astron. J.* 83:1437
Jones, T. W., Merrill, K. M. 1976. *Ap. J.* 209:509
Jura, M. 1975. *Astron. J.* 80:227
Jura, M., Jacoby, G. 1976. *Astrophys. Lett.* 18:5
Kafatos, M., Michalitsianos, A. G. 1979. *Ap. J. Lett.* 228:L115
Kenyon, S. J., Cahn, J. H. 1980. Preprint
Kleinmann, S. G., Payne-Gaposchkin, C. 1979. *Earth Extraterr. Sci.* 3:161
Kleinmann, S. G., Sargent, D. G., Moseley, H., Harper, D. A., Loewenstein, R. F., Telesco, C. M., Thronson, H. A. 1978.
Astron. Astrophys. 65:139
Knapp, G. R., Kuiper, T. B. H., Zuckerman, B. 1979. *Ap. J.* 233:140
Knapp, G. R., Phillips, T. G., Huggins, P. J., Leighton, R. G., Wannier, P. G. 1980. Preprint
Kobayashi, Y., Kawara, K., Maihara, T., Okuda, H., Sato, S., Noguchi, K. 1978. *Publ. Astron. Soc. Jpn.* 30:377
Kudritzki, R. P., Reimers, D. 1978. *Astron. Astrophys.* 70:227
Kuiper, T. B. H., Knapp, G. R., Knapp, S. L., Brown, R. L. 1976. *Ap. J.* 204:408
Kwan, J., Hill, F. 1977. *Ap. J.* 215:781
Kwok, S. 1975. *Ap. J.* 198:583
Kwok, S. 1976. *J. R. Astron. Soc. Can.* 70:49
Kwok, S. 1980. *Ap. J.* 236:592
Kwok, S., Purton, C. R. 1979. *Ap. J.* 229:187
Kwok, S., Purton, C. R., Fitzgerald, P. M. 1978. *Ap. J. Lett.* 219:L125
Lada, C. J., Reid, M. J. 1978. *Ap. J.* 219:95
Lambert, D. L. 1980. *Mem. Soc. R. Sci. Liège.* In press
Lambert, D. L., Snell, R. L. 1975. *MNRAS* 172:277
Lambert, D. L., Vanden Bout, P. A. 1978. *Ap. J.* 211:854
Landstreet, J. D., Angel, J. R. P. 1977. *Ap. J.* 211:825
Lefèvre, J. 1979. *Astron. Astrophys.* 72:61
Lepine, J. R. D., Rieu, N. Q. 1974. *Astron. Astrophys.* 36:469
Lewis, J. S., Ney, E. P. 1979. *Ap. J.* 234:154
Linsky, J. L., Haisch, B. M. 1979. *Ap. J. Lett.* 229:L27
Lo, K. Y., Bechis, K. P. 1976. *Ap. J. Lett.* 205:L21
Lo, K. Y., Bechis, K. P. 1977. *Ap. J. Lett.* 218:L27
Low, F. J., 1979. *High Angular Resolution Stellar Interferometry, IAU Colloq. No. 50*, ed. J. Davis, W. J. Tango, p. 15-1. Univ. Sydney
Lucy, L. B. 1976. *Ap. J.* 205:482
Lutz, J. H. 1978. *Planetary Nebulae, IAU Symp. No. 76*, ed. Y. Terzian, p. 185. Dordrecht: Reidel
Maciel, W. J. 1976. *Astron. Astrophys.* 48:27
Maciel, W. J. 1977. *Astron. Astrophys.* 57:273
Materne, J. 1976. *Astron. Astrophys.* 47:53
Mathis, J. S. 1978. See Lutz 1978, p. 281
McCabe, E. M., Smith, R. C., Clegg, R. E. S. 1979. *Nature* 281:263
McCarthy, D. W. 1979. See Low 1979, p. 18-1
McCarthy, D. W., Howell, R., Low, F. J. 1978. *Ap. J. Lett.* 223:L113
McCarthy, D. W., Howell, R., Low, F. J. 1980. *Ap. J. Lett.* 235:L27
McCarthy, D. W., Low, F. J., Howell, R. 1977. *Ap. J. Lett.* 214:L85
McLean, I. A. 1979. *MNRAS* 186:21
McLean, I. A., Clarke, D. 1977. *MNRAS* 179:293

McMillan, R. S., Tapia, S. 1978. *Ap. J. Lett.* 226:L87
Mengel, J. G. 1976. *Astron. Astrophys.* 48:83
Merrill, K. M., Ridgway, S. T. 1979. *Ann. Rev. Astron. Astrophys.* 17:9
Merrill, K. M., Stein, W. A. 1976. *Publ. Astron. Soc. Pac.* 88:294
Merrill, P. W. 1950. *Ap. J.* 112:514
Michalitsianos, A. G., Kafatos, M. 1978. *Ap. J.* 226:430
Michalitsianos, A. G., Kafatos, M., Hobbs, R. W. 1980. *Ap. J.* In press
Moran, J. M., Ball, J. A., Predmore, C. R., Lane, A. P., Huguenin, G. R., Reid, M. J., Hansen, S. S. 1979. *Ap. J. Lett.* 231:L67
Morris, M. 1975. *Ap. J.* 197:603
Morris, M. 1980. *Ap. J.* 236:823
Morris, M., Alcock, C. 1977. *Ap. J.* 218:687
Morris, M., Bowers, P. F. 1980. Preprint
Morris, M., Knapp, G. R. 1976. *Ap. J.* 204:415
Morris, M., Redman, R., Reid, M. J. Dickinson, D. F. 1979. *Ap. J.* 229:257
Mufson, S. L., Liszt, H. S. 1975. *Ap. J.* 202:183
Mufson, S. L., Lyon, J., Marionni, P. A. 1975. *Ap. J. Lett.* 201:L85
Mullan, D. J. 1978. *Ap. J.* 226:151
Mutel, R. L., Fix, J. D., Benson, J. M., Webber, J. C. 1979. *Ap. J.* 228:771
Ney, E. P., Merrill, K. M., Becklin, E. E., Neugebauer, G., Wynn-Williams, C. G. 1975. *Ap. J. Lett.* 198:L129
Nguyen-Q-Rieu, Laury-Micoulant, C., Winnberg, A., Schultz, G. V. 1979. *Astron. Astrophys.* 75:351
O'Brien, G., Lambert, D. L. 1979. *Ap. J. Lett.* 229:L33
Oemler, A., Tinsley, B. M. 1979. *Astron. J.* 84:985
Phillips, J. P. 1979. *Astron. Astrophys.* 71:115
Phillips, J. P., Reay, N. K. 1977. *Astron. Astrophys.* 59:91
Puetter, R. C., Russell, R. W., Soifer, B. T., Willner, S. P. 1978. *Ap. J. Lett.* 223:L93
Rank, D. M., Geballe, T. R., Wollman, E. R. 1974. *Ap. J. Lett.* 187:L111
Reid, M. J., Dickinson, D. F. 1976. *Ap. J.* 209:505
Reid, M. J., Muhleman, D. O. 1978. *Ap. J.* 220:229
Reid, M. J., Moran, J. M., Leach, R. W., Ball, J. A., Johnston, K. J., Spencer, J. H., Swenson, G. W. 1979. *Ap. J. Lett.* 227:L89
Reimers, D. 1975a. *Problems in Stellar Atmospheres and Envelopes*, ed. B. Baschek et al., p. 229. New York: Springer
Reimers, D. 1975b. *Mem. R. Soc. Liège 6 Ser.* 8, p. 369
Reimers, D. 1977a. *The Interaction of Variables Stars with Their Environment, IAU Colloq. No. 42*, ed. R. Kippenhahn et al., p. 559. Publ. Bamberg Obs.

Reimers, D. 1977b. *Astron. Astrophys.* 61:216
Renzini, A. 1977. See Reimers 1977a, p. 589
Ridgway, S. T., Carbon, D. F., Hall, D. N. B. 1978. *Ap. J.* 225:138
Ridgway, S. T., Hall, D. N. B. 1980. See Betz & McLaren 1980
Ridgway, S. T., Hall, D. N. B., Kleinmann, S. G., Weinberger, D., Wojslaw, R. S. 1976. *Nature* 264:345
Rodriguez-Kuiper, E. N., Kuiper, T. B. H., Zuckerman, B., Kakar, R. K. 1977. *Ap. J.* 214:394
Rosen, B. R., Moran, J. M., Reid, M. J., Walker, R. C., Burke, B. F., Johnston, K. J., Spencer, J. H. 1978. *Ap. J.* 222:132
Russell, R. W., Soifer, B. T., Willner, S. P. 1978. *Ap. J.* 220:568
Sagan, C., Khare, B. N. 1979. *Nature* 227:102
Salpeter, E. E. 1974. *Ap. J.* 193:579, 585
Salpeter, E. E. 1977. *Ann. Rev. Astron. Astrophys.* 15:267
Sanner, F. 1976. *Ap. J. Suppl.* 32:115
Savage, B. D., Mathis, J. S. 1979. *Ann. Rev. Astron. Astrophys.* 17:73
Scalo, J. M. 1976. *Ap. J.* 206:215
Scalo, J. M., Ross, J. E. 1976. *Astron. Astrophys.* 48:219
Scalo, J. M., Shields, G. A. 1979. *Ap. J.* 228:521
Scargle, J. D., Strecker, D. W. 1979. *Ap. J.* 228:838
Schmidt, G. D., Angel, J. R. P., Beaver, E. A. 1978. *Ap. J.* 219:477
Schwarzschild, M. 1975. *Ap. J.* 195:137
Shawl, S. J. 1975. *Astron. J.* 80:602
Shawl, S. J., Zellner, B. 1970. *Ap. J. Lett.* 162:L19
Shields, G. A. 1980. Preprint
Slovak, M. H. 1978. *Astron. Astrophys.* 70:L75
Smith, D., Adams, N. G. 1978. *Ap. J. Lett.* 220:L87
Smith, R. L., Rose, W. K. 1972. *Ap. J.* 176:395
Smith, S. E. 1980. Preprint
Snyder, L. E. 1980. See Betz & McLaren 1980
Spencer, J. H., Johnston, K. J., Moran, J. M., Reid, M. J., Walker, R. C. 1979. *Ap. J.* 230:449
Stencel, R. E. 1978. *Ap. J. Lett.* 223:L37
Stencel, R. E., Ionson, J. A. 1979. *Publ. Astron. Soc. Pac.* 91:452
Stephenson, C. B. 1973. *Publ. Warner Swasey Obs.* 1, No. 4
Stephenson, C. B. 1976. *Publ. Warner Swasey Obs.* 2, No. 2
Stothers, R. 1975. *Ap. J. Lett.* 197:L25
Sutton, E. C. 1979a. PhD Thesis. Univ. Calif., Berkeley
Sutton, E. C. 1979b. See Low 1979, p. 16-1
Sutton, E. C., Betz, A. L., Storey, J. W. V., Spears, D. L. 1979. *Ap. J. Lett.* 230:L105

Sutton, E. C., Storey, J. W. V., Betz, A. L., Townes, C. H., Spears, D. L. 1977. *Ap. J. Lett.* 217:L97
Sutton, E. C., Storey, J. W. V., Townes, C. H., Spears, D. L. 1978. *Ap. J. Lett.* 224:L123
Taylor, J. H., Manchester, R. N. 1977. *Ann. Rev. Astron. Astrophys.* 15:19
Terzian, Y. 1980. *Q.J.R. Astron. Soc.* In press
Tinsley, B. M. 1975. *Publ. Astron. Soc. Pac.* 87:837
Tinsley, B. M. 1977. In *Supernovae*, ed. D. N. Schramm, p. 117. Dordrecht: Reidel
Toombs, R. I., Becklin, E. E., Frogel, J. A., Law, S. K., Porter, F. C., Westphal, J. A. 1972. *Ap. J. Lett.* 173:L71
Torres-Peimbert, S., Peimbert, M., Daltabuit, E. 1980. *Ap. J.* In press
Treffers, R., Cohen, M. 1974. *Ap. J.* 188:545
Tsuji, T. 1973. *Astron. Astrophys.* 23:411
Tsuji, T. 1978a. *Astron. Astrophys.* 62:29
Tsuji, T. 1978b. *Astron. Astrophys.* 68:L23
Tsuji, T. 1978c. *Publ. Astron. Soc. Jpn.* 30:435
Tsuji, T. 1979. *Publ. Astron. Soc. Jpn.* 31:43
Tuchman, Y., Sack, N., Barkat, A. 1978. *Ap. J. Lett.* 225:L137
Tuchman, Y., Sack, N., Barkat, A. 1979. *Ap. J.* 234:217
van der Hucht, K. A., Bernat, A. P., Kondo, Y. 1980. *Astron. Astrophys.* 82:14
Vardya, M. S. 1966. *MNRAS* 134:347
Wallerstein, G. 1975. *Ap. J. Suppl.* 29:375
Wallerstein, G. 1977. *Ap. J.* 211:170
Wallerstein, G. 1978a. *Changing Trends in Variable Star Research, IAU Colloq. No. 46*, p. 177
Wallerstein, G. 1978b. *Publ. Astron. Soc. Pac.* 90:36
Wannier, P. G., Leighton, R. B., Knapp, G. R., Redman, R. O., Phillips, T. G., Huggins, P. J. 1979. *Ap. J.* 230:149
Wannier, P. G., Linke, R. A. 1978, *Ap. J.* 225:130
Webster, A. 1978. *IAU Circ. No. 3245*
Werner, M. W., Beckwith, S., Gatley, I., Sellgren, K., Berriman, G., Whiting, D. L. 1980. *Ap. J.* In press
Westbrook, W. E., Becklin, E. E., Merrill, K. M., Neugebauer, G., Schmidt, M., Willner, S. P., Wynn-Williams, C. G. 1975. *Ap. J.* 202:407
Weymann, R. J. 1977. See Reimers 1977a, p. 577
Wheeler, J. C. 1978. *Ap. J.* 225:212
Whittaker, A. G. 1978. *Science* 200:763
Willner, S. P., Jones, B., Puetter, R. C., Russell, R. W., Soifer, B. T. 1979. *Ap. J.* 234:496
Willner, S. P., Puetter, R. C., Russell, R. W., Soifer, B. T., 1980. See Betz & McLaren 1980
Willson, L. A. 1978. See Wallerstein 1978a, p. 199
Willson, L. A., Hill, S. J. 1979. *Ap. J.* 228:854
Winnberg, A. 1977. See Reimers 1977a, p. 495
Wood, P. R. 1978. See Wallerstein 1978a, p. 163
Wood, P. R. 1979. *Ap. J.* 227:220
Wood, P. R., Cahn, J. H. 1977. *Ap. J.* 211:499
Wyckoff, S., Clegg, R. E. S. 1978. *MNRAS* 184:127
Yorka, S. B., Wing, R. F. 1979. *Astron. J.* 84:1010
Zappala, R. R., Becklin, E. E., Matthews, K., Neugebauer, G. 1974. *Ap. J.* 192:109
Zellner, B. H., Serkowski, K. 1972. *Publ. Astron. Soc. Pac.* 84:619
Zuckerman, B. 1978. See Lutz 1978, p. 305
Zuckerman, B. 1979. *Ap. J.* 230:442
Zuckerman, B., Gilra, D. P., Turner, B. E., Morris, M., Palmer, P. 1976. *Ap. J. Lett.* 205:L15
Zuckerman, B., Palmer, P., Gilra, D. P., Turner, B. E., Morris, M. 1978. *Ap. J. Lett.* 220:L53
Zuckerman, B., Palmer, P., Morris, M., Turner, B. E., Gilra, D. P., Bowers, P. F., Gilmore, W. 1977. *Ap. J. Lett.* 211:L97
Zuckerman, B., Terzian, Y., Silverglate, P. 1980. *Ap. J.* In press

COSMIC-RAY CONFINEMENT IN THE GALAXY   ×2168

Catherine J. Cesarsky
Section d'Astrophysique, Centre d'Etudes Nucléaires de Saclay,
91 Gif sur Yvette, France

1 MAIN ISSUES

Most reviews on cosmic rays address themselves to the important question of the origin of these relativistic particles; we will, for the most part, sidestep this problem by assuming that cosmic rays are emitted by galactic sources of known distribution, and instead worry about their subsequent destiny: Are they free to wander around in the Universe? or Are they compelled to remain close to their sources? Are they confined to the galactic disk? Why? What do we know about the distribution of cosmic rays in local regions, within one kiloparsec from the Sun; in the Galaxy; in extragalactic space?

Since the last general review on cosmic rays published in these pages, in 1969, by P. Meyer, a considerable amount of work has been spent in gathering data on cosmic-ray elemental and isotopic composition, at various energies; on measuring electron and positron spectra; on detecting antiprotons and ultraheavy nuclei; on measuring anisotropies over an energy range covering eight decades. Cosmic-ray interactions with interstellar matter and fields give rise to the emission of gamma rays; thanks to the satellites OSO-3, SAS-2, and COS-B, the large scale features of the gamma-ray sky have been revealed. While I would like to make the reader aware of the particular difficulty of these measurements, I will not attempt here to give a complete account of recent experimental results on cosmic rays. This review is geared to the theoretical interpretation of some of the data. Apart from the two excellent monographs by Ginzburg & Syrovatskii (1964) and Hayakawa (1969), there are also the recent reviews by Daniel & Stephens (1975), Hillas (1975), Ginzburg & Ptuskin (1976) and Waddington (1977).

The interstellar spectrum of cosmic rays of energy $E \leq 1$ GeV is not well known, because the effects of the interactions of such particles with the expanding solar wind have not been disentangled. In the energy range 10–10^6 GeV the integral spectrum of nuclear cosmic rays is a power law of index $\simeq 1.7$ (with some possible structure, see Hillas 1975); then it steepens sharply at $E \simeq 3.10^6$ GeV. We will consider in most of the paper only particles with $E < 3.10^6$ GeV. The difficulties in interpretation at the higher energy range, $3 \cdot 10^6$–10^{11} GeV, are of a different nature, and will be mentioned briefly at the end, in Section 5.

The majority of the papers dealing with the propagation of galactic cosmic rays attempt to interpret the available experimental data on spectra, composition, and anisotropy in terms of phenomenological models, which we list in this section. In Section 2, we discuss their relationship with the basic underlying physics: the interaction of cosmic rays and complex magnetic field structures; in Section 3 we see how these models fare with regard to presently available observations of local cosmic rays, and analyze how future observations can help to distinguish between them. Section 4 summarizes our information on the cosmic-ray density in various regions of our and other galaxies.

In the following models, it is generally assumed that the Galaxy contains cosmic-ray sources, which emit nuclear particles and/or electrons whose spectrum, at least in the energy range 10–10^6 GeV, is a power law; a variety of source distributions are assumed. The aim of the propagation models is to account for a number of observational facts which we detail in Section 3: the power-law spectrum; the low anisotropy; the observed abundances of primary and of secondary nuclei; the fact that, at energies between a few GeV and ~ 100 GeV, the mean amount of matter traversed by cosmic rays before leaving the Galaxy, or escape length λ_e, decreases as the energy increases ($\lambda_e \propto E^{-a}$, with $a = 0.3$–0.6); the observed age $\geq 2 \cdot 10^7$ years for 1 GeV particles; and the electron spectrum.

(a) *The Closed Model* (Rasmussen & Peters 1975)   Cosmic rays cannot leave the Galaxy; they remain trapped in it and consequently they lose all their energy first by inelastic collisions with interstellar medium particles and then by Coulomb interactions.

(b) *The Leaky-Box Model*   Cosmic rays are trapped within reflecting boundaries surrounding the Galaxy; at each encounter with the boundary, they have a finite probability of escaping into extragalactic space. In the leaky-box model, as in the closed model, the cosmic-ray density is uniform throughout the volume of confinement.

(c) *The Nested Leaky-Box Model* (Cowsik & Wilson 1973)   Before pervading the Galaxy, cosmic rays are trapped near their sources, with a finite probability of escape per unit time.

(d) *Diffusion Models*   The irregularities of the galactic magnetic field allow cosmic rays to leave the Galaxy by a random walk process. A number of parameters can be incorporated in such models; for instance, it is possible to vary the distribution of the sources, to have different diffusion coefficients parallel and perpendicular to the galactic disk, etc. (e.g. Le Guet 1977). Obviously, both the leaky-box and the diffusion models can incorporate a halo, either spherical or flat, with a suitable choice of boundary conditions.

(*e*) *The Dynamical Halo Model* (Jokipii 1976) The Galaxy is assumed to be constantly expelling gas, at a given rate. The magnetic field and cosmic rays are linked to this gas, and so they are constantly being convected away; in addition the cosmic rays may diffuse with respect to the gas.

(*f*) *Very Inhomogeneous Models* Diffusion models predict cosmic-ray gradients on a galactic scale or at least on scales much larger than the inhomogeneities in the source distribution. Some authors assume or infer that there are small scale cosmic-ray gradients, owing to cosmic-ray preferential trapping in clouds (Rengarajan et al. 1973) or in interstellar "tunnels" (Scott 1975), or to cosmic-ray exclusion from molecular clouds (Skilling & Strong 1976).

(*g*) *Continuous Acceleration Models* An important alternative to the basic assumptions of all these models, and which has far-reaching consequences for the interpretation of composition data, is to postulate that cosmic rays are gradually accelerated in their passage through the interstellar medium (Fermi 1954, Blandford & Ostriker 1978) so that the acceleration and the nuclear interactions occur simultaneously.

(*h*) *The Extragalactic Model* Finally, there is the interesting suggestion that the whole Universe is pervaded by a cosmic-ray flux comparable to that received at the earth (Brecher & Burbidge 1972); we will see in Section 4 that the present gamma-ray data seems to rule out this hypothesis.

2 PHYSICAL MECHANISMS GOVERNING COSMIC-RAY CONFINEMENT AND ESCAPE

2.1 *Physical Background for Diffusion and Leaky-Box Models*

The radius of gyration of a cosmic-ray proton of energy E (in GeV), in the galactic magnetic field of strength $\sim 3\mu$ G, is $r_g \simeq 10^{12} E_{\text{GeV}}$ cm; the scale height of the galactic gaseous disk is $\sim 5 \times 10^{20}$ cm. Thus, the cosmic rays we are considering are closely attached to the magnetic field lines, and their galactic density must necessarily depend crucially on the configuration of the galactic magnetic field. Unfortunately, only partial information on the large and small scale characteristics of the galactic magnetic field can be extracted from the various types of measurements used (Heiles 1976, Simard-Normandin & Kronberg 1979). In the solar neighborhood, the field lines tend to run parallel to the galactic plane; the field strength is $\simeq 2.5\mu$ G, and there is a very important large scale ($L \simeq 100$ pc) component of fluctuation. There also seem to be small regions with highly nonuniform orientations. The large scale inhomogeneous component generates curvature and gradient drift velocities of order ($cr_g/L \simeq 10^3$ cm s^{-1}); such velocities are insufficient to transport cosmic rays to the border of the galactic disk, or even to significantly displace them across flux tubes, and can be neglected (Jokipii & Parker 1969).

Fermi (1949) had pointed out that large scale ($\gg r_g$), moving inhomogeneities in the magnetic field would reflect particles of large pitch angle, and accelerate them; but his model had difficulties in satisfying the energy requirements and was abandoned. As we mentioned, it has now become customary to decouple the problems of origin and of propagation. For the past ten years, with a few exceptions (e.g. Flewelling & Coroniti 1976), the work on cosmic-ray propagation has concentrated on another process: the resonant scattering of cosmic rays by Alfvén waves whose scale is comparable to their radius of gyration. Such waves can be generated by the cosmic rays themselves; their presence leads to particle diffusion along the magnetic field, with very little displacement across the lines. The review by Wentzel (1974) on "collective effects" gives the main formulae and outlines their derivations; here, we only give some approximate formulae, which are, nevertheless, sufficient for our purposes.

2.1.1 COSMIC-RAY SELF-CONFINEMENT TO THE GALAXY   In this and the following section (2.1.2), we adopt a "working model" analogous to that used by Wentzel (1974): cosmic rays are emitted at the galactic plane and propagate along a magnetic tube of force that is open at both ends, far away from the plane. Let us consider cosmic-ray propagation in a gas of uniform density. When cosmic rays stream at a velocity in excess of the Alfvén velocity, they excite resonant hydromagnetic waves which in turn scatter them. In the moving frame in which cosmic rays are isotropic, the growth rate of the waves of wavenumber k is approximately given by

$$\Gamma_{CR}(k) \simeq \Omega_0 \frac{n_{CR}(r_g > 1/k)}{n^*} (-1 + |v_R|/v_A) \qquad (1)$$

where Ω_0 is the Larmor frequency of a proton, $n_{CR}(>r_g)$ is the number density of cosmic rays having gyration radius larger than $(1/k)$; n^* is the density of charged particles in the medium; v_R the streaming velocity of cosmic rays with respect to the gas; and v_A the Alfvén velocity in the ionized part of the medium [$v_A = B/(4\pi n^* m)^{1/2}$, where B is the magnetic field strength and m is the proton mass]. In a neutral medium, the collisions between the charged and neutral particles are very effective in dissipating the wave energy (Kulsrud & Pierce 1969); the damping rate, for a temperature $T = 10^3 (10^4)$ K is $\Gamma^* = \Gamma_0 n_H = 3.3 (8.4) \times 10^{-9} n_H$ s^{-1}, where n_H is the number density of neutral hydrogen atoms. At energies of a few GeV, cosmic rays are numerous enough to overcome the wave damping and limit their streaming in the neutral intercloud medium (e.g. with $n_H = 0.2$ cm^{-3}, $n^* = 0.03$ cm^{-3}, $T = 5 \cdot 10^3$ K, $v_A = 3.6\ 10^6$ cm s^{-1}) which fills a sizeable portion of the interstellar space (Myers 1978). But it is obvious from the expression for Γ_{CR} and Γ^* that, if the amplitude of the waves were determined only by cosmic-ray collective effects in situ and by ambipolar collisional damping, the cosmic-ray streaming velocity would increase as $E^{1.7}$. Consequently particles of energy > 100 GeV would not be confined at all (Kulsrud & Cesarsky 1971).

Taking advantage of the large scale density gradients existing at the borders of the galactic disk, Skilling (1971) proposed a solution for the self-confinement of cosmic rays, which can be viewed as a physical realization of a leaky box. Assume that the cosmic rays only scatter off waves that they themselves have generated, and which propagate away from the plane. The propagation equation of the cosmic rays can then be written (Skilling 1975a)

$$\frac{\partial f}{\partial t} + \frac{\partial}{\partial p^3}(p^3 \mathbf{w}) \cdot \nabla f = 1/3(\nabla \cdot \mathbf{w})p\frac{\partial f}{\partial p} + \nabla \cdot (K\mathbf{nn} \cdot \nabla f), \qquad (2)$$

where $f(p, x)$ is the cosmic-ray distribution function, x is the coordinate along the field line; $\mathbf{n}$ is the unit vector parallel to the field line; with respect to a "rest" frame, the background plasma flows at a velocity $\mathbf{v}_0$, and the waves at $\mathbf{w} = \mathbf{v}_0 + v_A\mathbf{n}$. The diffusion coefficient along field lines, $K(p)$, is inversely proportional to the energy in the corresponding resonant waves: $3K(E)/c \simeq \Omega(E)kF[k = 1/r_g(E)]/B^2$ where $F(k)$ is the energy in waves of wavenumber between k and $(k+dk)$, and $\Omega(E) = \Omega_0 mc^2/E$. The first term on the right-hand side of Equation (2) represents convection of the cosmic-ray gas at a speed w; the second term accounts for the diffusion along field lines. For $v_0 = 0$, we have $K\mathbf{nn} \cdot \nabla f \simeq (v_R - v_A)f$. It is interesting that, if cosmic rays are self-confined, $(v_R - v_A)$ can be calculated by setting $\Gamma_{CR} = \Gamma$, where Γ is the appropriate wave damping rate. The diffusion term in Equation (2) can then be written simply as $p^{-3}\nabla \cdot (v_A n^*\Gamma/\Omega)$: cosmic-ray transport is dominated by diffusion or convection, depending on the effectiveness of the damping mechanism. In the case of weak damping, the cosmic rays are almost locked to the gas; they convect at v_A with respect to it.

Let us consider the consequences of Equation (2) for cosmic-ray transport along a field line. In first approximation, we assume that the galactic magnetic field strength is uniform, while the densities n_H and n^* decrease monotonically away from the plane. Then cosmic rays of a given energy E_1 are able to limit their streaming in all the regions further away from the plane than a critical height defined by $\Omega_0 n_{CR}(E_1) = \Gamma_0[n_H n^*(E_1)]_{crit}$. In the galactic disk, below the critical height, cosmic rays of energy E_1 are not scattered at all ("free zone"), but they are reflected at the critical height just often enough that their drift velocity is $v_{A_{crit}}(E_1) = B/[4mn^*_{crit}(E_1)]^{1/2}$. Beyond the critical height, the damping rate Γ^* decreases; if it is the dominant damping mechanism, the cosmic rays must soon reach a pure convective behavior, and flow at $\sim$ the local Alfvén velocity. If $n_H \propto n^*$, the mean amount of matter λ_e traversed by cosmic rays before escape has a power-law energy dependence $\lambda_e(E) \propto [v_{A_{crit}}(E)]^{-1} \propto E^{-0.4}$.

In fact, the scale height of ionized matter appears to be greater than that of neutral matter by a factor ≥ 2 (Falgarone & Lequeux 1973). But then the regions high above the plane are fully ionized, and nonlinear damping caused by interactions of hydromagnetic waves with sound waves (Chin & Wentzel 1972) becomes dominant. The inclusion of this effect leads to a gradually steeper dependence of the cosmic-ray pathlength, tending to $\lambda_e \propto E^{-0.75}$ at the highest energies (Holmes 1975).

Self-confinement of cosmic rays in fully ionized regions of the interstellar medium, with $T = 10^4$ K, has been shown by Wentzel (1974) and Skilling (1975b,c) to encounter similar problems as self-confinement in neutral regions: because of sound-wave cascade damping of hydromagnetic waves (Chin & Wentzel 1972), cosmic rays of energy ≥ 1000 GeV can stream unimpeded.

Since these papers were written, it has been realized that a substantial part of the interstellar volume is occupied by a hot plasma characterized by $T \simeq 10^6$ K, $n^* \simeq 0.003$ cm^{-3} (Cox & Smith 1974, McCray & Snow 1979 and references therein). The magnetic field strength in these regions has not been measured directly, but is generally expected from flux conservation arguments to be somewhat lower than in the intercloud medium; this would account for the observed anticorrelation between radio-continuum emission and soft X-ray emission (Hayakawa 1979). Thus the Alfvén velocity, which is still $\sim 3.6 \times 10^6$ cm s^{-1}, is lower than the sound velocity $c_s \simeq 10^7$ cm s^{-1}. In that case, the selection rules governing the decay of an Alfvén wave into another Alfvén wave and a sound wave cannot be obeyed, so that the damping of Chin & Wentzel (1972) is not operative. Holman et al. (1979), however, argue that in this case the instability cannot lead to cosmic-ray diffusion, because the very short waves, needed to scatter cosmic rays of pitch angle very close to 90°, and to turn them around, are rapidly destroyed by ion-cyclotron resonances with the thermal background protons. In the coronal medium, the problem arises only for particles with $(1 - |\cos\theta|) < c_s mc/E \simeq 3 \times 10^{-4}\ mc^2/E$; such particles are mirror-reflected by magnetic field fluctuations of amplitude $\delta B/B \gtrsim 10^{-7}(mc^2/E)^2$. If cosmic rays, propagating initially at the velocity of light, transferred all their momentum to linear Alfvén waves which are compressive only to second order, we would have $\delta B/B \lesssim [\varepsilon_{CR}/(B^2/8\pi)]v_A/c \simeq 10^{-4}$ (where ε_{CR} is the cosmic-ray energy density, taken to be comparable to $B^2/8\pi$); thus, cosmic rays (or other wave sources) may well be able to provide the required mirrors in the hot plasma.

Waves propagating at an angle are more or less effectively damped by collisionless processes (Foote & Kulsrud 1979); on the other hand, the circularly polarized waves traveling around the magnetic field are only known to be vulnerable to one kind of nonlinear decay process, which is analogous to the nonlinear Landau damping of plasma oscillations (Lee & Völk 1973, Kulsrud 1978). The damping rate of waves of wavenumber k, in first approximation, is proportional to the energy density of the waves, $kF(k)$. However, when the wave energy is higher than a value $(kF)_{crit}$, some of the thermal particles are trapped by the wave packets (Kulsrud 1978); as a consequence, the damping process becomes less effective.

In the coronal medium, $8\pi(kF)_{crit}/B^2 \simeq 4 \cdot 10^{-8}$. If cosmic rays have a bulk speed comparable to the Alfvén speed, the energy in waves is of the order of $(kF)_{k^{-1}=r_g(E)} \simeq (B^2/8\pi)(E/3 \cdot 10^5 L_1)$, where E is the energy in GeV of particles resonating with waves of wavenumber k, and L_1 is the length in kpc of the local flux tube in our "working model"; thus, the relevant wave energies are above the critical value. Then we estimate the nonlinear damping rate $\Gamma_{nl} \simeq 5 \cdot 10^{-12}/(EL_1)^{0.5}\ s^{-1}$ (if the trapping effect emphasized by Kulsrud is neglected, the damping rate of the waves is even higher). The growth rate of the waves, for a streaming velocity v_R comparable to v_A, is $\Gamma_{CR} \simeq 6 \cdot 10^{-10}\ E^{-1.7}\ s^{-1}$. It appears that, once again, only cosmic rays of energy ≤ 100 GeV can constrain themselves to stream at $v_R \simeq v_A$; and particles of energy higher than $\sim 10^4$ GeV are not confined at all in the coronal medium (C. J. Cesarsky and R. M. Kulsrud, in preparation).

2.1.2 DIFFUSION DUE TO AN INTERSTELLAR SPECTRUM OF HYDROMAGNETIC WAVES An interesting alternative to self-confinement is the possibility that cosmic-ray diffusion in interstellar space is due, as in interplanetary space, to waves generated by other sources. If a spectrum of hydromagnetic waves exists, it could force cosmic rays to stream at a velocity less than v_A, so that the instability discussed in Section 2.1.1 would never occur.

Hydromagnetic waves of the appropriate wavelength could be emitted by a number of discrete sources, such as high velocity stars with stellar winds, rotating white dwarfs or magnetic stars, etc. However, if waves of different wavelengths were emitted by different types of sources, the spectrum of waves would probably be very irregular, making it difficult to account for the simple cosmic-ray spectrum. On the other hand, it may be possible to obtain a smooth wave spectrum if only very long waves, with scales of several parsecs or more, are being generated directly by turbulent gas motions, supernova explosions, or expanding HII regions. In the case of homogeneous, isotropic hydrodynamic turbulence, nonlinear effects cause the energy in large eddies to be transmitted to eddies of successively smaller scales; Kraichnan (1965) asserts that a similar cascade process could operate in the case of hydromagnetic turbulence.

Let us first review the observational situation. Scintillations in pulsar signals bring some information on irregularities of scale $10^{11.5 \pm 1.5}$ cm in the thermal electron distribution of the interstellar medium. Rickett (1970), assuming a Gaussian spectrum of irregularities, derived from his data $\delta n^* \simeq 10^{-4 \pm 1}$ cm^{-3}. Lee & Jokipii (1976) have shown that the data can equally well be interpreted in terms of a Kolmogorov spectrum, i.e. $F(k) \propto k^{-5/3}$. The intensity of this spectrum, assuming that the turbulence is homogeneous, is such that it can be extrapolated to give $kF(k) \simeq B^2/8\pi$ at $1/k \sim$ a few tens of parsecs. As magnetic inhomogeneities are probably associated with the electron density fluctuations, Lee & Jokipii (1976) speculate that a Kolmogorov spectrum of hydromagnetic waves, extending over the range 10^{11}–10^{19} cm, may exist in the interstellar medium. The cosmic-ray diffusion coefficient would then be $K \simeq 3\Omega k F(k)/cB^2 \propto E^{1/3}$. [In the quasi-linear formulation (e.g. Kulsrud & Pierce 1969), K diverges if the wave spectrum has an index <2; the problem is again caused by the difficulty of scattering particles of pitch angle close to $90°$. If magnetosonic waves are also present, the problem is automatically solved by mirroring effects (Cesarsky & Kulsrud 1973); otherwise, a nonlinear theory has to be introduced (see review by Völk 1975). The energy dependence of K may be somewhat different because of these effects.]

The hypothesis of Lee & Jokipii (1976) meets with some unsolved difficulties. On the observational side, Rickett (1977) finds that the "turbulence" associated with pulsar scintillations is very inhomogeneous. There are also theoretical problems. When the wave frequency is higher than the ion-ion collision frequency, magnetosonic waves are Landau damped very rapidly. In clouds of density $n = 1$ cm^{-3}, $n^* = 0.01$ cm^{-3}, $T = 100$ K, this process affects waves with wavelength greater than 4×10^{11} cm. The damping time is $\simeq 8 \times 10^6$ s; if the level of fluctuations is $\delta n^* = 10^{-4}$ cm^{-3}, the energy input in the medium is ~ 100 times the cooling rate (Scheuer & Tsytovitch 1970); in a hotter medium, the energy expenditure is even higher.

The existence of an interstellar spectrum of waves at lower frequencies encounters similar difficulties (Cesarsky 1971, 1975). Let $t_t(k)$ be the time in which the energy present in waves of wavenumber k is transmitted to shorter waves. A lower limit to t_t is the time for a turbulent eddy to cross its length at its own velocity, or *turn-over* time (Heisenberg 1948): $t_H = 1/[k(kF)^{0.5}]$. If $t_t = t_H$, in the inertial range, the spectrum adopts the Kolmogorov form. In the presence of a magnetic field, known to be turbulent at the longest scales, it is probably more appropriate to use Kraichnan's (1965) theory[1]; then the transfer time is $t_K = v_A/Fk^2$, and the inertial range spectrum is $\propto k^{-1.5}$. Let $t_d(k)$ be the damping time of the waves. Assume that turbulence is fed to the medium at scales $\geq (1/k_0)$; if, at any $k > k_0$, $t_d(k) < t_t(k)$, the power input is dissipated as heat before having a chance to produce shorter waves. Thus, a spectrum of waves develops only if $t_d > t_t$ over the whole range of interest; if t_d is determined by linear damping mechanisms, to impose $t_d > t_t$ over a range of scales is equivalent to putting a lower bound on the turbulent power fed at long wavelengths. This power will finally be dissipated into heat. For the intercloud medium defined earlier, the power needed to maintain an inertial spectrum of turbulence from scales of 10^{19} cm to scales $\leq 10^{12}$ cm against the dissipative effect of the collisions between charged and neutral particles is six orders of magnitude more than needed to keep the medium at 10,000 K, and about four orders of magnitude more than required to collisionally ionize it. Infalling high-velocity clouds could generate parsec-sized waves in the regions bordering the Galactic disk; the energy dissipated by these waves would be sufficient to keep two slabs of material of density 0.02 cm^{-3} and width 200 pc, just above and under the Galactic disk, at 40,000 K. The corresponding degree of ionization would be so high that the damping of waves by ambipolar diffusion is negligible. Magnetoacoustic waves would be damped by thermal conduction effects, and would disappear for $(1/k) \leq 5 \cdot 10^{16}$ cm. A spectrum of waves will still exist if, over the range of k where the magnetoacoustic waves are damped, a sizeable fraction of the power input is transferred to the Alfvén modes; the mirroring of cosmic rays with pitch angles close to 90° would still be mainly due to the remaining magnetoacoustic waves. The slabs would act as partially reflecting boundaries for the disk cosmic rays, as in the leaky-box model.

In the "coronal" (hot plasma) phase of the interstellar medium, the sound velocity is higher than the Alfvén velocity; the properties of hydromagnetic waves propagating in such a medium have been studied by Foote & Kulsrud (1979). Using their results on wave damping, it appears that a Kolmogorov or Kraichnan spectrum of waves may develop over scales

---

[1] In fact, the Kolmogorov and the Kraichnan theories refer to incompressible turbulence; equivalent results for the case of compressible turbulence are not available.

$\geq 10^{19}$ to 10^{12}–10^{13} cm; at shorter wavelengths, the spectrum would be cut off. The heat input from such waves would be comparable to the cooling rate of the medium if, at $k_0 = 2\pi/100$ pc^{-1}, the energy in the waves is about half the energy in the field (Cesarsky 1975). If particles of pitch angle $\simeq 90°$ are easily mirrored, the pathlength of cosmic rays would have an energy dependence of $\lambda_e \propto E^{-(0.33-0.5)}$.

2.1.3 COSMIC-RAY REDISTRIBUTION IN THE GALAXY   So far, we have only discussed cosmic-ray propagation along a field line. This is because the rate of scattering perpendicular to field lines is very low; the maximum possible value of the perpendicular diffusion coefficient, $K_\perp = r_g c$, which corresponds to a displacement r_g across field lines every gyration period, is at least six orders of magnitude lower than the value of the parallel diffusion coefficient K needed to explain GeV cosmic-ray observations. Jokipii & Parker (1969) pointed out that the field lines of the turbulent galactic magnetic field must random-walk throughout the Galaxy, helping to spread cosmic rays around. This idea has led Lingenfelter et al. (1971) to the concept of *compound diffusion* (see also Getmantsev 1962) in which cosmic rays are assumed to always remain attached to the same field line: their motion is then the combination of cosmic-ray diffusion along the field lines, with random walk of the field lines in the Galaxy. A difficulty with this picture is that, if cosmic rays stay on the same field line on which they were initially injected, it would seem that adjacent field lines could contain significantly different amounts of relativistic particles, depending on the number of sources they happened to traverse. In fact, this simplistic view of the three-dimensional transport of cosmic rays is not widely held, as it has been shown that, in a turbulent magnetic field, temporarily adjacent lines are separated so fast that cosmic rays rapidly lose touch with their neighbors.

Let us assume, with Ptuskin (1979), that the galactic magnetic field is completely turbulent on a scale L. A line of force is specified by one of the points through which it passes, and its wandering is described by a Lagrangian variable representing the length along the line. It can then be shown that the lines random-walk with a characteristic step length $L_{\text{eff}} = L/\sqrt{2}$. Ptuskin shows that two particles that are initially at a distance r_g from each other will find themselves on separate tubes of force once they have diffused a distance $S_c \simeq 10L \ln(L/r_g)$ along the field. If instead the magnetic field is supposed to consist of a uniform component, plus a power-law spectrum of irregularities at $k > k_0$, of spectral index <2, the distance S_c is reduced to $(1/k_0) \sim L$ (Jokipii 1973, Skilling et al. 1974). Suppose that some scattering mechanism exists so that the mean free path of cosmic rays along the field is l_1; the time needed for a particle to diffuse a distance S_c

along the field is $(S_c l_1)^{0.5}/c$; if $L = 100$ pc, $l_1 = 1$ pc, cosmic rays forget their original field line in tens of years for a power-law spectrum of irregularities, and hundreds of years for a large scale turbulent magnetic field. Consequently, cosmic rays issued by a given source can be found over a significant volume of interstellar space.

As is the case for the *compound diffusion* theory, the three-dimensional diffusion coefficient of cosmic rays can be larger than $K = l_1 c/3$ by a factor as large as 30; the requirement on the energy in waves necessary to scatter cosmic rays can be relaxed by the same factor. Also, the problem remains that the anisotropy of cosmic rays, as measured at the earth, is determined only by the characteristics of cosmic-ray transport along field lines (Allan 1972).

2.2 *Cosmic-Ray Escape from the Galaxy; Static and Dynamical Halos*

2.2.1 MAGNETIC FIELD CONFIGURATION AND COSMIC-RAY BUBBLES We have considered the problem of cosmic-ray propagation in the Galaxy; let us now consider the boundary conditions. Piddington (1972) argues that the Galactic magnetic field is the result of the compression of an extragalactic field of $\sim 10^{-8}$ gauss; the two fields would still be connected. In that case, the question to be asked is, Why are cosmic rays retained in the Galaxy? and we have just presented theories leading to cosmic-ray confinement by diffusion processes taking place in the disk or at the border of the disk. In models of the Skilling (1971), Holmes (1974, 1975) type, cosmic rays would continue convecting out of the disk at essentially the local Alfvén velocity, which is thought to increase with distance from the disk. In models assuming decay of large scale turbulence, it seems reasonable to infer that the sources of turbulence die out at a distance from the galactic plane; thus the diffusive region is naturally bound.

Parker (1973) strongly disagrees with Piddington's views. He argues that the possible configurations of a galactic field connecting fully to a much weaker extragalactic field are violently unstable. If the magnetic field is mostly perpendicular to the disk, neutral sheets are formed at the borders, which he concludes would lead to field reconnection in times much shorter than the age of the Galaxy. If instead the field is mainly parallel to the Galactic disk, it is excluded from all but a thin surface layer in a time of the order of the characteristic turbulent diffusion time, which Parker estimates at $3 \cdot 10^8 – 3 \cdot 10^9$ years. He concludes that the present magnetic field has been generated by the cyclonic turbulence and the differential rotation in the disk, and that it is closed. The problem then becomes, Why do cosmic rays escape?

The answer proposed by Parker (1969, 1976) is well known: Jokipii &

Parker (1969) argue that magnetic field lines encounter the boundary of the galactic disk every ~ 600 pc if they random-walk throughout the Galaxy. There, the cosmic-ray pressure would overwhelm the magnetic field, and cosmic rays should then inflate a *bubble* at the disk boundary. Such cosmic rays can be considered as having escaped if they find themselves in a bubble that is expanding so fast that they cannot reenter in the disk (the minimum expansion velocity required is v_A if cosmic rays are self-limiting their streaming), or if they are in a bubble that detaches itself from the Galaxy after its field has reconnected to form a closed loop. The rate of inflation of the bubbles depends on the rate of generation of cosmic rays. Two limits can be distinguished.

1. If the bubbles inflate slowly, most cosmic rays entering them eventually return to the galactic disk; they only escape when they find themselves in a bubble that detaches itself or on a field line that is temporarily open. In this limit, we have a physical representation of a leaky-box model; pitch-angle scattering is not required to explain the long lifetime and the isotropy of cosmic rays. The relativistic particles have a better chance to sample different regions of the Galaxy than in a diffusion model, and they are expected to fill uniformly the galactic disk and a slab halo formed by the expanding bubbles. The mean pathlength traversed is energy independent, so that the energy dependence of λ_e observed must be attributed to particle trapping in the sources (*nested leaky-box model*).

2. If bubble inflation proceeds at a rapid pace, cosmic rays that cross the border of the disk do not, in general, reenter. Rapid bubble inflation would allow magnetic fields and cosmic rays to rise high above the plane, before resistive instabilities detached them completely. There would be a halo of cosmic rays, and a radio-halo, but since the halo would be made up of nonreentrant particles, the properties of the cosmic rays observed at the earth would be as predicted by a disk diffusion model.

2.2.2 GALACTIC WIND AND DYNAMICAL HALO Should the escape of cosmic rays from the Galaxy be accompanied by an outflow of thermal matter? Parker's bubble picture leaves this question open. Johnson & Axford (1971) suggested that the emergence of relativistic particles and magnetic fields from the Galaxy could give rise to an expanding galactic corona; the halo would be the dense, subsonic part, and the cosmic rays would escape with the supersonic outward flow. Ipavich (1975) has tackled this problem, using a simple, one-dimensional geometry (i.e. radial field lines, as for a monopole magnetic field, and a gravitational acceleration inversely proportional to the square of the distance to the center). He finds that the cosmic rays could transfer sufficient outward momentum to the gas to drive a galactic wind, even if the gas is cold. In all his models, cosmic rays are assumed to be constrained by self-generated waves to stream at the Alfvén velocity with respect to the background; their momentum would be transmitted to the unstable waves before being deposited into the medium. The waves are assumed to be dissipated slowly enough that the cosmic rays are essentially in the "convective" regime [i.e. the first term on the right-hand side of Equation (2) dominates over the second] and fast enough that the momentum transfer from the cosmic rays to the gas can be considered to take place in situ.

Jokipii (1976) has proposed a phenomenological model of cosmic-ray propagation, the *dynamical halo model*, whose goal is to combine the main features of the Parker bubbles and the Ipavich winds, and to allow a comparison with cosmic-ray observations. In this one-dimensional model, cosmic rays are generated in the disk and diffuse along the galactic magnetic field with a diffusion coefficient K. In addition, in the halo, the scattering centers responsible for the diffusion of cosmic rays are convected outwards with a velocity V. The cosmic-ray density goes to zero at a boundary situated at a distance H from the galactic plane. K and V are assumed to be uniform up to H. The main parameter of the model, $X = HV/K$, measures the importance of diffusion as compared to convection; the particles seen in the disk have not gone further from it than a distance $H^* = H/X$ (Jones 1978). At present, the gap separating the *dynamical halo model* from a self-consistent model of cosmic-ray–powered galactic wind is wide open. Ipavich's model was for a geometry very different from that envisaged by Jokipii and his co-workers; his results may be applicable to a galactic nuclei, but not to a region of the disk. In particular, as noted by Ginzburg (1979), in the solar neighborhood, the gravitational acceleration away from the plane does not decrease as $1/z^2$, where z is the distance to the plane; instead, it increases up to $z \simeq 1$ kpc, and only starts decreasing for $z \gtrsim 3$ to 4 kpc (Oort 1965). Also, in Ipavich's work, convection dominates the cosmic-ray transport through the halo, so that automatically $X \ll 1$; to drop that assumption and try to calculate K requires a good knowledge of the wave damping mechanisms. Moreover, if X is finite, the evaluation of the amount of momentum transferred to the gas at each point becomes much more complicated (e.g. Wentzel 1977a). Finally, we note that the transport equation used by Jokipii and his co-workers is that used for propagation in the solar wind; it is appropriate when describing a situation where cosmic rays are scattered by an isotropic distribution of waves. If cosmic rays can really power a wind, we expect most waves to go away from the disk; then a different equation must be used (Skilling 1975a).

Hayakawa (1979) proposed that gas and cosmic rays are convected away from the galactic plane by the expansion of supernova blast waves into the coronal interstellar medium; this scenario, where the scattering of cosmic rays may be due to preexisting microturbulence, can provide a better background for the *dynamical* halo equations. However, Chevalier & Oegerle (1979) argue that the presence of clouds in the halo indicates that the energy input from supernovae is lost to radiation, rather than used to power a galactic wind.

2.2.3 HYDROSTATIC EQUILIBRIUM OF THE GALACTIC DISK; PARKER INSTABILITY AND ENHANCED DIFFUSION OF THE MAGNETIC FIELD THROUGH THE GAS Let us forget, for a while, the existence of cosmic-ray sources and of supernova explosions, and consider the possible static equilibrium configurations of an interstellar medium composed of gas, magnetic fields and cosmic rays, immersed in the galactic gravitational field due to the presence of stars. The equilibrium equation is particularly simple when the lines of force of the magnetic field are assumed to be parallel to the galactic plane. Parker (1966, 1969) studied the stability of the static equilibrium obtained when the pressures of the gas, of the magnetic field, and of the cosmic rays vary with z in the same way; he found that the system is unstable if the adiabatic index γ of the gas is smaller than some critical value, γ_c. In the solar neighborhood, $\gamma_c \simeq 2.5$, while Parker argues that the interstellar gas responds to small perturbations as an isothermal gas, with $\gamma = 1$; thus, the system is unstable.

Let us take a coordinate axis such that the galactic plane is the xy plane, and the magnetic field is parallel to the y axis. Using a normal-mode analysis, Parker shows that the application of two-dimensional perturbations, proportional to $\exp(ik_y y + ik_z z)$, can lead to instability; if $(1/k_y) \simeq h$, the scale height of the system, the instability grows in a time comparable to the free-fall time of the gas, $h/c_s \simeq 10^7$ years. The effect of the instability is to allow the interstellar gas to slide down along the magnetic field lines, to form cloud complexes on the galactic plane, while the raised portions of the field lines, unburdened of the gas, can expand and form *bubbles* at the border of the disk. [Zweibel & Kulsrud (1975) find that the growth rate of the instability can be slower by a factor of order 10 if the magnetic field has also a component turbulent on a scale $\ll h$; they note that if the rate of bubble inflation is determined by the instability, cosmic rays would be confined to the Galaxy for too long times. In fact, as we have seen, the rate of bubble inflation is probably governed by the random walk of the field lines and the continuous production of cosmic rays, rather than by the instability.] Mouschovias (1974) has constructed numerically two-dimensional equilibrium states for the interstellar gas and field system, which would represent the end result of the two-dimensional Parker instability; he finds that these states are stable to two-dimensional perturbations. Cosmic rays are not included in his calculations, but in a subsequent paper (Mouschovias 1975), he conjectures that a similar result may be obtained if they are present. [However, he does not take into account the possible presence of sources continuously emitting new cosmic rays.]

In the equilibrium considered by Parker, the fastest growing instabilities are those triggered by three-dimensional perturbations, of the form $\exp(ik_x x + ik_y y + ik_z z)$, with $(1/k_x) \ll h$, and $(1/k_y) \simeq h$; thus, the first effect of the instability is to generate small scale turbulence. Using a variational method based on an energy principle, it is possible to show that this type of Rayleigh-Taylor instability affects all the equilibrium states that have been proposed to date for the interstellar medium gas and fields system. The horizontal equilibria calculated by Badhwar & Stephens (1977), which take into account the observational fact that the scale height of cosmic rays and magnetic fields is higher than that of the neutral gas, are all unstable at some height above the disk (Lachièze-Rey et al. 1980); so are the curved two-dimensional equilibria proposed by Mouschovias (1974). The presence of curved field lines with radius of curvature $R \lesssim h$ brings in an additional instability, whose effects complement those due to gravity. In addition, with the presence of a curved magnetic field, the cosmic-ray pressure P in the raised portions of the lines can exceed considerably the gas pressure p; then, the instability can grow in a time $\sim (h/c_s)(pR/Ph)^{1/2}$, which is shorter than the free-fall time (Asséo et al. 1978, 1980).

Mathematically, the fastest growth rates are obtained for $k_x \to \infty$; physically, k_x is limited by the effects of the diffusion of the magnetic field with respect to the gas, which tend to erase the perturbation (Parker 1967). The maximum k_x at which instabilities can grow is k_m, such that the growth time is equal to the diffusion time: $\eta k_m^2 \simeq c_s/h$, where η is the diffusion coefficient of the magnetic field through the gas. In HI regions, η is determined by ambipolar diffusion, and $k_m \simeq 1$ pc^{-1}; in ionized regions, the coefficient η is much smaller and k_m is several orders of magnitude larger.

The combined effect of two-dimensional and three-dimensional instabilities is to enhance the rate of separation of the magnetic field and the gas (Parker 1967): in a growth time, the gas in a field line "valley" can diffuse out a distance $1/k_m$ to a field line top, where it promptly falls a distance $1/k_y \sim h$. Thus, the gas drifts down with respect to the field at a velocity independent of k_m and comparable to c_s; the unburdened field lines rise continuously carrying the cosmic rays with them. In this case the magnetic field of the halo could not be formed by loops anchored in cloud complexes in the disk but would consist of "curly" field lines, with curls of characteristic size R. The instability here would lead to vertical transport of the gas at a distance $\sim R$ in a growth time, and the relative drift velocity of the magnetic field and the cosmic rays with respect to the gas is $\sim c_s(RP/hp)^{1/2}$. In this picture the escape of cosmic rays from the Galaxy is not accompanied by an outflow of thermal gas, but by infall of gas on the galactic plane.

2.4 Small Scale Cosmic-Ray Gradients

When interpreting cosmic-ray composition data or galactic gamma-ray observations, it is generally taken for granted that the cosmic-ray density is uniform on small scales ($\sim$ tens of parsecs). However, the possibility that cosmic rays are partially excluded from dense clouds deserves some consideration.

Let us consider a dense cloud that is threaded by the intercloud magnetic field and assume that, far from the cloud, the interstellar cosmic-ray distribution is isotropic. Then, in a stationary, static situation and in the absence of energy losses, the Liouville theorem implies that the cosmic-ray density is uniform along field lines. The column of matter through a typical molecular cloud is only $\sim 0.1\%$ of the stopping length of relativistic protons and electrons. Nevertheless, cosmic-ray depletion in dense regions could occur, if the particles were retained in them for sufficiently long times, either by the presence in the cloud of small or large scale magnetic turbulence (Nakano & Tademaru 1972), or because of the presence of self-excited hydromagnetic waves at the border of the cloud (Skilling & Strong 1976); a static but inhomogeneous magnetic field could achieve the same result (Cesarsky & Völk 1978). A close look at these three processes led the latter authors to conclude that none of them is likely to prevent cosmic-ray penetration in dense clouds. This conclusion is supported by a detailed study of the gamma-ray emission of a nearby cloud complex (Lebrun & Paul 1978). Scott (1975) has suggested that cosmic rays may remain trapped in the low density "tunnels" of coronal gas, carved by supernova remnants into the neutral interstellar medium (Cox & Smith 1974). McKee & Ostriker (1977) argue that in fact the hot plasma invades most of the interstellar volume, leaving the clouds of neutral gas disconnected from each other. In either case, as long as the magnetic fields from the hot and the cold regions are connected, the results of the study by Cesarsky & Völk (1978) indicate that cosmic rays permeate all the interstellar matter.

2.5 Continuous Acceleration by Supernova Shocks

Bell (1978), Axford et al. (1977), and Blandford & Ostriker (1978) have shown that shock waves induced by supernova remnants are capable of accelerating cosmic rays, producing a power-law spectrum of index close to, but somewhat higher than, 2. The mechanism relies on efficient resonant scattering of cosmic rays in the vicinity of the shock; if the waves are excited by the cosmic rays, the maximum energy reached is only ~ 300 GeV; much higher energies, up to $\sim 10^{18}$ eV, can be attained if an interstellar spectrum of waves, of the type discussed in Section 2, exists. If the interstellar medium is as described by McKee & Ostriker (1977), the "cross-section" of supernova remnants is large, and the average cosmic ray encounters several shocks in its galactic lifetime; then, the relationship between the cosmic-ray composition, the observed spectra, and the propagation characteristics becomes particularly involved (Ostriker 1979).

3 INTERPRETATION OF THE OBSERVATIONS OF COSMIC RAYS AT THE EARTH

3.1 *Implications of The Elemental Composition and the Spectra of Nuclei*

As cosmic-ray nuclei travel through interstellar space, they suffer inelastic collisions with interstellar medium nuclei; in this way "primary" cosmic-ray nuclei emitted by sources break up into lighter "secondary" nuclei. The amount of interstellar matter traversed by cosmic rays can be estimated by measuring the abundances of certain species expected to be absent in the primary spectrum.

3.1.1 ESCAPE PATHLENGTH AT MEDIUM AND HIGH ENERGIES At energies greater than a few GeV/nucleon, the effects of solar modulation and of Coulomb interactions in the interstellar medium are negligible and the cross sections of the spallation reactions affecting the cosmic-ray composition are energy-independent. In the *leaky-box* model, assuming an interstellar medium of pure hydrogen, the flux f_i of a species i (where i is the atomic number) is simply related to the source term Q_i (cm^{-3} s^{-1}) and the mean escape length λ_e (g cm^{-2}) through

$$f_i = \frac{Q_i/n_H m + \sum_{j>i} \sigma_{j,i} f_j/m}{(\lambda_{d_i})^{-1} + (\lambda_e)^{-1}} \quad (3)$$

where $\sigma_{j,i}$ is the cross section of the spallation reaction $j(p,\ )i$ induced by the element j on interstellar hydrogen atoms, and λ_{d_i} the pathlength for nuclear destruction of nuclei i on hydrogen; λ_{d_i} decreases when i increases [e.g. $\lambda_d(\text{He}) = 17$ g cm^{-2}, $\lambda_d(\text{C}) = 7$ g cm^{-2}, $\lambda_d(\text{Fe}) = 2.5$ g cm^{-2}]. For purely secondary species, such as the light elements lithium, beryllium, and boron, $Q_j \equiv 0$ and the knowledge of the fluxes f_i and of the nuclear cross sections involved is sufficient to determine the mean escape length λ_e; at $E = 5$ GeV/nucleon, the values of λ_e obtained are in the range (3.5–6) g cm^{-2}. [In the *leaky-box* model, the distribution of pathlengths around the mean is exponential; in comparison, the distributions predicted by the *nested leaky-box* model (Cowsik & Wilson 1975) and the *diffusion* models (especially at a distance from the cosmic-ray sources; Owens 1976, Lezniak & Webber 1979) are deficient in short pathlengths. At present, the spallation and destruction cross sections are not known with sufficient accuracy to permit a really precise determination of the pathlength distribution (Raisbeck 1979). With the parameters at hand, it appears that the exponential distribution is certainly a good first approximation to the actual distribution; however, it is often found that the best exponential fit for the CNO secondaries has a tendency to underproduce Fe secondaries, possibly implying that short pathlengths are indeed deficient (Shapiro et al. 1970, 1973).]

Several balloon measurements of cosmic-ray composition at energies up to 150 GeV/n have shown that, at energies greater than 5 GeV/n, the ratio of secondary to primary abundances decreases as the energy increases (Juliusson et al. 1972, Smith et al. 1973, Balasubrahmanyan & Ormes 1973, Webber et al. 1973, Juliusson 1974, Caldwell 1977, Ormes & Freier 1978, Lezniak & Webber 1978, Orth et al. 1978, Simon et al. 1979); also, the spectra of heavy primary species are flatter than those of lighter ones. In the framework of the *leaky-box model* these results simply imply that the probability of escape of cosmic rays increases with the energy. Assume, with Audouze & Cesarsky (1973), that the escape pathlength can be written $\lambda_e = \lambda_0 E^{-a}$. Then, if the source spectrum were $Q = Q_0 E^{-\alpha}$, the spectrum of primary nuclei would be approximately $E^{-\alpha}/(1/\lambda_{d_i} + E^a/\lambda_0)$. For protons, $\lambda_{d_i} \gg \lambda_e$, and the spectrum would be proportional to $E^{-\alpha-a}$; for heavier nuclei, it is flatter at first and only attains this slope at higher energies. As mentioned in the introduction, the quoted observations imply that a is in the range 0.3–0.6. A possible difficulty with this model is that, under the plausible assumption that the injection spectrum is a power law, $\lambda_e(E)$ must also be a power law that extends at least up to $3 \cdot 10^6$ GeV; otherwise, the proton spectrum would have a break at lower energies. In disk models, this condition imposes $a \leq 0.6$.

In the *nested leaky-box model* (Cowsik & Wilson 1973, 1975, Meneguzzi 1973a,b) cosmic rays traverse an amount of matter λ_s in their sources, before pervading the Galaxy, and it is the probability of escape from the source that is supposed to be energy dependent. For $E \leq 150$ GeV, the composition and the spectra of primaries and secondaries are essentially undistinguishable from those obtained with the energy-dependent *leaky-box model*, but in this case the galactic proton spectrum is identical to the injection spectrum, independent of the form of $\lambda_s(E)$. Several authors have shown (e.g. Ptuskin 1974) that the elemental composition of cosmic-ray nuclei in *diffusion models* is also determined almost exclusively by one parameter, λ_e, related to the mean amount of matter traversed by the particles before escape; in general, λ_e is inversely proportional to the diffusion coefficient K (in one-dimensional models, or in three-dimensional models with scalar diffusion) or to the component of the diffusion tensor perpendicular to the galactic plane. The constant of proportionality contains all the information on the distribution of the sources and on the boundaries of the containment region. We have seen that the model of Holmes (1975) can account for the presently available data, but cannot be extended to higher energies. Models with diffusion due to an interstellar spectrum of hydromagnetic waves, either in the disk or in slabs on each side of the disks (Cesarsky 1971, 1975) predict $\lambda_e \propto 1/K \propto E^{-a}$ for relativistic particles up to very high energies; the break at $E = 3 \cdot 10^6$ GeV is due to the fact that for high energy particles (of gyration radius $r_g > 1$ pc) the rate of escape from the Galaxy is no longer controlled by diffusion along the magnetic field lines.

In the original *closed model* (Rasmussen & Peters 1975), cosmic-ray sources do not emit protons. Of all the nuclei received at the earth, 15% would originate from a nearby source, and the rest, which is composed mostly of protons, then must be a very old population. This model cannot fit adequately the cosmic-ray composition at a few GeV, nor explain the energy-dependent composition. A modified version was proposed by Peters & Westergaard (1977) where cosmic rays are generated in galactic arms and trapped by them in an energy-dependent way. Since they assume the sun to be inside a spiral arm, cosmic-ray observations at energies less than 1000 GeV would not distinguish between this model and the leaky-box model; at very high energies this modified closed model predicts a much higher proportion of secondary nuclei. [Note that good fits of the galactic synchrotron radiation to a spiral structure are only obtained when the sun is supposed to be *outside* a spiral arm (Price 1974, Paul et al. 1976, Brindle et al. 1978).]

Models in which the cosmic-ray acceleration takes place during propagation in the interstellar medium generally predict a positive correlation between the mean age or the escape length of cosmic rays and the energy, in contrast with the observations (Cesarsky 1977). For instance, for the Fermi model (1949), where the confinement time in the Galaxy is energy independent, the ratio of secondary to primary nuclei increases as $\ln E$; it can be made to decrease with the energy by assuming that higher energy particles escape more rapidly from the Galaxy, but then the Fermi acceleration mechanism does not anymore lead to a power-law spectrum. If cosmic rays are accelerated by successive encounters with interstellar shocks (Blandford & Ostriker 1978, Ostriker 1979), the spectrum would be readjusted to a power law of index ~ 2 at each shock and the correspondence between age and energy becomes less clear-cut. Even so, the observational trend can only be reproduced if, in addition, the diffusion coefficient is energy dependent; and, if several shocks are involved, the fit to the data is not very good (Ostriker 1979). However, the combination of acceleration by a *single shock*, which accelerates particles with a power-law spectrum of index 2, and energy-dependent propagation with a containment time proportional to E^{-a}, accounts for all the observations at $E \leq 100$ GeV, including the slope of the proton spectrum.

3.1.2 ESCAPE PATHLENGTH AT LOW ENERGIES At energies less than 1 GeV/n, the interpretation of the data on the elemental composition of cosmic rays and on the spectrum is much less straightforward, owing to three effects: (a) the relevant spallation cross sections become energy dependent, (b) Coulomb interactions in the interstellar medium cease to be negligible for heavy nuclei, and (c) cosmic rays suffer adiabatic losses while being modulated by the expanding solar wind. Consequently, at low energies, despite the wealth of cosmic-ray data, the behavior of λ_e is not yet well established (Raisbeck 1979 and references therein).

Let us rapidly review the predictions of some of the models considered. If low energy cosmic rays are self-constrained to streaming at the Alfvén velocity in the galactic disk, the corresponding escape length λ_e is energy independent (Holmes 1974). If cosmic-ray escape is controlled by interactions with an interstellar spectrum of hydromagnetic waves $F \propto k^{-b}$, the diffusion coefficient is proportional to $r^a v$, where $a \simeq 2-b$, r is the particle rigidity $r = pc/eZ \propto r_g$, and v is the particle speed. If the containment region is the same for particles of all energies, the escape length λ_e is proportional to r^{-a}.

In the *dynamical halo* model, the high energy observations are accounted for if the diffusion coefficient is energy dependent, e.g. $K = K_0 r^{0.5} v/c$ (Jones 1979) [or $K \propto T^a$, where T is the kinetic energy per nucleon (Jokipii & Higdon 1979)]. Then the parameter X is also energy dependent; the transport of high energy particles is dominated by diffusion ($X < 1$) and that of low energy particles by convection ($X > 1$); the rigidity dependence of the escape length λ_e turns over at a rigidity r_0 such that $X = 1$. Jones (1979, see also Giler et al. 1979) gives an approximate expression for the escape length in this model; $\lambda_e(r)$ only involves two parameters, r_0 and L_0, where $L_0 \equiv \rho_0 dc/V$. ρ_0 is the disk gas density, and d is the half-length of the disk region traversed by cosmic rays; the outflow velocity V is that of the scattering centers acting on cosmic rays. If $2\rho_0 d$ is the surface density of the galactic disk in the solar environment, estimated at 5.16 $M_\odot$ pc^{-2} (Gordon & Burton 1976), and $r_0 = 2$ GV, V is only equal to 8 km s^{-1}; it is even less if r_0 is smaller. Note that if the waves are generated by the cosmic rays,

$V = v_{gas} + v_A$, and v_A is probably ~ 30 km s^{-1}! The simplest type of *dynamical halo model* appears to be in conflict with the observations.

3.2 Implications of The Isotopic Composition of Cosmic Rays

3.2.1 AGE OF COSMIC RAYS AND MEAN DENSITY OF THE MATTER TRAVERSED Measurements of the abundances of unstable secondary nuclei, such as ^{10}Be (with a decay period at rest of $\tau_d = 22 \cdot 10^6$ yr), ^{26}Al ($\tau_d = 0.85 \cdot 10^6$ yr), and ^{36}Cl ($\tau_d = 0.45 \cdot 10^6$ yr), can bring some information on the mean age of cosmic rays, and/or help to determine the parameters characterizing the different models. In the framework of the *leaky-box model*, such measurements, combined with the determination of λ_e through the elemental composition, permit us in principle to estimate the mean escape time of cosmic rays and the mean gas density in the box. However, because most measurements are done at low energies, solar modulation again complicates the interpretation of the data. Assuming that λ_e is energy independent, and using their own estimates of solar modulation effects, Garcia-Munoz et al. (1977) derive from their data at 30–150 MeV/nucleon a mean age of $17(+24, -8) \cdot 10^6$ yr, and a mean density $\langle n_H \rangle = 0.18 (+0.18, -0.11)$ cm^{-3}. Higher energy data on beryllium (Buffington et al. 1978, Webber & Kish 1979), chlorine (Meyer et al. 1977, Webber et al. 1979a), and aluminium (Webber et al. 1979b, Young et al. 1979), interpreted in the same way, lend support to the upper limit $\langle n_H \rangle < 0.3$–0.4 cm^{-3}. [While the results of Garcia-Munoz et al. make it highly probable that cosmic rays propagate in a low density medium, they are not completely inconsistent with a mean density n_H as high as 1 cm^{-3}, in view of the errors in the composition data and the combined uncertainties on both the low energy escape length and the solar modulation (Garcia-Munoz et al. 1977, Meyer 1975, Westergaard 1979).]

Since, in the solar neighborhood, the interstellar density (averaged over ~ 1 kpc in the disk) is estimated at 1–2 cm^{-3}, these results are generally interpreted as implying that galactic cosmic rays circulate in a low density halo which is at least 5 times wider than the disk.

In *diffusion models*, most of the ^{10}Be nuclei formed in the disk would decay while passing through the halo; in that case, the "mean age" derived from isotopic abundances is in fact the geometric mean of the diffusion time out of a halo of height H, H^2/K, and the decay time (Prishchep & Ptuskin 1975); the average confinement time of the particles in the Galaxy may be much longer. The composition data permit an estimate of some parameters for each of the models chosen; for instance, a one-dimensional diffusion model (where the cosmic-ray sources are concentrated in the disk) fits the data at 1 GeV if $K = 5 \times 10^{28}$ cm^2 s^{-1} and the halo size is $H = 5$ Kpc (Prishchep & Ptuskin 1979). In the *dynamical halo* model, the radioactive isotopes would behave approximately as in a diffusion model with a halo of size $H^* = H/X$ and their abundances can also be used to estimate the diffusion coefficient (Jones 1979, Jokipii & Higdon 1979).

3.2.2 ELECTRON-CAPTURE SECONDARY ISOTOPES AS DENSITY INDICATORS Secondary isotopes such as ^{7}Be, ^{49}V, and ^{53}Mn, which decay through electron-capture, can, in principle, be used to directly measure the density of the medium of propagation (Raisbeck et al. 1973, 1975). The most promising may be ^{53}Mn, for which restripping prevents decay unless the density is less than 10^{-2} cm^{-3}, and the energy lower than 300 MeV/n. [For some surprising ^{7}Be results, see Buffington et al. (1978).]

3.3 Relevance of the Electron Spectrum

According to several recent measurements, the electron spectrum is parallel to the proton spectrum in the energy range 2–10 GeV; in this range, the electron flux amounts to $\sim 1\%$ of the proton flux. At higher energies, the spectrum steepens to an index greater than 3, and possibly as large as 3.4 at several hundred GeV; however, in some cases the discrepancies between different sets of results are still large [Figure 1 from Prince (1979); see also

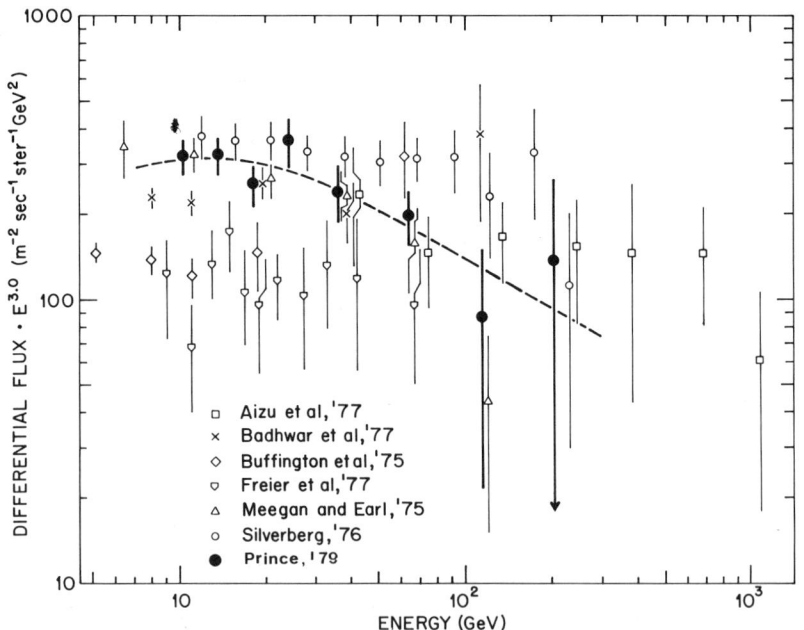

Figure 1 Measurements of the high energy electron spectrum (from Prince 1979).

discussions by Müller (1977) and Buffington (1978)]. A steepening of the high energy spectrum is expected, since the lifetime of a 30 GeV electron against radiation losses in the interstellar medium is $\sim 10^7$ years.

The observed electron spectrum does not impose strong constraints on the models proposed to explain the cosmic-ray composition. It is important to remember that the equations describing the behavior and the energy changes of electrons diffusing through the interstellar medium cannot be approximated by results obtained using the leaky-box model (Jones 1970). In *diffusion models*, the distribution of the sources plays an important, or even a predominant role (Shen 1970, Cowsik & Lee 1979). In addition, the injection spectrum of electrons is not known, and can generally be adjusted to ensure that a given model fits the data.

For lack of space, we will not attempt to describe here the very numerous models that can account for the observed electron spectrum; we refer the interested reader to Daniel & Stephens (1975), Ginzburg & Ptuskin (1976), and to some recent papers (Giler et al. 1978, Jokipii & Higdon 1979, Webber et al. 1979c).

3.4 Positrons and Antiprotons: A Test for the Nested Leaky-Box Model

Relativistic positrons and antiprotons are by-products of the inelastic collisions of cosmic-ray nuclei with interstellar medium particles. Because they are secondaries produced by particles with energy much higher than their own, their flux in the GeV range depends both on their propagation conditions and on those of their primaries, with energy ~ 20–150 GeV. As a result, the *nested leaky-box model* predicts lower intensities at a few GeV, and a flatter spectrum, than the energy-dependent *leaky-box model* (Orth & Buffington 1976, Lachièze-Rey & Cesarsky 1975). The antiproton data are potentially a better test; the positron spectrum may be partly shaped by radiative losses close to the cosmic-ray sources and/or in the Galaxy (Dilworth et al. 1974); also, the brightness of the annihilation line observed in the galactic center direction (Leventhal et al. 1978) raises the possibility that some of the positron cosmic rays are primary (Lingenfelter & Ramaty 1979).

Results of superconducting magnet observations of high energy positrons are available (Buffington et al. 1975, Golden et al. 1979a); but the uncertainties on this data and on the electron spectrum are still too large for definite conclusions to be reached (Müller 1977). The first detection of antiprotons has been reported at the most recent cosmic-ray conference (Golden et al. 1979b); the flux at 5.6–12.5 GV is as predicted by the energy-dependent *leaky-box* model.

3.5 *Relevance of the Anisotropy Measurements*

The distribution of the arrival directions of cosmic rays observed at the earth is very nearly isotropic. The existing measurements indicate that the anisotropy in the energy range 100–10^6 GeV is only $\simeq 0.1\%$ (Hillas 1975, Wolfendale 1977 and references therein); this can be interpreted as implying that cosmic rays are isotropic in a frame drifting at a velocity ~ 65 km s^{-1} with respect to the solar system.

Qualitatively, these observations are of crucial importance; they have led cosmic-ray physicists to describe cosmic-ray propagation in terms of diffusion models. Quantitatively, however, the constraints imposed by this type of observations on the models discussed throughout this paper are not clearly defined, for the following reasons:

(a) At energies less than 500 GeV, where the statistics are good (Marsden et al. 1976), the arrival directions may be mainly determined by cosmic-ray interactions with the solar wind (Parker 1976); even at energies as high as 1000 GeV, cosmic rays may be isotropized by magnetic field discontinuities in the solar cavity (Wentzel 1977b).

(b) Of course, all the measurements only relate to the component of the cosmic-ray streaming velocity that is perpendicular to the rotational axis of the earth; thus, even a considerable streaming toward the pole may go virtually undetected (Davies et al. 1977).

(c) In the leaky-box model, the anisotropy at the Sun is strongly dependent on the details of the large scale structure of the local magnetic field lines (see discussion in Wentzel 1974).

(d) In most diffusion models, the predicted anisotropy varies linearly with the distance $Z_\odot$ of the Sun to the center or plane of symmetry of the cosmic-ray source and/or confinement region; anisotropy measurements would then only serve to determine $Z_\odot$, and since the Sun is only ~ 12 pc above the galactic plane (Allen 1973), $Z_\odot$ is expected to be much smaller than the containment region dimensions. Still, if $K \propto E^a$, the anisotropy also increases rapidly with the energy; and with a as high as 0.6, the anisotropy limits at 10^6 GeV would impose upper limits on $Z_\odot$ that may be uncomfortably low.

4 COSMIC RAYS ELSEWHERE: SYNCHROTRON AND GAMMA-RAY DIFFUSE GALACTIC EMISSION

Studies of the synchrotron and gamma radiation from our, and, if possible, other galaxies, are a useful complement to the direct observations of cosmic rays at the earth.

Radio-continuum maps of our and of several other galaxies, at various frequencies, are available (see reviews by van der Kruit & Allen 1976, and by Wielebinski 1978); part of this radiation is emitted by relativistic electrons spiraling around the magnetic field lines. In our Galaxy, the synchrotron emission has been known for a long time to be partially correlated with spiral arms; this is also true in external galaxies of similar characteristics. In spiral galaxies, the synchrotron radiation from the disk peaks either at the center or at a ring surrounding the center, then falls off exponentially like the blue light of the Galaxy. Generally the synchrotron disk has a smaller radius than the disk of HI gas emitting 21-cm-line radiation. These observations give us some indications of the distribution of cosmic-ray electrons in the galactic plane, but their interpretation is heavily hindered by our lack of knowledge of the magnetic field distribution, and sometimes by the difficulty in separating the thermal and nonthermal components of the continuum radiation. Some external galaxies seen edge on have more or less flat radio halos, which are such that the spectral index increases with the distance to the galactic plane. The clearest and best-studied halo is that of NGC 891 (Allen et al. 1978, Beck et al. 1979). The observations can be fitted very well by models where relativistic electrons are generated in the disk, and diffuse outwards through the halo (Strong 1978, Dogiel 1979). In our Galaxy, the presence of a radio halo is difficult to ascertain: the task of separating out the disk, halo, and extragalactic components is complex (Baldwin 1976). Still, the existence of a radio-halo is suggested by a careful analysis of the data, under the plausible assumption that the halo spectrum is steeper than the disk spectrum (Webster 1975); and Bulanov et al. (1976) and Strong (1977) have shown that the present high latitude radio-continuum data can be interpreted in terms of a diffusion halo, with a diffusion coefficient compatible with that required to explain the cosmic-ray composition.

The radio-continuum data relates to the galactic distribution of relativistic electrons; it was hoped that the gamma-ray data would help to trace the main component of the cosmic radiation, the relativistic nuclei. Gamma rays in the 100 MeV range are emitted through the decay of neutral pions, generated by collisions of nuclear cosmic rays of energy > 700 MeV with interstellar medium particles. The spectrum of the radiation thus emitted exhibits a characteristic hump at 70 MeV (Stecker 1970). Two other processes contribute to the diffuse galactic gamma-ray emission in this energy range, and they both involve cosmic-ray electrons: bremsstrahlung emission, which involves electrons of energy comparable to that of the gamma rays, and Compton scattering of starlight photons by electrons in the GeV range and of blackbody photons by electrons in the 100 GeV range. In the galactic plane, the inverse-Compton component is generally expected to be the smallest of the three (Shukla & Paul 1976, Piccinotti & Bignami 1976; however see Hayakawa & Matsumoto 1979), but it may play a role at the border of the disk, or in the halo (Schlickeiser & Thielheim 1977). Solar modulation forbids a direct determination of the electron interstellar spectrum in the 100 MeV range; the spectrum derived from radio-continuum data was uncertain by a large factor (Cummings et al. 1973), making the prediction of the bremsstrahlung gamma-ray emission equally uncertain (Shukla & Cesarsky 1977), and most authors assumed that the π^0 decay component was dominant. Detailed observations of the galactic gamma radiation have now been obtained by instruments on board of two satellites, SAS-2, which operated during the period from November 1972 to June 1973, and COS-B, which is on orbit since August 1975. These satellites have measured the spectrum of the galactic radiation, received from various directions, SAS-2 in the energy range 35–100 MeV (Hartman et al. 1979) and COS-B at energies greater than 50 MeV (Paul et al. 1978). The observed spectra do not exhibit the characteristic π^0 decay hump. The spectra observed at medium latitudes by SAS-2, as well as the COS-B spectra from regions of the galactic plane external to the solar circle, presumably refer to diffuse radiation of local (within 1 kpc) origin: they can only be fitted if the electron contribution to the integral galactic background radiation at $E > 100$ MeV is at least as important as the nucleonic one. Thus, gamma-ray astronomy has failed in proving beyond doubt the presence of relativistic nuclei throughout the galactic disk; but it provides useful information on the local interstellar electron spectrum (Cesarsky et al. 1978).

Following early results by OSO-3 (Kraushaar et al. 1972), the SAS-2 observations have revealed the main characteristics of the gamma radiation from the Galaxy (Fichtel et al. 1975). The data of COS-B, which has longer exposure times, have permitted the construction of a gamma-ray map of the Milky Way (Figure 2). Both satellites find that the radiation is confined to a thin disk. The longitude distribution has a broad maximum in the inner Galaxy ($310° < l < 45°$); features of the interstellar gas distribution, such as the HI spiral arms, the Gould belt, and the warping of the plane, have gamma-ray counterparts.

The longitude distribution of gamma rays was interpreted by many authors as implying the presence of enhanced fluxes of cosmic rays in the inner Galaxy (e.g. Cesarsky et al. 1977, Hartman et al. 1979 and references therein). In fact, this conclusion is, at present, not firmly established: the galactic distribution of interstellar matter, or at least that of molecular hydrogen, is not known with sufficient precision, and the presence of numerous unresolved ($< 2°$) galactic "sources" of elusive nature (Wills et al. 1979), whose total contribution to the galactic gamma radiation is unknown, confuses the picture even more (for a more detailed discussion, see Cesarsky 1980).

Outside the solar circle, most of the interstellar gas is in the form of atomic hydrogen, and the situation is much clearer; the gamma-ray data require that the cosmic-ray density decreases substantially in the first

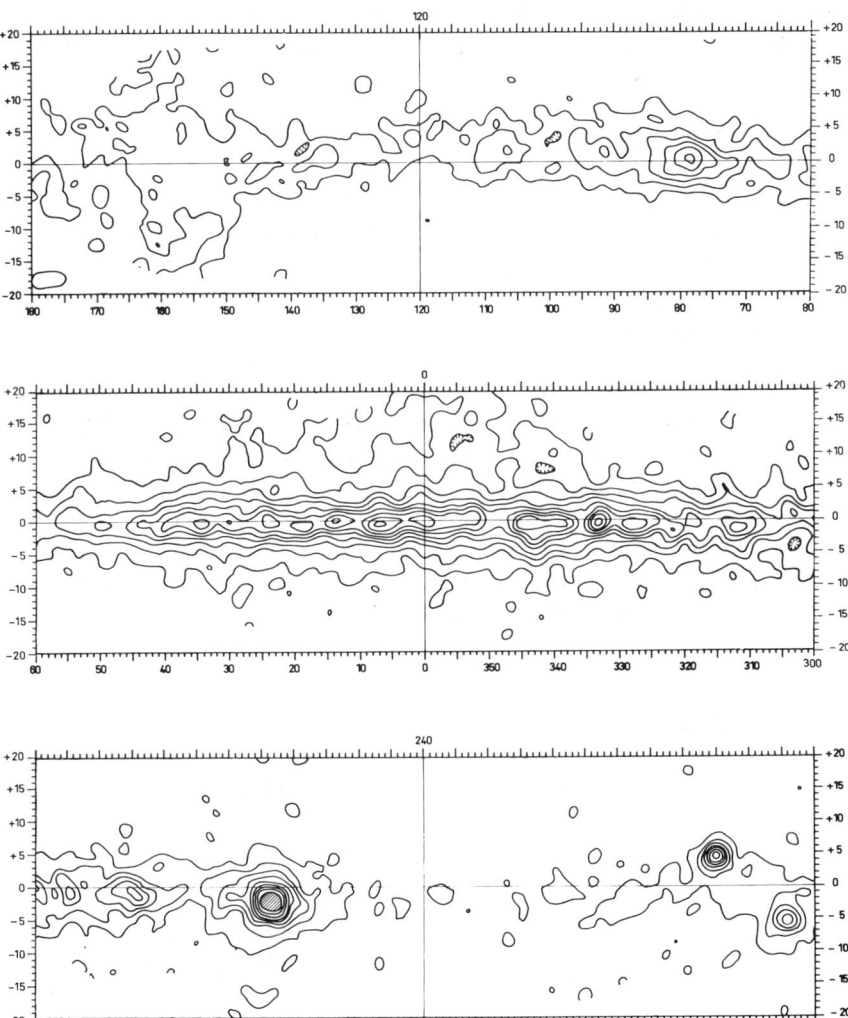

Figure 2 Gamma-ray map of the galactic plane, as seen by the COS-B satellite. The energy range is 70–5000 MeV; the contour lines are separated by $4 \cdot 10^{-3}$ "on axis" counts s^{-1} (Mayer-Hasselwander et al. 1980). Courtesy of the Caravane Collaboration.

kiloparsec beyond the Sun. This is the observational result mentioned in Section 1, which tends to rule out the hypothesis that cosmic-ray nuclei in the GeV range are universal (Dodds et al. 1975).

Fichtel et al. (1978) have shown that the Galaxy is not surrounded by a spherical gamma-ray halo; a flat gamma-ray halo may be present (Stecker 1978b), but cannot be too bright. Given the strength of the radiation losses suffered by high energy electrons, this result is not incompatible with the possible presence of a radio-halo and of an extended confinement region for the cosmic-ray nuclei.

5 VERY HIGH ENERGY COSMIC RAYS

Cosmic rays of very high energy, which trigger extensive air showers in the earth atmosphere, pose intriguing puzzles (Hillas 1975, Berezinsky 1977, Stecker 1978a and references therein); even their nature is open to speculation, and various classes of particles have been proposed: protons, iron nuclei, neutrons, neutrinos, relativistic dust grains.

Their origin is also discussed. If they were universal, their spectrum should have a break in the 10^{10} GeV range; instead, it appears to be flattening! Also, significant anisotropies have been measured at 10^8 GeV and at 10^9 GeV, while the highest energy particles tend to arrive nearly perpendicular to the galactic plane (Edge et al. 1978). Thus, in this case too, the universal hypothesis has been generally discarded: very high energy cosmic rays are now believed to originate and/or to be confined "locally", but what is meant by "local"—a galactic halo, the local cluster of galaxies, the super-cluster—is open to lively debate.

Acknowledgments

I thank my colleagues at Saclay for their continuous interest in my endeavor, and in particular Jean-Paul Meyer, Thierry Montmerle, and our guest Curtis Michel for useful comments on the manuscript.

Literature Cited

Aizu, E., Hiraiwa, H., Fujii, M., Nishimura, J., Taira, T., Kobayashi, T., Niu, K., Lord, J. J., Wilkes, R. J., Golden, R. 1977. *Int. Cosmic Ray Conf., 15th, Plovdiv* 1:372
Allan, H. R. 1972. *Astrophys Lett.* 12:237–41
Allen, C. W. 1973. *Astrophysical Quantities.* Univ. London: Athlone. 310 pp. 3rd ed.
Allen, R. J., Baldwin, J. E., Sancisi, R. 1978. *Astron. Astrophys.* 62:397–409
Asséo, E., Cesarsky, C. J., Lachièze-Rey, M., Pellat, R. 1978. *Ap. J.* 225:L21–L25

Asséo, E., Cesarsky, C. J., Lachièze-Rey, M., Pellat, R. 1980. *Ap. J.* In press
Audouze, J., Cesarsky, C. J. 1973. *Nature Phys. Sci.* 241:98–100
Axford, W. I., Leer, E., Skadron, G. 1977. *Int. Cosmic Ray Conf., 15th, Plovdiv* 11:132–37
Badhwar, G. D., Stephens, S. A. 1977. *Ap. J.* 212:494–506
Badhwar, G. D., Daniel, R. R., Cleghorn, T., Golden, R. L., Lacy, J. L., Stephens, S. A., Zipse, J. E. 1977. *Int. Cosmic Ray Conf.,*

15th, Plovdiv 1:404
Balasubrahmanyan, V. K., Ormes, J. F. 1973. Ap. J. 186:109–22
Baldwin, J. 1976. The structure and content of the Galaxy and galactic gamma rays. In NASA Symposium, Maryland, pp. 206–221. 392 pp.
Beck, R., Biermann, P., Emerson, D. T., Wielebenski, R. 1979. Astron. Astrophys. 77:25–30
Bell, A. R. 1978. MNRAS 182:147–56
Berezinsky, V. 1977. Int. Cosmic Ray Conf., 15th, Plovdiv 10:84–107
Blandford, R. D., Ostriker, J. P. 1978. Ap. J. 221:L29–L32
Brecher, K., Burbidge, G. R. 1972. Ap. J. 174:253–91
Brindle, C., French, D. K., Osborne, J. L. 1978. MNRAS 184:283–302
Buffington, A. 1978. Paper presented at The Topical Conference on Cosmic-Ray Astrophysics, Durham, New Hampshire
Buffington, A., Orth, C. D., Smoot, G. F. 1975. Ap. J. 199:669–79
Buffington, A., Orth, C. D., Mast, T. S. 1978. Ap. J. 226:355–71
Bulanov, S. V., Syrovatskii, C. I., Dogiel, V. A. 1976. Astrophys. Space Sci. 44:255–66
Caldwell, J. H. 1977. Ap. J. 218:269–85
Cesarsky, C. J. 1971. Interactions of cosmic rays with hydromagnetic waves in the Galaxy. PhD thesis. Harvard Univ., Cambridge. 124 pp.
Cesarsky, C. J. 1975. Int. Cosmic Ray Conf., 14th, Munich 12:4166–171
Cesarsky, C. J. 1977. Int. Cosmic Ray Conf., 15th, Plovdiv 10:196–216
Cesarsky, C. J. 1980. Proc. Texas Symp. 9th, Munich, Ann. NY Acad. Sci. 336:223–33
Cesarsky, C. J., Kulsrud, R. M. 1973. Ap. J. 185:153–65
Cesarsky, C. J., Cassé, M., Paul, J. A. 1977. Astron. Astrophys. 60:139–45
Cesarsky, C. J., Paul, J. A., Shukla, P. G. 1978. Astrophys. Space Sci. 59:73–83
Cesarsky, C. J., Völk, H. J. 1978. Astron. Astrophys. 70:367–77
Chevalier, R. A., Oegerle, W. R. 1979. Ap. J. 227:398–406
Chin, Y., Wentzel, D. G. 1972. Astrophys. Space Sci. 16:465–77
Cowsik, R., Wilson, L. W. 1973. Int. Cosmic Ray Conf. 13th, Denver 1:500–5
Cowsik, R., Wilson, L. W. 1975. Int. Cosmic Ray Conf., 14th, Munich 2:659–65
Cowsik, R., Lee, M. A. 1979. Ap. J. 228:297–301
Cox, D. P., Smith, B. W. 1974. Ap. J. 189:L105–L108
Cummings, A. C., Stone, E. C., Vogt, R. E. 1973. Int. Cosmic Ray Conf., 13th, Denver 1:335–39
Daniel, R. R., Stephens, S. A. 1975. Space Sci. Rev. 17:45–158
Davies, S. T., Elliot, H., Thambyahpillai, T. 1977. Int. Cosmic Ray Conf., 15th, Plovdiv 4:105–10
Dilworth, C., Maraschi, L., Perola, G. C. 1974. Astron. Astrophys. 33:43–48
Dodds, D., Strong, A. W., Wolfendale, A. W. 1975. MNRAS 171:569–77
Dogiel, V. A. 1979. Int. Cosmic Ray Conf., 16th, Kyoto 2:143–47
Edge, D. M., Pollock, A. M. T., Reid, R. J. O., Watson, A. A., Wilson, J. G. 1978. J. Phys. G.: Nucl. Phys. 4:133–57
Falgarone, E., Lequeux, J. 1973. Astron. Astrophys. 25:253–60
Fermi, E. 1949. Phys. Rev. 75:1169–74
Fermi, E. 1954. Ap. J. 119:1–6
Fichtel, C. E., Hartman, R., Kniffen, D. A., Thompson, D. J., Bignami, G. F. Ogelman, H., Ozel, M. E., Tümer, T. 1975. Ap. J. 198:163–82
Fichtel, C. E., Simpson, G. A., Thompson, D. J. 1978. Ap. J. 222:833–49
Flewelling, R. F., Coroniti, F. V. 1976. Ap. J. 205:L135–L138
Foote, E. A., Kulsrud, R. M. 1979. Ap. J. 233:302–16
Freier, P., Gilman, C., Waddington, C. J. 1977. Ap. J. 213:588–98
Garcia-Munoz, M., Mason, G. M., Simpson, J. A. 1977. Ap. J. 217:859–77
Getmantsev, G. G. 1962. Sov. Astron. 6:477–79
Giler, M., Wdowczyk, J., Wolfendale, A. W. 1978. J. Phys. G.: Nucl. Phys. 4:L269–L273
Giler, M., Kearsey, S., Osborne, J. L., Freedman, I. 1979. Int. Cosmic Ray Conf., 16th, Kyoto 2:131–36
Ginzburg, V. L. 1979. Int. Cosmic Ray Conf., 16th, Kyoto 2:148–152
Ginzburg, V. L., Syrovatskii, S. I. 1964. The Origin of Cosmic Rays. New York: Pergamon. 425 pp.
Ginzburg, V. L., Ptuskin, V. S. 1976. Rev. Mod. Phys. 48:161–89
Golden, R. L., Lacy, J. L., Stephens, S. A., Daniel, R. R. 1979a. Int. Cosmic Ray Conf., 16th, Kyoto 1:470
Golden, R. L., Horan, S., Mauger, B. G., Badhwar, G. D., Lacy, J. L., Stephens, S. A., Daniel, R. R., Zipse, J. E. 1979b. Int. Cosmic Ray Conf., 16th, Kyoto 12:76–81
Gordon, M. A., Burton, W. B. 1976. Ap. J. 208:346–53
Hartman, R. C., Kniffen, D. A., Thompson, D. J., Fichtel, C. E., Ogelman, H. B., Tümer, T., Ozel, M. E. 1979. Ap. J. 230:597–606
Hayakawa, S. 1969. Cosmic Ray Physics. N.Y., London: Wiley. 774 pp.
Hayakawa, S. 1979. Int. Cosmic Ray Conf., 16th, Kyoto 2:177–80

Hayakawa, S., Matsumoto, T. 1979. *Int. Cosmic Ray Conf., 16th, Kyoto* 1:85-90
Heiles, C. 1976. *Ann. Rev. Astron. Astrophys.* 14:1-22
Heisenberg, W. 1948. *Z. Phys.* 124:628-57
Hillas, A. M. 1975. *Phys. Rep. (Phys. Lett. C)* 2:59-136
Holman, G. D., Ionson, J. A., Scott, J. S. 1979. *Ap. J.* 228:576-81
Holmes, J. A. 1974. *MNRAS* 166:155-63
Holmes, J. A. 1975. *MNRAS* 170:251-60
Ipavich, F. M. 1975. *Ap. J.* 196:107-20
Johnson, H. E., Axford, W. I. 1971. *Ap. J.* 165:381-90
Jokipii, J. R. 1973. *Ap. J.* 183:1029-36
Jokipii, J. R. 1976. *Ap. J.* 208:900-2
Jokipii, J. R., Parker, E. N. 1969. *Ap. J.* 155:799-806
Jokipii, J. R., Higdon, J. C. 1979. *Ap. J.* 228:293-96
Jones, F. C. 1970. *Phys. Rev. D.* 2:2787-2802
Jones, F. C. 1978. *Ap. J.* 222:1097-1103
Jones, F. C. 1979. *Ap. J.* 229:747-52
Juliusson, E. 1974. *Ap. J.* 191:331-48
Juliusson, E., Meyer, P., Müller, D. 1972. *Phys. Rev. Lett.* 29:445-48
Kraichnan, R. H. 1965. *Phys. Fluids* 8:1385-87
Kraushaar, W. L., Clark, G. W., Garmire, G. P., Borken, R., Higbie, P., Leong, V., Thorsos, T. 1972. *Ap. J.* 177:341-63
Kulsrud, R. M., Pierce, W. P. 1969. *Ap. J.* 156:445-69
Kulsrud, R. M., Cesarsky, C. J. 1971. *Astrophys. Lett.* 8:189-91
Kulsrud, R. M. 1978. *Astronomical Papers Dedicated to Bengt Stromgren*, pp. 317-25. Copenhagen: Copenhagen Univ. Obs. 428 pp.
Lachièze-Rey, M., Cesarsky, C. J. 1975. *Int. Cosmic Ray Conf., 14th, Munich* 2:489-94
Lachièze-Rey, M., Asséo, E., Cesarsky, C. J., Pellat, R. 1980. *Ap. J.* In press
Lebrun, F., Paul, J. A. 1978. *Astron. Astrophys.* 65:187-91
Lee, L. C., Jokipii, J. R. 1976. *Ap. J.* 206:735-43
Lee, M. A., Völk, H. J. 1973. *Astrophys. Space Sci.* 24:31, 39
Le Guet, F. 1977. *Int. Cosmic Ray Conf., 15th, Plovdiv* 2:208-12
Leventhal, M., MacCallum, C. J., Stang, P. D. 1978. *Ap. J.* 225:L11-L14
Lezniak, J. A., Webber, W. R. 1978. *Ap. J.* 223:676-96
Lezniak, J. A., Webber, W. R. 1979. *Astrophys. Space Sci.* 63:35-56
Lingenfelter, R. E., Ramaty, R., Fisk, L. A. 1971. *Astrophys. Lett.* 8:93-97
Lingenfelter, R. E., Ramaty, R. 1979. *Int. Cosmic Ray Conf., 16th, Kyoto* 2:501-6
Marsden, R. G., Elliot, H., Hynds, R. J., Thambyahpillai, T. 1976. *Nature* 260:491-95
Mayer-Hasselwander, H. A., Bennett, K., Bignami, G. F., Buccheri, R., D'Amico, N., Hermsen, W., Kanbach, G., Lebrun, F., Lichti, G. G., Masnou, J. L., Paul, J. A., Pinkau, K., Scarsi, L., Swanenburg, B. N., Wills, R. D. 1980. *Proc. Texas Symp., 9th, Munich, Ann. NY Acad. Sci.* 336:211-22
McCray, R., Snow, T. P. 1979. *Ann. Rev. Astron. Astrophys.* 17:213-40
McKee, C. F., Ostriker, J. P. 1977. *Ap. J.* 218:148-69
Meegan, C. A., Earl, J. A. 1975. *Ap. J.* 197:219-33
Meneguzzi, M. 1973a. *Nature Phys. Sci.* 241:100-1
Meneguzzi, M. 1973b. *Int. Cosmic Ray Conf., 13th, Denver* 1:378-83
Meyer, J. P. 1975. *Int. Cosmic Ray Conf., 14th, Munich* 11:3698-3745
Meyer, J. P., Cassé, M., Goret, P. 1977. *Int. Cosmic Ray Conf., 15th, Plovdiv* 2:213-18
Meyer, P. 1969. *Ann. Rev. Astron. Astrophys.* 7:1-38
Mouschovias, T. 1974. *Ap. J.* 192:37-49
Mouschovias, T. 1975. *Astron. Astrophys.* 40:191-94
Müller, D. 1977. *Int. Cosmic Ray Conf., 15th, Plovdiv* 10:474-88
Myers, P. C. 1978. *Ap. J.* 225:380-89
Nakano, T., Tademaru, E. 1972. *Ap. J.* 173:87-101
Oort, J. H. 1965. *Galactic Structure*, ed. A. Blaauw, M. Schmidt, pp. 455-99. Univ. Chicago Press. 606 pp.
Ormes, J., Freier, P. 1978. *Ap. J.* 222:471-83
Orth, C. D., Buffington, A. 1976. *Ap. J.* 206:312-32
Orth, C. D., Buffington, A., Smoot, G. F., Mast, T. S. 1978. *Ap. J.* 226:1147-61
Ostriker, J. P. 1979. *Int. Cosmic Ray Conf., 16th, Kyoto* 2:124-29
Owens, A. J. 1976. *Astrophys. Space Sci.* 40:357-67
Parker, E. N. 1966. *Ap. J.* 145:811-33
Parker, E. N. 1967. *Ap. J.* 149:535-52
Parker, E. N. 1969. *Space Sci. Rev.* 9:651-712
Parker, E. N. 1973. *Astrophys. Space Sci.* 24:279-88
Parker, E. N. 1976. The structure and content of the Galaxy and galactic gamma rays. In *NASA Symposium, Maryland.* pp. 320-40. 392 pp.
Paul, J., Cassé, M., Cesarsky, C. J. 1976. *Ap. J.* 207:62-77
Paul, J. A., Bennett, K., Bignami, G. F., Buccheri, R., Caraveo, P., Hermsen, W., Kanbach, G., Mayer-Hasselwander, H. A., Scarsi, L., Swanenburg, B. N., Wills, R. D. 1978. *Astron. Astrophys.* 63:L31-L33
Peters, B., Westergaard, N. J. 1977. *Astrophys. Space Sci.* 48:21-46

Piccinotti, G., Bignami, G. F. 1976. *Astron. Astrophys.* 52:69–75
Piddington, J. 1972. *Cosmic Electrodyn.* 3:129–46
Price, R. M. 1974. *Astron. Astrophys.* 33:33–38
Prince, T. 1979. *Ap. J.* 227:676–93
Prishchep, V. L., Ptuskin, V. S. 1975. *Astrophys. Space Sci.* 32:265–71
Prishchep, V. L., Ptuskin, V. S. 1979. *Int. Cosmic Ray Conf., 16th, Kyoto* 2:137–42
Ptuskin, V. S. 1974. *Astrophys. Space Sci.* 28:17–30
Ptuskin, V. S. 1979. *Astrophys. Space Sci.* 61:259–65
Raisbeck, G. M. 1979. *Int. Cosmic Ray Conf., 16th, Kyoto* 14:146–58
Raisbeck, G., Perron, C., Toussaint, J., Yiou, F. 1973. *Int. Cosmic Ray Conf., 13th, Denver* 1:534–39
Raisbeck, G. M., Comstock, G., Perron, C., Yiou, F. 1975. *Int. Cosmic Ray Conf., 14th. Munich* 2:560–63
Rasmussen, I. L., Peters, B. 1975. *Nature* 258:412–13
Rengarajan, T. N., Stephens, S. A., Verma, R. P. 1973. *Int. Cosmic Ray Conf., 13th, Denver* 1:384–89
Rickett, B. J. 1970. *MNRAS* 150:67–91
Rickett, B. J. 1977. *Ann. Rev. Astron. Astrophys.* 15:479–504
Scheuer, P. A., Tsytovitch, V. N. 1970. *Astrophys. Lett.* 7:125–27
Schlickeiser, R., Thielheim, K. O. 1977. *Astrophys. Space Sci.* 47:415–21
Scott, J. S. 1975. *Nature* 258:58
Shapiro, M. M., Silberberg, R., Tsao, C. H. 1970. *Int. Cosmic Ray Conf., 11th, Budapest, Acta Phys. Hung.* 29 (Suppl. I):471–84
Shapiro, M. M., Silberberg, R., Tsao, C. H. 1973. *Int. Cosmic Ray Conf., 13th, Denver* 1:578–83
Shen, C. S. 1970. *Ap. J.* 162:L181–L186
Shukla, P. G., Paul, J. A. 1976. *Ap. J.* 208:893–99
Shukla, P. G., Cesarsky, C. J. 1977. *Proc. ESLAB Symp., 12th, Frascati*, pp. 131–36
Silverberg, R. F. 1976. *J. Geophys. Res.* 81:3944–52
Simard-Normandin, M., Kronberg, P. P. 1979. *Nature* 279:115–18
Simon, M., Spiegelhauer, H., Schmidt, K. M., Siohan, F., Ormes, J. F., Balasubrahmanyan, V. K., Arens, J. F. 1979. *Int. Cosmic Ray Conf., 16th, Kyoto* 1:352–57
Skilling, J. 1971. *Ap. J.* 170:265–73
Skilling, J. 1975a. *MNRAS* 172:557–66
Skilling, J. 1975b. *MNRAS* 173:245–54
Skilling, J. 1975c. *MNRAS* 173:255–69
Skilling, J., McIvor, I., Holmes, J. A. 1974. *MNRAS* 167:87P–91P
Skilling, J., Strong, A. W. 1976. *Astron. Astrophys.* 53:253–58
Smith, L. H., Buffington, A., Smoot, G. F., Alvarez, L. W. 1973. *Ap. J.* 180:987–1010
Stecker, F. W. 1970. *Astrophys. Space Sci.* 6:377–89
Stecker, F. W. 1978a. *Comments Astrophys.* 7:129–37
Stecker, F. W. 1978b. *Proc. IAU Symp.* 84, pp. 475–82
Strong, A. W. 1977. *MNRAS* 181:311–22
Strong, A. W. 1978. *Astron. Astrophys.* 66:205–9
van der Kruit, P. C., Allen, R. J. 1976. *Ann. Rev. Astron. Astrophys.* 14:417–46
Völk, H. J. 1975. *Rev. Geophys. Space Phys.* 13:547–66
Waddington, C. J. 1977. *Fundam. Cosmic Phys.* 3:1–85
Webber, W. R., Lezniak, J. A., Kish, J. C., Damle, S. V. 1973. *Nature Phys. Sci.* 241:96–98
Webber, W. R., Kish, J. 1979. *Int. Cosmic Ray Conf., 16th, Kyoto* 1:389–94
Webber, W. R., Kish, J. C., Simpson, G. A. 1979a. *Int. Cosmic Ray Conf., 16th, Kyoto* 1:430–35
Webber, W. R., Kish, J., Simpson, G. 1979b. *Int. Cosmic Ray Conf., 16th, Kyoto* 1:424–29
Webber, W. R., Goeman, R. A., Yushak, S. M. 1979c. *Int. Cosmic Ray Conf., 16th, Kyoto* 2:495–500
Webster, A. 1975. *MNRAS* 171:243–57
Wentzel, D. G. 1974. *Ann. Rev. Astron. Astrophys.* 12:71–96
Wentzel, D. G. 1977a. *J. Geophys. Res.* 82:714–16
Wentzel, D. G. 1977b. *Ap. J.* 216:L59–L62
Westergaard, N. J. 1979. *Ap. J.* 233:374–82
Wielebinski, R. 1978. *Proc. IAU Symp.* 84, pp. 113–18
Wills, R. D., Bennett, K., Bignami, G. F., Buccheri, R., Caraveo, P., D'Amico, N., Hermsen, W., Kanbach, G., Lichti, G. G., Masnou, J. L., Mayer-Hasselwander, H. A., Paul, J. A., Sacco, B., Swanenburg, B. N. 1979. *Proc. COSPAR Symp. Non-Solar Gamma Rays, Bangalore, Adv. Space Explor.* 7:43–47
Wolfendale, A. W. 1977. *Int. Cosmic Ray Conf., 15th, Plovdiv* 10:235–51
Young, J. S., Freier, P. S., Waddington, C. J. 1979. *Int. Cosmic Ray Conf., 16th, Kyoto* 1:442–47
Zweibel, E. G., Kulsrud, R. M. 1975. *Ap. J.* 201:63–73

OPTICAL AND INFRARED POLARIZATION OF ACTIVE EXTRAGALACTIC OBJECTS

✷2169

J. R. P. Angel and H. S. Stockman
Steward Observatory, University of Arizona,
Tucson, Arizona 85721

I Introduction

In bringing together the material for this review our attention was drawn repeatedly to the remarkably similar characteristics of virtually all strongly polarized extragalactic objects, which are found as the nuclei of giant elliptical galaxies or as quasi-stellar sources. These are violent variability, a compact, flat-spectrum radio source, and a very smooth continuum extending at least to $10\,\mu$. These properties are common to the polarized sources in BL Lac objects over a wide range of luminosity and to some QSOs and radio galaxies. In view of the similarities, which suggest a common process of energy release close to the central core of these objects, we will treat them all as a single group. In a memorable banquet speech at the Pittsburgh meeting on BL Lac objects (the only words spoken not faithfully reported in the proceedings) Ed Spiegel suggested the name "blazar" for this class of object. A combination of BL Lac object and quasar, with a strong feeling of the characteristic violent optical flaring, blazar seems an excellent name, one which we will adopt throughout the review. As we shall discuss in Sections V and VI, blazars may not be a different type of object from most quasars or active elliptical galaxies. These normal objects may have jets whose emission is beamed by relativistic bulk motion, and show blazar characteristics only when pointed at us. In reviewing the observational data we will be especially conscious of properties that could help distinguish an isotropic source from a beamed one.

The blazars form only a small portion of active extragalactic objects. Optical emission from most QSOs and Seyfert nuclei shows only very small polarization and little optical variability, and is perhaps the result of thermal or scattering processes. For these objects we will try to assess the information about source geometry and emission that can be derived from polarization data, particularly from the spectrum of polarization.

Polarization of active extragalactic objects has not been comprehensively reviewed in the past except by Visvanathan (1974) and Hagen-Thorn (1974). Both authors include rather complete discussions of the polarization by scattering in normal galaxies and the dusty halo of M82. In the interest of brevity we will not discuss these types of extended objects, except to point the reader to recent discussions of M82 by Bingham et al. (1976) and by Schmidt, Angel & Cromwell (1976). In this review we consider first the compact extragalactic sources for which strong polarization has been measured. The number of these is small enough, about 60, that we are able to at least mention them all. In Section II we first list the blazars, which form the large majority of the strongly polarized objects, and summaries of the observational data of some individual objects are given. The optical-infrared polarimetric properties and correlations of polarization with other properties of blazars are discussed in Section III. The remaining compact objects that show high polarization are PHL 5200 and knots in the jet of M87, whose properties are reviewed in Section II, and several Seyfert nuclei. The polarization properties of Seyfert nuclei, nearly all of which seem to be polarized by dust, and the majority of QSOs, which have only weak polarizations of unknown origin, are reviewed in Section IV. Theoretical models for the origin of optical-infrared polarization are reviewed in Section V, while in Section VI we consider how the observations of polarization can be accommodated by relativistically beamed jets. We finish in Section VII with some directions for future work.

II *Strongly Polarized Objects*

In this section, we summarize the optical-infrared and radio observations for all extragalactic objects that appear to be blazars and for which strong polarization has been measured, $\gtrsim 3\%$. We believe that the study of line-free, BL Lac objects has suffered from being too removed from the study of other polarized active objects. To redress that imbalance, the objects appearing in Table 1 have been included without regard to intrinsic luminosity or the presence of emission lines. A few faint objects, for which only single measurements of poor accuracy are available, have been omitted. Following Table 1, we present individual summaries of those interesting objects that bridge the compartments or subclasses of BL Lacs, QSOs, radio galaxies, etc. M87 and PHL 5200, which are strongly polarized but do not fall into the blazar class, are also summarized. A discussion of the only other compact extragalactic objects known, the strongly polarized Seyfert nuclei, is deferred until Section IV.

The information on blazars summarized in Table 1 is as follows. Coordinate designation and names are given in the first two columns, the redshifts where known and their sources are given in columns 3 and 4. If no emission lines are detected, and the redshift is that of the host galaxy, a "g" superscript is given. If absorption lines in the continuous spectrum are the only indication of distance, z is given as greater than that of the absorption lines. Rough visual magnitudes and their sources are given in columns 5 and 6. In the next three columns we summarize the optical polarization values with the appropriate reference(s) in column 10. Column 9 gives the number of separate measurements. The ranges of percentage polarization and position angle of the electric vector appear in columns 7 and 8. In column 11, we note whether the sources are known to have strong (s), weak (w), or no detectable emission lines (no entry). References to spectra are in column 12. The optical spectral index, α (column 13), and the known range of B magnitude (in magnitudes, column 15) are derived from the references in columns 14 and 16 respectively. In the next eight columns (17–24), we list radio flux measurements at frequencies of 178 MHz, 408 MHz, 1.4 GHz, 2.7 GHz, 5 GHz, 8.08 GHz, 22.2 GHz, and 90 GHz. This may seem rather clumsy, but there is no simple parameter that can describe the spectrum shape over nearly three decades in frequency. References to the radio spectra are given in column 25, and to additional data in column 26.

In the following discussion of individual objects and throughout this paper we will assume that the redshifts are cosmological with $H_0 = 75$ km s^{-1} Mpc^{-1}, and adopt $q_0 = 1/2$ when needed for objects at high z. The spectral index α will be defined by $F_\nu \alpha \nu^{-\alpha}$ for both optical and radio spectra.

0316+413 = NGC 1275 = 3C 84 This source is of special interest in being the closest listed in Table 1, at $z = 0.016$. It is the nucleus of the dominant central galaxy of the Perseus cluster, and is distinguished by the presence of moderate strength emission lines (Table 2) and a radio halo of steep spectrum around the compact flat-spectrum core.

Optical polarization of the nucleus was first reported by Dibaj & Shakhovskoy (1966), and it has since been repeatedly measured by Walker (1968) and by Babadzhanyants, Hagen-Thorn & Dombrovsky (references in Babadzhanyants & Hagen-Thorn 1975). This work showed the polarization to vary from month to month in strength up to 3% (26″ aperture), while nearly all points lie between 100° and 150° in position angle. More recently Martin, Angel & Maza (1976) and Angel et al. (1978) found erratic changes in position angle of up to 20° from night to night. Through a small aperture (4″) centered on the nucleus the polarization can be as strong as 6%

Table 1 Strongly polarized compact extragalactic objects

(1)	Object (2)	z (3)	Ref.[b] (4)	V (5)	Ref.[b] (6)	P (7)	θ (8)	N (9)	Ref.[b] (10)	Emission lines (11)	Ref.[b] (12)	α (13)	Ref.[b] (14)
0048−097	OB-081	—	128	16	105	7−14	—	3	53,143		15	1.8	112
0109+224	GC	—	128	15.5	128	3−6	55−85	2	0,90		85		
0215+015	PKS	≥1.345	37	18.3	128	20	—	1	37		37	1.9	37
0219+428	3C 66A	.444?	69	15.5	128	6−15	170−45[p]	41	2,53,142	w	69	1.1[γ]	69,112
0235+164	AO	≥.852	93	16.0	128	6−25	15−175	10	0,2,66,93		93	4.0	93
0300+470	4C 47.08	—	128	18.0	128	12−24	70−85	2	0,90	s	132		
0316+413	NGC 1275	.0172	102	11.9	102	1−6	100−160[p]	50	discussion	s	126		
0403−132	PKS	.571	13	17.2	13	0−4	170−195	6	0,106	s			13
0420−014	PKS	.915	13	18.0	13	8−20	150−175	7	0,68			1.1	
0422+004	OF 038	—	128	16.0	128	6−22	140−210[p]	36	2				
0521−365	PKS	.055	128	15.0	128	6	155	1	0	w	132	1.0[v]	28,53
0548−322	PKS	.069[a]	105	15.5	105	1.5−2	0−15	8	0		28	1.8	53
0735+178	PKS	≥.424	13	15.5	13	3−31	0−175	90	2,16,53,74, 90,94		69	1.3[v]	53,69 112
0736+017	PKS	.191	13	16.5	13	.5−6	25−135	9	0,106,118	s		1.1	6
0752+258	OI 287	.446	13	17.0	13	8	145[p]	8	0,106	s			
0754+100	OI 090.4	—	128	14.5	128	3−26	0−140	65	2,27,90,94,113		113		53
0808+019	OJ 014	—	128	17.5	128	4−14	—	5	53			0.9	
0818−128	OJ-131	—	128	15.5	128	8−36	60−115[p]	45	2,27,113		113		
0829+046	OJ 049	—	128	16.5	128	12	110−115	2	0	w	100	2.3	53
0851+202	OJ 287	.306?	69	14.0	69	1−32	0−180[p]	220	2,30,42,49, 53,54,73,90, 108,121,130		69	1.4	53,69
0906+430	3CR 216	.670	13	18.5	13	3−21	—	3	53,0	s	104	1.8	53,104
0912+297	OK 222	—	128	16.0	128	4−13	—	10	53		132	1.3	53,112
0957+227	4C 22.25	—	128	18.0	128	2−4[a]	—	2	53		95	.9	53
1057+100	HM	—	128	17.5	128	1−10	—	7	53,143		109	1.3	112
1101+384	Mkn 421	.030[a]	69	13.5	128	0−7	150−185[p]	25	53,65,66,90, 94,114		69	1.1	69
1133+704	Mkn 180	.044[a]	69	15.0	128	1−4	120−145	3	0		69		
1147+245	OM 280	—	128	16.0	128	3−13	5−155[p]	17	2,53,143	s	109	1.9[v]	112
1150+497	4C 49.22	.334	13	16.1	13	0.4[a]	20−180	2	0,106	s			
1156+295	4C 29.45	.729	13	15.6	13	1−9	0−120	6	0			.93	91
1215+303	ON 325	—	128	15.5	128	4−17	120−180[p]	62	2,53,74,108			1.8	112

POLARIZATION OF EXTRAGALACTIC OBJECTS 325

Table 1—continued

(1)	Object (2)	Δm (15)	Ref.[b] (16)	0.18 (17)	0.41 (18)	1.40 (19)	2.70 (20)	5.00 (21)	8.08 (22)	22.2 (23)	90.0 (24)	Ref.[b] (25)	Comments[b] (26)
0048−097	OB-081	2.7	116	<2	0.6	1.1	1.4	2.0[x]	2.4[y]		2.0[x]	1,32,36,47,124	VLBI (45)
0109+224	GC	3.1	85	<2	0.4[u]	0.7	1.9	1.2[y]	0.5[w]	0.5	0.8[v]	23,80,81,86,100	
0215+015	PKS	3.5	37		0.8[u]	0.5	0.4	0.4	<1.2			9,23,39,83, 122,123	
0219+428	3C 66A	1.0	105		5.0	1.3					<0.2	61,81	
0235+164	AO	5.2	87,93	<2	1.5	2.6	1.5[v]	2.8	2[x]	2.3	2.2	1,39,81,83,110	Compact radio (88)
0300+470	4C 47.08			2.2		2.1	2.3[x]	2.2	2.8[x]	3.3	2.2	1,11,39,81	Core-halo radio (10, 46)
0316+413	NGC 1275			63	29	13	16[v]	30[v]	50[v]	36[v]	36[v]	1,39,44,61,81 127,133,141	X-ray source
0403−132	PKS	.8	40	7.8[s]	7.2	3.2	2.9	2.8[v]	2.1[y]			32,36,97,101,124	Compact radio (24)
0420−014	PKS	2.8	87	1.8	1.2	1.5[v]	1.6[v]	1.8[v]	1.0[v]	3.3	4.1	9,32,81,82,123	VLBI (24, 45, 46)
0422+004	OF 038	1.5	135	1.5	1.2[u]	1.3[v]	1.0[v]	1.2[v]		1.6	1.7	9,22,36,81, 83,123	
0521−365	PKS	1.4	117	67[t]	26	15	11	10				32,36,98,101	Ext. radio (28), X-ray (56, 96)
0548−322	PKS				<2		0.3						32,99	X-ray source (72, 92, 96)
0735+178	PKS	2.5	87	2.9[s]	2.3[u]	2.2[v]	2.0[v]	2.0	2[v]	2.2	2.0	1,10,23,32,81, 101	VLBI (43, 45)
0736+017	PKS	1.0	67	1.1	2.6	2.6[v]	2.2[v]	2.0[v]	2.0[v]	2.6	3.4	9,32,81,98,123	VLBI (24, 45)
0752+258	OI 287			<2	1.4	0.5						20,75,86	Constant polarization (0)
0754+100	OI 090.4	1.0	27	<2	<2	0.7	0.8	0.9				32,39,75,100	
0808+019	OJ 014			<0.5	0.5[u]	0.5	0.7[v]	0.9[v]	<1.2			9,22,36,98, 122,123	
0818−128	OJ-131	2.9	27	3.1[t]	<2	1.1	0.9	0.8	0.7[v]			8,32,75,101	
0829+046	OJ 049	5.0	117	<2	<2	0.6	0.6	0.7	5[v]		0.5	9,32,39,81,100	
0851+202	OJ 287	4.0	87	<2	0.6[u]	1.5	2.8[v]	2.6		2.8	7[v]	1,23,31,82, 86,141	
0906+430	3CR 216			20	12	3.7	2.4	1.8	1.4[w]	1.0[v]		33,39,44,84,133	
0912+297	OK 222	1.9	117	<2	0.5	0.6						19,75,86	Extended radio (98)
0957+227	4C 22.25			3.7	2.7	1.1	0.7	0.4				23,82,86,136	
1057+100	HM			<2	1.2	0.6						32,39,136	
1101+384	Mkn 421	4.6	105	<2	1.1	0.6	0.6	0.7	0.5	0.4	0.5	59,81,86,136	X-ray source (72, 96)
1133+704	Mkn 180			<2			0.2	0.2	0.1	0.4[x]	0.2	39,59,115	
1147+245	OM 280			<2	0.8	0.6		1.0		0.8[x]	0.7	20,75,80,82,86	
1150+497	4C 49.22	2.0	106	5.6	3.2	1.4	1.6	1.1	1.5[w]	0.4[y]		5,84,86,133	Extended radio (24)
1156+295	4C 29.45	2.6	0	2.8	2.8	1.7		0.9				19,75,82,86	
1215+303	ON 325	2.1	117	<2	0.8	0.3		0.4				19,75,82,86	Compact radio (24)

Table 1—continued

(1)	Object (2)	z (3)	Ref.[b] (4)	V (5)	Ref.[b] (6)	P (7)	θ (8)	N (9)	Ref.[b] (10)	Emission lines (11)	Ref.[b] (12)	α (13)	Ref.[b] (14)
1219+285	W Com	—	128	16.5	128	2–10	30–95	14	53,74,108	s	108	2.3	112
1253−055	3C 279	.538	13	17.7	13	4–19	10–180	14	0,34,51,120	s	69	1.6	76
1308+326	B2	.996	69	19.0	128	0–25	25–160	44	70,89,90	w	7,69	1.6	69,89
1400+162	MC 3	.244	7	16.5	13	4–14	80–100[p]	6	0,7,74		27	1.5[v]	7,69
1418+546	OQ 530	—	128	15.0	128	2–19	50–105	5	27,90				
1514+197	GC	—	128	18.5	128	7–9	—	2	53			2.3	53
1514−241	AP Lib	.049	69	15.0	128	2–7	145–205[p]	18	2,14,53,108	w	69	2.7	53
1522+155	MC 3	.628	13	17.3	13	3–13	10–105	2	0	s		.20	103
1538+149	4C 14.60	—	128	15.5	128	22	—	1	53		131,132	1.9[v]	112
1641+399	3CR 345	.595	13	16.0	13	2–16	10–170	49	0,3,34,51,52, 55,58,107,120	s		1.1	76,120
1652+398	Mkn 501	.034[g]	69	13.8	105	2–4	125–145[p]	36	2,53,65,90,114		69	2.5	112
1717+178	OT 129	—	128	18.5	128	27	—	1	27		132	1.9[v]	112
1727+502	I Zw 186	.055[g]	69	16.0	128	4–6	—	6	53		69	2.2[v]	112
1749+096	OT 081	—	128	17.0	105	3–9	—	3	53		109	1.3	69
1807+698	3CR 371	.050	102	14.8	102	0–12	65–100[p]	68	0,14,53	w	69		
1845+797	3CR 390.3	.056	102	14.5	102	1–4	155–165[p]	30	0,4	s	79		
2032+107	MC	—	128	18.6	128	12	130	1	134				
2155−304		.17?	17	14.0	128	3–7	150–170	4	41	w	17	1.0	41
2200+420	BL Lac	.069	69	14.5	105	2–23	0–180	>500	0,2,53,57, 74,90,119,120	w	69	1.6[v]	69,105
2201+171	MC 3	1.080	13	18.8	13	9,5	30	1	134	s			
2208−137	PKS	.392?	13	17.0	13	5–9	100–170	3	0	s			
2223−052	3C 446	1.404	13	18.4	13	4–17	10–160	16	0,50,68,107,120	s		1.8	76,120
2225−055	PHL 5200	1.981	13	17.7	13	4	160[p]	3	0,107	s			
2230+114	CTA 102	1.037	13	17.3	13	1–11	100–170	6	0	s		1.0	76
2251+158	3CR 454.3	.859	13	16.1	13	0–16	0–170	24	0,107,118,120	s	109,132	1.5	76,120
2254+074	OY 091	—	128	16.5	128	14–21	—	6	53,143			2.1[v]	112
2345−167	PKS	.600	13	18.0	13	3–19	70–160	3	0	s			

Table 1—continued

(1)	Object (2)	Δm (15)	Ref.[b] (16)	0.18 (17)	0.41 (18)	1.40 (19)	2.70 (20)	5.00 (21)	8.08 (22)	22.2 (23)	90.0 (24)	Ref.[b] (25)	Comments[b] (26)
1219+285	W Com	4.0	117	0.8[u]	0.3	1.4	1.5[v]	0.7	1.6[v]	1.2	1.0	1,20,23,81,82	VLBI (26,45,46)
1253−055	3C 279	6.7	31	21	11	10	12	16		9[v]	6.6[v]	32,39,47,97,141	X-ray source (111); Superluminal (138)
1308+326	B2	5.6	38	<2	1.2	1.0	0.6	1.5	1.8[w]	3.0	2.7	19,81,83,86	
1400+162	MC 3			2.2	1.9	0.8	0.9	0.4		0.3[x]		7,36,39,110	Double radio (7)
1418+546	OQ 530	4.8	27	<2		0.8	0.5	1.1	1.4[w]	0.7	0.7	39,75,81,84	
1514+197	GC			<2	0.5[u]		2.2[v]	0.5				39,100	
1514−241	AP Lib	2.5	37		3.8	2.1		1.9	2.8[v]			1,32,36	Core-halo radio (48)
1522−155	MC 3			<2	0.6							39,110	
1538+149	4C 14.60	>2.8	117	3.6	2.3	2.1	2.0	2.0			0.5	39,44,71,81,82	D2 (29), VLBI (24,29,45,46,140)
1641+399	3CR 345	2.0	40	12	9.0	6.6	9.0[v]	8.0[v]	9.0[v]	9.0	7.7[v]	1,21,44,81,84, 86,141	Superluminal (138,139)
1652+398	Mkn 501			2.0	1.8	1.5	1.4	1.4	1.2	0.9	0.7	20,59,81,84, 86,136	X-ray source (72,96)
1717+178	OT 129			<2	<0.3[v]	0.5	0.7	0.7[v]				23,39,75,100	
1727+502	I Zw 186	1.9		<2		0.4		<0.5				39,75,84	X-ray source (56)
1749+096	OT 081		117	<2		1.3	1.0	1.8	1.5			1,23,36,39,82	
1807+698	3CR 371	2.0	117	5.3	1.3[v]	2.6	1.9	2.0	1.8[w]	2.1	1.4	39,44,81,84	
1845+797	3CR 390.3	1.8	117	47	34	12	6.8	4.5				44,60,102	N gal. X-ray source (96,63) Extended radio? (12,18,33,35)
2032+107	MC			<2	1.4	1.2	0.9	0.8				39,75,100,110	N gal. Double radio (60) X-ray source (63)
2155−304		1.4	41		2	0.3	0.3	0.3				32,75,136	X-ray source (41,96)
2200+420	BL Lac	4.0	117	<2	3.1	5.3[v]	5.0[v]	5.5[v]	7.0[v]	10[v]	11[v]	1,10,39,46,47, 81,84,133,136,141	VLBI (46,48); X-ray upper limit (96)
2201+171	MC 3			<2	1.1	0.8	0.6	0.7				39,75,83,100,110	
2208−137	PKS				<2		0.7	0.5				8,32	VLBI (45,46);
2223−052	3C 446	3.4	87	17	8.5	5.8	4.6	4.3		4.8[v]	3.0	32,39,47,97	X-ray source (111)
2225−055	PHL 5200	0?	40		<0.1		<0.4					8,137	Radio quiet; constant polarization
2230+114	CTA 102	1.0	0,40	5.5	8.0[v]	6.5[v]	4.5[v]	3.5	2.5[v]	1.3[v]	0.6	1,10,32,36,39, 47,97,129	VLBI (43,45,46)
2251+158	3CR 454.3	2.3	117	13	14	12[v]	12[v]	16[v]	12	6.9[v]	5.4[v]	1,32,36,39,47, 97,98,141	
2254+074	OY 091	1.6	117	<2	0.4[u]	0.4	0.7	0.8[v]	1.9[w]			23,39,75,83,100	D2 (29) VLBI (24,43,45,46,48)
2345−167	PKS	2.5	87	2.0[t]	2.1	2.5[v]	2.5[v]	3.6				32,36,101	VLBI (45,46)

Footnote to Table 1

[a] A general description is given in the text. Table notes are as follows—a: the polarization maximum is uncertain, g: redshift derived from the host galaxy absorption lines, p: polarization has preferred angle, t: flux measurement at 160 MHz, u: 318 MHz, w: 10.7 GHz, x: 15.1 GHz, y: 31.5 GHz, z: 86.0 GHz, v denotes that the radio flux is variable, and a mean value of reported values is given. v against a spectral index indicates that different values are reported in the cited references, and the lower value is given.

We should caution that the radio fluxes in this table are mostly obtained at different occasions at different frequencies, and so the shapes are not reliable for variable sources. Simultaneous observations are given by Owen & Mufson (1977), Owen, Spangler & Cotton (1980), and O'Dell et al. (1978). Jones & Rudnick (1980) find that fractional variations at 90 GHz are probably greater than at lower frequencies. Absence of a v superscript should not be taken as an indication of stability, only that variability is not established from the limited published observations.

[b] References:

0 Steward Obs, unpublished data
1 Altschuler & Wardle 1976
2 Angel et al. 1978
3 Babadzhanyants et al. 1972
4 Babadzhanyants & Hagen-Thorn 1975
5 Bailey & Pooley, 1968
6 Baldwin 1975
7 Baldwin et al. 1977
8 Bolton, Shimmins & Wall 1975
9 Brandie & Bridle 1974
10 Bridle et al. 1972
11 Bridle & Fomalont 1974
12 Broderick et al. 1972
13 Burbidge, Crowne & Smith 1977
14 Capps & Knacke 1978
15 Carswell et al. 1973
16 Carswell et al. 1974
17 Charles, Thorstensen & Bowyer 1979
18 Cohen et al. 1971
19 Colla et al. 1970
20 Colla et al. 1972
21 Colla et al. 1973
22 Condon & Jauncey 1974a
23 Condon & Jauncey 1974b
24 Conway et al. 1974
25 Cooke et al. 1978
26 Cotton et al. 1979
27 Craine, Duerr & Tapia 1978
28 Danziger et al. 1979
29 Davis, Stannard & Conway 1977
30 Dyck et al. 1971
31 Eachus & Liller 1975
32 Ekers 1969
33 Elsmore & Mackay 1969
34 Elvius 1968
35 Fomalont & Moffet 1971
36 Gardner, Whiteoak & Morris 1975
37 Gaskell 1978
38 Gottlieb & Liller 1976
39 Gower, Scott & Wills 1967
40 Grandi & Tifft, 1974
41 Griffiths et al. 1979
42 Hagen-Thorn 1972
43 Jauncey et al. 1970
44 Kellermann et al. 1969
45 Kellermann et al. 1970
46 Kellermann et al. 1971
47 Kellermann & Pauliny-Toth 1971

48 Kellermann et al. 1977
49 Kikuchi et al. 1976
50 Kinman, Lamla & Wirtanen 1966
51 Kinman 1967
52 Kinman et al. 1968
53 Kinman 1976
54 Kinman et al. 1974
55 Kinman 1977
56 Kinzer 1978
57 Knacke, Capps & Johns 1976
58 Knacke, Capps & Johns 1979
59 Kojoian et al. 1976
60 MacDonald, Kenderine & Neville 1968
61 Mackay 1971
62 Mackay 1969
63 Marshall et al. 1978
64 Martin, Angel & Maza 1976
65 Maza, Martin & Angel, 1978
66 Maza 1979
67 McGimsey et al. 1975
68 Miller & French 1978
69 Miller, French & Hawley 1978
70 Moore et al. 1980
71 Munro 1972
72 Mushotzky et al. 1978
73 Nordsieck 1972
74 Nordsieck 1976
75 Ohio 1415 MHz survey
76 Oke, Neugebauer & Becklin 1970
77 Oke 1966
78 Osterbrock & Miller 1975
79 Osterbrock, Koski & Phillips 1976
80 Owen & Mufson 1977
81 Owen et al. 1978
82 Pauliny-Toth & Kellermann 1972
83 Pauliny-Toth et al. 1972
84 Pauliny-Toth et al. 1978
85 Pica 1977
86 Pilkington & Scott 1965
87 Pollock et al. 1979
88 Porcas, Treverton & Wilkinson 1974
89 Puschell et al. 1979
90 Puschell & Stein 1980
91 Richstone & Schmidt 1980
92 Riegler, Agrawal & Mushotzky 1979
93 Rieke et al. 1976
94 Rieke et al. 1977
95 Schmidt 1974

96 Schwartz et al. 1979
97 Shimmins, Manchester & Harris 1969
98 Shimmins & Bolton 1972
99 Shimmins & Bolton 1974
100 Shimmins, Bolton & Wall 1974
101 Slee 1977
102 Smith, Spinrad & Smith 1976
103 Smith et al. 1977
104 Smith 1978
105 Stein, O'Dell & Strittmatter 1976
106 Stockman 1978
107 Stockman & Angel 1978
108 Strittmatter et al. 1972
109 Strittmatter et al. 1974
110 Sutton et al. 1974
111 Tananbaum et al. 1980
112 Tapia, Craine & Johnson 1976
113 Tapia et al. 1977
114 Ulrich et al. 1975
115 Ulrich 1978
116 Usher, Kolpanen & Pollock 1974
117 Usher 1975
118 Visvanathan 1968
119 Visvanathan 1969
120 Visvanathan 1973a
121 Visvanathan 1973b
122 Wall, Shimmins & Merkelijn 1971
123 Wall 1972
124 Wall, Wright & Bolton 1976
125 Wampler 1967
126 Wampler 1971
127 Wardle 1971
128 Weiler & Johnston 1979
129 Williams, Kenderdine & Baldwin 1966
130 Williams et al. 1972
131 Wills & Wills 1974
132 Wills & Wills 1976
133 Witzel et al. 1978
134 Zotov & Tapia, 1979
135 Kinman 1976, *IAU Circ. No.* 2908
136 From *Ohio Master List of Radio Sources*
137 Mills & Little 1970
138 Seilestad et al. 1979
139 Cohen et al. 1979
140 Readhead et al. 1979
141 Hobbs & Dent 1977
142 Puschell 1980
143 Serkowski & Tapia 1975

in the blue (Maza 1979). The position angle during 1976–1978 showed again nearly all points in the range 100°–150°. Interstellar polarization from our own galaxy does not affect the results for NGC 1275 appreciably, the polarization of three nearby galaxies in the cluster being 0.4% at 101°. In addition to the polarization variability, rapid variability of the optical continuum strength also on a time scale of a day, has now been observed (Geller, Turner & Bruno 1979).

The structure of the very bright compact radio core of NGC 1275, which is not polarized, has been explored by VLBI measurements. At centimeter wavelengths the emission is concentrated in three very small components in a line about 3 pc in length, at position angle 170°. The relative motion of the components is not greater than 0.05 pc/yr, or 0.15 c.

PKS 0521−36 and 1807+698 = 3C 371 These two sources are very similar and are distinguished by their combination of forbidden emission lines and strong, steep-spectrum extended components to their radio emission. Both are strongly polarized variable nuclei located in giant elliptical galaxies at about the same distance ($z \sim 0.05$). In both sources, the narrow forbidden lines of [O III] are about 10 times stronger than in BL Lac (Table 2). PKS 0521−36 has been studied most recently in the optical by Danziger et al. (1979). After subtracting the galaxy, they estimate the spectral index in the optical to be ~ 1.0. No polarization data are reported in the literature, but we recently obtained one measurement through a 4 arcsec aperture, giving $P = 6\%$ at angle 155°. The low-frequency radio emission from 0521−36 is strong (67 Jy at 178 MHz), or about the same

Table 2 Absolute luminosity of emission lines in units of 10^{41} ergs s^{-1} in the rest frame of the source, computed using $H_0 = 75$ km s^{-1} Mpc^{-1}, $q_0 = \frac{1}{2}$

	Object	MgII 2798	[OII] 3727	Hβ 4861	[OIII] 5007	Hα 6563	Ref.
0316+413	NGC 1275	—	1.5	0.8	3.0	8	Wampler 1971
0521−365	PKS	—	3	—	1.0	1.1	Danziger et al. 1979
0736+017	PKS	—	—	50	—	130	Baldwin 1975
0851+202	OJ 287	—	—	—	3.3	—	Miller et al. 1978
1308+326	B2	100	—	—	—	—	Miller et al. 1978
1400+162	MC3	—	0.6	0.3	0.6	—	Miller et al. 1978
1514−241	AP Lib	—	—	—	0.3	—	Miller et al. 1978
1641+399	3C 345	440	—	—	—	<0.5	Visvanathan 1973a
1807+698	3C 371	—	0.4	0.1	0.6	—	Miller et al. 1978
1845+797	3C 390.3	—	0.7	11	11	70	Osterbrock et al. 1976
1958+407	Cyg A	—	1.8	0.7	10	5	Osterbrock & Miller 1975
2200+420	BL Lac	—	—	—	0.08	0.06	Miller et al. 1978
2230+114	CTA 102	1200	—	—	860	—	Oke 1966
2251+158	3C 454.3	380	—	—	—	—	Visvanathan 1973a

luminosity as 3C 390.3. The structure is known to be extended over 15 arcseconds or 15 Kpc (Fomalont 1967), while all the low frequency emission in 3C 390.3 is from the double lobes with 200 Kpc separation.

3C 371 does not have such a strong low-frequency component as 0521 −36 (5 Jy at 178 MHz, and a flat 2 Jy component from 1.4–90 GHz) but its polarimetric properties have been extensively studied. Visvanathan (1967) found it to be polarized, and it has since been measured repeatedly by Dombrovsky et al. (1971), and Babadzhanyants & Hagen-Thorn (1975). Miller (1975) has recorded a high polarization of $\sim 10\%$, using a small 2.4 × 4 arcsec aperture. The bulk of the available data obtained through a 26 arcsec aperture rarely exceeds 6%, presumably because of dilution by the galaxy, though Dombrovsky et al. did measure 10–12% through a 26 arcsec aperture in September 1970. At that time the source flared for a period of about a week by 0.6 magnitude to $m_B = 14.5$. The direction of polarization is certainly not constant, but nearly all measurements taken in each of seven years lie within 35° of position angle 85°.

Optical spectrophotometry of 3C 371 has been obtained by Miller (1975) who gives reference to earlier work. He identifies three optical components: a power-law continuum of index 1.35 and $m_v = 15.4$ at the time of observation; the absorption line spectrum of the galaxy; and the emission line spectrum given in Table 2.

PKS 0736+017 This is a source with strong permitted lines at $z = 0.192$, recently found to show variable optical polarization by Moore & Stockman (1980). The strength of polarization varies between 0 and 6%, with sharp changes from night to night. The optical continuum shows optical variability of more than a magnitude (McGimsey et al. 1975) and has a spectral index of 1.1 over the range 0.3–0.7 μ in the rest frame (Baldwin 1975). Line strengths by Baldwin are given in Table 2. The radio spectrum is nearly flat at ~ 2.5 Jy from 0.41–90 GHz.

B2 0752+258 = OI 287 = VRO 25.07.04 This object at $z = 0.446$ has recently been discovered by Moore & Stockman (1980) to show essentially constant polarization of 8% at position angle 143°. Wills & Wills (1976) found the optical spectrum to show strong sharp forbidden lines of [OII] and [OIII] but no Balmer lines, and comment that from spectroscopic evidence alone it could be classified as a galaxy, though the appearance is stellar. The only permitted line detected is the resonance doublet of Mg II. The radio spectrum known only from 0.5–2 GHz appears steep, though there is no detection at 178 MHz (<2 Jy).

1156+295 = 4C 29.45 = Ton 599 This object at $z = 0.728$ was found by Moore & Stockman (1980) to show strongly variable polarization, up to

10% in magnitude and with no preferred position angle. The optical spectrum shows strong Mg II and CIII $\lambda 1909$ (Schmidt 1974), has a spectral index of 0.93 (Richstone & Schmidt 1980), and varies in brightness by at least 2 magnitudes (Moore & Stockman 1980). The radio spectrum drops slowly from 2.8 Jy at 178 MHz to 0.89 Jy at 5 GHz.

1228+127 = M87 Both the jet of polarized optical emission and the starlike nucleus of this central galaxy of the Virgo cluster are of particular interest. The optical jet is in the form of several unresolved knots out along a line from the nucleus, extending out to 25 arcsec (1.3 Kpc). Knot A, the brightest at $B = 16.8$, is about half way along the jet. Polarization was first studied photographically by Baade (1956) and photoelectrically by Hiltner (1959). Schmidt, Peterson & Beaver (1978) obtained a one-dimensional map of polarization and intensity along the jet with the 200-element Digicon detector, which allowed accurate determination of the diffuse galactic background. The knots themselves are polarized typically between 10% and 25%, with position angles apparently randomly oriented from knot to knot. Recently, Sulentic, Arp & Lorre (1979) have repeated Baade's photographs under similar conditions and find slight but significant changes in intensity and polarization of the knots, over a 22 year baseline.

In Turland's (1975) map of M87 made at 5 GHz, the radio emission from several knots is resolved, knot A having a flux of 1 Jy at 5 GHz and a spectral index of ~ 0.5. Comparing the optical and 5 GHz data, Schmidt et al. find the radio and optical polarization of individual knots agree closely in both strength and position angle, once a constant correction of $75°$ in angle is made for Faraday rotation. There is no detectable Faraday rotation, $\gtrsim 20°$, within the knots themselves.

Sulentic et al. argue that the knots appear to be a type of BL Lac object that is ejected from the galactic nucleus. We note, though, that there are some characteristics that appear rather different: the optical variability is slower and the absolute optical flux weaker than for any known BL Lac; the radio emission is relatively steep; and no compact cores are detected in the knots in VLBI observations. As in the Crab Nebula, the optical and radio emission can be interpreted as coming from a relatively large region of optically thin synchrotron radiation. In the absence of any evidence of compact, nuclear activity within individual knots, it seems likely that the energy is being supplied in a jet from the nucleus, where there is a compact core of flatter radio spectrum with $\alpha = 0.3 \pm 0.2$ and of strength ~ 3 Jy at 5 GHz.

Turning to the nucleus, it appears that optical polarization is present intermittently. In 1971-1972, Heeschen (1973) and Kinman (1973) found variable polarization of up to 6%, measured through a 3.2 arcsec diaphragm centered on the nucleus. This would appear to be associated with the stellar object at the nucleus which is distinct from the normal stellar core of the galaxy. In 1977, Schmidt et al. found no detectable polarization, obtaining an upper limit (3σ) of 0.9% in a 1×3 arcsec aperture centered on the nucleus. In 1979–1980, R. L. Moore (private communication) obtained several measurements consistent with $0.3 \pm 0.1\%$ at 120° position angle through a 4 arcsec aperture. This last measurement was with a red-sensitive detector and would be subject to considerable dilution by the galaxy. The featureless continuum of the stellar object (Sulentic et al. 1979) and its apparent variability (de Vaucouleurs & Nieto 1979) are all consistent with weak blazar activity, but since it is not active now we shall not group it with the nearby objects of much more consistent activity. Obviously, continued polarimetric observations of the nucleus are needed. M87 is of particular interest because of the dynamical and photometric evidence of a large, central mass of $5 \times 10^9 \, M_\odot$ (Young et al. 1978, Sargent et al. 1978), similar to the black hole masses inferred for BL Lac objects from the time scale of variability (Angel et al. 1978).

B2 1308 + 326 and AO 0235 + 164 These two distant ($z \sim 1$) and extremely luminous sources have many properties in common. Both are flat- or inverted-spectrum compact radio sources with no steep, low-frequency components. Both have shown outbursts in optical and radio emission lasting months, with strong, rapidly variable optical polarization. The emission lines in both cases are weak or undetectable relative to the polarized optical continuum.

Detailed studies of the polarimetric behavior of B2 1308 + 326 during the 1978 outburst have been made by Moore et al. (1980) and Puschell et al. (1979). The strength of optical polarization ranged up to 20%, with large variations in the strength and angle of polarization ($\Delta P \sim 7\%$, $\Delta\theta = 30°$), on a time scale of one day in the rest frame of the source. Repeated accurate measurements (typical errors of 0.25% in polarization) by Moore et al. over baselines of 2–5 hours showed only small variations consistent with the night-to-night trends. A recent reanalysis by Puschell (1980) of his data is also consistent with this result.

Moore et al. (1980) report infrared polarimetry obtained simultaneously with the optical on three different nights. On each of the three nights the strength and position angle at $2\,\mu$ and $0.6\,\mu$ are equal, indicating a common emission process. The angle of polarization is not always independent of wavelength though, since a rotation of 15° between $0.4\,\mu$ and $0.8\,\mu$ was observed on another night when only optical measurements were obtained. At centimeter wavelengths, the flux remained relatively steady during the 1978 outburst, and the polarization of a few percent was roughly constant in angle, with no Faraday rotation above 10 GHz (Puschell et al. 1979).

Only occasional polarization measurements have been made for AO 0235

+164 but these show similar variation over all position angles and sometimes great strength (20%), even when the source was faint (Rieke et al. 1976, Angel et al. 1978).

1400+162 = 4C 16.39 This object is important because it and 3C 390.3 are the only extended radio sources with central compact nuclei showing strong optical polarization known to have classic double lobe structure. The optical and radio properties have been extensively studied by Baldwin et al. (1977). The redshift of the object from weak emission lines (Table 2) is $z = 0.245$, the same as that of the brightest galaxy in a small adjacent group. The optical continuum from 2.3 to 0.5 μ is well represented by a power law of index 1.3, but then steepens to the ultraviolet. In June 1976 the V magnitude was 17.4; the object was not studied for flux variability. Optical polarization measured at Lick Observatory with the Nordsieck device over three months in 1974 was approximately constant with strength 12% and position angle 96°. Measurements in June 1978 and January 1980 by R. L. Moore (private communication) both gave 12% at 85°. A single measurement in January 1976 by Anderson and Garrison (quoted by Baldwin et al. 1977) is discrepant, 4.4% at 81°. The polarization is thus nearly steady in position angle and could perhaps be constant in strength, if the last point were in error.

Coincident with the stellar optical object is a compact radio source with a spectrum flat from 1.67 GHz (0.15 Jy) to 15 GHz (0.14 Jy). The extended structure shows approximately symmetric lobes lying along a line at position angle of approximately 115°, inclined some 20° to the direction of optical polarization. The total extent at 5 GHz is 25 arcsec, or 100 Kpc (projected on the plane of the sky), and the measured flux at 178 MHz is 2.55 Jy, extending to higher frequencies with a spectral index of 0.7.

1641+399 = 3C 345 As will be discussed in Sections V and VI, Blandford, Rees, and others have suggested that the polarized optical emission in extremely luminous polarized objects may originate in a relativistic jet directed toward us. The same type of relativistic beaming is invoked to explain superluminal expansion in compact radio sources. Thus the violently variable polarized object 3C 345, with strong emission lines at $z = 0.595$, is of exceptional interest since it shows superluminal expansion very clearly. VLBI measurements reviewed by Kellerman (1978) show two components whose separation has increased linearly from 1969–1977, at a rate of 0.17 milli arcsec/yr, corresponding to a velocity of 5 c. Only one other source in Table 1 (3C 279) is known to show superluminal expansion. In 3C 345 the position angle of the expanding double is 106°. As is often the case in other similar sources (Readhead et al. 1978), larger scale structure is not collinear.

The jet of low frequency emission lying 1–3 arcsec distant found by Davis, Stannard & Conway (1977) is at position angle 142°.

Polarization was discovered in 3C 345 by Kinman (1967), after it was found to be a radio (Dent 1965) and optical (Goldsmith & Kinman 1965) variable. The source has since been the subject of several polarimetric studies. Kinman et al. (1968) found the position angle to be about 80° during a period of high luminosity, with large changes in angle apparently correlated with short (10 day) bursts. Recently Kinman (1977) has examined all the data available over an eight-year period, and finds that the yearly polarization averages lie at approximately the same position angle (80–110°) when the source is bright. This indicates there is memory in the emission process and the angle may be related to the VLBI jet axis (105°). It is of interest that the polarization is not noticably weaker when the source is faint.

The wavelength dependence of polarization of 3C 345 was explored by Visvanathan (1973a), who found strength and angle to be constant at any given time within errors. The same result was found by R. L. Moore (private communication) using filters centered at 4500 Å and 7500 Å. Recently Knacke, Capps & Johns (1979) have reported 22–32% polarizations at 2.2 μ over a three-day period in April 1978, with no polarization detected at 1.6 μm and $P < 6\%$ at 0.44 μ. While two-component models may account for this extreme behavior, we note that the variability at 2.2 μ and in the optical is well correlated (Neugebauer et al. 1979). The spectrum from 0.3–10 μ obtained by Neugebauer et al. is close to being a single power law of index 1.4, with only a trace of the excess emission at ~ 0.3 μ and little of the complex structure common in QSOs. However, 3C 345 does appear to have some episodes when the spectrum changes, shown by Visvanathan's (1973a) multichannel spectrophotometry in which the index across the optical spectrum is seen to flatten when the source brightens.

1845 + 797 = 3C 390.3 This is a second key object in making the bridge between double lobe radio sources and those with variable stellar polarized nuclei. It is a classical radio double lying at $z = 0.056$ whose central galaxy contains a violently variable compact optical source (Cannon, Penston & Brett 1971) with substantial polarization ($\sim 3.5\%$). Optical polarization data over the years 1968–1973 have been published by Dombrovsky et al. (1971) and Babadzhanyants & Hagen-Thorn (1975). We will not here consider data by Efimov & Shakhovskoy (1972) which are not accurate enough to be useful. The strength of polarization measured through a 26 arcsec aperture increases with increasing brightness of the source, while remaining reasonably constant in position angle. Babadzhanyants & Hagen-Thorn (1975) make a least-squares fit to the yearly averages over

1968–1973, and obtain a best fit by superposing a source of variable intensity but constant polarization (3.7±0.6% at 155±3°) with an unpolarized galaxy component of B magnitude ~ 16.8. In 1979, we measured the nuclear polarization in a 4 arcsec aperture to be 2.1%±0.4% at 165° ±5°, consistent with the parameters given above since the remaining galaxy contribution has not been removed. An independent estimate of the underlying galaxy magnitude of $B \sim 16.6$, obtained by Penston & Penston (1973) from photometry with different aperture sizes, is also consistent with their separation model. In the radio map of 3C 390.3 given by Harris (1972) the central compact component had strength of 0.35 Jy at both 2.7 and 5.0 GHz, and is <0.4 Jy at 1.4 GHz. Hine & Scheuer (1980) have found variability in this source on a time scale of a year. The north and south outer components are separated by 2 and 1.5 arcminutes (20 and 90 Kpc) and are unusually compact. Their strengths are respectively 7 and 17 Jy at 408 MHz and their spectral indices are both 0.85. The position angle of the source axis is 143°, quite close to the mean polarization angle of 155°. This alignment is the same as that seen in the much more luminous double lobe QSOs (see Sections III and IV).

The nucleus of 3C 390.3 shows strong, extremely broad Balmer emission lines. Spectrophotometry by Osterbrock, Koski & Phillips (1976) shows these lines having complex profiles with a full width at half-maximum of 13,000 km s^{-1} (Table 2). Polarimetric measurements of these lines are now being undertaken and will clearly be of great value in locating the origin of the optical continuum polarization.

2200+420 = BL Lac Not only was this object the first blazar type with no emission lines to be recognized but it exemplifies the most rapid variability of flux and polarization in its optical, infrared, and radio flux (Angel et al. 1978, Aller & Ledden 1978). We will not give here a review of the general properties of BL Lac since this is already available in the literature (e.g. Stein, O'Dell & Strittmatter 1976, Miller 1978). It lies at the center of a giant elliptical galaxy of redshift 0.067, and Miller, French & Hawley (1978) find extremely weak emission lines (Table 2).

The variability of the optical polarization has recently been studied in some detail, and has been found to rotate in position angle at rates of 1–2°/hour during a night of observing, with changes of up to 30° from night to night (Angel et al. 1978, Angel & Moore 1980). The position angle has no preferred direction. In 1979 a group of observers in America, Europe, and Israel observed the optical polarization and intensity of BL Lac for one week, with continuous coverage for more than 12 hours on most days (R. L. Moore et al., in preparation). During this week the source was being measured sometimes by four different observers simultaneously. All the data lie on a curve that varies smoothly hour by hour, but where again one day is the time scale for substantial changes. Infrared polarization measurements at 2.2 μ were obtained by Rieke and Lebofsky on two nights during this week, and also exactly tracked the optical in strength and angle. During one night when the polarization was $\sim 3\%$ a rotation of $30°$ during 6 hours was seen at both wavelengths. This shows that there cannot be much significant interstellar polarization arising in our own galaxy, despite its low galactic latitude. The strength and angle of polarization are not always independent of wavelength, as discussed in Section III.

2225−055 = PHL 5200 PHL 5200 cannot be classified as a blazar since it is radio-quiet (Mills & Little 1970) and shows no variation in its high polarization or optical flux. PHL 5200 is the prototype for a class of radio-quiet QSOs whose spectra show broad, deep absorption troughs blueward of strong resonance emission lines (Lynds 1967, Burbidge 1968). Spectropolarimetric observations show the continuum to be polarized with the lines essentially unpolarized (Stockman, Angel & Hier 1978). Thus the polarization must originate within the emission line region and is not due to dust or resonance scattering outside this region. High polarizations are not a general property of QSOs with intrinsic, broad absorption lines. We have observed three PHL 5200 type objects discussed by Turnshek et al. (1979) and all three show weak polarization, $\lesssim 2\%$.

III *Properties of Blazars*

In this section, we describe the general properties of the blazar class represented by Tables 1 and 2 in Section II. We begin with the characteristic strength and variability of the optical-infrared polarization followed by the wavelength dependence of polarization. The strong correlation of the optical polarization with photometric variability and a smooth optical-infrared continuum is discussed as are the general radio characteristics of the blazar class. We end this section pointing out correlations of the X-ray and radio properties with the presence of strong line emission.

GENERAL CHARACTERISTICS OF THE OPTICAL-INFRARED POLARIZATION

Strength and timescale of variability A primary property of the polarization is its variability; this adds richness to what can be learned from polarimetric studies, and provides job security for polarimetrists. The variations can be substantial and erratic on every time scale from a few hours on up. In addition, even for a single source the general type of variability can change. For instance, for the first year or so after its identification, OJ 287 showed polarization varying wildly in strength and angle. Since 1972, however, its polarization, although still strong, has nearly constant position angle, with the radio emission polarized at the same angle (e.g. Rudnick et al. 1978). In another case, 3C 454.3, the strong variable polarization observed a decade ago has currently vanished, and the source has settled to only 1–2% polarization (Moore & Stockman 1980). The best we can do then is to characterize the behavior of specific objects from the known data, and recognize that it may not persist.

The strength of polarization, whose maximum and minimum recorded values are given in column 7 of Table 1, has exceeded 30% in three objects (PKS 0735+178, OJ-131, OJ 287) and 20% in another nine. While it is often the most extensively measured objects that yield the highest polarization values, there are also some well-studied examples that never show very high polarization. Another polarimetric distinction that can be made among the blazars is by the range over which the position angle varies. Angel et al. (1978) found that among 12 BL Lac objects monitored polarimetrically, 5 appeared to have a definite preferred angle. Making use of all the published data referenced in Table 1, we can distinguish two types of position angle variability. One group shows no tendency for preferred angle, and data span all points of the compass. These objects are nearly all distant and extremely bright: the 3C sources 279, 345, 446, and 454.3 (during outburst), AO 0235+164, B2 1308+326, PKS 0735+178, OM 280, OJ 287 (during outburst), BL Lac. The group of objects for which repeated measurements show a restricted range of angles are 3C 66A, NGC 1275, 0422+004, 0752+258, OJ-131, OJ 287 (recent years), Mkn 421, ON 325, 1400+162, AP Lib Mkn 501, 3C 390.3, 3C 371. These are generally less luminous objects. In Section VI, we consider the possibility that the two types of variability may reflect the difference between sources with jets pointed straight at us and those that are inclined to the line of sight.

We next turn to the time scale for changes in the polarization. Polarization measurements afford a good method for exploring variability on time scales less than a day since the linear polarization Stokes parameters Q and U change by amounts comparable to the total intensity parameter I. Rather small changes of Q/I and U/I are detected differentially to a precision limited only by photon statistics. Small changes of the total intensity I over periods of hours can only be detected by reference to standard stars, and require excellent conditions and careful treatment of extinction. Thus photometry is rarely limited by photon statistics except for the faintest objects. Variations on a time scale of a few days were recognized from the first discovery of strong polarization in the violently variable QSOs, and substantial night to night changes in the polarization of 3C 345 were measured in 1967–1968 by Kinman et al. (1968) and Visvanathan (1973a). OJ 287 also showed large nightly fluctuations in the early data of Dyck et al. (1971) and Kinman & Conklin (1971). Fluctuations on these time scales were also well established from photometric data. The synoptic study of 12 BL Lac objects made by Angel et al. (1978) with repeated measurements of the same objects during each night found that erratic variations on a time scale of a day are not uncommon in several objects but that variations from hour to hour are small. When significant hourly changes are detected they are monotonic, with dP/dt no larger than required to explain the daily variations. Thus it appears that the power spectrum of variability falls sharply for frequencies higher than $1\,d^{-1}$. These results have been confirmed by more recent work on BL Lac (see Section II). The variability of B2 1308 + 326, a source like AO 0235 + 164 that reaches extremely high brightness in outbursts of months duration, has recently been studied and is also discussed in Section II. The polarization changes very strongly (more even than BL Lac) on a time scale of 1 day in the rest frame of the source, but again there is no compelling evidence of much more rapid changes.

Not all strongly polarized sources show detectable variability; a few appear to be essentially constant in both strength and angle. OI 287 has been virtually constant with 8% polarization at 145° over a two year baseline (Moore & Stockman 1980), 3C 66A shows slight changes in polarization, but virtually all data points lie within $12 \pm 2\%$ in strength and $30 \pm 16°$ in angle. The double lobe radio source 1400 + 162 (see Section II) has polarization 12% at close to 90° for all but one measurement.

Wavelength dependence of polarization Another property of the polarization of considerable interest is its dependence with wavelength across the optical-infrared spectral range. It has long been known, particularly from the work of Visvanathan (1973a), that the polarization is usually wavelength independent at least across the optical spectrum. Sometimes, though, some objects do show small but significant rotations in position angle of up to 15° (Rieke et al. 1977, Moore et al. 1980), or changes in strength of polarization from blue to red (Kikuchi et al. 1976, Nordsieck 1976). The smoothness of the continuous emission from the optical to the infrared suggests that the infrared emission (where most energy is released) should share the polarimetric properties of the visible spectrum. This has recently been found to be the case. Simultaneous or nearly simultaneous observations of polarization over the range 0.4–2.2 μ have been reported for AO 0235, PKS 0735 + 178, OI 090.4, OJ 287, B2 1308 + 326, 3C 345, and BL Lac, in papers by Knacke, Capps & Johns (1976, 1979), Rieke et al. (1977), Puschell et al. (1979), Moore et al. (1980), and Puschell & Stein (1980). Generally the polarization is found to be the same in both strength and angle, but there are observations of substantial rotation of position angle (e.g. 35° in

OI 090.4) from the optical to the infrared. Until recently, no marked color dependence in the strength of polarization had been found, but some recent observations of BL Lac (Puschell & Stein 1980) and 3C 345 (Knacke et al. 1979; Section II) showed a strong increase to the ultraviolet in the former, to the red in the latter.

CORRELATION OF STRONG OPTICAL POLARIZATION WITH OPTICAL-INFRARED CONTINUUM PROPERTIES There seems to be virtually a one-for-one correspondence between the occurrence of polarization and large fluctuations in the optical flux. Essentially every object in Table 1 that has been monitored extensively shows variations larger than a magnitude. Conversely, almost all objects known to exhibit large fluctuations in brightness are found to be strongly polarized. For instance, in Usher's (1975) table of variable QSOs and BL Lac objects, there are 13 objects with $\Delta B > 2$ magnitudes. Eleven of these are in common with Table 1, one is in the south and has not been measured for polarization, and only one, 3C 323.1, is not strongly polarized. The correspondence is not perfect and a few variable objects are found not to be polarized (e.g. 3C 120 with $\Delta B = 1.7$). Selection effects play some role in this correlation, as some outstanding examples were originally picked out for polarimetric study after their variability was discovered. 3C 446, the first QSO found to be polarized, was measured by Kinman et al. (1966) after variability was discovered; the identification of BL Lac as an inverted-spectrum radio and optical variable (MacLeod & Andrew 1968, Schmitt 1968) led to the discovery of optical polarization. Nevertheless, many high polarization objects were identified from radio surveys before their optical variability was studied.

A second striking correlation of optical polarization is with the shape of the optical-infrared continuum. The fact that the optical spectrum of polarized objects tends to be steeper than that of most emission line QSOs has been recognized for some time (e.g. Stein, O'Dell & Strittmatter 1976). Now that good photometry from 0.3–10 μ has been obtained for many Seyfert nuclei, BL Lac objects, and QSOs, we find a division can be made between those that can be well represented by a simple power law or spectral shape, and those that have complex shapes (Rieke & Lebofsky 1979). The former class is virtually coincident with the class of highly polarized objects we are considering. The recent spectra of bright QSOs by Neugebauer et al. (1979) illustrate this very well. Picking out the straight spectra from the sample of 30 objects, one finds they are just the group of classical violent variable QSOs (3C 279, 3C 345, 3C 446, 3C 454.3). 3C 68.1 has the only other smooth spectrum in the sample, and its polarization has not yet been measured. Puschell (1980) remarks that there are no objects known showing correlated variability in the optical and infrared that are not highly polarized. Thus 3C 120 fails as a blazar not only in lacking polarization, but also in showing infrared radiation which does not track the optical on short time scales (Rieke & Lebofsky 1979).

RADIO EMISSION OF BLAZARS   An outstanding property we see from Table 1 is that all entries except PHL 5200 show radio emission, a correlation supporting the idea that nonthermal processes are responsible for the polarization. Historic interest in radio sources has biased most polarization surveys toward the discovery of radio-loud objects. Nevertheless the correlation of radio emission with the optical properties of polarization, strong variability, and smooth spectra is not a result of selection effects. Among emission line objects large numbers of radio-quiet QSOs searched by Stockman & Angel (1978) and Stockman (1978) yielded PHL 5200 as the only strongly polarized radio-quiet QSO. Seyferts, which are nearly all very weak radio sources, in the study by Maza, Martin & Angel (1980), showed exclusively dust type polarization, with only one or two possible exceptions (see Section IV).

Since nearly all known BL Lac objects are from radio surveys (Condon 1978) the observational bias against radio-quiet BL Lacs is severe. Certainly in one case, I Zw 186, the unusual bright featureless nuclear continuum was discovered by Zwicky (1966) while observing compact galaxies, and optical variability was then found by Oke et al. (1967) and Sandage (1967), still before weak radio emission was detected (Altschuler & Wardle 1976). Mkn 501 is another object noticed first by its unusual featureless spectrum, in a survey of Markarian galaxies by Khachikian & Weedman (1974). It then proved to have a flat radio spectrum and strong variable polarization. X-ray emission is not well correlated with radio emission at the lower frequencies of most radio surveys (see below), so one might hope that X-ray surveys would find radio-quiet objects. All-sky X-ray surveys have discovered at least one source (2155−304) not previously recognized, but this again has proven to be a radio source. From these data, it seems likely that if radio-quiet BL Lac objects exist, then they must be considerably less numerous than radio-loud ones. This situation is of course the reverse of that for emission line QSOs, where the radio-quiet QSOs far outnumber the radio-loud.

Having noted the common feature of radio emission for the objects in Table 1, we must go on to point out that the character of the emission varies widely among different sources. Nearly all have a flat-spectrum component, but there is a large range in luminosity. Five objects are among the intrinsically brightest of all known sources at 2700 MHz. 3C sources 279, 345, 446, 454.3, and CTA 102 all have apparent luminosities (assuming isotropic emission) of $\sim 10^{35}$ ergs Hz^{-1} s^{-1}, and constitute about 20% of all sources known to be this luminous. By contrast the faintest objects such as Mkn 180 have luminosities of 10^{31} ergs Hz^{-1} s^{-1}. It is interesting that the most direct evidence for bulk relativistic motion is only for the intrinsically brightest objects. The two certain cases of superluminal expansion that are known for objects in Table 1 are from this group, namely 3C 345 and 3C 279. Also, the only examples in Table 1 of low-frequency variability (which suggests bulk relativistic motion—see Section V) are 3C 454.3, CTA 102, and PKS 0420−014, which is also very bright. AO 0235+164 is so bright and rapidly variable at high radio frequencies that, again, brightness temperatures in excess of 10^{12} K are indicated. For the fainter objects with absent or weak emission lines (BL Lacs), VLBI data is reviewed by Shaffer (1978). Changes in structure are seen, but there is no strong case of superluminal expansion.

Radio structure on scales of arcseconds or larger, generally with steep spectra, has been measured in the following objects: NGC 1275 (Miley & Perola 1975), PKS 0521−365 (Fomalont 1967, Mills, Slee & Hill 1960), AO 0827+24 (Hazard, Gulkis & Sutton 1968), Mkn 180 (Kojoian et al. 1976), 1400+162 (Baldwin et al. 1977), AP Lib (Conway & Stannard 1972), 3C 345 and 3C 454.3 (Davis, Stannard & Conway 1977), 3C 371 (Fomalont & Moffet 1971), and 3C 390.3 (Harris 1972). Extended structure of 10–200 arcsec extent, generally of steep spectrum, is also known to be present in about half the sample of 27 BL Lac objects studied by Wardle (1978). Details of this structure are not known because the interferometer was used at only one position angle. Since symmetric extended radio emission, like the optical line emission, is almost surely isotropic, the geometry and strength of the extended sources can aid in distinguishing the degree to which relativistic beaming may be present, and will be considered further in Section VI.

Variability and polarization at radio wavelengths have quite different characteristics for different objects in Table 1. From the four year monitoring program by Altshuler & Wardle (1976) one finds that the optically violently variable quasars 3C 454.3, 3C 446, 3C 345, and CTA 102 generally have rather steady flux and polarization. By contrast, the BL Lac objects 0048−09, 0235+16, 0300+47, OJ 287, W COM, 1749+096, 2155+304, and BL Lac are strongly variable in both flux and polarization. Unfortunately there is little data for the nearby objects with $z < 0.1$, except for BL Lac, NGC 1275, whose radio emission is essentially unpolarized, and AP Lib, which has relatively steady polarization. Coordinated observations of blazars from radio through optical wavelengths by Rudnick et al. (1978) show the polarization is generally weaker at radio wavelengths. With the exception of OJ 287, for most blazars there is little correlation between the angles of optical and radio polarization.

CORRELATION OF RADIO AND X-RAY EMISSION WITH OPTICAL EMISSION LINES IN BLAZARS There is a strong correlation between the radio emission of blazars and their optical emission lines. Consider the group of all eleven objects in Table 1 with $z < 0.1$, all of which are found at the center of giant elliptical galaxies. The five of them (Mkn 180, 421, 501, I Zw 186, 0548 − 322) that have no detectable emission lines all show the weakest radio emission, all of similar strength with flat spectra at ~0.5 Jy scaled to a common z of 0.05. The remaining six objects that do show emission lines are 3C 84 (NGC 1275), 3C 371, 3C 390.3, PKS 0521 − 36, BL Lac, and AP Lib. All show stronger radio emission, at least 2 Jy at 1400 MHz, again scaled to $z = 0.05$, and all but BL Lac have a low-frequency steep-spectrum component.

The correlation between the presence of detectable emission lines and of a steep component to the radio spectrum holds for nearly all objects in Table 1, without regard to redshift. Another general correlation we have mentioned above is that the emission line objects tend to have relatively stable radio polarization and flux, in contrast to the line-free objects.

In contrast to the radio correlation, the X-ray emission observed from some of the same 11 objects is not stronger for those with lines, indeed there is a suggestion that the line-free objects may have stronger X-ray fluxes. The highest X-ray luminosities of > 5 μJy at 3.6 keV, referred to $z = 0.05$, have been observed in Mkn 521 and PKS 0548 − 322, both line-free objects. All the line-free objects except Mkn 180 have been detected at the ~1 UHURU count level, making the X-ray and optical luminosities comparable.

IV *Active Objects with Low Polarization*

As mentioned in the introduction, most active extragalactic objects show optical and infrared polarization of a percent or less. At these low levels, the case for any detected polarizations being intrinsic to the nuclear emission is much weaker than for the blazar class. The polarization may also be due to scattering by dust or free electrons or due to transmission through aligned grains either in the host galaxy or in our own. In the first half of this section, we review the polarimetric observations of Seyfert galaxies, where there is good evidence that the polarization is due to dust. Thus, rather than include the highly polarized Seyferts with sources of intrinsically polarized emission in Sections II and III, we discuss them here. For the majority of QSOs, whose polarizations are quite low, the source of the residual polarization is unknown. We review recent survey work of QSOs in the second half of this section.

SEYFERT GALAXIES Since the most detailed polarimetric observations to date have concentrated on two of the brightest Seyferts, NGC 1068 and NGC 4151, these are reviewed separately below. The first general polarization surveys of Seyferts were undertaken by Dombrovsky, Hagen-Thorn, and other collaborators and are summarized in Dombrovsky & Hagen-Thorn (1968), Dombrovsky et al. (1971), and Babadzhanyants & Hagen-Thorn (1975). This work has also been reviewed by Hagen-Thorn (1974) and in Maza's dissertation (1979). Although most of these observations utilized a rather large aperture (26″) which resulted in severe galaxy dilution of the nuclear light, these authors correctly attributed the observed polarization to the nuclear light, and found the strong wavelength dependence of polarization in NGC 1068 and NGC 4151, and detected weak polarization in NGC 3227 and NGC 7469 and several others. Because of the variable polarization observed in NGC 1275 (Section II) and NGC 4151 along with the nonthermal radio emission observed in NGC 1275 and other radio galaxies classified as Seyferts, the polarization was generally attributed to optical synchrotron radiation. However, a more recent polarization survey by Maza, Martin & Angel (1980) and spectropolarimetric work of several groups indicate that this is not the case.

Maza, Martin & Angel's survey of 47 Seyferts found only 8 with polarizations greater than 1.5% in a 4″ aperture: 5 Type 1's—Mkn 231, Mkn 486, Mkn 376, NGC 6814, IC 4329A; and 3 Type 2's—NGC 3227, Mkn 348, and Mkn 3. [Note: the polarization in NGC 6814 is probably local interstellar polarization (Maza 1979).] Additional multicolor polarimetry showed that most highly polarized Seyferts have stronger polarization in the blue than in the red with little rotation of position angle with wavelength. Spectropolarimetry of Mkn 376, IC 4329A, Mkn 231, NGC 3227, Mkn 3, and NGC 3516 (Stockman, Angel & Beaver 1976, Thompson et al. 1980) show that the emission lines and continuum in these sources have similar polarizations. Together with the strong wavelength dependence of polarization, this argues for dust scattering (see NGC 1068 below) or in at least one case, IC 4329A, an edge-on spiral, for transmission through aligned grains as the origin of the optical polarization. The case for polarization due to dust is not as strong for those few sources with a polarized continuum but unpolarized emission lines (e.g. NGC 4151 below, and our preliminary results for Mkn 486).

We must emphasize that observations of generally weak polarization in Seyferts and the absence of strong radio emission do not rule out the possibility of nonthermal emission in the nuclei that has been reprocessed inside the emission line region or diluted by thermal emission. Indeed, the high X-ray luminosities associated with most of the bright Seyferts (with the notable exception of NGC 1068) argue for a non-stellar origin for much of the optical infrared emission (Mushotzky et al. 1980, Dower et al. 1980).

NGC 1068 NGC 1068, a Type 2 Seyfert (narrow emission lines, FWOI ~ 1000 km s^{-1}), has had its optical polarization repeatedly measured by

Walker (1964, 1968), Dibaj & Shakhovskoy (1966), Kruszewski (1968), and by Dombrovsky and collaborators (see above). Apart from the early, rather crude measurements, the optical polarization and luminosity of the nucleus appear constant. The polarization measured through a $2''$ aperture increases toward the blue, roughly as $P \propto \lambda^{-3}$ from $\sim 0.6\%$ at $0.8\,\mu$ to $\sim 11\%$ at $0.32\,\mu$. The position angle increases gradually from $\sim 94°$ to $102°$ over the same spectral range (Angel et al. 1976).

Using data obtained through a large aperture ($\sim 10''$) Visvanathan & Oke (1968) and Kruszewski (1968) found the polarized flux to be essentially constant with wavelength, suggesting a flat, nonthermal component to the nuclear light. However, spectropolarimetric observations by Angel et al. (1976) showed the polarization of the permitted emission lines to be similar to that of the surrounding continuum. This evidence, together with the strong wavelength dependence and the observed large ellipticity, strongly point to dust scattering as the polarizing agent (see Section V). Infrared observations by Knacke & Capps (1974) and Lebofsky, Rieke & Kemp (1978) indicate strong wavelength dependence on degree and position angle of polarization in the spectral range 1.2–$10\,\mu$ which may be due to an added nonthermal component in the region 1–$5\,\mu$ or a complex cloud geometry around the nucleus (Jones et al. 1977, Elvius 1978). While NGC 1068 is the best-studied Seyfert, Maza (1979) points out that in visible light its core luminosity is very weak relative to the host spiral galaxy. If it were at redshifts typical of the majority of known Seyferts, its polarization would be practically undetectable, being less than 0.5% with the $4''$ aperture used in the survey.

NGC 4151 The spectrum of NGC 4151 shows strong narrow and broad emission line components and is generally classified as a Type 1 Seyfert (FWOI $\gtrsim 5000$ km s^{-1}). Unlike NGC 1068, the nuclear luminosity and degree of polarization of NGC 4151 does vary on a time scale of years, though with little change in the position angle (Dombrovsky & Hagen-Thorn 1968, Babadzhanyants, Hagen-Thorn & Lyutyi 1972, and Kruszewski 1977). The degree of polarization is small, $\sim 1\%$, and is roughly independent of wavelength into the near infrared, where it falls to $\sim 0.1\%$ at $2.2\,\mu$ (Kemp et al. 1977). Spectropolarimetric observations by Thompson et al. 1979 and Schmidt & Miller (1979) show the polarization of both the broad and narrow emission line components are much weaker than that of the surrounding continuum. These observations are consistent either with nonthermal continuum diluted by thermal emission and starlight or with scattering within the broad-line emission region.

While the weak but variable polarized component of the optical continuum in NGC 4151 resembles that seen in the blazars, it should be noted that the radio emission is far weaker than any object in Table 1. In addition, the radio source is not compact but is extended over ~ 300 pc, with a steep spectrum. De Bruyn & Willis (1974) observed a flux of 0.135 Jy at 5 GHz with a spectral index of 0.74.

QSOs Polarimetrists who first searched quasi-stellar sources for the high polarizations expected from synchrotron radiation were rewarded with the strong and rapidly variable polarization of the optically violent variables 3C 345, 3C 446, 3C 454.3, 3C 279 (Section II). Other QSOs such as 3C 273 (e.g. Whiteoak 1966, and Liller 1969) had disappointingly low polarizations. More extensive polarimetric surveys by Appenzeller & Hiltner (1967) and Visvanathan (1968) failed to discover any additional strongly polarized QSOs. The sensitivity of these surveys was such, however, that polarizations of a few percent could not be unambiguously detected.

Recently, Stockman & Angel (1978) have reported preliminary results of a large polarimetric survey of bright QSOs, $V \gtrsim 17$ (see also Stockman 1978). They find that $\sim 90\%$ of this sample have low polarizations, $< 2\%$. Much of the average polarization, $\sim 0.6\%$, is due to local interstellar polarization. With the exception of the unusual object PHL 5200 (Section II), none of the radio-quiet QSOs are strongly polarized. Since the vast majority of QSOs are believed to be radio quiet, the blazar type must be extremely rare. Indeed, the gap in polarimetric properties between the few strongly polarized ones and the remaining QSOs (even radio emitters) suggests that the blazar QSOs are a qualitatively distinct subclass of objects (Sections II and V).

The origin of the small polarizations seen in the majority of QSOs is unknown. Multicolor observations indicate that, while there is a statistical tendency for the polarizations to increase to the blue, the optical polarization is essentially wavelength independent (Moore, Stockman & Angel 1980). The weak polarizations and their lack of marked wavelength dependence suggests that, unlike Seyfert nuclei, QSOs have little surrounding dust ($\tau \ll 1$).

Stockman, Angel & Miley (1979) found that in those QSOs with double radio lobes the position angle of polarization was roughly aligned with the radio structure. Thus, unlike the most variable blazars, the position angle must be relatively constant for time scales of 10^6–10^7 yrs.

As with the Seyfert galaxies, the lack of strong polarization does not preclude a luminous nonthermal central source. Tananbaum et al. (1980) find many QSOs with an X-ray luminosity comparable to their optical-infrared luminosity, thus indicating a powerful, nonstellar engine. However, the very weak polarizations do argue for efficient reprocessing of any nonthermal optical emission or for dilution by an unpolarized, isotropic source which is probably thermal (Section V). More detailed study of the low polarization sources (wide baseline multicolor and spectropolarimetric observations; continued monitoring) will give valuable information concerning the origin of the polarization and should be capable of distinguishing between reprocessing and thermal dilution models.

V Theoretical Models and the Origin of Polarization

Of all active extragalactic objects, the blazar class exhibits the most extreme behavior in variability and polarization and places the greatest burden on current theoretical models of the central powerhouse. In this section, we review current theories with an emphasis on those areas pertaining to optical polarization and rapid variability. We devote the majority of the section to discussing basic emission mechanisms and, in particular, the canonical incoherent synchrotron model. After reviewing its success at describing extragalactic radio sources, we examine the difficulties of extending the theory into the optical, in addition to the well-known problems with the low-frequency variables, superluminal expansion, and short particle lifetimes—all of which are associated with the blazar class. Although the incoherent synchrotron theory and the alternative emission mechanisms are intimately connected to specific models for the central source of energy, we postpone discussing models for the central powerhouse and their common features until later in this section. Finally, we briefly review the processes by which dust scattering can cause linear and circular polarization.

INCOHERENT SYNCHROTRON RADIATION Because of its success in explaining the radio emission from the Crab Nebula, incoherent synchrotron emission was proposed by Shklovsky (1955) and others to be the source of the diffuse extragalactic radio emissions whose power-law spectra at high frequencies (optically thin regime) resembled that of the Crab. The synchrotron theory was soon extended to the compact sources, whose flat spectra are generally considered to be the superposition of many components of varying optical depth (see, for example, Kellermann & Pauliny-Toth 1968). One of the successes of this canonical theory of incoherent synchrotron theory was its "prediction" of the upper limit of observed brightness temperatures, implied by the observed flux and VLBI measurements or limits on angular size, $T_b \lesssim 10^{12}$ K. Such temperatures indicate that relativistic electron energies, coherent plasma effects, or both must be present. In the canonical theory, the maximum brightness temperature of $\sim 10^{12}$ K corresponds to the onset of significant Compton upscattering of the synchrotron radiation by the relativistic electrons. At higher brightness temperatures, the upscattered photons are themselves upscattered in energy leading to a "Compton catastrophe" and rapid cooling of the relativistic electron distribution (Hoyle, Burbidge & Sargent 1966, Kellerman & Pauliny-Toth 1969, Jones, O'Dell & Stein 1974). In addition to explaining the observed power-law spectra and surface brightness limits, the canonical synchrotron theory also predicts high polarizations (60–80%; Korchakov & Syrovatskii 1962) in the optically thin regions. Although radio sources generally show much lower polarizations, these could be understood as resulting from chaotic field geometries and self-absorption effects (Jones & O'Dell 1977a).

Despite the success of the canonical theory in describing the compact- and steep-spectra radio emission, there is little evidence that the observed optical emission from the large majority of QSOs or Seyfert nuclei is a simple extension of the observed radio emission (Section IV) or is even nonthermal in origin. For the blazar class, however, their smooth spectra, high polarization, and rapid variability make the incoherent synchrotron theory look more promising (Section III). In particular, rather simple models for the source can explain the observed polarization properties. For example, the presence of many misoriented emission regions with equal strength magnetic fields and particle distributions will produce a reduced net polarization that is independent of wavelength. Another model was explored by Nordsieck (1976) in which the field is preferentially stronger in one direction. In this case, the polarization depends on the electron energy distribution with the result that strong polarization should be associated with steep optical spectra and a convex curved spectrum will show polarization increasing to the blue. While indeed the blazars do tend to have steep spectra, recent observations do not support these specific predictions. Simultaneous infrared and optical measurements, summarized above, show polarization that is usually wavelength independent, even in BL Lac which falls steeply in the optical. We also find that several of the sources in Table 1 that show strong polarization have relatively flat spectra (Sections II and III).

Beyond explaining the spectral and polarization properties of the optical emission, the extension of the incoherent synchrotron spectrum into the optical region places several severe constraints on the emission source. For rapidly varying ($t_{var} \sim 1$ day), luminous sources ($\nu F_\nu \sim 10^{46}$ ergs s^{-1} or $L_{46} \sim 1$) the corresponding brightness temperature in the optical ($\nu \sim 10^{15}$) is $T_b \gtrsim 1 \times 10^5$ K. For a spectral index of 0.5–1, we would expect the corresponding radio spectrum to be self-absorbed ($T_b \sim 10^{12}$ K) in the frequency range 2–5×10^{12} Hz or $\lambda \sim 60$ μm in the far infrared. Thus the variable optical-infrared sources must be smaller than the observed radio components. Although relativistic beaming or lower luminosities may weaken this argument, in general we would expect little or no correlation between the optical and radio spectral bands on time scales of a week or less. As well as being more compact, the optical region must also have stronger fields. To avoid the Compton catastrophe, the lower limit to the magnetic field is given by Blandford & Rees (1978) $B_c \sim 125\ L_{46}^{1/2}\ t_{var}^{-1}$ gauss (here t_{var} is in units of one day). Blandford & Rees also derive a cooling time due to synchrotron radiation followed by mildly relativistic cyclotron radiation $t_{cyc} \lesssim L_{46}^{-1}\ t_{var}^2$ hr. Since $t_{cyc} < t_{var}$ for the rapid variations observed in most blazars, the electrons must be reaccelerated many times on a variability time scale if the density of thermal electrons is not to exceed the number of relativistic electrons. To avoid any significant Faraday rotation, the reacceleration may need to be almost continuous in the most variable sources. In the radio emission regions, the absence of Faraday rotation places an even more stringent upper limit on the number of nonrelativistic electrons, $n(\gamma < 100) \ll n(\gamma > 100)$ (Wardle 1977, Jones & O'Dell 1977b, see also Noerdlinger 1978).

BEAMING AND RELATIVISTIC JETS Two types of behavior found among the blazars, low-frequency variability with brightness temperatures inferred from variability arguments far exceeding 10^{12} K (Condon et al. 1979) and superluminal expansion (Cohen et al. 1979), present serious difficulties for the canonical incoherent synchrotron theory. Many remedies have been offered: non-cosmological distances (Hoyle, Burbidge & Sargent 1966); anisotropic electron distributions (Woltjer 1966); coherent emission processes (see below); and various phase and absorption effects (Rees & Sciama 1965, Jones, O'Dell & Stein 1974). One promising group of theories employs the relativistic Doppler effects present when a relativistic jet of plasma is viewed "end-on" (Lovelace 1976, Rees 1978b, Blandford & Königl 1979, Scheuer & Readhead 1979, Marscher 1980). Not only can these effects explain apparent superluminal velocities but the forward-beaming and time-dilation effects can increase the apparent radio brightness temperatures by very large factors ($T_{obs} \sim 10^2$–$10^3\ T_{true}$ for $\gamma \sim 5$–10; Blandford & Königl 1979). Recent VLBI observations indicate that many radio doubles have well-collimated structures on scales of 1 pc–1 Mpc (e.g. Readhead, Cohen & Blandford 1978). In addition, the lack of interstellar scintillation in the low-frequency variables provides indirect evidence for relativistic bulk motions (Condon & Dennison 1978). In the comoving frame of the relativistic jet, the dominant emission mechanism is still thought to be incoherent synchrotron emission. For observers essentially on axis ($\theta \lesssim \gamma^{-1}$), special relativistic effects will shift the spectrum blueward, enhance the luminosity, and decrease the observed time scales for variability (Rees & Simon 1968, Burbidge, Jones & O'Dell 1974). In addition, small variations in the beam direction may appear as quite large changes in the position angle of polarization, with 180° variations expected for end-on views. Blandford & Königl (1979) and Blandford & Rees (1978) suggest these relativistic effects may be responsible for the optical as well as radio properties of AO 0235+164 and other BL Lac objects.

COHERENT RADIATION   S. A. Colgate and collaborators have suggested a coherent emission process that is capable of describing the qualitative features of the radio, optical, and X-ray spectra and permits brightness temperatures far in excess of 10^{12} K (e.g. Petschek, Colgate & Colvin 1976). Coherent emission at $v \sim 2v_p$ is produced and frequency-scattered by nonthermal plasma oscillations in a mildly relativistic, thermal gas ($kT_e \sim 1/2 m_e c^2$). Coherent and incoherent Compton scattering then produce the low-frequency and high-frequency (optical) radiation. Thermal bremsstrahlung is chiefly responsible for the observed hard X-ray spectrum.

Because the optical photons are created through multiple Compton scatterings, this model has great difficulty explaining the high polarization and variability of polarization that characterizes the blazar class (as does the electron scattering model suggested by Katz 1976). An important constraint on these scattering models is the very low upper limit on the net magnetic field set by the lack of observed Faraday rotation in the radio, $B \leq 10^{-7}$ G (Colgate & Petschek 1978). Jones, O'Dell & Stein (1974) review the energetic problems of other coherent emission models.

THERMAL PROCESSES   There is a great amount of evidence, especially in Seyfert galaxies and in the unpolarized QSOs, that the nonthermal optical-infrared radiation has been diluted by thermal emission or reprocessed. Infrared studies of the variability and 10 μ features of nearby Seyfert galaxies indicate that most of the luminosity in both Type 1 and 2 Seyferts is due to thermal reradiation by dust (Rieke & Lebofsky 1979). Polarimetric evidence in support of dust is discussed in Section IV and reviewed by Maza (1979) and the high hydrogen column densities derived from recent X-ray data (Mushotzky et al. 1980) also indicate that absorption and scattering by dust may be important in these objects. For most QSOs, the nature of the infrared emission is unclear. Recent infrared-optical observations by Neugebauer et al. (1979) indicate significant structure and changes in the spectral slope in those QSOs that are not known polarized variables (see Section III). The complex spectra may be due to either thermal continuum emission from the photoionized regions responsible for the strong line emission seen in the objects (see the recent review by Davidson & Netzer 1979), reradiation by dust, or optically thick radiation from an accretion disk (Shields 1978).

A series of authors have suggested that Compton scattering from a hot gas ($kT_e \sim 5$–100 keV) may play an important role in reprocessing the nonthermal synchrotron spectrum (Katz 1976, Stockman 1978, Eardley et al. 1978). The scattering will tend to destroy any intrinsic polarization and variability and will harden the emergent spectrum. Thus these models are also more relevant for the nonpolarized sources.

MODELS OF THE CENTRAL POWERHOUSE   The tremendous luminosities of bright QSOs, $L \sim 10^{46}$–10^{48} ergs s^{-1}, are generally thought to be gravitational in origin. Models that invoke thermonuclear burning are usually inefficient and are incompatible with the high polarizations and rapid variability displayed by the blazar class. These blazar properties, the observed radio jets, and Eddington luminosity arguments suggest a central aligned supermassive object, $M \sim 10^8$–$10^{10} M_\odot$, either a magnetically and rotationally supported "spinar" (Pacini & Salvati 1978) or accretion onto a black hole (Rees 1978a).

Estimates of the ages for the double radio lobes and their alignments with the central VLBI sources and the optically polarized QSOs (Stockman, Angel & Miley 1979) require alignment of the central source over time scales of $\gtrsim 10^7$ years. For accretion onto black holes, the required accretion rate is $\dot{M} \sim 2\eta^{-1} L_{47} M_\odot$/year where η is the efficiency. For steady luminosities less than the Eddington limit, the lower limit on the mass of the black hole is $M \gtrsim L_{47}\, 10^9\, M_\odot$. The alignment of the rotation/jet/polarization axis will be maintained for time scales of $\sim M/\dot{M} \gtrsim 4 \times 10^8\, \eta$ years (Rees 1978b). It is interesting to note that an *upper* limit on the mass of the central source can be obtained from equating the variability time scale (t_{var} (days) ~ 1) with the light travel time across a Schwarzschild radius, $M \lesssim t_{\text{var}} 10^{10} M_\odot$ (Elliot & Shapiro 1974, Moore et al. 1980). Relativistic beaming effects can raise this upper limit by approximately the Lorentz factor.

The relativistic particles required to explain the observed nonthermal radiation are generated by strong electromagnetic fields in spinar or some accretion disk models (e.g. Lovelace 1976) or by nonequilibrium processes (turbulence, relativistic shocks, Fermi acceleration, etc.) in the region near the black hole where $v_{\text{infall}} \sim 1/2c$ ($E \sim 100$–200 MeV/nucleon). In this regime, the emission is likely to be dominated by nonthermal processes rather than bremsstrahlung.

Details of the accretion process and the formation of a relativistic jet are extremely uncertain. Disk models have been suggested by Lynden-Bell & Pringle (1974), Blandford (1976), Eardley et al. (1978) and others. Electromagnetic and hydrodynamic models for the formation of relativistic jets are discussed by Blandford & Rees (1974), Lovelace (1976), and Blandford (1976).

POLARIZATION BY DUST SCATTERING   The production of polarization by dust scattering is a well-known phenomenon that has been well studied in our own galaxy and is recently discussed by Martin (1979). Light transmitted through grains aligned by a magnetic field becomes linearly polarized, the effect responsible for the interstellar polarization of reddened stars. As in our own galaxy, the magnetic field of external galaxies appears to be in the direction of orbital motion (Elvius 1972). If light from a galactic nucleus were polarized by transmission through aligned grains in a spiral host galaxy, viewed nearly equatorially, we would expect the polarization to be parallel to the galaxy's major axis. If the grains are of a similar size to those causing polarization in the Milky Way, then polarization peaking in the visible part of the spectrum is a characteristic signature.

Polarization is also produced in the light that is scattered off small grains, regardless of their orientation. In some nuclei the very strong thermal radiation in the infrared shows the presence of optically thick dust, and most of the visible light must be scattered before leaving. Any departure from spherical symmetry will then result in net polarization of the source. This process is likely responsible for most of the polarization of Seyfert nuclei, including the strongly polarized ones where emission lines and continuum all share the same polarization. The very strong increase of polarization to the ultraviolet in most of these is probably indicative of scattering by very small particles (i.e. Rayleigh scattering).

When the scattering optical depth exceeds unity, multiple scattering in a skew geometry can produce significant circular polarization if the grains are not too small. In NGC 1068, the observed ellipticity, amounting to nearly 5% in the red, is probably due to this mechanism (Angel et al. 1976). For most active extragalactic objects with low linear polarization, circular polarization due either to scattering or synchrotron self-compton effects (Sciama & Rees 1967) is expected to be very small in the optical, $\lesssim 0.1\%$. This is consistent with current observational limits (Landstreet & Angel 1972, Kemp, Wolstencroft & Swedlund 1972, Maza 1979).

VI *Relativistic Jets and Optical Polarization*

We have reviewed in Sections III and V the suggestion by Blandford & Rees (1978) that in AO 0235+164 and similar objects relativistic jets may be responsible for beamed polarized optical radiation, as well as the very high radio brightness temperatures and apparent superluminal expansion. If the superluminal sources are beamed with cone angle of $\sim 1/\gamma$, and $\gamma \gtrsim 5$ is required for the more more rapidly separating doubles, then for every source beamed toward us there must be $\gtrsim 100$ sources at the same distance with misdirected beams. Analyzing the source counts for QSOs, Scheuer & Readhead (1979) identify these two classes as radio-loud flat-spectrum quasars, and the much larger number of radio-quiet objects. The radio emission from double-lobe or extended steep-spectrum sources is not beamed, so the statistics of occurrence of central flat-spectrum sources in such sources can be used to measure independently the degree of beaming. Scheuer & Readhead find $\gamma \lesssim 2$ by this method.

In this section we consider the arguments that can be made for the rapidly variable, highly polarized optical emission characteristic of blazars being produced in relativistic jets. The situation is closely related to but somewhat different from the radio emission. As we have seen, the ratio of polarized optical flux to high-frequency radio emission varies by over a thousand from flat-spectrum QSOs with no polarized optical emission, through violent QSOs like 3C 345, to weak radio sources like Mkn 421.

From the objects listed in Table 1 we can distinguish two types that must be inherently different, irrespective of any possible beaming. These are the strong emission line objects, and the generally weaker BL Lac (weak-lined or line-free) types. The first group contains the violent variable QSOs that are apparently strongly beamed, 3C 345 with superluminal expansion $v = 5\,c$ and 3C 279 with $v \simeq 10\,c$ (Cotton et al. 1979). There are also another dozen or so strong-lined sources in Table 1 for which there is no direct evidence of relativistic beaming, but which have similar luminosity and variability of polarized radiation. The space density out to $z = 0.7$ of these objects is $\sim 1/\text{Gpc}^3$, allowing for incomplete coverage in the south. If we assume that these sources are all beamed towards us with γs of ~ 5 and corresponding (half) cone angles of 0.2, then the space density of misdirected sources would be $\sim 40/\text{Gpc}^3$. This is close to the density of $140/\text{Gpc}^3$ found for local QSOs down to the lowest luminosity class of $L_{\text{opt}} = 10^{46}$ ergs s^{-1} considered by Schmidt (1978). Thus we reach a conclusion similar to that of Scheuer & Readhead. If relativistic beaming is responsible for the strong polarization seen in some QSOs, then most QSOs, including radio-quiet ones, must be emitting misdirected beams of polarized light.

If this picture is correct, then there are some interesting consequences. All these essentially identical strong line objects would be emitting most of their energy in the optical-infrared spectrum as isotropic unpolarized emission with the characteristic complex spectral structure, very possibly from thermal emission in an accretion disk (Shields 1978). The energy in the beam, since it does not swamp the emission lines even when directed toward us, is a small fraction of the QSOs energy budget when Doppler enhancement is accounted for. It may be significant that the maximum luminosity of extended radio sources, $\sim 10^{46}$ ergs s^{-1} (Miley 1980), is roughly equal to the power that would be required in a beamed component of a bright blazar. This supports the idea that the same jet is responsible for both phenomena. The scenario of a polarized, low-luminosity source that is relativistically beamed is quite different from that envisioned by Stockman (1978). If the polarized emission were *isotropic*, then at its brightest it would have a total luminosity equal to that of the brightest known QSOs ($\sim 10^{48}$ ergs s^{-1}). Unpolarized objects could then be explained as objects whose light was produced initially in the same way but was then depolarized and stabilized by scattering.

The situation for the BL Lac objects is not clear. The many similarities in the observed properties suggest we are seeing something closely related to the polarized quasars but without the strong emission lines and associated unpolarized continuum. In terms of the relativistic beam models, it may be that in the strongest BL Lac's, like AO 0235 + 164 or OJ 287, we are viewing directly into a similar beam, and the weaker ones are the same thing viewed from an oblique angle. Alternatively, or in addition, we could be viewing a different population of weaker objects head on.

Direct evidence that our view is sometimes oblique is provided by the two objects that lie at the center of double-lobe radio sources, 1400 + 162 and 3C 390.3. It is hard to guess what the angle between the lobe axis and the line of sight might be, but at least in 3C 390.3, which has an unusually small ratio of the lobe size to separation (Harris 1972), it would seem that it cannot be small. The fact that in both these objects the polarization is stable in angle and is aligned with the radio axis suggests that the property of stable position angle in many of the fainter BL Lac objects (Section III) may be related to their being viewed obliquely.

In a relativistic jet theory the misdirected jets that we argue must also be present in radio-quiet quasars could be emitting at the same strength of radio emission and polarized optical radiation that we see in BL Lac objects, and would pass unnoticed. The luminosity of the weaker BL Lacs of 10^{31} ergs s^{-1} Hz^{-1} would be detected at less than 1 mJy for $z > 0.3$, and the optical polarization could easily be diluted to less than 1% by the unpolarized continuum associated with the emission lines. It is tempting to speculate that the weakly aligned polarization found in double-lobe radio quasars could be of this type. If this were correct it would imply that the optical flux is not extremely dependent on angle, and that relatively low values of γ (<2) are typical.

A case that at least some BL Lac objects are viewed at small angle to the jet axis can be made if we identify the steep-spectrum extended component in some objects as radio doubles seen end on (M. J. Rees, private communication). This idea can be developed as follows. Three of the objects in Table 1 at $z < 0.1$ show such a component to their radio emission, namely 3C 84, 3C 371, and PKS 0521 − 36. A lower limit to the typical angle of view of these objects can be derived if we assume they emit the steep-spectrum component isotropically. The space density of all radio sources of similar luminosity is some 30 times greater than that of these objects. Thus if all radio galaxies would show polarized nuclei when viewed end on, then we deduce the typical angle of view for the three objects is ~ 0.25. Larger angles will be needed if only a subset of radio galaxies are involved. If the strong extended component of PKS 0521−36 and 3C 371 were shown to be a halo like 3C 84 and not two lobes closely spaced by foreshortening, this would be strong evidence that these objects were being viewed at close to random angles, since halo sources are rare. On the other hand, if these two sources showed core-jet structure it could indicate that even the low-frequency component is relativistically beamed. Such beaming is required for the strong, steep-spectrum emission of the classic OVV quasars in the statistical analysis of Scheuer & Readhead (1979).

One prediction made by relativistic beam models is that the fluctuation time scale for relativistic jets seen head on will be more rapid than when viewed obliquely. The fact that the time scale of one day for the extremely bright source B2 1308+326 is not markedly different from much fainter sources is given as an argument against relativistic beams by Moore et al. (1980). However, we note that rather modest values of γ of 2 or 3 give extremely large intensity enhancement in the forward over oblique directions, while causing modest changes in time scale. Given our present knowledge of variability, such changes could easily have escaped notice. In fact B2 1308+326 has as large an amplitude variation from day to day (allowing a factor two for cosmological redshift) as any known object. A good test of these ideas would be a comparison of the time scales for changes in the polarization in the most luminous sources with the time scales in sources where a restricted range of position angle or the presence of a double radio source indicates an oblique view. In general, these latter sources are not as well studied as the wildly varying ones.

VII Conclusions

The use of optical and infrared polarimetry to study active extragalactic objects is a powerful method to obtain information about the innermost core of the central engines. As Rees (1978a) has stressed, the spatial scale of 10^{15} cm indicated by the variability of the polarized optical emission is a thousand times smaller than the compact radio structures of $\sim 10^{18}$ cm studied by VLBI. Until recently there were few detailed predictions that could drive specific polarimetric tests of quasar models. The development of black hole accretion theories and relativistic jet models now suggests a wide range of studies.

One general study will be a quantitative analysis of the statistics of occurrence of polarization and large optical variability in objects ranging from the ellipticals with weak compact sources to the bright quasars. Correlating the polarization with on the one hand properties known to be isotropic such as galaxy luminosity, emission line strengths, and symmetric extended radio emission, and on the other with compact flat-spectra, low-frequency variability, and so on, should give a much better understanding of the role of relativistic beaming. Another area that should also be rewarding will be synoptic studies of the rapid variations in polarization and flux, ideally measurements correlated in the X-ray, infrared, and optical spectra. Because the time scales are around a day, coordinated measurements from satellites or several observatories around the world are needed. With modern instrumentation, modest telescopes of ~ 1 m aperture are adequate for this work. These data should shed much light on the primary mechanisms for energy release deep in the central powerhouse.

Additional studies of weakly polarized or dusty nuclei would be extremely valuable. For the majority of QSOs, high quality, multicolor polarimetry and spectropolarimetry should discriminate between scattering and nonthermal origins of the observed polarization. In the Seyfert nuclei, imaging polarimetry from space and spectropolarimetry similar in quality to the beautiful spectrum of NGC 4151 obtained by Schmidt & Miller (1979) could establish the spatial and kinematic relationship of the dust clouds within the narrow and broad line emission regions. The general polarization surveys should be extended into the ultraviolet to study the characteristics of nuclear dust and to smaller apertures to search for dust or nonthermal sources in active and normal galaxies.

ACKNOWLEDGMENTS

We are grateful to many colleagues for letting us have preprints in advance of publication, and especially to Richard Moore for preparing the two tables. This review has benefitted also from the help of Drs. Jones, Miley, Moore, Puschell, Rieke, Rudnick, Scheuer, Stein, Tapia, and Wardle, who pointed out errors and suggested improvements in the first draft. This work is supported by the NSF under grant AST-78-22714.

Literature Cited

Aller, H. D., Ledden, J. E. 1978. *Pittsburgh Conference on BL Lac Objects*, ed. A. M. Wolfe, p. 53. Univ. Pittsburgh Press

Altschuler, D. R. Wardle, J. F. C. 1976. *Mem. R. Astron. Soc.* 82:1

Angel, J. R. P., Stockman, H. S., Woolf, N. J., Beaver, E. A., Martin, P. G. 1976. *Ap. J. Lett.* 206:L5

Angel, J. R. P. et al. 1978. *Pittsburgh Conference on BL Lac Objects*, ed. A. M. Wolfe, p. 117. Univ. Pittsburgh Press

Angel, J. R. P., Moore, R. L. 1980. *Ann. NY Acad. Sci.* 336:55

Appenzeller, I., Hiltner, W. A. 1967. *Ap. J. Lett.* 149:L17

Baade, W. 1956. *Ap. J.* 123:550

Babadzhanyants, M. K., Hagen-Thorn, V. A., Ljutyi, V. M. 1972. *Astrophysics* 8:300

Babadzhanyants, M. K., Hagen-Thorn, V. A., Semenova, E. V. 1972. *Astron. Tsirk.* 701:1

Babadzhanyants, M. K., Hagen-Thorn, V. A. 1975. *Astrophysics* 11:259

Bailey, J. A., Pooley, G. G. 1968. *MNRAS* 138:51

POLARIZATION OF EXTRAGALACTIC OBJECTS 357

Baldwin, J. A. 1975. *Ap. J.* 201:26
Baldwin, J. A., Wampler, E. J., Burbidge, E. M., O'Dell, S. L., Smith, H. E., Hazard, C., Nordsieck, K. H., Pooley, G., Stein, W. A. 1977. *Ap. J.* 215:408
Bingham, R. G., McMullan, D., Pallister, W. S., White, C., Axon, D. J., Scarrott, S. M. 1976. *Nature* 259:463
Blandford, R. D., Rees, M. J. 1974. *MNRAS* 169:395
Blandford, R. D. 1976. *MNRAS* 176:465
Blandford, R. D., Rees, M. J. 1978. *Pittsburgh Conference on BL Lac Objects*, ed. A. M. Wolfe, p. 328. Univ. Pittsburgh Press
Blandford, R. D., Königl, A. 1979. *Ap. J.* 232:34
Bolton, J. G., Shimmins, A. J., Wall, J. V. 1975. *Aust. J. Phys. Suppl.* 34:1
Brandie, G. W., Bridle, A. H. 1974. *Astron. J.* 79:903
Bridle, A. H., Davis, M. M., Fomalont, E. B., Lequeux, J. 1972. *Astron. J.* 77:405
Bridle, A. H., Fomalont, E. B. 1974. *Astron. J.* 79:1000
Broderick, J. J., Kellermann, K. I., Shaffer, D. B., Jauncey, D. L. 1972. *Ap. J.* 172:299
Burbidge, E. M. 1968. *Ap. J. Lett.* 152:L111
Burbidge, G. R., Jones, T. W., O'Dell, S. L. 1974. *Ap. J.* 193:43
Burbidge, G. R., Crowne, A. H., Smith, H. E. 1977. *Ap. J. Suppl.* 33:113
Cannon, R. D., Penston, M. V., Brett, R. A. 1971. *MNRAS* 152:79
Capps, R. W., Knacke, R. F. 1978. *Astrophys. Lett.* 19:113
Carswell, R. F., Strittmatter, P. A., Disney, M. J., Hoskins, D. G., Murdock, H. S. 1973. *Nature* 246:89
Carswell, R. F., Strittmatter, P. A., Williams, R. E., Kinman, T. D., Serkowski, K. 1974. *Ap. J. Lett.* 190:L101
Charles, P., Thorstensen, J., Bowyer, S. 1979. *Nature* 281:285
Cohen, M. H., Cannon, W., Purcell, G. H., Shaffer, D. B., Broderick, J. J., Kellermann, K. I., Jauncey, D. L. 1971. *Ap. J.* 170:207
Cohen, M. H., Pearson, T. J., Readhead, A. C. S., Seielstad, G. A., Simon, R. S., Walker, R. C. 1979. *Ap. J.* 231:293
Colgate, S. A., Petschek, A. G. 1978. *Pittsburgh Conference on BL Lac Objects*, ed. A. M. Wolfe, p. 345. Univ. Pittsburgh Press
Colla, G., Fanti, C., Fanti, R., Ficarra, A., Formiggini, L., Gandolfi, E., Grueff, G., Lari, C., Padrielli, L., Roffi, G., Tomasi, P., Vigotti, M. 1970. *Astron. Astrophys. Suppl.* 1:281
Colla, G., Fanti, C., Fanti, R., Ficarra, A., Formiggini, L., Gandolfi, E., Lari, C., Marano, B., Padrielli, L., Tomasi, P. 1972. *Astron. Astrophys. Suppl.* 7:1

Colla, G., Fanti, C., Fanti, R., Ficarra, A., Formiggini, L., Gandolfi, E., Gioia, I., Lari, C., Marano, B., Padrielli, L., Tomasi, P. 1973. *Astron. Astrophys. Suppl.* 11:291
Condon, J. J., Jauncey, D. L. 1974a. *Astron. J.* 79:437
Condon, J. J., Jauncey, D. L. 1974b. *Astron. J.* 79:1220
Condon, J. J., Dennison, B. 1978. *Ap. J.* 224:835
Condon, J. J. 1978. *Pittsburgh Conference on BL Lac Objects*, ed. A. M. Wolfe, p. 21. Univ. Pittsburgh Press
Condon, J. J., Ledden, J. E., O'Dell, S. L., Dennison, B. 1979. *Astron. J.* 84:1
Conway, R. G., Stannard, D. 1972. *MNRAS* 160:31P
Conway, R. G., Haves, P., Kronberg, P. P., Stannard, D., Vallee, J. P., Wardle, J. F. C. 1974. *MNRAS* 168:137
Cooke, B. A., Ricketts, M. J., Maccacaro, T., Pye, J. P., Elvis, M., Watson, M. G., Griffiths, R. E., Pounds, K. A., McHardy, I., Maccagni, D., Seward, F. D., Page, C. G., Turner, J. J. L. 1978. *MNRAS* 182:489
Cotton, W. D., Counselman, C. C., Geller, R. B., Shapiro, I. I., Wittels, J. J., Hinteregger, H. F., Knight, C. A., Rogers, A. E. E., Whitney, A. R., Clark, T. A. 1979. *Ap. J. Lett.* 229:L115
Craine, E. R., Duerr, R., Tapia, S. 1978. *Pittsburgh Conference on BL Lac Objects*, ed. A. M. Wolfe, p. 99. Univ. Pittsburgh Press
Danziger, I. J., Fosbury, R. A. E., Goss, W. M., Ekers, R. D. 1979. *MNRAS* 188:415
Davidson, K., Netzer, H. 1979. *Rev. Mod. Phys.* 51:715
Davis, R. J., Stannard, D., Conway, R. G. 1977. *Nature* 267:596
deBruyn, A. G., Willis, A. G. 1974. *Astron. Astrophys.* 33:351
Dent, W. A. 1965. *Science* 148:1458
de Vaucouleurs, G., Nieto, J.-L. 1979. *Bull. Am. Astron. Soc.* 10:629
Dibaj, E. A., Shakhovskoy, N. M. 1966. *Astron. Tsirk.* 375:1
Dombrovsky, V. A., Hagen-Thorn, V. A. 1968. *Astrophysics* 4:163
Dombrovsky, V. A., Babadzhanyants, M. K., Hagen-Thorn, V. A., Houtkevich, S. M. 1971. *Astrophysics* 7:246
Dower, R. G., Griffiths, R. E., Bradt, H. V., Doxsey, R. E., Johnston, M. D. 1980. *Ap. J.* 235:355
Dyck, H. M., Kinman, T. D., Lockwood, G. W., Landolt, A. U. 1971. *Nature* 234:71
Eachus, L. J., Liller, W. 1975. *Ap. J. Lett.* 200:L61
Eardley, D. M., Lightman, A. P., Payne, D. G., Shapiro, S. L. 1978. *Ap. J.* 224:53

Efimov, Y. S., Shakhovskoy, N. M. 1972. *Izv. Krymskoj Astrofiz. Obs.* 46:3
Ekers, J. A. 1969. *Aust. J. Phys. Suppl.* 7:1
Elliot, J. L., Shapiro, S. L. 1974. *Ap. J. Lett.* 192:L3
Elsmore, B., Mackay, C. D. 1969. *MNRAS* 146:361
Elvius, A. 1968. *Lowell Obs. Bull. No. 142* 7:5
Elvius, A. 1972. *Interstellar Dust and Related Topics*, ed. J. M. Greenberg, H. C. van de Hulst. Dordrecht: Reidel
Elvius, A. 1978. *Astron. Astrophys.* 65:233
Fomalont, E. B. 1967. *Ap. J. Suppl.* 15:203
Fomalont, E. B., Moffet, A. T. 1971. *Astron. J.* 76:5
Gardner, F. F., Whiteoak, J. B., Morris, D. 1975. *Aust. J. Phys. Suppl.* 35:1
Gaskell, C. M. 1978. *Bull. Am. Astron. Soc.* 10:662
Geller, M. J., Turner, E. L., Bruno, M. S. 1979. *Ap. J. Lett.* 230:L141
Goldsmith, D. W., Kinman, T. D. 1965. *Ap. J.* 142:1693
Gottlieb, E. W., Liller, W. 1976. *IAU Circ. No. 2939*
Gower, J. F. R., Scott, P. F., Wills, D. 1967. *Mem. R. Astron. Soc.* 71:49
Grandi, S. A., Tifft, W. G. 1974. *Publ. Astron. Soc. Pac.* 86:873
Griffiths, R. E., Tapia, S., Briel, U., Chaisson, L. 1979. Preprint
Hagen-Thorn, V. A. 1972. *Astron. Tsirk.* 714:5
Hagen-Thorn, V. A. 1974. *Astrophysics* 10:79
Harris, A. 1972. *MNRAS* 158:1
Hazard, C., Gulkis, S., Sutton, J. 1968. *Ap. J.* 154:413
Heeschen, D. S. 1973. *Ap. J. Lett.* 179:L93
Hiltner, W. A. 1959. *Ap. J.* 130:340
Hine, R. G., Scheuer, P. A. G. 1980. Preprint
Hobbs, R. W., Dent, W. A. 1977. *Astron. J.* 82:257
Hoyle, F., Burbidge, G. R., Sargent, W. L. W. 1966. *Nature* 209:751
Jauncey, D. L., Bare, C. C., Clark, B. G., Kellermann, K. I., Cohen, M. H. 1970. *Ap. J.* 160:337
Jones, T. W., O'Dell, S. L., Stein, W. A. 1974. *Ap. J.* 192:261
Jones, T. W., Leung, C. M., Gould, R. J., Stein, W. A. 1977. *Ap. J.* 212:52
Jones, T. W., O'Dell, S. L. 1977a. *Ap. J.* 215:236
Jones, T. W., O'Dell, S. L. 1977b. *Astron. Astrophys.* 61:291
Jones, T. W., Rudnick, L. 1980. Private communication
Katz, J. I. 1976. *Ap. J.* 206:910
Kellermann, K. I., Pauliny-Toth, I. I. K. 1968. *Ann. Rev. Astron. Astrophys.* 6:417
Kellermann, K. I., Pauliny-Toth, I. I. K. 1969. *Ap. J. Lett.* 155:L71
Kellermann, K. I., Pauliny-Toth, I. I. K., Williams, P. J. S. 1969. *Ap. J.* 157:1
Kellermann, K. I., Clark, B. G., Jauncey, D. L., Cohen, M. H., Shaffer, D. B., Moffet, A. T., Gulkis, S. 1970. *Ap. J.* 161:803
Kellermann, K. I., Pauliny-Toth, I. I. K. 1971. *Astrophys. Lett.* 8:153
Kellermann, K. I., Jauncey, D. L., Cohen, M. H., Shaffer, B. B., Clark, B. G., Broderick, J., Rönnäng, B., Rydbeck, O. E. H., Matveyenko, L., Moiseyev, I., Vitkevich, V. V., Cooper, B. F. C., Batchelor, R. 1971. *Ap. J.* 169:1
Kellermann, K. I., Shaffer, D. B., Purcell, G. H., Pauliny-Toth, I. I. K., Preuss, E. Witzel, A., Graham, D., Schilizzi, R. T., Cohen, M. H., Moffet, A. T., Romney, J. D., Niell, A. E. 1977. *Ap. J.* 211:658
Kellermann, K. I. 1978. *Phys. Scripta* 17:257
Kemp, J. C., Wolstencroft, R. D., Swedlund, J. B. 1972. *Ap. J. Lett.* 173:L113
Kemp, J. C., Rieke, G. H., Lebofsky, M. J., Coyne, G. V. 1977. *Ap. J. Lett.* 215:L107
Khachikian, E. Y., Weedman, D. W. 1974. *Ap. J. Lett.* 189:L99
Kikuchi, S., Mikami, Y., Konno, M., Inoue, M. 1976. *Publ. Astron. Soc. Jpn.* 28:117
Kinman, T. D., Lamla, E., Wirtanen, C. A. 1966. *Ap. J.* 146:964
Kinman, T. D. 1967. *Ap. J. Lett.* 148:L53
Kinman, T. D., Lamla, E., Ciurla, T., Harlan, E., Wirtanen, C. A. 1968. *Ap. J.* 152:357
Kinman, T. D., Conklin, E. K. 1971. *Astrophys. Lett.* 9:147
Kinman, T. D. 1973. *Ap. J. Lett.* 179:L97
Kinman, T. D., Wardle, J. F. C., Conklin, E. K., Andrew, B. H., Harvey, G. A., Macleod, J. M., Medd, W. J. 1974. *Ap. J.* 79:349
Kinman, T. D. 1976. *Ap. J.* 205:1
Kinman, T. D. 1977. *Nature* 267:798
Kinzer, R. L. 1978. *Bull. Am. Astron. Soc.* 10:503
Knacke, R. F., Capps, R. W. 1974. *Ap. J. Lett.* 192:L19
Knacke, R. F., Capps, R. W., Johns, M. 1976. *Ap. J. Lett.* 210:L69
Knacke, R. F., Capps, R. W., Johns, M. 1979. *Nature* 280:215
Kojoian, G., Stramek, R. A., Dickinson, D. F., Tovmassian, H., Purton, C. R. 1976. *Ap. J.* 203:323
Korchakov, A. A., Syrovatskii, S. I. 1962. *Sov. Astron.-AJ* 5:678
Kruszewski, A. 1968. *Astron. J.* 73:852
Kruszewski, A. 1977. *Acta Astron.* 27:319
Landstreet, J. D., Angel, J. R. P. 1972. *Ap. J. Lett.* 174:L127
Lebofsky, M. J., Rieke, G. H., Kemp, J. C. 1978. *Ap. J.* 222:95
Liller, W. 1969. *Ap. J.* 155:1113
Lovelace, R. V. E. 1976. *Nature* 262:649

Lynden-Bell, D., Pringle, J. E. 1974. *MNRAS* 168:603
Lynds, C. R. 1967. *Ap. J.* 147:396
MacDonald, G. H., Kenderdine, S., Neville, A. C. 1968. *MNRAS* 138:259
Mackay, C. D. 1969. *MNRAS* 145:31
Mackay, C. D. 1971. *MNRAS* 154:209
Macleod, J. M., Andrew, B. H. 1968. *Astrophys. Lett.* 1:243
Marscher, A. P. 1980. *Ap. J.* 235:386
Marshall, F. E., Mushotzky, R. F., Boldt, E. A., Holt, S. S., Rothschild, R. E., Serlemitsos, P. J. 1978. *Nature* 275:624
Martin, P. G., Angel, J. R. P., Maza, J. 1976. *Ap. J. Lett.* 209:L21
Martin, P. G. 1979. *Cosmic Dust*. Oxford: Clarendon
Maza, J., Martin, P. G., Angel, J. R. P. 1978. *Ap. J.* 224:368
Maza, J. 1979. *Polarization of Seyfert Galaxies and Related Objects*. PhD. thesis. Univ. Toronto, Toronto
Maza, J., Martin, P. G., Angel, J. R. P. 1980. In preparation
McGimsey, B. Q., Smith, A. G., Scott, R. L., Leacock, R. J., Edwards, P. L., Hackney, R. L., Hackney, K. R. 1975. *Ap. J.* 80:895
Miley, G. K., Perola, G. C. 1975. *Astron. Astrophys.* 45:223
Miley, G. K. 1980. *Ann. Rev. Astron. Astrophys.* 18:165
Miller, J. S. 1975. *Ap. J. Lett.* 200:L55
Miller, J. S. 1978. *Comments Astrophys.* 7:175
Miller, J. S., French, H. B. 1978. *Pittsburgh Conference on BL Lac Objects*, ed. A. M. Wolfe, p. 228. Univ. Pittsburgh Press
Miller, J. S., French, H. B., Hawley, S. A. 1978. *Pittsburgh Conference on BL Lac Objects*, ed. A. M. Wolfe, p. 176. Univ. Pittsburgh Press
Mills, B. Y., Slee, O. B., Hill, E. R. 1960. *Aust. J. Phys.* 13:676
Mills, B. Y., Little, A. G. 1970. *Astrophys. Lett.* 6:197
Moore, R. L., Angel, J. R. P., Rieke, G. H., Lebofsky, M. J., Wisniewski, W. Z., Mufson, S. L., Vrba, F. J., Miller, H. R., McGimsey, B. Q., Williamson, R. M. 1980. *Ap. J.* 235:717
Moore, R. L., Stockman, H. S. 1980. *Ap. J.* In press
Moore, R. L., Stockman, H. S., Angel, J. R. P. 1980. In preparation
Munro, R. E. B. 1972. *Aust. J. Phys. Suppl.* 22:1
Mushotzky, R. F., Boldt, E. A., Holt, S. S., Pravdo, S. H., Serlemitsos, P. J., Swank, J. H., Rothschild, R. H. 1978. *Ap. J. Lett.* 226:L65
Mushotzky, R. F., Marshall, F. E., Boldt, E. A., Holt, S. S., Serlemitsos, P. J. 1980. *Ap. J.* 235:361

Neugebauer, G., Oke, J. B., Becklin, E. E., Matthews, K. 1979. *Ap. J.* 230:79
Noerdlinger, P. D. 1978. *Phys. Rev. Lett.* 41:135
Nordsieck, K. H. 1972. *Astrophys. Lett.* 12:69
Nordsieck, K. H. 1976. *Ap. J.* 209:653
O'Dell, S. L., Puschell, J. J., Stein, W. A., Owen, F., Porcas, R. W., Mufson, S., Moffett, T. J., Ulrich, M.-H. 1978. *Ap. J.* 224:22
Oke, J. B. 1966. *Ap. J.* 145:668
Oke, J. B., Sargent, W. L. W., Neugebauer, G., Becklin, E. E. 1967. *Ap. J. Lett.* 150:L173
Oke, J. B., Neugebauer, G., Becklin, E. E. 1970. *Ap. J.* 159:341
Osterbrock, D. E., Miller, J. S. 1975. *Ap. J.* 197:535
Osterbrock, D. E., Koski, A. T., Phillips, M. M. 1976. *Ap. J.* 206:898
Owen, F. N., Mufson, S. L. 1977. *Astron. J.* 82:776
Owen, F. N., Porcas, R. W., Mufson, S. L., Moffett, T. J. 1978. *Astron. J.* 83:685
Owen, F., Spangler, S. R., Cotton, W. 1980. Preprint
Pacini, F., Salvati, M. 1978. *Ap. J. Lett.* 225:L99
Pauliny-Toth, I. I. K., Kellermann, K. I. 1972. *Astron. J.* 77:797
Pauliny-Toth, I. I. K., Kellermann, K. I., Davis, M. M., Fomalont, E. B., Shaffer, D. B. 1972. *Astron. J.* 77:265
Pauliny-Toth, I. I. K., Witzel, A., Preuss, E., Kuhr, H., Kellermann, K. I., Fomalont, E. B., Davis, M. M. 1978. *Astron. J.* 83:451
Penston, M. V., Penston, M. J. 1973. *MNRAS* 162:109
Petschek, A. G., Colgate, S. A., Colvin, J. D. 1976. *Ap. J.* 209:356
Pica, A. J. 1977. *Astron. J.* 82:935
Pilkington, J. D. H., Scott, J. F. 1965. *Mem. R. Astron. Soc.* 69:183
Pollock, J. T., Pica, A. J., Smith, A. G., Leacock, R. J., Edwards, P. L., Scott, R. L. 1979. *Astron. J.* 84:1658
Porcas, R. W., Treverton, R. M., Wilkinson, A. 1974. *MNRAS* 167:41P
Puschell, J. J., Stein, W. A., Jones, T. W., Warner, J. W., Owen, F., Rudnick, L., Aller, H., Hodge, P. 1979. *Ap. J. Lett.* 227:L11
Puschell, J. J., Stein, W. A. 1980. Preprint
Puschell, J. J. 1980. Private communication
Readhead, A. C. S., Cohen, M. H., Blandford, R. D. 1978. *Nature* 272:131
Readhead, A. C. S., Cohen, M. H., Pearson, T. J., Wilkinson, P. N. 1978. *Nature* 276:768
Readhead, A. C. S., Pearson, T. J., Cohen, M. H., Ewing, M. S., Moffet, A. T. 1979. *Ap. J.* 231:299

Rees, M. J., Sciama, D. W. 1965. *Nature* 207:738
Rees, M. J., Simon, M. 1968. *Ap. J. Lett.* 152:L145
Rees, M. J. 1978a. *Phys. Scripta* 17:193
Rees, M. J. 1978b. *Nature* 275:516
Richstone, D., Schmidt, M. 1980. *Ap. J.* 235:361
Riegler, G. R., Agrawal, P. C., Mushotzky, R. F. 1979. Preprint
Rieke, G. H., Grasdalen, G. L., Kinman, T. D., Hinzen, P., Wills, B. J., Wills, D. 1976. *Nature* 260:754
Rieke, G. H., Lebofsky, M. J., Kemp, J. C., Coyne, G. V., Tapia, S. 1977. *Ap. J. Lett.* 218:L37
Rieke, G. H., Lebofsky, M. J. 1979. *Ann. Rev. Astron. Astrophys.* 17:477
Rudnick, L., Owens, F. N., Jones, T. W., Puschell, J. J., Stein, W. A. 1978. *Ap. J. Lett.* 225:L5
Sandage, A. 1967. *Ap. J. Lett.* 150:L177
Sargent, W. L. W., Young, P. J., Boksenberg, A., Shortridge, K., Lynds, C. R., Hartwick, F. D. A. 1978. *Ap. J.* 221:731
Scheuer, P. A. G., Readhead, A. C. S. 1979. *Nature* 277:182
Schmidt, G. D., Angel, J. R. P., Cromwell, R. H. 1976. *Ap. J.* 206:888
Schmidt, G. D., Peterson, B. M., Beaver, E. A. 1978. *Ap. J. Lett.* 220:L31
Schmidt, G. D., Miller, J. S. 1979. Preprint
Schmidt, M. 1974. *Ap. J.* 193:505
Schmidt, M. 1978. *Phys. Scripta* 17:135
Schmitt, J. L. 1968. *Nature* 218:663
Schwartz, D. A., Doxsey, R. E., Griffiths, R. E., Johnston, M. D., Schwarz, J. 1979. *Ap. J. Lett.* 229:L53
Sciama, D. W., Rees, M. J. 1967. *Nature* 216:147
Seilestad, G. A., Cohen, M. H., Linfield, R. P., Moffet, A. T., Romney, J. D., Schilizzi, R. T. 1979. *Ap. J.* 229:53
Serkowski, K., Tapia, S. 1975. *Bull. Am. Astron. Soc.* 7:499
Shaffer, D. B. 1978. *Pittsburgh Conference on BL Lac Objects*, ed. A. M. Wolfe, p. 68. Univ. Pittsburgh Press
Shields, G. A. 1978. *Nature* 272:706
Shimmins, A. J., Manchester, R. N., Harris, B. J. 1969. *Aust. J. Phys. Suppl.* 8:1
Shimmins, A. J., Bolton, J. G. 1972. *Aust. J. Phys. Suppl.* 23:1
Shimmins, A. J., Bolton, J. G. 1974. *Aust. J. Phys. Suppl.* 32:1
Shimmins, A. J., Bolton, J. G., Wall, J. V. 1974. *Aust. J. Phys. Suppl.* 34:63
Shklovsky, I. S. 1955. *IAU Symp. No. 4*, p. 205
Slee, O. B. 1977. *Aust. J. Phys. Suppl.* 43:1
Smith, H. E., Spinrad, H., Smith, E. O. 1976. *Publ. Astron. Soc. Pac.* 88:621
Smith, H. E., Burbidge, E. M., Baldwin, J. A., Tohline, J. E., Wampler, E. J., Hazard, C.,
Murdoch, H. S. 1977. *Ap. J.* 215:427
Smith, H. E. 1978. *Pittsburgh Conference on BL Lac Objects*, ed. A. M. Wolfe, p. 211. Univ. Pittsburgh Press
Stein, W. A., O'Dell, S. L., Strittmatter, P. A. 1976. *Ann. Rev. Astron. Astrophys.* 14:173
Stockman, H. S., Angel, J. R. P., Beaver, E. A. 1976. *Bull. Am. Astron. Soc.* 8:495
Stockman, H. S., Angel, J. R. P. 1978. *Ap. J. Lett.* 220:L67
Stockman, H. S., Angel, J. R. P., Hier, R. 1978. *Bull. Am. Astron. Soc.* 10:689
Stockman, H. S. 1978. *Pittsburgh Conference on BL Lac Objects*, ed. A. M. Wolfe, p. 149. Univ. Pittsburgh Press
Stockman, H. S., Angel, J. R. P., Miley, G. K. 1979. *Ap. J. Lett.* 227:L55
Strittmatter, P. A., Serkowski, K., Carswell, R., Stein, W. A., Merrill, K. M., Burbidge, E. M. 1972. *Ap. J. Lett.* 175:L7
Strittmatter, P. A., Carswell, R. F., Gilbert, G., Burbidge, E. M. 1974. *Ap. J.* 190:509
Sulentic, J. W., Arp, H., Lorre, J. 1979. *Ap. J.* 233:44
Sutton, J. M., Davies, I. M., Little, A. G., Murdoch, H. S. 1974. *Aust. J. Phys. Suppl.* 33:1
Tananbaum, H., Avni, Y., Branduardi, G., Elvis, M., Fabbiano, G., Feigelson, E., Giacconi, R., Henry, J. P., Pye, J. P., Soltan, A., Zamorani, G. 1980. *Ap. J. Lett.* 234:L9
Tapia, S., Craine, E. R., Johnson, K. 1976. *Ap. J.* 203:291
Tapia, S., Craine, E. R., Gearhart, M. R., Pacht, E., Kraus, J. 1977. *Ap. J. Lett.* 215:L71
Thompson, I., Landstreet, J. D., Angel, J. R. P., Stockman, H. S., Woolf, N. J., Martin, P. G., Maza, J., Beaver, E. A. 1979. *Ap. J.* 229:909
Thompson, I., Landstreet, J. D., Stockman, H. S., Angel, J. R. P., Beaver, E. A. 1980. *MNRAS*. In press
Turland, B. D. 1975. *MNRAS* 170:281
Turnshek, D. A., Weymann, R. J., Liebert, J. W., Williams, R. E., Strittmatter, P. A. 1979. Preprint
Ulrich, M. H., Kinman, T. D., Lynds, C. R., Rieke, G. H., Ekers, R. D. 1975. *Ap. J.* 198:261
Ulrich, M. H. 1978. *Ap. J. Lett.* 222:L3
Usher, P. D., Kolpanen, D. R., Pollock, J. T. 1974. *Nature* 252:365
Usher, P. D. 1975. *Ap. J. Lett.* 198:L57
Visvanathan, N. 1967. *Ap. J. Lett.* 150:L149
Visvanathan, N. 1968. *Ap. J. Lett.* 153:L19
Visvanathan, N., Oke, J. B. 1968. *Ap. J. Lett.* 152:L165.
Visvanathan, N. 1969. *Ap. J. Lett.* 155:L133
Visvanathan, N. 1973a. *Ap. J.* 179:1
Visvanathan, N. 1973b. *Ap. J.* 185:145
Visvanathan, N. 1974. *Planets, Stars and*

Nebulae, ed. T. Gehrels. Tucson: Univ. Arizona Press
Walker, M. F. 1964. *Astron. J.* 69:682
Walker, M. F. 1968. *Ap. J.* 151:71
Wall, J. V., Shimmins, A. J., Merkelijn, J. K. 1971. *Aust. J. Phys. Suppl.* 19:1
Wall, J. V. 1972. *Aust. J. Phys. Suppl.* 24:1
Wall, J. V., Wright, A. E., Bolton, J. G. 1976. *Aust. J. Phys. Suppl.* 39:1
Wampler, E. J. 1967. *Ap. J.* 147:1
Wampler, E. J. 1971. *Ap. J.* 164:1
Wardle, J. F. C. 1971. *Astrophys. Lett.* 8:183
Wardle, J. F. C. 1977. *Nature* 269:563
Wardle, J. F. C. 1978. *Pittsburgh Conference on BL Lac Objects*, ed. A. M. Wolfe, p. 39. Univ. Pittsburgh Press
Weiler, K. W., Johnston, K. J. 1979. Preprint
Whiteoak, J. B. 1966. *Z. Astrophys.* 64:181
Williams, P. J. S., Kenderdine, S., Baldwin, J. E. 1966. *Mem. R. Astron. Soc.* 70:53
Williams, W. L., Rich, A., Kupferman, P. N., Ionson, J. A., Hiltner, W. A. 1972. *Ap. J. Lett.* 174:L63
Wills, D., Wills, B. J. 1974. *Ap. J.* 190:271
Wills, D., Wills, B. J. 1976. *Ap. J. Suppl.* 31:143
Witzel, A., Pauliny-Toth, I. I. K., Geldzahler, B. J., Kellermann, K. I. 1978. *Astron. J.* 83:475
Woltjer, L. 1966. *Ap. J. Lett.* 146:597
Young, P. J., Westphal, J. A., Kristian, J., Wilson, C. P., Landauer, F. P. 1978. *Ap. J.* 221:721
Zotov, N. V., Tapia, S. 1979. *Ap. J. Lett.* 229:L5
Zwicky, F. 1966. *Ap. J.* 143:192

WHITE DWARF STARS ✶2170

James Liebert

Steward Observatory, University of Arizona, Tucson, Arizona 85721

1 INTRODUCTION

Strange objects, which persist in showing a type of spectrum out of keeping with their luminosity, may ultimately teach us more than a host which radiate according to rule.

From the President's Address of Arthur S. Eddington (1922), on the occasion of the Centenary of the Royal Astronomical Society

The violated rule to which Professor Eddington referred was the mass-luminosity relation for dwarf stars. The "strange objects" had the exceedingly low luminosities of the faint red dwarfs, yet their colors were quite bluish. Thus, they came to be called *white dwarfs*. The high surface emissivities required very small radii—of order $10^{-2} R_\odot$—compared with known stars. Yet, at least the companion star to Sirius was known from the orbital solution to have a mass of $\sim 1 M_\odot$. This implied interior densities and surface gravities orders of magnitude higher than for dwarf stars. The physical state of such superdense matter could not be understood until the quantum-statistical theory of the electron gas was worked out by E. Fermi and P. Dirac in the mid-1920's. R. H. Fowler then showed that the enhanced pressure of a degenerate electron gas could support an object of stellar mass against its own self-gravitation at precisely the radii of white dwarfs. Chandrasekhar (1939) modeled the basic interior structure and determined the basic mass-radius-density-composition relationships. Most importantly, Chandrasekhar established a critical mass ($\sim 1.44 M_\odot$) above which stable degenerate configurations cannot exist.

In the wake of their successes, theoreticians called the configurations *degenerate stars* or *degenerate dwarfs*, though observers still used *white dwarfs*. The stars are certainly not main-sequence *dwarfs*, and most are too cool to properly be called *white*. However, the term *degenerate stars* encompasses neutron stars as well. Since all of the above terms are in widespread use, it is futile to argue which of them is the biggest misnomer; we shall use them interchangeably in this discussion.

To this writer's knowledge, there has not been an extensive review of the general properties of single degenerate dwarfs written in the last several years. For exploration of problems in greater depth than is possible here, however, the reader is referred to various specialized reviews and contributions found in the *Proceedings of IAU Colloquium No. 53* on *White Dwarfs and Variable Degenerate Stars*. Earlier comprehensive reviews include those of Weidemann (1968, 1975) and Ostriker (1971, 1972); Bessell (1978) and Van Horn (1979) have provided shorter summaries. Degenerate stars in cataclysmic binary systems are reviewed by Robinson (1976) and Gallagher & Starrfield (1978). For an assessment of the probable role of accreting white dwarfs in higher luminosity X-ray sources, Kylafis et al. (1980) is a useful starting point.

2 OBSERVATIONAL CONSTRAINTS

2.1 *Methods of Discovery and Selection Effects*

White dwarfs have generally been identified in three ways: (*a*) as faint stars in proper motion catalogues, especially the bluer objects; (*b*) in surveys for faint blue stars, generally at high galactic latitudes; (*c*) as faint companions to brighter proper motion or parallax stars. Slit spectroscopy and/or photometric color measurements have generally been required to demonstrate that a given candidate is a degenerate star. Until recently, the majority of established degenerates were culled from the lists of proper motion surveys of W. J. Luyten and the Lowell Observatory (H. Giclas, N. Thomas, and associates). Luyten (1970, 1977) has published extensive lists of likely white dwarfs from the Luyten Palomar and his earlier proper motion surveys. Now the completion of Green et al.'s (1980) sky survey of blue stellar objects, all with spectroscopic classifications, has more than doubled the sample of spectroscopically confirmed degenerate stars. A catalogue of spectroscopically identified white dwarfs has been published by McCook & Sion (1977).

The discovery techniques have inevitably resulted in biasing the known sample in favor of (*a*) hotter stars, (*b*) those with above-average space motions, (*c*) stars with larger than average radii, and presumably lower than average masses and gravities (Shipman 1979a), and (*d*) stars with brighter, common-proper-motion (but well-separated) companions. The search for cool degenerates has also been frustrated by the relatively lower luminosities and effective search volumes for these stars and the prevalence of late-type subdwarfs among the redder stars of high proper motion. Little velocity information is yet available for the stars in Green's sample. However, most do not appear in the proper motion lists, including the Lowell "GD" lists of bluish objects showing only slight motion. Green's discoveries thus should include the low-velocity tail of the distribution of hot white dwarfs, a sample that is missing for the cooler stars.

2.2 White Dwarfs in Binaries and Clusters

While the presence of a distant, bright motion companion has resulted in the discovery of many degenerate stars in the known sample, the striking examples of Sirius B and Procyon B next to us in the sky alert us to the fact that most close degenerate companions could go undetected. Since the hottest degenerates stand the best chance of being found in the glare of brighter but redder companions, most of the known close-binary degenerates are hot stars. Of necessity, the studies of the luminosity function and stellar properties of white dwarfs must address the non-close binary. stars only. This is desirable also because close binary evolution will lead to different stellar parameters than for single stars.

However, the desired decoupling of the close binaries from the isolated stars is difficult to achieve. For example, G5-28 was first classified as degenerate M (DM), in part because it had an ultraviolet excess relative to other dwarf M stars. Yet, it is probably a composite dM plus cool white dwarf (Harrington et al. 1975, Liebert 1975). Likewise G107-70 seemed to be a very cool DC degenerate, but the accurate trigonometric parallax placed it 0.75 magnitude brighter than the mean for several other stars with similar photometric color. Then Mrs. B. Riepe noticed that the image was elongated on many Naval Observatory parallax plates; the star was shown to be two barely separated DC white dwarfs (Strand, Dahn & Liebert 1976).

One wonders how many peculiar spectra or atypical parameters found for white dwarfs might be explainable by hidden duplicity. In Section 3.4 we must consider seriously the possibility that some helium atmosphere white dwarfs have very low masses and are products of binary cannibalism. Greenstein (1979a) noted an ubiquitous pairing of very hot white dwarfs with companion stars. The unexplained X-ray emissions from Sirius B (Section 3.1) and the ultraviolet abundance anomalies in V471 Tauri (Guinan & Sion 1979) may yet owe their existence to some kind of particle flow in detached systems.

White dwarfs have been found in several open clusters, including some with turnoff masses $\gtrsim 5\ M_\odot$. The implications of these for pre–white dwarf mass loss and progenitor masses are discussed in Section 4.2. Richer (1978) has claimed the probable discovery of white dwarfs in the southern globular cluster NGC 6752. Richer (1979) has obtained low dispersion spectra showing that at least one of his candidates has a photometric parallax and Balmer line strength consistent with its identification as a DA white dwarf at the cluster distance. Clearly the resolution and limiting magnitudes offered by the space telescope may permit the discovery of large numbers of globular cluster white dwarfs.

2.3 Colors, Spectral Types, and Kinematic Properties

White dwarfs have been discovered ranging from the colors of O stars through K stars; one or two with early M colors have recently been found. These divide spectroscopically into 1. the "DA" sequence with hydrogen lines, covering nearly the full range in color, and 2. a group of non-DA or generally helium-rich types, also covering the full temperature range of white dwarfs. In Figure 1, we group these by temperature and approximate helium-to-hydrogen atmospheric composition, according to the analyses discussed in Section 3.

The spectra of white dwarfs generally appear monoelemental. The definitions of the basic types appear in Greenstein (1960). However, the recent discovery of some showing both helium and hydrogen lines has complicated the already unwieldy historical system of spectral types. Traditionally, the DA, F designation has been used for stars showing strong Balmer lines and weak Ca II. Analogously, DBA has recently been used for DB stars showing a trace of hydrogen. We now must also introduce other combinations where the first letter after "D" designates the primary spectroscopic type, and the next letter a secondary type. Thus DAO describes a hot star showing primarily Balmer lines but with weak He II $\lambda4686$. For convenience we omit the comma between letters.

Notwithstanding the biases in the known sample, it is clear that most local white dwarfs are members of the old disk population. There are small admixtures of young disk and halo stars (Eggen & Greenstein 1967). Attempts to find kinematic differences among the various spectral groupings have been frustrating and inconclusive (cf Sion & Liebert 1977). One reason is the difficulties with obtaining and interpreting radial velocities

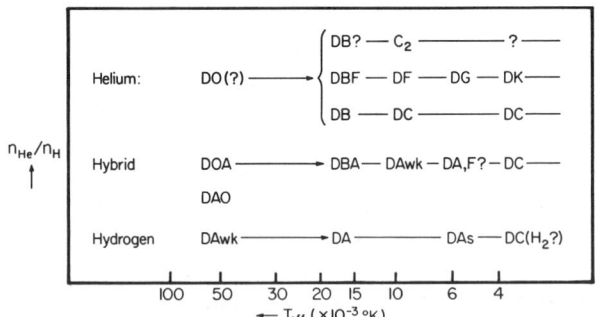

Figure 1 White dwarf spectroscopic $T_{\rm eff}$—composition sequences. Stars with an actual $n_{\rm He}/n_{\rm H}$ determinations are plotted in a similar diagram in Liebert et al. (1979c). The basic spectral types were defined in Greenstein (1960); hybrid types used here have the lead symbol (after D) representing the dominant atmospheric constituent (He or H), and the last symbol the secondary constituent appearing in the spectrum.

WHITE DWARF STARS 367

(see Section 3.2), which has largely limited the exercise to examining distributions of tangential velocities. Trimble & Greenstein (1972) and Wegner (1974) have studied the total space motions of the mostly-DA objects for which they measured radial velocities. A second limitation is that the total stellar ages of white dwarfs may blur any kinematic distinctions of their progenitors. Finally, the mean space motions of white dwarfs may increase with their cooling ages—aside from the selection effects—so that comparison must be made using samples having similar temperature distributions. There is currently no convincing evidence for 1. any significant difference between the kinematic properties of hot DA and non-DA stars, or that 2, the relative numbers of DA and non-DA depend on kinematic properties.

Most authors have assumed that the subgroup of non-DA's showing C_2 bands do come predominantly from a halo population, and hence could be the exceptions to the above statements. However, the previous use of UBV colors to estimate absolute magnitudes apparently resulted in the distances and velocities of several C_2 stars being greatly overestimated. The nearby C_2 stars having reasonable trigonometric parallaxes show a mean tangential velocity not significantly different from other groupings of cool white dwarfs (Humphreys et al. 1979; see also Sion, Fragola & O'Donnell 1979).

3 STELLAR AND ATMOSPHERIC PARAMETERS

3.1 *Astrometric Mass Determinations and Radii*

Mass estimates obtained through atmospheric analyses of colors, energy distributions, and line profiles (i.e. through gravity and temperature determinations) are only of statistical value; the reasonably accurate mass determinations for individual white dwarfs are those obtained from binary orbit solutions. Two of these (Sirius B and Procyon B) are in close binary systems with massive companions; the others are in more distant, lower mass systems, and hence are likely to be more useful for comparison with field white dwarfs.

SIRIUS B The most recent mass determination is 1.053 ± 0.028 $M_\odot$ (Gatewood & Gatewood 1978). Its proximity to the brightest star in the sky has complicated the determination of its temperature and radius. The Balmer line profiles and detected soft X-ray flux are consistent with the Greenstein, Oke & Shipman (1971) values of 32,000 K and ~ 0.0078 $R_\odot$. However, the recent far ultraviolet flux measurements and limits on extreme ultraviolet (EUV) radiation suggest $T_{\text{eff}} \simeq 27,000$ K and 0.009 ± 0.002 $R_\odot$ (Savedoff et al. 1976; see also Cash et al. 1978, Koester 1979a, Böhm-Vitense et al. 1979). The soft X-ray flux, localized primarily to the "B" component by HEAO-2 imagery (Giacconi 1979), must apparently be attributed to nonthermal processes such as a corona, for which a convincing theoretical justification is lacking (Section 4.3).

PROCYON B   The mass of 0.63 $M_\odot$ may be well determined, but the temperature, radius, and even the spectral type are uncertain due to the proximity of the companion.

40 ERI B   Heintz' (1974) astrometric solution for the BC pair yielded 0.43 ±0.02 $M_\odot$ for the white dwarf. This value seems preferable to the 0.372 $M_\odot$ advocated by Moffett, Barnes & Evans (1978), who forced the "C" component to fit a mass-luminosity relationship for late M dwarfs. Shipman (1979a) fits $T_{\text{eff}} = 16{,}900$ and 0.0124 ± 0.0005 $R_\odot$ for 40 Eri B, which is in good agreement with the analysis of Koester, Schulz & Weidemann (1979); the new radius determination is nearly 20% lower than the original Matsushima & Terashita (1969) value.

STEIN 2051B   This white dwarf is of great potential interest because it is a cooler, helium-atmosphere star. However, the astrometric mass is uncertain and remains "double-valued." Strand (1977) now argues that the "A" component is itself double. On the basis of the mass ratio determined between the red dwarf "A" system and the white dwarf "B", the derived mass for the latter is either 0.50 $M_\odot$ or 0.72 $M_\odot$, depending on whether "A" includes a low mass (0.02 $M_\odot$) companion or two components of nearly equal masses. The temperature has been estimated by Liebert (1976) as 7050 ±400 K, but the radius determination is much less certain (see Section 3.4) than for the DA stars.

G107-70AB   This barely resolved cool white dwarf pair (Section 2.2) now has a preliminary orbit determination by Harrington et al. (1980); this yields a total mass of 0.92 ± 0.22 $M_\odot$ for both stars, implying a low average mass.

MASS-RADIUS RELATION   The current "best" estimates for 40 Eri B and Sirius B are not in serious disagreement with the predictions of Hamada-Salpeter (1961) models for interior compositions lighter than iron. For Sirius B, the match to theory requires a radius (~ 0.007 $R_\odot$) at the low end of the range allowed by the ultraviolet data. If the mass were significantly below 0.4 $M_\odot$ for 40 Eri B, the recent Shipman radius estimate becomes a better match to the curve for iron than for lighter elements.

3.2   *Temperatures, Radii, and Gravities of DA Stars*

Several major observational and theoretical projects have recently improved our understanding of the distributions of radii, gravities, and masses among hydrogen-atmosphere degenerates. Larger problems remain with the interpretation of DA stars near the extremes in temperature.

The temperature of a DA star can be rather precisely estimated from careful measurements of the Balmer jump and the Paschen continuum. A basic temperture scale for DA stars is fairly well established for the range 30,000 K $\gtrsim T_{\rm eff} \gtrsim$ 7000 K. Greenstein & Oke's (1979) IUE ultraviolet scans for DA stars are generally consistent with the optically determined scale. Multichannel energy distributions by Greenstein and Strömgren colors have been widely used in recent analyses by Greenstein (1976, 1979c), Shipman (1979a), Koester, Schulz & Weidemann (1979 = KSW), Bessell & Wickramasinghe (1978), and Wegner (1979b). The Strömgren colors offer great photometric accuracy with telescopes smaller than the Hale 5-m, but systematic differences among observers' filter sets and techniques have led to some divergence in the Strömgren results. Since there is some contamination of the Strömgren b bandpass by the Hβ wing in strong-lined stars, Tapia's (1978) color system—originally proposed by Wickramasinghe and Strittmatter—offers some advantages in measuring the Paschen continuum slope. However, these colors have not yet been compared with recent model atmospheres (e.g. by Shipman or the Kiel group).

Radii may be directly estimated for the several dozen DA stars with reasonable distance estimates. While this may be done after determining $T_{\rm eff}$ and a bolometric correction, Shipman (1972), KSW, and Shipman (1979a), for example, compute the radius directly from the surface brightness method (Gray 1967), an approach that permits more explicit error analysis. The uncertainty in the radius determination for any object is usually dominated by the uncertainty in its parallax; for the small sample with distance determinations, it is possible to assess the mean radius for the sample, and the dispersion of radii independent of the surface gravity determination. KSW found a mean radius of 0.0120 $R_\odot$ (± 0.002) for 64 parallax DA stars; Shipman reported 0.01237 $R_\odot$ for his parallax sample. However, Shipman argued that the available stars may be enough like an apparent-magnitude-limited sample that we are biased towards discovering those with larger than average luminosities and radii. He suggests that the true mean is closer to 0.0111 $R_\odot$, implying also a larger mean mass.

Surface gravities affect the strengths and profiles of the Stark-broadened Balmer lines, especially above ~12,000 K. However, the wings are so extensive in hot DA stars that an entire theoretical synthetic spectrum should be fit to the energy distribution to avoid errors in defining the true continuum. Accurate, high resolution spectrophotometry is now available for a large sample of DA stars (Greenstein & Vauclair 1979, Greenstein 1979c, Weidemann & Koester 1980). Previously most efforts to derive surface gravities from atmospheric analyses used colors only. The Balmer jump has maximum gravity sensitivity in the range 10,000–15,000 K, but uncertainties in the model energy distributions due to treatment of convection are important in most of this region. Photometric and multichannel flux calibration errors are also a major problem and lead to systematic uncertainties probably exceeding 0.2 in the mean $\overline{\log g}$ for DA stars determined in this way. Observational errors, and discrepancies among different Strömgren observers, are a major uncertainty in the determination of the dispersion of surface gravities as well. The corresponding multichannel results from KSW, Greenstein (1979c), and Shipman & Sass (1980) agree tolerably well on $\overline{\log g} \sim 8.0 \pm 0.2$. However, KSW favor a smaller intrinsic dispersion than Shipman, leading to a disagreement on the implied mass dispersion of DA stars.

Surface gravities for DA stars have been independently determined from measurements of the redshifts of the Balmer lines (Greenstein & Trimble 1967, Trimble & Greenstein 1972, Wegner 1974); the agreement in $\overline{\log g}$ between the two methods has been less than satisfying. The problem of extracting the gravitational effect from the space velocity and pressure shifts is a nontrivial one and the line centers are ill-defined on early photographic spectra. Only a dozen or so DA stars have independent radial velocities from companions. However, higher resolution coudé spectra (Greenstein & Peterson 1973, Greenstein et al. 1977) revealed that DA stars have remarkably sharp non-LTE cores. This permitted line centers for the brighter stars to be more accurately measured. Red-sensitive image tube photographic plates covering Hα were later compared with Boksenberg spectrophotometry of Hβ and Hγ lines. These results and those of Wegner (1974) led Greenstein et al. to a mean gravitational redshift for DA stars of 45 ± 6 km s^{-1}. This implies $\overline{\log g} \simeq 8.25$, appreciably higher than the result obtained from the energy distributions. With the uncertainties discussed previously in the atmospheric $\overline{\log g}$ value, the difference should perhaps be regarded as only marginally significant. The discrepancy propagates into the derived mean masses for DA stars as well.

Masses and mass dispersions may then be indirectly determined from either the surface gravity or radius results, using the theoretical mass-radius or equivalent mass-gravity relation. As discussed in Section 4.1, it is generally assumed that the interior mean molecular weight in this relation is that for carbon, oxygen, or elements of similar weight; the composition dependence is small unless the core were composed of iron. KSW found mean masses of $0.46 M_\odot$ and $0.63 M_\odot$ computed from the theoretical mass-radius and mass-gravity relations of Hamada & Salpeter (1961), respectively. However, the mass required to fit the mean gravitational redshift estimate ($+45$ km s^{-1}) is significantly higher still at $\simeq 0.75 M_\odot$. Moffett et al. (1978) also recently emphasized the discrepancy between the gravitational and atmospheric results. Shipman's (1979a) selection effect could mean that the true mean mass is higher than that for the observed samples, but the selection should apply to the gravitational redshift sample as well. The dispersion in the mass distribution is also controversial. KSW conclude that a standard deviation of $\pm 0.10\ M_\odot$ is appropriate, relying mostly on their "weighted" sample of surface gravity determinations, though values higher than $0.15\ M_\odot$ were found for some of their data subsets. KSW placed most emphasis on the narrow "log g" dispersion for the modest sample of stars observed by Graham (1972) and Green (1977) in the most gravity-sensitive 10,000–15,000 K region. Accurate observations of more DA stars in this region are highly desirable, since other methods suggest a 1 σ mass dispersion nearer to ± 0.2 (Shipman & Sass 1980, Wegner 1979b, Bessell & Wickramasinghe 1978). Shipman relies heavily on his large dispersion of radii, $\sigma(R)/R \sim 0.23$, which he argues is three times the expected observational uncertainty. This transforms to $\sigma(M)/M \gtrsim 0.20$. In any case, none of the authors find convincing evidence for DA stars with masses well above $1.0\ M_\odot$ or below $0.4\ M_\odot$.

THE HOTTEST DA STARS   Temperature determinations for the hottest DA stars ($> 40,000$ K) require flux measurements nearer to the intensity maxima in the far ultraviolet (~ 1000 Å) or even the EUV (100–1000 Å). HZ43 and Feige 24 were among the first extrasolar EUV objects detected (Lampton et al. 1976, Margon et al. 1976); the former was identified as a soft X-ray source (Hearn et al. 1976). Recent model atmosphere analyses by Auer & Shipman (1977) and Wesselius & Koester (1978) indicate temperatures of 50,000–60,000 K for these objects. Because of the uncertainties, there is little useful information concerning the radii or masses of the hottest hydrogen stars. Shipman (1979b) outlines the particular pitfalls of calculating models for degenerates at high effective temperatures.

THE COOLEST DA STARS   Hydrogen lines have been seen in white dwarfs at temperatures as low as ~ 5500 K—see the analysis of BPM 4729 by Wickramasinghe & Bessell (1979). The onset of convection and other problems greatly increase the uncertainties in the $T_{\rm eff}$ and radii determinations below $\sim 10,000$ K. Furthermore, stars below 8,000 K are too cool for either the Balmer jump or Balmer profiles to serve as useful surface gravity indicators, independent of abundance effects (e.g. Shipman 1972, Wehrse 1977). This is most unfortunate, given that the mean parameters might be expected to change if some DA atmospheres are mixed with helium-rich envelopes during the cooling process (Section 4.3). There is the suggestion in the analysis of G128-7 ($T_{\rm eff} \sim 5800$ K) by Wehrse & Liebert

(1980a) that this cool object may have a higher than normal surface gravity. This tentative result was possible because the Hα profile fits a $T(\tau)$ atmospheric run indicative of the onset of H_2 molecule formation, which *is* gravity sensitive. Furthermore, there is the possibility of identifying cooler degenerates with hydrogen in their atmospheres through the detection of the pressure-induced H_2 dipole bands in the infrared, as attempted by Mould & Liebert (1978). The onset of this opacity source is gravity sensitive, but the available opacity calculations are too crude to be of much use.

3.3 *Abundances of DA Stars*

HELIUM ABUNDANCES Most hydrogen-atmosphere degenerates are both helium-poor and metal-poor; where detailed limits have been possible, the atmospheres are shown to be composed of nearly pure hydrogen. The high energy fluxes measured for the hottest stars HZ43 and Feige 24 were used to determine $n(He)/n(H) \gtrsim 10^{-6}$ for the former and perhaps as high as 10^{-3} for the latter (Auer & Shipman 1977, Wesselius & Koester 1978, Malina 1979). However, there are hot "DAO" stars like HZ34 with detectable He II $\lambda 4686$; Koester, Liebert & Hege (1979) found $n(He)/n(H) \sim 10^{-1.7}$ for this star. These analyses assumed homogeneous atmospheric compositions, but one must also consider the possibility that the hot atmospheres are layered (see Section 4.3), as Heise & Huizenga (1980) first proposed for HZ43.

HZ34 has no known spectroscopic counterparts among cooler stars with hydrogen-dominated atmospheres. He I lines have not yet been found in DA stars above $\sim 12{,}000$ K; this permits limits of $< 10^{-2}$ for the best cases. Below the He I line detection limit, it is difficult to infer $n(He)/n(H)$ limits because the effect of enhanced helium on Balmer profiles and energy distributions is small and difficult to distinguish from surface gravity effects.

METAL ABUNDANCES Wehrse's (1972) synthetic spectrum for a white dwarf with solar composition and temperature dramatized how easy it should be to see neutral metal lines in a cool DA atmosphere. While the limits are not as stringent as for lower-opacity helium atmospheres, the absence of Ca II H & K at 5500 K yields a calcium upper limit of only $\sim 10^{-4}$ of solar (Wickramasinghe & Bessell 1979). Since Stark broadening becomes unimportant below ~ 7000 K, the hydrogen lines become considerably narrower than for hotter stars and one expects any metallic lines to be narrow as well. Hintzen & Strittmatter (1975) argued that the gravity insensitivity of these profiles might result in some alleged metal-rich DA degenerates being misclassified as subdwarfs. However, the detailed model analysis of Wickramasinghe et al. (1977) separates the surface gravities successfully using narrow-band colors and the profiles of high Balmer lines. Indeed, no cool hydrogen-rich degenerate showing metal lines other than

"K" of Ca II has ever been established. Now Bessell & Wickramasinghe (1979) maintain that most of the stars photographically classified DA, F are really either 1. cool subdwarfs, 2. DA degenerates, the calcium lines coming from the night sky spectrum, or 3. helium-atmosphere degenerates with some hydrogen and calcium. It is important to demonstrate whether such a well-observed degenerate DA,F as Ross 627 truly belongs to the last category. However, very few if any new DA,F classifications have been claimed since the advent of sky subtraction spectrophotometry.

The higher continuum opacities of hotter DA stars preclude such stringent limits on composition from the absence of heavy element lines in these stars. However, at the highest temperatures, trace abundances of metal could be very influential on the EUV and far ultraviolet energy distributions. More accurate high energy flux distributions may permit a test of how rapid the onset of gravitational and thermal diffusion is in a hydrogen-atmosphere degenerate beginning its cooling evolution.

3.4 Temperatures, Radii, and Gravities for Helium-Atmosphere Stars

The atmospheric and stellar parameters for helium-atmosphere degenerates are not nearly so well determined as for the DA stars, and for a number of reasons. There are greater uncertainties in the continuous helium opacities and in the line broadening for the various He I transitions. Atmospheric convection is generally important. For the cooler stars in particular, the absence of electrons in the metal-poor atmospheres leads to very low He$^-$ continuous opacities, very high atmospheric transparencies and correspondingly high atmospheric pressures and densities; this in turn leads to greater uncertainties for cool helium stars due to the onset of non-ideal gas effects, and a dependence on unknown trace abundances of unseen metallic electron donors. Thus, the model atmosphere calculations for non-DA stars have not been as comprehensive as for DA stars. However, new grids relevant to DB and DO temperatures are being completed by Wickramasinghe (1979), Koester (1979b), and Wesemael & Van Horn (1979). These models include both line blanketing and convection. For the cooler stars in which the physical uncertainties are greater, available calculations have generally been aimed at fitting individual objects of types DF, DG, DK, C$_2$, or DC.

Pending the conclusions from the new series of DB models, the best available determinations of temperature, radii, and gravities for DB and somewhat cooler helium stars are published in Shipman (1979a). Though the uncertainties in the model fluxes are considerable, Shipman finds a similar mean radius for 28 helium stars (0.0111 $R_\odot$) as for the DA sample. He finds a remarkably similar relationship between the monochromatic flux and the (g-r) color of Greenstein's (1976) data for non-DA and DA stars. This is similar to the completely empirical claim of Barnes, Evans & Parsons (1976) and Moffett, Barnes and Evans (1978) that a similar relationship between surface brightness and (V-R) broad-band color exists for white dwarfs and a variety of lower gravity stars. However, the uncertainties in Shipman's flux-color relation may correspond to a 10% uncertainty in the mean radius. Furthermore, Koester's (1979b) preliminary conclusions from fitting colors and equivalent widths of DB stars to his new models are that $\overline{\log g} \sim 7.42$ (vice ~ 8.0 for DA stars), corresponding to only $M \sim 0.27\ M_\odot$ for a theoretical interior composed of carbon. For this $\overline{\log g}$ to be compatible with Shipman's radii, the interior of DB stars would have to be composed of much lighter elements (e.g. helium). There is currently no theoretically sensible way to produce core helium degenerates with $M \lesssim 0.4\ M_\odot$ from single star evolution (see Section 4.1). It is of course possible that the discrepantly low mean mass requirement will disappear when more accurate temperatures, radii, and gravities are obtained. However, Alcock (1979) and other envelope theorists argue that all DB masses need to be below $0.4\ M_\odot$ anyway in order to allow the small amount of matter accreted from the interstellar medium to be mixed in a sufficiently deep convective envelope so that the purity of observed DB atmospheres may be maintained.

Nather, Robinson & Stover (1979) have recently suggested a way in which close binary evolution might produce a single degenerate star with a helium core. They argue that the ultrashort period variable stars showing only helium lines—AM CVn and G61-29—are binaries consisting of helium degenerates with the low-mass secondary likely to be completely dissolved before the process finishes. The total masses of these systems are highly uncertain. However, Conti, Dearborn & Massey (1980) find that the short-period sdO system LB3459 clearly has a total mass $< 0.4\ M_\odot$ and may be a detached evolutionary precursor to the AM CVn stage. Thus, the completion of the evolution could result in a single helium white dwarf or subdwarf of very low mass, devoid of hydrogen, and having probably a high enough temperature to be at or above the "DB" range. Since the luminosities and lifetimes of the AM CVn variables are uncertain, it is difficult to assess what fraction of existing DB stars could be accounted for by such precursors.

Model atmosphere calculations for cooler helium-atmosphere stars have generally been aimed at fitting specific stars with abundance anomalies. The typical cool star is DC, with no spectral features to use. The electron supply, which controls the amount of He^- continuum opacity, is dominated by elements heavier than helium, many of which do not have strong transitions in the optical spectrum from which useful abundance limits may be inferred. The temperature uncertainties for cool DC stars may exceed

10%. Likewise, temperature and surface gravity inferences from fitting line profiles in DF-DG-DK stars are of dubious value.

For helium-atmosphere stars cooler than about 6000 K, greater uncertainties occur in the opacities, equations of state, and line broadening physics due to the proximity of third bodies on what are treated as two-body interactions. Simple model energy distributions currently predict such low temperatures for the reddest known DC-DK stars with good parallaxes that many of these stars would be required to have abnormally large radii or low masses, if they are assumed to have helium atmospheres (Mould & Liebert 1978, Shipman 1979a). Hence, Shipman argues that these stars too cool to show Balmer lines generally have hydrogen atmospheres; it is of course also possible that the helium-atmosphere temperature fits are simply $\sim 20\%$ lower than they should be.

Böhm, Carson, Fontaine & Van Horn (1977) have published the first exploratory calculations of the nature of pure helium envelopes at temperatures below 5000 K; they used a more detailed Thomas-Fermi equation of state and other refinements, but they allowed no metallic electron donors to the opacity. At temperatures below 4000 K, it appears that the atmospheres of such stars (if they exist) could be characterized by such "strange" effects as coulomb interactions, efficient convection, pressure ionization, degeneracy, and electron conduction; they could have extremely flat temperature $T(\tau)$ profiles, making it difficult for absorption features to be seen—see also Böhm (1979).

3.5 Abundances in Helium-Atmosphere Stars

HELIUM-TO-HYDROGEN RATIOS Because of the reduced continuum opacities at most temperatures, more stringent upper limits on hydrogen and heavy elements may be established for the helium-atmosphere white dwarfs. Virtually all well-studied helium-atmosphere cases have nearly pure helium atmospheres, despite the appearance of strong hydrogen and neutral metal lines in the spectra of some.

A few of the hottest known degenerates have helium atmospheres and "DO" spectra dominated by He II lines. The first abundance analysis of the prototype HZ21, assuming a homogeneous composition, indicated significant hydrogen abundance with $\log n_{\rm He}/n_{\rm H} \sim +1$ (Koester, Liebert & Hege 1979). The hydrogen abundance in such a star is difficult to derive given the coexistence of He II Brackett lines at essentially the same wavelengths as the Balmers. However, the large ratio of He I/He II $\lambda 4686$ strengths required a temperature below 50,000 K; then, the He II transitions alone could not account for the strength of the absorption at Brackett/Balmer wavelengths. Wray, Parsons & Henize (1979) argue that HD149499B, discovered from the *Skylab* UV experiment, may also have a significant hydrogen abundance but at a higher T_{eff} than HZ21. Wegner (1979a) found its optical spectrum to be DO.

Figure 5 in Liebert et al. (1979c) summarizes recently published data on degenerates having helium-to-hydrogen abundance determinations. It is curious that the analyzed stars hotter than 45,000 K like HZ21 cover an extensive range in n_{He}/n_H, while the known cooler stars showing hydrogen lines are still very helium-rich objects near the DB-DC lower limit $\sim 10^4$ (for no hydrogen lines). However, as discussed in the reference, selection might have worked against the discovery of any cooler stars with intermediate n_{He}/n_H abundances.

HEAVIER ELEMENTS IN HELIUM ATMOSPHERES   The great majority of helium-atmosphere degenerates show no optical absorption features due to elements heavier than helium; the exceptions belong to either the metallic-line (DF-DG-DK) or C_2 ($\lambda 4670$) spectral groups. Most of the former show *only* lines of Ca II H&K (DF stars) over a wide range in temperatures. A handful have three or more identified elements from good quality optical or ultraviolet (IUE) spectra and have had detailed atmospheric analyses; these have inspired a lively debate as to the origins of the surface heavy elements (Section 4.3). They range in temperature from the DBF star GD40 at 14,000 K to the heavily blanketed LP701-29 at ~ 4000 K. The dominant atmospheric constituent of LP701-29 is controversial (Dahn et al. 1978, Cottrell et al. 1977). Calcium, magnesium, iron, silicon, sodium, and hydrogen are among the trace elements that have been detected in various objects. Unfortunately, no CNO elements have been directly measured —except in the $\lambda 4670$ stars—though these may control the electron supply for the He$^-$ continuum opacity above ~ 7000 K.

Table 1   Some interesting abundance ratios for helium-rich white dwarfs

Name (WD #)	Ca/H	Mg/Ca	Si/Ca	Fe/Ca	Ref.[a]
GD40 (0300−013)	$\gtrsim 3 \times 10^{-3}$	—	—	—	SGB
GD401 (2215+388)	$\gtrsim 10^{-4}$	$\gtrsim 0.4$	—	—	CG
R640 (1626+369)	$\sim 2.5 \times 10^{-6}$	89	42	2.6	CG, L
G165-7 (1328+308)	$\gtrsim 3 \times 10^{-4}$	13:	37:	44:	WL
VMa2 (0046+052)	$\gtrsim 10^{-7}$	20:	$\gtrsim 100$:	20:	G
Solar	2.0×10^{-6}	13	16	20	A

[a] References:
  A—Allen (1973)
  CG—Cottrell & Greenstein (1980)
  G—Grenfell (1974)
  L—Liebert (1977)
  SGB—Shipman, Greenstein & Boksenberg (1977)
  WL—Wehrse & Liebert (1980b)

More detailed discussions of abundance determinations in specific metallic line stars may be found in Vauclair et al. (1979), Greenstein (1979a), Vauclair (1979), and Böhm (1979). Some of the abundance ratios most relevant to the issues discussed in Section 4.3 are summarized in Table 1.

3.6 Rotation

Much theoretical attention has been devoted to the implications of rapid rotation in degenerate dwarfs (cf Ostriker 1971); after all, such configurations could greatly exceed the Chandrasekhar mass limit and be eventual supernova candidates. However, the simple reality is that—save for those in close binaries—most white dwarfs rotate slowly, if at all. Various investigators have worried that rapidly rotating degenerates might be disguised as simple "DC" stars. Wickramasinghe & Strittmatter's (1970) and Kuzma's (1979) calculations showed that hydrogen lines should still be visible, except perhaps under rather extreme circumstances; Milton (1974) still argued that some differentially rotating hydrogen-atmosphere stars might appear with DAwk or even DC spectra. In any case, it is clear that DA stars with hydrogen lines weak for their color are in reality quite rare: Weidemann & Koester (1980) have disproved most of the prior claims. Any rotating "DC" group should include some at blue colors, where their lack of lines would quickly be noticed by the spectroscopic surveyor. Furthermore, the narrow DA distributions in the two-color diagrams, and the sharp line cores in DA and probably DB stars do not indicate any of the effects of moderate rotation rates calculated by Milton.

Slow rotation periods have been determined or inferred for some of the ZZ Ceti variable DA stars and for the polarization/spectrum-variable magnetic stars—see references given in Sections 3.7 and 3.8. These periods imply rotational velocities of only 0.4–8 km s^{-1}; the nonvariable strongly magnetic stars may not be rotating at all!

The best observational limits on nonvariable white dwarfs are those obtained from coudé-resolution line profiles of the brighter DA stars (Greenstein & Peterson 1973, Bessell & Wickramasinghe 1975, Greenstein et al. 1977). These Hα and Hβ profiles have revealed surprisingly sharp line cores, probably formed under non-LTE conditions high in the atmospheres. Greenstein et al. (1977) report a possible emission reversal in the Hα line of one cool DA star. Projected rotation velocities for eight well-observed stars do not exceed 40 km s^{-1}; several cases with somewhat broader cores could have $v \sin i \simeq$ 50–60 km s^{-1}—see also Kuzma (1979). Comparable observations for DB stars are not yet available. However, several cooler DB stars observed at ~ 2 Å Cassegrain resolution (at Steward Observatory and elsewhere) also appear to have sharp line centers.

3.7 Variability

ZZ CETI STARS   Aside from those known to be in binary systems, the only white dwarfs that are established photometric variables are the ZZ Ceti class of DA stars in the 10,000–14,000 K temperature range. Several recent reviews (e.g. McGraw 1979, Robinson 1979, Dziembowski 1979) describe well the properties and puzzles posed by these complicated pulsators. The light curves show variation amplitudes of only 0.01–0.30 magnitudes and periods ranging over $\sim$ 200–1200 seconds; no variation has yet been found at or near the fundamental (1–10^s) pulsation periods of degenerate stars. Multiple pulsation modes occur in all known stars, and the period structure and amplitudes may vary. The light curves are generally thought to be explicable as nonradial g-mode pulsations, as developed from the suggestion by Warner & Robinson (1972), though high radial overtones may also be excited (Starrfield, Cox & Hodson 1979). The curves may also show an effect on the derived modes attributable to rotation of the star; the inferred rotation periods range from a few hours to a few days, similar to those found for magnetic stars which are spectrum variables, and not inconsistent with the DA spectroscopic limits. The ZZ Ceti stars appear spectrophotometrically normal at optical and ultraviolet wavelengths (Greenstein 1979b).

While the nonradial g-mode interpretation has enjoyed some qualitative success in matching ZZ Ceti period spectra, the physical basis for the pulsations has yet to be developed. Many authors have noted that the ZZ Ceti position in the H-R diagram is not inconsistent with an extrapolation of the Cepheid instability strip to $\log g \sim 8$. Hence, most investigators have explored the possibilities that the pulsational driving is due to an underlying He II ionization layer at $\sim$ 50,000 K—the primary source of instability in classical Cepheids—or the Stellingwerf (1978) He II opacity edge near 150,000 K—which may power pulsations in higher gravity Beta Cephei and Delta Scuti stars. Dziembowski (1979) has now demonstrated that theoretical models may be unstable to g-mode pulsations, assuming that there is some helium at the required depths. However, Starrfield et al. (1979) have also found radial instabilities for nearly pure hydrogen compositions. Whatever the mechanism is, the theory must also explain why only specific high overtones appear in certain stars while other overtones, particularly lower ones, are not excited. It is clear that more realistic envelope structures must be incorporated into these calculations, models which attempt to account for the prior evolution in surface abundances of the cooling white dwarf. Hot DA white dwarfs are extremely helium deficient at their surfaces; the theoretical diffusion calculations (Section 4.3) have demonstrated that most surface helium should have sunk quite deep in the envelope, leaving an outer layer mass (M_H) of almost pure hydrogen. One may speculate that the M_H parameter holds the key to what fraction of DA stars cooling to 12,000 K becomes variables; the slow rotations inferred for some stars suggest that rotation is not a key property.

More detailed study of the pulsation properties should enhance our knowledge of the physical structure and evolution of DA degenerates. The detection of period changes would even afford a direct measurement of the cooling time (McGraw 1977), since Osaki & Hansen (1973) predict a linear period-luminosity relation. The slope of the period increase should depend inversely on the mean molecular weight of the interior. However, the stars have proven to be among the universe's better clocks: Stover, Robinson & Nather (1978) report a limit of $|\dot{P}|^{-1} > 10^{11}$ for Ross 548.

OTHER DEGENERATE VARIABLES Variability has been reported for a few non-DA degenerates, including the DC star G44-32 and the helium-rich suspect AM CVn = HZ 29. The latter is almost certainly a low-mass binary (Section 3.4), while the variability of the former is not well established (McGraw 1977). An updated list of degenerate suspects searched for variability is given by Hesser, Lasker & Neupert (1979). These authors emphasize that very little is known about variability in most white dwarfs on time scales $\gtrsim$ 30 minutes.

PG1159-035 The first stellar observations with the Arizona/Smithsonian 4.5-m multiple mirror telescope resulted in the discovery of a new kind of rapid variable star (McGraw et al. 1979, 1980). The object was selected from the Palomar Green faint blue stars because of its very blue color and unique spectrum (Green & Liebert 1979). Despite the drastically higher temperature, the object showed a power spectrum somewhat similar to some ZZ Ceti stars, with strong periodicities near 8.9^m and 7.6^m. Thus, pulsation is the most likely cause of the variability. Further study of the spectrum resulted in the identification of sharp lines due to such ions as He II and C IV, often blended together. Hence, the object is likely to be of lower surface gravity than the DO degenerates, though it may be a helium-rich precursor.

3.8 Magnetism

Excellent recent reviews by Angel (1977, 1978) and Landstreet (1979) preclude the need for much attention to this subject here. Thirteen isolated magnetic degenerates are now known, with surface field strengths ranging from 3 to over 100 megagauss. At least three have hydrogen-dominated atmospheres, and at least two have helium-rich atmosphere with some hydrogen. The temperature range (5600–22,000 K) appears representative. It is not clear that there are any significant differences in the space velocities from the sample of nonmagnetic white dwarfs. Stringent limits on surface magnetic field strengths as low as a few kilogauss are available for some DA stars whose line centers have been observed at coudé dispersion. However, Angel, Landstreet & Borra (1980) argue that the distribution of magnetic field strengths and limits for the available sample are not inconsistent with the hypothesis that the stars occur with an equal distribution ($\simeq 0.5\text{–}1\%$) of field strengths per octave to ~ 300 megagauss. These authors and Angel (1979) argue that magnetic degenerates may be descended from main-sequence magnetic A stars, whose fossilized core fields must presumably survive subsequent evolutionary stages; this idea might imply higher than average progenitor masses for the magnetic degenerates and lower than average space motions for a sufficient sample of hot magnetics.

4 THE ORIGIN AND EVOLUTION OF DEGENERATE DWARFS

4.1 *The Birthrate of Hot White Dwarfs*

The rate at which degenerate stars are currently being produced is best estimated from the empirical density and luminosity function of hot white dwarfs. The result is insensitive to uncertainties in cooling theory, which are more important at lower temperatures, and to the time dependence of the stellar birthrate function, since hot white dwarfs have not cooled very long. Early attempts to derive the birthrate were reviewed in Weidemann (1968). By far the best determination is due to the comprehensive survey of faint blue stars by Green (1977, 1980). His 18-inch Schmidt plate survey of the sky resulted in the discovery of some 3000 blue stellar objects with $\delta \geq -10°$ and galactic $|b| \geq 30°$; a subset of some 89 objects was used in the initial estimate of the luminosity function. The sample was argued to be complete for B (Magnitude) < 15.7 and for photographic colors corresponding to white dwarf with $M_v \gtrsim 12.75$. Green's (1977) luminosity function corresponds to a white dwarf birthrate of 1.4×10^{-12} pc^{-3} yr^{-1}, which compares with earlier estimates in the range of $1\text{–}5 \times 10^{-12}$. Green also discussed the rather modest constraints the hot white dwarf birthrate imposes on galactic star formation models. His modeling seems consistent with a scale height of ~ 250 pc for the young white dwarf distribution. Chiu's (1978) counts to much larger distances ought to provide a more accurate scale height determination for blue degenerates, though his value of 400–500 pc seems surprisingly large for what is argued to be an old disk population.

4.2 *Links to Prior Evolution*

The birthrate and other information determined for white dwarfs holds implications for identifying progenitor stars and tracing their prior evolution. The kinematical properties (Section 2.3) indicate that most progenitors were originally 1–2 $M_\odot$ Population I stars, but the existence of white dwarfs in young clusters may mean that some stars were initially much more massive. The low remnant masses of most DA's (Section 3.2) indicate that extensive envelope mass loss is a widespread phenomenon of post–main-sequence evolution; their slow rotation rates imply that most of the angular momentum of main-sequence stars is lost in subsequent evolution. It is desirable to match the birthrates with the formation rates of evolved giants, Miras, hot subdwarfs, planetary nebulae, and any other suspected progenitor stages, and to determine whether the atmospheric composition groupings can be identified with specific progenitor candidates. Forging the link with what has generally been learned about post–main-sequence stellar evolution can also improve our understanding of white dwarf parameters.

EVOLUTIONARY CONSTRAINTS ON INTERIOR PARAMETERS   The advances in stellar evolution theory since the advent of Henyey-method calculations constrain the core compositions of white dwarfs more accurately than is currently possible from the mass-radius relation (Section 3.1). The calculations have established that any star developing a helium core with $\gtrsim 0.45$ $M_\odot$ will ignite helium producing a core of carbon and/or oxygen. Single star evolution cannot produce a helium remnant with $<0.45\ M_\odot$ within the lifetime of the universe unless mass loss were so efficient as to strip away the entire hydrogen envelope before the helium core accumulates $0.45\ M_\odot$; only mass exchange in close binaries may produce such a low mass helium remnant (Section 3.4). Likewise, the expected cooling of the carbon-oxygen core due to neutrinos implies that the core mass should approach or exceed the Chandrasekhar limit before carbon burning—either explosive or nonexplosive—would commence (Paczynski 1970, Mazurek & Wheeler 1980). Below this mass limit, remnant degenerates should result from gravitational contraction following the end of the double-shell-source burning stage; they should generally have cores composed of carbon and oxygen, with a thin outer layer of helium ($\sim 10^{-2}\ M_\odot$) on top of which an even thinner hydrogen layer might survive a final envelope ejection.

ACCOUNTING FOR THE MASS LOSS   The relatively low masses and narrow mass dispersions derived for isolated DA degenerates (Section 3.2) greatly limit the buildup of mass in the carbon-oxygen core during late nuclear-burning stages. This is especially true if the calculations must account for some progenitor masses of $\gtrsim 5\ M_\odot$. Extensive mass loss prior to any planetary nebula ejection must occur, presumably by a "stellar wind" on the giant and asymptotic giant branches. The Reimer's (1975) expression

$$\dot{M} = 4 \times 10^{-13} \eta \frac{(L/L_\odot)(R/R_\odot)}{(M/M_\odot)} M_\odot \text{yr}^{-1} \qquad (1)$$

has the right dimensions for work done by the stellar luminosity in pushing mass away from the surface, and has a crude empirical calibration for the efficiency η. Other quasi-empirical expressions, e.g. Fusi-Pecci & Renzini (1975), make rather similar predictions. Still, it also appears that some stars have significant winds even on the subgiant branch (Dearborn et al. 1976) at luminosities too low for the Reimer's formula to predict any significant effect.

Several investigators have combined assumed formulae for mass loss with normal stellar evolution theory in order to derive theoretical relations between the progenitor and remnant masses. Following the stellar wind losses, Wood & Cahn (1977) assumed that any remaining hydrogen envelope may be ejected as a planetary nebula when the more massive stars reach the high luminosity side of the Mira instability region. The incorporation of an assumed stellar birthrate and initial mass function then permitted the mass distribution of remnants to be predicted. In general, these theoretical distributions (Weidemann 1977a, Wood & Cahn 1977) 1. have a sharp cutoff at the lower mass limit, generally near 0.5–0.6 $M_\odot$, and 2. predict that only $\sim 50\%$ of all remnants should be within 0.1 $M_\odot$ of this lower mass edge, while the remainder should form a skewed distribution of higher masses.

The theoretical distributions thus predict too many high mass remnants with $\not< 1$ $M_\odot$ compared to the empirical results. In fact, the known DA distribution offers little evidence for a high mass non-gaussian "tail" (cf Wegner 1979b). Even the well-studied degenerates in the Hyades and Pleiades clusters, with apparent progenitor masses of $\not< 2$ $M_\odot$ and $\not< 5$ $M_\odot$ respectively, appear to fit only marginally higher surface gravities and masses than the mean values (Weidemann 1975, Sweeney 1976, Shipman 1979a), and are spectroscopically normal. While the upper mass limit for white dwarf progenitors remains controversial (see below), the problem with accounting for the high mass progenitors underscores our lack of understanding of mass loss processes in late stages of evolution.

THE PROGENITOR MASS LIMIT The faint blue stars found by Romanishin & Angel (1980) in several open clusters with 3–6 $M_\odot$ turnoff masses indicate that the Pleiades case is not an isolated example and that some stars initially with $\gtrsim 5$ $M_\odot$ evolve into white dwarfs. Nevertheless, there is some evidence favoring an upper mass limit for progenitors (M_u) less than $5M_\odot$. The possibility of non-coeval star formation in young clusters introduces a potentially serious uncertainty. Landolt (1979) and Stauffer (1980) have shown that some low mass stars in the Pleiades fit pre–main-sequence contraction ages of $\gtrsim 10^8$ years, or more than twice the turnoff age. While there may prove to be some explanation other than differing ages for these objects, a high mass counterpart of ~ 4 $M_\odot$ could become a white dwarf in

~ 10^8 years. Additionally, the white dwarf counts in the Hyades apparently support a lower M_u value (van den Heuvel 1975), though this estimate is subject to uncertainties in the cluster initial mass function at high masses, the small numbers of stars, the difficulties with assessing membership, and the problem of escaped members. A statistically lower M_u is also favored by Taylor & Manchester (1977) in accounting for the pulsar/neutron star birthrate and scale-height evidence. However, the pulsar birthrate they wish to satisfy may be uncertain by as much as 600 (Arnett & Lerche 1980). Furthermore, Shipman & Green (1980) find that recent revisions in the stellar birthrate function permit them to accommodate the pulsar progenitors with a much higher M_u value than Taylor & Manchester used.

Finally, various authors have argued that differing angular momenta or magnetic field configurations for evolved stars should result in differential mass loss rates anyway—that is, M_u should vary with these properties (Weidemann 1977a). These possibilities are difficult to evaluate given the lack of a physical understanding of the mass loss mechanisms for the simpler assumptions of zero angular momenta and zero magnetic fields! The mean value of M_u thus remains uncertain despite a great deal of recent research, but is likely to be $\gtrsim 5\ M_\odot$.

PLANETARY NEBULAE AND HOT SUBDWARFS The white dwarf birthrate ($\dot{N}_{WD}$) may be compared with those for earlier phases of stellar evolution in order to determine which stars may generally be precursors. Planetary nebulae stars have long been assumed to be immediate progenitors, given that they are thought to be in the final stage of gravitational contraction and that the hottest white dwarfs approach the region of the Harman-Seaton sequence in the H-R diagram. In fact, Liebert, Greenstein & Green (1980) have now found that the central star of the low-surface-brightness planetary Abell 7 actually has the spectrum of a DAO white dwarf. With the older distance scale (Seaton 1968), the estimated birthrate for planetaries was substantially higher (Cahn & Kaler 1971) than Green's determination of $\dot{N}_{WD}$. However, there are persuasive arguments in favor of increasing the distance scale of planetaries by 30–45%, which would result in a nebula birthrate $\gtrsim \dot{N}_{WD}$ (Cudworth 1974, and especially Weidemann 1977b). The birthrates are therefore consistent with the assumption that most white dwarfs pass through the planetary nebula stage.

Other groups of stars suspected of being white dwarf progenitors are the subdwarf O and B stars, particularly those found to have relatively high surface gravities. Since most of these stars have been found at high galactic latitudes (e.g. Greenstein & Sargent 1974), they are known to be evolved stars from an old stellar population, though many may be disk rather than halo population objects. They are generally too hot and/or too low in luminosity to be plausibly associated with horizontal branch (helium burning) evolution, yet some exist at temperatures much too low to be fit by theoretical models for nonbinary gravitationally contracting stars of reasonable masses ($\gtrsim 0.4\ M_\odot$). Indeed, planetary nebulae stars are generally much hotter. Though Green's (1977) preliminary estimate implies that the space densities of both sdB and sdO stars were too low for them to be precursors for most white dwarfs, the uncertainties in the luminosities may be great enough for this to remain an open question.

Since stars classified sdO are generally believed to have helium-rich atmospheres, it is tempting to suppose that these may be progenitors for some helium-atmosphere white dwarfs. Schönberner (1977) has evolved a series of asymptotic branch giant models stripped of outer hydrogen envelopes, following the evolution to the white dwarf stage. He finds that configurations with $\sim 0.7\ M_\odot$ may cross through the regions of R CrB and other helium-rich stars (cf Hunger 1975), though the evolutionary tracks again do not account for the lower temperature sdO stars having high surface gravities. Schönberner finds that the number densities and time scales for evolved helium-rich stars suggest a lower limit death rate of $\gtrsim 4 \times 10^{-14}\ pc^{-3}\ yr^{-1}$, within an order of magnitude of the birthrate inferred for helium-atmosphere white dwarfs. It may be possible that the cooler helium sdO's require consideration of much lower masses and binary evolution (see Section 3.4).

4.4 Envelope Evolution of White Dwarfs

A variety of physical processes may change the surface abundances over the cooling time of a white dwarf. There is currently very extensive theoretical work on these problems, and the imminent availability of ultraviolet and X-ray spectra and fluxes for some hot stars promises to test some of the ideas concerning diffusion, mixing, stellar winds, and coronae. The author recently compared the empirical numbers and abundance distributions for cool DA, metallic-line and carbon-band white dwarfs with those that might be expected if convective mixing or interstellar accretion were taking place (Liebert 1979). Likewise, Vauclair (1979) has reviewed various envelope processes in detail—see also Böhm (1979), Vauclair, Vauclair & Greenstein (1979) and individual papers by Alcock, Michaud, Fontaine, and others referenced below. Here we attempt only a brief resumé of the potentially important physical processes.

DIFFUSION AND ACCELERATION PROCESSES   Since the work of Schatzman (1958), it has been recognized that diffusion processes can operate very quickly to separate elements in quiescent white dwarf envelopes. It was apparent that helium and the heavier elements could sink completely out of hot white dwarf atmospheres—leaving an essentially pure hydrogen upper layer—on time scales of < 100 years. Recent comprehensive calculations by Fontaine & Michaud (1979), Michaud & Fontaine (1979), Vauclair, Vauclair & Greenstein (1979), and by Alcock & Illarionov (1980) generally confirm and greatly extend this conclusion. The downward diffusion is caused by both pressure and temperature gradients, but radiative acceleration tries to drive material outwards. The net downward diffusion velocity may be expressed (adapting from the Vauclair et al. paper):

$$v = \frac{\bar{D}m}{kT}(g_{GT} - g_{rad}) \qquad (2)$$

where $\bar{D}$ is a diffusion coefficient and m the mass for a given ion; whether the ion sinks or rises thus depends on the relative sizes of the gravitational plus thermal acceleration (g_{GT}) and the radiative acceleration (g_{rad}). In the case of elements whose line transitions are not saturated, the latter may be approximated (Michaud et al. 1976) by

$$g_{rad} = 1.7 \times 10^{-4} \frac{T_{eff}^4 R^2}{ATr^2} \text{ cm s}^{-2} \qquad (3)$$

where T_{eff} is the effective temperature, T is the temperature at radius r, R is the stellar radius, and A is the atomic mass number of the element being considered. The ions must have transitions with large f-values near the peak wavelength of the flux distribution. Thus g_{rad} may become competitive for light CNO ions with ultraviolet resonance transitions near the surfaces of very hot stars. Vauclair et al. expect that some ions having $g_{rad} \gtrsim g_{GT}$ could be suspended in equilibrium or even expelled from very hot photospheres. However, when $g_{rad} < g_{GT}$ the element will sink rapidly and permanently downwards. Whether ions are suspended or ejected from the atmosphere depends on details of the outer boundary conditions of the envelope problem (see the subsection on accretion).

Saturation effects can reduce g_{rad} far below the estimates provided by Equation (3). The Vauclair et al. calculations suggest that saturation could effectively prevent any enhancements of CNO ions for hydrogen envelopes. Indeed, no elements other than hydrogen have yet been detected in early IUE ultraviolet spectra of hot DA stars (Greenstein & Oke 1979). However, in helium-rich envelopes, higher pressures produce relatively larger line widths thus reducing the effects of saturation. The calculations suggest that significant CNO abundances may be supported for a while at the tops of hot helium stars. Still, the onset of convection below about 60,000 K introduces a different deterrent. Due to the T^{-1} dependence in Equation (3), diffusion of heavy elements at the lower boundary of the homogeneous convective region could restrict the operation of the radiative acceleration process to very hot helium stars (Fontaine & Michaud 1979, Alcock & Illarionov

1980). Hence, radiation effects without accretion should not be relevant to the observed Ca II in a DB star such as GD40 at $\sim 15{,}000$ K (from Shipman, Greenstein & Boksenberg 1977). Again, the best chances for detecting heavy ions rest with ultraviolet spectrophotometry of hot helium-rich degenerates. Green & Liebert (1980) found C IV $\lambda 1550$ in at least two Palomar Green (PG) stars classified DO.

THE QUESTION OF CORONAE Numerous investigators have suggested that hot white dwarfs may have coronae which 1. inhibit accretion of hydrogen onto DB stars (Strittmatter & Wickramasinghe 1971), 2. contribute to the galactic soft X-ray background (Strittmatter, Brecher & Burbridge 1972), or 3. account for the soft X-ray flux from Sirius B (see Section 3.1). The first authors suggested that a hot corona-stellar wind might be generated for stars having a convective outer envelope—helium stars above $\gtrsim 20{,}000$ K and hydrogen stars below $\lesssim 12{,}000$ K. This idea was independently suggested and developed by Böhm & Cassinelli (1971), who calculated large acoustic fluxes for helium white dwarf envelopes.

Now Muchmore & Böhm (1978) have attempted preliminary modeling of the temperatures, pressures, and energy distributions for hot non-DA coronae, based on the main assumptions 1. that a corona is powered by turbulence (acoustic noise) generated in the envelope, 2. that acoustic fluxes may be estimated by the Lighthill-Proudman theory, and 3. that the corona structure adjusts itself to minimize the total flux emitted, i.e. the Hearn (1975) minimum flux theory. In contrast to the solar corona, this model produces coronae that have temperatures of $\sim 5 \times 10^6$ K and extend only tens of kilometers above the photosphere. The Muchmore-Böhm coronae lose energy by direct radiation, electron thermal conduction downwards, and by material outflow. However, the input energy estimate in assumption 2. is very uncertain because of the dependence on the eighth power of the maximum convective velocity; furthermore, the accuracy of the simple Hearn formalism 3. has been widely questioned.

Thus, the Böhm modeling offers only a qualitative physical foundation, but its predicted high energy fluxes may soon be tested from space observations, and it carries implications for other envelope problems. The total stellar EUV and soft X-ray radiation would be significantly enhanced by coronal radiation, particularly shortwards of atmospheric helium opacity edges (see also Böhm 1979, Böhm & Kapranidis 1980). However, a simple calculation suggests that not enough high energy photons would be created to produce detectable emission lines at He II $\lambda 4686$. Likewise, we note that the predicted corona is not nearly hot enough nor extended enough to selectively accelerate and eject protons from a DB atmosphere according to the mechanism of Michaud & Fontaine (1979). Still, it is implicit that there would be some outward mass flow, and its properties in inhibiting interstellar accretion have not yet been explored.

There remains no strong physical basis for postulating a corona around DA stars as hot as Sirius B. Lampton & Mewe (1979) suggest that Sirius B has a subsurface ionized helium layer at $\tau \gtrsim 4$ that provides the required acoustic power for a minimum flux corona. The outer hydrogen layer must be exceedingly thin yet cannot be mixed by the mechanical energy transport, unless it is resupplied by accretion from the primary. Still, Shipman (1979b) reminds us that the solar corona is neither static nor homogeneous in character. It is difficult to assess such an ad hoc, layered DA atmosphere model, given the difficulties with obtaining accurate line profile observations of Sirius B. The HEAO-1 steep X-ray energy distribution does not seem readily compatible with the thermal bremsstrahlung expected from a corona (Lampton et al. 1978). Hence, as of this writing, the Sirius X-ray question remains unanswered.

Not surprisingly, searches for soft X-ray emission from white dwarfs have been directed towards hot objects (Lampton & Kahn 1978). Yet it is in cooler DA stars at about the T_{eff} of the ZZ Ceti variables that deep convective envelopes develop which might readily provide coronal heating. The Starrfield et al. work (see Section 3.7) even suggests that acoustic modes are involved in the pulsations, which might therefore support a corona, as originally suggested by Strittmatter et al. (1972). However, the Böhm (1979) model suggests that such cool DA coronae would have temperatures well below 10^6 K, so that X-ray emission would be minimal.

CONVECTION AND ENVELOPE MIXING The occurrence of significant convective envelopes in cool DA stars and in non-DA stars of most surface temperatures holds several potential implications for envelope evolution:

1. the possible formation of coronae;
2. the dredging of gravitationally diffused heavy elements from deeper in the envelope;
3. penetration and mixing of an underlying carbon-oxygen layer for helium-envelope stars;
4. penetration and mixing of an underlying helium layer into hydrogen atmosphere stars; and
5. the mixing and dilution of accreted surface material.

Coronae (1) were discussed in the previous subsection. The reappearance of diffused heavy elements (2) at the surface of the star during any stage of the cooling process cannot be caused by convective mixing, according to the recent calculations of Fontaine & Michaud (1979) and Vauclair et al. (1979). These authors and Alcock (1979) insist that the appearance of metals in some cool helium degenerates must be caused at least in part by accretion.

Since outer helium convection zones should extend quite deep at cooler T_{eff}, it has been suggested that the sequence of non-DA stars showing C_2 bands covering $6000 \lesssim T_{eff} \lesssim 10{,}000$ K may result from some dredging of material out of the underlying carbon-oxygen core (3, above). D'Antona & Mazzitelli (1979, 1980) argued that pre–white dwarf evolution should not leave a helium envelope thin enough for the convection to penetrate. Calculations by these authors and by Vauclair & Fontaine (1979) indicate that, if this helium layer *were* small enough, penetration of the μ-discontinuity would result in complete mixing anyway and a nearly pure carbon surface composition for most stars. This would contradict the abundance analyses showing that several $\lambda 4670$ degenerates have helium-dominated atmospheric compositions (Bues 1973, Grenfell 1974) with C/He only $\sim 10^{-3}$. For these stars, a mechanism may exist that inhibits the diffusion of carbon in helium envelopes, regardless of the origin of the surface carbon. On the other hand, the predicted spectra for any stars with carbon-dominated atmospheres are not well established, and might even be DC-like; this leaves open the possibility that real carbon stars exist and have yet to be recognized!

A number of recent investigations (Koester 1976, Vauclair & Reisse 1977, D'Antona & Mazzitelli 1979) have explored the mixing of an outer hydrogen DA layer into an underlying helium envelope (4, above). The treatment of the μ-barrier and the possibilities for convection zones in both layers are serious complications to the evolutionary code. The actual surface T_{eff} at which mixing occurs should vary according to the hydrogen layer mass (M_H) and the total stellar mass. Given the theoretical uncertainties and the wide range of possible M_H values, it is unclear as to what fraction of stars should undergo mixing. However, Wehrse & Liebert (1980a), Liebert (1979), and Sion (1979) discuss empirical evidence that mixing *has* taken place in some of the stars reaching 6000–7000 K. Whether all DA stars get converted to helium atmospheres at still lower temperatures may be determined from further theoretical work and infrared observations of very cool stars.

THE QUESTION OF ACCRETION   The existence of nearly pure helium atmosphere degenerates over a wide range of temperatures has long been a puzzle, given the physical arguments that these stars should accrete gas as they pass through the interstellar medium. Strittmatter & Wickramasinghe (1971) argued that, with the assumption that a tail shock forms to destroy particle momentum perpendicular to the relative motion of the star, the Bondi-Hoyle rate should apply:

$$\dot{M} = 3 \times 10^{21} \alpha \frac{M^2 n}{v^3} \; (\text{g yr}^{-1}) \tag{4}$$

where $n =$ the interstellar gas density, v is the relative velocity of the star through the gas, α is an efficiency factor, and M the mass in solar units. Taking $n \sim 1 \text{ cm}^{-3}$, $v \sim 30 \text{ km s}^{-1}$, they showed that the DB atmosphere should be replenished by interstellar material in $\sim 10^{-1}$ years! If one assumed instead the "minimal" Eddington accretion rate

$$\dot{M} = 10^{14} n \bar{R} \bar{M}/v \, (\text{g yr}^{-1}) \tag{5}$$

the time scale of $\sim 10^3$ years is still far shorter than the cooling age of a DB star.

A realistic assessment, however, must take into account the inhomogeneous distribution of the interstellar gas. Using $n \gtrsim 10^{-2} \text{ cm}^{-3}$ for the typical "intercloud medium," the time scales are increased a few orders of magnitude. Koester (1976) argued that the Bondi-Hoyle rate should not apply at such low densities. Truran et al. (1977) suggested that most accretion occurs when a white dwarf encounters an interstellar cloud, as it is likely to do every 10^6–10^7 years. This idea has been developed by Wesemael (1979) and Alcock (1979), who argued that the Bondi-Hoyle mechanism must surely apply to a star moving slowly through a dense gas clump. Hence the more recent analyses still leave us with the same conclusion reached by Strittmatter & Wickramasinghe: either 1. a mechanism to inhibit accretion, especially the accretion of hydrogen, must be identified, or 2. the accreted material must be sufficiently diluted by an envelope mixing process. In the former category, a number of ideas have been suggested. The formation of a hot corona or "mini" H II region might deter accretion of protons (cf Vauclair et al. 1979). However, it is unclear that a low-lying, low temperature Böhm corona would prevent protons from reaching the surface. If electric fields create an ion wind in helium envelopes (Michaud & Fontaine 1979), then protons could be selectively expelled. However, the existence of such a wind is challenged by Alcock (1979). Finally, we note that the combination of a kilogauss magnetic field and a modest rotation rate provides another possibility for "batting away" charged particles from the magnetospheric environment, a possibility that has not been explored. Given the modest likely rotation rates of DB stars (Section 3.6), attention has focussed on convection zones as a means of diluting the accreted hydrogen if it does reach the surface. However, the outer convection layers are relatively thin for the hotter stars; Alcock (1979) argues that the DB stars must have masses $< 0.4 \, M_\odot$ in order to have deep enough layers for sufficient dilution to occur. Given that convective envelopes deepen with decreasing temperature, Alcock's argument may imply that the hottest helium white dwarfs generally have the lowest masses (see Section 3.4).

For the cooler helium-rich stars—the most numerous kind of white dwarf—there is less justification for assuming that accretion of interstellar material may be prevented. These stars have relatively less ultraviolet flux to ionize the approaching gas, relatively lower envelope acoustic fluxes to power a corona, and relatively few free electrons and ions for hypothesized acceleration processes (Liebert 1979). They do offer deeper convective envelopes in which to mix away the pollutants, but now a question arises specifically with regard to those cool helium stars showing metallic lines. The consensus view from the recent theoretical work requires these surface metals to be supplied by accretion, rather than by convective dredging from deeper layers. Vauclair et al. (1979) and Cottrell & Greenstein (1980) argue that the larger-than-cosmic abundance ratios of Mg/Ca in some metallic-line stars are consistent with the accretion hypothesis, though GD401 may be a counterexample (Table 1). Likewise, the metals in the accreted material should diffuse downwards (Alcock 1979), while hydrogen should remain in the convective layer. Thus, the predicted metals-to-hydrogen ratios would be *at or below* solar (interstellar) values, yet real DF-DG-DK stars have calcium-to-hydrogen abundance ratios ranging from about solar to well above solar (Table 1). This would imply that some cooler stars must have an effective screening mechanism that blocks hydrogen accretion in favor of heavier elements.

4.4 *Cooling Evolution and the Luminosity Function*

Extensive theoretical work on degenerate stellar interiors in the last fifteen years supersedes the simple Mestel cooling relation ($t \propto L^{-5/7}$) to describe the cooling of degenerate stars. Mestel (1952) assumed only that the luminosity of a white dwarf is due entirely to the thermal energy of ions, that Kramer's law opacity is applicable with radiative diffusion in a non-degenerate envelope, and that the degenerate core is isothermal. These assumptions grossly oversimplify the physics of the stars—Van Horn's (1971) review is still timely. Yet the sophisticated calculations show that the Mestel relation is a surprisingly good approximation for luminosities that are neither very high nor very low.

NEUTRINO COOLING   For the highest luminosity stars the principal uncertainty concerns the role of neutrinos in cooling. Koester (1978) estimated the expected luminosity function of very hot stars from modern calculations with and without the universal Fermi reaction. The predictions differ by factors of ~ 3 in the 50–70,000 K range. Gravitational contraction may also be significant for low mass stars at high luminosities. The heat capacity of the electrons is also incorporated into recent calculations. Unfortunately, the completeness problem, paucity of stars, and uncertainties in assigning temperatures and surface gravities have precluded a clear observational test of the function for $M_v \lesssim +10$ ($\gtrsim 40,000$ K). Wesemael (1978) argues that the X-ray background evidence supports the lower luminosity function derived from including neutrinos in the calculation. The combination of the PG star sample, space ultraviolet observations, and accurate model atmospheres for hot stars may soon permit a numerical test.

COLOUMB EFFECTS, CRYSTALLIZATION, AND CONVECTION   At lower luminosities, numerous complications in calculating the cooling rate are caused by coulomb interactions and the onset of envelope convection. Plasma interactions complicate the interior conductivity, the radiative opacities in the envelope and transition zone, and the particle ionization in the envelope. As the stars cool, the increasing interactions lead first to a liquid state, then to a crystallization of the ions, with the release of latent heat. During this state the extra energy input lengthens the cooling time. However, this is followed by rapid Debye cooling, due to the drop in the specific heat of the latticed ions. Lamb & Van Horn (1975) have provided a clear discussion of the evolutionary effects, while Shaviv (1979) gives a comprehensive treatment of the coulomb physics.

The onset of crystallization is a gradual process, beginning first in the deep interior and spreading slowly outwards as the degenerate dwarf cools. Hence, the latent heat is deposited over a range in luminosity (cf the 1 $M_\odot$ curve of Lamb & Van Horn 1975) and no pileup of stars at a single luminosity is expected. It is thus unlikely that crystallization sequences for different interior compositions (should they exist) could be recognized. The recent calculations (Lamb & Van Horn 1975, Shaviv & Kovetz 1976) also indicate that the onset of the rapid cooling phase—that is, the turndown from the peak of the luminosity function—should not occur until the stars have cooled below $\sim 10^{-4}$ $L_\odot$. The mass and envelope composition dependence of the peak position is not well established, though the above calculations are for reasonable carbon-oxygen core configurations covering 0.6–1.0 $M_\odot$.

The onset of extensive envelope convection offers a faster means for interior energy to escape the star, thus speeding up the cooling process at lower temperatures (Böhm 1968, Van Horn 1970). A comprehensive series of envelope calculations by Fontaine & Van Horn (1976) does not suggest that the uncertainties here could upset the basic cooling results discussed in the previous paragraph. However, envelope coulomb effects complicate the physics, and the basic relation between the core temperature and the surface effective temperature may remain uncertain by up to a factor of two.

THE EMPIRICAL LUMINOSITY FUNCTION   Green (1977, 1980) has provided the best determination of the observational luminosity function for $+10 \gtrsim M_v \gtrsim +13$. The data are in good agreement with the Mestel cooling relation, assuming that the white dwarf birthrate has remained constant for $\lesssim 10^9$ years.

At lower luminosities, there is no available sample free of motion selection and other biases. Early difficulties with finding cooler degenerates in proper motion surveys (e.g. Greenstein 1971) led to the suggestion that there might be a deficit of "yellow" degenerates at $M_v \gtrsim +13$ relative to the predictions of the Mestel relation. However, Hintzen & Strittmatter (1974) and especially Greenstein subsequently identified many new cool degenerates. Weidemann (1967), Kovetz & Shaviv (1976), and Sion & Liebert (1977) attempted to correct for the volume-dependence of the surveys at different absolute magnitudes, for successively larger spectroscopic samples. They found that the empirical distribution continues to rise to at least $M_{bol} = +14$. In Figure 2, the method is repeated for Sion's (1979) updated spectroscopic catalogue sample and yields very similar results; based on the result discussed in the next paragraph, stars near $M_{bol} \sim +15$ or $10^{-4} L_\odot$ may in fact be the most numerous kind of white dwarf! Green's function with a more secure absolute scale is also shown in Figure 2; that

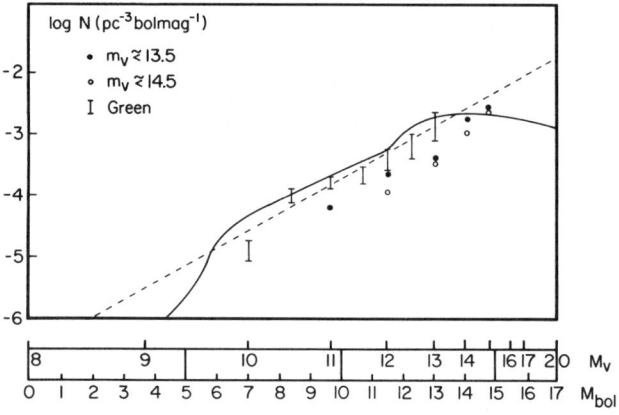

Figure 2 The empirical luminosity function for the known spectroscopic sample of white dwarfs (Sion 1979), updated from Sion & Liebert (1977) employing the technique of Weidemann (1967). Filled circles show values assuming relative completeness to $m_v \simeq 13.5$ (154 stars), open circles to 14.5 (200 stars). Numbers are converted to unit bolometric magnitude intervals using a blackbody bolometric correction. Green's (1977, 1980) empirical luminosity function is converted to the bolometric scale with 1σ error bars. The solid curve is the theoretical function of Lamb & Van Horn (1975) for $1 M_\odot$; the dashed line is the Mestel 2/7 M_{bol} cooling law, which is closer to the shape of the Shaviv & Kovetz (1976) $0.6 M_\odot$ calculation. In the region overlapping with Green, the spectroscopic sample appears to be less than 50% complete, though the slope offers satisfactory agreement. The technique basically hinges on the assumption that the fractional completeness remains essentially independent of absolute magnitude within a constant apparent magnitude limit. While 115 degenerates are now known within the interval $13.5 \gtrsim M_v \gtrsim 15.5$, too few stars with $M_v \gtrsim 15.5$ have been discovered for this method to be readily applicable. Other arguments indicate that the empirical luminosity function peaks near $M_{bol} \sim +15$.

the Weidemann technique matches tolerably well with Green in the overlap region is reassuring.

Recent observations of very faint Luyten (1976) proper motion stars and comprehensive surveys of brighter stars have failed to find any stars clearly having $M_{bol} \geq +16$, despite the abundance of cases near $+15$ (Liebert et al. 1979b, Liebert 1979). The reality of the shortfall is strongly supported by the absence of any such stars in proper motion binaries; the apparent magnitude difference would clearly earmark a low luminosity companion to any of the thousands of nearby stars having distance estimates. Yet the theoretical cooling curves predict that the stars spend at least as much time at $+16 \leq M_{bol} \leq +17$ as in each of the preceding two magnitude intervals. Hence, the cutoff implies 1. serious errors in the cooling calculations at $\lesssim 10^{-4} L_\odot$, and/or 2. a decreased white dwarf birthrate $\gtrsim 10^{10}$ years ago.

It is important to establish whether *any* such very low luminosity stars exist—e.g. such as LP131-66 (Liebert et al. 1979a); this will require parallax work, further investigations of the discrepant optical and infrared colors (Liebert 1979), and the calculation of more realistic energy distributions for stars below 5000 K (Böhm et al. 1977, Bessell, Wickramasinghe & Cottrell 1979). A cutoff would be expected if the galactic disk were effectively much younger than the halo, as some recent evidence suggests (e.g. Demarque & McClure 1977).

OTHER LUMINOSITY SOURCES AT LOW LUMINOSITY? Two effects which could help explain the observed paucity of degenerates at $< 10^{-4} L_\odot$ are 1. a physical separation of the elements when the core freezes, and 2. accretion of matter from the interstellar medium. Stevenson (1977) argued that the heavy elements in the core should become insoluble at the time the main constituent, i.e. carbon, crystallizes. Stevenson (1979) expects even the separation of oxygen from carbon to be an important energy source. The diffusion could release enough gravitational energy to slow the cooling process greatly, perhaps even suspending the star at the crystallization luminosity for the age of the galaxy!

A similar pileup of stars somewhere near $\simeq 10^{-4} L_\odot$ might be expected if interstellar accretion becomes a significant, gravitational energy source. The accretion rate necessary to provide $10^{-4} L_\odot$ is $\simeq 10^{-14} M_\odot$/year; such a mean rate is precluded in general for non-DA stars above ~ 6000 K, but one may speculate that any barriers to accretion which depend on surface temperature (Section 4.3) might be inoperative at very low T_{eff}. Starrfield & Sparks (1979) argue that such stars could even burn the accreted hydrogen quiescently with little increase in luminosity, though it seems unlikely that very cool stars could have sufficiently high internal temperatures.

Either mechanism discussed above might be expected to produce an excess of stars near the peak of the empirical luminosity function. A substantial excess near $M_{bol} = +14$–15, relative to the cooling predictions, is certainly not ruled out. However, there is still a limitation on the early stellar birthrate function (Liebert et al. 1979b).

THE SPACE DENSITY OF REMNANT DEGENERATE STARS The sum of the observed luminosity function and the numbers of any black degenerates below the detection limit is an important contributor to the local galactic mass. A number of authors have previously pointed out that the numbers of undiscovered cool and invisible degenerates might account for the missing Oort mass (insofar as there remains a "missing mass"). However, the number of single stars accounted for to $M_{bol} = +16$ accounts for only about 0.015 $M_\odot$ pc^{-3}, though a sizable increment may be necessary to include the contribution from close binaries. The use of a conventional stellar birthrate function with a disk age of 15×10^9 years would predict a local mass density for non-close binaries as high as 0.06 $M_\odot$ pc^{-3} in total remnants (Green 1977, Liebert et al. 1979b); however, the paucity of very low luminosity degenerates argues that the last number is a serious overestimate. The empirical numbers thus offer fair agreement with theoretical expectations based on global galaxy properties, such as those of Hills (1977).

ACKNOWLEDGMENT

I thank Drs. Roger Angel, K. H. Böhm, Dave Dearborn, Richard Green, Jesse Greenstein, Mike Lampton, Roger Malina, John McGraw, Harry Shipman, Ed Sion, Sumner Starrfield, and Francois Wesemael for results in advance of publication and/or useful discussions. The author acknowledges support from National Science Foundation grant No. AST 78-21071.

Literature Cited

Alcock, C. 1979. In *White Dwarfs and Variable Degenerate Stars, IAU Colloq. No. 53*, ed. H. M. Van Horn, V. Weidemann, p. 202. New York: Univ. Rochester

Alcock, C., Illarionov, A. 1980. *Ap. J.* 235: 534

Allen, C. W. 1973. *Astrophysical Quantities.* London: Athlone

Angel, J. R. P. 1977. *Ap. J.* 216: 1

Angel, J. R. P. 1978. *Ann. Rev. Astron. Astrophys.* 16: 487

Angel, J. R. P. 1979. In *White Dwarfs and Variable Degenerate Stars, IAU Colloq. No. 53*, ed. H. M. Van Horn, V. Weidemann, p. 313. New York: Univ. Rochester

Angel, J. R. P., Landstreet, J. D., Borra, E. F. 1980. Preprint

Arnett, W. D., Lerche, I. 1980. Preprint

Auer, L., Shipman, H. L. 1977. *Ap. J. Lett.* 211: L103

Barnes, T. G., Evans, D. S., Parsons, S. B. 1976. *MNRAS* 174: 503

Bessell, M. S. 1978. *Proc. Astron. Soc. Aust.* 3: 220

Bessell, M. S., Wickramasinghe, D. T. 1975. *MNRAS* 171: 11

Bessell, M. S., Wickramasinghe, D. T. 1978. *MNRAS* 182:275
Bessell, M. S., Wickramasinghe, D. T. 1979. Preprint
Bessell, M. S., Wickramasinghe, D. T., Cottrell, P. L. 1979. In *White Dwarfs and Variable Degenerate Stars, IAU Colloq. No. 53*, ed. H. M. Van Horn, V. Weidemann, p. 179. New York: Univ. Rochester
Böhm, K. H. 1968. *Astrophys. Space Sci.* 2:375
Böhm, K. H. 1979. In *White Dwarfs and Variable Degenerate Stars, IAU Colloq. No. 53*, ed. H. M. Van Horn, V. Weidemann, p. 223. New York: Univ. Rochester
Böhm, K. H., Carson, T. R., Fontaine, G., Van Horn, H. M. 1977. *Ap. J.* 217:521
Böhm, K. H., Cassinelli, J. 1971. *Astron. Astrophys.* 12:21
Böhm, K. H., Kapranidis, S. 1980. Preprint
Böhm-Vitense, E., Dettmann, T., Kapranidis, S. 1979. *Ap. J. Lett.* 232:L189
Bues, I. 1973. *Astron. Astrophys.* 28:181
Cahn, J. H., Kaler, J. B. 1971. *Ap. J. Suppl.* 22:319
Cash, W., Bowyer, S., Lampton, M. 1978. *Ap. J. Lett.* 221:L87
Chandrasekhar, S. 1939. *An Introduction on the Study of Stellar Structure.* Univ. Chicago Press
Chiu, G. 1978. PhD thesis. Univ. Calif., Berkeley
Conti, P. S., Dearborn, D., Massey, P. 1980. Submitted to *MNRAS*
Cottrell, P. L., Bessell, M. S., Wickramasinghe, D. T. 1977. *Ap. J. Lett.* 218:L133
Cottrell, P. L., Greenstein, J. L. 1980 *Ap. J.* In press
Cudworth, K. M. 1974. *Astron. J.* 79:1384
Dahn, C., Hintzen, P., Liebert, J., Stockman, H. S., Spinrad, H. 1978. *Ap. J.* 219:976
D'Antona, F., Mazzitelli, I. 1979. *Astron. Astrophys.* 74:161
D'Antona, F., Mazzitelli, I. 1980. *Ap. J.* 237: In press
Dearborn, D. S. P., Kozlowski, M., Schramm, D. 1976. *Nature* 261:210
Demarque, P., McClure, R. D. 1977. In *The Evolution of Galaxies and Stellar Populations*, ed. B. M. Tinsley, R. B. Larson. New Haven: Yale Univ. Press
Dziembowski, W. 1979. In *White Dwarfs and Variable Degenerate Stars, IAU Colloq. No. 53*, ed. H. M. Van Horn, V. Weidemann, p. 359. NY: Univ. Rochester
Eddington, A. S. 1922. *MNRAS* 82:436
Eggen, O. J., Greenstein, J. L. 1967. *Ap. J.* 150:927
Fontaine, G., Michaud, G. 1979. *Ap. J.* 231:826

Fontaine, G., Van Horn, H. M. 1976. *Ap. J. Suppl.* 31:467
Fusi-Pecci, F., Renzini, A. 1975. *Astron. Astrophys.* 39:413
Gallagher, J., Starrfield, S. G. 1978. *Ann. Rev. Astron. Astrophys.* 16:171
Gatewood, G. D., Gatewood, C. V. 1978. *Ap. J.* 225:191
Giacconi, R. 1979. Review paper presented at the 154th (Wellesley) A.A.S. Meeting
Graham, J. 1972. *Astron. J.* 77:144
Gray, D. F. 1967. *Ap. J.* 149:317
Green, R. F. 1977. PhD thesis. Calif. Inst. Tech., Pasadena
Green, R. F. 1980. *Ap. J.* 238: In press
Green, R. F., Liebert, J. 1979. In *White Dwarfs and Variable Degenerate Stars, IAU Colloq. No. 53*, ed. H. M. Van Horn, V. Weidemann, p. 118. New York: Univ. Rochester
Green, R. F., Liebert, J. 1980. In preparation
Green, R. F., Schmidt, M., Liebert, J., Green, J. 1980. *The PG Catalogue.* In preparation
Greenstein, J. L. 1960. In *Stars and Stellar Systems*, Vol. 6, ed. J. L. Greenstein, p. 676. Univ. Chicago Press
Greenstein, J. L. 1971. In *White Dwarfs, Proc. IAU Symp. No. 42*, ed. W. J. Luyten, p. 46. Dordrecht: Reidel
Greenstein, J. L. 1976. *Astron. J.* 81:323
Greenstein, J. L. 1979a. In *White Dwarfs and Variable Degenerate Stars, IAU Colloq. No. 53*, ed. H. M. Van Horn, V. Weidemann, p. 1. New York: Univ. Rochester
Greenstein, J. L. 1979b. In *White Dwarfs and Variable Degenerate Stars, IAU Colloq. No. 53*, ed. H. M. Van Horn, V. Weidemann, p. 374. New York: Univ. Rochester
Greenstein, J. L. 1979c. *Ap. J.* 233:239
Greenstein, J. L., Boksenberg, A., Carswell, R.; Shortridge, K. 1977. *Ap. J.* 212:186
Greenstein, J. L., Oke, J. B. 1979. *Ap. J. Lett.* 229:L141
Greenstein, J. L., Oke, J. B., Shipman, J. L. 1971. *Ap. J.* 169:563
Greenstein, J. L., Peterson, D. M. 1973. *Astron. Astrophys.* 25:29
Greenstein, J. L., Sargent, A. 1974. *Ap. J. Suppl.* 28:157
Greenstein, J. L., Trimble, V. 1967. *Ap. J.* 149:283
Greenstein, J. L., Vauclair, G. 1979. *Ap. J.* 231:491
Grenfell, T. C. 1974. *Astron. Astrophys.* 31:303
Guinan, E. F., Sion, E. M. 1979. *Bull. Am. Astron. Soc.* 11:683
Hamada, T., Salpeter, E. E. 1961. *Ap. J.* 134:683

Harrington, R., Christy, J., Dahn, C., Strand, K., Liebert, J. 1980. In preparation
Harrington, R., Dahn, C., Behall, A., Priser, J., Christy, J., Riepe, J., Ables, H., Guetter, H., Hewitt, A., Walker, R. 1975. *Publ. US Naval Obs.* 24: Part I
Hearn, D. R. 1975. *Astron. Astrophys.* 40: 355
Hearn, D. R., Richardson, J. A., Bradt, H. V. D., Clark, G. W., Lewin, W. H. G., Mayer, W. F., McClintock, J. E., Primini, F. A., Rappaport, S. A. 1976. *Ap. J. Lett.* 203: L21
Heintz, W. D. 1974. *Astron. J.* 79: 819
Heise, J., Huizenga, H. 1980. *Astron. Astrophys.* In press
Hesser, J. E., Lasker, B. M., Neupert, H. E. 1979. *Ap. J. Suppl.* 40: 577
Hills, J. G. 1977. *Ap. J.* 219: 550
Hintzen, P., Strittmatter, P. A. 1974. *Ap. J. Lett.* 193: L111
Hintzen, P., Strittmatter, P. A. 1975. *Ap. J. Lett.* 201: L37
Humphreys, R. M., Liebert, J., Romanishin, W., Strittmatter, P. A. 1979. *Publ. Astron. Soc. Pac.* 91: 107
Hunger, K. 1975. In *Problems in Stellar Atmospheres and Envelopes*, ed. B. Baschek, W. Kegel, G. Traving, p. 57. Berlin-New York: Springer
Koester, D. 1976. *Astron. Astrophys.* 52: 415
Koester, D. 1978. *Astron. Astrophys.* 65: 449
Koester, D. 1979a. *Astron. Astrophys.* 72: 376
Koester, D. 1979b. In *White Dwarfs and Variable Degenerate Stars, IAU Colloq. No. 53*, ed. H. M. Van Horn, V. Weidemann, p. 130. New York: Univ. Rochester
Koester, D., Liebert, J., Hege, E. K. 1979a. *Astron. Astrophys.* 71: 163
Koester, D., Schulz, H., Weidemann, V. 1979b. *Astron. Astrophys.* 76: 232
Kovetz, A., Shaviv, G. 1976. *Astron. Astrophys.* 52: 403
Kuzma, T. J. 1979. *Ap. J.* 227: 548
Kylafis, N. D., Lamb, D. Q., Masters, A. R., Weast, G. J. 1980. Ninth Texas Symposium on Relativistic Astrophysics, Munich, Germany, 1978. *Ann. NY Acad. Sci.* 336: 520
Lamb, D. Q., Van Horn, H. M. 1975. *Ap. J.* 200: 306
Lampton, M., Kahn, S. 1978. *Bull. Am. Astron. Soc.* 10: 511
Lampton, M., Mewe, R. 1979. *Astron. Astrophys.* 78: 104
Lampton, M., Margon, B., Paresce, F., Stern, R., Bowyer, S. 1976. *Ap. J.* 203: L71
Lampton, M., Tuohy, I., Garmire, G., Charles, P. 1978. COSPAR Meeting, Innsbruck, Austria
Landolt, A. 1979. *Ap. J.* 231: 468
Landstreet, J. D. 1979. In *White Dwarfs and Variable Degenerate Stars, IAU Colloq. No. 53*, ed. H. M. Van Horn, V. Weidemann, p. 297. New York: Univ. Rochester
Liebert, J. 1975. *Ap. J. Lett.* 200: L95
Liebert, J. 1976. *Ap. J.* 210: 715
Liebert, J. 1977. *Astron. Astrophys.* 56: 427
Liebert, J. 1979. In *White Dwarfs and Variable Degenerate Stars, IAU Colloq. No. 53*, ed. H. M. Van Horn, V. Weidemann, p. 146. New York: Univ. Rochester
Liebert, J., Dahn, C., Gresham, M., Hege, E. K., Moore, R. L., Romanishin, W., Strittmatter, P. A. 1979a. *Ap. J.* 229: 196
Liebert, J., Dahn, C., Gresham, M., Strittmatter, P. A. 1979b. *Ap. J.* 223: 226
Liebert, J., Greenstein, J. L., Green, R. F. 1980. In preparation
Liebert, J., Gresham, M., Hege, E. K., Strittmatter, P. A. 1979c. *Astron. J.* 84: 1612
Luyten, W. J. 1970. *White Dwarfs*. Minneapolis: Univ. Minn. Press
Luyten, W. J. 1976. *LHS Catalogue*. Minneapolis: Univ. Minn. Press
Luyten, W. J. 1977. *White Dwarfs—II*. Minneapolis: Univ. Minn. Press
Malina, R. 1979. PhD thesis. Univ. Calif., Berkeley
Margon, B., Lampton, M., Bowyer, S., Stern, R., Paresce, F. 1976. *Ap. J. Lett.* 210: L79
Matsushima, S., Terashita, Y. 1969. *Ap. J.* 156: 219
Mazurek, T. J., Wheeler, J. C. 1980. *Fundamentals of Cosmic Physics*. In press
McCook, G. P., Sion, E. M. 1977. *Villanova Univ. Obs. Contrib. No. 2*
McGraw, J. T. 1977. PhD thesis. Univ. Texas, Austin
McGraw, J. T. 1979. *Ap. J.* 229: 203
McGraw, J. T., Starrfield, S. G., Liebert, J., Green, R. F. 1979. In *White Dwarfs and Variable Degenerate Stars, IAU Colloq. No. 53*, ed. H. M. Van Horn, V. Weidemann, p. 377. New York: Univ. Rochester
McGraw, J. T., Starrfield, S. G., Liebert, J., Green, R. F. 1980. In preparation
Mestel, L. 1952. *MNRAS* 112: 583
Michaud, G., Charland, Y., Vauclair, S., Vauclair, G. 1976. *Ap. J.* 210: 447
Michaud, G., Fontaine, G. 1979. *Ap. J.* 229: 694
Milton, R. L. 1974. *Ap. J.* 189: 543
Moffett, T. J., Barnes, T. G., Evans, D. S. 1978. *Astron. J.* 83: 820
Mould, J., Liebert, J. 1978. *Ap. J. Lett.* 226: L29
Muchmore, D. O., Böhm, K. H. 1978.

WHITE DWARF STARS

Astron. Astrophys. 69:113
Nather, R. E., Robinson, E. L., Stover, R. J. 1979. In *White Dwarfs and Variable Degenerate Stars, IAU Colloq. No. 53*, ed. H. M. Van Horn, V. Weidemann, p. 453. New York: Univ. Rochester
Osaki, Y., Hansen, C. J. 1973. *Ap. J.* 185:277
Ostriker, J. O. 1971. *Ann. Rev. Astron. Astrophys.* 9:353
Ostriker, J. O. 1972. In *Stellar Evolution*, ed. H. Chiu, A. Muriel, p. 211. Cambridge, Mass.: MIT Press
Paczynski, B. 1970. *Acta Astron.* 20:47
Reimers, D. 1975. *Mem. Soc. Roy. Sci. Liège, 6te ser.* 8:369
Richer, H. 1978. *Ap. J. Lett.* 200:L95
Richer, H. 1979. In *White Dwarfs and Variable Degenerate Stars, IAU Colloq. No. 53*, ed. H. M. Van Horn, V. Weidemann, p. 259. New York: Univ. Rochester
Robinson, E. L. 1976. *Ann. Rev. Astron. Astrophys.* 14:119
Robinson, E. L. 1979. In *White Dwarfs and Variable Degenerate Stars, IAU Colloq. No. 53*, ed. H. M. Van Horn, V. Weidemann, p. 343. New York: Univ. Rochester
Romanishin, W., Angel, J. R. P. 1980. *Ap. J.* 235:992
Savedoff, M. P., Van Horn, H. M., Wesemael, F., Auer, L. H., Snow, T. P., York, D. G. 1976. *Ap. J. Lett.* 207:L45
Schatzman, E. 1958. *White Dwarfs*. Amsterdam: North Holland
Schönberner, D. 1977. *Astron. Astrophys.* 57:437
Seaton, M. J. 1968. *Astrophys. Lett.* 3:55
Shaviv, G. 1979. In *White Dwarfs and Variable Degenerate Stars, IAU Colloq. No. 53*, ed. H. M. Van Horn, V. Weidemann, p. 11. New York: Univ. Rochester
Shaviv, G., Kovetz, A. 1976. *Astron. Astrophys.* 51:383
Shipman, H. L. 1972. *Ap. J.* 177:723
Shipman, H. L. 1979a. *Ap. J.* 228:240
Shipman, H. L. 1979b. In *White Dwarfs and Variable Degenerate Stars, IAU Colloq. No. 53*, ed. H. M. Van Horn, V. Weidemann, p. 86. New York: Univ. Rochester
Shipman, H. L., Green, R. F. 1980. *Ap. J. Lett.* In press
Shipman, H. L., Greenstein, J. L., Boksenberg, A. 1977. *Astron. J.* 82:480
Shipman, H. L., Sass, C. 1980. *Ap. J.* 235:177
Sion, E. M. 1979. In *White Dwarfs and Variable Degenerate Stars, IAU Colloq. No. 53*, ed. H. M. Van Horn, V. Weidemann, p. 245. New York: Univ. Rochester
Sion, E. M., Fragola, J. L., O'Donnell, W. C. 1979. *Publ. Astron. Soc. Pac.* 91:460
Sion, E. M., Liebert, J. 1977. *Ap. J.* 213:468
Starrfield, S. G., Cox, A. N., Hodson, S. W. 1979. In *White Dwarfs and Variable Degenerate Stars, IAU Colloq. No. 53*, ed. H. M. Van Horn, V. Weidemann, p. 382. New York: Univ. Rochester
Starrfield, S. G., Sparks, W. M. 1979. *Bull. Am. Astron. Soc.* 11:663
Stauffer, J. 1980. Submitted to *Ap. J.*
Stellingwerf, R. F. 1978. *Astron. J.* 83:1184
Stevenson, D. J. 1977. *Proc. Astron. Soc. Aust.* 3:167
Stevenson, D. J. 1979. Paper presented at the C.N.R.S. International Colloquium on the Physics of Dense Matter, Paris, Sept. 17–22
Stover, R. S., Robinson, E. L., Nather, R. E. 1978. *Publ. Astron. Soc. Pac.* 89:912
Strand, K. 1977. *Astron. J.* 82:745
Strand, K., Dahn, C., Liebert, J. 1976. *Bull. Am. Astron. Soc.* 8:506
Strittmatter, P. A., Wickramasinghe, D. T. 1971. *MNRAS* 152:47
Strittmatter, P. A., Brecher, K., Burbidge, G. R. 1972. *Ap. J.* 174:L91
Sweeney, M. A. 1976. *Astron. Astrophys.* 49:375
Tapia, S. 1978. PhD thesis. Univ. Ariz., Tucson
Taylor, J. H., Manchester, R. N. 1977. *Ap. J.* 215:885
Trimble, V., Greenstein, J. L. 1972. *Ap. J.* 177:441
Truran, J. W., Starrfield, S. G., Strittmatter, P. A., Wyatt, S. P., Sparks, W. M. 1977. *Ap. J.* 211:539
Van den Heuvel, E. P. J. 1975. *Ap. J. Lett.* 196:L121
Van Horn, H. M. 1970. *Ap. J.* 160:L53
Van Horn, H. M. 1971. In *White Dwarfs, Proc. IAU Symp. No. 42*, ed. W. J. Luyten, p. 97. Dordrecht: Reidel
Van Horn, H. M. 1979. *Physics Today* 32:23
Vauclair, G. 1979. In *White Dwarfs and Variable Degenerate Stars, IAU Colloq. No. 53*, ed. H. M. Van Horn, V. Weidemann, p. 165. New York: Univ. Rochester
Vauclair, G., Fontaine, G. 1979. *Ap. J.* 230:563
Vauclair, G., Reisse, C. 1977. *Astron. Astrophys.* 61:415
Vauclair, G., Vauclair, S., Greenstein, J. L. 1979. *Astron. Astrophys.* 80:79
Warner, B., Robinson, E. L. 1972. *Nature* 239:2
Wegner, G. 1974. *MNRAS* 166:271
Wegner, G. 1979a. *MNRAS* 187:17
Wegner, G. 1979b. *Astron J.* 84:1384
Wehrse, R. 1972. *Astron. Astrophys.* 19:453

Wehrse, R. 1977. *Mem. Soc. Astron. Italiana* 48:13
Wehrse, R., Liebert, J. 1980a. *Astron. Astrophys.* 83:184
Wehrse, R., Liebert, J. 1980b. *Astron. Astrophys.* In press
Weidemann, V. 1967. *Z. Astrophys.* 67:286
Weidemann, V. 1968. *Ann. Rev. Astron. Astrophys.* 6:351
Weidemann, V. 1975. In *Problems in Stellar Atmospheres and Envelopes*, ed. B. Baschek, W. Kegel, G. Traving, p. 173. New York-Berlin: Springer
Weidemann, V. 1977a. *Astron. Astrophys.* 59:411
Weidemann, V. 1977b. *Astron. Astrophys.* 61:L27
Weidemann, V., Koester, D. 1980. Submitted to *Astron. Astrophys.*
Wesemael, F. 1978. *Astron. Astrophys.* 65:301
Wesemael, F. 1979. *Astron. Astrophys.* 72:104
Wesemael, F., Van Horn, H. M. 1979. In *White Dwarfs and Variable Degenerate Stars, IAU Colloq. No. 53*, ed. H. M. Van Horn, V. Weidemann, p. 130. New York: Univ. Rochester
Wesselius, P., Koester, D. 1978. *Astron. Astrophys.* 70:745
Wickramasinghe, D. T. 1979. In *White Dwarfs and Variable Degenerate Stars, IAU Colloq. No. 53*, ed. H. M. Van Horn, V. Weidemann, p. 35. New York: Univ. Rochester
Wickramasinghe, D. T., Bessell, M. S., Cottrell, P. L. 1977. *Ap. J. Lett.* 217:L65
Wickramasinghe, D. T., Bessell, M. S. 1979. *MNRAS* 186:399
Wickramasinghe, D. T., Strittmatter, P. A. 1970. *MNRAS* 147:123
Wood, P. R., Cahn, J. H. 1977. *Ap. J.* 211:499
Wray, J. D., Parsons, S. B., Henize, K. G. 1979. *Ap. J. Lett.* 234:L187

NUCLEAR ABUNDANCES AND EVOLUTION OF THE INTERSTELLAR MEDIUM

✳2171

P. G. Wannier

Owens Valley Radio Observatory, California Institute of Technology, Pasadena, California 91125

1 INTRODUCTION

It is taken for granted that nucleosynthesis in stars is responsible for the production of heavy elements in the Galaxy. The study of nuclear abundances involves virtually every major branch of theoretical and observational astronomy. Observed interstellar abundances form a record of the past events of nuclear production, with different nuclei sensitive to different types of events. Thus, for example, the isotopes of hydrogen and helium are considered to be important indicators of conditions in the Big Bang. Certain isotopes of heavier elements are thought to be produced and expelled by metal-poor stars and can be used to follow nuclear evolution in the early Galaxy. Other isotopes, which are formed (or destroyed) primarily in metal-rich stars, provide information about recent nuclear production and mass loss, primarily in the galactic disc. Elements of the CNO group are affected by stars of intermediate mass while heavier elements are produced and expelled by more massive stars. Thus, by measuring the abundances of many nuclear species, we are able to reconstruct many details about the history of star formation and nucleosynthesis in the Galaxy.

An extensive review, "The Origin and Abundances of the Chemical Elements" (Trimble 1975), provides a bibliography for work prior to 1975. Since that time, the observation of interstellar abundances has developed dramatically. Millimeter-wave observations now provide the sensitivity to measure nine independent isotope ratios in dense clouds. Cosmic ray experiments now provide the mass resolution to distinguish among adjacent isotopes of elements up to and including those in the iron group.

Ultraviolet satellite observations such as Copernicus and the I.U.E. ultraviolet telescope have made accessible a large number of new atomic and molecular lines.

The interstellar medium (ISM) is very diverse, containing both the enriched material that is ejected from old stars and material that is about to be incorporated into new ones. Different components of the ISM provide information on various aspects of nucleosynthesis and mass loss. Millimeter-wave measurements have been used to study giant molecular clouds whose compositions are probably representative of a long-term average of the nuclear enrichment. Planetary nebulae and their precursors can be observed by means of optical infrared and millimeter-wave lines, yielding information about the nuclear contribution of mass lost from late-type stars. Cosmic ray analysis and observations of supernova remnants fulfill a similar role for the contribution from supernovae. For each of these components, isotope measurements hold the promise of providing accurate (better than $\pm 10\%$) abundance determinations which may be able to illuminate some of the details of nucleosynthesis. The measurement of isotope ratios neatly sidesteps some particularly knotty problems which result from the different physical properties of the elements. For example, different ionization potentials will affect recombination line strengths, particle acceleration, and chemical reactivity. Chemical bond strengths will determine the formation of molecules and depletion onto grains. Such problems persist for all determinations of chemical abundances whether in cosmic rays, meteors, or stellar atmospheres. This review presents observations of interstellar abundances with special emphasis given to isotope ratios.

2 MOLECULAR ISOTOPE ABUNDANCES AND CHEMICAL FRACTIONATION

Molecules have very rich spectra which allow for observations ranging from the radio to the far UV. Rotational and vibrational levels may be observed at their fundamental frequencies in the millimeter-wave and IR, respectively, or in the optical or UV by means of the structure imposed on otherwise simple electronic transitions. The splitting of rotational levels can give rise to a variety of centimeter-wave transitions.

Rotational and vibrational energy levels can be excited in many interstellar clouds: the rotational levels in dense clouds and the vibrational levels in a variety of infrared emitting regions, especially in and around late-type stars. Visible and UV lines can be observed in absorption along lines of sight to early-type stars. Rotational (or vibrational) energy levels vary inversely with atomic mass (or its square root) so that the isotopic splitting is resolvable and isotope abundances are measurable. Element abundances, on the other hand, cannot be measured in molecular clouds because of many imponderables of cloud chemistry.

The logical progression from a measured spectral line to a derived interstellar isotope abundance involves two important steps. First, it must be ascertained that measured lines accurately measure molecular abundances. Line formation problems vary greatly with observing technique and are discussed individually along with the results. Second, it must be determined that the isotope composition of molecules is the same as that prevailing in the ISM. This second problem, which is the subject of the present section, potentially involves all isotope abundance determinations.

The isotopes of a given element are very similar, but they are not identical. In certain circumstances the difference in nuclear mass can cause a difference in chemical behavior. The separation of isotopes by chemistry is referred to as chemical fractionation, an effect which is useful for industrial applications but of little value to astronomers who wish to learn about interstellar abundances. Fortunately, fractionation does not appear to be important for the interesting elements, nitrogen, oxygen, sulphur, and silicon. For these elements, the fractional ionization in molecular clouds is too small for ion-molecule fractionation reactions to occur and the neutral fractionation reactions proceed too slowly to be important (Langer 1979, Herbst & Klemperer 1973). Fractionation is, however, important for hydrogen and can be important for carbon in certain circumstances.

The fractionation reactions for hydrogen and carbon involve ionized species because ion-molecule reactions are known to proceed with very small activation energies. For example, C^+ and H_3^+ can react via[1]

$$^{13}C^+ + CO \leftrightarrows C^+ + ^{13}CO + \Delta E \qquad (1)$$

and

$$H_2D^+ + HCN \leftrightarrows H_3^+ + DCN + \Delta E. \qquad (2)$$

Even with a suitable reaction, fractionation cannot proceed unless the values of ΔE are large enough to drive the reaction forward at the temperatures (~ 10–100 K) of interstellar clouds. For fractionation reactions, ΔE results from a slight difference in the molecular binding energy of different isotopes and is easy to calculate.

---

[1] Henceforth I shall adopt the usual convention that any molecule, unless otherwise specified, is composed of its most abundant nuclear species. Thus $^{13}C^{16}O$ is written ^{13}CO and $^{1}H^{12}C^{14}N$ is HCN. All abundance ratios, unless otherwise specified, are by number, not by mass.

In interstellar clouds, the kinetic temperature is usually low enough (and the collision rate slow enough) that molecules exist in their ground vibrational state ($v = \frac{1}{2}$) with an associated energy, $\frac{1}{2}hv_{vib}$. This energy depends on nuclear mass since $v_{vib} \, \alpha \mu^{-1/2}$, where μ is the appropriate reduced mass. The relative advantage in binding energy of incorporating heavy isotopes in a given molecule is the difference of ground state energies for the two isotopes in question. In a case where one molecule must fractionate relative to another, this advantage must be calculated for each molecule and the difference becomes the quantity ΔE.

Hydrogen Fractionation

The case for D/H fractionation is clear. Significant interstellar fractionation is both predicted and observed. The values of ΔE associated with deuterating reactions are large due to the small hydrogen mass and the large D/H mass ratio. For the transfer of D from atomic form to H_2, $\Delta E/k$ = 404 Kelvins. In very dense clouds the transfer of deuterium from HD to HCN liberates an additional energy corresponding to 456 K. The potential for D/H fractionation is clear. In interstellar space, equilibrium is not actually achieved and the predicted enhancements of deuterium in both H_2 and in several more complex molecules result from a series of competing reactions which are thoroughly explored by Watson (1976). The observations of D/H fractionation are also very extensive. Since the initial discovery of DCN by Jefferts et al. (1973), the deuterated species of seven other molecules have been observed, and they all show a large D/H enhancement. An elegant set of measurements is provided from the ultraviolet lines of H, D, H_2, and HD toward several nearby stars (York & Rogerson 1976). Because these measurements provide a look at almost all of the hydrogen, the real interstellar D/H ratio (1.8×10^{-5} by number) can be measured, as can the relative fractionation of deuterium in molecular hydrogen. A recent survey of DNC, HNC, and $HN^{13}C$ in sources representing a variety of temperatures even provides an astronomically determined value of $\Delta E/k (\sim 240$ K) for the deuteration of this molecule (Snell & Wootten 1980). In large part, the observed D/H fractionation is in very satisfactory agreement with the predictions of Watson.

Carbon Fractionation

On theoretical grounds, carbon isotope fractionation is generally expected to be a much weaker effect, but the problem is very important due to the key role of carbon in the study of molecular isotopic abundances. Reaction (1) was first introduced by Watson et al. (1976) and its potential astrophysical importance was emphasized especially by Langer (1976) and Watson

(1977). In contrast to the D/H fractionation reaction, the ΔE associated with reaction (1) is rather small, $\Delta E/k \sim 35$ K. An understanding of the actual role of carbon fractionation has emerged gradually as a result of new and sensitive observations and of increasingly sophisticated model calculations. The conclusion is that significant fractionation is likely to be important mainly at the outer boundaries of clouds. Fortunately, most isotope ratio determinations refer to dense cores of clouds, which are less likely to be affected by fractionation. Both the observations and the model calculations indicate that fractionation is not very significant for the observed abundances in the dense cores of cool clouds. This conclusion also applies to the dense cores of giant active clouds where, in addition, the kinetic temperature may be large compared to $\Delta E/k$. A look at the evidence is certainly worthwhile.

The actual operation of reaction (1) involves the formation and destruction of CI, CII, and CO, the dominant reservoirs of gas-phase carbon. If we are interested in the ^{13}C enhancement of CO (rather than the ^{13}C depletion of atomic or ionized carbon) then it is important that CO be a relatively minor constituent of the total gas-phase carbon. This condition is met in cloud exteriors, where CI and CII are thought to dominate. Ultraviolet ionization of CI is very important since it provides a link between CO and atomic carbon. The inverse reaction is the radiative recombination of C^+. In the outer regions of clouds where there is a significant ($>1\%$) fractional ionization of CI by UV photons, the time scale for operation of the fractionation reaction is $\lesssim 10^6$ years. The H_3^+ abundance, depressed by the rich supply of electrons, is probably too low to feed CI into the rich chemical pathways starting with CH^+. All these conditions favor the operation of $C/^{13}C$ fractionation. The time-dependent models of Liszt (1978) clearly show a significant enhancement of $^{13}CO/CO$ relative to the $^{13}C/C$ abundance ratio. This enhancement is expressed as $R(CO) \equiv (\xi_{12C}/\xi_{13C})[x(^{13}CO)/x(^{12}CO)]$, where $\xi \equiv$ the total isotope abundance, $x(\text{mol}) \equiv n(\text{mol})/n$ and $n \equiv$ total equivalent H density $\approx 2n(H_2)$. Liszt derives values of R up to 8 or so in the outer region of cool clouds, in general agreement with the previous, but less extensive calculations of Langer (1977).

The situation is different in cloud cores where the UV flux is low and the fractional ionization is reduced. In the cores neutral carbon is largely processed into molecules and most of the carbon probably exists in the form of CO. Under these circumstances, fractionation is not only very difficult to initiate, but it is even difficult to maintain. If most of the carbon is processed into CO, then, by necessity, CO must reflect the isotope abundance of most of the carbon. The model calculations predict a degree of fractionation which depends on the physical conditions in the cloud and on the molecule being considered. Such model calculations immediately suggest two types of observational tests.

The first test involves determining one ratio such as $C/^{13}C$ using one molecule, but made in a variety of locations with different physical conditions. The model described above suggests the mapping of $C/^{13}C$ in a single cool cloud including both the core and cloud edge. Alternatively, determination from many clouds can be used to seek a systematic variation of apparent isotope abundance with temperature or density. Because cloud temperatures range from about 10 to 100 K, the Boltzman factor $e^{-35/T}$ can take on very different values. Whereas carbon fractionation is seldom expected to reach equilibrium, the Boltzman factor does provide an upper limit to the extent of fractionation. An elegant set of observations by Langer et al. (1980) provides a test of carbon fractionation by its thorough measurement of carbon monoxide line emission at several locations in each of three isolated dark clouds. These observations include $J = 1$–0 spectra of CO, ^{13}CO, and $C^{18}O$ as well as $J = 2$–1 spectra of CO and ^{13}CO. In addition, a $J = 1$–0 observation of the very rare $^{13}C^{18}O$ was made at the center of each cloud. The $J = 2$–1 and $J = 1$–0 lines together provide a determination of density and temperature, thus allowing for the calculation of reliable column densities for each of the isotopic species. The results provide definite evidence for a ^{13}C enhancement (fractionation) by a factor of ~ 2 at the cloud edges, a result in agreement with model predictions. In the cloud cores, the $C/^{13}C$ molecular abundance ratio is provided by the measurement of $C^{18}O/^{13}C^{18}O$ and yields an average value of 60 ± 5. This value is in excellent agreement with $C/^{13}C = 60 \pm 8$, a result obtained in giant molecular clouds by a variety of techniques (see below).

A second, and independent test of the importance of carbon fractionation is the simultaneous measurement of $^{13}C/^{12}C$ in one location, but using several different molecules. The idea here is that if C^+ fractionates relative to CO according to reaction (1), then any molecule forming from C^+ must appear fractionated relative to CO (and the molecules which form from CO). Fortunately, for cloud cores such tests are readily available from the data. HCN and CS fall into the former category (downstream from C^+) in contrast to CO (and probably HCO^+ and H_2CO). The observational results, which refer to cloud cores, are presented in the next section and are not discussed here. As we shall see, the evidence is consistent with a lack of relative fractionation among the molecules, at least within the accuracy of the present abundance determinations.

Thus, the conclusion to be drawn is that the model predictions and the observational tests are in satisfactory agreement. There appears to be carbon fractionation in certain circumstances, namely in the outer boundaries of cool clouds. Fortunately, abundance measurements seldom probe these locations. The results presented in the following section are from the dense cores of giant clouds where fractionation effects should be negligible. The fractionation results do, however, warn of potential pitfalls in the handling of the data. Langer et al. stress especially the danger of using the spectral line wings to escape from problems of line saturation at the line centers. According to the cloud models that espouse a large-scale gravitational collapse, the line wings select material from the cloud edges. Such selection applies especially to sources, such as small cold clouds with no prominent cores. However, most of the giant clouds display distinctly broader spectral lines ($\Delta v \gtrsim 10$ km s^{-1}) in the direction of their dense cores, definitely indicating that, in these cases, the line width is due to some local phenomenon such as turbulence (cf. Linke & Goldsmith 1980). In such a case, even the line wings should be safe for abundance measurements. With one or two possible exceptions, the sources in Tables 1–3 are of the broad line type which should be dependable even in the line wings. A possible exception is NGC2024 which has a value of $\Delta v \lesssim 5$ km s^{-1}, rather small by the standards of giant clouds (Wannier et al. 1976a).

3 GIANT CLOUDS: MILLIMETER-WAVE AND CENTIMETER-WAVE SPECTRA

Isotope abundances have been determined in over thirty dense clouds scattered throughout the galactic disc. Most of these determinations have used two or more of the nineteen observed isotopic species of CO, H$_2$CO, HCN, CS, and SiO. Several of the most extensive surveys provide double isotope ratios rather than single ones. The difficulty of obtaining single isotope ratios is an experimental one, resulting from the fact that many elements are dominated by one isotopic form. Thus, the solar value of O/^{17}O is 2675 or of N/^{15}N is 272. These large ratios create an experimental bind. Spectral lines of the more abundant isotopic species must be weak enough to ensure that the line suffers no serious optical depth effects (see below). This requirement often makes the line of the less abundant species impossible to detect. Hence, many experiments yield double isotope abundance ratios by comparing the line strengths of two molecules having isotope substitutions of two different elements [for example, |C^{18}O|/|^{13}CO| which yields (C/^{13}C)/(O/^{18}O)]. The observations thus provide a set of single ratios as well as a number of interlocking double ratios. A fruitful approach to unraveling the results is to focus attention on those clouds where several isotope experiments have yielded significant results. This approach is also the best one to control possible errors, either experimental or resulting from possible effects of optical depth or chemical fractionation. The results of many experiments have been collected by Penzias (1980a) for his talk at the 87th IAU Symposium. This collection contains most of the recent results, including those from eleven papers which are unpublished as of this writing. The results are contained in Tables 1–3, which can be viewed in several ways to yield useful information. First, the rows are arranged in order of ascending distance from the galactic center. Therefore, a glance down the columns can indicate a possible radial abundance gradient similar to those reported in external galaxies (cf Smith 1975, or Jensen et al. 1976). Second, the measured abundances at ~ 10 kpc can be compared to the terrestrial abundances listed at the bottom. A difference might indicate an evolution of the ISM since the time (~ 5 billion years ago) when the protosolar nebula condensed. Third, consistent cloud-to-cloud variations at comparable galactocentric radii may indicate recent (local) changes, possibly from some single event or set of events such as a supernova or extensive mass loss from a cluster of stars. These three views are considered for the various reported ratios.

Carbon Isotopes

Carbon figures heavily in many of the measured isotope abundances, in large part because of its central role in interstellar chemistry. All the data in Table 1 measure the $C/^{13}C$ ratio, either singly or in combination with a second isotope ratio. To ease the job of isolating the $C/^{13}C$ results, each of the double ratios (columns 7–11) is multiplied by the terrestrial value of the companion ratio to $C/^{13}C$. Thus, $C^{34}S/^{13}CS$ is multiplied by 23, the terrestrial value of $S/^{34}S$. If the interstellar $S/^{34}S$ ratio were actually equal to the terrestrial one, then the resulting value would equal the interstellar $C/^{13}C$ abundance ratio.

The several different surveys together provide tests for consistency as well as for possible effects of fractionation and high optical depth. One of the two tests for fractionation discussed above involves a comparison of the isotope ratio of CO to the ratio in HCN or CS. It would be expected that if fractionation were significant, then the ratios for CO (columns 2 and 8) ought to be systematically smaller than the ratios for HC_3N, CS, or HCN (columns 4, 7, and 11). Yet, to a small extent, the opposite is in fact the case!

Another type of check is for possible systematic errors resulting from insufficient allowance for opacity. Each survey contains its own internal checks for the effects of line saturation and these are discussed in the original references. A crosscheck is possible by noting that of the seven most extensive surveys, the ^{13}C-bearing isotopic form is the less abundant one in three cases (columns 1, 2, and 7) and is the more abundant form in the four other cases (columns 8, 9, 10, and 11). If there were a general tendency to

Table 1 $C/^{13}C$ ratio in giant clouds[a]

Source	R (kpc)	(1) H_2CO / $H_2^{13}CO$	(2) $C^{18}O$ / $^{13}C^{18}O$	(3) HCO^+ / $H^{13}CO^+$	(4) HC_3N / $H^{13}CCCN$	(5) NH_2CHO / $NH_2^{13}CHO$	(6) OCS / $O^{13}CS$	(7) $C^{34}S$ / ^{13}CS (×23)	(8) $C^{18}O$ / ^{13}CO (×500)	(9) $H_2C^{18}O$ / $H_2^{13}CO$ (×500)	(10) $HC^{18}O^+$ / $H^{13}CO^+$ (×500)	(11) $HC^{15}N$ / $H^{13}CN$ (×272)
Sgr A	0.0	20±10(2)	26±5	21±2				33±2	43±5	60±9		<12(3)
Sgr B	0.1	25±10(2)	23±3	26±4				35±2	32±11	77±4	>45	<14(3)
W43	5.5	42±6 (1)			22±1	24±3	21±5	49±5	61±8			
W33	5.7	74±11(1)	42±4					68±4	52±40	79±10		~64(3)
W51	7.6	70±11(1)	32±4					74±5	62±1	96±15		~48(3)
W31	8(?)	37±6 (1)							54±25			
M17	8.0	85±14(2)	91±20					62±3	49±7		91±10	
NGC 6334	9.3	34±9 (2)	91±28					44±3	74±3		63±7	35±7(13)
W49	9.4	53±8 (1)						79±8	39±18			
DR21 (DR21OH)	9.9	73±11(1)	143±43					115±12	34±9		62±6	~58(3)
Ori A	10.9		>74		50±5			64±3	50±4		24±5	~48(3)
NGC 2024	11.0	69±7 (2)	56±8	83±24				96±20	27±4	<75		~51(3)
NGC 2264	11.1		83±31					72±7	60±3			~38(3)
W3 (OH)	12.2	86±13(1)	111±40						37±3			~63(3)
NGC 7538	12.7		77±21					53±4	42±7			~51(3)
Solar System	10	89	89	89	89	89	89	89	89	89	89	89
Ref.[b]		1,2	4	5	6	7	8	9	10	11	5,12	3,13

[a] The values in columns 1–6 are line ratios, occasionally corrected for effects of spectral line formation. The values in columns 7–11 are actually double isotope ratios of carbon with sulphur (7), oxygen (8–10), and nitrogen (11) and have been multiplied by the solar system values of $S/^{34}S$, $O/^{18}O$ and $N/^{15}N$, respectively. All uncertainties are $\pm\sigma$ and the inequalities are always given with 2σ confidence. R is galactocentric radius.

[b] References
1. Henkel et al. 1980
2. Gardner & Whiteoak 1979
3. Wannier et al. 1980a
4. Linke 1980
5. Stark 1980
6. Wannier & Linke 1978a
7. Lazareff et al. 1978
8. Goldsmith & Linke 1980
9. Frerking et al. 1980
10. Wannier et al. 1980b
11. Kutner et al. 1980
12. Guélin & Thaddeus 1979
13. Linke et al. 1977

overlook optical depth effects, then the first set of surveys might produce apparent ratios systematically smaller than the second. Again, such is not the case.

THE GALACTIC PLANE VS THE GALACTIC CENTER   *There is no evidence for a $C/^{13}C$ gradient in the Galaxy, except that the galactic center sources appear to be different than those in the remainder of the disc.* For the remainder of this review, the galactic disc without the galactic center region is referred to as the galactic "ring." Another reason to consider the ring ($\gtrsim 3$ kpc) separately from the galactic center ($\lesssim 1$ kpc) is the conspicuous gap in molecular gas in an annulus of ~ 2 kpc, as deduced from CO surveys (see Burton 1976).

The radial variation (or lack thereof) is most easily seen in Figure 1 which contains the data from three $C/^{13}C$ surveys, namely $H_2CO/H_2{}^{13}CO$, $C^{34}S/^{13}CS$, and $C^{18}O/^{13}CO$ (columns 1, 7, and 8). One other extensive survey ($C^{18}O/^{13}C^{18}O$) has been omitted from the figure because of its very large experimental uncertainties and one ($HC^{15}N/H^{13}CN$) is omitted because uncertainties have not yet been estimated. It is apparent that the three surveys, taken together or considered individually, are consistent with a uniform ^{13}C abundance. The same conclusion results from the two surveys not included in the figure. The existence of a gradient has been

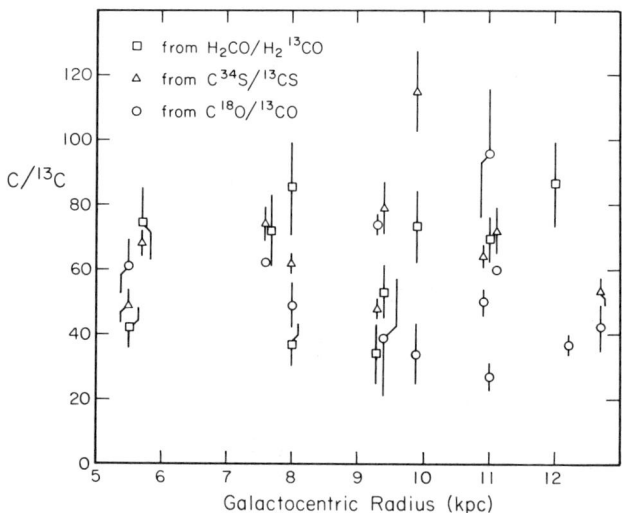

Figure 1 $C/^{13}C$ ratios in giant clouds, from Table 1, plotted according to galactocentric radius. The results of the two double ratio experiments have been multiplied by the solar values of $S/^{34}S$ ($\triangle$) and $O/^{18}O$ ($\bigcirc$), respectively. There is no apparent radial gradient, nor is there any obvious systematic variation of $C/^{13}C$ either by source or by molecule (see text).

suggested on the basis of formaldehyde data (Gardner & Whiteoak 1979) but may result from a systematic effect emphasized by Henkel et al. (1980). In their important paper, Henkel et al. have shown that the observed K-doubling transitions of formaldehyde are not equally excited in the two isotopic species, H_2CO and $H_2^{13}CO$. This difference results from the high opacity in the rotational transitions of H_2CO. Optical trapping of the rotational line would decrease the depth of the K-doubling absorption line of the more abundant ^{12}C species and would thus decrease its apparent abundance. Allowance for this effect in the very dense clouds toward the inner part of the galactic ring has virtually eliminated what would otherwise appear to be a $C/^{13}C$ gradient.

It is interesting to note that the lack of a gradient is apparent in two surveys that measure $C/^{13}C$ alone and in three that simultaneously measure the $N/^{15}N$, $S/^{34}S$, and $O/^{18}O$ gradients. A more complete discussion of these results will be given in Section 5, which also presents other methods of estimating local abundance gradients.

SOLAR VS GALACTIC ABUNDANCES *The $C/^{13}C$ ratio appears to be 60 ± 8 in the ring, an overabundance of ^{13}C by a factor 1.5 relative to the solar system value.* Table 1 contains an assortment of $C/^{13}C$ determinations in dense clouds and the calculation of an average can be approached in several ways. If it is assumed that a single isotope ratio prevails in the ring and if the formal uncertainties are accepted at face value, then a statistically weighted average of the $C/^{13}C$ determinations is best. This approach emphasizes those results that have small formal errors. As is argued below, the observational uncertainties appear to be somewhat larger than those reported. If unreported uncertainties are dominant, then a straight unweighted average of values in Table 1 would be the more valid approach. In this second case, the uncertainty in the calculated mean will be dominated by the most uncertain of the individual determinations. This could be especially awkward in the case of a weak spectral line that is barely detected. Consider, for example, the effect of averaging 60 ± 1 with 300 ± 300 if the second ratio merely indicates very poor signal-to-noise of the less abundant isotopic species. A straight (unweighted) average of the reciprocal ratio ($^{13}C/C$) would be more appropriate for the above example and would also produce a different average value.

The most conservative approach is to calculate the average in several ways (weighted or unweighted; $C/^{13}C$ or $^{13}C/C$) and adopt a mean value and uncertainty which are consistent with any of the methods described above. The value in the ring is derived as follows. The single $C/^{13}C$ isotope experiments (columns 1–6) indicate a ^{13}C overabundance by a factor 1.2– 1.8 depending on whether the values are (1.8) or are not (1.2) weighted according to their reported uncertainties. The overabundance of ^{13}C is also a general feature of all the double isotope results (columns 7–11) if terrestrial values are assumed for the interlocking isotope ratio. These interlocking ratios include S/^{34}S (column 7), O/^{18}O (columns 8–10), and N/^{15}N (column 11). The C^{34}S/^{13}CS experiment yields an apparent overabundance of ^{13}C by a factor 1.3 (unweighted average) to 1.5 (weighted). The HC^{15}N/H^{13}CN results in four sources from the results of Linke et al. (1977) yield 1.7 (unweighted) and 1.6 (weighted). A more recent survey of eight sources by Wannier et al. (1980a; column 11) does not yet report a statistical uncertainty and yields 1.69 (unweighted). Taken together, all of the results are consistent with a ^{13}C overabundance by a factor 1.3–1.7, yielding a formal abundance ratio of C/^{13}C = 60 ± 8.

LOCAL VARIATIONS There is no compelling evidence for any cloud-to-cloud variation of the C/^{13}C abundance ratio. This conclusion results from a rather detailed examination of Table 1. Any *single* isotope experiment does, in fact, yield significant cloud-to-cloud differences. It is edifying to follow a particular example. DR21 and W51 are both well-studied sources. In column 8, we find that the C/^{13}C ratio in W51 exceeds that in DR21 by 3σ, apparently a significant result. However, in column 7, exactly the opposite is the case, also a 3σ result. In column 2, DR21 is again larger while the two sources agree perfectly in column 1. The question is, do these several differences represent real molecular abundance differences or is there some additional source of experimental error that is not reflected by the formal uncertainties? It has been suggested that some apparent differences may result from isotope fractionation. Two lines of evidence point against the suggestion. First, fractionation is unlikely to be significant for these dense clouds (see above). Second, if fractionation were the explanation, then there should be some pattern to the observed differences. However, an exhaustive search of Table 1 yields no apparent systematic effect. The source-to-source differences do not correlate with temperature, column density, galactic radius, or even with each other. There may yet be a good explanation for these differences. However, for the moment the conclusion that can be drawn is that real observational uncertainties, for one reason or another, are larger than those reported.

Oxygen Isotopes

The oxygen isotope results, along with those of nitrogen, silicon, and sulphur, are given in Table 2. The single isotope ratios are presented as they were in Table 1. However, the double isotope ratios have been treated in a different manner. Since all of the double ratios include C/^{13}C, the results from the previous section have been applied to find the apparent single ratios listed at the top of each column. For example, the ^{13}CS/C^{34}S survey

Table 2 N, O, Si, and S isotope abundances in giant clouds[a]

Source	R (kpc)	(1) ($^{18}O/^{17}O$) $\frac{C^{18}O}{C^{17}O}$	(2) ($^{18}O/^{17}O$) $\frac{^{18}OH}{^{17}OH}$	(3) ($O/^{18}O$) $\frac{^{13}CO}{C^{18}O}$ [× 26 center × 60 ring]	(4) ($O/^{18}O$) $\frac{H_2^{13}CO}{H_2C^{18}O}$ [× 26 center × 60 ring]	(5) ($O/^{18}O$) $\frac{H^{13}CO^+}{HC^{18}O^+}$ [× 26 center × 60 ring]	(6) ($O/^{18}O$) $\frac{OH}{^{18}OH}$	(7) ($N/^{15}N$) $\frac{H^{13}CN}{HC^{15}N}$ [× 26 center × 60 ring]	(8) ($Si/^{29}Si$) $\frac{^{28}SiO}{^{29}SiO}$	(9) ($^{29}Si/^{30}Si$) $\frac{^{29}SiO}{^{30}SiO}$	(10) ($^{34}S/^{33}S$) $\frac{C^{34}S}{C^{33}S}$	(11) ($S/^{34}S$) $\frac{^{13}CS}{C^{34}S}$ [× 26 center × 60 ring]
Sgr A	0.0	3.4 ± 0.4	4.2 ± 0.4	302 ± 35	217 ± 33		275 ± 100	> 600	8.4 ± 0.5	1.3 ± 0.1		18.1 ± 1.1
Sgr B	0.1	3.2 ± 0.2		406 ± 140	168 ± 9	< 288	240 ± 100	> 500	10.8 ± 6.0	1.6 ± 0.2	5.3 ± 1.0	17.1 ± 1.0
W43	5.5			491 ± 64								28.1 ± 2.9
W33	5.7	3.6 ± 0.2		577 ± 444	379 ± 48			253				20.3 ± 1.2
W51	7.6	2.8 ± 0.1		483 ± 16	312 ± 49			341	11.8 ± 1.5			18.6 ± 1.3
W31	8(?)			555 ± 257								
M17	8.0	3.5 ± 0.3		612 ± 87		329 ± 36						22.2 ± 1.1
NGC 6334	9.3	3.3 ± 0.2		405 ± 16		476 ± 53		460 ± 94				31.3 ± 2.1
W49	9.4			769 ± 355								17.5 ± 1.8
DR21 (DR21OH)	9.9	3.4 ± 0.3		882 ± 233		484 ± 47		281				12.0 ± 1.3
Ori A	10.9	3.6 ± 0.3		600 ± 48	> 400	1250 ± 260	900 ± 300	342	9.0 ± 0.5	1.4 ± 0.1	4.8 ± 0.5	21.6 ± 1.0
NGC 2024	11.0	3.5 ± 0.2		1100 ± 163				319				14.4 ± 3.0
NGC 2264	11.1	3.5 ± 0.3		500 ± 25				428				19.2 ± 1.9
W3	12.2	3.2 ± 0.3		811 ± 66				259				
NGC 7538	12.7	3.5 ± 0.3		714 ± 119				318				26.0 ± 2.0
Solar system	10	5.5	5.5	500	500	500	500	272	20	1.5	5.5	23
Ref.[b]		16, 26	17	10	11	5, 12	14, 15	3, 13	18	18	19	9

[a] Table 2 is similar to Table 1 except that the double isotope ratios (columns 3, 4, 5, 7, 11) have been multiplied by adopted values of the galactic $C/^{13}C$ ratio. These values are 26 for Sgr A and Sgr B, and 60 for all other sources (see text). The desired abundance ratio is indicated at the top of each column and the corresponding solar system ratio is given at the bottom of the table.

[b] References:
1–13. Listed in Table 1.
14. Whiteoak & Gardner 1975
15. Whiteoak & Gardner 1978
16. Penzias 1980c
17. Gardner & Whiteoak 1976
18. Wolff 1980
19. Wilson et al. 1976
26. Wannier et al. 1976b

which measures $(S/^{34}S)/(C/^{13}C)$ has been made to yield the $^{34}S/S$ by multiplying by the appropriate $C/^{13}C$ ratio, namely 26 in the galactic center and 60 in the galactic ring sources. If the $C/^{13}C$ results are valid, then the resulting number can be compared directly to other observations or to the corresponding solar system value.

$^{18}O/^{17}O$ *This ratio is ~ 3.2 throughout the galactic disc, significantly different from the solar value of 5.5.* There are three stable oxygen isotopes, namely ^{16}O, ^{18}O, and ^{17}O with relative solar abundances of 2750/5.5/1. Predictably, the large $O/^{18}O$ and $O/^{17}O$ ratios are very difficult to measure directly, and only approximate or indirect ratios are available (see below). Happily, the $^{17}O/^{18}O$ ratio is not only easy to measure, but it provides a very consistent and unambiguous result. The extensive carbon monoxide survey in column 1 of Table 2 is striking for its relative lack of source-to-source variation, a result that applies equally to the galactic center sources. It is thought that the oxygen isotopes cannot fractionate in interstellar conditions, a prediction that is borne out by the lack of variability and by the consistent $^{18}OH/^{17}OH$ result in Sgr A (column 2). Indeed, the surprise is that all of the galactic measurements are significantly lower than the solar system abundance ratio, a result that strongly implies an evolutionary increase in the ratio if the solar system is taken to be typical of prevailing abundances 5 billion years ago. This implication is strengthened by the large $^{17}O/^{18}O$ abundance measured in a protoplanetary nebula (see below).

$O/^{18}O$ *The measurements are somewhat uncertain, but indicate that there is an overabundance of ^{18}O by a factor ~ 2 in the galactic center, but not in the galactic ring.* The chief uncertainty results from the spatially constant $^{17}O/^{18}O$ ratio (above). While the constant $^{17}O/^{18}O$ does not contradict a spatial variation of $O/^{18}O$, it does indicate that $^{17}O/O$ and $^{18}O/O$ must vary in the same manner, a surprising coincidence in view of the very different nucleosynthesis of ^{17}O and ^{18}O (below).

The case for an overabundance of ^{18}O in the galactic center rests largely on double ratio measurements (columns 3–5). If the adopted $C/^{13}C$ values are the correct ones, then the galactic center sources have an $O/^{18}O$ value that varies among experiments, but is consistently lower than the solar value. The one direct $O/^{18}O$ measurement (column 6) also indicates an overabundance of ^{18}O, but Penzias (1980b) reports an argument by Cernicharo & Guélin (1980) that casts some doubt on the OH results. It is argued that the rotational excitation of the common OH species can be significantly affected by radiative trapping of its saturated rotational lines. ^{18}OH is not affected in the same way because of its much smaller column density. The calculated trapping corrections could possibly make the observed line strengths in the galactic center consistent with the terrestial value of 500.

The $O/^{18}O$ data in the ring are consistent with the solar value although there is some significant scatter. Because of the large fractional uncertainties in some cases, the data are best averaged in their reciprocal form, namely $^{18}O/O$. When this is done, the data in columns 3–6 yield an average $O/^{18}O$ ratio which ranges from 450 (weighted) to 550 (unweighted), or from 0.9 to 1.1 times the solar value. It may be of interest that the largest $(O/^{18}O)$ value in column 3 (Table 2) comes from NGC 2024, the one giant cloud with a less-than-prominent broad line core (Section 2).

Nitrogen: $N/^{15}N$

^{15}N is underabundant in the galactic center by a factor of ≥ 2 relative to the solar system value. Also, there may be an underabundance by a factor ~ 1.2 in the galactic ring, but this conclusion is sensitive to the adopted value of the $C/^{13}C$ ratio in the ring and consequently is a bit uncertain.

The case for the galactic center result is quite strong. The relative underabundance of ^{15}N (or the overabundance of ^{14}N!) was first noted by Linke et al. (1977), a paper which includes a complete discussion of excitation and line formation as applied to hydrogen cyanide. A most important point is that because $H^{13}CN$ is the more abundant of the two species, it is the more likely to be underestimated from the spectra. Therefore, an apparent relative *overabundance* of $H^{13}CN$ is strengthened by additional corrections for line formation. The more abundant $H^{13}CN$ is clearly saturated in portions of the Sgr A and Sgr B spectra and the reported nitrogen abundance ratio is calculated at those velocities where the opacity is lowest. The measurements of $HC^{15}N$ are therefore regarded only as providing a lower limit (2σ) to the $N/^{15}N$ ratio, already significantly higher in the galactic center than in the ring or in the solar system.

The case for an underabundance of ^{15}N in the ring is weaker than for the galactic center. It is clear from column 11 of Table 1 that the double isotope ratio $(C/^{13}C)/(N/^{15}N)$ is very much smaller than the terrestrial one. However, a large part of the small value is accounted for by the adopted $C/^{13}C$ ratio in the ring. The unweighted average of $^{15}N/^{14}N$ after accounting for the carbon ratio (column 7 of Table 2) is 0.82 times the corresponding solar system value. The apparent underabundance of ^{15}N in the ring is also suggested by the recent detection of $^{15}NH_3$ in Orion A. After correcting for appreciable opacity in the $^{14}NH_3$ line, Wilson & Pauls (1979) report $^{14}N/^{15}N > 300$ with 2σ confidence. More secure measurements of the nitrogen isotope ratio are still needed.

Silicon and Sulphur: ^{28}Si, ^{29}Si, ^{30}Si, ^{32}S, ^{33}S, ^{34}S

These two elements are considered together, in part because of their similar nucleosynthetic origin and in part because of similar properties of their measured abundance ratios in the Galaxy. All the abundance results are derived from the CS and SiO data which appear in columns 8–11 of Table 2. Both of these molecules are chemically similar to the very abundant CO molecule, silicon being electronically similar to carbon, and sulphur being electronically similar to oxygen. SiS, the fourth similar molecule, is also observed, but no isotopic measurements have been forthcoming. Both silicon and sulphur have three observed isotopes and therefore each yield two independent isotope ratios.

THE LACK OF SPATIAL VARIATION   *A property of all the silicon and sulphur isotopes is that the measured abundances are similar in all clouds, including those near the galactic center.* The four measured abundance ratios each include sources both in the galactic ring and in the galactic center (columns 8–11). When the statistically weighted averages of the ratios are calculated separately for the center sources and for the ring sources, the resulting values agree to within 1σ for all four of the isotope ratios. (In the case of $S/^{34}S$, the only isotope ratio that is part of a double ratio, the statistical uncertainty necessarily includes the $\sim 10\%$ uncertainty in the adopted $C/^{13}C$ abundances of the ring and the center.)

COMPARISON TO SOLAR SYSTEM ABUNDANCES   *The $Si/^{29}Si$ ratio is significantly (0.5 $\times$) smaller than the solar system value, whereas both of the ratios of rare isotopes, namely $^{28}Si/^{29}Si$ and $^{34}S/^{33}S$, are perfectly consistent with solar system abundances.* $S/^{34}S$ also appears to be slightly ($\sim 0.85 \times$) smaller than the solar value, but the effect is only marginally significant in view of the $\sim 15\%$ uncertainty of the interlocking $C/^{13}C$ adopted galactic ratios. For the $S/^{34}S$ result, it is of interest to note that any correction for line saturation would exaggerate the difference from the solar value. In contrast, the $Si/^{29}Si$ differs from solar in the direction that might result from line saturation effects so that, in this case, the internal checks for line saturation are particularly important. A detailed discussion is given in the original reference (Wolff 1980). Briefly stated, the SiO brightness temperature is very small (<0.7 K in all cases), even considering the small apparent source sizes. Confirmation of the sub-solar $Si/^{29}Si$ ratio must await the results of new observations now in progress (Penzias 1980b).

Deuterium

Deuterium occupies a unique nitch in the study of isotope abundances; see Table 3. The D/H ratio is of central importance to cosmologists who wish to determine the conditions in the Big Bang. However, the D/H ratio is notoriously hard to measure and misinterpretation of observations can lead to estimated abundances that are incorrect by factors of ten or more. Part of the problem was discussed in the previous section on chemical fractionation, namely that the D/H mass ratio is large compared to other isotope mass ratios. As a result D and H have significantly different chemical and physical properties. Symptomatic of the problems associated with the interstellar D/H ratio is the great difficulty in even determining a reliable solar system value. D is destroyed in the sun and H preferentially escapes from the Earth. Fractionation is certain to play a large role in dense interstellar clouds. Yet, the results in dense clouds are important, so that serious efforts have been made to study molecular D/H ratios and understand their significance. The D/H observations and the problems associated with their interpretation are reviewed in a comprehensive way by Penzias (1978) and in a way that relates specifically to the DCN results by Penzias et al. (1977).

EVIDENCE FOR A GALACTIC GRADIENT Measurements have been made of the deuterated and non-deuterated species of the three molecules HCN, HCO^+, and NH_3. While all three are thought to be heavily fractionated, so that the apparent D/H ratios in themselves are only upper limits, the significant result is the consistent appearance of a radial D/H gradient in the sense that deuterium appears relatively depleted toward the galactic center. The most important problem in interpreting the results of Table 3 is

Table 3 D/H in giant clouds[a]

Source	R (kpc)	(18) DCN/HCN (× 1000)	(19) DCO^+/HCO^+ (× 1000)	(20) NH_2D/NH_3 (× 1000)
Sgr A	0.0	1.1 ± 0.4	0.4	
Sgr B	0.1	0.8 ± 0.5	0.4	17
W33	5.7	2.3 ± 0.4		
W51	7.6	1.8 ± 0.7		
M17	8.0	1.8 ± 0.6	0.7 ± 0.3	
DR 21 (DR21OH)	9.9	1.3 ± 0.3	5.3 ± 0.5	
Ori A	10.9	4.8 ± 1.0	1.7 ± 0.6	50
NGC 2264	11.1		10.2 ± 0.7	
W3(OH)	12.2	4.3 ± 0.9		
NGC 7538	12.7	4.6 ± 1.0		
Ref.[b]		20, 21	21	22

[a] Table 3 is similar to Table 1. The values are the measured D/H ratios in molecules, which are thought to be proportional to the true D/H ratio (see text).
[b] References:
20. Penzias et al. 1977
21. Penzias 1979
22. Turner et al. 1978

the large but unknown extent of D/H fractionation. The hope is clearly that there is no systematic radial dependence of deuterium fractionation.

4 CURRENT ENRICHMENT OF THE ISM: MASS LOSS FROM EVOLVED STARS

For studies of nuclear evolution it is important to know the composition of material being returned to the ISM from evolved stars. Circumstellar clouds and planetary nebulae (PN) provide information about static nuclear processing in stars of intermediate mass (2–6 $M_\odot$), while supernova remnants (SNR) and cosmic rays play a similar role for the products of explosive burning from supernovae. Such observations, combined with model calculations of nuclear burning, provide the input to models of nuclear evolution on a galactic scale.

Planetary nebulae (PN) play two important roles in the study of interstellar abundances. The first and more unique role is to provide a look at actual enrichment of the ISM by the products of stellar nucleosynthesis. The second role is to act as an element of material passively reflecting the general interstellar abundances at the time of formation of the parent star. This second role is discussed in Section 5.

The task is to clearly distinguish the two roles, that is, to separate the initial stellar abundances from those that result from nucleosynthesis in the parent star. This separation has been accomplished by careful studies of chemical composition, making use of the forbidden optical lines and several UV transitions such as those of CIII and CIV. One such study by Peimbert (1978 and references therein) has produced four distinctive types of PN which from their composition as well as their kinematics (low $|\Delta v|$ to high $|\Delta v|$), spatial distribution (low $|\Delta z|$ to high $|\Delta z|$) and mass (high mass to low mass) have been identified as ranging from Population I to extreme Population II. It is the type I PN and to a lesser extent the type II PN that provide a direct look at stellar nucleosynthesis.

Circumstellar clouds also probe the details of stellar nucleosynthesis. The study of mass loss from late-type stars is relatively new because of the exacting demands placed on instruments in order to measure abundances in the freshly expelled material. To date, the abundances in only one mass loss object have been studied with the same thoroughness applied to giant molecular clouds, namely the cloud surrounding the carbon star IRC +10216. Objects available to mm/IR techniques are necessarily nearby. As a result they almost certainly belong to the same population as the type I PN, if they are not in fact actual precursors to them. The great advantage of the mm/IR observations comes from the ability to study a number of isotope abundance ratios with the reliability apparent from similar studies of giant clouds. The interpretation of circumstellar measurements in terms of stellar evolution is limited mainly by the difficulty in obtaining reliable mass estimates for the parent star.

IRC + 10216: Evidence for Stellar Processing

Studies of the late-type carbon star IRC + 10216 indicate that the C, N, and O isotope ratios are all not only nonsolar, but also vary significantly from the ratios observed in giant clouds. A complete discussion of the millimeter-wave isotope results is given by Wannier & Linke (1978b) and is summarized in Table 4 along with two infrared results. Although the millimeter and IR measurements probe the same source, it must be remembered that different layers of the circumstellar cloud might also represent different layers in the parent star. It is thus possible that the abundances in the two layers, and hence from the two techniques, could differ.

The nitrogen and oxygen results are especially interesting, implying $(N/^{15}N) \gg \odot$ and $(O/^{18}O) \gg \odot$. These results apparently rest on solid ground because any optical depth correction, if applicable, would strengthen the case for underabundant ^{15}N and ^{18}O. Likewise, it was argued that fractionation is especially unlikely in this object because of the youth of the molecular gas (<5000 years) and because of the lack of ion-molecule chemistry (Wannier & Linke 1978b).

Table 4 Isotope ratios in IRC + 10216[a]

	Isotope ratio	Observed value	Solar value	Method	Ref.[b]
(1)	$C/^{13}C$	40 ± 8	89	$(CS)/(^{13}CS)$	23
		(~ 20 in IR)	89	$(CO)/(^{13}CO)$	24
(2)	$S/^{34}S$	22 ± 2	23	$(CS)/(C^{34}S)$	23
(3)	$O/^{18}O$	>2700	500	$(^{13}CO)(CS)/(C^{18}O)(^{13}CS)$	23
(4)	$N/^{15}N$	>515	272	$(H^{13}CN)(CS)/(HC^{15}N)(^{13}CS)$	23
(5)	$^{18}O/^{17}O$	<1	5.5	$(C^{18}O)/(C^{17}O)$	23
		(0.2–1.0 in IR)	5.5	$(C^{18}O)/(C^{17}O)$	24, 25

[a] Table 4 is similar to Table 1 except that the double isotope ratios (rows 3 and 4) use the measured $C/^{13}C$ ratio in IRC + 10216. The errors are $\pm \sigma$ and the inequalities are at the 2σ confidence level. The IR results are in absorption and sample the inner ~ 2 arcsec. All other results are from millimeter emission lines and sample the more extended material (~ 2 arcmin).

[b] References:
23. Wannier & Linke 1978b
24. Rank et al. 1974
25. Barnes et al. 1977

Also of interest is the very small $^{18}O/^{17}O$ ratio, measured both in the extended ($\sim 2'$) envelope by the millimeter emission lines and measured in the compact ($\sim 2''$) source by two independent measurements of the $\sim 5\,\mu m$ vibration-rotation absorption lines of carbon monoxide. All the results yield $^{18}O/^{17}O < 1$, whereas the solar value is 5.5.

The $C/^{13}C$ ratio is most accurately measured by the millimeter emission lines. The measured value of 40 ± 8 is certainly less than solar, but not out of keeping with the value of 60 adopted for the galactic ring. It was argued that the CS opacity is $\ll 1$, a conclusion strengthened by $CS/C^{34}S = 22 \pm 2$. The sulphur result is in perfect accord with the solar value as well as with that in the galactic plane.

Type I Planetary Nebulae: He and N Enrichment

Results from a survey of type I PN indicate that He/H and N/O are significantly enhanced by the products of nucleosynthesis in the parent star. These objects are of Population I and presumably have masses of $\gtrsim 2\,M_\odot$. As a group, they have very filamentary structure and have lines from very different ionization states of given elements. A particularly dramatic instance of helium enrichment comes from the study of A30, an extended PN which contains a number of filaments close to the central star. Sargent (1980) has shown that the ratio of He (4686 Å)/Hβ in the filaments is greater than 20, indicating a hydrogen depletion (relative to helium) by a factor of at least 50. The same enrichment is not seen in the extended nebula, indicating an evolutionary change in the composition of ejected material.

Another interesting result is from the study of NGC2818, a type I PN with likely membership in the open cluster of the same name. The cluster has a main-sequence turnoff corresponding to $\sim 2.1\,M_\odot$, implying a slightly larger mass for the parent star of the PN (Peimbert 1978, Tifft et al. 1972). Obviously, cluster membership cannot be proven in a single instance, and other similar objects should be sought out. The association of stellar mass with abundance enhancements is clearly important to models of galactic enrichment.

Cassiopeia A: Explosive Burning in a Supernova

Cas A is a particularly interesting object for abundance studies. Observations of this SNR provide strong evidence for two distinct episodes of mass loss, one associated with the supernova event and the other associated with stellar mass loss, presumably from the pre-SN star.

The supernova event is associated with a series of fast-moving knots displaying a complete lack of H and He lines (Baade & Minkowski 1954). These fast-moving knots are of special interest for the understanding of nucleosynthesis in SN because they appear to be variously overabundant in the products of advanced stages of burning and they may well represent uncontaminated material from deep within the pre-SN (Peimbert & van den Bergh 1971; Chevalier & Kirshner 1978). Recent observations by Chevalier & Kirshner (1979) demonstrate significant compositional differences among the fast-moving knots, suggesting that the knots have originated from different layers within the pre-SN. The line strengths of S, Ar, and Ca appear to be correlated with each other and, as a group, they can vary significantly relative to the oxygen lines. Carbon and nitrogen are notable, along with H and He, by their absence, implying O/C and O/N significantly larger than the cosmic abundance ratios.

In contrast to the fast-moving knots, the quasistationary flocculi (QSF) have H and He lines. Kamper & van den Bergh (1976) found that the outward radial motion of the QSF's is approximately 150 km s^{-1}, implying an age of about 10^4 years. This time scale should be comparable to that between carbon core burning and core collapse in massive stars (Chevalier & Kirshner 1978). An overabundance of N in the QSF's provides evidence for the presence of products of CNO processing, in agreement with the above observations of IRC + 10216. These results are discussed below.

5 CHEMICAL ABUNDANCE GRADIENTS: HII REGIONS AND PLANETARY NEBULAE

Measured abundance ratios include O/H, N/H, N/S, He/H, and C/H. There is observational evidence for strong radial gradients of the first three ratios. The results for He/H and C/H are less certain, but may also indicate measurable gradients. All five gradients are in the sense of decreasing with increasing distance, R_c, from the galactic center.

In Section 3, it was concluded that none of the measured isotope ratios in interstellar clouds indicate a significant radial gradient in the Galaxy. Yet, the study of chemical abundances in PN and HII regions tells a somewhat different story. The subject of element abundances in galactic PN and HII regions was recently reviewed by Peimbert (1979, and references therein). The road from observed lines to derived abundance ratios is not a straight one and the abundance determinations depend on the nebular temperature and density and, therefore, to a lesser degree, on their variability within a source. The possible roles of the ionizing spectra have also been discussed (e.g. Balick & Sneden 1976, or Shields & Tinsley 1976). Yet, for the measurement of a radial abundance gradient, none of these effects are of central importance in the absence of a radial gradient of the physical properties of ionized regions. The measurement of absolute abundance ratios in ionized regions may be difficult, but a measurement of gradients of abundance ratios is more straightforward.

HII Regions

Galactic HII regions are valuable for studies of radial gradients because they are disc population objects, which are unlikely to be strongly affected by local nucleosynthesis (but see Tinsley 1979). On the other hand, studies are rendered difficult by the limited numbers of visible HII regions and by the large range of their physical parameters such as density, temperature, size, and the spectrum of ionizing radiation. The positive and negative virtues of HII regions for the study of abundances contrast to those of galactic PN which are discussed below. The results for five different surveys of HII regions are given in Table 5. The bottom row of the table includes all those HII regions from the five surveys for which electron temperatures had been observationally determined.

The significance of Table 5 is that the derived abundance gradients are very large. Using the definition that $\gamma(X/Y) \equiv d[\log (X/Y)]/d[R_c(\text{kpc})]$, we see that $\gamma(C/H)$, $\gamma(O/H)$, and $\gamma(N/H)$ are about -0.1, amounting to factors of 10 or so if extrapolated to a distance of 10 kpc. Because of the high extinction in the Galaxy, one might speculate about possible systematic effects of surveying distant HII regions. In order to test for such effects, the results for the Galaxy can be profitably compared to observed abundance gradients in external galaxies where extinction is no problem. The presence of O/H, N/H, and N/S gradients in Sc galaxies seems well established

Table 5 Solar neighborhood abundance gradients[a]

Object	He/H	C/H	O/H	N/H	N^+/S^+	Ref.[b]
PN	$-0.02\pm.01$		$-0.06\pm.02$	$-0.18\pm.04$		1
H II					$-.04$	2
H II	$-0.02\pm.01$		$-0.13\pm.04$	$-0.23\pm.06$	$-0.09\pm.05$	3
H II	0.00		-0.04	-0.10	-0.05	4
H II			-0.11	-0.11	-0.06	5
H II		-0.09	-0.09	-0.12	-0.05	6
H II			-0.10	-0.14	-0.05	6

[a] From a tabulation by Peimbert (1979), the abundance gradients are given as $\gamma = d \log (X/Y)/dR(\text{kpc}^{-1})$.
[b] Original references:
1. Torres-Peimbert & Peimbert 1977
2. Sivan 1976
3. Peimbert et al. 1978
4. Hawley 1978
5. Talent & Dufour 1978
6. Peimbert 1979

(Searle 1971, Smith 1975, Shields & Searle 1978). The abundance gradients, especially those from Shields & Searle in M101, are in close agreement with those derived for the Galaxy.

Another abundance gradient which has received attention in the literature is that of N/O. Significant N/O gradients were reported, for example, by Searle (1971) and by Smith (1975). More recent observations have used improved measured values of the electron temperatures and have avoided possible blending of the lines of [NII] with Hα at 6563Å by improved spectral resolution (Alloin et al. 1979, Pagel et al. 1979, Shields 1980). The more recent conclusion is that N/O does not vary significantly (γ(N/O) $\gtrsim -0.04$) either in our Galaxy (Table 5) or in external galaxies. This small limit is particularly interesting in light of the millimeter-wave result in dense clouds and is discussed below.

Planetary Nebulae

In contrast to the case for HII regions, the number of well-studied PN is large ($\gtrsim 100$) and they represent a smaller range of physical conditions. On the other hand, the PN originate from different stellar populations and they may be appreciably affected by nucleosynthesis in the parent stars. In order to determine radial abundance gradients in the disc, Peimbert (1978) has classified PN into four types according to their observed properties. Type I PN are disc objects, but their abundances are significantly affected by nucleosynthesis in the central star. The type II PN are intermediate Population I objects, with $\langle |z| \rangle \sim 150$ pc and they have orbits that are circular enough that they have probably not wandered far from their original galacto-centric radii. Type II PN also contain some products of stellar production, but Peimbert (1978) argues that the He/H, O/H, and N/H abundance ratios truly measure initial abundance gradients and that the results in Table 5 indicate a significant gradient for each. The significance of the gradients in Table 5 is the general agreement between results for the PN and for the HII regions. The two sets of objects are very dissimilar and should not suffer from the same type of systematic errors.

There is, however, room for disagreement when interpreting results for the PN. Kaler (1978, 1979, 1980) has analyzed the observations in a somewhat different manner. He has also separated the PN according to population but considers that enrichment by the parent star can be generally ignored. Kaler's analysis indicates a strong variation of O/H, N/O, and He/H with population type in the sense of being larger in the disc objects and smaller in the halo objects (large $|z|$ and large radial velocity). However, he sees no significant radial variation of the observed abundance ratios and he concludes that the gradients measured by Peimbert and others may result from observational selection.

6 COSMIC RAYS

Cosmic ray astronomy holds the unique distinction of directly sampling a component of the ISM. Instruments now available permit nuclei up to and including those of the iron group to be readily distinguished so that the "arriving" nuclear abundances (above the earth's atmosphere) can be measured. Available mass resolution is as small as 0.2 amu (atomic mass unit) with several instruments capable of a resolution of 0.4 amu (see, for example, Althouse et al. 1978, or Greiner et al. 1978). Indeed, the chief uncertainties in deriving "source" abundances of several common nuclei result from questions of interpretation rather than from instrumental error.

For purposes of this review, it is convenient to distinguish the cosmic ray results that measure the relative abundances of the elements from those that focus on the isotopic composition of a given element. Much information can be derived from the elemental abundances and this subject has been well discussed in the review by Trimble (1975). Briefly, the cosmic ray abundances have many of the same features present in all abundance data, namely a preponderance of H and He, and a gradual decline in abundance with increasing atomic number starting from the CNO nuclei except for a distinct peak at the iron group. Effects of propagation are evident from many of the details of the measured abundances, such as the enhanced presence of spallation products like Li, Be, and B, or the smoothing of the saw-tooth appearance (odd-even effect) of the abundances (Shapiro & Silberberg 1974). The first step in recovering source abundances from arriving abundances is to correct for the effects of propagation through the intervening material. Models of the intervening material are calculated to fit the many observed abundance anomalies which are particularly sensitive to propagation effects. This procedure is remarkably successful in the sense that many such anomalies are eliminated by very simple models. The most common model is a simple exponential probability distribution of intervening column densities (path lengths): $p(N) = (1/N_0) \exp(-N/N_0)$ with N_0 taking on a value of $\sim 4\text{--}6$ g cm^{-2}. The straight exponential is justified (somewhat) on the basis of cosmic ray confinement by galactic magnetic fields. If the cosmic rays are produced more or less uniformly and escape by a random walk process, then the resulting model (called the "leaky box" model) produces the exponential. Shapiro & Silberberg (1974) discuss a similar two-parameter model which improves the fit with existing data and eliminates the obvious overemphasis of zero path length appearing in the straight exponential. They use

$$dp/dN = [1 - e^{-(N/0.60)^2}] e^{-N/4.3}$$

and results for the most abundant arriving nuclei are given in Figure 2 which is discussed below.

Isotope data will play an important role in future cosmic ray research. Despite the great successes achieved with the element abundances alone, a very worrisome problem is that the cosmic ray acceleration mechanism may discriminate among the different elements. Such a selective effect is well documented from solar flare experiments where relative abundance enhancements as large as factors of ten are seen especially in the low-energy particles (<15 MeV/amu). High-energy solar particles seem to be less affected by the preferential acceleration. As for galactic cosmic rays, their acceleration mechanism is not known, so that some uncertainty remains regarding the possible role of the different ionization potentials and ionization cross sections of the elements. In contrast, the isotope results should be almost entirely free of such effects. The only selection among the isotopes would come from their different charge/mass ratios. For the heavy nuclei (≥ 20 amu) which are discussed below this possible effect should be negligible.

The nature of isotope abundance measurements is best illustrated in Figure 2, which compares the abundances of (a) the cosmic rays leaving a

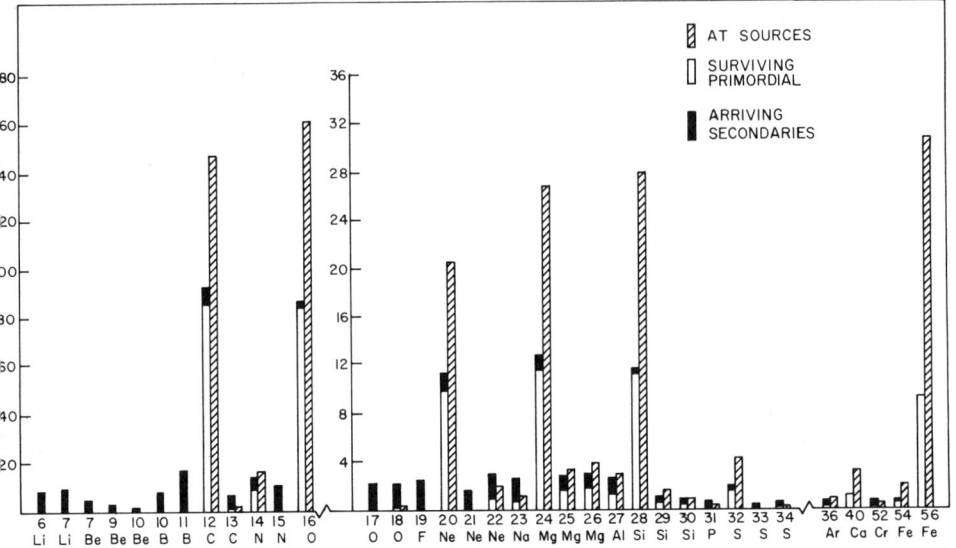

Figure 2 The composition of cosmic rays leaving the source is compared to the composition of arriving cosmic rays at the top of the Earth's atmosphere. The data are those of Silberberg et al. (1976) who discuss ways in which corrections were made for spallation in the interstellar medium.

hypothetical source, (b) the surviving source nuclei, and (c) the arriving secondary nuclei produced as spallation products from heavier nuclei. The results are restricted to the most abundant species likely to be observed. It is apparent that all nuclei lighter than ^{11}B and all the rare isotopes up through ^{18}O are dominated by secondary production resulting from collisions with interstellar H and He. As a result, the most valuable and dependable cosmic ray results have come from the heavier elements Ne, Mg, Si, and Fe, each of which have at least two isotopes for which the "surviving primordial" abundances represent a sizeable fraction of the total arriving abundances. For these nuclear species a propagation model such as that of Shapiro & Silberberg is likely to be dependable since the correction factors are in any event not very large. Selected relative abundances are presented in Figure 3 for the rare isotopes of Ne, Mg, and Si. Also shown are the values calculated by Mewaldt et al. (1980a) for the case of a source with solar system isotopic abundances propagated through 5.5 g cm^{-2} of intervening material.

Neon

The best results available are those for ^{22}Ne, which indicate a ^{22}Ne/^{20}Ne abundance ratio substantially larger than the solar value. Only two instruments are capable of clearly distinguishing among the three neon isotopes (see Greiner et al. 1979, Mewaldt et al. 1980a). However, it is generally assumed (and it appears to be the case) that the observed ^{21}Ne is a result of secondary production (an arriving ^{21}Ne/Ne value of $\sim 12\%$ results from the propagation calculations). With this simplifying assumption, several experiments can measure the single remaining ^{20}Ne/^{22}Ne abundance ratio, including those which yield only the mean atomic weight of neon. Six direct observations of ^{22}Ne/(Ne + ^{22}Ne + ^{21}Ne) are presented in Figure 3. After correcting for propagation effects, the weighted average of the observations yields ^{22}Ne/Ne = 0.34 ± 0.04 at the cosmic ray source. This value is very significantly larger than the corresponding solar system value of ~ 0.12. The solar system value is itself of some interest because of the presence of more than one distinct component of neon. The component thought by some to be representative of the solar average is a meteoritic component, "Neon-A", which yields ^{22}Ne/Ne = 0.12 in agreement with solar flare experiments (Dietrich & Simpson 1979, Mewaldt et al. 1979). Direct observations of the solar wind show a relative depletion in ^{22}Ne(^{22}Ne/^{20}Ne ~ 0.07) (Geiss et al. 1970). It should be noted that if this smaller value is truly representative of the solar system abundance, then the apparent overabundance of cosmic ray ^{22}Ne is even more dramatic. A full discussion of the solar isotope ratios is given by Podosek (1978).

Magnesium and Silicon

In contrast to Ne, the elements Mg and Si each have two rare isotopes that are comparably abundant, namely ^{25}Mg, ^{26}Mg, ^{29}Si, and ^{30}Si. Hence, successful measurements of these rare isotopes can only be made with equipment having good mass resolution. The results presented in Figure 3 are those of Mewaldt et al. (1980a) obtained with a resolution of ~0.2 amu and the indicated errors are due to counting statistics. These results suggest an overabundance of both ^{25}Mg and ^{26}Mg relative to ^{24}Mg. Each abundance ratio is only about 1 σ larger than the solar system value but, taken together, the two rare magnesium isotopes certainly appear overabundant. The silicon results indicate that both ^{29}Si/^{28}Si and ^{30}Si/^{28}Si are consistent with the solar system values and there is no indication of an overabundance for the two rare isotopes taken together.

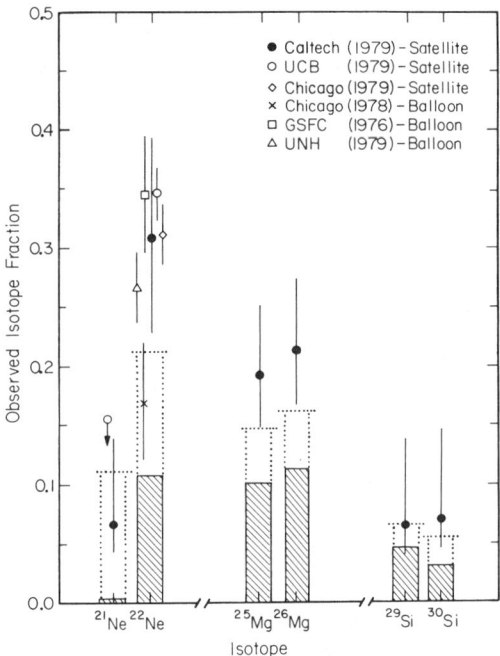

Figure 3 Cosmic ray isotope abundance measurements compared to results expected on the basis of a solar system source composition (modified from Mewaldt et al. 1980a). All abundances are relative to the total elemental abundances consisting primarily of ^{20}Ne, ^{24}Mg, and ^{28}Si, respectively. The shaded portions of the bars represent actual solar system abundances while the dotted bars give the expected fraction after propagation through 5.5 g cm^{-2} (see text). The references are Caltech (Mewaldt et al. 1980a), UCB (Greiner et al. 1979), Chicago 1979 (Garcia-Munoz et al. 1979), Chicago 1978 (Dwyer 1978), GSFC (Fisher et al. 1976), UNH (Webber et al. 1979a).

Iron and Nickel

The relatively abundant nuclei of the iron peak are of special interest for the study of pre-SN. These nuclei form in the very last stages of evolution and very close to the physical boundary separating a forming neutron star from its expelled envelope. The iron isotopes are also unique in that they behave very differently when propagated through the ISM. ^{56}Fe, ^{57}Fe, and ^{58}Fe can only be depleted because there are no abundant nuclei heavier than ^{56}Fe that could contribute such spallation products. On the other hand, ^{54}Fe and ^{55}Fe can be produced from ^{56}Fe so that propagation can produce a significant secondary component.

Clearly, an independent measurement of all five Fe isotopes requires an adequate mass resolution such as the ~ 0.37 amu resolution achieved by the Heavy Isotope Spectrometer Telescope (Mewaldt et al. 1980b). It is apparent that the results (Table 6) are consistent with a solar system abundance distribution of the iron isotopes. Confirmation of these results is available from other experiments with poorer mass resolution under the simplifying assumption that the source abundances of ^{55}Fe and ^{57}Fe (and possibly ^{58}Fe) are negligible. (Since ^{55}Fe decays to ^{55}Mn with $t_{1/2} = 2.6$ yr, the low assumed abundance of this isotope is not out of line.) Such confirmation is available from two experiments that indicate the dominance ($>90\%$) of ^{56}Fe and by implication indicate upper limits on the potentially important ^{54}Fe and ^{58}Fe species (Tarlé et al. 1979, Webber et al. 1979b).

The only result for nickel comes from Tarlé et al. who, with the simplifying assumption that there is no significant source abundance of ^{59}Ni, report that the nickel isotopes, ^{60}Ni, ^{58}Ni, ^{56}Ni, are also consistent with the solar system values.

Table 6 Fe isotope composition[a]

Isotope	Observed (%)	Cosmic ray source	Solar abundance
^{54}Fe	10^{+8}_{-4}	9^{+8}_{-5}	5.8
^{55}Fe	≤ 10	≤ 7	0
^{56}Fe	90^{+0}_{-15}	91^{+5}_{-11}	91.7
^{57}Fe	≤ 8	≤ 8	2.2
^{58}Fe	≤ 6	≤ 6	0.3

[a] All abundances are given as percentage of total iron. The errors are $\pm 1\sigma$, and the inequalities are also 1σ limits (84% confidence). The source abundances are the observed abundances after correcting for propagation effects. In the case of ^{55}Fe, the "source" upper limit is calculated without regard for the ^{55}Fe lifetime of 2.6 years for decay to ^{55}Mn. The actual source abundance could be higher than indicated. All data are from Mewaldt et al. (1980b).

7 CONCLUSION

Discussion

EVOLUTION OF THE ISM   The study of nuclear abundances in the ISM is far-reaching. A buildup of the elements, starting with the Big Bang some eighteen billion years ago, is still going on with contributions from different populations of stars at different times and places. Comprehensive models of nuclear evolution incorporate many assumptions about galactic structure, stellar evolution, and stellar mass loss. A review of such comprehensive models is given by Audouze & Tinsley (1976). Since the time of that review, one change in the modeling has been a careful treatment of the actual lifetimes of stars responsible for nucleosynthesis, replacing the instant recycling approximation. This has caused a de-emphasis of the primary ($p-p$ or 3α) or secondary (CNO processing) nature of different nuclei. [For up-to-date calculations, see Dearborn et al. (1978) or Tinsley (1979).] The observations reported in Sections 3–6 are very diverse in their technique and in their view of the ISM. Yet there is a common theme: each one of the reported measurements deals in a direct way with the chemical evolution of the ISM. Several observations sample the local ISM. These results can be compared to the solar abundances, which we take to typify the ISM as it was five billion years ago. The observations of PN and stellar mass loss provide a direct look at actual enrichment of the ISM by freshly synthesized material, and thus help to identify the production of specific nuclei with specific stellar types. The cosmic rays and SNR observations fulfill the same role for pre-SN and SN production. Finally, spatial variations of individual nuclear abundances can be used to further identify the sites of production, or they may be used, with a knowledge of the production sites, to draw conclusions about the distribution of stellar types.

The nuclear species discussed in this review are listed in Table 7, and their probable production mechanisms are given in Table 8. [For complete discussions of nucleosynthesis see Fowler et al. (1975) or Trimble (1975) and references therein.] It can be seen that these nuclei represent most of the basic nuclear processes. The processes are of one of two types, namely those that proceed hydrostatically in stars and those that proceed explosively. The latter occur in such various sites as the Big Bang, supernovae, novae, and possibly a speculative early population of supermassive objects. The spallation of cosmic rays plays a relatively minor role in chemical evolution, but is of obvious importance to the observations of cosmic rays.

BIG BANG: THE DESTRUCTION OF D AND DETECTION OF ^{3}HE   The implication of the D and ^{3}He abundance to cosmology is well known (see discussions by

Ostriker & Tinsley 1975 and Penzias 1978). Of interest to this review is the evidence for (or against) galactic evolution of D/H. It is thought that D cannot be produced in significant quantities in stellar sites (see, for example, Epstein 1977). If it could, then it would be a disappointment to cosmologists. On the other hand, D burns very easily and significant cycling of material through stars ought to lead to its depletion. The evidence for a depletion of D toward and in the galactic center (Section 3), where astration has been much larger, is therefore significant. The observed depletion provides observational evidence against galactic production of D, and supports a cosmological interpretation. The best interstellar determination

Table 7 Observed isotopic species

Isotopes				Observing techniques
^{1}H	^{2}D			millimeter, UV
^{4}He	^{3}He			optical: (PN, HII regions)
^{12}C	^{13}C			radio, IR, UV, optical: (PN, HII regions)
^{14}N	^{15}N			millimeter (PN, HII regions)
^{16}O	^{17}O	^{18}O		millimeter, centimeter, IR (PN, HII regions)
^{20}NE	(^{21}Ne)	^{22}Ne		cosmic rays
^{24}Mg	^{25}Mg	^{26}Mg		cosmic rays
^{28}Si	^{29}Si	^{30}Si		cosmic rays, millimeter
^{32}S	^{33}S	^{34}S		millimeter
^{54}Fe	^{56}Fe	(^{57}Fe)	(^{58}Fe)	cosmic rays
^{56}Ni	^{58}Ni	(^{59}Ni)	^{60}Ni	cosmic rays

Table 8 Major production processes[a]

		Quiet-burning		Explosive-burning				
Isotope	Cosmological	H	He	H	He	C-burning	O-burning	e-process
H, ^{3}He	+							
D	+	−						
^{4}He	+	+						
^{12}C, ^{16}O, ^{22}Ne			+					
^{13}C, ^{17}O		+	+					
^{14}N		+						
^{15}N		−		+	+			
^{18}O			+		+			
^{20}Ne, ^{29}Si, ^{30}Si, ^{33}S, Mg						+		
^{28}Si, ^{32}S, ^{34}S							+	
Fe, Ni								+

[a] Minus sign signifies a major destruction process.

of D/H is that of York & Rogerson (1976) which yields D/H = $1.8 \pm 0.4 \times 10^{-5}$.

In a similar vein, Rood et al. (1979) have reported the probable detection of the 3.5 cm hyperfine line of ^{3}He in W51 at a level comparable to the protosolar value. ^{3}He is thought not to be significantly destroyed in the envelopes of normal stars, so that the approximately solar value also argues against the production of ^{3}He in red giants.

HYDROSTATIC HYDROGEN BURNING: THE CNO CYCLES *The CNO process implies evolutionary trends which find support in the observed isotope abundance ratios.* The accepted scenario for the production of the CNO elements is as follows. H and He from the Big Bang form a first generation of stars in which, after p–p H burning, form ^{12}C by the 3α process and subsequently ^{16}O (but probably not ^{20}Ne) by He burning. These products (^{4}He, ^{12}C, and ^{16}O) are generally regarded as *primary* products since they can form in first generation, metal-poor stars. There is some evidence (Section 5) that in some stars deep mixing may bring primary carbon and oxygen up to the hydrogen-burning zone. In such a case, products of CNO processing such as ^{14}N may also be primary in nature. When primary CNO nuclei are incorporated into second generation stars H burning may be accomplished by the catalytic CNO process. The important mass range is 3–5 $M_\odot$, too small for supernovae but massive enough for extensive CNO processing to occur. In such stars the mass lost to the ISM comes from the outer envelope and produces the PN and dense circumstellar clouds described above. The exact nuclear processes that affect the observed interstellar abundances depend on the stellar mass, the mass of the expelled envelope (the depth to which it penetrates in the star), and the amount of mixing occurring within the stars prior to expulsion of the envelope. It seems certain that a significant portion of the mass expelled will have been exposed to hydrogen processing in the CNO cycle. Equilibrium CNO hydrogen burning proceeds at a rate slow compared to the several β decays, so that the bottlenecks in the cycle (which determine the "equilibrium" CNO abundances) are always the preceeding (p, γ) reaction. The tricycle is represented schematically below:

^{12}C(p,γ)^{13}N(,e$^+$v)^{13}C
^{13}C(p,γ)^{14}N
^{14}N(p,γ)^{15}O(,e$^+$v)^{15}N
^{15}N(p,α)^{12}C   or ^{15}N(p,γ)^{16}O
                    ^{16}O(p,γ)^{17}F(,e$^+$v)^{17}O
                    ^{17}O(p,α)^{14}N   or ^{17}O(p,γ)^{18}F(,e$^+$v)^{18}O
                                         ^{18}O(p,α)^{15}N.

Of the two important branching points, the first (at ^{15}N) heavily favors the (p, α) reaction over the (p, γ) reaction while the second branching point (at ^{17}O) shows no such great preference. The exact branching ratios are of obvious interest for the implied abundances. However, these ratios are difficult to measure because of the small energies of the reacting particles compared to the typical run of energies in nuclear laboratories.

The measured nuclear cross sections for the CNO cycle imply the production, in equilibrium, of an overwhelming amount of ^{14}N. Such material, in combination with ^{12}C and ^{16}O from primary processing, explains the abundances of the isotopes of C, N, and O. The protonation of ^{12}C, ^{14}N, or ^{16}O are each followed by a β decay which occurs at a fixed rate. Thus, if hydrogen burning could be interrupted then the odd nuclei ^{13}C, ^{15}N, or ^{17}N could be enhanced. Such nonequilibrium or incomplete burning may occur in novae by quick quenching of burning material or in supernovae in the supernova shock. A similar enrichment occurs in the envelopes of giant stars (Dearborn et al. 1978). The odd nuclei, though rarer than the even nuclei, have a cosmic abundance that implies a significant enrichment by incompletely processed material.

Regardless of the site of CNO processing or of the experimental values of the branching ratios, a few trends are predictable. The abundances of those isotopes that are secondary products of nucleosynthesis are expected to increase in time relative to primary products, which can be produced in first-generation, low-metal stars. Assuming an initial CNO composition that is approximately solar, the CNO nuclei will get rearranged by equilibrium burning to produce ^{14}N (increasing N/H) especially at the expense of oxygen (increasing N/O). Although the total abundance of carbon is decreased by equilibrium processing, ^{13}C is enhanced and, in the steady-state operation, yields C/^{13}C ≈ 4 (decrease of C/^{13}C).

To the extent that nonequilibrium CNO processing is important ^{13}C is also enhanced, as are ^{15}N and ^{17}O. The predicted evolutions of N/^{15}N and ^{18}O/^{17}O are somewhat obscure. Initial ^{15}N certainly forms ^{14}N by the equilibrium CN cycle (in giants), but ^{15}N forms explosively from ^{14}N in novae. The fate of ^{15}N/N thus depends on the relative importance of novae (or SN) and of equilibrium burning in stars. The time evolution of ^{15}N/N may therefore reflect evolution in the stellar population (see the discussion in Wannier et al. 1980). ^{17}O/^{18}O is also difficult to model because of the somewhat uncertain values of the second branching ratio and because ^{18}O is formed mostly by helium burning in very different conditions from those which apply to CNO processing.

THE OBSERVED CNO ISOTOPES: EVIDENCE FOR EVOLUTION Table 9 summarizes the CNO isotope abundances from molecular cloud observations. The four astronomical sites are arranged according to their nucleosynthetic "age," where the solar system is taken to represent the interstellar abundances about five billion years ago (but see discussion of non-uniformities below). The galactic ring sources, which include many clouds at approximately the solar galactocentric radius, have about twice the solar age, or ~ 10 billion years. The galactic center sources have the same chronological age, but are part of a region in which stellar processing has been much more rapid, as evidenced by the relative depletion of interstellar gas. The last entry in Table 9 is IRC+10216, a carbon star currently undergoing mass loss and containing newly processed material. Of course, the material in the parent star may be stratified and the deepest layers have not yet been shed.

The point of interest in Table 9 is that $C/^{13}C$ is largest for the sun, is smaller for the ring, and yields smaller, but comparable values for IRC +10216 and for the galactic center. In no case is the value as small as the CNO equilibrium value of 4. $N/^{15}N$ behaves in a like manner. The observed ratio in the ring is larger than solar, and IRC+10216 and the galactic center are larger yet. The equilibrium CNO value is truly large ($> 10^5$). $S/^{34}S$, to within the scatter of the observations, shows no significant change among the four sites and the CNO process would certainly not induce one. $^{18}O/^{17}O$ yields observations with very small scatter and the decrease of $^{18}O/^{17}O$ for the ring and center sources is significant. Again, the decrease with age is monotonic, with IRC+10216 displaying a remarkably small ratio. The value of $O/^{18}O$ indicates a dramatic change in IRC+10216 which will be discussed below.

The comparison in Table 9 only refers to one example of observed enrichment. Correct predictions of the CNO isotope abundances come from models that are based on the cycling of the ISM through realistic stellar populations and that correctly keep track of the production, mass loss, and turnover time for each stellar type. A discussion of the most recent models is given by Guélin & Lequeux (1980).

THE RADIAL ABUNDANCE GRADIENT (OR THE LACK THEREOF) The molecular isotope abundances of C, N, O, Si, and S never show a gradient in the

Table 9 Observed abundances

Source	$C/^{13}C$	$O/^{18}O$	$N/^{15}N$	$S/^{34}S$	$^{18}O/^{17}O$
Solar system	89	500	272	23	5.5
Galactic ring	60	500	333	20.3	3.2
Galactic center	26	300	>550	17.6	3.3
IRC+10216	40 mm 20 IR	>2700	>515	22	$\lesssim 1$

galactic ring. The element abundances of C, N, and O show large radial gradients in the neighborhood of the sun. These results stand in apparent contradiction.

The best way to measure large scale (> 5 kpc) abundance variations in our Galaxy is by sampling the isotope ratios in giant clouds. The results, discussed in Section 3 above, yield an apparent step function, that is, one set of abundance ratios for the galactic center region and one set for the galactic ring ($R \gtrsim 3$ kpc). The observations include five surveys, namely $C^{17}O/C^{18}O$, $H_2CO/H_2^{13}CO$, $C^{34}S/^{13}CS$, $C^{18}O/^{13}CO$, and $HC^{15}N/H^{13}CN$. The first two surveys measure single ratios, namely $^{18}O/^{17}O$ and $C/^{13}C$. The next three measure $C/^{13}C$ in combination with $S/^{34}S$, $O/^{18}O$, $N/^{15}N$. None of the surveys display a significant gradient! The simple explanation is that all five of the isotope abundance ratios are constant. More quantitatively, the five sets of data are each absolutely inconsistent with a factor of two variation between 5.5 and 12.5 kpc. Indeed the $C^{17}O/C^{18}O$ survey is incompatible with even a factor of 1.3. Expressed in terms of $\gamma(A/B)$, the logarithmic gradient per kpc, this implies that all five isotope ratios are incompatible with values of γ exceeding 0.04 in absolute value. In contrast, the chemical abundance gradients in Table 5 yield four significant gradients, namely C/H, O/H, N/H, and N^+/S^+. The errors are hard to assess, because of the complexity of the data analysis. However, all the gradients have the same sign and magnitude as a series of similar element abundance gradients in external galaxies with values of γ exceeding 0.04 in absolute value.

An abundance gradient that has received attention in the literature is that of N/O. As indicated above, ^{14}N is usually considered to be a secondary product of stellar nucleosynthesis. As a result, simple models would predict that N/O = O/H, so that the N/H gradient might be expected to be twice as steep as that of O/H. In fact, the indications are that there is only a very small N/O abundance gradient whether in our Galaxy or in other galaxies (Section 5). This result has been interpreted by several authors as indicating the existence of a primary production process for ^{14}N (Pagel et al. 1979, Alloin et al. 1979, Kaler 1979). However, *several* of the isotope abundance gradients discussed above show no radial gradient including $^{13}C/C$, $^{15}N/N$, and $^{17}O/O$. If a lack of radial variation indicates a primary origin, then one would be forced to conclude that ^{13}C, ^{17}O, and ^{15}N are *all* primary nuclei (Wannier 1980). Such a conclusion seems untenable. Instead, it seems that a more complete revision of current ideas is necessary in order to correctly interpret the observed abundance ratios in the galactic disc.

The problem of gradients is touched upon by Audouze et al. (1975), who conclude that their model is just able to produce small chemical gradients with no $C/^{13}C$ gradient. However, it seems less likely that such a feat can be repeated for the array of local gradients that have now been measured (or not detected). Because of the importance of composition gradients to the discussion of galactic evolution, it is most important that this problem be successfully resolved.

NUCLEOSYNTHESIS OF Ne, Mg, Si, S, Fe, AND Ni   For nucleosynthesis beyond the CNO elements, He burning plays a key role. By the 3α process, ^{12}C can be formed directly and, in turn, undergoes an (α, γ) reaction to form ^{16}O (hydrostatic He burning). The ^{14}N from equilibrium CNO cycling undergoes He burning to produce ^{18}O by way of ^{14}N$(\alpha,\gamma)^{18}$F$(,e^+\nu)^{18}$O. The ^{18}O, in turn, can form ^{22}Ne by further He burning. In contrast, the corresponding reaction of ^{16}O to form ^{20}Ne is not important and ^{20}Ne is thought to form in more advanced stages of nuclear burning. Because ^{18}O and ^{22}Ne are both formed from ^{14}N, they are both secondary products of nucleosynthesis in the sense that they only form in stars with an initial concentration of the CNO nuclei. The e^+ decay of ^{18}F is significant in that ^{18}O provides excess neutrons for subsequent explosive C and O burning (Table 8). Elements that have obvious potential for showing evolutionary trends are those having isotopes formed by different nuclear processes. Neon is one such element since ^{22}Ne can form by quiet He burning in massive stars in contrast to ^{20}Ne. Silicon is another such element since ^{28}Si is formed during explosive carbon burning, but not ^{29}Si or ^{30}Si. In fact, these two elements may both show evolutionary effects in their observed abundances. Even when the isotopes of a given element share a common production process, it is by no means a foregone conclusion that the isotopes always come out in the same ratio. In explosive C burning and explosive O burning, as well as in the e-process, the neutron excess $[\eta \equiv (n-Z)/(n+Z)]$ is an important parameter because the burning proceeds too rapidly to allow for β decays to change protons into neutrons. If a star began its life with 2% CNO, then, after conversion to ^{14}N, ^{18}O, and ^{22}Ne, the neutron excess would be $\eta \sim 0.0033$. η, in turn, affects the production of the neutron-rich isotopes. Thus, a possible evolutionary trend is in favor of the isotope favored by high η, generally the neutron-rich ones.

NEON AND SILICON: EVIDENCE FOR EVOLUTION   Much of the evidence for production of elements beyond the CNO group comes from the observations of cosmic rays and of giant molecular clouds. However, the observations of Cas A provide direct measurements of nucleosynthesis in the final stages of pre-SN processing and the large observed abundances of Si, S, Ar, and Ca indicate that the fast-moving knots contain material from different layers of the evolved star (Chevalier & Kirshner 1979). The cosmic ray data (Section 6) imply a source ratio of ^{22}Ne/^{20}Ne which is several times higher than the solar value. ^{22}Ne probably results from hydrostatic He burning while ^{20}Ne is a product of explosive carbon burning (Table 8). ^{22}Ne is a so-called secondary product, made in stars needing seed CNO nuclei, while ^{20}Ne is a primary product. Thus, it is reasonable to expect an evolutionary increase in ^{22}Ne/Ne just as is found for ^{13}C/C.

Silicon is the one element that allows for a comparison of the cosmic ray isotopes and the molecular isotope abundances. From Section 6, we see that the cosmic ray abundances are entirely consistent with the solar value. However, from the reported molecular data, ^{28}Si seems to be underabundant by a factor 2 relative to ^{29}Si and ^{30}Si, with the underabundance persisting in both the galactic center and the galactic ring. ^{28}Si is also thought to be somewhat different in terms of its production (Table 8), forming by carbon burning rather than closely related oxygen burning. The relative overabundance is, however, consistent with a galactic evolution toward increasing abundances of the more neutron-rich nuclei (see above).

ABUNDANCES OF S, Fe, Ni, Mg: A SOLAR MIX (ALMOST) Iron, nickel, and magnesium share the property that all of their nuclei are produced by a common process. The case for sulphur is less clear. Carbon burning rather than oxygen burning may dominate its production (see Pardo et al. 1974). If the stellar conditions for nucleosynthesis are constant, then these elements might behave in a lock-step manner, increasing or decreasing together, but never producing isotope anomalies. On the other hand, it is very unlikely that every instance of oxygen burning (or of carbon burning or of the e-process) will occur under identical conditions. Different initial abundances will certainly affect η if not the temperature and density of the burning zone during a supernova.

The observed interstellar abundances of S, Fe, and Ni appear to be consistent with the solar values. In the case of sulphur, the mm-wave results indicate that the sulphur isotope abundances are constant throughout the Galaxy, slightly surprising in view of the very different metallicity and stellar population of the galactic center region. The case of Mg is an interesting exception, indicating an overabundance of the neutron-rich ^{25}Mg and ^{26}Mg relative to the solar ratios. That this overabundance represents an evolutionary effect is supported by recent observations of the magnesium isotopes in G and K dwarf stars (Tomkin & Lambert 1980). Using the ^{24}Mg, ^{25}Mg, and ^{26}Mg isotopic variants of MgH, they find a significant *underabundance* of ^{25}Mg and ^{26}Mg in the old, metal-deficient star, Gmb 1830, but abundances consistent with the solar values for younger stars. Again, as for the cosmic ray results, the indication is for an evolutionary increase of the neutron-rich isotopes. This indication is consistent with our understanding of nucleosynthesis. An increased metallicity implies a higher η, which forms the neutron-rich isotopes. However, the isotopes of iron, sulphur, and nickel show no such evolutionary trend.

UNIFORMITY OF THE ISM: A WORRY ABOUT SOLAR ABUNDANCES The ISM is by no means homogeneous and it would be naive to assume that local samples of the ISM are necessarily representative of the average interstellar abundances. One of the ways to trace evolutionary effects is to use the solar abundances as a five billion year benchmark. Thus, abundances in the solar system are often taken to represent the average composition of the ISM as it existed when the protosolar nebula condensed some five billion years ago. A potential weakness of such comparisons is that they could be in error if the protosolar nebula was not, in fact, representative of the ISM. Several analyses of meteoritic material have produced evidence for isotope abundance anisotropies in the protosolar nebula (see, for example, Lee et al. 1979 and Clayton et al. 1973). These results, based on the isotopes of oxygen (Clayton et al.) and aluminum (Lee et al.), are taken to indicate that two or more poorly mixed reservoirs of material may have contributed to the protosolar mix.

One test for abundance inhomogeneities in the ISM comes from detailed observations in the giant clouds thought to be the sites of star formation. From the molecular isotope data, the isotopic abundances in giant clouds seem to have a modest cloud-to-cloud scatter, but there are no systematic differences that persist from the observations of several molecular species. There are also no convincing instances of abundance variations within giant clouds. The only observed inhomogeneities in the ISM are those directly associated with mass loss from an evolved stellar object.

Concluding Remarks

The quality and quantity of interstellar abundance data are growing very rapidly. Many of the observed results point to a significant evolution of the ISM, in ways supported by (and supportive of) our knowledge of past and present nucleosynthesis in stars. As is often the case, the problems seem to loom larger than the successes. One area of theoretical and observational research that deserves more attention deals with the way in which freshly synthesized material is mixed into the ISM. Large scale abundance variations in the Galaxy are particularly puzzling. The molecular isotope data in dense clouds show no radial abundance gradients, a result which is at odds with chemical abundance gradients determined in HII regions. Until the problem of radial abundance variations is understood in our own Galaxy, it seems premature to discuss the scantier results from external galaxies.

Acknowledgments

It is a pleasure to thank Arno Penzias who suggested this review and who has provided many helpful suggestions. My wife, Louise, typed the final manuscript and provided welcome encouragement. Richard Mewaldt helped greatly with the material on cosmic rays. Beatrice Tinsley, Peter Goldreich, and William Fowler reviewed a preliminary manuscript and made several useful suggestions. Thanks also go to Manuel Peimbert, Jim Kaler, Leonard Searle, and Greg Shields who discussed with me various aspects of the observations of planetary nebulae and HII regions. Finally, my thanks to to the editorial staff of the Annual Reviews for their patience.

Literature Cited

Alloin, D., Collin-Souffrin, S., Joly, M., Vigroux, L. 1979 *Astron. Astrophys.* 78:200
Althouse, W. E., Cummings, A. C., Garrard, T. L., Mewaldt, R. A., Stone, E. C., Vogt, R. E. 1978. *Geosci. Elec.* 16:204
Audouze, J., Lequeux, J., Vigroux, L. 1975. *Astron. Astrophys.* 43:71
Audouze, J., Tinsley, B. M. 1976. *Ann. Rev. Astron. Astrophys.* 14:43
Baade, W., Minkowski, R. 1954. *Ap. J.* 119:206
Balick, B., Sneden, C. 1976. *Ap. J.* 208:336
Barnes, T. G., Beer, R., Hinkle, K. H., Lambert, D. L. 1977. *Ap. J.* 213:71
Burton, W. B. 1976. *Ann. Rev. Astron. Astrophys.* 14:275
Cernicharo, J., Guélin, M. 1980. In preparation
Chevalier, R. A., Kirshner, R. P. 1978. *Ap. J.* 219:931
Chevalier, R. A., Kirshner, R. P. 1979. *Ap. J.* 233:154
Clayton, R. N., Grossman, L., Mayeda, L. K. 1973. *Science* 182:485
Dearborn, D., Tinsley, B. M., Schraam, D. N. 1978. *Ap. J.* 223:557
Dietrich, W. F., Simpson, J. A. 1979. *Ap. J. Lett.* 231:L91
Dwyer, R. 1978. *Ap. J.* 224:691
Epstein, R. I. 1977. *Ap. J.* 212:595
Fisher, A. J., Hagen, F. A., Maehl, R. C., Ormes, J. F., Arens, J. F. 1976. *Ap. J.* 205:938
Fowler, W. A., Caughlan, G. R., Zimmerman, B. A. 1975. *Ann. Rev. Astron. Astrophys.* 13:69
Frerking, M. A., Wilson, R. W., Linke, R. A., Wannier, P. G. 1980. *Ap. J.* In press
Garcia-Munoz, M., Simpson, J. A., Wefel, J. P. 1979. *Ap. J. Lett.* 232:L95
Gardner, F. F., Whiteoak, J. B. 1976. *MNRAS* 176:57P
Gardner, F. F., Whiteoak, J. B. 1979. *MNRAS.* In press
Geiss, J., Eberhardt, P., Bühler, F., Meister, J., Signer, P. 1970. *J. Geophys. Res.* 75:5972
Goldsmith, P. F., Linke, R. A. 1980. Submitted to *Ap. J.*
Greiner, D. E., Bieser, F. S., Heckman, H. H. 1978. *IEEE Trans. Geosci. Elec.* GE-16, 163
Greiner, D. E., Wiedenbeck, M. E., Bieser, F. S., Crawford, H. J., Heckman, H. H., Lindstrom, P. J. 1979. *16th Int. Cosmic Ray Conf., Kyoto* 1:418
Guélin, M., Lequeux, J. 1980. *Interstellar Molecules*, ed. B. H. Andrew. Boston: Reidel
Guélin, M., Thaddeus, P. 1979. *Ap. J. Lett.* 227:L139
Hawley, S. A. 1978. *Ap. J.* 224:417
Henkel, C., Walmsley, C. M., Wilson, T. L. 1980. *Astron. Astrophys.* In press
Herbst, E., Klemperer, W. 1973. *Ap. J.* 185:505
Jefferts, K. B., Penzias, A. A., Wilson, R. W. 1973. *Ap. J. Lett.* 179:L57
Jensen, E. B., Strom, K. M., Strom, S. E. 1976. *Ap. J.* 209:748
Kaler, J. B. 1978. *Ap. J.* 226:947
Kaler, J. B. 1979. *Ap. J.* 228:163
Kaler, J. B. 1980. Submitted to *Ap. J.*
Kamper, K., van den Bergh, S. 1976. *Ap. J. Suppl.* 32:351
Kutner, M. L., Machnik, D. E., Tucker, K. D., Massano, W. 1980. In press
Langer, W. D. 1976. *Ap. J.* 210:328
Langer, W. D. 1977. *Ap. J. Lett.* 212:L39
Langer, W. D. 1979. Private Communication
Langer, W. D., Goldsmith, P. F., Carlson, E. R., Wilson, R. W. 1980. *Ap. J. Lett.* 235:L39
Lazareff, B., Lucas, R., Encrenaz, P. 1978. *Astron. Astrophys.* 70:L77

Lee, T., Russell, W. A., Wasserburg, G. J. 1979. *Ap. J. Lett.* 228:L93
Linke, R. A. 1980. In preparation
Linke, R. A., Goldsmith, P. F. 1980. *Ap. J.* 235:437
Linke, R. A., Goldsmith, P. F., Wannier, P. G., Wilson, R. W., Penzias, A. A. 1977. *Ap. J.* 214:50
Liszt, H. S. 1978. *Ap. J.* 222:484
Mewaldt, R. A., Spalding, J. D. Stone, E. C., Vogt, R. E. 1979. *Ap. J.* 231:L97
Mewaldt, R. A., Spalding, J. D., Stone, E. C., Vogt, R. E. 1980a. *Ap. J. Lett.* 235:L95
Mewaldt, R. A., Spalding, J. D., Stone, E. C., Vogt, R. E. 1980b. *Ap. J. Lett.* 236:L121
Ostriker, J. P., Tinsley, B. M. 1975. *Ap. J.* 201:L51
Pagel, B. E. J., Edmunds, M. G., Blackwell, D. E., Chun, M. S., Smith, G. 1979. *MNRAS*, 189:95
Pardo, R. C., Couch, R. G., Arnett, W. D. 1974. *Ap. J.* 191:711
Peimbert, M. 1978. In *Planetary Nebulae, Observations and Theory*, ed. Y. Terzan, p. 215. Boston: Reidel
Peimbert, M. 1979. *The Large Scale Characteristics of the Galaxy*, ed. W. B. Burton. Dordrecht: Reidel
Peimbert, M., Torres-Peimbert, S., Rayo, J. F. 1978. *Ap. J.* 220:516
Peimbert, M., van den Bergh, S. 1971. *Ap. J.* 167:223
Penzias, A. A. 1977. *Ap. J.* 214:50
Penzias, A. A. 1978. *Am. Sci.* 66:291
Penzias, A. A. 1979. *Ap. J.* 228:430
Penzias, A. A. 1980a. *Interstellar Molecules*, ed. B. H. Andrew. Boston: Reidel. In press
Penzias, A. A. 1980b. Submitted to *Science*
Penzias, A. A. 1980c. In preparation
Penzias, A. A., Wannier, P. G., Wilson, R. W., Linke, R. A. 1977. *Ap. J.* 211:108
Podosek, F. A. 1978. *Ann. Rev. Astron. Astrophys.* 16:293
Rank, D. M., Geballe, T. R., Wollman, E. R. 1974. *Ap. J. Lett.* 187:L111
Rood, R. T., Wilson, T. L., Steigman, G. 1979. *Ap. J. Lett.* 227:L97
Sargent, W. L. W. 1980. In preparation
Searle, L. 1971. *Ap. J.* 168:327
Shapiro, M. M., Silberberg, R. 1974. *Philos. Trans. R. Soc. London, A.* 277:319
Shields, G. A. 1980. Private communication
Shields, G. A., Searle, L. 1978. *Ap. J.* 222:821
Shields, G. A., Tinsley, B. M. 1976. *Ap. J.* 203:66
Silberberg, R., Tsao, C. H., Shapiro, M. M. 1976. *Spallation Nuclear Reactions and their Applications*, ed. B. S. P. Shen, M. Merker, p 49. Dordrecht: Reidel

Sivan, J. P. 1976. *Astron. Astrophys.* 49:173
Smith, H. E. 1975. *Ap. J.* 199:591
Snell, R., Wootten, A. 1980. Submitted to *Ap. J.*
Stark, A. A. 1980. In preparation
Talent, D. L., Dufour, R. J. 1978. Private communication, reported in Peimbert 1979
Tarlé, G., Ahlen, S. P., Cartwright, B. G., Solarz, M. 1979. *16th Int. Cosmic Ray Conf., Kyoto* 1:455
Tifft, W. G., Connolly, L. P., Webb, D. F. 1972. *MNRAS* 158:47
Tinsley, B. M. 1979. *Ap. J.* 229:1046
Tomkin, J., Lambert, D. L. 1980. *Ap. J.* 235:925
Torres-Peimbert, S., Peimbert, M. 1977. *Rev. Mexicana Astron. Astrofiz.* 2:181
Trimble, V. 1975. *Rev. Mod. Phys.* 47:877
Turner, B. E., Zuckerman, B., Morris, M., Palmer, P. 1978. *Ap. J. Lett.* 219:L43
Wannier, P. G. 1980. In preparation
Wannier, P. G., Linke, R. A. 1978a. *Ap. J.* 226:817
Wannier, P. G., Linke, R. A. 1978b. *Ap. J.* 225:130
Wannier, P. G., Frerking, M. A., Carlson, E. R. 1980b. In preparation
Wannier, P. G., Linke, R. A., Penzias, A. A. 1980a. In preparation
Wannier, P. G., Lucas, R., Linke, R. A., Encrenaz, P. J., Penzias, A. A., Wilson, R. W. 1976b. *Ap. J. Lett.* 205:L169
Wannier, P. G., Penzias, A. A., Linke, R. A., Wilson, R. W. 1976a. *Ap. J.* 204:26
Watson, W. D. 1976. *Rev. Mod. Phys.* 48:513
Watson, W. D. 1977. *CNO Isotopes in Astrophysics*, ed. J. Audouze, p. 105. Dordrecht: Reidel
Watson, W. D., Anicich, V. G., Huntress, W. T. Jr. 1976. *Ap. J. Lett.* 205:L165
Webber, W. R., Kish, J. C., Simpson, G. A. 1979a. *16th Int. Cosmic Ray Conf., Kyoto* 1:424
Webber, W. R., Kish, J. C., Simpson, G. A. 1979b. *16th Int. Cosmic Ray Conf., Kyoto* 1:430
Whiteoak, J. B., Gardner, F. F. 1975. *Proc. Astron. Soc. Aust.* 2:360
Whiteoak, J. B., Gardner, F. F. 1978. *MNRAS* 183:67
Wilson, T. L., Pauls, T. 1979. *Astron. Astrophys.* 73:L10
Wilson, R. W., Penzias, A. A., Wannier, P. G., Linke, R. A. 1976. *Ap. J. Lett.* 204:L135
Wolff, R. S. 1980. *Ap. J.* In press
York, D. G., Rogerson, J. B. 1976. *Ap. J.* 203:378

STELLAR CHROMOSPHERES[1] ✖2172

Jeffrey L. Linsky[2]
Joint Institute for Laboratory Astrophysics, National Bureau of Standards and University of Colorado, Boulder, Colorado 80309

I Introduction

Important progress in our understanding of stellar chromospheres has occurred in the past few years as a result of new observations, developments in spectral line formation theory, and the application of this theory to constructing detailed model chromospheres. Significant trends are beginning to emerge from such analyses, and we are on the threshold of a meaningful confrontation between purely theoretical models and the data. The range of stars thought to possess chromospheres may be widening, and we now have a better understanding of the enigmatic problem of why the Wilson-Bappu relation between Ca II emission core widths and stellar absolute visual magnitudes actually works.

An important element in this progress has been the realization that much can be learned by studying the outer atmospheres of the Sun and stars in the same context, and that such an approach is a two-way street. Not only are the theoretical techniques for analyzing spectra, modeling atmospheric structures, and computing the consequences of different heating processes the same, but also the wide variety of structures and phenomena seen on the Sun with high spatial and spectral resolution may be useful prototypes for stellar atmospheric structures and phenomena that we cannot hope to resolve, but whose existence is implied by indirect evidence. In a sense, we use the Sun as a plasma physics laboratory to develop insight into the physical processes occurring in the outer atmospheres of stars. Needless to say, our understanding of the Sun can be strengthened by studying phenomena in stars that have values of gravity, rotational velocity, chemical composition, and luminosity very much different from those of the Sun.

---

[1] The US Government has the right to retain a nonexclusive, royalty-free license in and to any copyright covering this paper.
[2] Staff Member, Quantum Physics Division, National Bureau of Standards.

Despite the importance of solar-stellar cross fertilization, it is unwise to pursue solar analogies too far. At some point most and perhaps all of the solar analogies will fail to explain observables for certain stellar chromospheres. When we run into such situations, as I think we have in several cases, we will make important advances in our understanding of underlying physical processes operating in stars.

The general topic of stellar chromospheres has been reviewed recently by Linsky (1977, 1979, 1980), Praderie (1977), Ulmschneider (1979), Snow & Linsky (1980), and Jordan (1980a). Important earlier reviews include those of Praderie (1973), Doherty (1973), and Kippenhahn (1973) in the proceedings of the IAU Colloquium on Stellar Chromospheres held in 1972 (Jordan & Avrett 1973). Recent reviews and monographs on the solar chromosphere include Athay (1976) and Withbroe & Noyes (1977). Because of the extensive review literature in this field and the rapid advances made most recently and presently under way, I will adopt a nonstandard approach here. In each subsequent section I will pose a broad, open-ended question as a means of focusing on recent observational and theoretical work in one aspect of the general problem of stellar chromospheres. I will not specifically discuss plasma diagnostics because this topic has been reviewed by Praderie (1973), Linsky (1977), and Dupree (1978). However, several interesting developments in our understanding and use of chromospheric diagnostics are described below under the relevant headings.

II What is a Stellar Chromosphere?

The broadening range of stellar phenomena that are commonly called "chromospheric" is taxing our commonsense idea of what should be called a chromosphere. For this reason and to clarify the physical processes responsible for chromospheres, it is productive to delve further into the question of what is and what is not a chromosphere.

Initially, the term chromosphere was coined to describe a region of the solar atmosphere extending some 10^4 km above the limb that is visible in emission lines of neutral and singly ionized atoms at the time of eclipse. Subsequently, it was recognized that this layer in the solar atmosphere gives rise to the cores of many strong Fraunhofer lines seen in projection against the disk and many emission lines and continua in the ultraviolet. Thomas & Athay (1961) pointed out that the anomalously large-scale heights (exponential decrease in intensity with height above the limb) of typical solar chromospheric emission lines requires either an energy source in addition to that supplied by radiation from the underlying photosphere (nonradiative heating) or a source of mechanical momentum. At the present time it is generally assumed that nonradiative heating is primarily responsible for the large extent of the solar chromosphere, although direct momentum transfer to the gas by mechanical waves or Lα radiation pressure may be important in cool giants and supergiants (Haisch et al. 1980, Hartmann & MacGregor 1980).

Thomas & Athay (1961) and Athay (1976) point out that the existence of the solar chromosphere and corona is a vivid manifestation of the failure of classical stellar atmospheres theory, with its assumptions of radiative equilibrium, hydrostatic equilibrium, and spherical symmetry, to predict the real Sun. Clearly the fundamental aspect of a chromosphere is the violation of these assumptions, particularly radiative equilibrium, and an important goal of stellar chromospheres research is to identify and quantify the nonradiative heating mechanisms responsible for the nonclassical behavior of the Sun's outer atmosphere.

Praderie (1973) proposed a tentative definition of a stellar chromosphere as that layer in which both mass flux and nonradiative energy dissipation occur. By contrast, she proposed that a stellar photosphere is that region in which no nonradiative energy deposition occurs. Such a definition is motivated by the prevailing belief that the dissipation of mechanical energy in acoustic or other types of waves (see Thomas & Athay 1961, Ulmschneider 1979) heats the chromosphere, and that turbulent convection generates the required mechanical energy flux. The nonradiative heating forces a rise in temperature over the monotonic outward decline otherwise expected in an LTE radiative equilibrium photosphere The temperature inversion then produces in a complex way the emission lines and continua that we call the chromospheric spectrum.

One problem with the scenario discussed by Praderie (1973, 1977) is that temperature inversions are possible even in a purely radiative equilibrium atmosphere. In the most extreme example of the effect originally described by Cayrel (1963, 1964), the local electron temperature in the outer layers can rise to the color temperature of the background photospheric radiation field. The Cayrel effect is opposed by surface line cooling (cf. Athay 1970a) in solar-type stars, but a Cayrel-type temperature inversion is a standard feature of non-LTE models of early-type stars (cf. Ulmschneider 1979). Thus our intuitive feeling that a chromosphere begins at the temperature minimum, where dT_e/dh changes from negative to positive values, may ignore much of the underlying physics.

I think that it is important to go beyond Praderie's definition of a chromosphere for the following reasons:

1. There is growing evidence for nonradiative heating in atmospheric layers that we intuitively characterize as photospheres. Theoretical calculations of the propagation of acoustic waves in solar-type and cooler stars by Ulmschneider et al. (1977), among others, predict that as much as 90% of the initial acoustic wave energy is dissipated in the photosphere (where

$dT_e/dh < 0$) below the temperature minimum. The temperature gradient remains negative despite the mechanical heating because the high density photosphere is an efficient radiator of energy by H^- and other continua, and the acoustic dissipation rate is small compared with the local radiative heating and cooling. (By comparison, the lower density chromosphere is an inefficient radiator of energy and a small amount of heating is sufficient to produce a temperature inversion.) Also, analyses of the Ca II and Mg II resonance line wings suggest that the outer portions of the photospheres of many main-sequence and giant stars (Kelch et al. 1978, 1979) and both quiet and active regions of the solar chromosphere (Ayres & Linsky 1976, Morrison & Linsky 1978) are hotter than predicted by radiative equilibrium considerations. Thus the assumption that nonradiative heating is unimportant in photospheres needs to be reconsidered.

2. Withbroe & Noyes (1977) have summarized the solar data that demonstrate the pervasive and fundamental role played by magnetic fields in defining the structure, mass balance, and energy flow of the chromosphere and corona. They point out that the differences in physical conditions in the chromospheric and coronal layers over different areas of the solar surface are intimately related to the strength and configuration of the local magnetic fields. A specific example is the excellent correlation of Ca II K-line intensity with magnetic field strength (Skumanich et al. 1975), which demonstrates the close connection between magnetic field concentrations and nonradiative heating. Heating mechanisms involving magnetic fields include MHD waves, ohmic dissipation of currents induced by evolving magnetic structures, and field reconnection. The latter two mechanisms do not involve mechanical waves directly. Furthermore, X rays from solar flares can heat the chromosphere (see, for example, Henoux & Nakagawa 1978). Consequently, the assumption that a mechanical energy source and a mass flux are necessary conditions for the existence of a chromosphere may not be valid.

3. It is important to differentiate between chromospheres and coronae in a meaningful way.

With these problems in mind, I propose new tentative definitions of different atmospheric layers. The defining characteristics are summarized in Table 1.

I use *outer atmosphere* as a generic term encompassing all of the layers described below. The specific quantity defining an outer atmosphere and separating it from the inner atmosphere of a star is the presence of significant nonradiative heating. I use the term nonradiative heating to include dissipation of mechanical waves, magnetic heating processes, thermal conduction, and even radiation from higher layers, for example X rays from flares and coronae. The important point is that the heating does not involve the emergent photospheric radiation field.

Table 1 Atmospheric layers

Parameter	Inner atmosphere	Outer atmosphere			
		Photosphere	Chromosphere	Transition region	Corona
Nonradiative heating	not present	present but not dominant	dominates the energy balance equation		
Temperature gradient		$\frac{dT}{dh} < 0$ except for Cayrel-type temperature inversions	$\frac{dT}{dh} > 0$ but gradual	$\frac{dT}{dh} > 0$ but steep	$\frac{dT}{dh} > 0$ but small
Geometrical extent (in units of the local P scale height)		many	several	much less than 1	many
Dominant cooling terms for Sun		radiation in continua and lines	radiation in resonance and some subordinate lines, H^- and H continua	line radiation in UV and EUV	X-ray and EUV radiation, thermal conduction, thermally driven wind
Range of temperatures in Sun		6000–4300	4300–25,000 K	25,000–1×10^6 K	1–3×10^6 K
Important structures seen in Sun		sunspots, faculae, granulation, flux tubes	network, plages, spicules, prominences	network, active regions	magnetic loops, coronal holes

By the term *photosphere*, I specify the atmospheric layers where nonradiative heating is present but not dominant. By this I mean that the nonradiative heating is not sufficiently large to force a temperature inversion. This definition is consistent with our intuitive feeling that a photosphere is a region of $dT_e/dh < 0$, such as usually occurs in radiative equilibrium. By this definition, layers in which temperature inversions occur within the radiative equilibrium constraint, such as by the Cayrel mechanism, are included within the term photosphere.

A necessary but not sufficient condition for the existence of a *chromosphere* is that nonradiative heating dominate the energy balance. By this I mean that the local nonradiative heating rate is sufficient to force a positive temperature gradient. Since this particular property does not uniquely distinguish a chromosphere from other layers, I will propose another condition based on our understanding of the solar example. The first empirical aspect of the solar chromosphere that called attention to its anomalous character was its geometrical thickness. That is, the chromosphere extends over many pressure scale heights, and the temperature gradients are generally small. This is presumably due to the large opacity of the Lyman continuum and the resonance lines of H I, Ca II, and Mg II. These lines, among others, and the Lyman continuum are relatively efficient radiators of energy and act as thermostats to produce a gradual temperature increase with increasing height and decreasing density. In fact, it is useful to define the upper extent of a chromosphere as that height where the last of these resonance lines, Lα, becomes very thin. At this point, typically above 20,000 K, the atmosphere loses an important cooling mechanism (cf. Thomas & Athay 1961, Athay 1976) and the temperature rises steeply with height.

Therefore, I propose that the necessary and sufficient conditions for the existence of a chromosphere are that nonradiative heating dominate the energy balance and that temperature gradients are small compared with the local pressure scale height. Mass flux need not be an important term in the momentum equation, although in late K and M giants and supergiants there is evidence for supersonic winds in the chromosphere (Stencel 1978, Mullan 1978).

In the Sun the region immediately above the chromosphere is characterized by steep temperature gradients. In fact, the temperature rises from about 30,000 K to 1×10^6 K in less than a pressure scale height, presumably owing to the absence of efficient cooling agents. It therefore seems reasonable to define stellar *transition regions* as those layers where nonradiative heating is dominant (in the same sense as defined above), but the cooling mechanisms are sufficiently weak that the geometrical extent of the layer is smaller than a pressure scale height. Jordan (1980b) has discussed the interrelationships between conductive heating, radiative cooling, and additional nonradiative heating terms in the context of the solar transition region. The role played by wave dissipation processes is unclear although there appears to be insufficient mechanical energy available to balance estimated cooling rates in either the transition region or the corona (Athay & White 1978). Since the lower boundary temperature is governed by the ionization of hydrogen, one might expect transition regions to begin near 30,000 K. (This argument also suggests that the hotter O and WR stars may not have chromospheres as defined above owing to the absence of appreciable neutral hydrogen.)

The solar corona was initially identified by the presence of 10^6 K emission lines and its large geometrical extent. The high temperatures are a consequence of nonradiative heating processes that are largely or entirely magnetic in character (Tucker 1973, Withbroe & Noyes 1977), and cooling processes involving X ray and EUV radiation, thermal conduction both down to the transition region and out to space, and a thermally driven wind, all of which require 10^6 K temperatures to operate efficiently. The geometrical extent of the corona is a consequence of the large pressure scale heights at high temperatures, conductive smoothing of temperature gradients, and perhaps also heating over large spatial scales.

It therefore seems natural to define a *corona*, in contradistinction to a chromosphere or a transition region, as a region heated by nonradiative processes that is characterized by small temperature gradients and sufficiently hot temperatures that the dominant cooling mechanisms include X-ray and EUV radiation, thermal conduction, or a thermally driven wind. For the solar corona these latter three terms are roughly comparable (Withbroe & Noyes 1977), but this need not be true in general as shown schematically by Hearn (1975). In addition, the interface temperature between a corona and the underlying transition region need not be at 10^6 K, but will depend on the balance of heating and cooling rates, which controls the local temperature gradient. Based on initial IUE observations, Linsky & Haisch (1979) have proposed that among G, K, and M stars the TR-corona interface temperature is not in the range 20–250,000 K and that stars dissipate the available nonradiative heating either by an outer atmosphere including a chromosphere, transition region, and corona (hotter than 250,000 K) or by a chromosphere that has a strong wind as an important cooling agent, but little or no plasma hotter than 20,000 K.

It is important to recognize that the outer atmosphere layers of a star are probably inhomogeneous. In particular, magnetic loops dominate the Sun's coronal geometry (Rosner et al. 1978), and emission from the solar transition region may arise from less than 1% of the solar surface (Nicolas et al. 1979, Feldman et al. 1979).

III   In What Regions of the H-R Diagram Do Chromospheres Exist?

This question is important, among other reasons because it is a means of testing the validity of theoretical models of stellar chromospheres and the assumed heating mechanisms that underlie such calculations.

(*a*) CHROMOSPHERES IN LATE-TYPE STARS (SPECTRAL TYPES F-M)   Important spectroscopic indicators of chromospheres in late-type stars include the Ca II H ($\lambda 3968$) and K ($\lambda 3934$) lines, Mg II h ($\lambda 2803$) and k ($\lambda 2796$), Lα ($\lambda 1216$), the H I Balmer series (particularly in M dwarfs), He I ($\lambda\lambda 10830$, 5876), O I ($\lambda\lambda 1305$, 1355), Si II ($\lambda\lambda 1808$, 1817), Fe II, and continua in the ultraviolet and infrared. The usefulness of these direct indicators as well as other less direct diagnostics has been surveyed by Praderie (1977), Linsky (1977), and Ulmschneider (1979), among others. The Ca II and Mg II features are among the most readily observed of these diagnostics, consequently surveys of stellar chromospheres generally are based on these lines.

A considerable literature now exists concerning observations of the stellar H and K lines, and Bidelman (1954) has compiled a very useful bibliography of the various stellar types that exhibit Ca II emission. H and K emission is occasionally seen in F stars, is usually seen in G stars, and is essentially ubiquitous in K and M stars. The earliest stars having Ca II emission include the F0 dwarf, γ Vir N (Warner 1968), and the F0 supergiant, α Car (Warner 1966). Occasionally Ca II is seen in the A7 III δ Scuti star γ Boo (Le Contel et al. 1970, Auvergne et al. 1979). *Copernicus* observations of the Mg II lines by Evans et al. (1975) confirm the existence of a chromosphere in α Car. Böhm-Vitense & Dettman (1980) have surveyed A and F stars with IUE to see if there is a boundary line in that portion of the H-R diagram between stars with and without chromospheres. They find that chromospheres occur in dwarfs and giants of spectral types F2 and later and in supergiants redward of the Cepheid instability strip. The stars that exhibit Mg II emission cores usually also show evidence for transition region material (e.g. C IV $\lambda 1548$, Si IV $\lambda 1394$, and He II $\lambda 1640$). The most luminous stars they observed, HR 8752 (F8 Ia0) and HR 4511 (G0 Ia0), show no emission lines, but both stars have hot companions that would have covered up any chromospheric emission otherwise present.

Chromospheres have been observed for some time in stars along the main sequence at least as cool as the dMe stars. Recent IUE observations include those of EQ Peg (M3.5e V + M4.5e V; Hartmann et al. 1979), YZ CMi (M4e V; Carpenter & Wing 1979), and Prox Cen (M5.5e V; Carpenter & Wing 1979, Haisch & Linsky 1980). Chromospheres also occur in M giants and supergiants as indicated by their bright Fe II spectra (see, for example, Linsky & Haisch 1979, Carpenter & Wing 1979, van der Hucht et al. 1979). However, for the coolest low gravity stars, Dyke & Johnson (1969), Jennings & Dyke (1972), and Jennings (1973) find an inverse correlation between K-line emission and polarization or infrared excess, two symptoms of grains in a cool outer atmosphere. This empirical result suggests that, when the grain density is high, the grains provide more effective plasma cooling than emission lines such as H and K. Chromospheric emission spectra also occur in RS CVn-type systems (Hall 1976), T Tauri stars (for example, Ulrich & Knapp 1979), and UV Ceti-type flare stars during flares and quiescent periods.

(b) CHROMOSPHERES IN EARLY-TYPE STARS (SPECTRAL TYPE O-A) The ample evidence for massive winds in early-type stars has been reviewed recently by Hearn (1979), Snow (1979), Cassinelli (1979), Snow & Linsky (1980), Hutchings (1979), and Conti (1978). In addition, recent X-ray observations by HEAO-1 and HEAO-2 indicate the presence of coronae in OB supergiants (Harnden et al. 1979) and A-type dwarfs (Topka et al. 1979, Cash et al. 1979). The anomalously strong O VI and N V emission lines in O stars and C IV emission lines in B supergiants detected by *Copernicus* may be due to a 10^5 K wind (cf. Lamers & Rogerson 1978), or as now appears more likely given the new X-ray data, to Auger ionization by X rays from a thin 10^6 K corona at the base of these winds (Cassinelli & Olson 1979). However, there is little hard evidence for chromospheres, as defined in Section II, among stars hotter than spectral type F0. Böhm-Vitense & Dettmann (1980) have searched without success for chromospheric emission lines in 16 A- and late B-type stars using IUE, but they are not certain whether the absence of emission lines indicates the lack of chromospheres, or instead is a consequence of the bright photospheric continuum against which such measurements must be made.

Using *Copernicus*, Kondo et al. (1975) have found evidence for Mg II emission in λ Ori (O8 III), γ Ara (B1 Vep), and possibly α Gru (B7 IV), but their results have not been confirmed. Similarly, 20-μm excess emission from Vega (Morrison & Simon 1973) has not been verified and emission satellites in the infrared Ca II and O I triplet lines in Vega (Johnson & Wisniewski 1978) have not been observed subsequently. Vega does not show emission in H and K (Freire et al. 1978) or in the ultraviolet C II and Si II lines (Freire 1979), but Freire concludes that these data do not preclude a chromospheric temperature rise at log $\tau_{5000} \leq -4$ in Vega and other early A-type stars. Renzini et al. (1977) and Lamers & de Loore (1976) predict that the acoustic energy generation rate is relatively small in A stars, consequently chromospheric emission lines may be faint. In any event, it seems unreasonable that Vega can have a corona, as indicated by X-ray emission (Topka et al. 1979), without also having a chromosphere (or transition region) that interfaces the warm photosphere and hot corona. Thus, I expect that chromospheres eventually will be discovered in A-type and possibly hotter stars.

(c) TRANSITION REGIONS The short wavelength spectrograph on IUE has, for the first time, permitted observations of emission lines formed in the transition regions of a large number of stars. Material at 50,000–250,000 K is indicated by strong lines of C III (λ1175, λ1909), Si IV (λ1394, λ1403), C IV (λ1548, λ1551), N V (λ1238, λ1242), and O V (λ1371). Lines of C II (λ1334, λ1335) and Si III (λ1206, λ1892) are likely formed near 16,500 K when a plateau exists at the top of the chromosphere, as appears likely for the Sun and ε Eri (see Section IV).

The initial IUE observations of late-type stars (Linsky et al. 1978) showed that the spectra of α Aur (G6 III + F9 III), HR 1099 (G5 V + K1 IV), λ And (G8 III-IV), and ε Eri (K2 V) contain most or all of the emission lines described above in roughly the same relative line strengths as in the quiet Sun spectrum. On this basis, these stars are thought to posses solar-like transition regions. Because no strong lines formed at temperatures hotter than 250,000 K fall in the IUE bandpass, it is not possible to say whether these stars also have solar-like coronae (as defined in Section II) on the basis of IUE observations alone. However, in analogy with the Sun it is reasonable to expect coronae when transition-region emission is prominent. In fact, the first three stars (all spectroscopic binaries) have been detected in the HEAO-1 A2 soft X-ray survey (Walter et al. 1980), and presumably therefore have hot and extensive coronae.

Linsky & Haisch (1979) observed 22 stars with IUE, and noted a sharp dividing line in the H-R diagram between stars showing evidence for transition regions, and by implication also coronae, and those stars showing evidence for chromospheres only. Their results are plotted in Figure 1, together with data from Böhm-Vitense & Dettmann (1980), Hartmann et al. (1979), Brown et al. (1979), Dupree et al. (1979a), Carpenter & Wing (1979), Haisch & Linsky (1980), and Blanco et al. (1979). The stars in Figure 1 fall into the following groups: 1. stars with chromospheric and transition region emission lines, 2. stars with chromospheric lines and C II λ1335 and/or Si III λ1892, 3. stars that show only chromospheric lines such as Si II $\lambda\lambda$1808, 1817 and Mg II h and k, and 4. stars that show no evidence of emission lines in the ultraviolet. Group 2 may not be distinct from group 3, because Jordan (1979, private communication) suggests, at least for α Tau, that the feature identified as Si III λ1892 may be S I λ1900 and Linsky & Haisch (1979) note that there is a feature near 1340 Å in spectra of α Boo and α Ori that is probably not C II λ1335 but could be Fe II λ1338 (Brown et al. 1979). Finally, some stars in group 4 may have chromospheres even though no Mg II emission is seen in the IUE spectra.

The data in Figure 1 should be viewed as preliminary: the spectra were exposed to very different levels and individual authors may have used different criteria in identifying weak spectral features in noisy data. Nevertheless, these data are consistent with the existence of solar-like transition regions in only the limited region of the H-R diagram outlined, with fairly sharp boundary lines between stars with and without prominent transition-region emission.

IV What Trends Are Emerging from Semiempirical Chromospheric Models of Single Stars?

During the past several years two important developments have greatly facilitated the computation of semiempirical models of stellar chromospheres. The first is the acquisition of absolute flux profiles of important diagnostics such as Ca II H and K, the Ca II infrared triplet, He I λ10830, and Hα from the ground; and Mg II h and k, Lα, and the resonance lines of C II, Si II, and Si III from space experiments, particularly IUE. The second development is the refinement in our understanding of optically thick resonance line formation. We now have increased confidence in the use of these diagnostics to provide some insight into the gross properties of stellar chromospheres.

The various diagnostics used in building semiempirical model chromospheres have been reviewed by Linsky (1977, 1979), and Ulmschneider

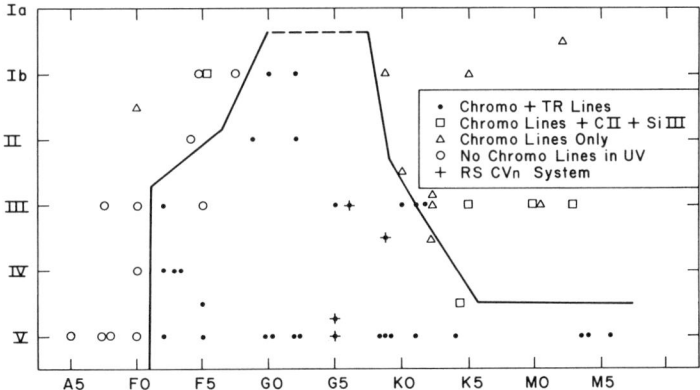

Figure 1 The location of stars that show spectroscopic evidence of chromospheres and/or transition regions in the H-R diagram. The lines tentatively outline the region of the H-R diagram in which stars generally have transition regions.

(1979). The usefulness of different spectral features for identifying chromospheres has been reviewed by Praderie (1973, 1977). Here I will describe some recent developments in the field, identify interesting trends that are emerging, and call attention to uncertainties in the conventional analyses.

First, I will mention important observational programs that are producing chromospheric line profiles, calibrated in absolute flux units, which are the backbone of model chromosphere studies. Linsky et al. (1979a) have obtained 120 mÅ spectra of the Ca II H and K lines in a wide variety of stars later than spectral type F0, using the Kitt Peak 4-m echelle spectrograph. These data were calibrated in absolute flux units at the stellar surface based on R. V. Willstrop's (1972, private communication) narrow band photometry and the Barnes & Evans (1976) relation for deriving stellar angular diameters. Giampapa (1980) has extended the program to dMe stars. Blanco et al. (1974, 1976) have published absolute K-line fluxes for F5-G8 dwarfs and G2-M5 giants, and Blanco et al. (1978) have presented absolute fluxes and brightness temperatures for the H_1 and K_1 features in 21 late-type stars. Echelle spectra of λ And obtained with the Mt. Hopkins Kron camera system have been discussed by Baliunas & Dupree (1979). Anderson (1974) and Linsky et al. (1979b) have obtained spectra of the Ca II infrared triplet lines, and A. Young (1979, private communication) is obtaining spectra of these features in RS CVn-type systems and other active chromosphere stars using the KPNO CID system. The Balmer lines in dMe flare stars have been studied by many observers, most recently by Worden et al. (1980). Zirin (1976) has obtained He I λ10830 equivalent widths for some 200 stars later than F5, and in several stars such as R Aqr, T Tau, and 12 Peg that have circumstellar envelopes. O'Brien & Lambert (1979) are studying λ10830 with an echelle-reticon system at McDonald Observatory. They find λ10830 emission in α Boo and α^1 Her and are presently studying nearly 60 F-M stars, with monitoring programs on several particularly interesting objects.

A number of important observing programs are under way in the ultraviolet. Surveys of Mg II emission fluxes include the *Copernicus* observations of 49 stars by Weiler & Oegerle (1979), BUSS observations with 0.1 Å resolution (e.g. Kondo et al 1979, de Jager et al. 1979, van der Hucht et al. 1979), and IUE observations by Pagel & Wilkins (1979), Basri & Linsky (1979), Carpenter & Wing (1979), Stencel & Mullan (1980), and several other groups. The 1175–2000 Å short wavelength spectral region of IUE permits the study of prominent chromospheric resonance lines of H I, O I, C I, Si II, C II, and Si III, and transition region lines of C III, Si IV, C IV, N V, and O V. Initial observations of cool stars in the shortwavelength region include Linsky et al. (1978), Linsky & Haisch (1979), Ayres & Linsky (1980a, b), Dupree et al. (1979a), Hartmann et al. (1979), Brown et al. (1979), Carpenter & Wing (1979), Böhm-Vitense & Dettman (1980).

Table 2 summarizes semiempirical chromospheric models that have been constructed to match various diagnostic features observed in quiet and active regions on the Sun and in stars cooler than spectral type F0 V. For the most part these models were constructed to fit the Ca II and Mg II emission cores and damping wings using partial redistribution (PRD) radiative transfer codes, although prior to 1975 only the less accurate complete redistribution (CRD) codes were available. The use of PRD diagnostics has resulted in computed Ca II line profiles which now exhibit steep sides beyond the K_2 emission features, in agreement with observations (Wilson 1973). Because the Ca II and Mg II lines are formed in chromospheric layers cooler than 8000 K, these models are meaningful only below that temperature. The Lyman and millimeter continua are useful for extending the models to 10,000 K, but suitable data are available only for the Sun. Other diagnostics of the 6500–8000 K temperature range include Si II $\lambda\lambda$1808, 1817, 1265, and 1553 and the damping wings of Lα. Tripp et al. (1978) have described the formation of the Si II lines in the quiet Sun, while Basri et al. (1979) have used Lα wing observations to construct quiet and active solar models. However, the accuracy of the Lα diagnostic is compromised by the uncertain amount of frequency redistribution beyond the Doppler core. Because the Si II lines and Lα wings can be observed in stars by IUE, these diagnostics are potentially very important. In dMe stars, Hα and other Balmer series members go into emission and are useful diagnostics of the 8,000–15,000 K temperature range (Fosbury 1974, Cram & Mullan 1979).

Vernazza et al. (1973) have proposed that the solar upper chromosphere has a plateau near 20,000 K with small temperature gradients, presumably produced when Lα becomes optically thin and an efficient radiator. Such a plateau can be studied by analyzing the Lα core. However, interstellar H I absorption prevents the observation of the Lα core in any star other than the Sun, or perhaps a few short period binary systems for which the orbital motions are large enough to unmask the stellar Lα core from the saturated interstellar absorption feature. Instead, one can study the C II $\lambda\lambda$1334, 1335 and Si III $\lambda\lambda$1206, 1892 lines which are also formed in the plateau (Lites et al. 1978, Tripp et al. 1978). These features, except Si III λ1206, can be observed even in faint stars by IUE at low dispersion, and chromospheric models of the 20,000 K region have now been constructed for several stars (e.g. Simon et al. 1980a, Basri 1979, Simon & Linsky 1980).

Before discussing the general trends emerging from these models, I should bring to your attention their inherent limitations:

1. All of the stellar models assume one-component atmospheres, whereas the Sun exhibits an embarrassingly rich variety of chromospheric structure. Inhomogeneities must also be important in many other stars as indicated by Wilson's (1978) observations of time variability in K-line

Table 2 Semiempirical chromospheric models

Stars or Solar Features	Diagnostics used	Approximations used	Chromospheric temperature range	References
Solar models				
Quiet Sun (one comp.)	H, H⁻, C I, Si I, IR continua	nonLTE	4100–25,000	Vernazza et al. 1973
Quiet Sun (one comp.)	Mg II h+k, Ca II H+K	PRD	4450–5320	Ayres & Linsky 1976
Quiet Sun (one comp.)	UV and IR continua	nonLTE	4150–5360	Vernazza et al. 1976
Quiet Sun (one comp.)	C II λ1334, λ1335; Lyman lines Lyman continuum	CRD	6884–57,000	Lites et al. 1978
Quiet Sun (one comp.)	Si II λ1816, λ1256, λ1533; Si III λ1206	CRD	VAL (1973)	Tripp et al. 1978
Plage	Ca II H+K, λ8498, λ8542, λ8662	CRD	4200–8000	Shine & Linsky 1974
Plage	Mg II h+k, Ca II H+K	PRD	4600–8000	Kelch & Linsky 1978
4 Component Sun	Lα, Lyman and mm continua	PRD	4460–25,000	Basri et al. 1979
6 Component Sun	UV and IR continua	PRD	4150–25,000	E. H. Avrett 1979, private communication
Flares	Ca II H+K	CRD	5000–8400	Machado & Linsky 1975
Flares	Ca II K, UV continua	PRD	4890–6100	Machado et al. 1978
Main-sequence and subgiant stars				
α CMi (F5 IV-V)	Ca II K, λ8542	CRD	4750–8000	Ayres et al. 1974
α Cen A (G2 V), α Cen B (K1 IV)	Ca II K	PRD	3650–8000	Ayres et al. 1976
70 Oph A (K0 V), ε Eri (K2 V)	Ca II K, Mg II h+k	PRD	3850–8000	Kelch 1978
γ Vir N (F0 V), θ Boo (F7 IV-V) 59 Vir (F8 V), HD 76151 (G4 V) 61 UMa (G8 V), ξ Boo A (G8 V) EQ Vir (dK 7e), 61 Cyg B (dM0)	Ca II H+K	PRD	3000–8000	Kelch et al. 1979
dMe stars	Hα, Hβ, Hγ	CRD	3000–15,000	Cram & Mullan 1979
ε Eri (K2 V)	Mg II h+k, C II, Si II, Si III	PRD, CRD	3850–30,000	Simon et al. 1980a

Table 2 (continued)

Stars or Solar Features	Diagnostics used	Approximations used	Chromospheric temperature range	References
Giants				
α Boo (K2 III)	Ca II H+K, λ8542; Mg II h+k	CRD, PRD wings	3200–8000	Ayres & Linsky 1975
β Gem (K0 III), α Tau (K5 III)	Ca II K, Mg II h+k	PRD	2700–8000	Kelch et al. 1978
Supergiants				
β Dra (G2 II)	Ca II K, Mg II k, C II	PRD	4000–16,000	Basri 1979
ε Gem (G8 Ib), α Ori (M2 Iab)	Ca II K, Mg II k	PRD	2730–7000	Basri 1979
RS CVn-type systems				
α Aur (G6 III + F9 III)	Ca II K, Mg II h+k	PRD	4700–8000	Kelch et al. 1978
λ And (G8 III–IV), α Aur (G6 III + F9 III)	Ca II K, Mg II k, Hα	PRD	3800–10,000	Baliunas et al. 1979
HR 1099 (G5 V + K1 IV), UX Ari (G5 V + K0 IV)	Mg II k, C II, Si II, Si III	PRD, CRD	4100–30,000	Simon & Linsky 1980

emission, which is likely produced in part by the appearance and temporal evolution of plage regions on the visible hemispheres of stars. The important question is whether a one-component analysis is sufficient for an assessment of critical auxiliary quantities, for example nonradiative heating rates, or for comparison with purely theoretical chromospheric models. This question will be deferred to the next section, but it is clear that the chromosphere of at least one cool giant—Arcturus (K2 III)—is structured enough to make models based on diagnostics such as Ca II or Mg II, which are representative of the hotter atmospheric components, inconsistent with observations of diagnostics such as the CO fundamental vibration-rotation bands, which are sensitive to the cooler components (cf. Heasley et al. 1978).

2. Models based on optically thick chromospheric resonance lines are uncertain to the extent that frequency redistribution in the line wings is not properly treated. The redistribution problem is most acute for estimating temperatures near the stellar temperature minimum. In these low density layers, scattering in the Ca II and Mg II resonance lines is nearly coherent at and beyond the K_1 minima; consequently, the monochromatic source functions are strongly decoupled from the Planck function. For the Ca II resonance lines, radiative transitions to the 3d ^{2}D metastable levels provide a lower limit of roughly 0.05 to the incoherence fraction. However, Mg II lacks analogous subordinate levels between the upper and lower states of the resonance lines, and the lower limit to the incoherence fraction Λ is only 10^{-4} (Basri 1979). Supergiants provide the most extreme examples of nearly pure coherent scattering in the Ca II and Mg II wings, owing to the low chromospheric densities and correspondingly reduced collisional redistribution. Basri (1979) finds that the largest sources of incoherence in such situations are radiative transitions to other levels. As described below, highly coherent scattering can even lead to self-reversed profiles with K_1 minimum features in *isothermal* models. Finally, the disagreement in solar chromospheric temperature structures based on the Ca II and the Mg II lines (Ayres & Linsky 1976) is a clear warning either that our understanding of the underlying atomic physics of the scattering process in resonance lines may be lacking in some important respect, or, most likely, that single-component, homogeneous models are not a satisfactory description of the solar outer atmosphere.

3. Baliunas et al. (1979) have shown that steep temperature gradients relatively deep in the chromospheres of active stars such as λ And (G8 III–IV) and Capella (G6 III + F9 III) can produce high pressures in the upper chromospheres and Ca II and Mg II resonance line cores with small K_2-K_3-K_2 contrasts or no central reversals at all, consistent with observations. Their approach takes advantage of properties of the line cores that were ignored in constructing earlier models (e.g. Kelch et al. 1978). While the

Baliunas et al. approach may be a reasonable way to model active chromosphere stars, it is important to recognize that other processes are equally effective in filling in the line core; in particular, rotation and intermediate scale turbulence (mesoturbulence; see, for example, Shine 1975, Basri 1979). Baliunas et al. (1979) also suggest that the usual approach of assuming a steep temperature rise beginning at 8000 K, as appears to be valid in quiet and active regions of the Sun, is an overly restrictive assumption.

Bearing these problems in mind and the nonunique aspects of the diagnostics, it is nonetheless important to consider the basic trends that are surfacing from the modeling effort. In some cases, such trends may not be overly sensitive to the uncertainties of the modeling process.

1. In most cases studied, temperatures in the stellar upper photosphere inferred from the K-line wings are hotter than predicted by radiative equilibrium models, implying nonradiative heating in the photosphere itself. This result is sensitive to uncertainties in the PRD theory, but such uncertainties are least for the Ca II lines in dwarfs, and solar active regions show temperature enhancements of 400–1500 K (Chapman 1977, 1979, Morrison & Linsky 1978) compared with quiet Sun models below the temperature minimum. If active chromosphere stars, that is stars having chromospheric line surface fluxes comparable to or brighter than solar plages, are at all analogous to solar plages, then such stars should also have photospheres with temperatures significantly hotter than radiative equilibrium models predict. In fact, enhanced photospheric temperatures may occur even for nonactive chromosphere stars (Kelch 1978, Kelch et al. 1978, 1979, Desikachary & Gray 1978). However, Schmitz & Ulmschneider (1980c) find that acoustic waves of large amplitude may explain much of the photospheric temperature enhancements determined using the Ca II and Mg II line wings as diagnostics.

2. Kelch et al. (1979) have presented evidence that the temperature minimum in dwarfs moves outward to smaller mass column densities with decreasing nonradiative heating. Furthermore, there is some evidence that the T_{min}/T_{eff} ratio decreases with age among main-sequence stars (Linsky et al. 1979a) as might be expected if the nonradiative heating rate decreases with age.

3. Active chromosphere dwarf stars have larger chromospheric radiative loss rates and steeper chromospheric temperature rises than quiet chromosphere stars. The latter result is depicted in Figure 2, where the chromospheric temperature gradients for plages (Shine & Linsky 1974, Kelch & Linsky 1978) and flares (Machado & Linsky 1975) are compared to those of 13 dwarf stars (Kelch et al. 1979). The correlation of increasing temperature gradients with increasing nonradiative heating rates is as expected. Kelch et

al. (1979) and Cram & Mullan (1979) have shown that Hα emission, which distinguishes dMe stars from normal M dwarfs, can be simply explained by the steep chromospheric temperature gradient computed to match the bright K-line emission characteristic of dMe stars.

4. For quiet chromosphere dwarfs and giants, Kelch et al. (1979) found a correlation between the mass column density at the 8000 K level of the chromosphere (m_0) and stellar gravity. However, Baliunas et al. (1979) argue that the Kelch et al. relation is not valid, at least for RS CVn stars, and that the Kelch et al. assumption that the rapid increase in temperature occurs near 8000 K may lead to inaccurate values of the mass column density at the top of a chromosphere. In addition, the relation does not hold for solar plages and flares, where m_0 can be several orders of magnitude larger than typical quiet Sun values. Furthermore, based on an analysis of the Mg II, Si II, Si III, and C II lines, Simon et al. (1980a) find that the active chromosphere star ε Eri (K2 V) has a value of m_0 a factor of 6 larger than predicted by the quiet chromosphere relation.

5. A chromospheric temperature plateau near 20,000 K in the Sun (Vernazza et al. 1973), or near 16,500 K as suggested by Lites et al. (1978), may be a real feature of other stars. For example, Simon et al. (1980a) propose that ε Eri (K2 V) has such a plateau. Furthermore, Basri et al. (1979) suggest that the plateau decreases in geometrical thickness with increasing brightness of Lα in the Sun.

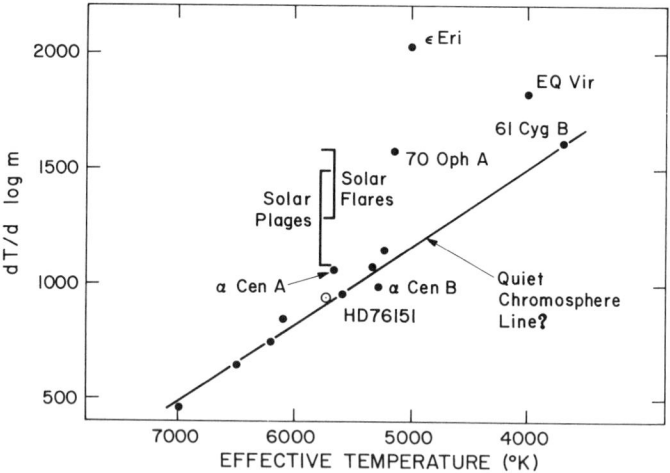

Figure 2 Mean temperature gradients in the chromospheres of main-sequence stars, the quiet Sun, solar plages, and solar flares (taken from Kelch et al. 1978). The temperature gradient, measured in log (mass column density) units is defined between the temperature minimum and 8000 K.

Clearly much work remains to be done to better understand the available chromospheric diagnostics and to extend chromospheric modeling to other groups of stars. In particular, further studies of supergiants, RS CVn stars, dMe stars, F stars, and T Tauri stars are needed, as well as the extension of chromospheric models to transition regions using IUE spectra. The study of transition regions and coronae per se is beyond the scope of this review, but they are relevant to chromospheres since hydrostatic equilibrium, if valid, requires vertical pressure continuity between the top of the chromosphere and the base of the transition region. In this regard I would like to stress the following points: 1. The most reliable means of deriving pressures at the top of a chromosphere is to analyze diagnostic lines formed near 20,000 K such as the Lα core (useful primarily for the Sun), C II $\lambda\lambda$1334, 1335 and Si III $\lambda\lambda$1206, 1892. The Ca II and Mg II resonance lines are formed deeper in the chromosphere where the pressures are typically much larger than at the top, consequently the emission cores of these lines are not unique diagnostics. 2. Doschek et al. (1978) have proposed that the Si III λ1892/C III λ1909 line ratio, which is easily observed by IUE, is a useful diagnostic for densities at the base of the transition region. However, Simon et al. (1980) find that if a chromospheric plateau is present near 20,000 K, then Si III λ1892 is formed primarily in the plateau. Under these circumstances the Si III line is sensitive to the plateau thickness and the Si III/C III line ratio is a poor density diagnostic. Also, charge-exchange can affect the temperature at which the Si III lines are formed (Baliunas & Butler 1980). Other prominent density diagnostics accessible to IUE include Si IV λ1403/C III λ1909 (Cook & Nicolas 1979) and C III λ1176/λ1909 (Raymond & Dupree 1978). However, one must keep in mind that such ratios can be sensitive to systematic flows (Raymond & Dupree 1978, Dupree et al. 1979b, Joselyn et al. 1979). Alternatively one can derive transition region pressures from emission measures, using an auxiliary relation such as constant conductive flux (e.g. Evans et al. 1975, Haisch & Linsky 1976, Brown et al. 1979). However, the validity of this approach is questionable. For example, Ayres & Linsky (1980b) have argued that the Capella B transition region is not conductively heated because this would require pressures of a factor of 30–50 times larger than those estimated from transition-region density diagnostics, coronal emission measures, and the Mg II lines.

V Are Theoretical Models of Chromospheres Becoming Realistic?

Despite the rapidly growing number of spectroscopic observations of stellar chromospheres and the semiempirical models computed to match these data, we cannot claim that we understand chromospheres without first identifying the important heating mechanism(s), and second computing ab initio theoretical chromosphere models based on these heating mechanisms which accurately match the available data and semiempirical models. This particular goal of understanding stellar chromospheres has not yet been achieved, but considerable progress has been made recently.

Stein & Leibacher (1974, 1980) have described the different types of hydrodynamic and magnetohydrodynamic waves that are thought to exist in the solar atmosphere and are candidates for heating stellar chromospheres. Ulmschneider (1979) argues that magnetic modes are likely to dominate the transition region and coronae, but that short-period acoustic waves are the best candidate for heating the chromospheres of the Sun and stars of spectral type A and later. He argues for the latter on the basis of 1. Deubner's (1976) measured short-period acoustic flux in the solar photosphere of 10^8–10^9 ergs cm^{-2} s^{-1}, which is ample enough to balance the total energy loss of about 6×10^6 ergs cm^{-2} s^{-1} in the solar outer atmosphere even with considerable wave damping in the upper photosphere; and 2. the rather good agreement between empirical and theoretical chromospheric heating rates, based on the short-period acoustic wave mechanism, for several late-type stars and the Sun.

I will consider the validity of these arguments in the context of a comparison between the predictions of the acoustic wave models and empirical data, but first it is important to state the approximations made in recent theoretical calculations. Prior to 1977, theoretical models of chromospheres (e.g. Kuperus 1965, 1969, Ulmschneider 1967, 1971, de Loore 1970) assumed weak shock theory and time-independent solutions. In more recent calculations (e.g. Ulmschneider & Kalkofen 1977, Ulmschneider et al. 1978, 1979, Schmitz & Ulmschneider 1980a, b) shocks are treated explicitly with time-dependent hydrodynamic codes. These numerical approaches typically assume mixing-length convection, the Lighthill-Proudman theory for acoustic wave generation, a single period for the acoustic waves, and grey opacity stellar atmospheres. I now consider four comparisons between these calculations and empirical measurements.

1. A fundamental test of the acoustic wave heating models is that they correctly predict the T_{eff} and gravity dependences of chromospheric radiative loss rates. This test is difficult because several important emission features (i.e. resonance lines of Ca II, Mg II and H I) must be measured to estimate the radiative loss rate in lines, and it is presently impractical to directly measure the chromospheric radiative loss rate in the H$^-$ and other important continua. As a first attempt at testing the theoretical models, Linsky & Ayres (1978) estimated that Mg II h and k account for 30% of the chromospheric line radiative losses in the Sun and other late-type stars for which the Ca II, Mg II, and H I resonance lines are effectively thick.

They then compared normalized radiative loss rates in the Mg II lines, $\mathscr{F}(\text{Mg II})/\sigma T_{\text{eff}}^4$, for 32 stars including the Sun and found a systematic trend of decreasing $\mathscr{F}(\text{Mg II})/\sigma T_{\text{eff}}^4$ with decreasing effective temperature, but essentially no dependence on stellar surface gravity. The computations by Ulmschneider et al. (1977) of the acoustic flux available to heat chromospheres exhibit the observed T_{eff} dependence, but predict an increase in chromospheric heating of 1–2 orders of magnitude between log $g = 4$ and log $g = 2$ that is not seen in the data. Subsequently, Basri & Linsky (1979) determined more accurate values of $\mathscr{F}(\text{Mg II k})/\sigma T_{\text{eff}}^4$ from their IUE data and the *Copernicus* spectra of Weiler & Oegerle (1979). These data are illustrated in Figure 3. The normalized Mg II fluxes are widely scattered, but they do show little if any dependence of $\mathscr{F}(\text{Mg II k})/\sigma T_{\text{eff}}^4$ on stellar luminosity in agreement with the previous data, and perhaps a slow decrease with decreasing T_{eff}.

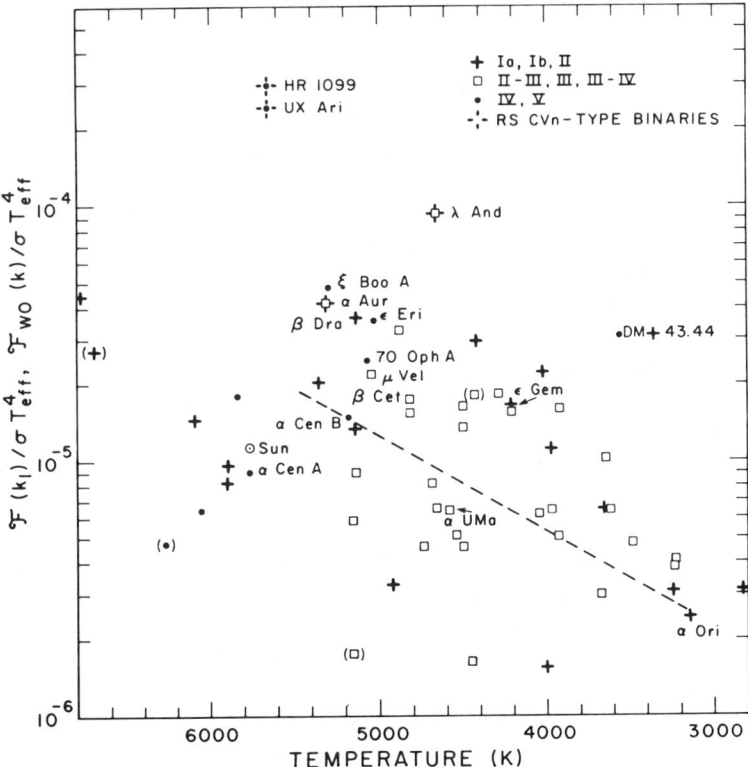

Figure 3 Ratios of the chromospheric radiative loss rates in the Mg II k line to the total surface luminosity (σT_{eff}^4) for stars observed by IUE (Basri & Linsky 1979) and by *Copernicus* (Weiler & Oegerle 1979).

An important question in this regard is the amount of chromospheric cooling in the H^- continuum and the extent to which the H^- cooling term depends on gravity. Schmitz & Ulmschneider (1980a,b) revised the previous work of Ulmschneider et al. (1977) in a way that has reduced the gravity dependence of the acoustic flux that survives radiative damping in the photosphere, and is therefore available to heat the low chromosphere. They find that the ratio of the computed H^- radiative loss rates to the empirical Mg II radiative loss rates may increase with decreasing gravity in a manner consistent with their acoustic flux calculations. However, unlike radiative losses in lines, H^- cooling cannot be measured directly. Schmitz & Ulmschneider instead estimated H^- cooling indirectly using chromospheric models constructed to fit the Ca II and Mg II lines. Not only are the derived H^- radiative loss rates extremely model-dependent, but the use of models constructed to match the Ca II and Mg II lines to calculate the H^- losses is itself questionable. The reason is that the Ca II and Mg II lines average over the intrinsically inhomogeneous stellar atmosphere in a much different way than does the H^- continuum. Since emission in the near ultraviolet resonance lines is strongly weighted toward the hotter components of an atmosphere, while the mainly visible and near infrared H^- emission is more evenly weighted over the thermal irregularities, the use of models constructed to fit the Ca II and Mg II lines will inevitably overestimate $\mathscr{F}(H^-)$. Finally there remains fundamental disagreement on the proper way to compute H^- radiative losses from a stellar chromosphere (see Section IX b).

2. The chromospheric k-line radiative loss rates illustrated in Figure 3 exhibit a wide range of values for stars of similar effective temperature and luminosity. Excluding the RS CVn-type systems, that range is typically an order of magnitude. The analogous plot of normalized chromospheric radiative losses in the H and K lines (Linsky et al. 1979a) shows a comparable spread. While acoustic wave heating may explain quiet chromospheres, it cannot explain the large diversity of chromospheric radiative loss rates among single stars of similar effective temperature and gravity (same luminosity). In solar plages, chromospheric radiative losses in the Ca II and Mg II resonance lines are commonly 10 times those of the quiet Sun (Kelch & Linsky 1978, Kelch et al. 1979). Since Ca II K-line intensities are well correlated with magnetic field strength in the Sun (Skumanich et al. 1975), it is generally presumed that magnetic field concentrations play an important role in enhancing the chromospheric nonradiative heating rates. In fact, the correlation of Ca II strength with the magnetic field must be viewed as evidence for a hydromagnetic origin of chromospheric heating. One likely explanation, then, for the factor of 10 range in $\mathscr{F}(k)/\sigma T_{\text{eff}}^4$ at each effective temperature is that stars with low

$\mathscr{F}(k)/\sigma T_{\text{eff}}^4$ ratios have few solar-type plages, while stars with large ratios are mainly covered with solar-type plages.

3. A third test is a comparison of the theoretical acoustic wave heating in specific stars with empirically determined radiative loss rates. Ulmschneider et al. (1977) and Schmitz & Ulmschneider (1980a) have computed theoretical chromosphere models for the same dwarfs, subgiants, and giants that Ayres et al. (1974), Ayres & Linsky (1975), Kelch (1978), and Kelch et al. (1978, 1979) had previously computed semiempirical models based on the Ca II and Mg II lines. For many of these stars the agreement between computed acoustic fluxes and empirical radiative loss rates (including the uncertain estimated H^- contribution) is within a factor of 6, which is not a large factor when one considers the gross uncertainty in the amount of acoustic flux generated in the convective zone (see Gough 1976) and the potential importance of inhomogeneities on determining the "empirical" H^- cooling. The acoustic wave theory has the most difficulty for the cooler dwarfs such as 70 Oph A (K0 V) and EQ Vir (dK5e). For the latter star, in particular, Schmitz & Ulmschneider (1980b) estimate that a factor of 145 times more energy is needed to balance the empirical chromospheric radiative loss rate than is predicted by the acoustic wave theory. Part of the missing flux may be attributed to inadequacies in the theory, for example, the treatment of line blanketing, molecular opacity, and the effects of atmospheric stratification on wave production (Schmitz & Ulmschneider 1980a). However, I feel that the principal reason is the dominance of *magnetic* heating mechanisms in these active chromosphere stars. The same is almost certainly true for the RS CVn-type systems, including Capella.

4. A final test is the location in mass column density of the temperature minimum, which is determined by the nonradiative heating rate at the base of the stellar chromosphere. Cram & Ulmschneider (1978) called attention to inconsistencies between observed and predicted widths of the Ca II K_1 features, which should be formed in the vicinity of T_{min} (but see Section VI concerning supergiants). In subsequent computations by Schmitz & Ulmschneider (1980a, b) the agreement between computed and empirical mass column densities at T_{min} has improved except for those stars (70 Oph A, EQ Vir, Capella) that show considerable deficiencies in the computed acoustic flux.

In summary, theoretical models based on the short-period acoustic wave theory show promise for explaining the heating of the lower chromospheres of quiet chromosphere stars, but they are clearly inadequate to explain active chromosphere stars and solar plages for which magnetic heating mechanisms are presumably dominant. Even for the quiet chromosphere stars the acoustic theory needs to be carefully examined. For example, H^-

radiative losses should be calculated for the several classes of thermal inhomogeneities known in the solar case, to establish the reliability of estimating H$^-$ cooling rates from single-component models. In addition, the acoustic wave theory should be extended to more realistic atmospheric models, including nongrey opacity sources and a more complete, nonlinear treatment of the propagation and damping of the sound waves. Finally, Bruner (1978), Athay & White (1978, 1979), and Bruner & McWhirter (1979) have shown, on the basis of OSO-8 observations of chromospheric transition line widths and oscillations, that there is insufficient acoustic wave energy available to heat the solar transition region and corona. Thus the acoustic heating theory can not be extended reliably to predict transition-region and coronal properties of other stars.

VI Why Does the Wilson-Bappu Relation Work?

Ever since Wilson & Bappu (1957) discovered a simple correlation between the widths of the Ca II H- and K-line emission cores and stellar absolute luminosity (M_v) extending over 15 magnitudes, many astronomers have expanded the data base and have attempted to explain the origin of this relation. Part of the fascination of this subject must be the inherent simplicity of the correlation and the prospect of obtaining valuable information about stellar chromospheres from simple measurements of line widths. Unfortunately, one can easily be deluded into feeling that one understands something of this topic. In particular, many authors have made back-of-the-envelope calculations and arrived at relations similar in functional form to that originally proposed by Wilson & Bappu (1957), whereas the wide variety of assumptions used are often very different and even contradictory. However, there is no substitute for careful analysis of this problem based on realistic radiative transfer calculations. Here I will briefly summarize past observational and theoretical work, and then discuss in detail two recent papers that may contribute significantly to our understanding of the underlying physics of the Wilson-Bappu effect.

Width-luminosity relations have now been found for several lines in addition to Ca II H and K, for example the analogous Mg II h and k resonance lines, Lα, and Hα. The width that Wilson & Bappu (1957) measured, W_0 (km s^{-1}), is the separation between the outer edges of the Ca II emission features on 10 Å mm^{-1} spectrograms. Lutz (1970) has shown that W_0 is very nearly the full width at half maximum (FWHM) of the emission core. Wilson & Bappu (1957) and Wilson (1959) proposed the following expression for the K-line width-luminosity relation

$$M_v = 27.59 - 14.94 \log W_0(K). \tag{1}$$

The most extensive compilation of Mg II k-line widths is that of Weiler &

Oegerle (1979), based on *Copernicus* observations of 49 late-type stars. Their expression,

$$M_v = 34.93 - 15.15 \log W(k), \tag{2}$$

has essentially the same slope as that for Ca II, even though the Mg II widths were measured at the base of the emission feature. IUE observations will add to this data set and provide widths at different portions of the line profile. Using a small sample of Lα widths (FWHM), McClintock et al. (1975) obtained the relation

$$M_v = (40.2 \pm 4.5) - (14.7 \pm 1.6) \log W(L\alpha). \tag{3}$$

Finally, Kraft et al. (1964), LoPresto (1971), and Fosbury (1973) have studied the dependence of the Hα line half-width, defined in several different ways, on M_v, but they have not proposed definite functional relationships.

Different characteristic features within the K line also show width-luminosity relations. For example, Ayres et al. (1975) and Engvold & Rygh (1978) find that the separation of the K_1 minimum features is correlated with M_v. Based on high resolution spectra of 26 late-type stars, Engvold & Rygh (1978) derive

$$M_v \simeq \text{const} - 15.2 \log W(K_1). \tag{4}$$

In addition, Lutz et al. (1973) find that the entire damping wings of the H and K lines broaden with increasing stellar luminosity. Finally, Cram et al. (1979) have found in a sample of 32 stars observed at high dispersion, that both the K_2 peak separation and the K_1 minimum separation exhibit essentially the same width-luminosity slope as the FWHM.

The agreement among the slopes of the four relations above is certainly suggestive of a common, presumably simple, origin. As a first step toward understanding that origin, several authors have expressed the line widths empirically in terms of fundamental stellar parameters. For example, Lutz & Pagel (1979) have proposed the relation

$$\log W_0 = -0.22 \log g + 1.65 \log T_e + 0.10 [\text{Fe/H}] - 3.69, \tag{5}$$

based on data from 55 stars. Lutz & Pagel argue that their relation is more accurate than those derived previously without an abundance term. For the K_1 width, Cram et al. (1979) find

$$\log W(K_1) = -(0.20 \pm 0.02) \log g + (1.1 \pm 0.2) \log T_e - 3.76, \tag{6}$$

whereas Engvold & Rygh (1978) derive a coefficient of -0.16 for the first term.

Physical interpretations of these width-luminosity relations fall into two distinct classes. The first assumes that the width, generally taken to be W_0

for Ca II K, is formed in the Doppler core of the line profile and therefore responds mainly to turbulent velocities in the chromosphere. For example, Scharmer (1976) followed Goldberg (1957) in assuming that $W_0 \simeq 6\bar{u}$, where $\bar{u}$ is the mean chromospheric turbulent velocity, and arrived at the inevitable result that such velocities are supersonic in giants and highly supersonic in supergiants. Using conservation equations and taking the mechanical energy flux proportional to $\bar{u}^3$, Scharmer (1976) was able to show that $W_0 \simeq 6\bar{u} \sim g^{-1/4}$ in agreement with Equation (5). However, the derived value of $\bar{u}$ for the Sun is 22 km s^{-1}, far larger than any estimates of turbulent velocities in the solar chromosphere. Fosbury (1973) has studied the Ca II and Hα widths together in order to better determine chromospheric turbulent velocities. He estimated chromospheric wave fluxes from the line widths and concluded that the amplitude of upward propagating acoustic waves increases with luminosity. Most recently, Lutz & Pagel (1979) have derived a relation similar to Equation (5) assuming slab geometry with a geometrical thickness equal to the K-line thermalization length, complete frequency redistribution (CRD), and Doppler control of W_0.

These three papers and previous studies of the same type share a number of difficulties. 1. They assume that W_0 is located in the Doppler core, but give no theoretical or empirical justification for this. 2. They either ignore radiative transfer altogether, or treat line transfer in a wholly unrealistic manner. 3. They rarely compute chromospheric densities and ionization self-consistently. 4. They do not attempt to explain high resolution solar observations or turbulent velocities measured in the solar chromosphere by various techniques. The ability to arrive at expressions similar to Equation (5) is often given as sufficient justification that the approaches are accurate enough to explain the underlying physical mechanism.

An alternative physical interpretation for the width-luminosity relations assumes that the width, generally taken to be $W(K_1)$, is formed in the damping wings of the line profile. In this scenario, the width is sensitive primarily to the mass column density above the temperature minimum and is relatively insensitive to chromospheric turbulent velocities. Engvold & Rygh (1978) have argued that $W(K_1) \simeq 7.8 \pm 1.7$ Doppler half-width units and thus the K_1 minimum features must be formed in the damping wings of the line profile. With this assumption and the approximation that the K_1 feature is formed at the temperature minimum, which empirical models suggest occurs at roughly the same continuum optical depth in late-type stars, Ayres et al. (1975) used the pressure-squared dependence of H$^-$ opacity to derive

$$\log W(K_1) = -0.25 \log g + 0.25 \log A_{\text{met}} + \text{const.} \tag{7}$$

The coefficient of the log g term above is close to that of Equation (6).

Thomas (1973) has proposed a somewhat different version of this explanation by showing that the location of the base of the chromosphere, which determines $W(K_1)$, may depend on the upward mass flux, which in turn is related to the chromospheric heating rate.

At this point it should be clear that a synthesis is needed to determine whether these very different explanations for W_0 and $W(K_1)$ are related, and which if either remain valid after a realistic study of spectral line formation in the K line. An important paper by Ayres (1979) goes far in addressing these questions. Ayres proposed simple scaling laws for the thickness and mean electron density of stellar chromospheres as functions of surface gravity and nonradiative heating, based on hydrostatic equilibrium and the assumption that the nonradiative heating is relatively constant with height. He also took into account the influence of ionization on the general structure of a chromosphere and on the plasma cooling functions. He argued that the K_1 features are formed in the damping wings of the line profile on the grounds that PRD calculations now match the limb darkening of the solar Ca II features (Shine et al. 1975, Zirker 1968), which in itself provides strong evidence for near coherent scattering (and thus little Doppler redistribution) beyond the emission peaks. Ayres found that

$$\log W(K_1) = -1/4 \log g + (7/4 \pm 1/2) \log T_{\text{eff}} + 1/4 \log \mathscr{F}$$
$$+ 1/4 \log A_{\text{met}} + \text{const}, \qquad (8)$$

where $\mathscr{F}$ is a scaled nonradiative heating rate that measures how "active" a star is and A_{met} is the metal abundance. The coefficients are close to the empirical relation [Equation (6)] and the derived width ratio $W(k_1)/W(K_1) \cong 2.5$ is consistent with the stellar value of 2.5 ± 0.3 estimated by Ayres et al. (1975) and the solar value of $\cong 2.3$.

Ayres (1979) further assumed that the K_2 emission peaks are formed just outside of the Doppler core and that the line source function is a maximum at one thermalization length below the top of the chromosphere. These approximations lead to

$$\log W(K_2) = -1/4 \log g - (5/4 \pm 1/2) \log T_{\text{eff}} - 1/4 \log \mathscr{F}$$
$$- 1/4 \log A_{\text{met}} + 1/2 \log \xi + \text{const}, \qquad (9)$$

where ξ is the turbulent broadening velocity. The Mg II–Ca II emission peak separation ratio based on the above relation is $W(k_2)/W(K_2) \cong 0.9$.

Several important conclusions can be drawn from this work.

1. Both $W(K_2)$ and $W(K_1)$ scale as $g^{-1/4}$ if the total nonradiative heating is independent of gravity; the former owing to the dependence of chromospheric electron density on gravity and the latter owing to the dependence of chromospheric thickness on gravity. Because both $W(K_2)$

and $W(K_1)$ scale with gravity in the same way, it is reasonable that every width between K_1 and K_2, in particular W_0, should also scale the same way. The theoretical gravity dependence is consistent with the Lutz-Pagel scaling law $W_0 \sim g^{-0.22}$, and explains why the width-luminosity laws for W_0, $W(K_1)$, and $W(K_2)$ have essentially the same slopes (Cram et al. 1979).

2. Ayres (1979) found that $W(K_2) \sim \xi^{1/2}$ not $\sim \xi^1$ as generally assumed. This is a result of placing the K_2 feature just outside the Doppler core, in the Lorentzian damping wings. The assumption of Lorentzian control at K_2 is based on the argument that the solar Ca II emission peaks limb-darken (Zirker 1968), which suggests significant coherent scattering, and thus little Doppler redistribution, where K_2 is formed. Engvold & Rygh (1978) estimate that on the average $W(K_2)/\Delta\lambda_D = 3.4 \pm 0.2$ for their sample of 26 stars, which suggests that K_2 is formed just outside the Doppler core. Consequently, the $W(K_2)$ and W_0 width-luminosity relations can be explained simply as a gravity effect without having to invoke highly supersonic turbulence in stellar chromospheres.

3. Ayres predicts that $W(K_2)$ should decrease with increasing $\tilde{\mathscr{F}}$ (i.e. increasing activity), while $W(K_1)$ should increase with $\tilde{\mathscr{F}}$. A good test of this prediction is found by comparing solar plage profiles of Ca II (e.g. Smith 1960, Shine & Linsky 1972) with those of the mean quiet Sun. In particular, the enhancement of the nonradiative heating rate is large and easily estimated in plages, and the stellar parameters of a plage are the same as those of the quiet Sun. The Ca II plage observations and corresponding data for the k line are consistent with the $\tilde{\mathscr{F}}$ dependence of the $W(K_2)$ and $W(K_1)$ scaling laws. In addition, the K-line FWHM (approximately equal to W_0) appears to be independent of $\tilde{\mathscr{F}}$, which is in accord with Wilson's (1966) result that W_0 appears to be independent of the strength of the emission feature. However, Glebocki & Stawikowski (1978) do claim that there is a dependence of W_0 on visually estimated line strengths, at least for very active stars. Furthermore, Linsky et al. (1979a) find rough agreement between measured K_1 widths of a large sample of stars and the gravity and chromospheric heating rate dependences given in Equation (8).

Finally, I comment on the calculations of Basri (1979), which cast the whole question of the interpretation of width-luminosity relations in a new light. Basri has made a number of prototype PRD calculations of line profiles for chromospheric models of late-type supergiants. His goal was to determine the factors involved in the formation of the emission widths of self-reversed chromospheric resonance lines under conditions of extreme coherency in the line wings. The PRD effects were expected to be particularly severe in supergiants owing to the very small collisional redistribution rates in the low density envelopes of such stars.

As shown in Figure 4, Basri finds that the emergent line profile in an *isothermal atmosphere* can have a self-reversed character for small values of the incoherence fraction Λ and r_0, the ratio of continuum to line center opacity, even though the atmosphere is isothermal and therefore has no temperature inversion. The origin of the phantom K_1 minimum feature is as follows: Outside the Doppler core the photon scattering becomes more coherent with increasing $\Delta x = \Delta\lambda/\Delta\lambda_D$, where $\Delta\lambda_D$ is the Doppler half-width, and the monochromatic source functions rapidly decouple from the line center source function (which itself is close to the local Planck function) such that the emergent intensity decreases with increasing Δx. The intensity then rises in the far wings as the monochromatic source function begins to include a significant contribution from the assumed purely thermal background continuum. The extent to which the phantom K_1 feature occurs depends on the relative importance of the competing terms. Increasing coherence (decreasing Λ) tends to emphasize this Schuster-type process (Mihalas 1970), while increasing collisional redistribution (increasing Λ) or increasing continuum absorption (increasing r_0) deemphasizes it. In addition, Doppler drifting—the frequency diffusion of photons in the observer's frame with each coherent scattering in the atom's frame owing to Doppler motions—tends to increase the probability of core photons wandering into the line wings. The redistribution of core photons increases

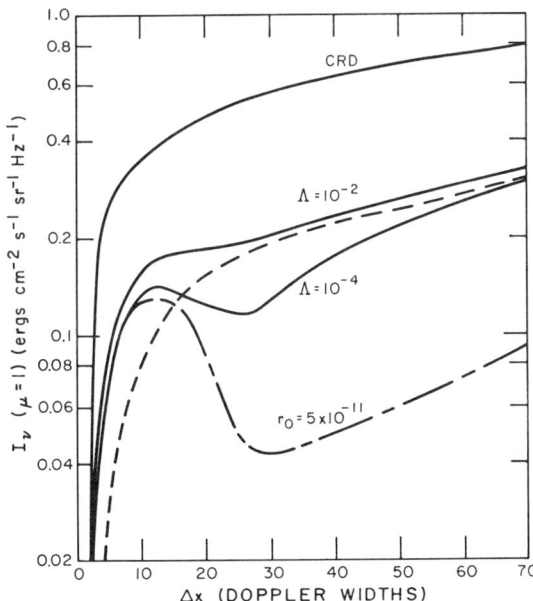

Figure 4 Mg II k-line profiles computed for an isothermal atmosphere and different line parameters (taken from Basri 1979).

the effective noncoherent scattering term, which in turn raises the K_1 minimum flux over the pure coherent case. In fact, Basri found that *microturbulence* can influence the location of the K_1 feature even though K_1 is formed far from the Doppler core, by enhancing the Doppler drifting mechanism. Basri's calculations demonstrate the critical importance of properly treating frequency redistribution for very coherent cases and the potential dangers of naïvely assuming that the K_1 minimum feature is formed at the temperature minimum in extremely low gravity stars.

As a test of what mechanisms affect line profile shapes in realistic supergiant chromospheres, Basri considered the Mg II k line for the not so extreme example of β Dra (G2 II), for which $W(k_1) = 2.5$ Å. Figure 5 depicts atmospheric parameters for solar-type temperature distributions (Models A and A') with T_{min} at log $m_R = -1$ g cm^{-2} and a supergiant-type temperature distribution (Model B) with T_{min} shifted inward in mass column density by a factor of 100. For Model A, with a maximum turbulent

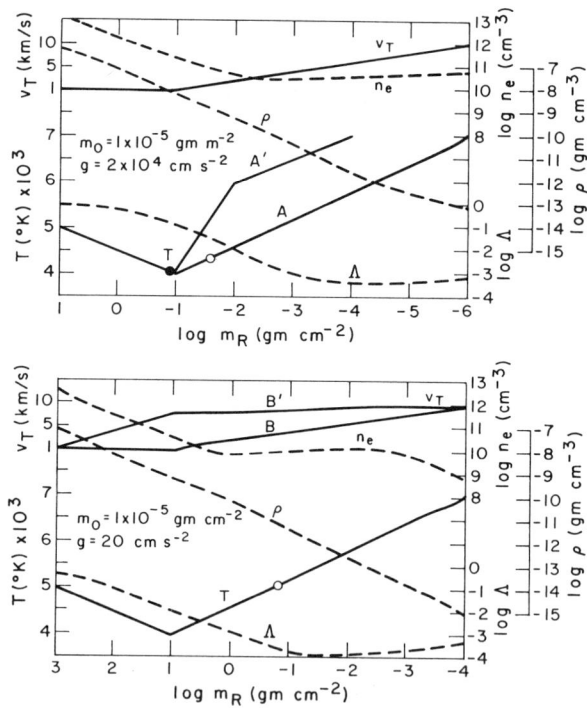

Figure 5 Atmospheric parameters for solar-type temperature distributions (Models A and A') and a supergiant-type temperature distribution (Model B). Two microturbulent velocity distributions (see text Section VI) are shown (taken from Basri 1979).

velocity in the chromosphere of 10 km s^{-1} and either the gravity of the Sun or β Dra, $W(k_1)$ occurs near 0.5 Å. By increasing the chromospheric temperature gradient (Model A') to produce a strong emission profile for β Dra, Basri finds that $W(k_1)$ is about 1.5 Å, even when the microturbulence is increased to the highly supersonic value of 30 km s^{-1}. However, the large values of microturbulence tend to wash out the k_2 emission features. Basri concludes that microturbulence by itself cannot be a viable explanation of the $W(k_1) - M_v$ or $W(K_1) - M_v$ relations in supergiants, although highly supersonic microturbulence can broaden W_0.

Alternatively, increasing the mass column density down to the temperature minimum (as is done in Model B) results in $W(k_1) = 1.5$ Å (see Figure 6), even when $m_{T_{min}}$ is increased without limit. The apparent lack of sensitivity of $W(k_1)$ to chromospheric column densities occurs because the k_1 minimum feature is strongly decoupled from the temperature structure (as was seen in the isothermal example). The location of k_1 is now determined by the balance of coherent and noncoherent processes. In particular, Doppler drifting is an important incoherence term which can be enhanced by increasing the microturbulence. Basri finds that $W(k_1)$ can be broadened to the observed value of 2.5 Å simply by using a barely supersonic microturbulence of 8 km s^{-1} (Model B' in Figures 5 and 6).

The prototype supergiant calculations leave us with the following

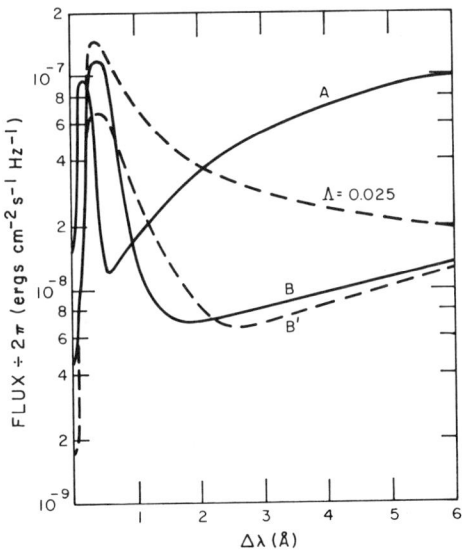

Figure 6 Computed Mg II line profiles for β Dra, using different microturbulent velocity distributions (Models B and B') (taken from Basri 1979).

unanticipated results: 1. k_1 and K_1 need not be formed at the temperature minimum. 2. Neither the turbulent velocity nor the mass column density explanation for the $W(K_1) - M_v$ relation is viable by itself. 3. Microturbulent velocities can play an important role in determining $W(k_1)$ and $W(K_1)$ via the Doppler drifting mechanism. Even though these conclusions refer to the perhaps extreme case of low density supergiant chromospheres, the most important message is that it is essential to solve the transfer equation properly before one can make meaningful statements concerning the physical basis of width-luminosity relations.

VII Are There Systematic Flow Patterns in Stellar Chromospheres?

Until now we have been concerned with the interpretation of chromospheric line intensities and widths. These data provide valuable information on thermal structure and random nonthermal velocities in chromospheres, but they do not contain useful information on systematic flow patterns. Line profile asymmetries may provide such information with proper interpretation, and they are the only means at present of studying chromospheric systematic velocity fields. Because the analysis of line asymmetries is complex, I first consider available theoretical calculations of profiles of optically thick chromospheric lines in the presence of systematic flow patterns, and then consider the extent to which solar and stellar data can be understood in terms of these models.

1. Athay (1976) has emphasized the caution with which one should approach the question of determining velocity fields from line asymmetries. He provides examples of velocity fields for which even absorption line profiles predict the wrong sign of the flow vector. Needless to say, the analysis of optically thick self-reversed emission cores is more complex and must be undertaken with care.

2. Using as a test case the Ca II K line formed in the solar chromosphere, Athay (1970b) and Cram (1972) have shown that asymmetries provide information on velocity *gradients* in the line-forming region but not on the magnitude or even the direction of the flow in any specific atmospheric layer. For example, Athay (1970b) showed that downward motions of 10–20 km s^{-1} in the region of K_3 formation shift K_3 to the red, thereby moving absorbing material (material with a smaller source function than the underlying material producing the K_2 emission peaks) so as to partially obscure the K_{2R} emission peak and uncover the K_{2V} emission peak. The net effect is to produce brighter emission at K_{2V} and weaker emission at K_{2R}, and $\Delta\lambda_{K_3} > 0$. Unfortunately, nearly the same asymmetries are produced by assuming upward motions of 3–7 km s^{-1} in the K_2-forming region and no systematic velocities where K_3 is formed. What I refer to as "blue

asymmetry", $I(K_{2V}) > I(K_{2R})$ and $\Delta\lambda_{K_3} > 0$, thus only provides information on the systematic vertical velocity gradient, specifically, that $dv/dh < 0$, but no unique information on absolute vertical velocities anywhere. Similarly "red asymmetry", that is $I(K_{2V}) < I(K_{2R})$ and $\Delta\lambda_{K_3} < 0$, implies only that $dv/dh > 0$.

3. Vertically propagating waves with scales between the microturbulent and macroturbulent limits ("mesoturbulence"), and which are nonsinusoidal in character, can also produce line asymmetries. Shine (1975) has synthesized Na D profiles for a solar chromosphere model permeated by shock waves (approximated by vertically propagating sawtooth waves). He finds that the basic asymmetry of the sawtooth function produces time-averaged absorption profiles with line center shifted to the red, and the blue wing brighter than the red wing. Shine's calculations suggest that vertically propagating shock waves can produce blue asymmetries in chromospheric emission cores that might be mistaken for the symptoms of downflows where K_3 is formed.

4. Heasley (1975) and Cram (1976) have synthesized the Ca II resonance and infrared triplet lines for models of the solar atmosphere including upward propagating acoustic pulses, perturbations in the local temperature and density induced by the pulses, and resultant changes in the line source functions, all self-consistently. They find that the passage of a pulse through the chromosphere produces first a blue asymmetry, owing to upward motion where K_2 is formed and no motion where K_3 is formed, and then a red asymmetry, owing to upward motion where K_3 is formed and no motion where K_2 is formed. The effect of including the density and temperature perturbations associated with the pulse is to enhance hydrogen ionization. The increased electron density in turn drives the collision-dominated line source function closer to the Planck function, thereby enhancing the K_2 emission. In fact, they note that the dominant influence on the Ca II line source functions is not the velocity field of the pulses, but rather the atmospheric perturbations. Furthermore, the pulses do not change the wavelengths of the K_2 peaks, only their strengths.

The mean quiet Sun K-line profile (see, for example, White & Suemoto 1968) and the integrated solar disk K-line profile (Beckers et al. 1976) both show blue asymmetry, as do the Mg II resonance lines (Lemaire & Skumanich 1973) and Lα (Basri et al. 1979). To interpret this asymmetry, we can take advantage of the high spatial resolution spectra that can be obtained from the Sun. Such spectra in the K line (see, for example, Pasachoff 1970, Wilson & Evans 1972) show a variety of single and double peak profiles with red and blue asymmetry, but relatively constant K_2 peak wavelengths, consistent with Heasley's (1975) calculations.

A clue to the cause of the blue asymmetry seen in the whole Sun K-line profile is provided by the high spatial resolution time-sequence spectra of Liu (1974). These data show intensity perturbations that propagate from the far wings toward line center and produce a strong blue asymmetry when the disturbance reaches the line core. Liu (1974) has interpreted this behavior in terms of local heating of the chromosphere by upward propagating waves. The important point is that the solar blue asymmetry is produced for these waves by something analogous to Heasley's acoustic pulses, that have the largest effect on the K-line profile at those phases when the temperature and density perturbations and the upward motions are all positive in the chromospheric layers where K_2 is formed.

However, Durrant et al. (1976) and Cram (1978) conclude that these simple upward propagating shock events are probably rare, and that line asymmetries are typically produced by 200 second period gravity waves propagating energy upward in the chromosphere. For such waves, Cram et al. (1977) argue that the predominant blue asymmetry is due to the temporal coincidence of maximum heating where K_2 is formed and maximum downward velocity where K_3 is formed. Thus the net blue asymmetry tells us nothing about the direction or magnitude of the solar wind or the nature of possible circulation patterns in the solar chromosphere.

With this background we can consider the stellar data. Profiles of the Ca II and Mg II lines in F-K main sequence stars (cf. Linsky et al. 1979a, Basri & Linsky 1979) typically have blue asymmetry, like the Sun, indicative of upward propagating gravity or acoustic waves in their chromospheres if our solar explanation is correct.

The G and K giants exhibit more complex behavior. Arcturus (K2 III), for example, shows Mg II line profiles with pronounced red asymmetry that did not change during 1973–1976 (McClintock et al. 1978), whereas Chiu et al. (1977) found that the K-line asymmetry is variable in their data and in previous work going back to 1961, with blue asymmetry perhaps more common. Chiu et al. (1977) have modeled the red asymmetry Ca II and Mg II line profiles with a mass flux conservative stellar wind; that is, the outflow velocity is inversely proportional to the density and therefore increases rapidly with height. Such a velocity field produces red asymmetries in both the Ca II and Mg II lines, as expected from Athay's (1970b) analysis for $dv/dh > 0$. The question remains, however, why Arcturus can have simultaneously a blue asymmetry Ca II profile and a strongly red asymmetry Mg II profile. McClintock et al. (1978) argue that the deep absorption feature at -38.3 km s^{-1} that produces or contributes to the red asymmetry of the Mg II profiles is very likely not interstellar absorption and must therefore be intrinsic to the star. Possible explanations for the opposite asymmetries include large systematic velocity gradients in the chromosphere (the Mg II lines should be formed slightly higher in the chromosphere than Ca II) or an expanding circumstellar envelope with large Mg II column density and small Ca II column density, presumably owing to ionization of Ca$^+$. The latter possibility may be correct, because the slightly cooler giant Aldebaran (K5 III) has Mg II line profiles very similar to Arcturus (van der Hucht et al. 1979), but Ca II lines which contain variable, weak circumstellar absorption features blue shifted by about 30 km s^{-1} (Reimers 1977, Kelch et al. 1978). Reimers (1977) has designated such circumstellar components "K$_4$".

Stencel (1978) found a statistical trend of K-line blue asymmetry for giants hotter than spectral type K3 and red asymmetry for giants cooler than K4. The location of Stencel's Ca II asymmetry dividing line in the H-R diagram is depicted in Figure 7. If blue asymmetry is symptomatic of upward propagating waves but no large wind in the chromosphere, and red asymmetry indicates the presence of a significant outward mass flux and possibly also a circumstellar shell, then the dividing line indicates the onset of massive winds in stellar chromospheres. Most recently, Stencel & Mullan (1980) (cf. *Copernicus* data of Weiler & Oegerle 1979) have determined a locus in the H-R diagram (see Figure 7) where the Mg II resonance lines change their asymmetry in a manner similar to that found by Stencel (1978) for H and K. The Ca II and Mg II dividing lines are located slightly to the right of that proposed by Linsky & Haisch (1979) to separate stars exhibiting high excitation emission lines characteristic of a solar-like transition region (material at 20–250×10^3 K) from stars showing only

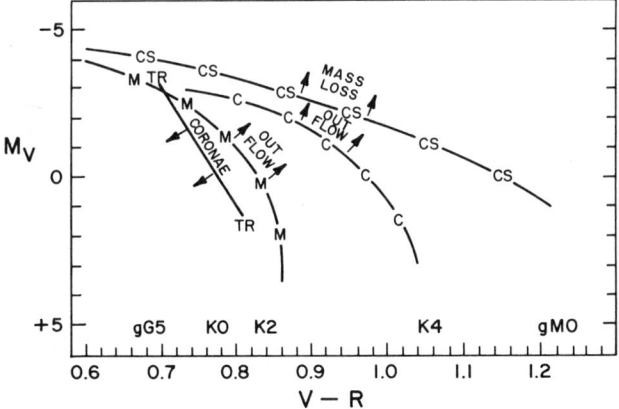

Figure 7 Location in the H-R diagram of dividing lines between stars with or without transition regions (Linsky & Haisch 1979), stars with Ca II red or blue asymmetry (Stencel 1978), stars with Mg II red or blue asymmetry (Stencel & Mullan 1980), and stars showing Ca II circumstellar features (Reimers 1977).

chromospheric emission lines (material at less than 10^4 K). Taken together these data suggest that cool stars fall into two main classes: those with outer atmospheres consisting of chromospheres, transition regions, and presumably also hot coronae; and those with chromospheres and massive cool winds.

Mullan (1978) has attempted to explain the apparent onset of massive winds in the early K stars. He proposes that the location of the point where the stellar wind becomes supersonic moves deeper into the stellar atmosphere as gravity and effective temperature decrease. Haisch et al. (1980) have shown that Lα radiation pressure may play an important role in initiating the cool stellar winds. The cooling effect of the wind may explain the following:

(a) For giants with color index $(V-R) > 0.80$ (about spectral type K0 III), the stellar wind is optically thin, but the energy associated with the mass flow is about equal to the nonradiative energy that would otherwise heat a hot corona. Such stars therefore do not have coronae and fall to the right of the Linsky-Haisch dividing line. However, the sonic point of the wind lies above the region where the Mg II lines are formed, consequently the Mg II lines are symmetric or show blue asymmetry. These stars therefore fall to the left of the Mg II asymmetry locus.

(b) For giants with $(V-R) > 0.85$ (about spectral type K2 III), the wind now affects the upper chromosphere where the Mg II lines are formed. The wind reverses the Mg II asymmetry and the mass loss rate rises because the sonic point has penetrated into the chromosphere where the density is high.

(c) For giants with $(V-R) > 1.00$ (about spectral type K4 III), the sonic point occurs well into the middle chromosphere where H and K are formed. The Ca II lines acquire red asymmetry and the wind now carries enough material to produce observable circumstellar features. The presence of a wind deep in the chromospheres of M III giants is confirmed by Boesgaard & Hagen's (1979) observations of Hα asymmetries and blue shifts of the H_2 and K_2 features.

In cool supergiants like ε Gem (G8 Ib), ε Peg (K2 Ib), ξ Cyg (K5 Ib), and α Ori (M2 Iab), the Ca II and Mg II line asymmetries (cf. Linsky et al. 1979a, Basri & Linsky 1979) are dominated by apparent circumstellar absorption and it is difficult to determine the asymmetry of the unmutilated chromospheric line. The problem of the intrinsic asymmetries of chromospheric emission cores in supergiants is further complicated by the following points: 1. ε Gem shows apparent circumstellar absorption features at both positive and negative velocities. 2. The Mg II k profile of α Ori and presumably other stars is mutilated by overlying circumstellar Fe II and Mn I absorption lines (cf. Bernat & Lambert 1976, de Jager et al. 1979). 3. Profiles of optically thick Fe II lines in α Ori show blue asymmetry, which Boesgaard (1979) and Boesgaard & Magnan (1975) interpret as indicating an inward accelerating circumstellar shell. However, van der Hucht et al. (1979) present evidence that the flow in α Ori accelerates outwards. 4. Basri (1979) has constructed a chromospheric model for α Ori to match the Ca II and Mg II line profiles. He finds that broad, flat-bottomed reversals in the Ca II lines can be easily predicted by the chromospheric model without including any circumstellar shell whatsoever. Consequently, deep absorption features in supergiant resonance lines do not necessarily imply the existence of an extended circumstellar shell.

The early G supergiants, of which α Aqr (G2 Ib) and β Aqr (GO Ib) are prototypes, are interesting in that they have both bright transition region emission lines and strong winds [their Ca II and Mg II lines exhibit absorption components blue-shifted up to ~ 125 km s^{-1} (Dupree & Hartmann 1980, Hartmann et al. 1980)]. Thus the outer atmospheres of these stars are hybrid in character. In addition, Dupree & Baliunas (1979) report that both the wind velocity and emission line fluxes in α Aqr are variable. One possible explanation for the hybrid character of the outer atmospheres of these stars is that they may consist of closed loops and open field structures similar to solar coronal holes. The closed loops contain the hot, transition region material while the coronal holes produce the supersonic wind. The supergiant wind is much cooler and has far greater mass loss than the high-speed streams from solar coronal holes probably as a result of the three orders of magnitude smaller stellar gravity.

Not all cool supergiants show "circumstellar" features. In particular, β Dra (G2 II) exhibits Ca II profiles with very strong blue asymmetry but nearly symmetric Mg II profiles. Basri (1979) has modeled these data using a comoving PRD code and a vertical velocity that increases with height rapidly where the Ca II lines are formed but slowly where the Mg II lines are formed. Such a velocity field produces computed Ca II and Mg II profiles similar to observed, but it is unphysical and, as noted above, upward propagating waves may produce the same kind of asymmetry at least for the Ca II lines.

Variable asymmetry in chromospheric emission lines may turn out to be common in late-type stars as specific stars are monitored for long periods. For example, Hollars & Beebe (1976) have noted changes in the K line of α Aqr (G2 Ib), and Dravins et al. (1977) have found K-line asymmetry changes in the δ Scuti star ρ Pup. O'Brien & Lambert (1979) have reported variable He I λ10830 emission from α^1 Her, which may result from a shock front created when high velocity gas from the secondary accretes onto the photosphere of the primary. G. O'Brien (1979, private communication) has also reported λ10830 observations of a number of F-M stars. Many show variable absorption or emission, and α Aqr shows evidence of enormous outflow velocities of nearly 200 km s^{-1}.

The interpretation of asymmetries in chromospheric lines (Balmer series,

Ca II, Na I) of T Tauri stars is a matter of dispute. Blue-shifted absorption components in these lines have generally been interpreted as symptoms of a strong stellar wind (see Herbig 1962). Nevertheless, Ulrich (1976) has demonstrated that nonspherically symmetric accretion can produce apparent blueshifted absorption components from an infalling postshock gas. Ulrich & Knapp (1979) find that absorption components in Hα are not reliable indicators of gas flow direction, contrary to the common presumption, but instead the Na D line absorption features are more reliable flow-vector diagnostics. They conclude that accretion occurs in the major fraction of T Tauri stars and that discrete clouds of ejected material, perhaps due to flares, continually pass through the generally infalling gas.

VIII How Are Chromospheres in Close Binary Systems Different from Chromospheres in Single Stars?

Until now we have considered the relatively simple case of chromospheres in single stars. The addition of a nearby star in a close binary system introduces possible additional complexity due to tidal forces, reduced chromosphere pressure scale heights, rotational-orbital synchronism, resonances, and interacting magnetic fields. In particular, the approximation of chromospheric thickness being small compared with stellar radius for dwarfs and giants may not be valid; one can anticipate large loop geometries, mass transfer between component stars, and even a common extended chromosphere envelope surrounding both stars for nearly contact systems.

There are a number of different types of close binary systems involving late-type stars, but I will discuss here only the RS Canum Venaticorum-type and BY Draconis-type systems as chromospheres in these systems are presently under active study. Hall (1976) and Eaton & Hall (1979) have reviewed the properties of the RS CVn and related systems with periods longer than or shorter than the 1–14 day period RS CVn group. Kunkel (1975) and Bopp (1979) have summarized the present status of the BY Dra systems. The RS CVn systems typically consist of a K0 IV star, which is generally the more chromospherically active, and a F-G star of luminosity class V or IV, which is usually the brighter component and often shows weaker chromospheric emission lines than the K0 IV secondary. The long-period RS CVn group, containing binaries with periods of 20–100 days or somewhat longer like Capella (104 days) often consist of luminosity class III stars. The BY Dra systems generally consist of 2 dMe stars, both of which are known to flare, with orbital periods in the range 1–10 days (cf. Busko & Torres 1978).

Both groups show evidence for strong magnetic fields and bright chromospheric emission line spectra with surface brightness larger than solar active regions. For example, both groups exhibit persistent, nearly sinusoidal wavelike distortion in their light curves outside of eclipse with low amplitude ($\approx 0\overset{m}{.}2$). The light minima tend to migrate slowly toward decreasing orbital phase for the RS CVn systems, with cycles of about 10 years, but for BY Dra the wave migration is very fast. These phenomena are generally explained by rotational modulation of large cool spots covering a substantial fraction of one hemisphere of the more active star (e.g. Eaton & Hall 1979). The migration toward decreasing phase is likely due to differential rotation in the same sense as the Sun (equatorial acceleration) and the concentration of active regions in bands near the equator. Since differential rotation apparently is an essential aspect of dynamo processes that are thought to generate large magnetic fields, the identification of migration toward decreasing phase with equatorial acceleration is important to our understanding of these systems. Additional indirect evidence for spots with large magnetic fields is the correlation of brightest Ca II and Hα emission with minimum light (spots on the visible hemisphere) in RS CVn systems (Weiler 1975) and flares with minimum light in BY Dra systems (Gershberg 1975). In addition, RS CVn systems exhibit many phenomena similar to solar activity (Eaton & Hall 1979), including spot cycles, spot evolution, and high speed winds on the hemisphere away from the spots. Also, RS CVn systems are strong X-ray sources (see, for example, Walter et al. 1978, 1980, Swank 1979) with coronal temperatures near 10^7 K, and emit strong nonthermal microwave bursts (Gibson et al. 1975, Mutel & Weisberg 1978), which are likely due to gyrosynchrotron emission (Spangler 1977, Owen et al. 1976). The circular polarization observed during these radio flares provides additional evidence for magnetic fields in these systems. BY Dra has also been detected in soft X rays (Ayres et al. 1979).

IUE has now obtained ultraviolet spectra of a number of these systems including Capella (Ayres & Linsky 1980b), UX Ari, HR 1099, and λ And (Linsky & Haisch 1979, Basri & Linsky 1979). In addition, UX Ari was observed near the peak of a major flare by Simon et al. (1980b). The surface fluxes of chromospheric and transition-region lines, obtained using angular diameters estimated by the Barnes & Evans (1976) relation, lie in the range 10^1–10^3 times the quiet Sun and 3–10^2 times a typical active region. Chromospheric models for λ And and Capella proposed by Baliunas et al. (1979) have top mass column densities 1–4×10^{-3} g cm^{-2}, and quiescent chromospheric models for the K0 IV stars in UX Ari and HR 1099 proposed by Simon & Linsky (1980) have top mass column densities of 2–5×10^{-4} g cm^{-2}. These top mass column densities are larger than the 10^{-4} g cm^{-2} value Shine & Linsky (1974) found for bright solar plages, and are consistent with the pressure of more material in the outer atmospheres of close binary systems than for solar plages.

The ultraviolet emission line spectra, chromospheric models, and indirect indicators of strong magnetic fields in close binary systems suggest that much can be learned by pursuing the analogy with solar plage chromospheres. In particular, the bright emission line spectra require enhanced chromospheric mechanical heating rates and significant variations in chromospheric emission line fluxes (see, for example, Baliunas & Dupree 1979) imply significant time variations in these heating rates. To get additional insight we will need detailed studies of specific stars. The best candidate is Capella, which is a 104 day long period RS CVn system consisting of a G6 III primary and F9 III secondary, because this system is bright enough to be studied in high dispersion by IUE. In their analysis of high dispersion spectra of Capella at times of velocity crossing and maximum velocity separation, Ayres & Linsky (1980b) found that most of the emission in chromospheric and transition-region lines is due to the rapidly rotating secondary star (the F9 III component), and propose that the very broad (FWHM ≈ 150 km s^{-1}) transition-region lines may be due to large corotating loops which extend several radii above the limb of the F9 III star.

A second argument for the existence of large flux tubes in RS CVn systems is the application of the expression

$$T_{cor} = 1.4 \times 10^3 \, (PL)^{1/3}, \tag{10}$$

derived by Rosner et al. (1978), where P is the (constant) loop pressure and L is the loop length. If we assume a quiescent UX Ari coronal temperature of $T_{cor} = 10^7$ K (Walter et al. 1980) and pressure $P = 0.18$–0.55 dynes cm^{-2} (Simon & Linsky 1980), then we derive a flux tube length $L = 4$–$11 \, R_\odot$. By comparison, the stellar radius for the K0 IV active star in UX Ari is approximately $3 \, R_\odot$, its Roche lobe distance is $7 \, R_\odot$, and the orbital semimajor axis is $16 \, R_\odot$ (Young & Koniges 1978). This calculation is very rough, but it does suggest large loop dimensions. During a flare P may be increased to 1.1 dynes cm^{-2}, but T_{cor} should also be larger. If T_{cor} is raised to 2–3×10^7 K, then $L = 6$–$20 \, R_\odot$ if the scaling law is valid.

A third argument follows from the observed bright wings of the Mg II resonance lines during the UX Ari flare of 1 January 1979 (Simon et al. 1980b) which extend redward to $+475$ km s^{-1}. Since the escape velocities from either component of UX Ari are 400–600 km s^{-1}, the observed red asymmetry of the Mg II lines suggests free-falling material from a distance far from one star, presumably along a flux tube. In view of this indirect evidence for flux tubes with dimensions approaching the separation of the two stars in UX Ari, Simon et al. (1980b) proposed the following very speculative scenario for major flare events in RS CVn-type systems. Large flux tubes from the K0 IV secondary star occasionally come close to smaller flux tubes of the G5 V star, perhaps due to spot migration in the K0

IV star. When this occurs, magnetic reconnection driven by the relative velocities of the two flux tubes may provide the energy necessary to explain the enhanced X-ray and ultraviolet emission that characterizes the flare. They have in mind the field annihilation processes generally assumed to occur in solar flares when convective motions force oppositely directed field lines together and some instability facilitates rapid reconnection. These processes occur for much longer time scales in RS CVn system flares than in solar flares due to the much larger geometrical scales that we expect in the RS CVn systems. As a result of magnetic reconnection, the two stars may be temporarily connected by a flux tube, facilitating mass exchange. Material falling down the flux tube to one or both stars could produce the large red asymmetries observed in the UX Ari Mg II lines. For the orbital circumstances of this particular flare, the flow is from the K0 IV star to the G5 V star. The up to ± 250 km s^{-1} velocity components seen by Weiler (1978) and by Weiler et al. (1978) in HR 1099 and UX Ari and by Baliunas & Dupree (1979) in λ And during quiescent times may be due to normal streaming motions along smaller flux tubes that do not interconnect the stars. This scenario predicts considerable hot material streaming along lines of force between the stars during flares, which may account for the huge radio emission fluxes observed. It also suggests that in long-period, and thus widely separated, systems, flux tubes are unlikely to extend from one star to another and large flare events should not occur or should occur rarely. Indeed, long-period systems such as Capella (104 day period) are not known to flare and are not strong radio emitters.

A final point concerns the importance of rotation. The 1–14 day period RS CVn systems and BY Dra stars are generally rapid rotators (Hall 1976, Kunkel 1975), presumably due to tidally induced synchronism of rotation with orbital revolution, which occurs on a rapid time scale for stars with periods less than 20 days (Zahn 1977). However, tidal synchronism is unimportant for the 104 day period Capella system, where the more rapidly rotating star is the more active star exhibiting a bright transition region emission line spectrum and bright X-ray spectrum (Cash et al. 1978, Holt et al. 1979) similar to shorter-period RS CVn systems. The difference between Capella A and B suggests that rapid rotation is the key to much of the chromospheric phenomena associated with the close binary systems and the close proximity of another star is important mainly because it can induce rapid rotation by tidal synchronism. If this is true, then the existence of a close binary is sufficient but not necessary to produce active chromosphere phenomena. Indeed, Bopp (1979) has argued that not all BY Dra stars are binaries! He points out that EQ Vir, which has the spectroscopic properties of BY Dra stars and is considered a member of this group, exhibits no radial velocity changes but does show rotational modulation of its spots. Consequently, the star is not viewed pole on, and if it is a binary with sufficiently short period to produce tidal synchronism, then the orbital plane should be close to the stellar equator, resulting in large orbital Doppler shifts. EQ Vir therefore is either a single star or a member of a very wide system; in either case EQ Vir is a BY Dra star owing to rapid rotation rather than duplicity. The visual binary system ξ Boo A may be another example of an effectively single RS CVn "star", like Capella B, whose similarity to the close binaries is a consequence of rapid rotation and not binarity. Clearly much theoretical and observational work remains to be done to understand the roles that binariness and rapid rotation play in determining the properties of stellar chromospheres.

IX Future Prospects

I would like to conclude this review by describing in some detail three general topics, which provide prospects for significant advances in our understanding of stellar chromospheres during the next decade and which should be pursued vigorously.

(*a*) CHROMOSPHERIC VARIABILITY Wilson (1978) has summarized the results of an 11 year program to monitor the H and K flux in 1 Å bandpasses with high precision in 91 main-sequence stars of spectral type F5-M2. Of this sample about a dozen stars appear to have completed a cycle of H and K flux variation, and a number of others show evidence of uncompleted cycles of duration greater than 11 years. In addition, most of the stars in his sample show real variability on time scales from 1 day to several months. Wilson argues that the observed cycles are likely to be analogous to the solar 11 years activity cycle (22 year full magnetic cycle) and that the incidence of cycles increases toward cooler stars, but the later result may be due in part to observational selection. Curiously, although the solar cycle has been extensively studied using sunspots, magnetic fields, plage area, and coronal indicators, the only efforts to systematically monitor the Ca II H and K lines for evidence of a cycle are Wilson's (1978) lunar observations and a program begun in 1974 by White & Livingston (1978) to monitor the Sun at disk center and in integrated light.

There are a number of important questions that can be addressed by well-planned monitoring programs.

1. The basic physics of dynamo processes, which presumably drive stellar activity cycles, can be probed by determining how the period and amplitude of the cycle depend upon fundamental stellar parameters such as mass, rotational velocity, chemical composition, luminosity, and depth of the convective zone. Because these parameters depend implicitly on stellar age, the study of statistical samples of stars of similar age, such as cluster members, may be a useful way to study the effects of important independent

parameters, for example the convective zone depth, that can not be easily measured. Furthermore, because magnetic fields likely control stellar activity, magnetic fields should be directly measured as now appears to be feasible (Robinson et al. 1980). The morphology of stellar cycles, that is the detailed shape of the variation in the measured cycle indicator, may provide important clues as to the basic physics of the dynamo processes operative.

2. An important question for terrestrial climatology is the probability of occurrence and duration of Maunder minima events (Eddy 1977), when solar cycles have reduced or zero amplitudes. Wilson points out that this question can be addressed statistically by determining what fraction of solar-type stars show no evidence for cycles. Phillips & Hartmann (1978) and Hartmann & Rosner (1979) have shown that the total luminosity of BY Dra stars varies with time, perhaps due to (transitory) inhibition of convection by magnetic fields. If this mechanism were operative in the Sun, then it would have important consequences for the terrestrial climate.

3. Short term variability could be due to rotational modulation of active regions, time evolution of active regions, or flares. The systematic monitoring of specific stars offers the prospect of determining some properties of stellar plages, such as their area and differences in chromospheric properties from quiet regions. In addition, the measurement of rotation rates is feasible with accuracies far higher than the Fourier deconvolution methods pioneered by Gray (1976) and Smith (1978). These data are particularly interesting because stellar rotational velocity may be the single most important parameter which determines whether stars of similar spectral type are chromospherically active or quiet (see Kraft 1967, Skumanich 1972, Bopp & Espenak 1977, and Ayres & Linsky 1980b), since dynamo processes for generating magnetic fields may depend critically on differential rotation, which in turn depends on rotational velocity and convective zone properties (cf. review by Gilman 1980).

(b) ENERGY BALANCE IN STELLAR CHROMOSPHERES At present we do not have comprehensive measurements of the cooling rates in stellar chromospheres, and there is no reliable theory to describe the mechanical heating process necessary to balance chromospheric energy losses. Athay (1976) has summarized the dominant radiative cooling transitions in the solar chromosphere; he concluded that the H^- continuum is the major cooling term for the temperature minimum region at the base of the chromosphere, resonance and subordinate lines of Ca II, Mg II, Fe I, and the Hα line are most important in the middle chromosphere (5–8000 K), and Lα dominates in the upper chromosphere (temperature above 8000 K).

The main problem at present is the extent to which H^- cools the base of the chromosphere. This point is controversial. One approach (e.g. Athay 1976, Ulmschneider 1970) is to evaluate a differential H^- radiative loss, by subtracting the H$^-$ cooling determined for the boundary temperature T_0 of a solar radiative equilibrium model from the cooling rate at chromospheric temperature $T > T_0$, as taken from an empirical model. Praderie & Thomas (1972, 1976) and Gebbie & Thomas (1970) have argued that departures from LTE in H$^-$ can strongly affect derived cooling rates based on Ulmschneider's (1970) LTE prescription. Ulmschneider & Kalkofen (1978) and Kalkofen & Ulmschneider (1979) conclude that the Praderie & Thomas (1972, 1976) criticisms are not valid in practical situations. Recently, Ayres (1980) has reexamined the question and has argued that the Praderie & Thomas objections are indeed legitimate, and that large overestimates of chromospheric cooling are in fact possible using Ulmschneider's empirical approach. Also, Avrett & Vernazza (1979) argue that H$^-$ cooling in the solar chromosphere is of secondary importance to that of Ca II and Mg II, and that cooling by the Hα line is effectively balanced by radiative heating in the Balmer continuum. The resolution of this fundamental disagreement is essential if we are ever to successfully determine the true magnitude of chromospheric radiative cooling, that in turn tells us the magnitude of chromospheric heating.

In their recent review of mechanical heating processes, Stein & Leibacher (1980) conclude that acoustic waves probably can heat the low chromosphere, but not higher layers or magnetic regions. Unfortunately, there is not yet available a physically self-consistent theory of convection and the acoustic wave energy density depends critically on turbulent velocities deep in the photosphere. Until such theory is available, all acoustic heating calculations must be viewed skeptically and alternative heating mechanisms should be investigated in detail. In particular, mechanical heating in regions of solar magnetic field concentrations and in stars much cooler than the Sun should be pursued more aggressively.

(c) PHYSICS OF CHROMOSPHERIC ACTIVITY "Solar activity" is a broad characterization that encompasses transient events, such as flares and flare-triggered phenomena, and longer term phenomena such as sunspots, chromospheric plages, and coronal active regions. Very likely, all of the solar phenomena characterized as "active", and corresponding phenomena in stars, are intimately related to magnetic fields. Since the spatial correlation of bright chromospheric emission features and chromospheric flow patterns with the small scale magnetic field is now generally accepted, the fundamental question of chromospheric activity is how do strong magnetic fields produce the observed phenomena. This complex question can be broken down into the following parts.

1. Why is the solar chromospheric and photospheric magnetic field not uniform, but instead filamentary on subarcsecond scales (cf. Frazier &

Stenflo 1973, Stenflo 1973)? Zwaan (1977) has argued that filamentary magnetic fields should exist in all main-sequence stars cooler than late A-type, and such stars as a consequence should be chromospherically active. Parker (1978, 1979) has proposed a hydraulic concentration mechanism driven by the suppression of convective heat transport by magnetic fields. If this mechanism is viable, then further work should be done to assess its importance in stars of different effective temperature, gravity, and total magnetic field.

2. Why are photospheric faculae (filigree) (cf. Dunn & Zirker 1973) bright near the solar limb? Is this simply a consequence of the Wilson depression associated with reduced opacity in magnetic regions (Spruit 1976, 1977, 1979, Stenholm & Stenflo 1977), or are there huge mechanical heating rates deep in the photosphere?

3. How does the existence of strong magnetic fields in the chromospheric network and plage regions lead to enhanced heating? One possibility is that the diverging field geometry permits the channeling of enthalpy flux or heat conduction from a large volume of the corona to a small volume of the transition region (see, for example, Gabriel 1976). Alternatively, strong magnetic fields permit a variety of mechanical wave modes that may be able to dissipate significantly more energy in the chromosphere than acoustic waves. Clearly detailed numerical calculations are needed. An important point to consider is that acoustic waves dissipate roughly 90% of their energy in the photosphere (Stein & Leibacher 1980). If there are magnetic wave modes that are not appreciably damped in the photosphere, then appreciably more energy would be available to heat the chromosphere. Another set of alternatives involves the direct conversion of magnetic to thermal energy either by current dissipation or field reconnection (cf. reviews by Withbroe & Noyes 1977, and Wentzel 1980). Although these mechanisms are generally discussed in the context of coronae, they may be operative to some extent in chromospheres as well.

4. What is the relation between chromospheric motions and mechanical heating? A particularly interesting point is that the kinetic flux of spicular motions is 10 times that needed to balance the energy losses of the chromosphere (Athay 1976).

ACKNOWLEDGMENT

I wish to thank Drs. T. R. Ayres, E. H. Avrett, L. Hartmann, D. Mihalas, and C. Zwaan for their comments on the text, Dr. G. S. Basri for permission to cite unpublished results from his thesis, and many colleagues for sending me preprints and unpublished data for inclusion in this paper. This work was supported in part by NASA through grants NAS5-23274 and NGL-06-003-057 to the University of Colorado.

Literature Cited

Anderson, C. M. 1974. *Ap. J.* 190:585
Athay, R. G. 1970a. *Ap. J.* 161:713
Athay, R. G. 1970b. *Solar Phys.* 11:347
Athay, R. G. 1976. *The Solar Chromosphere and Corona: Quiet Sun.* Dordrecht: Reidel
Athay, R. G., White, O. R. 1978. *Ap. J.* 226:1135
Athay, R. G., White, O. R. 1979. *Ap. J. Suppl.* 39:333
Auvergne, M., Le Contel, J.-M., Boglin, A. 1979. *Astron. Astrophys.* 76:15
Avrett, E. H., Vernazza, J. E. 1979. *Bull. Am. Astron. Soc.* 11:657
Ayres, T. R. 1979. *Ap. J.* 228:509
Ayres, T. R. 1980. *Solar Phys.* In press
Ayres, T. R., Linsky, J. L. 1975. *Ap. J.* 200:660
Ayres, T. R., Linsky, J. L. 1976. *Ap. J.* 201:212
Ayres, T. R., Linsky, J. L. 1980a. *Ap. J.* 235:76
Ayres, T. R., Linsky, J. L. 1980b. *Ap. J.* In press
Ayres, T. R., Linsky, J. L., Garmire, G., Cordova, F. 1979. *Ap. J. Lett.* 232:L117
Ayres, T. R., Linsky, J. L., Rodgers, A. W., Kurucz, R. L. 1976. *Ap. J.* 210:199
Ayres, T. R., Linsky, J. L., Shine, R. A. 1974. *Ap. J.* 192:93
Ayres, T. R., Linsky, J. L., Shine, R. A. 1975. *Ap. J. Lett.* 195:L121
Baliunas, S. L., Avrett, E. H., Hartmann, L., Dupree, A. K. 1979. *Ap. J. Lett.* 233:L129
Baliunas, S. L., Butler, S. E. 1980. *Ap. J.* In press
Baliunas, S. L., Dupree, A. K. 1979. *Ap. J.* 227:870
Barnes, T. G., Evans, D. S. 1976. *MNRAS* 174:489
Basri, G. S. 1979. PhD thesis. Univ. Colorado, Boulder, Colorado
Basri, G. S., Linsky, J. L. 1979. *Ap. J.* 234:1023
Basri, G. S., Linsky, J. L., Bartoe, J.-D. F., Brueckner, G., Van Hoosier, M. E. 1979. *Ap. J.* 230:924
Beckers, J. M., Bridges, C. A., Gilliam, L. B. 1976. *A High Resolution Spectral Atlas of the Solar Irradiance from 380 to 700 Nanometers*, Vol. 1. *AFGL-TR-76-0126(I)*
Bernat, A. P., Lambert, D. L. 1976. *Ap. J.* 204:803
Bidelman, W. P. 1954. *Ap. J. Suppl.* 1:214
Blanco, C., Catalano, S., Marilli, E. 1976. *Astron. Astrophys.* 48:19
Blanco, C., Catalano, S., Marilli, E. 1978. In *High Resolution Spectrometry*, ed. M. Hack, p. 501. Trieste: Obs. Astron.
Blanco, C., Catalano, S., Marilli, E. 1979. *Nature* 280:661
Blanco, C., Catalano, S., Marilli, E., Rodono, M. 1974. *Astron. Astrophys.* 33:257
Boesgaard, A. M. 1979. *Ap. J.* 232:485
Boesgaard, A. M., Hagen, W. 1979. *Ap. J.* 231:128
Boesgaard, A. M., Magnan, C. 1975. *Ap. J.* 198:369
Böhm-Vitense, E., Dettmann, T. 1980. *Ap. J.* 236:560
Bopp, B. W. 1979. Paper presented at the IAU Joint Meeting on Close Binaries and Stellar Activity, Montreal, 18 August 1979
Bopp, B. W., Espenak, F. 1977. *Ap. J.* 82:916
Brown, A., Jordan, C., Wilson, R. 1979. *Proc. Symp. "The First Year of IUE."* ESA and SRC
Bruner, E. C. Jr. 1978. *Ap. J.* 226:1140
Bruner, E. C. Jr., McWhirter, R. W. P. 1979. *Ap. J.* 231:557
Busko, I. C., Torres, C. A. O. 1978. *Astron. Astrophys.* 64:153
Carpenter, K. G., Wing, R. F. 1979. *Bull. Am. Astron. Soc.* 11:419
Cash, W., Bowyer, S., Charles, P. A., Lampton, M., Garmire, G., Riegler, G. 1978. *Ap. J. Lett.* 223:L21
Cash, W. C., Snow, T. P., Charles, P. 1979. *Ap. J. Lett.* 232:L11
Cassinelli, J. P. 1979. *Ann. Rev. Astron. Astrophys.* 17:275
Cassinelli, J. P., Olson, G. L. 1979. *Ap. J.* 229:304
Cayrel, R. 1963. *Comp. Rend. Acad. Sci. Paris* 257:3309
Cayrel, R. 1964. *SAO Spec. Rep.* 167:169
Chapman, G. A. 1977. *Ap. J. Suppl.* 33:35
Chapman, G. A. 1979. *Ap. J.* 232:923
Chiu, H. Y., Adams, P. J., Linsky, J. L., Basri, G. S., Maran, S. P., Hobbs, R. W. 1977. *Ap. J.* 211:453
Conti, P. S. 1978. *Ann. Rev. Astron. Astrophys.* 16:371
Cook, J. W., Nicolas, K. R. 1979. *Ap. J.* 229:1163
Cram, L. E. 1972. *Solar Phys.* 22:375
Cram, L. E. 1976. *Astron. Astrophys.* 50:263
Cram, L. E. 1978. *Astron. Astrophys.* 70:345
Cram, L. E., Brown, D. R., Beckers, J. M. 1977. *Astron. Astrophys.* 57:211
Cram, L. E., Krikorian, R., Jefferies, J. T. 1979. *Astron. Astrophys.* 71:14
Cram, L. E., Mullan, D. J. 1979. *Ap. J.* 234:579
Cram, L. E., Ulmschneider, P. 1978. *Astron. Astrophys.* 62:239
de Jager, C., Kondo, Y., Hockstra, R., van

der Hucht, K. A., Kamperman, T., Lamers, H. J. G. L. M., Modisette, J. L., Morgan, T. 1979. *Ap. J.* 230:534
de Loore, C. 1970. *Astrophys. Space Sci.* 6:60
Desikachary, K., Gray, D. F. 1978. *Ap. J.* 226:907
Deubner, F.-L. 1976. *Astron. Astrophys.* 51:189
Doherty, L. R. 1973. In *Stellar Chromospheres*, ed. S. D. Jordan, E. H. Avrett, p. 99. *NASA SP-317*
Doschek, G. A., Feldman, U., Mariska, J. T., Linsky, J. L. 1978. *Ap. J. Lett.* 226: L35
Dravins, D., Lind, J., Sarg, K. 1977. *Astron. Astrophys.* 54:381
Dunn, R. B., Zirker, J. B. 1973. *Solar Phys.* 33:281
Dupree, A. K. 1978. In *Advances in Atomic and Molecular Physics*, Vol. 14, ed. D. R. Bates, B. Bederson. New York: Academic
Dupree, A. K., Hartmann, L. 1980. In *Stellar Turbulence*, ed. D. F. Gray, J. L. Linsky, p. 279. New York: Springer
Dupree, A. K., Baliunas, S. 1979. *IAU Circ. No. 3435*
Dupree, A. K., Black, J. H., Davis, R., Hartmann, L., Raymond, J. C. 1979a. *Proc. Symp. "The First Year of IUE."* ESA and SRC
Dupree, A. K., Moore, R. T., Shapiro, P. R. 1979b. *Ap. J. Lett.* 229: L101
Durrant, C. J., Grossmann-Doerth, U., Kneer, F. J. 1976. *Astron. Astrophys.* 51:95
Dyke, H. M., Johnson, H. R. 1969. *Ap. J.* 156:389
Eaton, J. A., Hall, D. S. 1979. *Ap. J.* 227:907
Eddy, J. A. 1977. In *The Solar Output and its Variation*, ed. O. R. White, p. 51. Boulder: Colo. Assoc. Univ. Press
Engvold, O., Rygh, B. O. 1978. *Astron. Astrophys.* 70:399
Evans, R. G., Jordan, C., Wilson, R. 1975. *MNRAS* 172:585
Feldman, U., Doschek, G. A., Mariska, J. T. 1979. *Ap. J.* 229:369
Fosbury, R. A. E. 1973. *Astron. Astrophys.* 27:129
Fosbury, R. A. E. 1974. *MNRAS* 169:147
Frazier, E. N., Stenflo, J. O. 1973. *Solar Phys.* 27:330
Freire, R. 1979. *Astron. Astrophys.* In press
Freire, R., Czarny, J., Felenbok, P., Praderie, F. 1978. *Astron. Astrophys.* 68:89
Gabriel, A. H. 1976. *Philos. Trans. R. Soc. London A* 281:339
Gebbie, K. B., Thomas, R. N. 1970. *Ap. J.* 161:229

Gershberg, R. E. 1975. In *Variable Stars and Stellar Evolution*, ed. V. E. Sherwood, L. Plaut, p. 47. Dordrecht: Reidel
Giampapa, M. S. 1980. *Ap. J.* In press
Gibson, D. M., Hjellming, R. M., Owen, F. N. 1975. *Ap. J. Lett.* 200: L99
Gilman, P. A. 1980. In *Stellar Turbulence*, ed. D. F. Gray, J. L. Linsky, p. 19. New York: Springer
Glebocki, R., Stawikowski, A. 1978. *Astron. Astrophys.* 68:69
Goldberg, L. 1957. *Ap. J.* 126:318
Gough, D. O. 1976. *Proc. IAU Colloq. No. 36*, ed. R. M. Bonnet, P. Delache, p. 3. Paris: Clermont-Ferrand
Gray, D. F. 1976. *The Observation and Analysis of Stellar Photospheres*, p. 392. New York: Wiley
Haisch, B. M., Linsky, J. L. 1976. *Ap. J. Lett.* 205: L39
Haisch, B. M., Linsky, J. L. 1980. *Ap. J. Lett.* 236: L33
Haisch, B. M., Linsky, J. L., Basri, G. S. 1980. *Ap. J.* 235:519
Hall, D. S. 1976. In *Multiple Periodic Variable Stars*, ed. W. S. Fitch, p. 287. Dordrecht: Reidel
Harnden, F. R., Branduardi, G., Elvis, M., Gorenstein, P., Grindlay, J., Pye, J. P., Rosner, R., Topka, K., Vaiana, G. S. 1979. *Ap. J. Lett.* 234: L51
Hartmann, L., Davis, R., Dupree, A. K., Raymond, J., Schmidtke, P. C., Wing, R. F. 1979. *Ap. J. Lett.* 233: L69
Hartmann, L., Dupree, A. K., Raymond, J. C. 1980. *Ap. J. Lett.* 236: L143
Hartmann, L., MacGregor, K. B. 1980. Preprint
Hartmann, L., Rosner, R. 1979. *Ap. J.* 230:802
Hearn, A. G. 1975. *Astron. Astrophys.* 40:355
Hearn, A. G. 1979. *Proc. IAU Symp.* In press
Heasley, J. N. 1975. *Solar Phys.* 44:275
Heasley, J. N., Ridgway, S. T., Carbon, D. F., Milkey, R. W., Hall, D. N. B. 1978. *Ap. J.* 219:970
Henoux, J. C., Nakagawa, Y. 1978. *Astron. Astrophys.* 66:385
Herbig, G. H. 1962. *Adv. Astron. Astrophys.* 1:47
Hollars, D. R., Beebe, H. A. 1976. *Pub. Astron. Soc. Pac.* 88:934
Holt, S. S., White, N. E., Becker, R. H., Boldt, E. A., Mushotzky, R. F., Serlemitsos, P. J., Smith, B. W. 1979. *Ap. J. Lett.* 234: L65
Hutchings, J. B. 1979. *IAU Symp. No. 83, Mass Loss and Evolution in O stars*, ed. C. de Loore, P. S. Conti. Dordrecht: Reidel.
Jennings, M. C. 1973. *Ap. J.* 185:197

Jennings, M. C., Dyke, H. M. 1972. *Ap. J.* 177:427
Johnson, H. L., Wisniewski, W. Z. 1978. *Publ. Astron. Soc. Pac.* 90:139
Jordan, C. 1980a. In *Highlights of Astronomy, Vol. 5*. Dordrecht: Reidel. In press
Jordan, C. 1980b. *Astron. Astrophys.* In press
Jordan, S. D., Avrett, E. H. 1973. *Stellar Chromospheres. NASA SP-317*
Joselyn, J. A., Munro, R. H., Holzer, T. E. 1979. *Ap. J. Suppl.* 40:793
Kalkofen, W., Ulmschneider, P. 1979. *Ap. J.* 227:655
Kelch, W. L. 1978. *Ap. J.* 222:931
Kelch, W. L., Linsky, J. L. 1978. *Solar Phys.* 58:37
Kelch, W. L., Linsky, J. L., Basri, G. S., Chiu, H. Y., Chang, S. H., Maran, S. P., Furenlid, I. 1978. *Ap. J.* 220:962
Kelch, W. L., Linsky, J. L., Worden, S. P. 1979. *Ap. J.* 229:700
Kippenhahn, R. 1973. *Stellar Chromospheres*, ed. S. D. Jordan, E. H. Avrett, p. 265. *NASA SP-317*
Kondo, Y., de Jager, C., Hoekstra, R., van der Hucht, K. A., Kamperman, T. M., Lamers, H. J. G. L. M., Modisette, J. L., Morgan, T. H. 1979. *Ap. J.* 230:533
Kondo, Y., Modisette, J. L., Wolf, G. W. 1975. *Ap. J.* 199:110
Kraft, R. P. 1967. *Ap. J.* 150:551
Kraft, R. P., Preston, G. W., Wolf, S. C. 1964. *Ap. J.* 140:237
Kunkel, W. E. 1975. In *Variable Stars and Stellar Evolution*, ed. V. E. Sherwood, L. Plaut, p. 15. Dordrecht: Reidel
Kuperus, M. 1965. *Rech. Astron. Obs. Utrecht* 17:1
Kuperus, M. 1969. *Space Sci. Rev.* 9:713
Lamers, H. J. G. L. M., Rogerson, J. B. 1978. *Astron. Astrophys.* 66:417
Lamers, H. J. G. L. M., de Loore, C. 1976. In *Physique des Mouvements dans les Atmospheres Stellaires*, ed. R. Cayrel, M. Steinberg, p. 453. Paris: Editions de CNRS
Le Contel, J. M., Praderie, F., Bijaoui, A., Dantel, M., Sareyan, J. P. 1970. *Ap. J. Lett.* 203:L71
Lemaire, P., Skumanich, A. 1973. *Astron. Astrophys.* 22:61
Linsky, J. L. 1977. In *The Solar Output and Its Variation*, ed. O. R. White, p. 477. Boulder: Colorado Assoc. Univ. Press
Linsky, J. L. 1979. In Report of Commission 36 in *Trans. IAU*. XVIIA, Part 2:197
Linsky, J. L. 1980. In *Stellar Turbulence*, ed. D. F. Gray, J. L. Linsky, p. 248. New York: Springer
Linsky, J. L., Ayres, T. R. 1978. *Ap. J.* 220:619
Linsky, J. L., Haisch, B. M. 1979. *Ap. J. Lett.* 229:L27
Linsky, J. L., Hunten, D. M., Sowell, R., Glackin, D. L., Kelch, W. L. 1979b. *Ap. J. Suppl.*, 41, 47
Linsky, J. L., Worden, S. P., McClintock, W., Robertson, R. M. 1979a. *Ap. J. Suppl.* 41:47
Linsky, J. L. et al. 1978. *Nature* 275:389
Lites, B. W., Shine, R. A., Chipman, E. G. 1978. *Ap. J.* 222:333
Liu, S. Y. 1974. *Ap. J.* 189:359
Lo Presto, J. C. 1971. *Pub. Astron. Sou. Pac.* 83:674
Lutz. T. E. 1970. *Ap. J.* 75:1007
Lutz, T. E., Furenlid, I., Lutz, J. H. 1973. *Ap. J.* 184:787
Lutz, T. E., Pagel, B. E. J. 1979. *MNRAS*. Submitted
Machado, M. E., Emslie, A. G., Brown, J. C. 1978. *Solar Phys.* 58:363
Machado, M. E., Linsky, J. L. 1975. *Solar Phys.* 42:395
McClintock, W., Henry, R. C., Moos, H. W., Linsky, J. L. 1975. *Ap. J.* 202:733
McClintock, W., Moos, H. W., Henry, R. C., Linsky, J. L., Barker, E. S. 1978. *Ap. J. Suppl.* 37:223
Mihalas, D. 1970. *Stellar Atmospheres*, p. 330. San Francisco: Freeman
Morrison, N. D., Linsky, J. L. 1978. *Ap. J.* 222:723
Morrison, D., Simon, T. 1973. *Ap. J.* 186:193
Mullan, D. J. 1978. *Ap. J.* 226:151
Mutel, R. L., Weisberg, J. M. 1978. *Ap. J.* 83:1499
Nicolas, K. R., Bartoe, J.-D. F., Brueckner, G. E., Van Hoosier, M. E. 1979. *Ap. J.* 233:741
O'Brien, G., Lambert, D. L. 1979. *Ap. J. Lett.* 229:L33
Owen, R. N., Jones, T. W., Gibson, D. M. 1976. *Ap. J.* 210:L27
Pagel, B. E. J., Wilkins, D. R. 1979. *MNRAS*. Submitted
Parker, E. N. 1978. *Ap. J.* 221:368
Parker, E. N. 1979. *Ap. J.* 232:291
Pasachoff, J. M. 1970. *Solar Phys.* 12:202
Phillips, M. J., Hartmann, L. 1978. *Ap. J.* 224:182
Praderie, F. 1973. *Stellar Chromospheres*, ed. S. D. Jordan, E. H. Avrett, p. 79. *NASA SP-317*
Praderie, F. 1977. *Mem. Soc. Astron. Italiana*, 553
Praderie, F., Thomas, R. N. 1972. *Ap. J.* 172:485
Praderie, F., Thomas, R. N. 1976. *Solar Phys.* 50:333
Raymond, J. C., Dupree, A. K. 1978. *Ap. J.* 222:379

Reimers, D. 1977. *Astron. Astrophys.* 57:395
Renzini, A., Cacciari, C., Ulmschneider, P., Schmitz, F. 1977. *Astron. Astrophys.* 61:39
Robinson, R. D., Worden, S. P., Harvey, J. W. 1980. *Ap. J.* In press
Rosner, R., Tucker, W. H., Vaiana, G. S. 1978. *Ap. J.* 220:643
Scharmer, G. B. 1976. *Astron. Astrophys.* 53:341
Schmitz, F., Ulmschneider, P. 1980a. *Astron. Astrophys.* In press
Schmitz, F., Ulmschneider, P. 1980b. *Astron. Astrophys.* In press
Schmitz, F., Ulmschneider, P. 1980c. *Astron. Astrophys.* In press
Shine, R. A. 1975. *Ap. J.* 202:543
Shine, R. A., Linsky, J. L. 1972. *Solar Phys.* 25:357
Shine, R. A., Linsky, J. L. 1974. *Solar Phys.* 39:49
Shine, R. A., Milkey, R. W., Mihalas, D. 1975. *Ap. J.* 199:724
Simon, T., Kelch, W. L., Linsky, J. L. 1980a. *Ap. J.* 237:72
Simon, T., Linsky, J. L. 1980. *Ap. J.* In press
Simon, T., Linsky, J. L., Schiffer, F. H. III. 1980b. *Ap. J.* In press
Skumanich, A. 1972. *Ap. J.* 171:565
Skumanich, A., Smythe, C., Frazier, E. N. 1975. *Ap. J.* 200:747
Smith, E. van P. 1960. *Ap. J.* 132:202
Smith, M. A. 1978. *Ap. J.* 224:584
Snow, T. P. 1979. *Proc. IAU.* In press
Snow, T. P., Linsky, J. L. 1980. In *Proc. Conf. Astron. Astrophys. from Spacelab,* ed. P. L. Bernacca, R. Ruffini. Dordrecht: Reidel. In press
Spangler, S. R. 1977. *Ap. J.* 82:169
Spruit, H. C. 1976. *Solar Phys.* 50:269
Spruit, H. C. 1977. *Solar Phys.* 55:3
Spruit, H. C. 1979. *Solar Phys.* 61:363
Stein, R. F., Leibacher, J. 1974. *Ann. Rev. Astron. Astrophys.* 12:407
Stein, R. F., Leibacher, J. W. 1980. In *Stellar Turbulence,* ed. D. F. Gray, J. L. Linsky, p. 225. New York: Springer
Stencel, R. E. 1978. *Ap. J. Lett.* 223:L37
Stencel, R. E., Mullan, D. J. 1980. *Ap. J.* In press
Stenflo, J. O. 1973. *Solar Phys.* 32:41
Stenholm, L. G., Stenflo, J. O. 1977. *Astron. Astrophys.* 58:273
Swank, J. 1979. Paper presented at the IAU Joint Meeting on Close Binaries and Stellar Activity, Montreal, 18 August 1979
Thomas, R. N. 1973. *Astron. Astrophys.* 29:297
Thomas, R. N., Athay, R. G. 1961. *Physics of the Solar Chromosphere.* New York: Interscience

Topka, K., Fabricant, D., Harnden, F. R., Gorenstein, P., Rosner, R. 1979. *Ap. J.* 229:661
Tripp, D. A., Athay, R. G., Peterson, V. L. 1978. *Ap. J.* 220:314
Tucker, W. H. 1973. *Ap. J.* 186:285
Ulmschneider, P. 1967. *Z. Ap.* 67:193
Ulmschneider, P. 1970. *Solar Phys.* 12:403
Ulmschneider, P. 1971. *Astron. Astrophys.* 12:297
Ulmschneider, P. 1979. *Space Sci. Rev.* In press
Ulmschneider, P., Kalkofen, W. 1977. *Astron. Astrophys.* 57:199
Ulmschneider, P., Kalkofen, W. 1978. *Astron. Astrophys.* 69:407
Ulmschneider, P., Schmitz, F., Hammer, R. 1979. *Astron. Astrophys.* 74:229
Ulmschneider, P., Schmitz, F., Kalkofen, W., Bohm, H. U. 1978. *Astron. Astrophys.* 70:487
Ulmschneider, P., Schmitz, F., Renzini, A., Cacciari, C., Kalkofen, W., Kurucz, R. L. 1977. *Astron. Astrophys.* 61:515
Ulrich, R. K. 1976. *Ap. J.* 210:377
Ulrich, R. K., Knapp, G. R. 1979. *Ap. J. Lett.* 230:L99
van der Hucht, K. A., Stencel, R. E., Haisch, B. M., Kondo, Y. 1979. *Astron. Astrophys. Suppl.* 36:377
Vernazza, J. E., Avrett, E. H., Loeser, R. 1973. *Ap. J.* 184:605
Vernazza, J. E., Avrett, E. H., Loeser, R. 1976. *Ap. J. Suppl.* 30:1
Walter, F. M., Cash, W., Charles, P. A., Bowyer, C. S. 1980. *Ap. J.* 236:212
Walter, F. M., Charles, P., Bowyer, C. S. 1978. *Ap. J.* 83:1539
Warner, B. 1966. *Observatory* 86:82
Warner, B. 1968. *Observatory* 88:217
Weiler, E. J. 1975. *IAU Inf. Bull. Var. Star,* No. 1014
Weiler, E. J. 1978. *Ap. J.* 83:795
Weiler, E. J., Oegerle, W. R. 1979. *Ap. J. Suppl.* 39:537
Weiler, E. J., Owen, F. N., Bopp. B. W., Schmitz, M., Hall, D. S., Fraquelli, D. A., Piirola, V., Ryle, M., Gibson, D. M. 1978. *Ap. J.* 225:919
Wentzel, D. G. 1980. In *The Sun as a Star. NASA-CNRS monogr. ser.* Vol. *VIA.* Preprint
White, O. R., Livingston, W. 1978. *Ap. J.* 226:679
White, O. R., Suemoto, Z. 1968. *Solar Phys.* 3:523
Wilson, O. C. 1959. *Ap. J.* 130:499
Wilson, O. C. 1966. *Science* 151:1487
Wilson, O. C. 1973. In *Stellar Chromospheres,* ed. S. D. Jordan, E. H. Avrett, p. 311. *NASA SP-317*
Wilson, O. C. 1978. *Ap. J.* 226:379

Wilson, O. C., Bappu, M. K. V. 1957. *Ap. J.* 125:661
Wilson, P. R., Evans, C. D. 1972. *Solar Phys.* 18:29
Withbroe, G. L., Noyes, R. W. 1977. *Ann. Rev. Astron. Astrophys.* 15:363
Worden, S. P., Schneeberger, T. J., Giampapa, M. S. 1980. *Ap. J.* In press
Young, A., Koniges, A. 1978. *Ap. J.* 211:836
Zahn, J. P. 1977. *Astron. Astrophys.* 57:383
Zirin, H. 1976. *Ap. J.* 208:414
Zirker, J. B. 1968. *Solar Phys.* 3:164
Zwaan, C. 1977. *Mem. Soc. Astron. Italiana* 48:525

MEASUREMENTS OF THE COSMIC BACKGROUND RADIATION

✽2173

Rainer Weiss[1]

Department of Physics, Massachusetts Institute of Technology, Cambridge, Massachusetts 02139

INTRODUCTION

This article presents a critical review of measurements of the attributes of the cosmic background radiation and indicates the prospects for improved measurements. Since the last review in these volumes (Thaddeus 1972), there has been substantial progress in the field and of course several other review articles (Alpher & Herman 1975, Danese & DeZotti 1977, Ulfbeck 1980). An up-to-date review of theoretical work on the background is presented by Sunyaev & Zel'dovich (1980) in this volume.

Whether one's taste runs to the presently favored "big bang" cosmology or to more heretical models, the background radiation has become an accepted and essential part of observational cosmology. The radiation has unique and remarkable properties: it is isotropic on both large and small angular scales to a higher degree than any other source in the universe; it exhibits no linear polarization; and it has a spectrum so close to that of a blackbody that one can talk only of deviations from a thermal spectrum. In short, the radiation satisfies beyond almost reasonable expectations the simple hypothesis that it is a remnant of a homogeneous and structureless primeval explosion. The observed blandness and apparent pristine state of the radiation beg for more detail. One hopes that precision observations would show imprints left by early cosmic processes and that observed deviations could be fruitfully interpreted. Studies of the background radiation have passed the discovery phase and are beginning an analytic one.

This review is dedicated to the analytic phase. How well have the

---

[1] Supported in part by NASA Grant NGR 22-009-526.

0066-4146/80/0915-0489$01.00

spectrum, intensity distribution on large angular scales, and polarization been measured? What limits these measurements and where is there a reasonable chance for improvement?

MEASUREMENTS OF THE SPECTRUM

Absolute measurements are always difficult. Measurements of the cosmic background radiation (CBR) spectrum are no exception and are further complicated by the fact that it is impossible to modulate the signal, the CBR. The CBR is therefore the residue, what is left over after one has accounted for everything else. The quality of this accounting, the precision and reliability of measurement of the individual terms, determines the value of the observation. The evolution of experiments to measure the CBR spectrum demonstrates this well. Future progress lies in improved experimental design or operation in more benign environments which permit reduction of the most uncertain elements in the accounting sum.

Direct measurements of the CBR spectrum fall crudely into two categories: low frequency observations in the Rayleigh-Jeans portion of the spectrum and observations at high frequencies in the region embracing the blackbody peak and into the Wien tail. This categorization results from the different technologies used in the two regions, and differences between the magnitudes of the terrestrial atmospheric emission. The low frequency measurements employ conventional microwave technology—coherent receivers—at sea-level and mountain observing sites while the high frequency observations have been carried out with less well developed infrared techniques—incoherent detectors, broad-band filters, Fourier transform spectrometers—on balloon- and rocket-borne platforms.

As the sub-millimeter technology of coherent receivers, in particular the development of efficient broad-band mixers, improves, the low frequency techniques will be applied to the high frequency region from balloon and aircraft platforms.

Figure 1 outlines the basis of all the low frequency measurements. The individual experiments listed in Table 1 may differ in details, even in some important ones, but nevertheless all consist of the same components: a

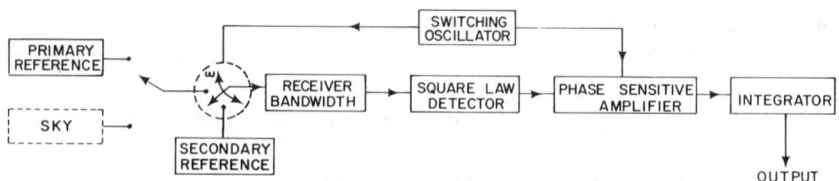

Figure 1 Schematic diagram of a generic absolute radiometer.

coherent receiver, a primary absolute reference calibrator, and an antenna The experimental strategy has been to compare the radiation entering the antenna with that of the primary calibrator, the gain of the receiver being determined by either varying the temperature of the primary reference or by injecting a known power into the receiver input.

(a) Receivers

The receiver sensitivity is often specified in terms of the smallest measurable change in temperature of a Rayleigh-Jeans thermal source that covers the entire antenna aperture, in a post-detection bandwidth of 1 Hz—an integration time of approximately 1/2 second. The minimum detectable temperature change is given by

$$T_{\min}(K/Hz^{1/2}) = gT_{\text{system}}/B^{1/2} \tag{1}$$

where g is a factor between 1 and 2 determined by the type of modulation and integrating filter. B is the receiver bandwidth determined by the mixer or the intermediate frequency amplifier. The system noise is characterized by T_{system}, the temperature of a fictitious Rayleigh-Jeans source placed over the antenna aperture that would produce the same *fluctuations* in a noise-free receiver output as are actually observed.

The early experiments had system temperatures of several thousand degrees and receiver bandwidths of tens of MHz, which permitted measurements to 1/10 degree in an integration time of about 1 minute. Inadequate receiver sensitivity has never been a real limit in these measurements except in the process of measuring some of the systematic noise sources, in particular the contribution from antenna side lobes, and in some of the experiments, the atmospheric radiation. The reasons for this will become clearer when each of these terms is discussed below.

The system output signal is proportional to the integral of the input power spectral density over the receiver bandwidth. The receivers, however, amplify the currents and voltages at the input terminals which are proportional to the incident electric and magnetic fields and are therefore sensitive to the phases of these fields. The receiver and antenna together are generally designed to accept one spatial mode (and often one polarization state) of the radiation field. For coherent systems like this the area—solid angle product (étendue) is λ^2. The total power accepted from an isotropic blackbody distribution is then

$$\Delta P = \frac{hf}{e^{hf/kT} - 1} B, \quad f \gg B. \tag{2}$$

The receiver systems are usually calibrated with blackbody sources for which $hf/kT \ll 1$ (Rayleigh-Jeans sources). For example at $\lambda = 1$ cm, $T = 300$ K, $hf/kT \sim 5 \times 10^{-3}$ so that the receiver power is just equal to

Table 1 Heterodyne measurements of the cosmic background spectrum

Reference	λ (cm)	f (GHz)	ν (cm^{-1})	Altitude (km)	Receiver noise (K/Hz$^{1/2}$)	Beam width (deg.)	Thermodynamic temperature of primary ref.	Antenna temperature of primary ref.	Properties of primary calibrator $\varepsilon_p T_p$ (K)	R	$\varepsilon_w T_w$ (K)
Howell & Shakeshaft 1967	73.5	0.41	0.014	0	—	15	4.2	4.19	1.7±0.2	—	—
Penzias & Wilson 1967	49.2	0.61	0.020	0	—	—	4.2	4.17	1.4±0.2	—	—
	21.2	1.415	0.047	0	—	—	4.2	4.17	0.6	—	—
Howell & Shakeshaft 1966	20.7	1.41	0.048	0	—	13×15	4.2	4.17	1.7±0.2	—	—
Penzias & Wilson 1965, Penzias 1968	7.35	4.08	0.136	0	—	—	4.2	4.10	1.3	—	—
Roll & Wilkinson 1966	3.2	9.37	0.313	0	—	20	4.2	3.98	2.6±0.35	—	—
Stokes et al. 1967	3.2	9.37	0.313	3.8	~1.5	4	3.77	3.55	0.16±0.10	6±3×10^{-4}	0.06±0.02
Stokes et al. 1967	1.58	19.0	0.633	3.8	~1.5	4	3.77	3.33	0.21±0.08	1±0.3×10^{-3}	0.04±0.01
Welch et al. 1967	1.5	20.0	0.666	3.8	~2	12	73.6	73.12	0.42±0.1	<4.10^{-4}	—
Ewing et al. 1967	0.924	32.5	1.08	3.8	~3	20	3.78	3.056	1.0±0.15	—	0.06±0.02
Wilkinson 1967	0.856	35.05	1.168	3.8	~1.5	4	3.77	2.99	0.28±0.11	0±3×10^{-4}	0.06±0.06
Puzanov et al. 1968	0.82	36.6	1.22	0	1.6	4	77.36	76.48	—	—	—
Kislyakov et al. 1971	0.358	83.8	2.79	3	2–4	10	~75.0	~73.0	—	—	—
Boynton et al. 1968	0.33	90.0	3.00	3.44	1.5	—	3.8	2.042	0.22±0.11	3×10^{-4}	0.27±0.04
Millea et al. 1971				3.1			3.85	2.087			
	0.33	90.4	3.00	2.8	0.5	6.6	3.88	2.114	1.1±0.11	3±0.6×10^{-4}	—
Boynton & Stokes 1974	0.33	90.0	3.00	14.9	0.1	—	4.2	2.405	0.22±0.11	8±4×10^{-4}	0.27±0.04

Table 1—*continued*

Reference	Properties of antenna			External sources[a]			Switch asymmetry	Cosmic background thermodynamic temperature	See note[b]
	$\varepsilon_{ant}T_{ant}$ (K)	$f_{gnd}T_{gnd}$ (K)		$\varepsilon_{atm}T_{atm}$ (K)	T_{gal} (K)				
Howell & Shakeshaft 1967	0.4±0.2	0.6±0.4		1.3±0.1	20.6±2.2		—	3.7±1.2	1,2
Penzias & Wilson 1967	0.4±0.2	0.6±0.4		1.95±0.1	6.7±0.7		—	3.2±1.0	—
Howell & Shakeshaft 1966	0.55			2.3	0.3		0.05	2.8±0.6	2
Penzias & Wilson 1965, Penzias 1968	1.3±0.2	≤0.1		2.2±0.2	0.5±0.2			3.3±1.0	3
	0.8±0.4	≤0.1		2.3±0.3	<0.2		0.05		
Roll & Wilkinson 1966	1.08±0.15	—		3.0±0.2	<0.0015		3.8±0.2	3.0±0.5	—
Stokes et al. 1967	0.08±0.06	<0.05		1.37±0.1	—		—	$2.69^{+0.16}_{-0.21}$	4
Stokes et al. 1967	0.15±0.05	<0.15		4±0.1	—		—	$2.78^{+0.12}_{-0.17}$	4
Welch et al. 1967	1.9±0.2	≤0.1		4±0.2	—		—	2.45±1.0	5
Ewing et al. 1967	0.21±0.06	0.03±0.01		4.6±0.2	—		0.5±0.1	3.09±0.26	6
Wilkinson 1967	0.12±0.04	<0.05		6.5±0.2	—		—	$2.56^{+0.17}_{-0.22}$	4
Puzanov et al. 1968	—	—		17.0	—		—	3.7±1.0	5
Kislyakov et al. 1971	5±0.4	—		14.7±1.2	—		—	2.4±0.7	—
Boynton et al. 1968	0.21±0.12	—		11.5±0.22	—		—	$2.46^{+0.40}_{-0.44}$	4
Millea et al. 1971	0.21±0.11	≤0.02		11.8±0.22	—		—	—	—
				12.1±~0.4				2.61±0.25	
Boynton & Stokes 1974	—	—		1.21±0.37	—		—	$2.48^{+0.50}_{-0.54}$	7

$\langle T \rangle = 2.74 \pm 0.087$ K normalized $\chi^2 = 0.44$ 14 degrees of freedom

See overleaf for footnotes.

FOOTNOTES FOR TABLE 1

[a] Atmospheric radiation entries are typical numbers for a data set.

[b] Notes:
  1. The data analysis assumed that the cosmic background has a Rayleigh-Jeans spectrum and that the galactic component can be characterized by a power law spectrum. The two points together produce one measurement of the background temperature and the magnitude of the galactic emission.
  2. Atmospheric contribution is calculated from other data, not measured in the course of the experiment.
  3. Penzias & Wilson (1965) is the discovery observation. Penzias (1968) gives a corrected value for the measured temperature.
  4. The four Princeton experiments, due to their small error bars, have the largest weight in determining the average temperature. Looking at the data from the four experiments together, one can see that the measured contribution of the emission by the reflector scales only marginally with the square root of the frequency. The emissivity of idealized metal surfaces in the case $E_{rf} \perp$ to the plane of incidence varies as $\varepsilon \sim 2 \cos\theta \sqrt{v/\sigma}$ where θ is the angle of incidence, v the rf frequency, and σ the conductivity (cgs units). An aluminium reflector at 290 K has a minimum theoretical emission of 0.09, 0.125, 0.170, 0.27 K_{ant} at 3.2, 1.58, 0.856, and 0.33 cm for $\theta = 30°$.
  5. The published result has been converted to thermodynamic temperature.
  6. The emission by the horn antenna in this experiment is the estimated difference between the radiative contributions of two similar horns, one pointed to the load, the other to the sky. The published result has been corrected for an error in the way atmospheric absorption enters the measurement of the background radiation.
  7. The only 3 mm experiment in which the atmospheric contribution is less than the cosmic background. The apparatus was not calibrated using a primary calibration source during the airborne measurements. It was calibrated before and after flight. The properties of the calibrator were not remeasured and assumed to be the same as in Boynton et al. (1968).

$\Delta P = KTB$. It becomes convenient to measure power in terms of a temperature. Measurements of the background radiation and liquid helium reference sources at wavelengths shorter than 3 cm show substantial deviations from the Rayleigh-Jeans spectrum; the power is not proportional to the thermodynamic temperature. To relate the received power to the calibration, it is useful to define the concept of an antenna temperature, T_{ant}. The temperature of a Rayleigh-Jeans source that radiates the same power as the actual source at a thermodynamic temperature T and frequency f:

$$kT_{ant} = hf/e^{hf/kT} - 1. \tag{3}$$

The general practice has been to measure all terms in units of antenna temperature and then to convert the CBR contribution to a thermodynamic temperature. Several of the references have not been consistent in this regard and the results in Table 1 have been adjusted accordingly.

(b) Antenna

The antenna design in background experiments is critical. The background radiation was discovered by Penzias & Wilson (1965) primarily because a very low side-lobe antenna had been constructed at Bell Laboratories for satellite communications. It is most unlikely that any general purpose radio antenna existing at that time could have been used to make the discovery with any degree of confidence. The remaining observations in Table 1 were carried out with special purpose horn antennae.

Since the background radiation covers the entire sky, there is no optimum antenna beam size to investigate the spectrum unless one wants to avoid discrete local sources. Most of the spectrum experiments have used beam widths large by radio astronomy standards, 4° to 20°, but still small enough to allow zenith angle scanning to measure the atmospheric emission. The primary concerns are antenna thermal emission and the reduction of antenna diffraction at large angles to the optic axis—the side lobes. The importance of controlling the side lobes is shown by a simple calculation. Let the main beam, observing the CBR at 3 K, include a solid angle of 0.02 steradians (10° beam). The 300 K earth, airplane, or spacecraft fills the 2π steradian backplane. If the contribution from the backplane is to be less than 10% of the CBR, the point source response of the antenna for angles greater than 90° to the optic axis must be less than 10^{-6} (-60 db) of that along the optic axis, and still smaller if the experiment is to measure the spectrum at high frequencies where the Rayleigh-Jeans approximation for the CBR is no longer valid. Since the off-axis response of the antenna is usually a steep function of the angle from the optic axis, the contribution from atmospheric emission, determined by scanning the zenith angle, can be confused by an increase in the ground radiation contribution from poorly suppressed side lobes.

Most of the antennas used in the low frequency experiments were single mode rectangular or cylindrical horns 10 to 15 wavelengths wide at their last diffracting surface. The field distribution at the horn mouth is close to cosinusoidal giving a far field angular response function at best varying as $\sim (2n \sin \theta)^{-4}$ at large angles where n is the aperture width in units of the wavelength and θ the angle to the optic axis. These horns were just barely good enough and most groups resorted to placing large metallic reflectors under and around the horn to reflect the "cold" sky into the antenna back lobes.

The high precision experiments pioneered by the Princeton group used a horn aimed at a mirror as the antenna. In this scheme the radiometer was held fixed throughout the experiment while the mirror was rotated to perform zenith scanning or removed for calibration. This technique eliminated a possible systematic error source arising from changes in the radiometer offset due to varying gravitational loading of the waveguide plumbing. However, the mirror is an additional polarization- and angle-sensitive emitting surface in the beam.

In the past few years substantial advances in low side-lobe horn designs have been made. The narrow-band corrugated horns used in the U2 experiment to measure the large angular scale anisotropy of the CBR (Gorenstein et al. 1978) have demonstrated a back-lobe rejection of 10^{-8} (-80 db) (Janssen et al. 1979). Flared broad-band multimode (apodizing) horns developed for balloon-borne spectrum measurements (Mather 1974,

Woody 1975) have been designed with the Keller theory of geometric diffraction (Levy & Keller 1959, Mather 1980). The flared horns, tested at JPL, also exhibit back-lobe rejection of 10^{-8} or smaller (Janssen & Weiss 1977).

The thermal emission by the antenna is not a serious problem at low frequencies and is measured by varying the temperature of the antenna while looking into a cold load or at the sky. In those experiments where the calibrator covered the entire antenna aperture, antenna emission does not enter the accounting procedure in first order. High frequency observations are more troubled by antenna emission both because the emissivity of the metal surfaces grows at least as fast as the square root of the frequency and because the CBR power is not rising as fast as a Rayleigh-Jeans spectrum. The best recourse here is to cool the antenna.

(c) *Calibration*

The basis of absolute receiver calibration is the blackbody law; it is the only radiation source that is calculable in actual practice. The implicit assumption is that an isothermal enclosure that does not reflect or become transparent radiates a blackbody spectrum. There are a host of problems in making a practical calibrator approach the idealized limit and furthermore, given a good calibrator, in being assured that the act of placing the calibrator at the radiometer input does not affect the performance of the receiver or alter the geometry at the input in an incalculable way. The calibrator should appear to the radiometer as similar to the measured source as possible.

Most of the low frequency and all the high frequency experiments have used cryogenic reference bodies to reduce the demands on receiver linearity and to minimize the change in operating point of the square law detector when switching between the calibrator and the sky. The temperature–vapor pressure relation of liquid helium is known to better than 0.1% and with reasonable care cryogenic thermometry can be performed reliably to this precision. The calibrators are constructed by terminating a waveguide or light pipe on a prismatic or conical absorber several wavelengths thick that is in good thermal contact with the cryogen. The termination is designed to trap the radiation through multiple reflections by the poorly reflecting absorber material. The central difficulty in the calibrator designs has been the thermal gradients at the transition from the cryogenic environment to the warm world outside, the problems being emission by the warmer sections of the waveguide and from windows used to avoid condensation of air in the liquid helium. These contributions are not negligible; in some experiments, in fact, they are comparable with the emission by the absorber. As the frequency increases this problem becomes more acute.

The emissivity of the calibrator is determined by reflection measurements, using $\varepsilon = 1 - R$, and is generally larger than 99% for narrow-band calibrators. Broad-band calibrators are more difficult to design. The reflectivity of the calibrator has to be known because, in use, emission by the warm receiver components (i.e. waveguides, local oscillators, horns) is reflected by the calibrator back into the receiver and this effect is not balanced out when the receiver looks into the sky.

Most of the detailed considerations involved with the interaction of the calibrator, receiver, and the thermal gradients are eliminated if the entire apparatus is maintained at cryogenic temperatures. This has been the practice in high frequency measurements and is clearly indicated for any new precision measurements at lower frequencies.

(d) *Atmospheric Emission and Absorption*

The profound role of the atmosphere in the measurements of the CBR is demonstrated in Figure 2 which shows the average atmospheric emission at four altitudes—0, 4, 14, 44 km—as well as the Planck spectra of 300, 3, and

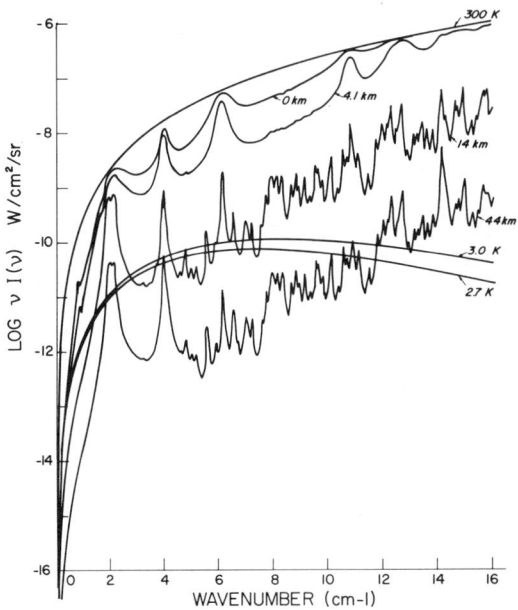

Figure 2 Zenith atmospheric emission at various altitudes drawn for an instrument with 0.04 cm^{-1} hwhm resolution. The parameters are: 0 km (H$_2$O, 3.7×10^{22} mol/cm^2, 300 K; O$_2$, 5×10^{24}, 300; O$_3$, 9×10^{18}, 220); 4.2 km (7×10^{21}, 250; 3×10^{24}, 250; 9×10^{18}, 220); 14 km (3.8×10^{19}, 215; 6.7×10^{23}, 215; 8×10^{18}, 215); and 44 km (2.7×10^{17}, 260; 1×10^{23}, 260; 1.7×10^{17}, 260).

2.7 K blackbodies for reference. The curves correspond to the emission observed from the ground on a good day or night (column density of 1 cm precipitable H_2O), a mountain site (2 mm), the typical altitude attained by available and instrumented jet aircraft (10μ), and balloons (0.05μ).

The atmospheric emission in this spectral region is due to the rich spectra of O_2, H_2O, and O_3. The contribution of aerosols, the other minor molecular constituents of the atmosphere, CO, N_2O, free radicals such as OH in the upper atmosphere, which emit in this spectral region, are assumed to be negligible. On the ground and mountain sites the O_2 and H_2O contributions are so overwhelming that O_3 radiation is never considered. However, at airplane and balloon altitudes this is no longer the case.

O_2 is assumed to be uniformly mixed in the atmosphere making up 23.14% of the atmospheric mass up to at least 60 km (proof positive of the uniform mixing hypothesis is missing but meteorologists argue it can't be any other way). H_2O column densities, as everyone knows, are highly variable up to altitudes of 14 km. The stratospheric concentration is a few parts per million by mass although this wasn't known until high altitude balloon experiments to measure the CBR were begun in the late 1960s. (Prior sampling experiments had measured much larger concentrations but were evidently measuring the water brought up in the apparatus.) O_3, again parts per million by mass, is concentrated in the lower stratosphere at about 30 km. The O_3 densities vary with season and latitude.

All of the emission lines come from rotational or fine structure transitions in the ground electronic and vibrational states of these molecules. The O_2 lines arise from two mechanisms—a cluster of lines at 2 cm^{-1} and a single one at 4 cm^{-1} originate from magnetic dipole transitions associated with the reorientation of the electronic angular momentum (2 Bohr magnetons from the unpaired electron spins) relative to the molecular rotation ($\Delta J = \pm 1$, $\Delta K = 0$). These lines have been extensively studied experimentally (Meeks & Lilley 1963, Liebe et al. 1977). The other lines, occurring in triplets spaced 2 cm^{-1} apart at 14, 25, 37... cm^{-1}, are magnetic dipole rotational transitions ($\Delta J = \pm 1$, $\Delta K = \pm 2$) predicted by Tinkham & Strandberg (1955a,b) and discovered by Gebbie et al. (1969). These lines play an important role in the recent and most precise measurement of the background spectrum at high frequencies (Woody & Richards 1979) and will be discussed in more detail later.

H_2O and O_3 are both asymmetric rotors with complicated spectra. H_2O, owing to its large electric dipole moment and relatively small partition sum at atmospheric temperature, is responsible for the bulk of the atmospheric emissivity in the spectral region (Benedict 1976). The influential H_2O lines are well separated in frequency. O_3, on the other hand, has a multitude of weak lines, no less than 300 in the 1 to 20 cm^{-1} band (Gora 1959). A much appreciated and used resource is the Air Force Cambridge Geophysical Laboratory compilation of atmospheric line parameters (McClatchey 1979, Rothman 1978), which is available on magnetic tape. The compilation gives the frequency, line strength, energy of the ground state, and pressure-broadening coefficient of all known atmospheric lines from the radio through optical region.

It is well worth reviewing the assumptions and outlining the steps involved in calculating the models for atmospheric emission and absorption that have been used in CBR observations [Goody (1964) is a useful reference]. The end result of the calculation is the absorption and emission integrated over the atmospheric column at each frequency. The input parameters of a complete model would include 1. the temperature and partial pressure of the constituents as a function of altitude, 2. the individual molecular line strengths as a function of temperature, 3. the line shapes at different pressures and temperatures. To be complete, the Zeeman effect in the earth's magnetic field should be considered, especially at high altitude where the lines are narrow. The Zeeman effect makes the atmospheric propagation anisotropic. The calculation is carried out layer by layer beginning at some maximum altitude above which the radiation is deemed negligible. Each layer contributes its radiation and absorbs some fraction of the radiation from the preceding layer. The integration is continued to the altitude of the observation.

The complete calculation would pose a substantial computing problem, especially in maintaining sufficient frequency resolution to resolve the narrow lines at high altitudes. More important, the calculation is not worth doing because many of the input parameters are poorly known. Except possibly for O_2, the partial pressure of the constituents as a function of altitude is the most uncertain element. The actual temperature distribution with altitude at the time of the measurement would be the next important unknown. Furthermore, the line profiles do not conform in detail to simple theoretical models because the collisional line-broadening mechanisms depend on the colliding species and their kinetic energies. At present the only reliable way to determine the line-broadening parameters is to measure them directly with high resolution instruments.

All this is not to say the atmosphere is impossible to model but rather to urge caution in the interpretation of absolute CBR measurements where the atmospheric contribution is large.

In those parts of the atmospheric spectrum where the total absorption is small, less than $\sim 10\%$, the atmospheric emission can be measured by zenith angle scanning, since the emission is linearly proportional to the total column density of emitters. The result depends only on the assumptions of a laminar atmosphere and homogeneity within each layer and not on the specific model. Temperature, pressure, and constituent inhomogeneities occur and in fact are the largest source of random noise in the ground-based experiments. However, they do not contribute systematic errors unless the particular observing site is anisotropic in a gross manner—because of a large lake or the ocean in the direction of the zenith scan, for example. The atmospheric and CBR contributions are separable in this case without further measurements or modeling. The total emittance B as a function of zenith angle θ is given by

$$B_{\text{total}}(f, \theta) = \alpha(f, \theta) B(f, T_{\text{atm}}) + [1 - \alpha(f, \theta)] B(f, T_{\text{CBR}}) \tag{4}$$

where $\alpha(f, \theta)$ is the atmospheric absorption coefficient (in this case also the emissivity) at frequency f which is proportional to the column density of emitters (absorbers) and therefore depends on $\sec \theta$. $B(f, T)$ is the blackbody emittance at the thermodynamic temperature T. Measurement of $B(f, \theta)$ at two angles gives $B(f, T_{\text{CBR}})$ to a theoretical precision of order α^2. The procedure implied by Equation (4) has been used in all of the low frequency ground-based measurements.

Since the atmosphere at low frequencies appears primarily as a random noise source rather than a systematic one, there is hope for improved measurements of the CBR at low frequencies from the ground. The strategy of using two frequencies, X (3 cm) and K band (1.5 cm), simultaneously with up-to-date low noise radiometers, would allow measurement of the atmospheric fluctuations during the course of zenith scanning. Furthermore, rapid zenith scanning with co-addition of scans would overcome some of the dominating low frequency components of the atmospheric fluctuations.

Once self-absorption becomes important, atmospheric modeling is inevitable; Equation (4) then looks more like

$$B(f, \theta) = [1 - e^{-\alpha(f, \theta, h)}] B[f, T_{\text{atm}}(h)] + e^{-\alpha(f, \theta, h)} B(f, T_{\text{CBR}}), \tag{5}$$

although this equation still does not represent the complete integration required.

The atmospheric absorption coefficient and the atmospheric blackbody emittance no longer appear as a simple product and must be calculated or determined separately as a function of altitude, h. Zenith scanning no longer gives a model-independent measure of the atmospheric contribution. The CBR observations at high frequency from balloons had to face up to this.

The multifilter MIT (Muehlner & Weiss 1973a,b) and the Berkeley spectrometer experiments (Woody et al. 1975, Woody 1975, Woody & Richards 1979) have used similar strategies to handle the atmospheric contribution. At high altitudes the individual molecular lines are narrower than the spectrometer or filter resolution widths. Pressure line-broadening parameters range around 0.1–0.3 cm^{-1} per atmosphere pressure so that, at an altitude of 44 km (pressure 2×10^{-3} atm), the line widths at the base of the column are $\sim 10^{-4}$ cm^{-1}. Doppler widths are typically a factor of 10 smaller (as a consequence, absorption of the CBR is neglected). The difference between Van Vleck–Weisskopf and Lorentzian line shapes becomes negligible for such narrow lines and the Lorentzian profile is adopted for ease of calculation. All the emitting constituents are assumed to be uniformly mixed with an exponentially decreasing density as a function of altitude. The emitting column is furthermore assumed to be isothermal with the temperature measured at its base.

Under these assumptions, the power in a line from the column, integrated over all frequencies, can be cast in an analytic form that depends on the *total column density*, the line strength, the pressure-broadening parameter, and the temperature (Goody 1964). The equivalent emission width of the line is given by

$$A(\text{cm}^{-1}) = 2\sqrt{\pi} \, \Delta v(T, P_0) \, \Gamma(x+1/2)/\Gamma(x), \tag{6}$$

where x is a dimensionless parameter indicating the degree of saturation of the line given by

$$x = S(T)\sigma/2\pi \, \Delta v(T, P_0). \tag{7}$$

$S(T)$ is the line strength in units of cm^{-1}/molecule/cm^2 and includes the matrix elements, multiplicity, fraction of molecules in the lower state, and the relative population difference of the two levels involved in the transition. $\Delta v(T, P_0)$ is the pressure-broadened Lorentzian line width at the base of the column, and σ is the column density in molecules/cm^2. The total emittance due to the line is

$$B(v) = B(v, T_{\text{atm}})A. \tag{8}$$

The limiting cases are

$$A = \begin{cases} S(T)\sigma & x \ll 1 \quad \text{unsaturated lines} \\ (2s(T) \, \Delta v(T)\sigma)^{1/2} & x \gg 1 \quad \text{saturated lines} \end{cases}$$

When the lines are unsaturated (the emissivity at line center is much less than 1), the emission is linearly proportional to the column density and therefore the atmospheric contribution could be measured directly with zenith scanning in an almost model-independent determination. On the other hand, if the line is saturated, the emission is proportional to the square root of the column density—this is due to the assumed Lorentzian line function—and depends on the pressure-broadened line width. The increase in emission with column density can only come from molecules emitting in the wings of the line. Zenith scanning would produce an atmospheric signal proportional to $(\sec \theta)^{1/2}$ and, providing that all lines falling into the instrument resolution width were saturated, could be used to aid in determining the atmospheric contribution. The real problem comes with a mixture of both saturated and unsaturated lines or partially saturated lines, which is the situation in the stratosphere; the O_2 lines are partially saturated, the strong H_2O lines are fully saturated, and all O_3 lines are unsaturated. Zenith scanning, with an instrument of low frequency resolution, can without knowledge of the individual constituent column densities only offer a model-independent lower limit to the atmospheric radiation and hence an upper limit to the CBR. This was the case with the MIT experiments (Muehlner & Weiss 1973a). A subsequent flight by this group (Muehlner & Weiss 1973b) used different narrow-band filters to isolate individual atmospheric lines of H_2O and a cluster of O_3 lines to determine the constituent column densities, but the signal to noise and flight duration were insufficient to improve substantially on their prior results. The spectrometer observations have enough resolution to isolate the lines of O_2 and H_2O and line clusters of O_3 so that it becomes possible, using the model outlines above, to solve for the individual column densities by fitting the observed spectrum.

A question that keeps recurring when the high frequency balloon measurements are discussed is the validity of the simple atmospheric model, which is inadequate at lower altitudes. In preparing this review I made a comparison over a restricted frequency band (10–20 cm^{-1}) of the model described above at 43 km with a 10 layered one that 1. used Doppler lines in the upper stratosphere and Van Vleck–Weisskopf lines at lower altitudes, 2. removed the isothermal restriction using the average atmospheric temperature profile with altitude (US 1966), 3. retained the uniform mixing hypothesis. The same line parameters were used in both calculations but the estimated temperature dependences of both the line strengths and line-broadening parameters were included. The result was that for the same column base temperature and pressure and equal column densities, the two models differed by less than 5%, not a remarkable result since the major emission occurs within a scale height. The difference between models is no worse than the "noise" of the uncertainties in the line parameter listings, which I estimate to be at least 10%.

Another recurring worry is the possibility that there exist *influential* atmospheric constituents that emit in this spectral region but are not now included in the line listings. The response is an equivocal "no." Finally, the notion of calibrating a CBR experiment using the atmospheric emission is a bad one; more on this later.

(e) Galactic Background Contribution

The next "atmosphere" is the solar system and the galaxy, the "molecules" being the astrophysical phenomena distinguished by their spectra and identified by their anisotropic distribution. The individual "molecules" are not indistinguishable and can only be characterized in a statistical sense. As a consequence, the success in modeling the "local" astrophysical background depends heavily on sampling much of the sky with extended spectral coverage.

Figure 3 shows the present best estimates of known astrophysical sources that emit in the frequency range important to CBR measurements. The CBR lies predominantly in a clear region of the local astrophysical spectrum embraced by synchrotron emission of cosmic ray electrons moving in galactic magnetic fields and free-free thermal emission by ionized gas surrounding hot stars (H II regions) at low frequencies and by the thermal emission of interstellar dust at high frequencies.

Full sky maps of synchrotron emission have been made at low frequencies, 200 MHz (Droge & Priester 1956) and 404 MHz (Pauliny-Toth & Shakeshaft 1962), where it is the major source of the galactic background and pervades the sky. There is no region of the sky which emits less than $\sim 5\%$ of the strongest sources contained in the galactic plane in a band $b \pm 10°$, $l \pm 40°$. The spectral index n, defined by $I(v) \propto v^{-n}$, ranges between 0.8 to 0.9 (Witebsky 1978) as determined by high frequency measurements in selected regions of the sky. The spectral index depends on the energy distribution of the radiating electrons and may well be different in discrete sources such as supernova remnants and in the general galactic background. In clearer regions, toward the galactic poles, the antenna temperature due to synchrotron emission is approximated by

$$T_{syn} \sim 1.5 f(GHz)^{-2.9} \text{ K}.$$

Strong H II regions lie primarily in the galactic plane although the Orion nebula is a notable exception. A compilation of emission by H II regions at 5 GHz is given by Lang (1978), from data of Reifenstein et al. (1970) and Wilson et al. (1970). A proper estimate of the contribution from discrete H II source emission requires knowledge of the specific H II regions in the observing beam. Typical parameters for the stronger H II regions are electron temperatures ~ 7000 K, emission measure, $N_e^2 d$, $\sim 10^6$ electron^2 × parsec per cm^6, and angular size, θ_S, $\sim 0.1°$. In the spectral region of interest to CBR measurements, $1 \text{ GHz} < f < 1000 \text{ GHz}$, quantum effects are negligible, and the optical depth in the source is much less than unity. The spectrum is almost flat, modified only by the classical Gaunt factor, giving a spectral index of approximately 0.1. The antenna tempera-

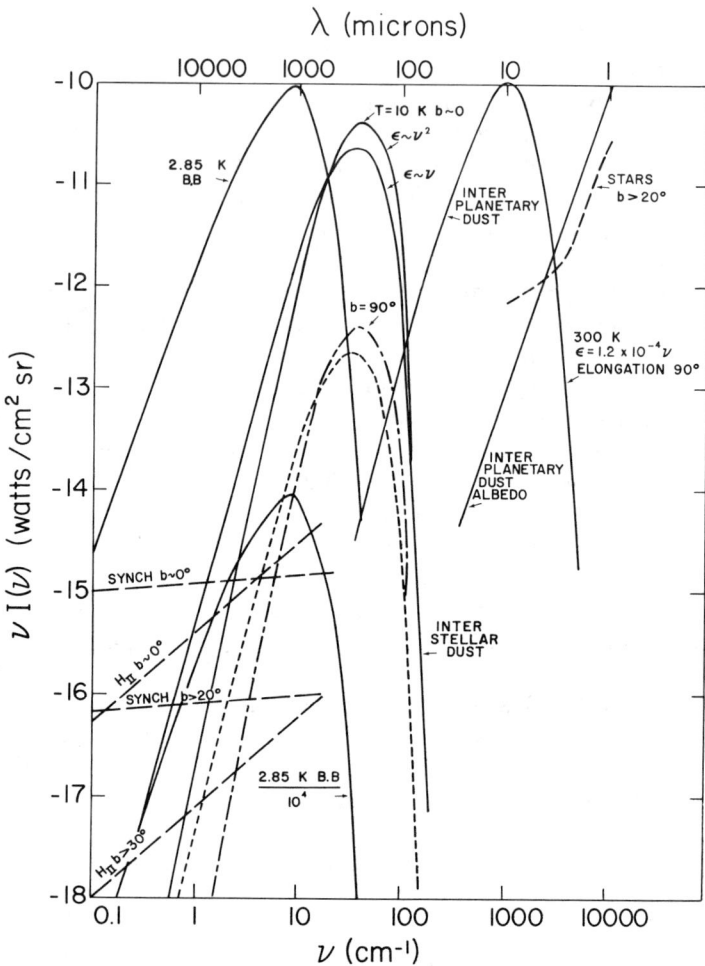

Figure 3 The astrophysical background.

ture of such a typical H II region, when diluted in a beam of angular size $\theta_B \sim 10°$, is given by (Lang 1978)

$$T_{\text{H II}} \sim (\theta_s/\theta_B)^2 8.2 \times 10^{-2} T_e(\text{K})^{-0.35} \nu(\text{GHz})^{-2.1} N_e^2 d \sim 0.4\nu(\text{GHz})^{-2.1} \text{ K}. \quad (9)$$

In addition to the discrete H II regions, Hirabayashi (1974) has observed a diffuse galactic background ascribed to free-free emission with an emission measure of 4000 pc cm^{-6} and an average electron temperature of 3000 K, directly in the galactic plane. Combining this observation with the model proposed by Ellis & Hamilton (1966), which assumes the ionized hydrogen is distributed uniformly ($N_e \sim 0.2$ cm^{-3}) in a disk 300 pc thick on either side of the plane, gives the following estimated antenna temperature (Smoot 1977) using Equation (9):

$$T_{H\,II} \sim 0.07\nu(\text{GHz})^{-2.1}/\sin b \text{ K for } b > 5 \text{ K.}$$

The estimates for thermal emission by interstellar dust grains shown in Figure 3 are merely suggestive since both the physical properties of the dust grains as well as the dust distribution throughout the Galaxy are still poorly known. Direct measurements of interstellar dust emission in the far infrared, $\nu < 50$ cm^{-1}, have been made solely in regions of strong emission such as in surveys of the galactic plane—at 50 cm^{-1} (Low et al. 1977), at 20 and 10 cm^{-1} (Owens et al. 1979)—and observations toward the galactic center at 28 cm^{-1} (Gezari et al. 1973 and Rieke et al. 1973), and at 18 cm^{-1} (Hildebrand et al. 1978). It has proved difficult to determine the spectrum by comparing measurements of the same objects made by instruments using different beam widths and spatial chopping techniques because many of the dusty regions are complex with both compact and diffuse emitting regions.

For $\nu \lesssim 50$ cm^{-1}, the radiating efficiency of the grains is small since the wavelength is much larger than the grain dimensions, assumed 10^{-5} cm and smaller, so that even for the densest regions of the Galaxy the optical depth at $\nu < 50$ cm^{-1} is less than 1. The emittance of a dust column is then

$$B(\nu) \sim \pi a^2 N_D Q_{\text{abs}}(\nu) B(\nu, T_D) \qquad (10)$$

where πa^2 is a typical cross sectional area of the grain, N_B the column density of grains, $B(\nu, T_D)$ the blackbody emittance at the dust temperature T_D, and $Q_{\text{abs}}(\nu)$ the emissivity per grain.

The most general form of the emissivity per grain is

$$Q_{\text{abs}} = a\nu f(\nu) \qquad (11)$$

where $f(\nu)$ is a function of the imaginary part of the grain dielectric constant. The observations in strongly emitting regions indicate that $f(\nu)$ varies as $\nu^{0.5}$ to ν^2 and implies that the far infrared emission is most likely due to the low frequency tails of solid state absorption bands in the grains (Aannestad 1975).

Combining Equations 10 and 11, the dust brightness is proportional to the volume of dust and therefore to the mass column density. The dust brightness alone, however, cannot give the grain emissivities because the mass column density is not known directly. In lieu of a dust model, estimates of the dust mass column densities are made from the observed correlation of dust with neutral gas. The dust column densities, estimated by optical extinction and reddening, related to H I column densities determined from 21 cm line profiles (see for example Knapp & Kerr 1974) indicate that the dust-to-hydrogen mass ratio is of the order of 1%.

In the past few years observation of CO emission in dusty regions, where it is a minor constituent, have provided estimates of the H_2 densities, the major constituent. These measurements again give dust-to-neutral-gas ratios of $\sim 1\%$ (Scoville & Solomon 1975) and furthermore yield independent measurements of the dust temperature.

Along the galactic plane $|l| < 50°$, plausible parameters that fit the measured emission are $T_D \sim 25$ K, dust mass column density 3×10^{-4} gm/cm^2, corresponding to 2×10^{22} gas atoms/cm^2, individual grain emissivity 2×10^{-7} v(cm^{-1})^2, $a = 10^{-5}$ cm, and a grain density of 2 gm/cm^3. These are the parameters used in estimating the emission at $b = 0$ in Figure 3 in a beam width of $16°$, the galactic plane filling about 1/15 of the beam.[2]

The estimates of dust emission out of the plane, especially toward the galactic poles, are even more uncertain. The steps involved in the estimate are the following: the dust type is assumed to be the same out of the plane as in the plane, the temperature of the dust is calculated on the basis of radiative equilibrium with the absorption of star light $T_D \sim 10$ K, the dust mass column density is estimated from H I column densities, and extinction and reddening data (Heiles & Jenkins 1976, Daltabuit & Meyer 1972). Various methods of calculating the dust mass density toward the galactic poles do not agree. Reddening data of Feltz (1972) gives an extinction due to dust of 0.000 ± 0.006 mag while Appenzeller (1975) infers an extinction ≥ 0.03 from stellar polarization due to dust grain alignment. The assumed ratio of gas density to visual extinction is 1.6×10^{21} atoms/cm^2/mag or $\sim 2.5 \times 10^{-5}$ gm dust/cm^2/mag so that the dust column density to the galactic poles from extinction data is close to zero or larger than 7.5×10^{-7} gm/cm^2 (4×10^{19} atoms/cm^2), while H column densities are measured directly as $1-2 \times 10^{20}$ atoms/cm^2. The dust mass column density adopted for Figure 3 is 8×10^{-7} gm/cm^2, at $b = 90°$, and to the same precision as any of these estimates, may follow a cosecant law with galactic latitude.

(f) *Summary of the Low Frequency Measurements*

The terms in the account have been discussed and are assembled in the following two equations. To first order in small quantities, the antenna

---

[2] The dust model isn't totally crazy; using simple Lorentz-Lorenz theory with one resonance, the emissivity per grain is $(2/3\pi)(\rho/m_a)(ar_e)(v^2\Delta v/v_0^4)$ where m_a is the mass per atom, assuming each atom contributes one oscillator, r_e the Thomson radius of the electron, and v_0 the resonance frequency in cm^{-1} with width Δv. Assuming $\Delta v \sim 1/10v_0$, $v_0 \sim 1000$ cm^{-1}, typical of silicates and carbonates, the emissivity/grain is of the same order of magnitude as used above.

temperature of the reference is

$$T_{ref} = \varepsilon_W T_W + (1 - R_W)[\varepsilon_P T_P + \varepsilon_L T_L(1 - \varepsilon_P)] + R_{cal} T_{rec} \qquad (12)$$

where ε_W, R_W, T_W are the emissivity, reflectivity, and temperature of the window, ε_P, T_P the emissivity and temperature of the warm parts of the calibrator, ε_L, T_L the emissivity and antenna temperature of the load and R_{cal} is the reflectivity of the reference taken as a whole. T_{rec} is an equivalent temperature of the receiver and antenna viewed as an emitter. The terms, when known, are listed in Table 1 for the various experiments.

The antenna temperature, looking at the sky, is

$$T_{sky} = (1 - \varepsilon_{ant})[\varepsilon_{atm}(\theta)T_{atm} + f_{ant}(\theta)\varepsilon_{grd}T_{grd} + (T_{Gal} + T_{CBR})(1 - \varepsilon_{atm}(\theta))]$$
$$+ \varepsilon_{ant}T_{ant}, \qquad (13)$$

where ε_{ant} and T_{ant} are the emissivity and temperature of the antenna, $\varepsilon_{atm}(\theta)$ and T_{atm} are the emissivity as a function of elevation angle and temperature of the atmosphere, $f_{ant}(\theta)$ is the side-lobe response of the antenna at the ground with emissivity and temperature, ε_{grd} and T_{grd}, T_{Gal} is the antenna temperature of the Galaxy, and finally T_{CBR} the antenna temperature of the cosmic background radiation. Equations (12) and (13) apply broadly to all the experiments. However, the analysis of any particular experiment may require modifications of the two equations. For example, if the reference is placed at the input of the antenna, Equation (12) would be multiplied by $(1 - \varepsilon_{ant})$ and the emission of the antenna, $\varepsilon_{ant}T_{ant}$, would have to be added.

Table 1 is intended to show the magnitudes and uncertainties of the individual terms. As can be seen from the table many of the systematic error sources could be substantially reduced by operating the entire input end of the receiver and calibrator at cryogenic temperatures. The atmosphere, although a dominant random noise source, can be handled as indicated before. There is therefore a future in performing improved observations of the spectrum at frequencies 1 to 90 GHz from the ground.

(g) *High Frequency Observations of the Spectrum*

The high frequency experiments have gone through two generations. The first were the rocket- and balloon-borne observations using cryogenic broad-band radiometers which laid the technological groundwork for the second and present generation of high resolution cryogenic spectrometer observations. Some of the broad-band experiments were discussed in the Thaddeus (1972) review and the final results of these observations are listed in Table 2. The earliest experiments in the late 1960s were attended by controversy; the first rocket observations of the Cornell group showed a large energy excess at high frequencies, while the first multifilter radiometer

Table 2 Summary of broad-band high frequency observations: final results

Reference	Band (cm^{-1})	Altitude (km)	Beam size (deg.)	P (side lobe + warm parts) P (total)	P (atmosphere) P (total)	CBR (K thermodynamic)
Houck et al. 1972	7.7–25	144	~2	?	—	≤4.1
Williamson et al. 1973	1.7–12.5	325	20	?	—	$3.4^{+1.4}_{-3.4}$
	1.7–16.7				—	$5.1^{+0.8}_{-1.5}$
	1.7–33.3				—	$3.8^{+0.8}_{-1.9}$
Muehlner & Weiss 1973a,b	1–5.4	44, 39	10	<0.04	0.13	$2.7^{+0.4}_{-0.6}$
	1–7.8			<0.04	0.09	2.8±0.2
	1–11.1			<0.05	≧0.53	≦2.7
	1–18.5			<0.02	≧0.73	≦3.4
Dall'oglio et al. 1976	7.1–11.1	3.5	2	?	≧0.8	≦2.7

balloon experiments of the MIT group showed an excess in one of three spectral channels. By the early 1970s the rocket experiments, in particular those of the Los Alamos group and a new set of balloon observations of the MIT group using 5 spectral channels, settled the controversy. Furthermore, with the improved understanding of the atmospheric emission by 1973, the balloon experiments had established that there was a peak in the CBR spectrum and that the total power was consistent with the extrapolation of the ground-based low frequency measurements. The detailed shape of the spectrum was, however, not measured.

Some of the underlying problems in the early experiments are still manifested in the second generation experiments. The worst and unavoidable problem is just the fact that a fussy absolute observation must be carried out by remote control; the effect of this is that in the course of the experiment what might have been an easy calibration or test for a systematic error, or iteration in the experiment design, turns into a substantial and often expensive engineering project. The next order difficulty is due to interaction of three factors: small observation times, inadequate test facilities, and low sensitivity detectors. The nonthermal equilibrium environment—a 3 K sky, a hot earth below, and in the case of a balloon experiment, a warm and condensable atmosphere surrounding the apparatus—are difficult to simulate in terrestrial test facilities. As a consequence the final testing for systematic errors and in principle the calibration under the actual observing conditions must (or should) be carried out during the course of the experiment. The total observing time in a rocket observation is measured in minutes while for balloons it is approximately 10 hours. At very high altitudes where the atmospheric emission is smallest, the stratospheric winds are irregular and strong and the observing times may only be a few hours. With time so limited, a premium is put on having sensitive detectors to test for systematic error sources such as the contribution from warm parts of the surroundings and side-lobe response, to mention just two. The detectors have improved considerably. In the early experiments the detector noise equivalent power (NEP) was $\sim 10^{-11}$ to 10^{-12} W/Hz$^{1/2}$, just barely good enough to measure the total power in the CBR in seconds of integration time with systems of 10% optical efficiency and an étendue of 0.1–0.3 cm^2 sr. Present day detectors (composite bolometers) are a factor of 100 to 1000 more sensitive and the second generation experiments have made good use of them. But the fact remains that a table, such as Table 1, developed for the low frequency experiments, cannot be assembled for the high frequency experiments under observing conditions, and further, *none* of the high frequency experiments have experienced a primary calibration during flight.

All the high frequency spectroscopic observations of the CBR have employed Fourier transform spectrometers—Michelson interferometers. In these devices the incident radiation is split between two paths and recombined after one path is delayed relative to the other. The output signal as a function of the delay is the autocorrelation function of the radiation which, Fourier-transformed, becomes the power spectrum.

Fourier transform spectrometers are the logical choice for measuring the small power in the CBR with present day detector technology since they offer a wide spectral range, typically 2 decades, multiplexed on one or two detectors, and a large étendue, 0.1 to 0.5 cm² sr. The optical efficiency of the spectrometer in the millimeter and sub-millimeter range can approach 100% by using polarizing Michelson interferometers (Martin & Puplett 1969) in which the beam division is performed by a polarizer, an almost ideal optical component at these wavelengths. The frequency resolution is adjustable and limited by the maximum delay for which beam divergence or wave front distortion and displacement within the instrument destroys the relative spatial coherence of the two recombined paths. Resolutions of a few tenths of a wavenumber are typical for the instruments that have been used.

Fourier transform spectroscopy is not without hazards, the most serious being the opportunities for frequency distortion of the measured spectrum. Total intensity variations due either to fluctuations in the input signal (random noise) or incurred by the scanning (systematic errors), transform into spurious frequency components in the spectrum. Systematic spectrum distortions can also result from periodic errors in the mechanism that produces the delay. Careful design reduces these errors and thorough calibration can uncover them. A safe way to reduce their effect is to limit the optical bandwidth of the system but this of course limits the spectral coverage. The high optical efficiency, étendue, and multiplexed frequency coverage also put burdens on the detector linearity. Consequently calibration of the spectrometer requires care so as not to saturate the detector and is best carried out with cryogenic thermal sources.

The emittance measurable with a Fourier transform spectrometer is given qualitatively by

$$B_{\text{input}}(v) - B_0(v) = (S/N)(NEP)/\varepsilon_{\text{opt}} T(v) A\Omega \, \Delta v_{\text{res}} t^{1/2} \tag{14}$$

where $T(v)$, $A\Omega$, and ε_{opt} are the overall optical transfer function, étendue, and optical efficiency of the system t is the total observing time using a detector with a given NEP to achieve a signal to noise S/N, in estimating the emittance in a frequency resolution element Δv_{res}. $B_0(v)$ is the thermal radiation by the interferometer enclosure or an absorbing surface entering the other port of the interferometer or incident on the detector during the alternate half of the chopping cycle when the instrument "looks" at itself. The contribution from this term produces an offset which, if constant throughout the measurement, can be determined from the external absolute calibration.

The design of the three high frequency spectroscopic instruments used in CBR measurements are shown in Figures 4, 5, and 6. The Queen Mary College balloon-borne instrument (Beckman & Robson 1972, Beckman et al. 1974, Robson et al. 1974), Figure 4, is a cryogenic polarizing Michelson interferometer operated at 1.4 K at 40 km altitude. The instrument parameters are 0.1 cm^2 sr étendue, 0.25 cm^{-1} unapodized resolution, 6° beam width maintained at a 50° zenith angle during flight. The beam was defined by the 45° off-axis paraboloid collimator at the entrance of the interferometer section after the beam had traversed a warm 50μ polyethlyne window (not shown in the figure) and a stationary entrance polarizer followed by a rotating polarizer used as a chopper. Two InSb hot electron bolometers detect the beams reflected and transmitted by the output polarizer yielding two interferograms 180° out of phase. The technique of chopping with a rotating polarizer eliminates the signal that is not modulated by the delay and thereby reduces one opportunity for frequency distortion of the spectrum. The overall system transfer function has not

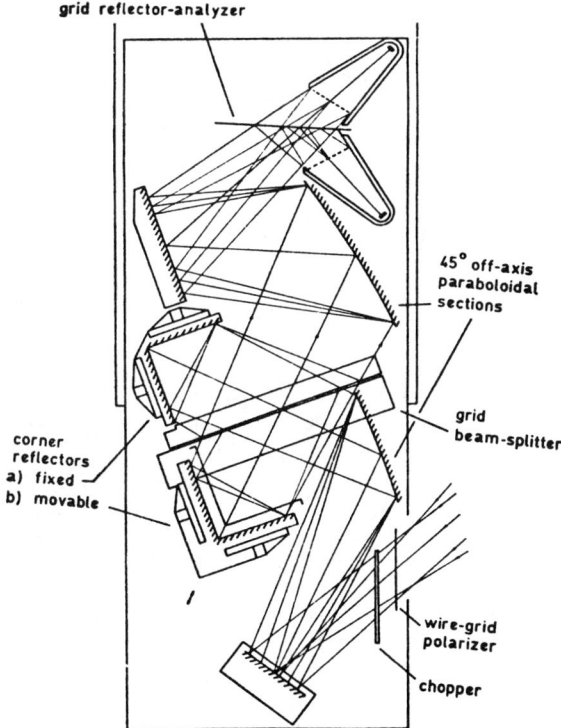

Figure 4 The Queen Mary College polarizing Michelson Interferometer.

been published but must be dominated by the high frequency roll off of the InSb detectors beginning near 10 cm^{-1} varying as $1/\nu$ and then further attenuated above 40 cm^{-1} by the reduced efficiency of the polarizers as the wavelength approaches the polarizer grid spacing (100μ).

I have difficulty in evaluating the results of the experiment, Figure 7, as there are inconsistencies in the data and the accounting of the radiation incident on the instrument is incomplete. The emission by the warm polyethlyne window, which is not removed during the observation, is unknown and may never be known as it is possible that frost condensed on it. An estimate of the atmosphere emission has not been published. However, an attempt (on my part) to model the atmosphere as observed in this experiment, especially in the critical minimum between 8 and 18 cm^{-1}, fails by a substantial margin. The contribution of the ground in the beam side lobes due to diffraction by edges at the entrance optics has not been published; finally, the calibration of the instrument in flight is not established.

The Berkeley balloon-borne experiment, Figure 5 (Mather 1974, Mather et al. 1974, Woody 1975, Woody et al. 1975, Woody & Richards 1979), has been flown twice successfully with improvements in the apparatus between

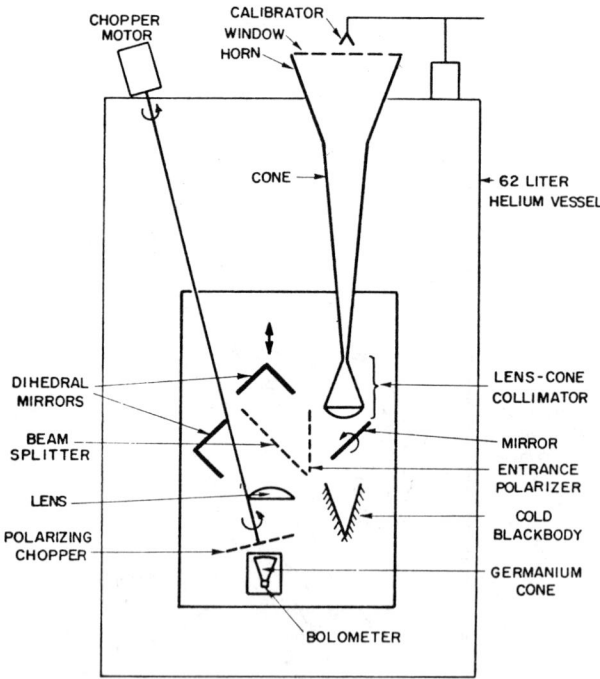

Figure 5 The Berkeley polarizing Michelson Interferometer.

flights. The instrument is again a polarizing Michelson using germanium bolometers. A composite bolometer maintained at 0.3 K (NEP $\leq 1 \times 10^{-15}$ W/Hz$^{1/2}$) was used in the second flight, which enhanced the sensitivity of the instrument by a factor of 10. The instrument parameters are 0.23 cm^2 sr étendue, maximum unapodized resolution 0.14 cm^{-1}, 6° beam width, and an overall optical efficiency in mid-band $\sim 1\%$. A significant feature of the design is the beam-forming cone, held at liquid helium temperature, which has a demonstrated and calculable off-axis rejection of 10^6 at 60° and greater for larger angles. In the second flight an even better cone was used in concert with ground shields. Zenith scanning measurements carried out during flight demonstrated that the large angle contributions from the earth were less than $5 \times 10^{-13} \nu^{1/2}$ W/cm^2 sr cm^{-1}. In flight, the window over the cone was removed. The primary calibration was carried out on the ground with cryogenic calibration sources placed in the beam-forming cone. Calibration in flight, however, was estimated with an ambient temperature

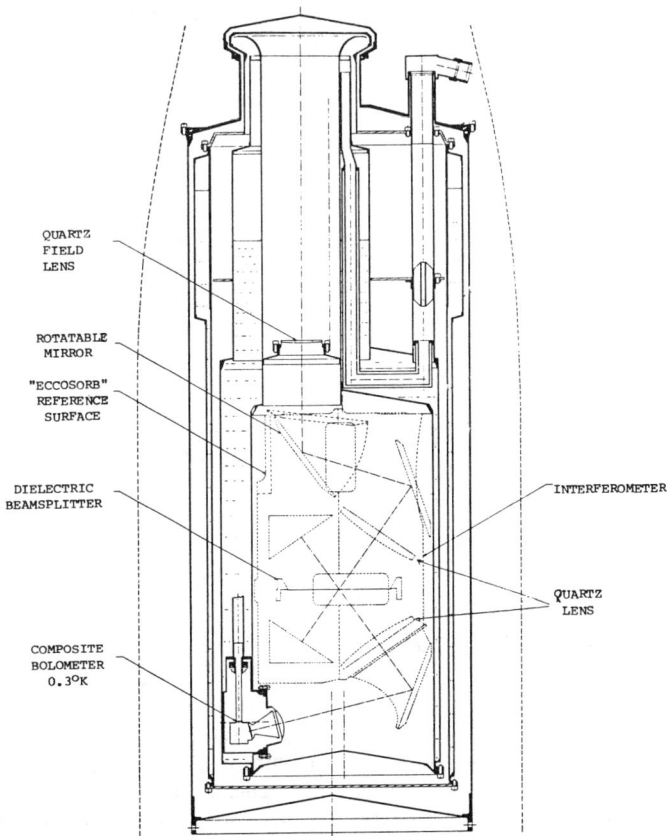

Figure 6 The rocket-borne Michelson interferometer of Gush.

secondary calibrator that could be swung into the beam. The cold blackbody source inside the interferometer at the temperature of the interferometer enclosure served to establish the offset.

The data of the second flight, Figure 8, are the most precise measurements of the CBR spectrum to date at high frequencies. (The data of the first flight is much the same with decreased signal to noise.) Panel a is the instrument transfer function which has been rolled off at high frequencies by a low pass filter at the detector. Panel b is the total measured spectrum uncorrected for the instrument transfer function while c is the fitted atmospheric emission spectrum using the techniques described previously. The scheme used by the group was to fit the interferogram rather than the spectrum to the atmospheric model. This streamlines the computations, eliminating the need to calculate the convolution of the instrument response function with the fitted spectrum. Panels d and e are the residuum, in principle the CBR at 0.28 and 1.8 cm^{-1} resolution. The data corrected for the instrument transfer function are given in numerical form in Table 3 (D. P. Woody, 1979, private communication).

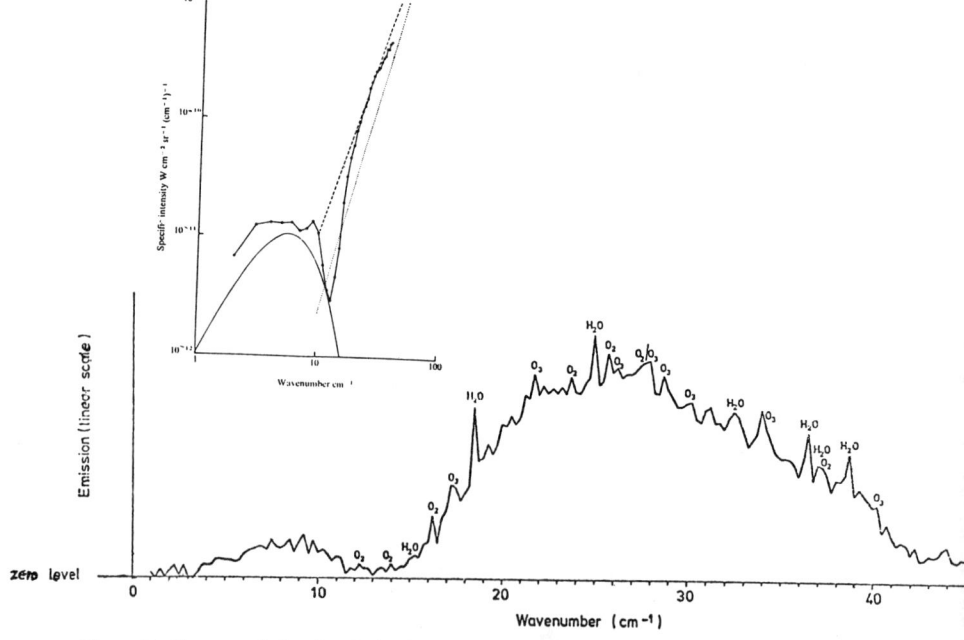

Figure 7 Spectra of the sky derived from the Queen Mary College instrument. The lower trace is the total measured spectrum including the CBR, an unknown window emission, and the atmosphere at 40 km (50° zenith angle) uncorrected for the instrument transfer function and without an absolute vertical scale. The unapodized resolution of the spectrum is 0.25 cm^{-1}. The insert is the corrected spectrum including several estimates of the window emission. The solid line is the difference between a 2.7 K and a 1.4 K spectrum.

The Berkeley data deserves close scrutiny because the major terms in the accounting are measured. Treating the errors in flux determination at each frequency as independent, the best fit to a Planck spectrum, with temperature as the only fitted parameter, is $T_{CBR} = 2.96$ K at only a 35% χ^2 contour—a 1 in 3 chance that data with the assumed random errors would fit as poorly as it does if the true spectrum were thermal at this temperature. In other words, the deviations from a thermal spectrum at 2.96 K are statistically significant. The question is, are these deviations real or artifacts of observation? Caution driven by experience dictates that we be hard on the experiment. The weakest link in all the high frequency CBR spectrum experiments is in the absolute calibration, not so much in the spectral

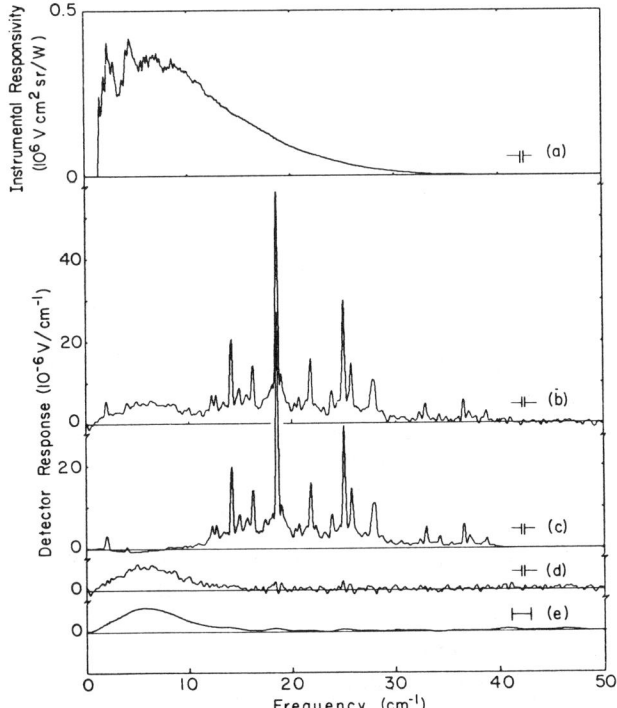

Figure 8 The Berkeley spectrum. Panel *a* shows the instrument transfer function. Panel *b* shows the total measured spectrum uncorrected for the instrument transfer function. The spectra are derived from asymmetric interferograms of 3.72 and 0.48 cm optical delay on either side of the central maximum. The apodization used is $(1-x^2/L^2)^2$. The resolution of the larger optical delay is 0.28 cm^{-1} but with wings from the shorter delay. Panel *c* is the best fit of *b* to the model atmosphere and instrument offset. Panel *d* shows the residual and *e* the same residual with 1.8 cm^{-1} resolution. The lines at 12, 14, and 16 cm^{-1} are a group of the O_2 pure rotational lines discussed in the text.

Table 3 Tabular results of the Berkeley experiments (D. P. Woody, 1979, private communication)

ν (cm^{-1})	Average flux (10^{-12} w/cm^2 sr cm^{-1})	1σ flux limits		$1\sigma\ T_{CBR}$ limits (K)	
		min.	max.	min.	max.
2.38	8.74	7.76	9.99	3.05	3.57
3.40	12.13	11.02	13.67	2.95	3.29
4.41	14.81	13.65	16.51	2.97	3.22
5.42	16.05	14.81	17.89	2.97	3.18
6.44	16.09	14.86	17.94	2.98	3.16
7.45	13.89	12.76	15.57	2.91	3.07
8.46	12.42	11.36	13.98	2.92	3.07
9.48	8.63	7.71	9.89	2.79	2.94
10.49	6.64	5.21	7.75	2.76	2.91
11.50	5.52	4.48	6.74	2.75	2.95
12.52	4.07	2.69	5.47	2.66	2.97
13.53	4.92	2.82	7.15	2.80	3.23
15.20	1.87	—	4.10	—	3.15
17.28	0.96	—	3.07	—	3.27
20.03	0.70	—	1.81	—	3.36
22.89	0.48	—	5.76	—	4.21

response of the instrument but rather in the overall frequency-independent scale factor that converts the measured signal into an absolute power incident on the instrument.

In the Berkeley experiment, primary calibration before and after flight agreed but the instrument was not directly calibrated with the same precision during flight, the properties of the secondary ambient temperature calibrator not being well enough known. If one then assumes (without specific justification) that the instrument behaved differently in flight, one might try a two-parameter fit to the data, fitting to a thermal spectrum and a frequency-independent overall calibration factor. The results of this procedure are $T = 2.79$ and a calibration factor of 1.27. The χ^2 contour is close to 90%, a very good fit. Woody & Richards (1979) point out this fact and argue that a 27% calibration error is inconsistent with the emission measured by the atmospheric O_2 lines. The argument hinges on the fact that emission by O_2 is the one atmospheric contribution that can be calculated directly. The calculation is built on the following assumptions: 1. the O_2 column density (uniformly mixed) is given by the barometric pressure at altitude, 2. the column temperature distribution is known, 3. the line shapes are known, and finally 4. the tabulated line parameters are correct.

In preparing this review I redid the O_2 calculations independently and agree with the Woody & Richards result using the current best theoretical estimates of the line strengths and line widths of the magnetic dipole rotational transitions ($\Delta K = \pm 2$) (Greenebaum 1975, Liebe et al. 1977, Weiss 1980). However, there is little direct experimental evidence of the line strengths, and in particular the pressure widths, except for measurements of insufficient absolute precision by Gebbie et al. (1969), which incidentally give smaller line strength than the theoretical values. Furthermore, the Zeeman effect of O_2 in the earth's magnetic field must be treated in detail and this is still in progress. The O_2 fine structure line, $\Delta K = 0$, at 4 cm^{-1} is far better known and would serve as a more reliable calibration but the instrument transfer function is changing rapidly at this frequency. The conclusion is that there are too many uncertainties in the calculation to allow an *absolute* calibration to a precision of 30% using the atmospheric emission.

The rocket-borne instrument of Gush (1979), Figure 6, performed the first spectrometer observations of the CBR above the atmosphere. The observations took place in the approximately 7 minutes while the payload was between 370 and 150 km altitude. The instrument is a conventional but cryogenically cooled Michelson interferometer using a dielectric pellicle beam splitter and a composite bolometer operated at 0.3 K. The interferogram is developed by varying the delay at a constant rate without additional beam chopping (rapid scan mode of operation). The instrument transfer function is determined by the reflection spectrum of the beam splitter at long wavelengths, becoming more inefficient the longer the wavelength, and by both the beam splitter and the high audio frequency roll-off of the detector and associated electronics at short wavelengths. The optical bandwidth of the interferometrically unmodulated component of the input signal is large, extending past 35 cm^{-1}, and this may have aggravated one of the difficulties experienced in the experiment. The instrument was calibrated before and after flight with a cryogenic reference source.

Unfortunately the results of the experiment, Figure 9, are tentative because the analysis of the data requires modeling of a significant time-varying radiative contribution from the rocket clam shell covers that moved unexpectedly to the edges of the spectrometer beam. Another time-varying and significant correction had to be made for the temperature variation of the entire instrument due to venting of helium efflux gas during the course of the flight.

Figure 10 summarizes the direct measurements of the CBR spectrum and includes the determinations using the excitation temperature of CN observed in molecular clouds at 3.79 and 7.58 cm^{-1} (Thaddeus 1972, Hegyi et al. 1974, Danese & DeZotti 1977). The figure gives the thermodynamic

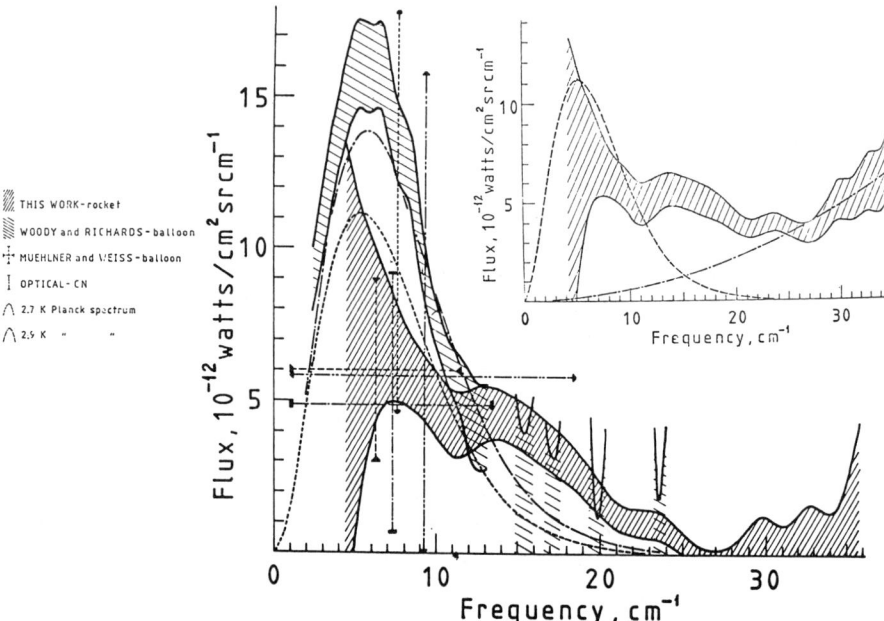

Figure 9 Rocket data of Gush. The inset shows the measured spectrum corrected for the instrument transfer function. The spectrum includes a contribution from hot objects at the edges of the beam modeled by the dot-dash line. The larger figure is a composite of various measurements of the CBR spectrum at high frequency.

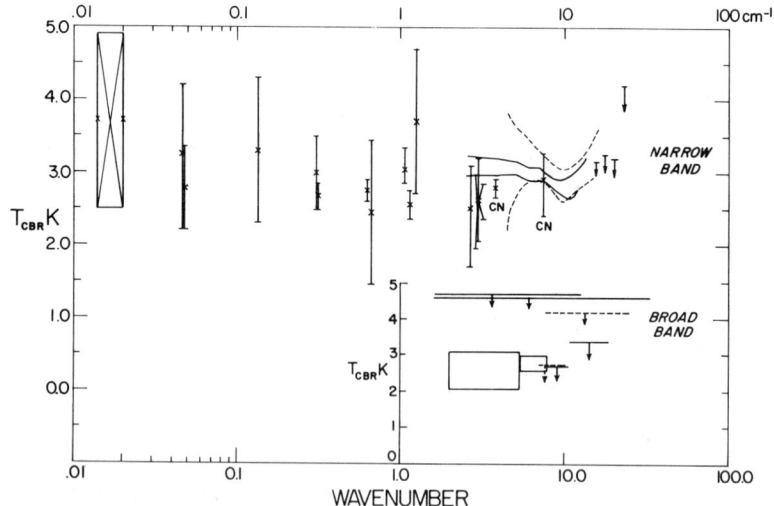

Figure 10 Summary of the measurements of the thermodynamic temperature of the CBR as a function of frequency.

temperature of a blackbody source with the same flux as that measured. The uncertainty in flux is related to the uncertainty in temperature by

$$\frac{\Delta T}{T} = \frac{x}{[\ln(1+x)](1+x)} \frac{\Delta B}{B},$$

$$x = (e^{h\nu/kT} - 1). \tag{15}$$

In summary, the overall shape of the spectrum is thermal, characterized by a single parameter, the temperature. Considering the moderate precision of the low frequency experiments and the uncertainties in the high frequency observations, it is, in my opinion, too early to fruitfully analyze the present data for the subtle distortions that may discriminate different cosmic histories. Another generation of observations is needed.

CBR measurements using optical spectroscopy to probe the state of excitation of interstellar molecules equilibrated to the background radiation have not advanced since 1974. It would be delightful to observe the CBR at different locations in the universe and at different times by in situ measurements using molecules in other galaxies. The idea may be fanciful but not entirely impossible.

MEASUREMENTS OF THE LARGE ANGULAR SCALE INTENSITY DISTRIBUTION OF THE CBR

The most convincing evidence that the background radiation is of cosmological origin—due to very distant sources if not the primeval explosion—is its isotropy, or more explicitly, that it does not share the anisotropic distribution associated with local sources such as the solar system or the Galaxy. The CBR, in fact, exhibits such a high degree of isotropy on both small ($<1°$) and large angular scales, short of the dipole term, as to pose a quandary. On small angular scales, no evidence has yet been found for granularity, thereby stringent limits are set on any discrete-source hypothesis for the origin of the background. On the other hand, at some level, still not observed, small scale anisotropies must occur as density perturbations in the early universe are believed to have preceded the presently observed aggregations. The status of these observations is not part of this review [see the article by Partridge in Ulfbeck (1980)]. On large angular scales ($\geq 10°$), the anisotropy of the CBR is less than 1/3000, limited only by measurement uncertainties, lending strong support to the naïve assumptions of homogeneity and isotropy made in developing the spherically symmetric cosmological models, but leaving the puzzle of how unconnected regions of the universe could have evolved so identically.

In addition to the search for evidence of intrinsic cosmic anisotropies, such as might be due to aspheric expansion or large scale anisotropies in the primeval matter distribution, an impetus for the large angular scale isotropy experiments has been the measurement of the kinematic anisotropy associated with the earth's motion relative to the distant sources of the CBR. This anisotropy had to exist at least on a level of 10^{-4} due to the earth's orbital motion around the sun in the unlikely case that the sun were at rest with respect to these sources. The anisotropy, now definitely observed at a 10^{-3} level is easily derived by a special relativistic calculation of the intensity measured by an observer moving relative to the walls of a blackbody cavity. The anisotropy retains a Planck spectrum but with an observation angle–dependent temperature (dipole term) given by

$$T(\theta) = T_0(1-v^2/c^2)^{1/2}/1 - v/c \cos \theta \sim T_0(1+v/c \cos \theta), \quad v/c \ll 1,$$

where T_0 is the temperature observed in a frame at rest with respect to the sources, v is the velocity of the observer, and θ is the angle between the observing direction and the velocity.

Kinematic anisotropies of higher order (quadrupole) could exist if the universe rotates (Collins & Hawking 1973) or is filled with long wavelength gravitational radiation (Burke 1975). In general, the intensity distribution expanded in spherical harmonics is a convenient method of interrelating observation and theory in this almost spherically symmetric system. To date, no positive evidence exists for moments higher than the dipole; however, as the measurement sensitivity improves and, in particular, the sky coverage becomes more complete, interesting limits (if not measurements) can be set on the higher moments. Both extended sky and spectral coverage will become essential to unscramble the effect of local sources which are expected (Figure 3) to become important at 10^{-4} anisotropy levels even in the clearest region of the CBR spectrum. The measured spectrum of the dipole anisotropy is close to thermal. If this holds for higher order anisotropies, it becomes a discriminant against the local sources with their own characteristic spectra. Sky coverage is also required to break the correlation in estimates of the amplitude of different multipole moments. Higher order moments, if they exist, corrupt estimates of the lower order moments in fitting the multipole expansion with a data set derived from partial sky coverage.

Isotropy experiments do not have the same character as those designed to measure the CBR spectrum. The major difference is, of course, that isotropy experiments make relative measurements. The quantity measured is the difference in intensity from different directions; absolute calibration, therefore, plays a secondary role. Furthermore, because one is looking at a fixed structure in the celestial sphere, the differential data can be tested for internal consistency. Apparatus- and environment-generated anisotropies can, in principle, be removed by observing the same celestial locations under different experimental conditions. In other words, the data alone, without further accounting as in the spectrum observations, can reveal whether or not the experiment is in trouble. The ultimate systematic errors come from local astrophysical sources, in particular from extended objects or regions of low surface brightness which may have eluded less sensitive sky surveys.

Receiver or detector sensitivity is at a premium in these experiments because the CBR anisotropies are small and tests for systematic noise in the instruments have to be carried out at the same low levels. Present day receivers at 20–30 GHz have sensitivities (Equation 1) ~ 50 mK/Hz$^{1/2}$. Observation of a systematic effect at the 10^{-3} anisotropy level requires ~ 10 minutes of integration time. In the near future, Maser preamplifiers will be used at 20 GHz, increasing the sensitivity by a factor of 10 to 20 (S. Gulkis and D. T. Wilkinson, private communication). Broadband mm and sub-mm incoherent radiometers using composite bolometers have achieved sensitivities ~ 1 mK/Hz$^{1/2}$ (2.7 K). However, at these high frequencies the galactic dust background (systematic errors) and the residual atmospheric fluctuations, even at balloon altitudes (random errors), intrude.

Table 4 presents the results and some of the characteristics of large scale anisotropy experiments. All experimets, except those of Penzias & Wilson (1965) and Wilson & Penzias (1967), who noted in their discovery paper that the background was isotropic to 10%, have been carried out by comparing the intensity from one part of the sky with that from another without reference to an absolute calibrator. The early ground-based experiments were plagued by the large galactic background at long wavelengths and atmospheric emission fluctuations at shorter wavelengths where the galactic background is smaller. In these experiments, the radiation from two directions at the same elevation angle (to equalize the average atmospheric contribution) is compared as the earth's rotation sweeps the beams over the sky. Various scanning strategies were used. Partridge & Wilkinson (1967) compared the radiation from the celestial equator with a fixed point at the celestial pole using a single beam switched by a large mirror. This experiment established that the CBR anisotropies were less than a few parts in a thousand. Conklin (1969, 1972) compared the radiation entering two horns set at 30° to the zenith (fixed declination 32°), one pointing toward the east, the other toward the west. The horns were periodically interchanged in order to search for systematic differences in the apparatus. This experiment was the first to report a 24 hour anisotropy but the result was tentative because the large galactic contribution had to be modeled by extrapolation with a poorly known spectral index from 400 MHz sky maps. The observation, which measured the East–West com-

Table 4 Results of large scale anisotropy experiments

Reference	λ (cm)	f (GHz)	ν (cm^{-1})	Altitude (km)	Receiver noise (mK/Hz$^{1/2}$)	Beam width (deg.)	ΔT_{rms} mK / Beam width	ΔT mK (24 hr)
Wilson & Penzias 1967	7.35	4.08	0.14	0	—	—	<100	—
Partridge & Wilkinson 1967	3.2	9.4	0.31	0	—	10	~10	1 ± 2.2
Conklin 1969	3.75	8.0	0.27	3.8	65	12	—	1.6 ± 0.8 (projected $\delta = 32$)
Conklin 1972								2.3 ± 0.9 equatorial
Boughn et al. 1971	0.86	35	1.16	0	1600	4	~50	7.5 ± 11.6
Henry 1971	2.96	10.1	0.34	24	~250	15	3	3.2 ± 0.8
Corey & Wilkinson 1976 Corey 1978	1.58	19	0.63	25	100	~10	3	2.9 ± 0.7
Muehlner & Weiss (Muehlner 1977)			3–10	39	90	18	3	<3.3
Smoot et al. 1977 Gorenstein 1978	0.9	33	1.11	20	44	7	1.5	3.5 ± 0.6
Cheng et al. 1979	1.21	24.8	0.83	27	51	8	1.5	2.99 ± 0.34
Analysis includes Corey 1978	0.955	31.4	1.05	27	45	6	1.5	
Smoot & Lubin 1979	0.9	33	1.11	20	44	7	1 (95% CF)	3.1 ± 0.4
Muehlner & Weiss 1980	—	—	3–10	39	90	18	3	2.8 ± 0.8

Note: Since this review was written the measurement of the large scale anisotropy in the 3–20 cm^{-1} band has been reported by Fabbri et al. (1980). The data imply a dipole anisotropy of amplitude $T_D = 2.9^{+1.3}_{-0.6}$ mK in the direction $\alpha_D = 11$ h 24 m ± 40 m, $\delta_D = 3 \pm 10°$. The data require a fit to a quadrupole-like term of amplitude $0.9^{+0.2}_{-0.2}$ mK.

COSMIC BACKGROUND RADIATION 523

Reference	ΔT mK (24 hr galactic correction)	RA_D (hr)	δ_D	ΔT mK (12 hr)	Sky coverage
Wilson & Penzias 1967	—	—	—	—	29 Locations distributed $70° > \delta > -20°$ $2\,hr < \alpha < 20\,hr$
Partridge & Wilkinson 1967	—	—	—	4.9 ± 2.0	Circle, $\delta = -8°$, vs celestial pole
Conklin 1969	2.9	$13 ^{+1.9}_{-2.3}$	—	—	Circle, $\delta = 32°$, East-West difference 60° apart
Conklin 1972	—	11	—	—	
Boughn et al. 1971	—	—	—	5.5 ± 6.6	Circle, $\delta = 0°$, beam switched $\pm 2.1\,hr$ RA Celestial pole reference
Henry 1971	<0.5	10.5 ± 4	-30 ± 25	—	North-South only $3\,hr < \alpha < 15\,hr\ -13° < \delta < 77°$
Corey & Wilkinson 1976 Corey 1978	<0.3	12.3 ± 1.4	-21 ± 21	—	$12\,hr < \alpha < 20\,hr$ $-13° < \delta < 77°$
Muehlner & Weiss (Muehlner 1977)	—	—	—	—	$4\,hr < \alpha < 6\,hr$ $15\,hr < \alpha < 23\,hr$ $-13° < \delta < 77°$
Smoot et al. 1977	<0.5?	11.0 ± 0.5	6 ± 10	<1	22 spots $0\,hr < \alpha < 18\,hr$ $5° < \delta < 65°$
Gorenstein 1978 Cheng et al. 1979 (Analysis includes Corey 1978)	<0.3	12.3 ± 0.4	-1 ± 6	$Q(\alpha,\delta)$ <2	$12\,hr < \alpha < 26\,hr$ $-13° < \delta < 77°$
Smoot & Lubin 1979	—	11.4 ± 0.4	9.6 ± 6	$Q(\alpha,\delta)$ <1 $Q(\alpha,\delta)$	14 spots $4\,hr < \alpha < 17\,hr$ $-50° < \delta < 25°$
Muehlner & Weiss 1980	<1	9.6 ± 1.5	-9 ± 20	$Q(\delta)$	$14\,hr < \alpha < 31\,hr$ $-13° < \delta < 77°$

ponent of the anisotropy, agrees with newer measurements, but it took courage to publish the result at the time. Boughn et al. (1971) attempted an isotropy measurement at 35 GHz, where the galactic contribution is small. The instrument was a differential radiometer with one beam (the reference) pointed to the zenith and the other switched periodically between two points on the celestial equator 2.1 hr either side of the meridian plane. Atmospheric emission fluctuations dominated the noise budget, partly aggravated by the experiment design, but nevertheless highlighting one of the main difficulties with ground-based large angular scale anisotropy experiments. The sensitivity to a dipole anisotropy increases as the sine of the beam separation. However, the effect of atmospheric emission fluctuations grows as well, since the fluctuations become more independent the larger the beam separation.

The definitive observations have all been carried out above the bulk of the atmospheric emission at balloon or jet airplane altitudes. The balloon-borne experiments by Henry (1971), Corey & Wilkinson (1976), Muehlner & Weiss (1977, 1980), and Cheng et al. (1979) have been similar in concept (Figure 11). The intensity in two beams 90° apart, bisected by the zenith and separated by 180° in azimuth, is differenced at a rapid rate (10–200 Hz). The beams are formed with low side-lobe horns and further protected from radiation by the ground and the lower atmosphere with a large ground shield that reflects the sky. The entire instrument is set into rotation about the zenith at about 1 revolution per minute. The rotation serves both to scan the sky and to allow measurement of the intrinsic anisotropy of the apparatus. The azimuth is determined with magnetometers using the earth's field as a reference. Provision is made to measure or eliminate the systematic noise terms that might be synchronous with the rotation. The most serious of these have been magnetic interactions in the ferrite components of the microwave receivers which perturbed the results of Henry (1971). Other effects are rotation-induced wobbles of the apparatus or pendulation of the instrument which cause an azimuth-dependent variation in elevation angle of both beams, thereby modulating the atmospheric emission. The wobbling motions are measured with tilt sensors and pendulation is determined from the magnetometers. The effect of these motions can also be determined by using a differential radiometer at a frequency where the atmospheric emission is stronger but not saturated (Muehlner 1977). Finally, although balloons follow the stratospheric winds, wind shears do occur and cause differential temperature variations in the instrument as it rotates under the balloon. These are measured with sufficient precision using differential thermometers. Typical balloon flights last 8–10 hours, the balloons traveling at almost constant declination. With the given beam configuration, about 1/4 of the sky can be covered in a single flight.

COSMIC BACKGROUND RADIATION   525

The signal is the difference in intensity measured in the two beams as a function of azimuth. The analysis is made by averaging this signal over a set of rotations occurring in the time it takes the celestial sphere to move about 1/2 a beam width. These averages are then expanded in a harmonic series in multiples of the rotation frequency. The fundamental component includes

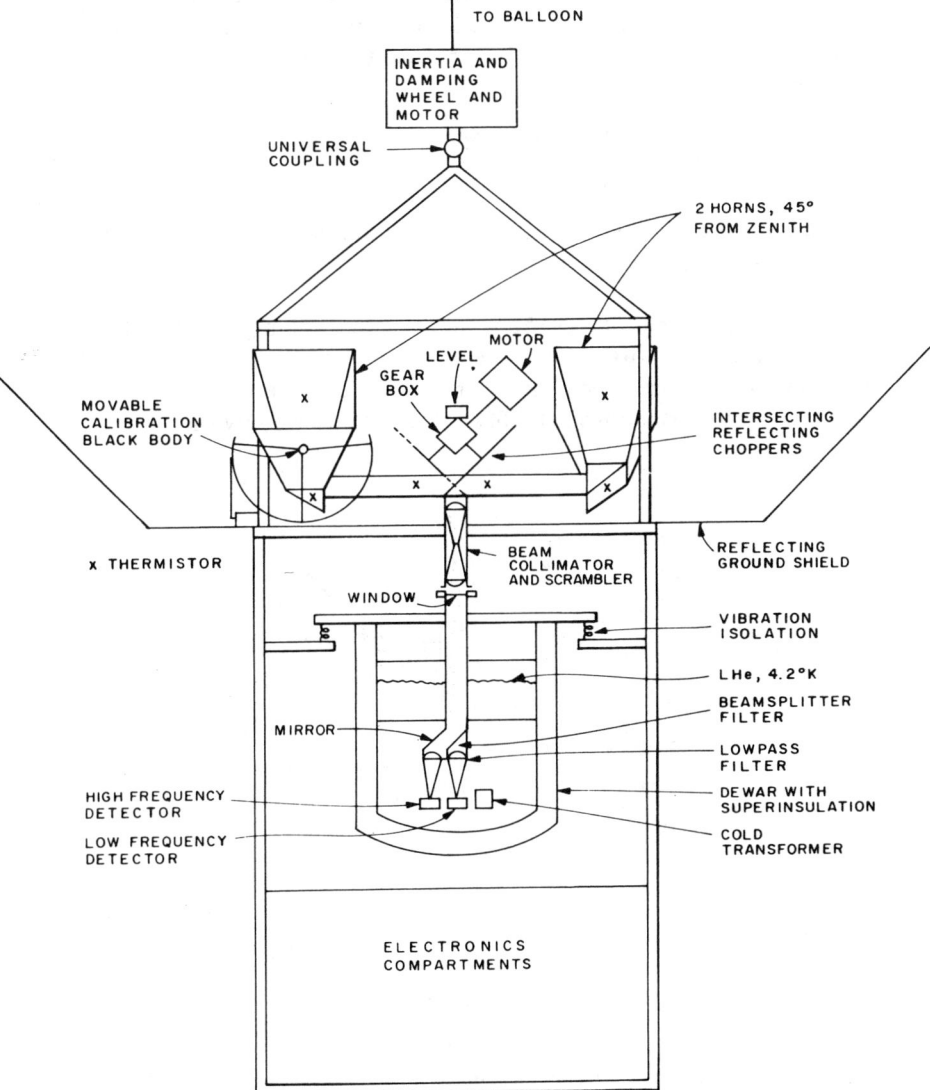

Figure 11 A typical balloon-borne large angular scale isotropy experiment. This one is the M.I.T. instrument. The more successful Princeton balloon-borne instrument is similar in concept.

information on all multipole moments of the intensity distribution; higher odd harmonics exclude the lower order anisotropy moments and are a diagnostic for discriminating discrete sources. With beams 180° apart in azimuth, the apparatus is insensitive to even harmonic components although signals at these frequencies are valuable in diagnosing internal difficulties such as radio frequency interference and microphonics.

A dipole anisotropy of amplitude T_d pointing along α_d (RA) and δ_d (dec) would produce a fundamental component with polar and equatorial projections given by

$$\Delta T_{\text{N-S}} = 2T_d \sin \text{EA} \,[\sin \delta_d \cos \delta - \cos \delta_d \sin \delta \cos (\alpha - \alpha_d)]$$
$$\Delta T_{\text{W-E}} = 2T_d \sin \text{EA} \,[\cos \delta_d \sin (\alpha - \alpha_d)], \tag{16}$$

where α and δ are the right ascension and declination of the zenith and EA is the elevation angle of the beams (45°). The best fit of Equation (16) to the data of Chang et al. (1979) is shown in Figure 12.

The balloon-borne observations were begun by Henry (1971) at 10 GHz. He measured the North–South component of the dipole anisotropy. The first definitive measurement (sufficient sky coverage and signal to noise) is that of Corey & Wilkinson (1976) at 19 GHz followed by the observation

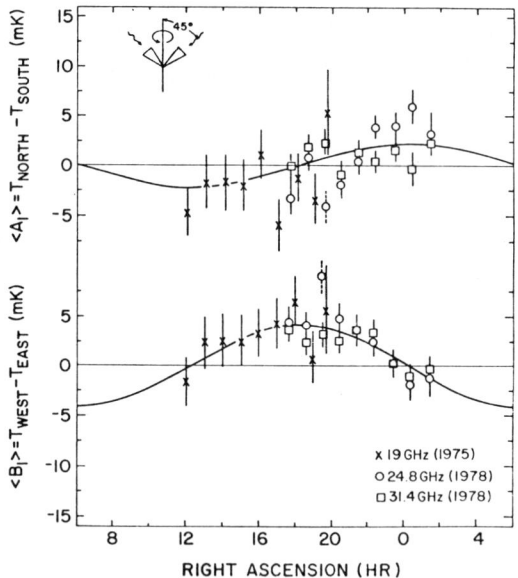

Figure 12 The Princeton anisotropy measurements using a balloon platform. The best fit to a dipole anisotropy is shown as the solid line.

using the U2 airplane (Smoot et al. 1977) to be discussed below. Isotropy measurements in the spectral band embracing the blackbody peak, 3–10 cm^{-1}, were carried out from balloons by Muehlner and Weiss (Muehlner 1977). The data, perturbed by galactic dust emission, could only be used to set an upper limit on the dipole anisotropy. With improved knowledge of the dust sources (Owens et al. 1979), the high frequency data has been reanalyzed, exhibiting a dipole anisotropy consistent with the lower frequency measurements (Muehlner & Weiss 1980). The spectrum of the dipole anisotropy appears to be close to Planckian.

The group at Berkeley (Smoot et al. 1977) has made good use of the U2 airplane as a platform to measure the large scale anisotropy. Although the airplane does not attain as high an altitude as the balloon, it offers some logistic advantages over the balloon, in particular in the relative ease of launching flights. The U2 instrument, Figure 13, is similar to the balloon-borne differential radiometers. It operates at 33 GHz and includes a 54 GHz radiometer to monitor the tilt of the airplane in the atmosphere. The radiometer beams are separated by 60° and are switched at 100 Hz. The entire apparatus is turned periodically within the airplane housing to measure intrinsic instrument anisotropies. Finally, the flights are arranged

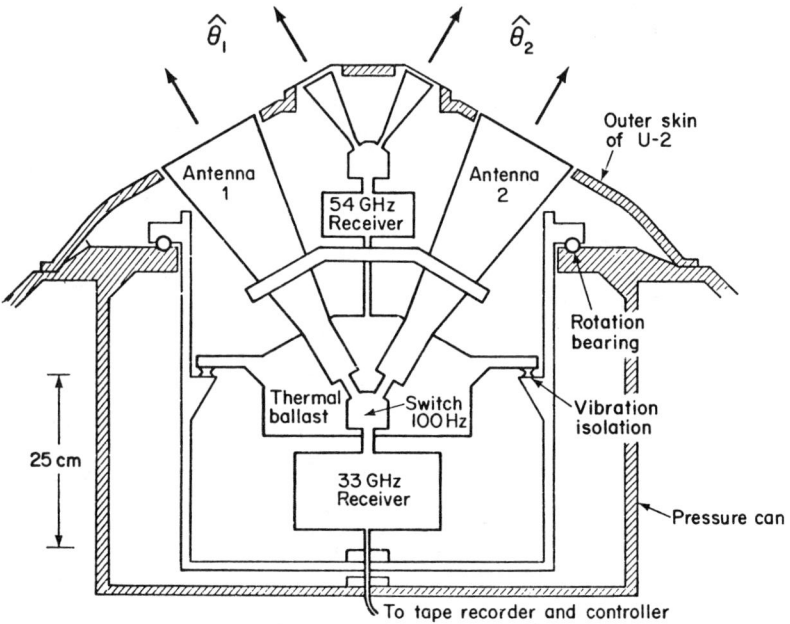

Figure 13 The U2 differential radiometer.

so as to observe the same regions of the sky with the airplane reversed in direction to account for anisotropies that might be fixed in the airplane coordinate system.

The observing and data analysis strategy (Gorenstein 1978, Gorenstein & Smoot 1980) is different than in the balloon experiments. Observation points in the sky are selected, Figure 14b and d, and the difference in intensity of the two beams is fit to a dipole distribution in a celestial coordinate system. The presentation of the data and the fit, Figure 14a and c, superposes all points that make the same polar angle with the dipole axis. Figure 14, as it stands, cannot be used to make a unique correspondence with a sky map.

With the enhanced precision and more extended sky coverage of both the U2 (Smoot & Lubin 1979) and balloon experiments (Cheng et al. 1979), limits have been set on the five independent terms of a quadrupole moment

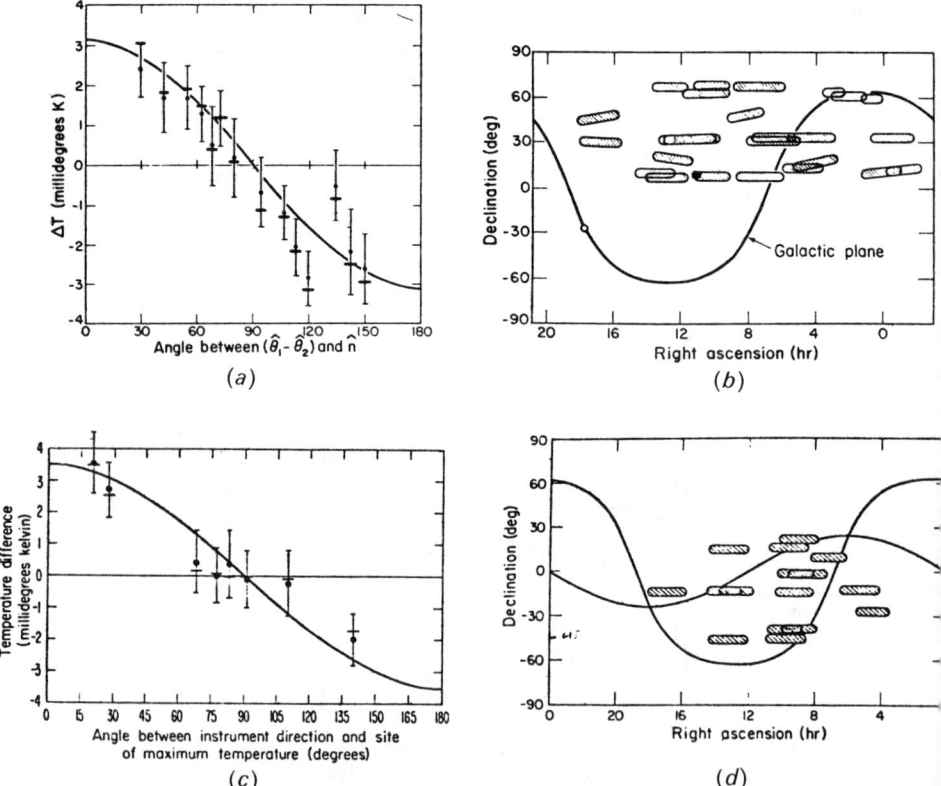

Figure 14 The Berkeley anisotropy data using the U2 airplane. Panel *a* and *c* show the fit to the same dipole anisotropy using the data derived from sky coverage shown in *b* and *d*.

Table 5 Multipole fits of Berkeley and Princeton data (Lubin & Smoot 1979, Cheng et al. 1979)

Quadrupole and dipole fit

Princeton:
$$T = \mathbf{T}\cdot\mathbf{n} + \sum_{m=-2}^{2}(a_{2m}+ib_{2m})Y_{2m}$$

Berkeley:
$$T(\alpha,\delta) = T_0 + T_x \cos\delta\cos\alpha + T_y \cos\delta\sin\alpha + T_z \sin\delta$$
$$+ Q_1(\tfrac{3}{2}\sin^2\delta - \tfrac{1}{2}) + Q_2 \sin 2\delta \cos\alpha + Q_3 \sin 2\delta \sin\alpha$$
$$+ Q_4 \cos^2\delta \cos 2\alpha + Q_5 \cos^2\delta \sin 2\alpha \quad \alpha = \text{RA},\ \delta = \text{DEC}$$
$$Q_1 = \sqrt{5/4\pi}\,a_{20},\quad Q_2 = -\sqrt{15/8\pi}\,a_{21},\quad Q_3 = \sqrt{15/8\pi}\,b_{21},$$
$$Q_4 = \sqrt{15/8\pi}\,a_{22},\quad Q_5 = -\sqrt{15/8\pi}\,b_{22}$$

	Berkeley	Princeton
T_x (mK)	-2.78 ± 0.28	-3.27 ± 0.57
T_y	$+0.66 \pm 0.29$	-0.17 ± 0.73
T_z	-0.18 ± 0.39	$-0.10 \pm 0.72\ (T_z + 1.125 Q_1)$
Q_1	$+0.38 \pm 0.26$	—
Q_2	$+0.34 \pm 0.29$	$+0.22 \pm 0.50$
Q_3	$+0.02 \pm 0.24$	$+0.26 \pm 0.67$
Q_4	-0.11 ± 0.16	-0.05 ± 0.36
Q_5	$+0.06 \pm 0.20$	-0.22 ± 0.38
Dipole fit only		
T_x	-3.01 ± 0.24	-2.98 ± 0.30
T_y	$+0.39 \pm 0.25$	-0.24 ± 0.30
T_z	$+0.52 \pm 0.23$	-0.06 ± 0.31
T mK	3.1 ± 0.4	2.99 ± 0.34
$\delta°$	9.6 ± 6	-1 ± 6
α hr	11.4 ± 0.4	12.3 ± 0.4
$v(2.7\text{ K})$ km s^{-1}	344 ± 44	332 ± 38
f(GHz)	33	19, 24.8, 31.4

of the CBR intensity distribution, Table 5. The balloon observations, all having been carried out at the same zenith declination, are not able to separate the polar dipole contribution, T_z, from the exclusively polar quadrupole term Q_1. Of the approximately 20 U2 flights that have been made, four were in the southern hemisphere specifically to break the correlation of these terms. The maximum magnitude of the quadrupole moment is less than 1 mK.

The large scale isotropy experiments are now at a level where further progress requires instruments of increased sensitivity or the use of observing platforms allowing substantial increase in integration time. More extended sky coverage is essential.

LIMITS ON THE LINEAR POLARIZATION OF THE CBR

The degree of polarization of the CBR is another attribute that can be directly measured. Polarization anisotropies may be produced by the same sources generating intensity anisotropies and could survive if the intergalactic medium does not randomize the polarization through a dispersive Faraday rotation (Rees 1968).

Penzias & Wilson (1965) noted that over a substantial portion of the sky, the background was less than 10% linearly polarized. Nanos (1973) carried out the first systematic search for linear polarization. The instrument was a single beam X band (3.2 cm) polarization sensitive radiometer aimed at the zenith. The output signal of the radiometer was the difference in intensity of two orthogonally polarized components. By rotating the entire apparatus about the optic axis, two Stokes parameters (Q or S_1 and U or S_2) as well as the instrumental polarization asymmetries were measured. As the same beam includes both polarization states, the atmospheric fluctuations cancel in the measurement, providing only that the polarization is switched at a rate fast compared to the time scale of the atmospheric fluctuations. The observation covered a celestial circle of fixed declination, 40°, sampled every beam width, 12°, in right ascension. Each sample consists of two differences in intensity of polarization, one difference being between the North–South plane and the East–West plane and another pair at $\pm 45°$. The rms noise per point is close to receiver noise and corresponds $\leq 0.03\%$ polarization. No statistically significant average, 24 hour or 12 hour periodic components were observed in the data at the same level of sensitivity.

A similar experiment, carried out at 9 mm ($\theta_B = 7°$) (Lubin & Smoot 1979) with increased sky coverage, $\delta = 38°$, 53°, and 63°, and a factor of about 2 improvement in sensitivity, produced a null result as well.

Caderni et al. (1978) have set upper limits on the polarization in the 3 to 20 cm^{-1} band with a balloon-borne polarimeter. The instrument consisted of a telescope with a choice of fields of view, $1/2°$ or 15°, coupled to a bolometer through a rotating linear polarizer. Such systems generally have intrinsic wavelength-dependent polarization anisotropies which must be calibrated by viewing an unpolarized source with a known spectrum. The atmospheric emission, by way of the instrument polarization anisotropy, produced a large systematic polarization offset which, although measured by secant scanning, still limited the observation. The observations, extending over a patch of sky $30° \times 60°$ in the vicinity of the galactic center, set limits of $\sim 1\%$ to $\sim 0.1\%$ on the degree of polarization on angular scales of $1.5°$ to $40°$. The limits could be improved if the experiment were done by rotating the entire apparatus rather than the polarizer as is done in the microwave experiments.

No experiments have measured the third Stokes parameter, V or S_3, to establish the degree of circular polarization of the CBR. At present all the polarization measurements are limited by instrument-derived random noise rather than by polarization anisotropies in the atmosphere or from galactic sources. It seems, therefore, that with improved instrumentation a 10 to 100 times better limit could be set before the intervention of local polarization anisotropies. The open question is on what angular scales is the search most interesting?

THE COBE PROJECT

It has long been recognized that measurements of the spectrum, especially at high frequency, and the large angular scale intensity distribution of the CBR would benefit greatly by being carried out from a long-lived space platform. Aside from the obvious and important advantages of freedom from atmospheric emission, fluctuations in the emission, and absorption, which limits the spectral range, a satellite platform offers full sky coverage with a single instrument and allows sufficient time to test for systematic errors as well as to integrate the weak CBR signals. Furthermore, a shielded apparatus in space offers a thermally controlled environment in vacuum, which facilitates absolute primary calibration with cryogenic sources.

Since 1974 a group consisting of S. Gulkis (Jet Propulsion Laboratory), M. Hauser (NASA Goddard Space Flight Center), J. Mather (NASA Goddard Space Flight Center), G. Smoot (University of California, Berkeley), R. Weiss (MIT), and D. Wilkinson (Princeton) has been involved in planning the COBE (Cosmic Background Explorer) mission. The advance possible by using a space platform should bring measurements of the CBR to the level where the dominant "noise" will be the contribution of the local astrophysical environment. The complement of instruments chosen and the mission specifications of full sky coverage with ~ 1 year lifetime are designed to discriminate the CBR from more local astrophysical sources by their spectra and anisotropic angular distribution.

The present NASA plan is to fly the COBE mission in early 1985. A schematic diagram of the instrument is shown in Figure 15. The instrument, in sun-synchronous polar orbit, points radially out from the earth and rotates at 1 revolution per minute. A liquid helium cryostat with a 1 year storage time sits in the center of a 5 meter diameter sun and rf shield. The cryostat contains a Fourier transform spectrometer (FIRAS) and an infrared photometer (DIRBE). FIRAS (Far Infrared Absolute Spectrophotometer) points along the axis of rotation while DIRBE (diffuse infrared background experiment) points at 30° to the rotation axis. Four differential microwave radiometers (DMR) are attached to the outside of the cryostat.

FIRAS is intended to measure the CBR spectrum. It consists of a rapid-scan polarizing Michelson interferometer divided into two bands 1–20 cm^{-1} and 20–100 cm^{-1} with a minimum resolution width of 0.2 cm^{-1} at long wavelengths and a 5% resolution at short wavelengths. The beam is defined to 7° by a trumpet-shaped horn. The expected sensitivity for each field of view, in a one year mission, is $\sim 10^{-13}$ W/cm^2 sr.

DIRBE is incorporated in the COBE mission to serve two purposes: first, to measure the interstellar dust emission which may contaminate the CBR

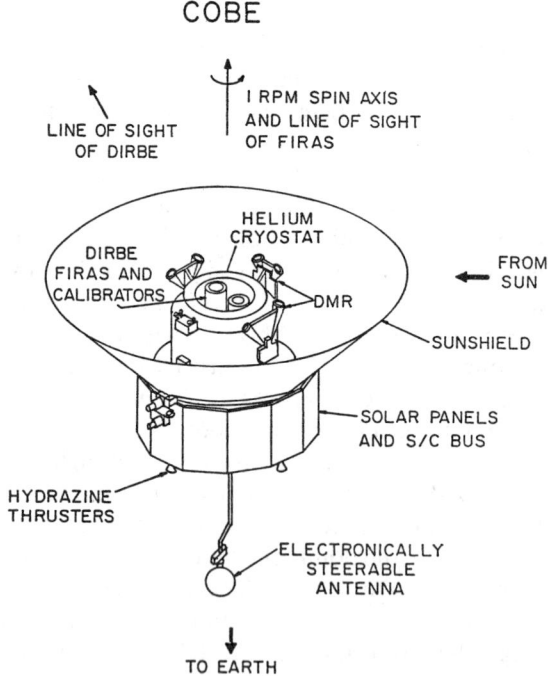

Figure 15 Schematic of the proposed COBE satellite.

spectrum at high frequencies and, second, to perform an all-sky survey of the diffuse infrared background from 1 to 250 microns which, although it is not a residue of the primeval radiation, may contain components that are the result of later epochs in cosmic history. The instrument is a multiband filter absolute photometer with a 1° field of view. The low background conditions in space and in particular in the COBE configuration should allow a sensitivity of $\sim 10^{-14}$ W/cm^2 sr per field of view in 1 year, $\lambda < 160\mu$, and about a factor of 10 worse $\lambda > 200\mu$.

The DMR experiment measures the large angular scale anisotropies of the CBR with a beam width of 7°. The differential radiometers operate at 23, 31, 53, and 90 GHz. Each radiometer compares the radiation of two patches of the sky 60° apart. The expected sensitivity in a one year mission is 0.3 mK in each of 1000 independent elements of the sky. The radiometer frequencies are chosen to measure both the CBR as well as the contamination by synchrotron and thermal emission by H II regions at low frequency and the interstellar dust at high frequencies.

The results of the COBE mission will be a set of sky maps from 1 to 10,000 cm^{-1}, which includes all the processes indicated in Figure 3, the galactic synchrotron emission, diffuse H II thermal emission, interstellar and interplanetary dust, as well as the integrated starlight. The residue will be the CBR and whatever other cosmological background might exist.

Literature Cited

Aannestad, P. A. 1975. *Ap. J.* 200:30
Alpher, R. A., Herman, R. 1975. *Proc. Am. Philos. Soc.* 119:235
Appenzeller, I. 1975. *Astron. Astrophys.* 38:313
Beckman, J. E., Robson, E. I. 1972. *Infrared Detection Techniques for Space Research*, ed. V. Manno, J. Ring, p. 63. Dordrecht: Reidel
Beckman, J. E., Clegg, P. E., Huizinga, J. S., Robson, E. I., Vickers, D. G. 1974. Paper presented at *IAU Conf.*, Sept. 3, 1976, Trieste
Benedict, W. S. 1976. *Infrared Spectra of Atmospheric Molecules.* AFGL-TR-76-0145. Air Force Geophysical Lab.
Boughn, S. P., Fram, D. M., Partridge, R. B. 1971. *Astrophys. J.* 165:439
Boynton, P. E., Stokes, R. A., Wilkinson, D. T. 1968. *Phys. Rev. Lett.* 21:462
Boynton, P. E., Stokes, R. A. 1974. *Nature* 247:528
Burke, W. L. 1975. *Astrophys. J.* 196:329
Caderni, N., Fabbri, R., Melchiorri, B., Melchiorri, F., Natale, V. 1978. *Phys. Rev. D* 17:1908

Cheng, E. S., Saulson, P. R., Wilkinson, D. T., Corey, B. E. 1979. *Astrophys. J. Lett.* 232:L139
Collins, C. B., Hawking, S. W. 1973. *MNRAS* 162:307
Conklin, E. K. 1969. *Nature* 222:971
Conklin, E. K. 1972. In *External Galaxies and Quasi-stellar Objects. IAU Symp. 44*, ed. D. S. Evans, p. 518. Dordrecht: Reidel
Corey, B. E., Wilkinson, D. T. 1976. *Bull. Am. Astron. Soc.* 8:351
Corey, B. E. 1978. PhD thesis. Princeton Univ.
Dall'oglio, G., Fonti, S., Melchiorri, B., Melchiorri, F., Natale, V., Lombardini, P., Trivero, P., Sivertsen, S. 1976. *Phys. Rev. D* 13:1187
Daltabuit, E., Meyer, S. 1972. *Astron. Astrophys.* 20:415
Danese, L., DeZotti, G. 1977. *Riv. Nuovo Cimento* 7:277
Droge, F., Priester, W. 1956. *Z. Astrophys.* 40:236
Ellis, G. R. A., Hamilton, P. A. 1966. *Ap. J.* 146:78

Ewing, M. S., Burke, B. F., Staelin, D. H. 1967. *Phys. Rev. Lett.* 19:1251

Fabbri, R., Guidi, I., Melchiorri, F., Natale, V. 1980. Submitted to *Phys. Rev. Lett.*

Feltz, K. A. 1972. *Publ. Astron. Soc. Pac.* 84:497

Gebbie, H. A., Burroughs, W. J., Bird, G. R. 1969. *Proc. R. Soc. London Ser. A* 310:579

Gezari, D. Y., Joyce, R. R., Simon, M. 1973. *Ap. J. Lett.* 179:67

Goody, R. 1964. *Atmospheric Radiation*. Oxford Univ. Press

Gora, E. K. 1959. *J. Mol. Spectrosc.* 3:78

Gorenstein, M. V. 1978. PhD thesis. Univ. Calif., Berkeley

Gorenstein, M. V., Muller, R. A., Smoot, G. F., Tyson, J. A. 1978. *Rev. Sci. Instrum.* 49:440

Gorenstein, M. V., Smoot, G. F. 1980. In preparation

Greenebaum, M. 1975. *Tech. Rep. T-1-306-3-14*. Riverside Res. Inst. NY

Gush, H. P. 1979. Preprint

Hegyi, D. J., Traub, W. A., Carleton, N. P. 1974. *Astrophys. J.* 190:543

Heiles, C., Jenkins, F. B. 1976. *Astron. Astrophys.* 46:333

Henry, P. S. 1971 *Nature* 231:516

Hildebrand, R. H., Whitcomb, S. E., Winston, R., Steining, R. F., Harper, D. A., Moseley, S. H. 1978. *Ap. J. Lett.* 219:101

Hirabayashi, H. 1974. *Publ. Astron. Soc. Jpn.* 26:263

Houck, J., Soifer, B., Harwit, M., Pipher, J. L. 1972. *Ap. J. Lett.* 178:29

Howell, T. F., Shakeshaft, J. R. 1966. *Nature* 210:1318

Howell, T. F., Shakeshaft, J. R. 1967. *Nature* 216:753

Janssen, M., Weiss, R. 1977. COBE Rep.

Janssen, M. A., Bednarczyk, S. M., Gulkis, S., Marlin, H. W., Smoot, G. F. 1979. Preprint

Kislyakov, A. G., Chernyshev, V. I., Lebskii, Y. V., Mal'tsev, V. A., Serov, N. V. 1971. *Sov. Astron.* 15:29

Knapp, G. R., Kerr, F. J. 1974. *Astron. Astrophys.* 35:361

Lang, K. R. 1978. *Astrophysical Formulae*. New York: Springer

Levy, D. R., Keller, J. B. 1959. *Comments Pure Appl. Math.* 12:159

Liebe, H. J., Gimmestad, G. G., Hopponen, J. D. 1977. *IEEE Trans. Antennas Propag.* AP-25:327

Low, F. J., Kurtz, R. F., Poteet, W. M., Nishimura, T. 1977. *Ap. J. Lett.* 214:L115

Lubin, P. M., Smoot, G. F. 1979. *Phys. Rev. Lett.* 42:129

Martin, D. R., Puplett, E. 1969. *Infrared Phys.* 10:105

Mather, J. C. 1974. PhD thesis. Univ. Calif., Berkeley

Mather, J. C., Richards, P. L., Woody, D. P. 1974. *IEEE Trans. Microwave Theory Tech. MTT-22*:1046

Mather, J. C. 1980. Preprint

McClatchey, R. A. et al. 1979. *Air Force Cambridge Research Laboratory Absorption Line Parameters Compilation*

Meeks, M. L., Lilley, A. E. 1963. *J. Geophys. Res.* 68:1683

Millea, M. R., McColl, M., Pedersen, R. J., Vernon, F. L. 1971. *Phys. Rev. Lett.* 26:919

Muehlner, D. J., Weiss, R. 1973a. *Phys. Rev. D* 7:326

Muehlner, D. J., Weiss, R. 1973b. *Phys. Rev. Lett.* 30:757

Muehlner, D. J. 1977. *Infrared and Submillimeter Astronomy*, ed. G. G. Fazio. Dordrecht: Reidel

Muehlner, D. J., Weiss, R. 1980. To be published

Nanos, G. P. 1973. PhD thesis. Princeton Univ.

Owens, D. K., Muehlner, D. J., Weiss, R. 1979. *Ap. J.* 231:702

Partridge, R. B., Wilkinson, D. T. 1967. *Phys. Rev. Lett.* 18:557

Pauliny-Toth, I. I. K., Shakeshaft, J. R. 1962. *MNRAS* 124:61

Penzias, A. A., Wilson, R. W. 1965. *Ap. J.* 142:419

Penzias, A. A., Wilson, R. W. 1967. *Astron. J.* 72:315

Penzias, A. A. 1968. *IEEE Trans. Microwave Theory Tech. MTT-16*:608

Puzanov, V. I., Salomonovich, A. E., Stankevich, K. S. 1968. *Sov. Phys. AJ* 11:905

Rees, M. J., 1968. *Ap. J.* 153:L1

Reifenstein, E. C., Wilson, T. L., Burke, B. F., Mezger, P. G., Altenhoff, W. J. 1970. *Astron. Astrophys.* 4:357

Rieke, G. H., Harper, P. A., Low, F. J., Armstrong, K. R. 1973. *Ap. J. Lett.* 183:67

Robson, E. I., Vickers, D. G., Huizinga, J. S., Beckman, J. E., Clegg, P. E. 1974. *Nature* 251:591

Roll, P. G., Wilkinson, D. T. 1966. *Phys. Rev. Lett.* 16:405

Rothman, L. S. 1978. *Appl. Opt.* 17:3517

Scoville, N. Z., Solomon, P. M. 1975. *Ap. J. Lett.* 199:105

Smoot, G. F. 1977. *COBE Rep. No. 5002*

Smoot, G. F., Gorenstein, M. V., Muller, R. A. 1977. *Phys. Rev. Lett.* 39:898

Smoot, G. F., Lubin, P. M. 1979. *Ap. J. Lett.* 234:L83

Stokes, R. A., Partridge, R. B., Wilkinson, D. T. 1967. *Phys. Rev. Lett.* 19:1199

Sunyaev, R. A., Zel'dovich, Ya. B. 1980. *Ann. Rev. Astron. Astrophys.* 18:537
Thaddeus, P. 1972. *Ann. Rev. Astron. Astrophys.* 10:305
Tinkham, M., Strandberg, M. W. P. 1955a. *Phys. Rev.* 97:937
Tinkham, M., Strandberg, M. W. P. 1955b. *Phys. Rev.* 97:951
Ulfbeck, O., ed. 1980. *Universe at Large Red Shifts.* Proc. Niels Bohr Institut, June 24–31, 1979. *Physica Scripta* 21:599
US 1966. *U.S. Standard Atmosphere Supplements.* Washington, DC: US Gov't. Print. Off.
Weiss, R. 1980 In preparation
Welch, W. J., Keachie, S., Thornton, D. D., Wrixon, G. 1967. *Phys. Rev. Lett.* 18:1068
Wilkinson, D. T. 1967. *Phys. Rev. Lett.* 19:1195
Williamson, K. D., Blair, A. G., Catlin, L. L., Hiebert, R. D., Loyd, E. G., Romero, H. V. 1973. *Nature Phys. Sci.* 241:79
Wilson, T. L., Mezger, P. G., Gardner, F. F., Milne, D. K. 1970. *Astron. Astrophys.* 6:634
Wilson, R. W., Penzias, A. A. 1967. *Science* 156:1100
Witebsky, C. 1978. *COBE Rep. No. 5006*
Woody, D. P. 1975. PhD thesis. Univ. Calif., Berkeley
Woody, D. P., Mather, J. C., Nishioka, N. S., Richards, P. L. 1975. *Phys. Rev. Lett.* 34:1036
Woody, D. P., Richards, P. L. 1979. *Phys. Rev. Lett.* 42:925

MICROWAVE BACKGROUND RADIATION AS A PROBE OF THE CONTEMPORARY STRUCTURE AND HISTORY OF THE UNIVERSE

×2174

R. A. Sunyaev and Ya. B. Zel'dovich

Space Research Institute, USSR Academy of Sciences, Moscow, USSR

1 Introduction

The discovery of the microwave background by Penzias & Wilson (1965) has been interpreted as confirming Gamow's (1948a,b) primordial-fireball hypothesis. The longwave part of the microwave background spectrum is quite well described by the Rayleigh-Jeans law. In the shortwave part the intensity is observed to decrease in accordance with the blackbody spectrum. The radiation is extremely isotropic on both large and small scales. There is no noticeable polarization.

Interest has now shifted to departures of the spectrum from that of a blackbody. Also of great interest are measurements of angular anisotropy, small-scale angular fluctuations, and polarization. What is known already is that these distortions and deviations (both spectral and angular) are small.

We believe the universe may never have been completely uniform. Even in the earliest stages there may have been small perturbations of matter and radiation densities, and of the velocity field. The fireball hypothesis requires such perturbations to account for the formation of galaxies and clusters of galaxies. Observations of the microwave background are a powerful method of investigating such perturbations. Even perturbations that are dissipated by viscosity, thermal conductivity, and nonlinear effects too early to influence the present structure of the universe could have distorted the microwave spectrum and might therefore be observable. The angular distribution of the radiation could give information about perturbation spectra on scales much larger than those of clusters of galaxies, where the amplitude of the perturbations is too small for the formation of gravitationally bound objects.

Microwave photons in the directions of clusters of galaxies have been Compton-scattered by electrons in the hot intergalactic gas bound in such clusters. Detailed observations of the microwave background in these directions could also give important information about the present-day universe. Together with X-ray observations, they offer the possibility of direct determinations of the Hubble constant and, possibly, of the deceleration parameter q_0, and even of the peculiar velocities of clusters of galaxies relative to the microwave background radiation.

The experimental data are reviewed by R. Weiss (present volume; see also reviews written by Longair & Sunyaev 1971, Sato et al. 1971, Thaddeus 1972, Danese & de Zotti 1977, Blair 1974). We discuss below the theoretical background of such observational studies. Our point of view on small-scale angular fluctuations of the microwave background is discussed in two papers (Sunyaev & Zel'dovich 1970a and Sunyaev 1978).

1.1 THE MAIN PARAMETERS AND EVOLUTIONARY STAGES OF THE HOT UNIVERSE The hot universe is characterized by three parameters: the present temperature of the microwave radiation $T_0 = 3\Theta$ K, the present value of the Hubble parameter $H_0 = 50h$ km s^{-1} = 1.62 $10^{-18}h$ s^{-1}, and the average mass density $\rho = \Omega \rho_{\text{crit}}$, where $\rho_{\text{crit}} = 3H_0^2/8\pi G = 4.7\ 10^{-30}h^2$ g cm^{-3}. The parameter Ω is of fundamental significance. According to the theory of A. A. Friedman, if $\Omega > 1$ the universe is closed and the observed expansion will someday be replaced by contraction. If $\Omega \leq 1$, the expansion must continue indefinitely. The experimental data are consistent with the inequalities

$$0.03 \leq \Omega \leq 2, \quad 0.8 \leq h \leq 2.5, \quad 0.88 \leq \Theta \leq 1.05.$$

In an expanding universe temperature and mass density decrease with time. The wavelength and frequency of a photon depend on the redshift z:

$$v = v_0(1+z), \quad \lambda = \lambda_0/(1+z).$$

The radiation temperature, radiation energy density, and photon number, are given by

$$T_r = T_0(1+z),$$
$$E_r = bT_r^4 = 6 \cdot 10^{-13}\ \Theta^4(1+z)^4\ \text{erg cm}^{-3},$$
$$N_\gamma = E_r/2.7kT_r = 550\Theta^3(1+z)^3\ \text{cm}^{-3}.$$

The baryon number density changes according to

$$N_b = \Omega \rho_{crit}(1+z)^3/m_p = N_1 \Omega h^2 (1+z)^3$$

where $N_1 = 2.82\ 10^{-6}\ cm^{-3}$.

The photon number greatly exceeds the baryon number (the entropy of the universe is high).

$$\frac{N_\gamma}{N_b} = 1.4\ 10^8\ \frac{\Theta^3}{\Omega h^2}.$$

For $\Omega z \gg 1$ the parameter Ω enters most formulae multiplied by h^2. Therefore it is convenient to introduce the parameter $\omega = \Omega h^2$.

The radiation energy density $E_r \sim (1+z)^4$ increases with z more rapidly than the rest-mass energy density $\rho_m c^2 \sim (1+z)^3$. They became equal at $z_1 = 5.3\ 10^3(\gamma \omega/\Theta^4)$; $E_r < \rho_m c^2$ for $z < z_1$ and $E_r > \rho_m c^2$ for $z > z_1$. The total energy density of massless particles is $E_{rt} = \gamma E_r$, where the factor γ, which takes into account the contribution of neutrinos, satisfies the inequalities $1.68 \leq \gamma \leq 4$.

The time t (reckoned from the initial singularity) and the redshift z are connected by the simple relations:

$$\frac{dt}{dz} = \begin{cases} -\dfrac{1}{H_0(1+z)^2\sqrt{1+\Omega z}} \underset{\Omega z \gg 1}{\approx} -\dfrac{6.3 \cdot 10^{17}}{\omega^{1/2} z^{5/2}}\ s & \text{for } z < z_1 \\ -\dfrac{6.4 \cdot 10^{19}}{z^2 \Theta^2 \sqrt{\gamma}}\ s & \text{for } z > z_1. \end{cases}$$

In Figure 1 the main stages in the evolution of the universe are presented as a function of redshift or time. We are interested mainly in $t > 30$ s, $z < 10^9$.

Before this time electron-positron pairs are in thermodynamical equilibrium. Their number density is of the order of N_γ. Then it drops rapidly. When $z \approx 10^8$ their density is of the order of the baryon density $\approx 10^{-9} N_\gamma$. The energy released by positron-electron annihilation increases the radiation energy density by the factor $(11/4)^{4/3} = 3.85$. The main energy release occurs when the density of positrons and electrons is so high that

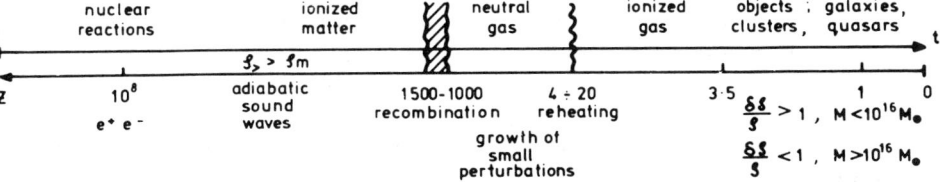

Figure 1 The history of the universe.

bremsstrahlung, Compton scattering, and other processes rapidly establish a blackbody spectrum with a higher temperature.

At $3 \cdot 10^8 < z < 10^9$ nuclear reactions lead to the synthesis of He^4, He^3, and deuterium. The energy released in nuclear reactions hardly changes the radiation energy density: $\Delta E/E_r \approx 2 \cdot 10^{-7} \omega/\Theta^3$.

From $z \sim 10^9$ to z_1 the universe was radiation-dominated. Matter was fully ionized up to $z \sim 1500$ when, according to the Saha formula, recombination of hydrogen occurred. Up to $z \sim 150$ scattering of photons by residual electrons keeps the matter temperature equal to the radiation temperature. Then, in the absence of heat sources, the radiation and matter temperatures decrease according to the laws $T_r = 3\Theta(1+z)$ K (adiabatic index $\gamma = 4/3$), and $T_m \approx 2 \cdot 10^{-2} (1+z)^2$ K (adiabatic index $\gamma = 5/3$). The ideal picture leads to a cold ($T_m \approx 0.02$ K), rarefied ($N_H \approx 10^{-6}$ cm^{-3}) gas uniformly filling the universe.

However, in reality some time after recombination but before $z = 2$ the formation of galaxies, quasars, and other objects occurred presumably owing to the growth of small density perturbations. The energy released by the gravitational contraction of protoclusters of galaxies, ultraviolet radiation of young galaxies and quasars, explosions of the nuclei of galaxies, and so on, reheats and ionizes intergalactic matter. The radiation energy density decreases more rapidly than $\rho_m c^2$. Therefore the energy released at small redshifts may strongly influence the spectrum of the microwave background. For example, the transformation of 30% of the hydrogen into helium increases the radiation energy density by the factor $(1+(16\omega/1+z)) = 2$ if $z = 15$ and $\omega = 1$. More detailed information about the history of the universe is given in reviews and books by Zel'dovich (1967), Peebles (1971), Weinberg (1972), Harrison (1973), and Zel'dovich & Novikov (1967, 1975).

2 Distortions of the Microwave Spectrum

2.1 IDEALIZED HOT BIG BANG MODEL
For equilibrium blackbody radiation the occupation number in the photon phase space

$$n = \frac{c^2 I_v}{2h v^3} \quad \text{is} \quad n = \left[\exp\left(\frac{hv}{kT_r}\right) - 1\right]^{-1},$$

where I_v is the intensity of radiation. Both hv and kT_r are proportional to $(1+z)$. Therefore, the blackbody spectrum remains blackbody during the expansion, because for any group of photons the ratio hv/kT_r does not depend on z.

In the idealized model only recombination of hydrogen leads to noticeable distortions of the blackbody spectrum. Recombination occurs at $z \approx 1500$, when $T_r \approx 4000$ K and $\chi/kT_r \approx 40$, where $\chi = 13.6$ eV is the ionization potential of hydrogen. Although the total number density of photons is 10^9 times the density of baryons N_b, the density of ionizing photons with $h\nu > \chi$ is less than N_b. The absorption optical depth of the universe for the ionizing photons and photons of the Lyman series is extremely high. Therefore, the intensity of the Ly-α line photons is in thermodynamical equilibrium with the $2p$ level. The $2p$ and $2s$ levels are overpopulated in comparison with $1s$. All higher levels and the densities of free electrons and ions are in thermodynamical equilibrium with $2p$ and $2s$ levels. Recombination occurs mainly through two-photon decay of the $2s$ level, with transition probability $W_{2s} = 8 \text{ s}^{-1}$. As a result, the degree of ionization exceeds that given by the Saha formula (Zel'dovich et al. 1969, Peebles 1968). Sunyaev & Zel'dovich (1970a) found a useful approximate formula for the degree of ionization $\alpha = N_e/N_e + N_H$ as a function of redshift:

$$\alpha = \frac{A}{z\sqrt{\omega}} e^{-B/z}, \quad \text{where} \quad A = \frac{(2\pi m_e k T_0)^{3/2}}{4 w_{2s} k T_0 h^3} = 6 \cdot 10^6 \theta^{1/2}$$

and

$$B = \frac{\chi}{4kT_0} = 1.62 \cdot 10^4/\Theta.$$

This formula is valid for $10^{-4} < \alpha < \frac{1}{2}$, $900 < z < 1500$. The density of electrons decreases by a factor e during the redshift interval $\Delta z \sim 70$, $\Delta z/z \sim \frac{1}{20}$. After $z \sim 800$ the degree of ionization is frozen at $\alpha \approx 3 \cdot 10^{-5}/\sqrt{\omega}$.

The density of recombining electrons and ions is very small compared with N_γ. Each recombination is followed by the emission of several photons. It is obvious that they cannot change the spectrum near its peak. Distortions might be noticeable only in the far Wien ($x = h\nu/kT_r > 20$) and far Rayleigh-Jeans ($x < 0.1$) regions of the spectrum, where the number of photons is small. Zel'dovich et al. (1969) and Peebles (1968) computed distortions in the region $x > 20$. These distortions are strong. Unfortunately the radiation intensity is very weak there, and other sources such as dust in galaxies and compact infrared sources may make greater contributions to the background. Dubrovich (1975) has discussed line radiation from cascade recombination and transitions between highly excited levels $n \sim 10 \div 20$, $\Delta n = 1$. These lines are redshifted to radio wavelengths. Detailed computations by Dubrovich et al. (1977) and Beigman & Sunyaev (1978) show that the intensity of these lines is not great

$$\frac{\Delta I_\nu}{I_\nu} \approx \frac{\Delta T}{T_r} = 10^{-6} \omega \left(\frac{\lambda}{10 \text{ cm}}\right)^2.$$

Their spectral widths are determined by the effective duration of recombination $\Delta\lambda/\lambda \sim \Delta z/z \sim 0.1$. The lines are overlapping. The recombination $He^{++} \to He^+$ and $He^+ \to He$ also must produce weak overlapping radiofrequency lines. Their intensities are an order of magnitude less than those of hydrogen, owing to the lower abundance of helium. During recombination $He^+ \to He$ the lines corresponding to transitions between fine-structure levels of the excited neutral helium atom also appear, both in emission and absorption. All these recombinational lines are very weak.

2.2 INTERACTION OF MATTER AND RADIATION IN THE UNIVERSE   Compton scattering is the most important elementary mechanism of interaction of radiation with matter in the early universe. The optical depth for Thomson scattering

$$\tau_T = \int_0^{z_{max}} \sigma_T N_e(z) c \, \frac{dt}{dz} \, dz$$

is $\tau_T = 0.024 \omega^{1/2} z_{max}^{3/2}$ if the primeval hydrogen-helium plasma is fully ionized during time interval corresponding to $0 \ll z \ll z_{max}$. Before recombination this optical depth was enormously high. The Rosseland optical depth of the universe for free-free absorption was very small up to the time of electron-positron annihilation. The absorption probability is small, but a typical photon experiences multiple scatterings. Under such conditions Compton scattering with change of frequency ("comptonization") is the most important mechanism of energy transfer between plasma and radiation.

2.3 ENERGY RELEASE   There are several possible energy sources. At redshifts $z < 10$ these include explosions in the galactic nuclei and quasars, ionization losses of subcosmic rays in intergalactic medium (Ginzburg & Ozernoy 1965), and shock waves formed during the contraction of protoclusters of galaxies (Sunyaev & Zel'dovich 1972a). In the early universe ($z > 1000$) the dissipation of acoustic and turbulent energy may be important (Silk 1968). Other sources of energy include matter-antimatter annihilation, evaporation of primordial black holes, and heavy unstable particle decay. The energy released by these processes increases the radiation energy density and distorts its spectrum.

2.4 LATE ENERGY RELEASE AND HOT INTERGALACTIC GAS   Observations of the Ly-α absorption band in the spectra of distant quasars showed the absence of neutral intergalactic hydrogen at redshifts $z \leq 3.5$ (Gunn & Peterson 1965). Therefore, we can assume that the intergalactic gas was at some time heated and ionized. If the gas is hot enough Compton scattering influences the spectrum of the microwave background and by observing it we can learn when secondary heating occurred and how it affects the intergalactic gas (Weymann 1967, Sunyaev 1968, Zel'dovich & Sunyaev 1969, Chan & Jones 1975a–c, Field & Perrenod 1977).

If the energy was released during the epoch $0 < z < 4 \times 10^4 \, \omega^{-6/5}$ it raises the electron temperature, so that $T_e > T_r$. For a wide range of physical conditions a typical spectrum arises which depends only on a single parameter,

$$y = \int_{t_{min}}^{t_{max}} \frac{k(T_e - T_r)}{m_e c^2} \sigma_T N_e(z) c \, dt.$$

The photons are redistributed over frequency. The number of photons is conserved but their average energy increases. For $y \leq 0.25$ the Rayleigh-Jeans slope of the spectrum is maintained in the low-frequency region, but the intensity and brightness temperature decrease: $T = T_r \exp(-2y)$. For $h\nu > 3.93 \, kT_r$ the intensity increases. Distortions of this type are shown in Figure 2. They are described by Equations (A7) and (A8) of the Appendix (Zel'dovich & Sunyaev 1969). The energy release is simply related to the parameter y:

$$\frac{\Delta E}{E_r} = e^{4y} - 1$$

where E_r is the initial radiation energy density. The resulting radiation energy density

$$E_{tot} = \int_0^\infty E_\nu \, d\nu = E_r e^{4y} = bT_{RJ} e^{12y}$$

where T_{RJ} is the brightness temperature at long wavelengths. Recent observations (Muehlner & Weiss 1970, Robson et al. 1974, Woody & Richards 1979) place upper limits on the parameter y, $y \leq 0.055$, and on the energy released during the time interval corresponding to $8 < z < 4 \, 10^4 \, \omega^{-6/5}$:

$$\frac{\Delta E}{E_r} = \int_{t_{min}}^{t_{max}} \frac{\dot{E}(z)}{E_r(z)} \frac{dt}{dz} dz \leq 0.25.$$

For $z < 8$ the plasma cooling time $t = \frac{3}{4}(m_e c / \sigma_T E_r)$ for comptonization exceeds the Hubble time and only a small part of the energy released is transferred to radiation.

Sunyaev (1968) and Zel'dovich & Sunyaev (1969) showed that neutral hydrogen remained neutral during some time after recombination at z_r. Otherwise, distortions of the blackbody spectrum would be observable.

It is interesting that the spectral distortions shown in Figure 2 and described by Equations (A7) and (A8) of the Appendix are predicted under very different conditions:

Case 1. The temperature of electrons exceeds that of the radiation. This case corresponds, for example, to plasma heating by shock waves arising in a nonlinear stage of the formation of clusters of galaxies, or to heating by subcosmic rays generated by explosions in quasars of galactic nuclei. In each scattering changes in the photon frequency result from the Doppler

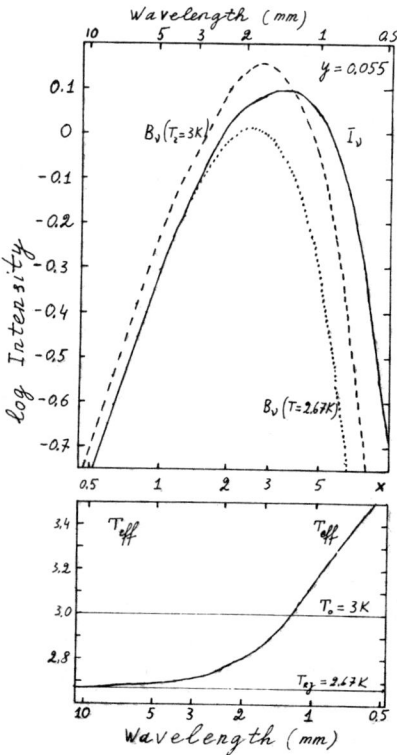

Figure 2 The effect of Compton scattering on the spectrum of the microwave background in the case of late energy release. The broken line corresponds to the initial blackbody spectrum, the solid line corresponds to the spectrum that results from Compton scattering, and the dotted line to a blackbody spectrum that mimics the spectrum produced by Compton scattering at long wavelengths. In the lower part of the figure the effective temperature of the Compton-scattered radiation is shown as a function of the wavelength. The effective temperature is defined as the temperature of blackbody radiation with the same intensity at a given wavelength.

effect. The first-order effect, proportional to v_e/c, vanishes because equal numbers of electrons are moving in opposite directions. Only a second-order effect proportional to $v^2/c^2 \approx kT_e/m_e c^2$ remains. More exactly, the effect is proportional to $k(T_e - T_r)/m_e c^2$, so that for $T_e = T_r$ the blackbody spectrum does not change. If the number of scatterings is large, the parameter y becomes appreciable and comptonization is important.

Case 2. *The energy released is stored in chaotic motions of optically thick plasma clouds together with their electrons, protons, and bottled-up radiation.* Perturbations of the matter and radiation energy density perturb the radiation temperature. Kinematic energy is transformed into thermal energy, and photon diffusion smoothes the radiation field. The same process dissipates adiabatic perturbations on small scales. Let us consider the limiting case of zero scattering. In nature this might occur after hydrogen recombines. The radiation field is now Planckian in any direction, but has different temperature in different directions. Let us average this radiation field over direction (this might be accomplished by the last few scatterings which change the directions of photons but not their energies). The resulting spectrum differs from a blackbody one. For the same value of $\Delta E/E_r$ it resembles that shown in Figure 2 (Zel'dovich et al. 1972).

Case 3 *A small fraction of the photons (5% for example) is scattered by very hot electrons ($T_e \sim 10^8$ K), and the other 95% are not scattered.* This might occur if the background photons interact with hot gas clouds in clusters of galaxies (see below). The spectrum in this case is very similar to that of case 2 (Sunyaev 1980a).

The radiation spectrum after energy release is defined by two parameters: the initial temperature of the blackbody radiation T_0 and the parameter y characterizing the energy release. There are also two physical quantities characterizing the photon distribution, the photon number density N_γ and the energy density E_r, with

$$N_\gamma = \frac{8\pi}{c^3} \int n v^2 \, dv \quad \text{and} \quad E_r = \frac{8\pi h}{c^3} \int n v^3 \, dv.$$

In equilibrium

$$N_0 = \frac{bT_0^3}{2.7} = aT_0^3, \quad E_r = bT_0^4.$$

After energy release, Compton scattering conserves $N_\gamma = N_0 = aT_0^3$, but increases $E_r = E_0 e^{4y} = bT_0^4 e^{4y}$. This gives a relation between the pairs N_γ, E_r and T_0, y. The Rayleigh-Jeans part of the spectrum is characterized by $T_{RJ} = T_0 e^{-2y}$ and $E_r = bT_{RJ}^4 e^{12y}$. Therefore, measurements of T_{RJ} and E_r give $y = \frac{1}{12} \ln (E_r/bT_{RJ}^4)$.

The parameter y may be defined by either of the formulas

$$y_1 = \int \frac{k(T_e - T_r)}{m_e c^2} d\tau_T, \quad e^{4y_2} = \int \frac{\dot{E}}{E_r} dt.$$

Given the energy release function $\dot{E}(t)$, conservation of energy ensures that $T_e(t)$ satisfies the identity $y_1 = y_2$.

This is true for every process, for example

$$p + \bar{p} \rightarrow \pi^+, \pi^0, \pi^- \rightarrow \mu^+, \mu^-, \gamma \rightarrow e^+, e^-, \gamma \rightarrow \gamma$$

which gives energetic electrons but also high-energy photons. The additional number of photons is small and unimportant compared with the change of the soft photon spectrum.

2.5 RELAXATION OF THE RADIATION SPECTRUM AFTER ENERGY RELEASE The distorted spectrum tends to relax to a blackbody spectrum with a new temperature. Relaxation occurs in stages characterized by different relaxation times. The relaxation times are proportional to the particle number density $t_e/t_\gamma = N_e/N_\gamma \sim 10^{-9}$.

First the electrons adjust their temperature to the nonequilibrium spectrum. This process gives (see Appendix) very rapidly

$$kT_e = \frac{h}{4} \frac{\int_0^\infty (n + n^2) v^4 \, dv}{\int_0^\infty n v^3 \, dv} = \frac{2\pi h^2}{E_r c^3} \int_0^\infty n(1 + n) v^4 \, dv.$$

For a y-distorted spectrum (as defined earlier), $T_e = T_0(1 + 5.4y)$ (Illarionov & Sunyaev 1975a,b).

The next step is an adjustment of the spectrum by multiple scatterings, without emission or absorption. During this process E_r and N_γ remain constant and N_γ is smaller than its equilibrium value for E_r. The result is given by the Bose-Einstein equilibrium formula

$$I_\nu = \frac{2h\nu^3}{c^2} \left[\exp\left(\frac{h\nu}{kT_r} + \mu\right) - 1 \right]^{-1}$$

with chemical potential $\mu \geq 0$. For $0 < \mu \ll 1$ the difference between this and the blackbody spectra is large at frequencies $h\nu/kT_r \ll \mu$, where $I_\nu \sim \nu^3$ instead of $I_\nu \sim \nu^2$, as it would be for the Planck spectrum.

The efficiency of free-free radiation depends strongly on frequency. It is important at low frequencies, where it rapidly establishes the Rayleigh-Jeans spectrum. The chemical potential also depends in this case on frequency. The nondimensional frequency $x_0 = h\nu_0/kT_e$ defined by equation (A10) is important. In the vicinity of this frequency Compton scattering and bremsstrahlung are comparable in importance (see Appendix). Sunyaev & Zel'dovich (1970b) found that $\mu = \mu_0 \exp\{-2x_0/x\}$. Illarionov & Sunyaev (1975a,b) gave a solution for $\mu > x_0$. The production of photons with $x > x_0$ causes μ to decrease. They are redistributed over a frequency by Compton scattering.

The final step is the establishment of full thermodynamical equilibrium, i.e. the formation of a blackbody radiation spectrum with $T_e = T_r$ and $N_\gamma = E_r/2.7kT_r$. Relaxation times can be found in Sunyaev & Zel'dovich (1970b) and Illarionov & Sunyaev (1975a). All traces of the released energy have now been lost. We observe blackbody radiation with a new temperature T_1 instead of the original temperature T_0, which we have no way of measuring. However, the intermediate situation is very informative.

2.6 EARLY ENERGY RELEASE; $z > 10^4$ In the early universe ($z \gg 10$) the time scale for the cooling of electrons by interaction with the radiation background, $t_e = \tfrac{3}{4}(m_e c/\sigma_T E_r)$, was very short compared with the Hubble time. Therefore T_e was very close to T_r independent of the rate of energy release.

Even if the energy release is local its influence is rapidly smoothed by photon diffusion. The energy release increases both T_r and T_e. Doppler and recoil effects redistribute the photons over frequency. The relaxation time for the radiation is much longer than t_e because the photon density is 10^9 times the electron density. Relaxation requires $m_e c^2/kT_e$ scatterings on the electrons. These scatterings conserve the number of photons. Free-free emission and absorption are weak. Therefore relaxation gives rise to an equilibrium with given values of the radiation energy density and number density of the photons. Multiple Compton scatterings lead to the Bose-Einstein spectrum $n = (e^{x+\mu}-1)^{-1}$. This spectrum is formed at high frequencies $x \gtrsim 1$ during the time corresponding to

$$u = \int \frac{kT_r}{m_e c^2} \sigma_T N_e c\, dt \geq 0.25.$$

This condition is satisfied for $z_\mu > 1.7\,10^4\,\omega^{-1/2}\theta^{1/2}$. However, for small x the condition is more complicated $u > \tfrac{1}{2}\ln 2.2/x$ (Zel'dovich & Sunyaev 1969, Illarionov & Sunyaev 1975a).

The frequency $x_0 \approx 70\omega^{1/2}z^{-3/4}$ (see Appendix and 2.5) depends on redshift z. The observed spectrum must be formed at the epoch when $u \approx 0.25$. Therefore, for $\mu < x_0(z_\mu)$ the minimum in the dependence of the effective temperature of the background radiation on frequency lies close to $x_0(z_\mu) = 5\cdot 10^{-2}\omega^{7/8}$ or $\lambda_0 \approx 10\omega^{-7/8}$ cm. The effective temperature T_{eff} is determined by the relation

$$n(v) = (e^{hv/kT_{\text{eff}}} - 1)^{-1} = \left[\exp\left(x \cdot \frac{T_r}{T_{\text{eff}}}\right) - 1\right]^{-1}. \quad (1)$$

It follows from (1) that

$$T_{\text{eff}} = \frac{T_r}{x \ln\left(1 + \dfrac{1}{n(v)}\right)} = \frac{T_r}{1 + \mu/x}. \quad (2)$$

Relaxation to a blackbody spectrum is so slow that the energy release might be detected up to $z = 5 \cdot 10^8 \mu^2 \omega^{-4}$. However, for $z > 10^7$ the double Compton effect (the radiation of an additional soft photon during Compton scattering) becomes important. This makes it impossible to observe any energy release occurring earlier than $z \approx 10^7$ (Sunyaev 1980b, see also Gould 1972, and Danese & de Zotti 1977). For small $\mu \ll 1$ one can easily find the relation between μ and $\Delta E/E_r = 0.714\mu$ expanding

$$E_r = \frac{8\pi h}{c^3} \int_0^\infty nv^3\, dv \quad \text{and} \quad N_\gamma = \frac{8\pi}{c^3} \int_0^\infty nv^2\, dv$$

into the series

$$E_1 = E_0 + \Delta E = bT_0^4 + \Delta E = bT_1^4(1 - 1.11\mu)$$

and

$$N_\gamma = \frac{bT_0^3}{2.7k} = \frac{bT_1^3}{2.7k}(1 - 1.368\mu).$$

Comparison of Figure 3 with existing long-wavelength experimental data (Howell & Shakeshaft 1966, 1967, Pelyushenko & Stankevich 1969, Penzias & Wilson 1967, Roll & Wilkinson 1966, Stokes et al. 1967) gives an upper limit to the energy release in the early universe $\Delta E/E_r < 0.03$ for $\omega = 1$ and $\Delta E_r/E_r < 7 \cdot 10^{-3}$ for $\omega \approx 0.1$ (see discussion in Illarionov & Sunyaev 1975b and Danese & de Zotti 1977).

These upper limits were used to obtain restrictions on

1. the spectrum of primeval adiabatic density perturbations and vortex motions dissipating at $10^4 < z < 10^7$ (Sunyaev & Zel'dovich 1970b),
2. the amount of antimatter in the early universe (Sunyaev & Zel'dovich 1970b,c, Stecker & Puget 1972),
3. the density of primordial black holes of small mass evaporating by the Hawking mechanism, and
4. the density of heavy unstable leptons (see review by Dolgov & Zel'dovich 1980).

2.7 THE NARROW SPECTRAL FEATURES   The theory of the spectral distortions formed during hydrogen recombination changes strongly if there was strong energy release before recombination $0.003 < y < 0.05$ (Sunyaev 1980b). Before and during recombination the plasma was in a non-equilibrium radiation field whose temperature depended on frequency. In the Rayleigh-Jeans region the brightness temperature $T_{RJ} = T_0 e^{-2y}$ was less than in the Wien region (see Figure 2) and the electron temperature $T_e = T_{RJ}(1+7.4y)$. Under such conditions radiation bands with quite sharp edges must appear, corresponding to transitions between highly excited levels.

2.8 ALTERNATIVE THEORIES   Early star formation ($z \sim 30 \div 100$) might lead to strong distortions of the background spectrum near and beyond its maximum. At redshift $z \sim 100$ the universe had matter density of the order of that in the disk of the Galaxy. The difference was only in dimensions. The horizon $ct \sim 1$ Mpc greatly exceeded the dimensions of the Galaxy disk.

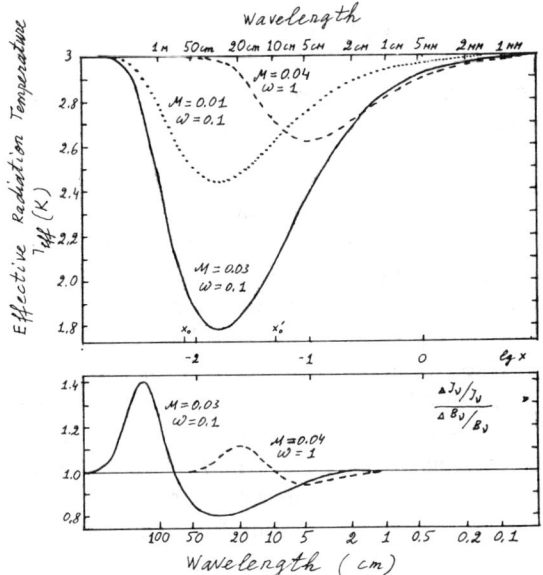

Figure 3   The distortions of the background-radiation spectrum by Compton scattering in the case of early energy release. Three cases are illustrated, computed for three different values of the chemical potential μ and two values of the parameter ω or x_0. At high frequencies ($x > x_0$) the spectrum has the Bose-Einstein form and the effective temperature of the radiation increases with frequency. At low frequencies $x < x_0$ free-free processes establish $T_{\text{eff}} = T_e$. The lower part of the figure illustrates the possibility of detecting distortions of the background spectrum through observations in the direction of a cluster of galaxies containing hot intergalactic gas (see discussion in Section 4.4).

If an early generation of stars (Rees 1978) ejected heavy elements and dust was formed, the universe was optically thick for optical light and even for $\lambda \sim 10\mu$, corresponding to $\lambda \sim 1$ mm today. Present 10-cm radiation corresponds to $\lambda \approx 1$ mm at $z = 100$. From observations of the center of our Galaxy we know that the Galactic plane is highly transparent to $\lambda \approx 1$ mm photons. The same must be true of the early universe. However, energy released by early stars might affect the background spectrum in the vicinity of its maximum. It hardly can change the R-J part of the spectrum which remains the witness of the hot Big Bang.

3 Intergalactic Gas in Clusters of Galaxies

Data obtained by X-ray satellites (Lea et al. 1973, Culhane 1978) show that the rich clusters of galaxies contain a large amount of hot ($T_e \sim 10^8$ K), rarefied ($N_e \sim 10^{-2} \div 10^{-3}$ cm^{-3}) intergalactic gas. Clusters are high-temperature plasma clouds with noticeable optical depth for Thomson scattering

$$0 < \tau_T \ll 1, \quad \tau_T = \int_{-\infty}^{+\infty} \sigma_T N_e(l) \, dl.$$

Bremsstrahlung by this gas is observed at X-ray frequencies. Compton scattering of microwave photons on free electrons leads to many important effects, discussed below.

3.1 THERMAL EFFECT   A fraction of order τ of the background photons observed in the direction of a cluster of galaxies has been scattered once by hot electrons. The remaining fraction $(1 - \tau)$ is not scattered. Scattering by electrons with $kT_e \gg hv$ changes the photon frequency through the Doppler effect. The left side of Figure 4 demonstrates the spectrum of once-scattered monochromatic line photons. The electrons have a Maxwellian distribution with temperature $kT_e = 5.1$ keV $\gg hv_0$. The line is broadened by the Doppler effect. The right wing is stronger than the left. The photons on the average increase their energies. The cusp in the distribution at $v = v_0$ is real. It results from the fact that Δv, which is proportional to $1 - \cos \alpha$, vanishes at small deflection angles α. The computations show that

$$\frac{\overline{\Delta v}}{v} = 4 \frac{kT_e}{m_e c^2}.$$

The right side of Figure 4 shows how one scattering by Maxwellian electrons influences a blackbody radiation spectrum with $T_r \ll T_e$. The frequency of the photons is increased. In the Rayleigh-Jeans ($hv \ll kT_r$) region the intensity and the brightness temperature decrease: $\Delta T/T_r = -2kT_e/m_e c^2$. In the Wien region ($hv > 3.83 \, kT_r$) the intensity increases. The effect is obviously proportional to the fraction of photons that have undergone one scattering in the cloud. In the Rayleigh-Jeans region $\Delta T/T_r = -(2kT_e/m_ec^2)\tau_T$. The exact computation (Sunyaev 1980a) shows that the blackbody spectrum after one scattering by electrons with $kT_e = 0.01m_ec^2$ coincides well with the solution [(A7), see Appendix] of the Kompaneets equation, which was obtained by Zel'dovich & Sunyaev (1969) (see also Gould & Raphaely 1978). The difference is important only for $hv > 10kT_r$.

As a result the effective temperature of the background observed in the direction of a cluster must differ from the temperature in other directions. The difference depends strongly on the wavelength (see Figure 5, case $v_r = 0$). In the Rayleigh-Jeans region the effect does not depend on wavelength: in the direction of the hot gas cloud the brightness temperature decreases (Figure 6; Sunyaev & Zel'dovich 1972b). This effect was investigated by Pariyskiy (1973), Gull & Northhover (1975), Lake & Partridge (1977), Birkinshaw et al. (1978); see also Schallwich & Wielebinsky (1979). In the cluster A2218 with redshift $z = 0.17$, according to the last three papers mentioned, the effect is quite large, $\Delta T = (-2.65 \pm 0.23) \cdot 10^{-3}$ K. Assuming that the gas temperature is close to $kT_e = 5$ keV and using the formula $\Delta T/T_r = -(2kT_e/m_ec^2)\tau_T$ one easily estimates the optical depth of the cloud

$$\tau_T = \frac{1}{2}\frac{\Delta T}{T_r}\frac{m_ec^2}{kT_e} \approx \frac{1}{20}.$$

It is possible that superclusters and bridges between clusters of galaxies filled by ionized gas have even larger optical depth. This increases even more the thermal effect in the direction toward them. The large angular

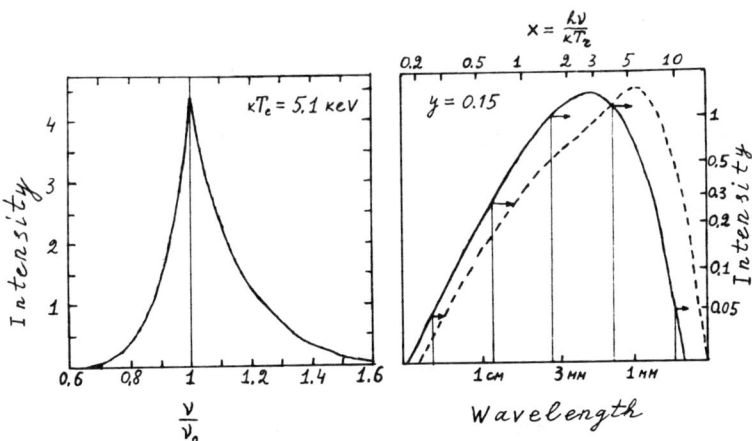

Figure 4 The monochromatic spectral line profile after one scattering by hot Maxwellian electrons (left). The spectrum of blackbody radiation (solid line) after multiple Compton scattering (broken line) is shown on the right.

dimensions may permit us to observe them both from the Earth's surface and from future radio satellites. This is important because observations in λ1–2-mm band may be possible where the effect is strongest.

Owing to the thermal effect a cluster of galaxies with hot intergalactic gas becomes a strong source of submillimeter radiation

$$L = 8 \cdot 10^{43}(1+z)^4 \left(\frac{M_{\text{IGG}}}{10^{14} M_\odot}\right) \left(\frac{kT_e}{5.1 \text{ keV}}\right) \left(\frac{T_r}{3 \text{ K}}\right) \text{ erg/s}$$

and a "negative" radiation source in millimeter and centimeter bands (Figure 7; Sunyaev 1980a). Here M_{IGG} is the mass of intergalactic gas in the cluster.

3.2 COSMOLOGICAL SIGNIFICANCE What could these observations contribute to astrophysics and cosmology? It may be possible to look for the hot intergalactic gas in clusters of galaxies and superclusters from the Earth's surface with radio telescopes. Large-scale irregularities in the gas might also be investigated in this way. The amplitude of the effect does not depend on the redshift. Only the angular dimension of the cloud depends on

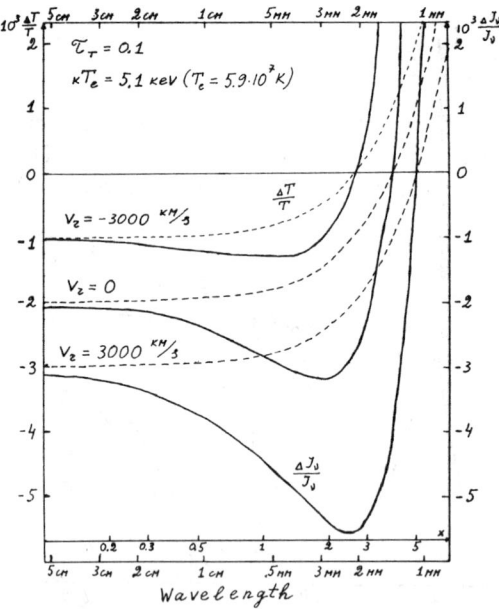

Figure 5 The dependence of changes in the background intensity (solid lines) and temperature (broken lines) in the direction of a cluster of galaxies for three different values of its peculiar velocity.

z. Therefore, it may be possible to "see" clusters and protoclusters of galaxies, even if they are at cosmological distances.

Radio observations combined with X-ray data would permit us to determine the gas density and temperature distribution inside clusters. Moreover we have the unique possibility of finding both the radial and angular dimensions of the cloud. Let us assume for simplicity that the cloud is isothermal and uniform. Then the surface brightness of the X-ray bremsstrahlung in the direction to the center of cloud is equal to

$$I_x = 2ARN_e^2 T_e^{-1/2} \exp\left\{-\frac{hv}{kT_e}\right\} g \qquad (3)$$

where A is constant, R is the radius of the cloud, and g is Gaunt factor. The background brightness temperature decrease in the same direction is equal to

$$\frac{\Delta T}{T_r} = -\frac{4kT_e}{m_e c^2} \sigma_T N_e R. \qquad (4)$$

X-ray observations at several frequencies corresponding to $hv < kT_e$ and $hv > kT_e$ would permit us to find kT_e. Using (3) and (4) one could easily find both N_e and R. Detailed X-ray and radio observations of the clusters would permit us, in principle (assuming that the cloud has spherical symmetry), to find $N_e(r)$, $T_e(r)$ and the effective radius R of the cloud (at a given angular distance from its center; Silk & White 1978).

Knowing the absolute and angular dimensions of the cluster we can easily find the distance to the cluster. The redshift of the lines in the spectra of the galaxies gives the recession velocity of the cluster. Therefore it is

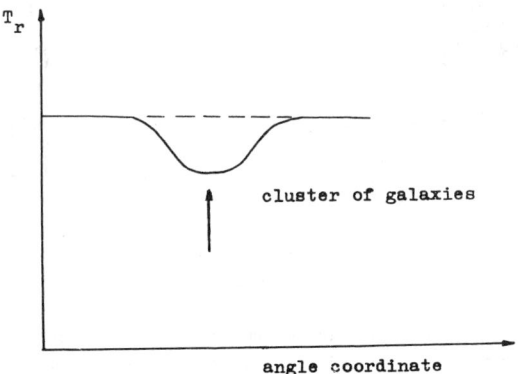

Figure 6 The decrease in the brightness temperature of the microwave background in the direction of a cluster of galaxies. This effect takes place only for $hv < 3.83kT_r$.

possible to find the Hubble constant (Cavaliere et al. 1977, 1979, Gunn 1978, Silk & White 1978, Birkinshaw 1979) without any intermediate steps and transitions from one distance indicator to another. Using data about clusters with high redshifts $z \gtrsim 0.5$ one could in principle also find the parameter $\Omega = 2q_0$, and hence the spatial curvature, all this without any assumptions about the evolution of clusters. X-ray data from HEAO-B-type satellites would be necessary for realization of this program. There are many difficulties. For example, bremsstrahlung by cold intergalactic clouds, compact radio sources in galaxies, and other sources of radio emission may influence the brightness of the background in the direction of the clusters (Bubukin 1974).

3.3 THE PECULIAR MOTIONS OF CLUSTERS Radio observations of hot gas in clusters of galaxies permit us to measure their peculiar velocities relative to the background in their vicinity (Sunyaev & Zel'dovich 1980).

Consider a cloud of cold electrons moving as a whole with radial peculiar velocity v_r. Scattering of the background photons by electrons leads, by virtue of the Doppler effect, to a change of the scattered radiation temperature in the direction to the cloud. Taking into account the small

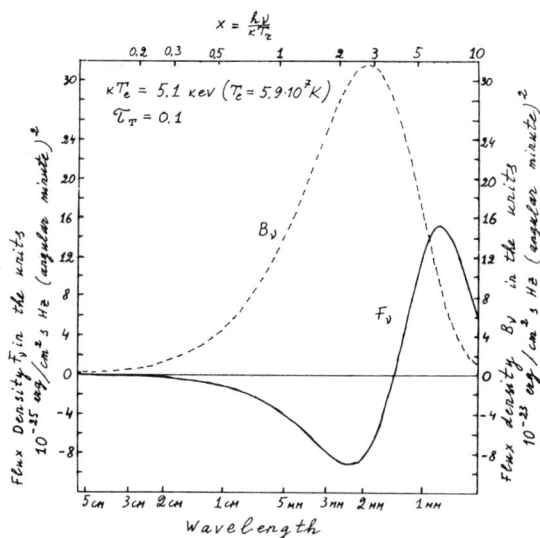

Figure 7 The change of the microwave-background energy density (solid line) in the direction of a cluster of galaxies as a function of the wavelength. The spectrum of blackbody radiation is shown on a different scale for comparison (broken line). The cluster becomes a "positive" radiation source in the submillimeter band, and a "negative" one at longer wavelengths. $T_r = 2.7$ K is assumed.

optical depth of the cloud we obtain

$$\frac{\Delta T}{T_r} = -\frac{v_r}{c}\tau_T.$$

The sign of the effect depends on the direction of the velocity (the positive sign corresponds to recession). The effect does not depend on the frequency. The small thermal and velocity effects add linearly. Figure 5 shows how the effective temperature and intensity of the background changes in the direction of the cloud with $\tau_T = 0.1$, $kT_e = 5.1$ keV for different peculiar velocities of the cluster through both thermal and motion effects. Measurements in two frequency bands (centimeter and millimeter) are needed to separate these two effects and find the radial component of peculiar velocity.

Measurements of polarization in the direction of the cloud would permit us to measure also the tangential component of its peculiar velocity.

Detailed measurements of brightness and polarization over the cluster surface would permit us to infer large-scale motions of gas in the cluster, including rotation. On scales $\lambda > 100$ Mpc present perturbations of the matter density are small $\delta\rho/\rho \ll 1$. If $\Omega \gtrsim 1$ they are growing $\delta\rho/\rho \sim t^{2/3}$. It follows from the continuity equation

$$\frac{\partial}{\partial t}\frac{\delta\rho}{\rho} + \text{div}\,\mathbf{v} = 0$$

that

$$\frac{v}{c} \approx \frac{1}{2\pi}\frac{\delta\rho}{\rho}\frac{\lambda}{ct}$$

in this case. Therefore, measurements of the peculiar velocities of clusters of galaxies would give important information about the spectrum of the large-scale matter density perturbations (Sunyaev & Zel'dovich 1980). If $\Omega < 1$ the perturbations grow only at $\Omega z > 1$ (see, for example, Sunyaev 1971). At smaller z they are frozen and peculiar velocities decrease with time.

3.4 THE SPECTRUM OF THE MICROWAVE BACKGROUND If the background spectrum differs from the blackbody one, the dependence of thermal effect on the frequency changes. Measurements of $\Delta I_v/I_v$ in the direction of a hot gas cloud on several frequencies permit us to find spectral distortions due both to interaction with the hot gas and to energy release at $z > 1$ (Sunyaev & Zel'dovich 1972b, Fabbri et al. 1978). Results of computations (Sunyaev 1980c) are presented in Figure 3, which shows the ratio of $\Delta I_v/I_v$ for the distorted spectrum to $\Delta B_v/B_v$ for a blackbody spectrum.

For $I_v \sim v^3$ (long-wavelength part of Bose-Einstein spectrum), $\Delta I_v/I_v$

= 0 in the direction to the hot gas cloud. For $I_\nu \sim \nu^{-\alpha}$ we easily obtain from the Kompaneets equation $\Delta I_\nu/I_\nu = \alpha(\alpha+3)y$. For $\alpha = 0.75$ (the spectral index of background radiation at meter wavelengths) we have

$$\frac{\Delta I_\nu}{I_\nu} = 2.8y, \quad \text{where} \quad y = \int_{-\infty}^{+\infty} \sigma_T N_e \, dl.$$

3.5 The anisotropy of the microwave background also changes its intensity in the direction of a cluster (Zel'dovich & Sunyaev 1980).

APPENDIX: ELEMENTS OF THE THEORY OF COMPTON SCATTERING

Thomson scattering has a cross section

$$\sigma_T = \frac{8\pi}{3}\left(\frac{e^2}{m_e c^2}\right)^2 = 6.65 \; 10^{-25} \text{ cm}^2$$

and angular dependence for unpolarized radiation

$$\frac{d\sigma}{d\Omega} = \tfrac{3}{8}\sigma_T(1+\cos^2\alpha)$$

where α is the scattering angle. A photon with frequency ν and wave vector $\nu\mathbf{\Omega}$ after being scattered by an electron with velocity $\mathbf{v}$ becomes a photon with $\nu'\mathbf{\Omega}'$. It follows from the momentum and energy conservation laws that

$$\frac{\nu'}{\nu} = \frac{1 - \dfrac{\mathbf{v}}{c}\mathbf{\Omega}}{1 - \dfrac{\mathbf{v}}{c}\mathbf{\Omega}' + \dfrac{h\nu}{\gamma m_e c^2}(1-\mathbf{\Omega}\mathbf{\Omega}')} \tag{A1}$$

where $\gamma = (1-v^2/c^2)^{-1/2}$, $\mathbf{\Omega}\cdot\mathbf{\Omega}' = \cos\alpha = \mu$. For hard photons in the limit $v/c \ll h\nu/m_e c^2$ the recoil effect plays the main role in the photon frequency change: $\Delta\nu/\nu = -h\nu(1-\cos\alpha)/m_e c^2$. In the opposite case the Doppler effect dominates $\Delta\nu/\nu = \mathbf{v}/c(\mathbf{\Omega}'-\mathbf{\Omega})$. The scattering of isotropic unpolarized radiation with $h\nu \ll m_e c^2$ in a thermal Maxwellian plasma with $kT_e \ll m_e c^2$ is of special interest for us. Kompaneets (1957; see also Weymann 1965) derived the kinetic equation describing the Compton interaction of electrons and isotropic unpolarized radiation field:

$$\frac{\partial n}{\partial t} = \frac{\sigma_T N_e h}{m_e c}\frac{1}{\nu^2}\frac{\partial}{\partial \nu}\nu^4\left(n+n^2+\frac{kT_e}{h}\frac{\partial n}{\partial \nu}\right) \tag{A2}$$

or

$$\frac{\partial n}{\partial y} = \frac{1}{\tilde{x}^2}\frac{\partial}{\partial \tilde{x}}\tilde{x}^4\left(n+n^2+\frac{\partial n}{\partial \tilde{x}}\right)$$

where

$$y = \int \frac{kT_e}{m_e c^2} d\tau_T \quad \text{and} \quad \tilde{x} = \frac{h\nu}{kT_e}, \quad \text{and} \quad n = \frac{c^2 I_\nu}{2h\nu^3}$$

is the occupation number in the photon phase space. This equation is valid for any relation between $h\nu \ll m_e c^2$ and $kT_e \ll m_e c^2$, if the spectrum is sufficiently smooth. The Equation (A2) conserves the number of photons. Its static solution given by $n + n^2 + (\partial n/\partial \tilde{x}) = 0$ is the Bose-Einstein equilibrium distribution $n = (e^{x+\mu} - 1)^{-1}$ with chemical potential $\mu \geq 0$. In the limiting cases we obtain the blackbody spectrum $n = (e^x - 1)^{-1}$ for $\mu = 0$ and the Wien distribution $n = e^{-x} \cdot e^{-\mu}$ for $\mu \gg 1$. Equation (A2) can be used to obtain the energy balance

$$\frac{dE_r}{dt} = -\frac{dE_e}{dt} \times N_e = \frac{8\pi h}{c^3} \int_0^\infty \frac{\partial n}{\partial t} \nu^3 \, d\nu = \frac{8\pi \sigma_T N_e h^2}{m_e c^4}$$

$$\times \int_0^\infty \nu \frac{\partial}{\partial \nu} \nu^4 \left(n + n^2 + \frac{kT_e}{h} \frac{\partial n}{\partial \nu} \right) d\nu = -w^+ + w^-.$$

After simple computations it is easy to find the plasma heating rate (Levich & Sunyaev 1971)

$$w^+ = \frac{8\pi h \sigma_T N_e}{m_e c^4} \int_0^\infty \nu^4 (n + n^2) \, d\nu \tag{A3}$$

and cooling rate (Weymann 1967)

$$w^- = \frac{4\sigma_T k T_e E_r}{m_e c} \tag{A4}$$

In a stationary state for electrons $w^+ = w^-$. Using (A3) and (A4) we find the stationary temperature of electrons in an arbitrary radiation field

$$kT_e = \frac{h}{4} \frac{\int_0^\infty (n + n^2) \nu^4 \, d\nu}{\int_0^\infty n\nu^3 \, d\nu} = \frac{2\pi h^2}{E_r c^3} \int_0^\infty n(1+n) \nu^4 \, d\nu \tag{A5}$$

(Zel'dovich & Levich 1970). For blackbody and Bose-Einstein distributions, $T_e = T_r$.

In the limiting case $T_e \gg T_r$ we can neglect terms n^2 and n in (A2), which then simplifies to the diffusion equation

$$\frac{\partial n}{\partial y} = \frac{1}{x^2} \frac{\partial}{\partial x} x^4 \frac{\partial n}{\partial x} \tag{A6}$$

where $x = h\nu/kT_r$ and $y = \sigma_T N_e ct$. Substituting into the right side of (A6) the initial blackbody spectrum $n_0 = (e^x - 1)^{-1}$ Zel'dovich & Sunyaev (1969) found the first iteration (Figure 2).

$$\frac{\Delta I_\nu}{I_\nu} = \frac{\Delta n}{n} = y \frac{xe^x}{(e^x - 1)} \cdot \left\{ x\left(\frac{e^x + 1}{e^x - 1}\right) - 4 \right\},$$
$$\frac{\Delta T_r}{T_r} = \frac{d \ln I_\nu}{d \ln T_r} \cdot \frac{\Delta I_\nu}{I_\nu} = y\left\{ x\left(\frac{e^x + 1}{e^x - 1}\right) - 4 \right\}. \tag{A7}$$

The exact solution for (A6) has the form (Zel'dovich & Sunyaev 1969)

$$n(x, y) = \frac{1}{\sqrt{4\pi y}} \int_0^\infty n_0(w) \exp\left\{ -\frac{(\ln x - \ln w + 3y)^2}{4y} \right\} \frac{dw}{w} \tag{A8}$$

where n_0 is the initial arbitrary radiation spectrum. It follows from (A8) that in the Rayleigh-Jeans region the radiation brightness decreases according to the law $T_{RJ} = T_r e^{-2y}$. Multiplying (A6) by $8\pi h\nu^3/c^3$ and integrating over the frequency, Kompaneets (1956) found that $E_r = E_0 e^{4y}$.

For the blackbody spectrum $n + n^2 = -\partial n/\partial x$. Therefore, for the initial blackbody spectrum and small deviations, Equation (A2) becomes identical to (A6) if we introduce a new variable

$$\tilde{y} = \frac{k(T_e - T_r)}{m_e c^2} \sigma_T N_e ct$$

(Weymann 1967).

In the early universe T_e is very close to T_r. The kinetic equation describing simultaneous action of the Compton and free-free processes was written by Kompaneets (1957)

$$\frac{\partial n}{\partial y} = \frac{1}{x^2} \frac{\partial}{\partial x} x^4 \left(\frac{\partial n}{\partial x} + n + n^2 \right) + \frac{K e^{-x}}{ax^3} [1 - n(e^x - 1)] \tag{A9}$$

where

$$a = \frac{dy}{dt} = \frac{kT_e}{m_e c^2} \sigma_T N_e c,$$

$$K = K_0 g(x) = \frac{8\pi}{3} \frac{e^6 h^2}{\sqrt{6\pi m_e^3}} N_e^2 (kT_e)^{-3.5} g(x)$$

$$= 1.25 \cdot 10^{-12} \cdot N_e^2 T_e^{-3.5} g(x)$$

and

$$g(x) \approx \begin{cases} 1 & \text{for } x \geq 1 \\ \frac{\sqrt{3}}{\pi} \ln \frac{2.35}{x} & \text{for } x < 1 \end{cases}$$

is the Gaunt factor. It is useful to introduce

$$x_0 = \sqrt{\frac{K(x_0)}{4a}} = 3 \cdot 10^5 N_e^{1/2} T_e^{-9/4} \sqrt{g(x_0)}. \quad (A10)$$

For $x \leq x_0 < 1$ free-free processes dominate, but for $x > x_0$ Compton scattering takes photons to the region with $x \sim 3$, forming the Bose-Einstein spectrum (see details in Sunyaev & Zel'dovich 1970b, Illarionov & Sunyaev 1975a,b; these papers are carefully reviewed by Danese & de Zotti 1977).

Literature Cited

Beigman, I. L., Sunyaev, R. A. 1978. *Preprint N 163*. Lebedev Physical Inst., Moscow
Birkinshaw, M., Gull, S. F., Northover, K. J. E. 1978. *MNRAS*. 185:245
Birkinshaw, M.1979. *MNRAS*. 187:847
Blair, A. G. 1974. *Confrontation of Cosmological Theories with Observational Data*, ed. M. S. Longair. *Proc. IAU Symp. No. 63*, p. 143
Bubukin, S. 1974. Thesis. Gorki State Univ.
Burbidge, G. 1967. Private Communication
Cavaliere, A., Danese, L., De Zotti, G. 1977. *Astron. Astrophys*. 217:6
Cavaliere, A., Danese, L., De Zotti, G. 1979. *Astron. Astrophys*. 75:322
Chan, K. L., Jones, B. J. T. 1975. *Ap. J*. 195:1
Chan, K. L., Jones, B. J. T. 1975. *Ap. J*. 200:454
Chan, K. L., Jones, B. J. T. 1975. *Ap. J*. 200:461
Culhane, J. L. 1978. In *Large Scale Structure of the Universe*, ed. Y. Einasto, M. S. Longair. Dordrecht: Reidel
Danese, L., De Zotti, G. 1977. *Riv. Nuovo Cimento* 7 (3):277
Dolgov, A., Zel'dovich, Ya. B. 1980. *Rev. Mod. Phys*. In press
Dubrovich, V. K. 1975. *Sov. Astron. Lett*. 1 (10):3
Dubrovich, V. K., Bernstein, I. N., Bernstein, D. N. 1977. *Sov. Astron*. 54:727
Fabbri, R., Melchiorri, F., Natale, V. 1978. *Astrophys. Space. Sci*. 59:223
Field, G., Perrenod, S. C. 1977. *Ap. J*. 215:717
Gamov, G. 1948a. *Phys. Rev*. 74:505
Gamov, G. 1948b. *Nature* 162:680
Ginzburg, V. L., Ozernoy, L. M. 1965. *Sov. Astron*. 42:943
Gould, R. J. 1972. *Ann. Phys. (NY)* 69:321
Gould, R. J., Raphaeli, Y. 1978. *Ap. J*. 219:12
Gull, S. F., Northover, K. J. E. 1975. *MNRAS* 173:535
Gunn, J. E. 1978. In *Observational Cosmology*, ed. Maeder, Tammann.

Geneva Obs.
Gunn, J. E., Peterson, B. A. 1965. *Ap. J*. 142:1633
Harrison, E. R. 1973. *Ann. Rev. Astron. Astrophys*. 11:155
Howell, T. F., Shakeshaft, J. R. 1966. *Nature* 210:1318
Howell, T. F., Shakeshaft, J. R. 1967. *Nature* 216:753
Illarionov, A. F., Sunyaev, R. A. 1975a. *Sov. Astron*. 18:413
Illarionov, A. F., Sunyaev, R. A. 1975b. *Sov. Astron*. 18:691
Kompaneets, A. S. 1957. *Sov. Phys. JETP* 4:730
Lake, G., Partridge, R. B. 1977. *Nature* 270:502
Lea, S. M., Silk, J., Kellogg, E., Murray, S. 1973. *Ap. J. Lett*. 184:L105
Levich, E. V., Sunyaev, R. A. 1971. *Sov. Astron*. 48:461
Longair, M. S., Sunyaev, R. A. 1971. *Sov. Phys. Usp*. 105:41
Muehlner, D., Weiss, R. 1970. *Phys. Rev. Lett*. 24:742
Pariyskiy, Yu. N. 1973. *Sov. Astron. AJ* (2) 50:453
Peebles, P. J. E. 1970. *Phys. Rev. D*. 1:397
Peebles, P. J. E. 1968. *Ap. J*. 153:1
Peebles, P. J. E. 1971. *Physical Cosmology*. Princeton Univ. Press
Pelyushenko, S. A., Stankevich, K. S. 1969. *Sov. Astron*. 13:223
Penzias, A. A., Wilson, R. W. 1965. *Ap. J*. 142:419
Penzias, A. A., Wilson, R. W. 1967. *Astron. J*. 72:315
Rees, M. J. 1978. *Nature* 275:35
Robson, E. I., Vickers, D. G., Huizinga, J. S., Beckman, J. E., Clegg, P. E. 1974. *Nature* 251:591
Roll, P. G., Wilkinson, D. T. 1966. *Phys. Rev. Lett*. 16:405
Sato, H., Matsuda, T., Takeda, H. 1971. *Prog. Theor. Phys. Suppl*. 49:11
Schallwich, D., Wielebinsky, R. 1979. *Astron. Astrophys*. 71:L15

Silk, J. 1968. *Ap. J.* 151:459
Silk, J., White, S. D. M. 1978. *Ap. J. Lett.* 226:L3
Smoot, G. F., Gorenstein, M. V., Muller, R. A. 1977. *Phys. Rev. Lett.* 39:14, 898
Stecker, F. W., Puget, J. L. 1972. *Ap. J.* 178:57
Stokes, R. A., Partridge, R. B., Wilkinson, D. T. 1967. *Phys. Rev. Lett.* 19:1199
Sunyaev, R. A. 1968. *Sov. Phys. Dokl.* 13:183
Sunyaev, R. A. 1971. *Astron. Astrophys.* 12:190
Sunyaev, R. A. 1974. *Confrontation of Cosmological Theories with Observational Data*, ed. M. S. Longair. *Proc. IAU Symp. No. 63*, p. 167
Sunyaev, R. A. 1978. In *Large Scale Structure of the Universe*, ed. Y. Einasto, M. S. Longair. Dordrecht: Reidel
Sunyaev, R. A. 1980a. *Sov. Astron. Lett.* 6:387
Sunyaev, R. A. 1980b, c. *Sov. Astron. Lett.* In press
Sunyaev, R. A., Zel'dovich, Ya. B. 1970a. *Astrophys. Space Sci.* 7:3
Sunyaev, R. A., Zel'dovich, Ya. B. 1970b. *Astrophys. Space Sci.* 7:20
Sunyaev, R. A., Zel'dovich, Ya. B. 1970c. *Astrophys. Space Sci.* 9:368
Sunyaev, R. A., Zel'dovich, Ya. B. 1972a. *Astron. Astrophys.* 20:189
Sunyaev, R. A., Zel'dovich, Ya. B. 1972b. *Comments Astrophys. Space Sci.* 4:173
Sunyaev, R. A., Zel'dovich, Ya. B. 1980. *MNRAS.* 190:413
Thaddeus, P. 1972. *Ann. Rev. Astron. Astrophys.* 10:306
Weinberg, S. 1972. *Gravitation and Cosmology.* New York, NY: Freeman
Woody, D., Richards, P. 1979. *Phys. Rev. Lett.* 42:925
Weymann, R. 1965. *Phys. Fluids* 8:2112
Weymann, R. 1967. *Ap. J.* 147:887
Zel'dovich, Ya. B. 1967. *Sov. Phys. Usp.* 9:602
Zel'dovich, Ya. B., Illarionov, A. F., Sunyaev, R. A. 1972. *Sov. Phys. JETP* 35:643
Zel'dovich, Ya. B., Kurt, V. G., Sunyaev, R. A. 1969. *Sov. Phys. JETP* 28:146
Zel'dovich, Ya. B., Levich, E. V. 1970. *Sov. Phys. JETP Lett.* 11:57
Zel'dovich, Ya. B., Novikov, I. D. 1967. *Relativistic Astrophysics.* Moscow: Nauka
Zel'dovich, Ya. B., Novikov, I. D. 1975. *The Structure and Evolution of the Universe.* Moscow, USSR: Nauka
Zel'dovich, Ya. B., Sunyaev, R. A. 1969. *Astrophys. Space Sci.* 4:301
Zel'dovich, Ya. B., Sunyaev, R. A. 1980. *Sov. Astron. Lett.* In press

AUTHOR INDEX

(Names appearing in capital letters indicate authors of chapters in this volume.)

A

Aannestad, P. A., 505
Aarseth, S. J., 193
Ables, H., 365
Achterberg, A., 257
Adachi, I., 98, 108
Adams, N. G., 280
Adams, P. J., 472
Adams, T. F., 22, 25
Adgie, R. L., 168, 169
Aggarwal, H. R., 105
Ahern, F. J., 284
Ahlen, S. P., 426
Aitken, D. K., 49, 283
Aizu, E., 310
Ake, T. B., 264
Alcock, C., 271, 374, 385, 387, 389, 390
Alexander, D. R., 278
Alfvén, H., 80, 104, 106
Allan, H. R., 299
Allègre, C. J., 79
Allen, C. W., 312, 376
Allen, L. R., 138, 145, 153, 169
Allen, R. J., 313
Aller, H. D., 326, 333, 336, 339, 340
Alloin, D., 421, 432
Alpher, R. A., 489
Altenhoff, W. J., 269, 503
Althouse, W. E., 422
Altschuler, D. R., 325, 327, 341, 342
Alvarez, L. W., 306
AMBARTSUMIAN, V. A., 1-13
Ammann, M., 129
Anderegg, M., 49
Andernach, H., 173, 180, 203
Anders, E., 79
Andersen, J., 134, 136, 137, 143, 160
Anderson, B., 168, 169
Anderson, C. M., 450
Anderson, D. L., 105
Andrew, B. H., 324, 340
Andrews, D. H., 134, 135
ANGEL, J. R. P., 321-61; 209, 234, 268, 275, 276, 322-24, 326, 333, 334, 336-39, 341, 344-46, 351, 352, 355, 379, 380, 382
Angione, R., 203
Anicich, V. G., 402
Annestad, P. A., 235, 245
Appenzeller, I., 346, 506

Arawal, P. C., 325
Arens, J. F., 306, 425
Argue, A. N., 172
Armstrong, K. R., 505
Arnett, W. D., 241, 383, 434
Arnold, C. N., 242
Arp, H., 332, 333
Arrhenius, G., 80, 106
Asbridge, J. R., 19, 228
Asséo, E., 303
Athay, R. G., 440, 441, 444, 445, 451, 452, 458, 462, 470, 472, 477, 481, 483
Audouze, J., 306, 427, 432
Auer, L. H., 367, 371, 372
Aumann, H. H., 49
Auriemma, C., 178
Auvergne, M., 446
Avedisova, V. S., 254
Avni, Y., 327, 346
Avrett, E. H., 440, 451-56, 477, 482
Axford, W. I., 169, 255, 300, 304
Axon, D. J., 322
Ayres, T. R., 442, 450, 452-54, 457, 461, 463-66, 477, 478, 481, 482

B

Baade, W., 16, 197, 332, 418
Baath, L., 194-96
Babadzhanyants, M. K., 323, 326, 331, 335, 344, 345
Babcock, H. W., 19
Badhwar, G. D., 303, 310, 311
Baggio, R., 189
Bahcall, J. N., 115
Bailey, J. A., 325
Baker, J. R., 173, 180, 203
Balasubrahmanyan, V. K., 306
Baldwin, B. J., 43, 66
Baldwin, J., 313
Baldwin, J. A., 324, 326, 327, 330, 331, 334, 342
Baldwin, J. E., 176, 198, 313, 327
Balick, B., 419
Baliunas, S. L., 450, 453-57, 475, 477-79, 482
Ball, J. A., 274, 276, 284
Balona, L. A., 16, 28, 34, 35
Baluteau, J. P., 49
Bame, S. J., 19, 228
Bappu, M. K. V., 462
Barber, D., 169

Bare, C. C., 168, 325, 327
Barkat, A., 271, 278, 280
Barker, E. S., 472
Barlow, D. J., 128
Barnes, J. V., 136, 156
Barnes, T. G., 116, 265, 266, 271, 368, 370, 374, 417, 450, 477
Barrel, S. L., 28, 29, 36
Barrell, J., 105
Barry, D. C., 119
Barshay, S. T., 55
Bartoe, J.-D. F., 445, 451, 452, 456, 471
Basri, G. S., 270, 442, 450-56, 459, 461, 466-69, 471-75, 477
Batchelor, R., 325, 327
Batten, A. H., 124, 126-31, 134, 136-40, 142-45, 156
Baud, B., 194-96
Beaver, E. A., 197, 203, 268, 275, 332, 344, 345, 352
Bechis, K. P., 265, 275, 282
Beck, R., 313
Beck, S. C., 250, 277
Becker, R. H., 479
Becker, S. A., 18, 20, 38
Beckers, J. M., 471, 472
Becklin, E. E., 250, 272, 275
Beckman, J. E., 49, 511, 543
Beckwith, S., 243, 250, 251, 258, 268, 273, 278, 280, 283
Beddoes, D. R., 144
Bednarczyk, S. M., 495
Beebe, H. A., 475
Beer, R., 49, 55, 266, 417
Behall, A. L., 148, 149, 151, 365
Beigman, I. L., 541
Bell, A. R., 229, 255, 256, 304
Bell, R. A., 121
Benedict, W. S., 498
Benford, G., 170
Bennett, K., 314, 315
Benson, J. M., 277
Benvenuti, P., 240, 241
Berezinsky, V., 316
Berezne, J., 49, 58
Berlin, A. B., 194
Bernat, A. P., 265, 266, 274, 275, 279, 474
Bernstein, D. N., 541
Bernstein, I. N., 541
Berriman, G., 273, 278, 280
Bertola, F., 209

AUTHOR INDEX

Bessell, M. S., 364, 369, 371-73, 376, 377, 393
Betz, A. L., 265, 269, 271, 272, 274-76, 280
Bezzerides, B., 224
Bidelman, W. P., 446
Biermann, P., 313
Bieser, F. S., 422, 424, 425
Bigeleisen, P. E., 89
Bignami, G. F., 314, 315
Bijaoui, A., 446
Bingham, R. G., 322
Binnendijk, L., 115
Birck, J. L., 79
Bird, G. R., 498, 517
Birkinshaw, M., 176, 181, 551, 554
Biskamp, D., 230
Black, D. C., 79
Black, J. H., 237, 241, 251, 252, 448, 450
Blackman, G. L., 252
Blackwell, D. E., 269, 421, 432
Blair, A. G., 508, 538
Blanco, C., 448, 450
Blanco, V. M., 81, 108, 203, 209
Blandford, R. D., 170, 187, 193, 195, 196, 200, 202, 204, 255, 291, 304, 307, 349-52
Blazit, A., 145
Bless, R. C., 116
Blitz, L., 252
Bloomer, R. H., 139
Blumenthal, G., 194
Bodenheimer, P., 79
Boesgaard, A. M., 270, 274
Boglin, A., 446
Bohm, H. U., 458
Böhm, K. H., 239, 243, 375, 377, 384, 386, 387, 391, 393
Böhm-Vitense, E., 367, 446-48, 450
Boksenberg, A., 209, 212, 333, 376, 386
Boldt, E. A., 190, 325, 327, 344, 350, 479
Bolton, J. G., 324, 325, 327
Bonneau, D., 145
Boone, R. L., 326, 331-33, 336, 338, 339, 346, 351, 355
Bopp, B. W., 476, 479, 481
Borken, R., 314
Boughn, S. P., 522-24
Bowers, P. F., 265, 269, 274-79, 282, 283
Bowyer, C. S., 477-79
Bowyer, S., 326, 367, 371
Boynton, P. E., 492-94
Bozkurt, S., 128

Bradt, H. V. D., 344, 371
Braes, L. L. E., 212
Brand, P. W. J. L., 243
Brandie, G. W., 208, 325
Branduardi, G., 327, 346, 447
Breakiron, L. A., 145
Brecher, K., 291, 386, 387
Bregman, J. D., 49, 58, 60, 66, 267
Bregman, J. N., 267
Brett, R. A., 335
Bridges, C. A., 471
Bridle, A. H., 176, 180, 185, 194, 195, 198, 202-5, 208, 210, 325, 327
Briel, U., 326, 327
Brindle, C., 307
Broadfoot, A. L., 240
Broderick, J. J., 325, 327
Broten, N. W., 168
Brown, A., 448-50, 457
Brown, D. R., 472
Brown, J. C., 452
Brown, R. H., 168
Brown, R. L., 243, 271
Brueckner, G., 445, 451, 452, 456, 471
Bruner, E. C. Jr., 462
Bruno, M. S., 329
Bubukin, S., 554
Buccheri, R., 314, 315
Budding, E., 127
Bues, I., 388
Buff, J., 225
Buffington, A., 306, 309-11
Bühler, F., 424
Bulanov, S. V., 313
Bunch, T. E., 43, 66
Burbidge, E. M., 203, 209, 324, 326, 327, 334, 342
Burbidge, G. R., 169, 193, 203, 209, 291, 324, 326, 348, 349, 386, 387
Burch, S. F., 172, 188-90, 192, 198, 202, 205
Burke, B. F., 277, 492, 493, 503
Burke, E. M., 139
Burke, J. R., 234
Burke, W. L., 520
Burns, J. A., 108
Burns, J. O., 198, 202, 204, 205, 210
Burroughs, W. J., 498, 517
Burton, W. B., 308, 408
Busko, I. C., 476
Bussoletti, E., 49
Butcher, H., 203, 209, 258
Butler, S. E., 236, 241, 457
Bychkov, K. V., 241
Byrd, G. G., 191
Bystedt, J., 172, 192, 198, 202

C

Cacciari, C., 441, 447, 459-61
Caderni, N., 530
Cahn, J. H., 268, 271, 280, 281, 283, 284, 382, 383
Caldwell, C. N., 156
Caldwell, J. H., 306
Callahan, P. S., 193
Cameron, A. G. W., 80, 82, 89
Cameron, M. J., 173
Cameron, R. C., 133
Campbell, B., 153
Campbell, M. F., 268
Cannon, R. D., 335
Cannon, W., 327
Capps, R. W., 326, 335, 339, 340, 345
Caraveo, P., 314
Carbon, D. F., 274, 275, 454
Carleton, N. P., 517
Carlson, E. R., 404, 405, 407
Carlson, R. W., 55
Carpenter, K. G., 446-48, 450
Carson, T. R., 18, 20, 375, 393
Carstensen, E., 130
Carswell, R. F., 324, 326
Carter, B., 16
Cartwright, B. G., 426
Cash, W. C., 367, 447, 477-79
Cassé, M., 307, 309, 314
Cassinelli, J. P., 278, 386, 447
Castor, J. I., 22, 25, 220, 221, 232
Catalano, S., 448, 450
Catlin, L. L., 508
Catura, R. C., 203
Caughlan, G. R., 427
Cavaliere, A., 554
Cayrel, R., 441
Cernicharo, J., 412
CESARSKY, C. J., 289-319; 293, 296-98, 303, 306, 307, 311, 314, 315
Cess, R. D., 56
Cester, B., 115, 122, 126-28, 130, 132, 136, 137, 139, 156
Chaiken, J., 105
Chaisson, L., 326, 327
Chambliss, C. R., 138
Chan, K. L., 543
Chandrasekhar, S., 94, 363
Chang, S. H., 442, 453-56, 461, 473
Chapman, C. R., 107
Chapman, E. A., 455
Charland, Y., 385
Charles, P., 326, 387, 447, 477-79
Chen, K.-Y., 128

AUTHOR INDEX   563

Cheng, E. S., 522-24, 526, 528, 529
Chernyshev, V. I., 492, 493
Chevalier, R. A., 220, 221, 225, 229, 230, 232, 234, 241, 253, 254, 302, 419, 433
Chin, G., 251
Chin, Y., 294
Chipman, E. G., 451, 452, 456
Chisholm, R. M., 168
Chiu, G., 380
Chiu, H. Y., 442, 453-56, 461, 472, 473
Christiansen, W. A., 169, 195, 200
Christiansen, W. N., 172, 193
Christy, J., 365, 368
Christy, R. F., 15, 22, 25, 26
Chu, C. K., 222, 224
Chun, M. S., 421, 432
Chu. S.-I., 235, 249-51
Ciatti, F., 284
Cipolla, J. W., 231
Ciurla, T., 326, 335, 338
Clark, B. G., 168, 195, 325, 327
Clark, G. W., 314, 371
Clark, R. N., 66
Clark, T. A., 327, 353
Clarke, D., 276
Clarke, R. W., 168
Clary, W. G., 147
Clausen, J. V., 123, 126, 128, 130, 136, 137
Claussen, M. J., 277
Clayton, M. L., 281
Clayton, R. N., 435
Clegg, P. E., 511, 543
Clegg, R. E. S., 264, 265, 267
Cleghorn, T., 310
Clements, G. L., 119
Cochran, W. D., 251
Code, A. D., 116, 136, 139
Cogan, B. C., 15-18, 20, 21, 25, 28, 29, 34, 36
Cohen, J. G., 275, 277
Cohen, M. H., 166, 168, 172, 181, 195, 196, 202, 204, 267, 273, 325, 327, 334, 349
Colacevich, A., 145
Cole, D. J., 172, 173, 180, 209
Colgate, S. A., 350
Colla, G., 325, 327
Collins, C. B., 520
Collins, G. W., 19
Collin-Souffrin, S., 421, 432
Colvin, J. D., 350
Combes, M., 49, 51, 55-58, 60, 65
Comstock, G., 310

Condon, J. J., 325, 327, 341, 349
Conklin, E. K., 324, 338, 521-23
Connes, J., 48
Connes, P., 48, 147
Connolly, L. P., 418
Conrath, B., 49, 53
Consolmagno, G. J., 72
Conti, P. S., 138, 374, 447
Conttrell, P. L., 383
Conway, R. G., 167, 172, 197, 205, 325, 327, 335, 342
Cook, J. W., 457
Cooke, B. A., 185
Cooper, B. F. C., 172, 173, 180, 209, 325, 327
Cordova, F., 477
Corey, B. E., 522-24, 526, 528, 529
Corliss, C. H., 55
Cornett, R. H., 251
Coron, N., 49
Coroniti, F. V., 292
Costain, C. H., 185
Costero, R., 241
Cotton, W. D., 327, 328, 353
Cottrell, P. L., 376, 390
Couch, R. G., 434
Counselman, C. C., 327, 353
Cousins, A. W. J., 21, 37, 128, 136
Cowie, L. L., 178, 191, 225, 233, 234, 241, 250, 251
Cowley, A. P., 138
Cowsik, R., 290, 306, 311
COX, A. N., 15-41; 16-22, 25-29, 34-37, 378
Cox, D. P., 235, 237, 240, 257, 294, 304
Cox, J. P., 16, 19, 22, 25-27, 39
Cox, L. P., 92, 100, 102, 104
Coyne, G. V., 324, 339, 345
Crabtree, D. R., 272
Craine, E. R., 324, 326, 329, 330
Cram, L. E., 451, 452, 456, 461, 463, 466, 470-72
Crane, P., 193
Crane, P. C., 194
Crawford, D. L., 125, 136, 138, 156
Crawford, H. J., 424, 425
Cromwell, R. H., 119, 322
Crowne, A. H., 324, 326
Crowther, J. H., 168, 169
Cruikshank, D. P., 43, 49, 58, 61, 66, 72
Crutcher, R. M., 247, 251
Cudworth, K. M., 383
Culhane, J. L., 233, 234, 550
Cummings, A. C., 314, 422

Curtis, H. D., 197
Czarny, J., 447
Czyzak, S. J., 267

D

Dachs, J., 126
Dahn, C. C., 365, 368, 376, 393, 394
Dainty, J. C., 144
Dalgarno, A., 227, 245, 248, 249
Dall'oglio, G., 508
Daltabuit, E., 235, 257, 283, 506
D'Amico, N., 314, 315
Damle, S. V., 309
Danese, L., 489, 517, 538, 548, 554, 559
Daniel, R. R., 289, 310, 311
Danielson, R. E., 53
Dantel, M., 446
D'Antona, F., 388
Danziger, I. J., 324-26, 329, 330
Darland, J. J., 156
Das Gupta, M. K., 169
David, D. S., 51, 58, 60
Davidson, K., 282, 284, 350
Davidson, R. C., 230, 231
Davies, I. M., 325, 327
Davies, S. T., 312
Davis, C. G., 22, 25
Davis, J., 116, 138, 145, 153
Davis, L. E., 277, 283
Davis, M., 233
Davis, M. M., 180, 198, 204, 210, 325, 327
Davis, R., 446, 448, 450
Davis, R. J., 172, 327, 335, 342
Davis, W. D., 122
Davison, D. K., 25
Dawson, J. M., 231
Dean, J. F., 21, 37
Dearborn, D., 374, 427, 430
Dearborn, D. S. P., 382
de Bergh, C., 49, 51, 56, 58, 60
de Bruyn, A. G., 165, 194, 346
DeCampli, W. M., 82
de Jager, C., 450, 474
de Jong, T., 245, 249, 252
Dekkers, N. H., 66
de Loore, C., 447, 458
Delvaille, J., 203
Demarque, P., 79, 393
Dennison, B., 349
DeNoyer, L. K., 250
Dent, W. A., 169, 325, 327, 335
Dermott, S. F., 80
Desikachary, K., 455
Dettmann, T., 367, 446-48, 450

AUTHOR INDEX

Deubner, F.-L., 458
Deupree, R. G., 18, 19, 21, 25, 27, 28, 35, 36
Deutsch, A. J., 277, 279
de Vaucouleurs, G., 203, 333
Devinney, E. J., 135, 139
De Young, D. S., 165, 169, 186, 190
DeZotti, G., 489, 517, 538, 548, 554, 559
Dibaj, E. A., 323, 345
Dickel, H. R., 251
Dickel, J. R., 251
Dickinson, D. F., 228, 265, 270, 274, 281, 325, 327, 342
Dietrich, W. F., 424
Dilworth, C., 311
Disney, M. J., 324
Dodds, D., 316
D'Odorico, S., 240, 241
Dogiel, V. A., 313
Doherty, L. R., 440
Dole, S. H., 92, 100
Dolgov, A., 548
Dombrovsky, V. A., 331, 335, 344, 345
Dominy, J. F., 143
Donahue, T. M., 109
Donaldson, W., 168, 169
Donnison, J. R., 82
Dopita, M. A., 234, 235, 239-43, 257
Doschek, G. A., 445, 457
Dower, R. G., 344
Downes, A., 189, 194, 198
Doxsey, R. E., 185, 325, 327, 344
Draine, B. T., 234, 240, 247
Dravins, D., 475
Drawin, H. W., 236
Dreher, J. W., 189, 192, 194, 195
Droge, F., 503
Dubrovich, V. K., 541
Duerr, R., 324, 326, 327
Dufour, R. J., 203, 209, 239, 241, 420
Dukes, R. J., 136
Dumont, P. J., 126-30, 136, 137, 139, 142
Dunn, R. B., 483
Dupree, A. K., 237, 241, 440, 446, 448, 450, 453-57, 475, 477-79
Durand, R. A., 18
Durrant, C. J., 472
Dwek, E., 267, 280
Dworetsky, M. M., 145
Dwyer, R., 425
Dyck, H. M., 324, 338, 447
Dziembowski, W., 19, 378

E

Eardley, D. M., 350, 351
Earl, J. A., 310
Eaton, J. A., 156, 476, 477
Eberhardt, P., 424
Eddington, A. S., 363
Eddy, J. A., 481
Edge, D. M., 316
Edgeworth, K. E., 83
Edmunds, M. G., 421, 432
Edwards, P. L., 321, 325, 327
Efimov, Y. S., 335
Egan, W. G., 267
Eggen, O. J., 129, 147, 149, 366
Eggleton, P. P., 143
Eichhorn, H., 147
Eichler, D., 187, 255-57
Eilek, J. A., 187
Ekers, J. A., 325, 327
Ekers, R. D., 166, 177, 178, 180, 181, 189, 190, 193-95, 324-26, 329, 330
Elgaroy, O., 168
Elias, J. H., 268
Elitzur, M., 245, 247-49, 252
Elliot, H., 312
Elliott, J. L., 351
Ellis, G. R. A., 504
Elmegreen, B. G., 228, 251, 252
Elmegreen, D. M., 251, 252
Elsmore, B., 325, 327
Elvis, M., 327, 346, 447
Elvius, A., 326, 345, 352
Emerson, D. T., 166, 313
Emslie, A. G., 452
Encrenaz, T., 49, 56-58, 60, 65
Encrenoz, P. J., 407, 411
Engels, D., 265, 276
Engvold, O., 463, 464, 466
Eoll, J. G., 18
Epstein, A., 229
Epstein, R. I., 428
Erickson, E. F., 43, 49, 53, 54, 58, 66
Erpenbeck, J. J., 253
Esepkina, N. A., 194
Espenak, F., 481
Etzel, P. B., 122
Evans, C. D., 471
Evans, D. S., 116, 143, 368, 374, 380, 450, 477
Evans, N. J. II, 251
Evans, N. R., 16, 34
Evans, R. G., 446, 457
Ewing, M. S., 195, 327, 492, 493

F

Fabbiano, G., 185, 229, 327, 346
Fabbri, R., 522, 523, 530, 555
Fabian, A. C., 203, 229, 276
Fabricant, D., 447, 448
Fachus, L. J., 325, 327
Falgarone, E., 294
Fanaroff, B. L., 169, 178
Fanti, C., 325, 327
Fanti, R., 178, 180, 189, 190, 194, 195, 325, 327
Faulkner, D. J., 17, 21, 27-29, 36
Feast, M. W., 137, 281
Fedel, B., 122, 126-28, 130, 132, 136, 137, 139
Feierberg, M. A., 72, 73
Feierman, B. H., 148
Feigelson, E., 203, 327, 346
Fekel, F. C., 143, 145
Feldman, P. A., 185
Feldman, U., 445, 457
Feldman, W. C., 19
Felenbok, P., 447
Feltz, K. A., 506
Fermi, E., 291, 292, 307
Fernie, J. D., 15, 17, 20
Ferrari, A., 187
Ficarra, A., 325, 327
Fichtel, C. E., 314, 316
Field, G. B., 235, 254, 543
Fink, U., 43, 44, 46, 49, 51, 55, 56, 58, 60-62, 64, 66, 68-70, 72, 73, 250
Finsen, W. S., 146-49, 152, 154
Fischel, D., 25
Fischer, P. C., 203
Fisher, A. J., 425
Fisher, W. A., 133, 139
Fisk, L. A., 255, 298
Fitch, W. S., 29
FitzGerald, M. P., 117, 284
Fitzgerald, P. M., 282, 284
Fix, J. D., 277, 278
Flasar, F. M., 193
Flasar, M., 49, 53
Fletcher, J. M., 124, 126-31, 134, 136-40, 142-45, 156
Flewelling, R. F., 292
Flower, P. J., 28, 36
Fogh Olson, H. J., 147
Fomalont, E. B., 167, 168, 171, 176, 180, 190, 194-96, 198, 202, 204, 205, 208, 210, 325, 327, 331, 342
Fontaine, G., 375, 385-89, 393
Fonti, S., 508
Foote, E. A., 295, 297
Forbes, J. E., 18
Ford, H. C., 258

AUTHOR INDEX 565

Ford, W. K., 209
Formiggini, L., 325, 327
Forrest, W. J., 58, 60, 67
Forslund, D. W., 224
Forster, J. R., 188, 194, 195, 203
Fosbury, R. A. E., 239, 257, 324-26, 329, 330, 451, 463, 464
Fourcade, S., 79
Fowler, W. A., 427
Fowles, G. R., 253
Fox, K., 55
Fragola, J. L., 367
Fram, D. M., 522-24
Franz, O. G., 147
Fraquelli, D. A., 479
Fraser, C. W., 203
Frater, R. H., 172, 193
Frawley, W. M., 270
Frazier, E. N., 442, 460, 482
Freedman, I., 308
Freier, P. S., 306, 309, 310
Freire, R., 447
French, D. K., 307
French, H. B., 324, 326, 330, 336
Frerking, M. A., 407
Fricke, K., 21, 25, 26, 35
Fridman, A. M., 84
Frogel, J. A., 272, 275, 277
Fujii, M., 310
Furenlid, I., 442, 453-56, 461, 463, 473
Furniss, I., 49
Fusi-Pecci, F., 280, 382

G

Gabriel, A. H., 483
Gallagher, J., 364
Galt, J. A., 168
Gamov, G., 537
Ganapathy, R., 79
Gandolfi, E., 325, 327
Garcia-Munoz, M., 309, 425
Gardner, C. S., 253
Gardner, F. F., 187, 188, 325, 327, 407, 409, 411, 503
Garmire, G. P., 233, 234, 314, 387, 477, 479
Garrard, T. L., 422
Garrison, R. F., 156
Gaskell, C. M., 237, 324, 325
Gatewood, C. V., 367
Gatewood, G., 145
Gatewood, G. D., 367
Gatley, I., 243, 268, 273, 278, 280, 283
Gautier, D., 49, 53
Gautier, T. N. III, 55, 66, 70, 250

Gavazzi, G., 176, 180
Gearhart, M. R., 324
Geballe, T. R., 49, 250, 266, 277, 417
Gebbie, H. A., 498, 517
Gebbie, K. B., 482
Gebel, W., 251
Gehrels, T., 80
Gehrz, R. D., 279
Geiss, J., 424
Geldzahler, B. J., 194, 325, 327
Geller, M. J., 329
Geller, R. B., 327, 353
Gent, H., 168, 169
Gerola, H., 252
Gershberg, R. E., 477
Getmantsev, G. G., 167, 298
Gezari, D. Y., 268, 505
Ghigo, F. D., 203
Giacconi, R., 203, 229, 327, 346, 368
Giampapa, M. S., 450
Giannone, P., 115
Giannuzzi, M. A., 115
Gibson, D. M., 208, 477, 479
Gierasch, P., 49, 53
Gilbert, G., 324, 326
Giler, M., 308, 311
Gillett, F. C., 48, 49, 58, 60, 67, 277
Gilliam, L. B., 471
Gilman, C., 310
Gilman, P. A., 481
Gilman, R. C., 267, 269
Gilmore, W., 265, 275, 278, 282, 283
Gilra, D. P., 265, 267, 273, 275, 278, 282, 283
Gimmestad, G. G., 498, 517
Ginzburg, V. L., 167, 289, 301, 311, 542
Gioia, I., 327
Gisler, G. R., 176, 191, 210
Giuli, R. T., 92
Giuliani, J. L., 254
Glackin, D. L., 450
Glebocki, R., 466
Gliese, W., 146, 148, 149, 152
Goeman, R. A., 311
Goirocom, G., 122, 126-28, 130, 132, 136, 137, 139
Goldberg, L., 275, 464
Golden, K. I., 231
Golden, R., 310
Golden, R. L., 310, 311
Goldreich, P., 84, 85, 89, 266, 274, 276, 280
Goldsmith, D. W., 251, 335
Goldsmith, P. F., 249, 404, 405, 407, 410, 413
Gol'nev, V. Ya., 194
Golson, J. C., 136, 156

Goody, R., 499, 501
Goorvitch, D., 49, 53, 54, 58
Gopal-Krishna, 189, 192, 194
Gora, E. K., 499
Gordon, H. A., 308
Gorenstein, M. V., 495, 522, 523, 527, 528
Gorenstein, P., 233, 447, 448
Goret, P., 309
Gosling, J. T., 19, 228
Goss, W. M., 172, 193, 209, 239, 252, 257, 324-26, 329, 330
Gottlieb, E. W., 327
Gough, D. O., 25, 461
Gould, R. J., 345, 548, 551
Gow, C. E., 272
Gower, J. F. R., 325, 327
Graham, D., 195, 327
Graham, J. A., 203, 209, 371
Grandi, S. A., 325, 327
Grasdalen, G. L., 78, 250, 324, 325, 334
Gray, D. F., 369, 455, 481
Green, J., 364
Green, R. F., 364, 371, 373, 379, 380, 383, 384, 391, 392, 394
Greenberg, R., 90, 91, 103, 104, 107
Greenbaum, M., 517
Greenstadt, E. W., 225, 229
Greenstein, J. L., 365-67, 369, 370, 373, 376-78, 383-87, 389, 390, 392
Greiner, D. E., 422, 424, 425
Greisen, E. W., 198, 202
Grenfell, T. C., 376, 388
Gresham, M., 366, 376, 393, 394
Griffiths, R. E., 325-27, 344
Grindlay, J. E., 185, 203, 447
Grønbech, B., 123, 126, 127, 130, 137
Gronenschild, E. H. B. M., 229, 233, 234
Gross, R. A., 222, 224
Grossman, L., 435
Grossmann-Doerth, U., 472
Grueff, G., 325, 327
Güdür, N., 128, 137
Guelin, M., 251, 407, 412, 431
Guetter, H., 365
Guidi, I., 522, 523
Guinan, E. F., 365
Guindon, B., 211
Gulkis, S., 325, 327, 342, 495
Gull, S. F., 551
Gull, T. R., 233, 244
Gülmen, O., 128
Gundermann, E. J., 166
Gunn, J. E., 542, 554

AUTHOR INDEX

Gurevich, L. E., 83
Gursky, H., 185, 190
Gush, H. P., 517
Gustaffson, B., 121
Guthrie, B. N. G., 208
Gwinn, W. D., 252
Gyldenkerne, K., 126, 127, 129, 130

H

Haber, I., 230, 231
Hackney, K. R., 325, 331
Hackney, R. L., 325, 331
Haddock, F. T., 169
Hagen, F. A., 425
Hagen, W., 270, 273, 274, 278, 279, 474
Hagen-Thorn, V. A., 323, 326, 331, 335, 344, 345
Haisch, B. M., 269, 270, 445-48, 450, 457, 473-75, 477
Halbig, M., 228
Hall, D. N. B., 265, 266, 269, 271, 274, 275, 454
Hall, D. S., 142, 447, 476, 477, 479
Hall, R. G., 150, 151
Halliwell, M. J., 145
Halran, E., 326, 335, 338
Hamada, T., 368, 370
Hamilton, P. A., 504
Hammer, D. A., 230, 231
Hammer, R., 458
Hanbury Brown, R., 116, 138, 145, 153, 169
Hanel, R., 49, 53
Hanisch, R. J., 176
Hanks, T. C., 105
Hans, E. M., 145
Hansen, C. J., 16, 22, 27, 276, 379
Hansen, S. S., 274, 284
Hanson, H. K., 128
Hanson, R. B., 21
Hargrave, P. J., 171, 180, 192
Härm, R., 278
Harnden, F. R., 233, 447, 448
Harper, D. A., 277, 505
Harrington, J. P., 276, 283
Harrington, R., 365, 368
Harris, A., 177, 180, 336, 342, 354
Harris, A. W., 79, 80
Harris, B. J., 325, 327
Harris, D. E., 166, 173, 176, 177, 185, 190
Harris, D. L., 146, 147, 150, 151, 153
Harrison, E. R., 540
Hartman, R. C., 314

Hartmann, L., 237, 241, 441, 446, 448, 450, 453-56, 475, 477, 481
Hartmann, W. K., 87, 90, 91, 104, 107
Hartquist, T. W., 248, 252
Hartsuijker, A. P., 172, 180, 195, 200, 208
Hartwick, F. D. A., 209, 212, 333
Harvel, C. A., 209
Harvey, G. A., 324
Harvey, J. W., 481
Harvey, P. M., 268, 284
Harwit, M., 508
Haslam, C. G. T., 166
Hauck, R., 126-30
Haves, P., 187, 205, 325, 327
Hawking, S. W., 520
Hawley, S. A., 324, 326, 330, 346, 420
Hayakawa, S., 289, 294, 301, 314
Hayashi, C., 20, 83, 88, 98, 106, 108, 109
Hayes, D. S., 116, 117, 119, 120
Hazard, C., 167, 198, 326, 327, 334, 342
Hearn, A. G., 445, 447
Hearn, D. R., 371, 386
Heasley, J. N., 454, 471
Heckman, H. H., 422, 424, 425
Heeschen, D. S., 332
Hege, E. K., 366, 372, 376, 393
Hegyi, D. J., 517
Heiles, C., 292, 506
Heintz, W. D., 115, 146-49, 151, 153, 368
Heise, J., 372
Heisenberg, W., 297
Henbest, S. N., 180
Henize, K. G., 375
Henkel, C., 407, 409
Henon, M., 104
Henoux, J. C., 442
Henry, J. P., 327, 346
Henry, P. S., 522-24, 526
Henry, R. C., 463, 472
Herbig, G. H., 275, 476
Herbison-Evans, D., 138, 145
Herbst, E., 401
Herman, R., 489
Hermsen, W., 314, 315
Hershey, J. L., 149, 150
Hertzsprung, E., 15
Hesser, J. E., 379
Hewish, A., 166, 167, 192
Hewitt, A., 365
Hiebert, R. D., 508
Hier, R., 337
Higbie, P., 314
Higdon, J. C., 308, 311

Hildebrand, R. H., 505
Hilditch, R. W., 126-30, 133, 137, 139
Hilgeman, T., 267
Hill, E. R., 342
Hill, F., 264, 266, 274, 280
Hill, G., 126-30, 133, 135, 137, 139
Hill, J. K., 235, 244
Hill, J. M., 169
Hill, S. J., 271, 278, 281
Hillas, A. M., 289, 290, 312, 316
Hills, J. G., 100, 394
Hiltner, W. A., 156, 197, 324, 332, 346
Hilton, W. B., 158
Hine, R. G., 207, 336
Hinkle, K. H., 265, 266, 271, 274, 417
Hinteregger, H. F., 327, 353
Hintzen, P., 189, 211, 372, 376, 392
Hinzen, P., 324, 325, 334
Hirabayashi, H., 504
Hiraiwa, H., 310
Hirshfeld, A., 251
Hirst, W. P., 148, 151
Hjellming, R. M., 477
Hobbs, R. W., 243, 284, 325, 327, 472
Hodge, P., 333, 339, 340
Hodson, S. W., 18, 19, 21, 25-29, 35-37, 378
Hoekstra, R., 450, 474
Hoffman, W. F., 268
Høg, E., 147
Högbom, J. A., 167, 171, 172, 180, 189, 190, 192, 202
Hogg, D. E., 190, 197
Hollars, D. R., 475
HOLLENBACH, D. J., 219-62; 222, 223, 225, 226, 235, 238, 242, 244, 245, 247, 249-51
Holman, G. D., 186, 256, 294
Holmes, J. A., 294, 298, 299, 307
Holt, S. S., 190, 325, 327, 344, 350, 479
Holzer, T. E., 232, 457
Honeycutt, R. K., 272
Hopponen, J. D., 498, 517
Horan, S., 311
Horedt, G. P., 80, 88, 101
Hoshi, R., 20
Hoskins, D. G., 324
Houck, J. R., 49, 67, 508
Horedt, G. P., 80, 88, 101
Houtkevich, S. M., 331, 335, 344
Howell, R., 272, 275, 276
Howell, T. F., 492, 493, 548
Hoyle, F., 267, 348, 349

AUTHOR INDEX 567

Huang, S.-S., 138, 145
Hudson, H. S., 268
Huggins, P. J., 264, 271, 276, 278, 280
Huguenin, G. R., 274, 284
Huizenga, H., 372
Huizinga, J. S., 511, 543
Humphreys, R. M., 269, 270, 282, 284, 367
Hunger, K., 384
Hunt, G. E., 56
Hunten, D. M., 109, 450
Huntress, W. T., 248
Huntress, W. T., Jr., 402
Hutchings, J. B., 135, 138, 447
Hyland, A. R., 275

I

Ibanoglu, C., 128
Iben, I., 18, 20, 22, 38
Iglesias, E. R., 245, 248
Illarionov, A., 385
Illarionov, A. F., 545-59
Ingham, W., 177
Inoue, H., 234
Inoue, M., 324, 339
Ionson, J. A., 186, 256, 276, 294, 324
Ip, W.-H, 104
Ipatov, A. V., 194
Ipavich, F. M., 300
Iriarte, B., 127, 144, 149, 156
Isaacman, R., 100
Itoh, H., 233, 234

J

Jacobson, T. S., 284
Jacoby, G., 268
Jaffe, W. J., 176, 178, 190, 191
Jaffrin, M. Y., 227
Janssen, M., 496
Janssen, M. A., 495
Jauncey, D. L., 168, 325, 327
Jeans, J. H., 81
Jefferies, J. T., 463, 466
Jefferts, K. B., 402
Jenkins, C. J., 171, 179, 180, 189, 192, 195
Jenkins, E. B., 233, 242, 506
Jenkins, L. F., 152, 154
Jenkner, H., 19
Jennings, M. C., 447
Jennings, R., 49
Jennison, R. C., 168, 169
Jensen, E. B., 406
Johns, M., 326, 335, 339, 340
Johnson, H. E., 300
Johnson, H. L., 127, 136, 144, 149, 156, 447
Johnson, H. M., 203
Johnson, H. R., 447

Johnson, K., 324, 326
Johnston, K. J., 172, 207, 274, 276, 277, 325, 326
Johnston, M. D., 185, 325, 327, 344
Jokipii, J. R., 256, 291, 292, 296, 298, 299, 301, 308, 310, 311
Joly, M., 421, 432
Jones, A. V., 43
Jones, B., 49, 267, 283
Jones, B. J. T., 543
Jones, F. C., 301, 308, 310, 311
Jones, T. J., 66
Jones, T. W., 178, 204, 273, 326, 328, 333, 338, 339, 342, 345, 348-50, 477
Jordan, C., 440, 444, 446, 448-50, 457
Jordan, S. D., 440
Jørgensen, H. E.,127, 129, 130, 160
Joselyn, J. A., 457
Joshi, M. N., 194
Josse, M., 145
Joyce, R. R., 250, 505
Judge, D. L., 55
Juliusson, E., 306
Jura, M., 249, 251, 268, 275

K

Kafatos, M., 254, 268, 279, 281, 284
Kahn, S., 387
Kaidanovskii, N. L., 194
Kakar, R. K., 266
Kalberla, P., 166
Kaler, J. B., 383
Kalkofen, W., 441, 458-61, 482
Kamper, K., 419
Kamper, K. W., 229
Kamperman, T. M., 450, 474
Kanbach, G., 314, 315
Kapahi, V. K., 167, 171, 177, 192
Kaplan, S. A., 235
Kapranidis, S., 367, 386
Karp, A. H., 25
Katgert-Merkelijn, J., 171, 193
Kato, T., 258
Katz, J. I., 350
Kaula, W. M., 87, 89, 94, 105
Kawara, K., 275, 277
Keachie, S., 492, 493
Kearsey, S., 308
Kelch, W. L., 442, 450-56, 460, 461, 473
Keller, J. B., 496
Kellermann, K. I., 165, 168, 194, 195, 325, 327, 334, 347, 348
Kellogg, E., 550

Kemp, J. C., 324, 339, 345, 352
Kenderdine, S., 169, 327
Kennel, C. F., 231
Kent, S. M., 258
Kenyon, S. J., 283, 284
Kephart, J. E., 272
Kerr, F. J., 274, 279, 505
Khachikian, E. Y., 341
Khaliullin, Kh. F., 137
Khare, B. N., 267
Kiess, C. C., 55
Kiess, H. K., 55
Kikuchi, S., 324, 339
King, D. S., 16-19, 21, 22, 25-29, 35-37
King, K. J., 49
Kinman, T. D., 324-26, 332, 334, 335, 338, 340
Kinzer, R. L., 325, 327
Kippenhahn, R., 440
Kirkpatrick, R. C., 239
Kirshner, R. P., 229, 232-34, 242, 419, 433
Kirsten, T., 79
Kish, J. C., 306, 309, 425, 426
Kislyakov, A. G., 492, 493
Kitamura, M., 120, 128
Klein, U., 166
Kleinmann, S. G., 274, 277
Klemperer, W., 401
Knacke, R. F., 49, 52, 53, 58, 60, 326, 335, 339, 340, 345
Knapp, G. R., 243, 251, 264, 266, 271, 276-80, 447, 476, 505
Knapp, S. L., 271
Kneer, F. J., 472
Kniffen, D. A., 314
Knight, C. A., 327, 353
Kobayashi, T., 310
Kobayashi, Y., 275, 277
Koch, R. H., 124, 126-28, 130, 136-39, 142, 156
Koechlin, L., 145
Koegler, C. A., 137
Koester, D., 367-69, 371-74, 377, 388-90
Kojoian, G., 325, 327, 342
Kollberg, E., 281
Kolpanen, D. R., 325
Kompaneets, A. S., 556, 558
Kondo, H., 127
Kondo, M., 120, 128
Kondo, Y., 115, 279, 447, 450, 473-75
Koniges, A., 478
Königl, A., 200, 202, 349, 350
Konno, M., 324, 339
Kopylov, I. M., 156
Korchakov, A. A., 348
Korol'kov, D. V., 194

Koski, A. T., 257, 326, 330, 336
Kotanyi, C. G., 209
Kovetz, A., 391, 392
Kozlowski, M., 382
Kraft, R. P., 19, 21, 38, 463, 481
Kraichnan, R. H., 296, 297
Krall, N. A., 222, 223, 230, 231
Kraus, J., 324
Kraushaar, W. L., 314
Krikorian, R., 463, 466
Kristian, J., 209, 212, 275, 333
Kríz, S., 115
Kronberg, P. P., 167, 187, 292, 325, 327
Kruskal, M. D., 253
Kruszewski, A., 345
Krymsky, G. F., 255
Kudritzki, R. P., 279
Kuhr, H., 325, 327
Kuiper, G. P., 48, 160
Kuiper, T. B. H., 266, 271, 279
Kulsrud, R. M., 293, 295-97, 302
Kumar, S., 49, 53
Kunde, V., 49, 53
Kundu, M. R., 269
Kunkel, W. E., 119, 152, 476, 479
Kuperus, M., 458
Kupferman, P. N., 324
Kurt, V. G., 541
Kurtz, R. F., 505
Kurucz, R. L., 120, 441, 452, 459-61
Kusaka, T., 83, 106
Kutner, M. L., 251, 407
Kuzma, T. J., 19, 377
Kwan, J., 245, 250, 264, 266, 274, 280
Kwok, S., 266, 268, 274, 282-84
Kylafis, N. D., 364

L

Labeyrie, A., 145
Lachiéze-Rey, M., 303, 311
Lacombe, C., 187
Lacroute, P., 147
Lacy, C. H., 115, 119, 121, 131, 133, 146, 148, 149, 153, 160
Lacy, J. H., 49, 250, 277
Lacy, J. L., 310, 311
Lada, C. J., 228, 248, 251, 252, 277
Laing, R., 181
Lake, G., 551
Lamb, D. Q., 364, 391, 396

Lambert, D. L., 158, 264-66, 269, 271, 272, 274, 275, 417, 434, 450, 474, 475
Lamers, H. J. G. L. M., 447, 450, 474
Lamla, E., 326, 335, 338, 340
Lampe, M., 231
Lampton, M., 367, 371, 387, 479
Landauer, F. P., 209, 212, 333
Landolt, A. U., 324, 338, 382
Landstreet, J. D., 276, 344, 345, 352, 379, 380
Lane, A. P., 274, 284
Lang, K. R., 503, 504
Langer, W. D., 249, 401-5
Lari, C., 171, 178, 180, 189, 190, 193, 210, 325, 327
Larrabee, A. I., 49
LARSON, H. P., 43-75; 43, 44, 46, 49, 51, 55, 56, 58, 60-62, 64, 66, 68-73, 250
Larson, R. B., 77
Lasker, B. M., 203, 209, 243, 379
Laurent, C., 233
Laury-Micoulant, C., 274
Lauterborn, D., 18
Lavrov, A. P., 194
Law, S. K., 272
Lawrence, A., 185
Lazareff, B., 407
Lea, S. M., 550
Leach, R. W., 276
Leacock, R. J., 325, 327, 331
Lebedinskii, A. I., 83
Lebofsky, L. A., 72
Lebofsky, M. J., 258, 324, 326, 333, 339-41, 345, 350, 351, 355
Lebrun, F., 304, 315
Lebskii, Y. V., 492, 493
Lecacheux, J., 49, 51, 56, 58
Lecar, M., 92, 100
Leckrone, D. S., 134, 160
Le Contel, J. M., 446
Ledden, J. E., 336, 349
Lee, L. C., 296
Lee, M. A., 295, 311
Lee, T., 435
Leer, E., 255, 304
LeFèvre, J., 268
Legg, T. H., 168
LeGuet, F., 291
Leibacher, J. W., 458, 482, 483
Leighton, R. B., 264, 271, 276, 278, 280
Lemaire, P., 471
Leong, V., 314
Lepine, J. R. D., 283
Lequeux, J., 294, 325, 327, 431, 432
Lerche, I., 225, 383

Lester, D. F., 58, 60
Leung, C. M., 250, 345
Leung, K.-C., 131, 135, 138, 156
Leventhal, M., 311
Levich, E. V., 557
Levin, B. J., 103
Levy, D. R., 496
Lewin, W. H. G., 371
Lewis, J. S., 53-55, 72, 92, 100, 265, 267
Lezniak, J. A., 306
Lichti, G. G., 314, 315
Liebe, H. J., 498, 517
LIEBERT, J. W., 363-98; 337, 364-68, 371, 372, 374, 376, 379, 383, 384, 386, 388, 390, 392-94
Lightman, A. P., 350, 351
Liller, W., 325, 327, 346
Lilley, A. E., 498
Lind, J., 475
Lindemann, E., 126-30
Lindman, E. L., 224
Lindstrom, P. J., 424, 425
Linfield, R. P., 327
Lingenfelter, R. E., 298, 311
Linke, R. A., 265, 266, 405, 407, 410, 411, 413, 415, 417, 430
Linnell, A. P., 127
LINSKY, J. L., 439-88; 269, 270, 440, 442, 445-61, 463-66, 471-74, 477, 478, 481
Lippincott, S. L., 149, 151
Liszt, H. S., 279, 403
Lites, B. W., 451, 452, 456
Little, A. G., 172, 325, 327, 337
Little, L. T., 166
Litvak, M. M., 252
Liu, S. Y., 472
Livingston, W., 480
Ljutyi, V. M., 345
Lo, K. Y., 265, 275, 282
Locke, J. L., 168
Lockhart, I. A., 172, 193
Lockwood, G. W., 324, 338
Loeser, R., 451, 452, 456
Loewenstein, R. F., 277
Lomb, N. R., 145
Lombardini, P., 508
London, R., 250
Longair, M. S., 169, 171, 183, 193, 207, 538
Lo Presto, J. C., 463
Lord, J. J., 310
Lorre, J., 332, 333
Lovelace, R. V. E., 170, 349, 351
Low, F. J., 48, 49, 272, 275, 276, 505

AUTHOR INDEX

Lowman, P., 49, 53
Loyd, E. G., 508
Lubin, P. M., 522, 523, 528-30
Lucas, R., 407, 411
Lucke, R. L., 234
Lucy, L. B., 274
Lutz, B. L., 56
Lutz, J. H., 283, 463
Lutz, T. E., 462-64
Luyten, W. J., 364, 393
Lynden-Bell, D., 84, 212, 351
Lynds, C. R., 209, 212, 324, 326, 333, 337
Lyon, J., 266
Lyttleton, R. A., 84

M

MacAlpine, G. M., 235, 257
MacCallum, C. J., 311
Macdonald, G. H., 169, 197, 327
MacGregor, K. B., 441
Machado, M. E., 452, 455
Machnik, D. E., 407
Maciel, W. J., 270
Mackay, C. D., 171, 193, 325, 327
Mackey, M. B., 167, 198
Macleod, J. M., 324, 340
Madore, B. F., 15, 28
Maehl, R. C., 425
Magalashvili, N. I., 132
Magnan, C., 474
Maguire, W., 49, 53
Maihara, T., 275, 277
Maillard, J. P., 48, 49, 51, 56, 58, 60
Malina, R., 372
Maltby, P., 169
Mal'tsev, V. A., 492, 493
Mammano, A., 138, 284
Manchester, R. N., 280, 325, 327, 383
Manfroid, J., 243
Mann, P. J., 124, 126-31, 136-40, 142, 143, 156
Mantz, A. W., 55
Maran, S. P., 243, 442, 453-56, 461, 472, 473
Marano, B., 325, 327
Maraschi, L., 311
Mardirossian, F., 122, 126-28, 130, 132, 136, 137, 139
Margolis, J. S., 56
Margon, B., 371
Margoni, R., 138
Marilli, E., 448, 450
Marionni, P. A., 266
Mariska, J. T., 445, 457
Marlin, H. W., 495
Marscher, A. P., 349
Marsden, R. G., 312

Marsh, K. A., 284
Marshall, F. E., 327, 344, 350
Marten, A., 49
Martin, D. R., 510
Martin, P. G., 272, 323, 324, 326, 341, 344, 345, 352
Martin, T. Z., 49, 58
Martins, D. H., 209
Martynov, D. Ya., 137
Masnou, J. L., 314, 315
Mason, G. M., 309
Mason, H. P., 56
Massano, W., 407
Massey, P., 374
Masson, C. R., 178, 198, 202
Mast, T. S., 306, 309, 310
Masters, A. R., 364
Materne, J., 268, 276
Mather, J. C., 495, 496, 500, 512
Mathis, J. S., 243, 267, 268
Matsuda, T., 538
Matsui, T., 105
Matsumoto, T., 314
Matsushima, S., 368
Matthews, K., 243, 272, 335, 340, 350
Matthews, T. A., 176
Matveyenko, L., 325, 327
Mauder, H., 129
Mauger, B. G., 311
Mayeda, L. K., 435
Mayer, C. J., 178
Mayer, W. F., 371
Mayer-Hasselwander, H. A., 314, 315
Maza, J., 324, 326, 329, 341, 344, 345, 350, 352
Mazurek, T. J., 381
Mazzitelli, I., 388
McAlister, H. A., 144, 145
McBride, J. B., 230
McCabe, E. M., 265
McCarthy, D. W., 272, 275-77
McCarthy, R., 67
McClatchey, R. A., 499
McClintock, J. E., 371
McClintock, W., 450, 455, 460, 463, 466, 472, 474
McClure, R. D., 79, 393
McCluskey, C. E., 115
McColl, M., 492, 493
McCook, G. P., 364
McCord, T. B., 66
McCray, R., 220, 221, 232, 235, 242, 250, 251, 254, 294
McCrea, W. H., 83
McCuskey, S. W., 81, 108
McEllin, M., 171, 179
McGimsey, B. Q., 325, 326, 331, 333, 339, 351, 355
McGraw, J. T., 378, 379

McHardy, I. M., 177, 210, 211
McIvor, I., 298
MCKEE, C. F., 219-62; 178, 191, 221-23, 225, 226, 230, 231, 233-39, 241, 242, 245, 249-51, 256, 304, 305
McKellar, A., 155
McKellar, A. R., 56
McLaren, R. A., 265, 271, 274, 275, 280
McLean, I. A., 276
McLeish, C. W., 168
McMillan, R. S., 268, 273, 275
McMullan, D., 322
McNamara, D. H., 127-29, 137, 156, 158
McVittie, G. C., 186
McWhirter, R. W. P., 462
Mebold, U., 239, 257
Medd. W. J., 324
Meegan, C. A., 310
Meeks, M. L., 498
Meister, J., 424
Melchiorri, B., 508, 530
Melchiorri, F., 508, 522, 523, 530, 555
Meloy, D. A., 180, 198, 210
Meneguzzi, M., 306
Mengel, J. G., 280
Merkelijn, J. K., 325
Merrill, J. E., 122
Merrill, K. M., 264, 267, 269, 273, 275, 280, 282, 284, 324, 326
Merrill, P. W., 284
Mestel, L., 390
Meuve, R., 387
Mewaldt, R. A., 422-26
Mewe, R., 229, 233, 234
Meyer, J. P., 309
Meyer, P., 289, 306
Meyer, S., 506
Meyerott, A. J., 203
Meyers, K. A., 192
Mezger, P. G., 503
Mezzetti, M., 122, 126-28, 130, 132, 136, 137, 139
Michalitsianos, A. G., 268, 279, 281, 284
Michaud, G., 19, 27, 28, 385-87, 389
Michel, G., 49
Mickelson, M., 55, 60
Mihalas, D., 465, 467
Mikami, Y., 324, 339
MILEY, G. K., 165-218; 165, 166, 168, 169, 171-73, 175-78, 180-82, 189-91, 194-98, 200-5, 208-10, 342, 346, 351, 353
Milione, V., 251
Milkey, R. W., 454, 465
Millea, M. R., 492, 493

AUTHOR INDEX

Miller, H. R., 326, 333, 339, 351, 355
Miller, J. M., 208
Miller, J. S., 239, 240, 324, 326, 330, 331, 336, 345, 356
Miller, S. C., 251
Millis, R. L., 139
Mills, B. Y., 327, 337, 342
Mills, D. M., 169
Milne, D. K., 503
Milone, E. F., 142, 159
Milton, R. L., 377
Minkowski, R., 257, 418
Mitchell, R. I., 127, 144, 149, 156
Mitton, S., 169
Mizuno, H., 109
Mizutani, H., 105
Modisette, J. L., 447, 450, 474
Moffet, A. T., 165, 168, 169, 182, 187, 195, 325, 327, 342
Moffett, T. J., 116, 325, 327, 328, 368, 370, 374
Moiseyev, I., 325, 327
Moore, R., 220, 232
Moore, R. L., 326, 331-33, 336, 338, 339, 346, 351, 355, 393
Moore, R. T., 234, 457
Moore, W. E., 233
Moorwood, A., 49
Moos, H. W., 463, 472
Moran, J. M., 251, 274, 276, 277, 284
Morbey, C. L., 145, 147
Morgan, B. L., 144
Morgan, J. G., 143
Morgan, T. H., 450, 474
Morgan, W. W., 136, 139
Morris, D., 168, 325, 327
Morris, M., 265, 266, 271, 273-78, 280-83, 415
Morrison, D., 72, 447
Morrison, N. D., 442, 455
Morrison, P., 177, 193
Morrison, P. Jr., 257
Morton, D. C., 160
Morton, W. A., 251
Moseley, H., 277
Moseley, S. H., 505
Mott-Smith, H., 227
Mould, J., 372, 374
Mouschovias, T., 227, 302, 303
Muchmore, D. O., 386
Muehlner, D. J., 500, 502, 505, 508, 522-24, 527, 543
Mufson, S. L., 254, 266, 279, 325-28, 333, 339, 351, 355
Muhleman, D. O., 277
Mullan, D. J., 227, 270, 277, 444, 451, 452, 456, 473, 474

Müller, D., 306, 311
Muller, R. A., 495, 527
Münch, G., 60
Munro, R. E. B., 325
Munro, R. H., 457
Murdoch, H. S., 324, 325, 327
Murray, S., 550
Murray, S. S., 229
Mushotzky, R. F., 190, 325, 327, 344, 350, 479
Mutel, R. L., 277, 477
Muzylev, V. V., 253
Myers, P. C., 293

N

Nadeau, D., 250
Nakagawa, Y., 107, 442
Nakano, T., 83, 106, 304
Nakazawa, K., 88, 98, 108, 109
Nanos, G. P., 530
Natale, V., 508, 522, 523, 530, 555
Nather, R. E., 374, 379
Neal, D. S., 172
Neff, J. S., 119
Neff, S. G., 195
Neidhöfer, J., 166
Nelson, B., 122
Netzer, H., 350
Neugebauer, G., 243, 250, 268, 272, 275, 326, 335, 340, 341, 350
Neupert, H. E., 379
Neville, A. C., 169, 327
Ney, E. P., 265, 267, 275
Nguyen-Q-Rieu, 274
Nicolas, K. R., 445, 457
Niehaus, R. J., 128
Niell, A. E., 195, 327
Nieto, J.-L., 333
Nishimura, J., 310
Nishimura, T., 505
Nishioka, N. S., 500, 512
Niu, K., 310
Niva, G. D., 28, 29
Noerdlinger, P. D., 349
Noguchi, K., 275, 277
Nordsieck, K. H., 324, 326, 327, 334, 338, 342, 348
Nordström, B., 126, 137, 143
Norman, C. A., 187, 190, 213, 251, 252, 257
Northover, K. J. E., 180, 192, 551
Nousek, J. A., 234
Novikov, I. D., 540
Noyes, R. W., 440, 442, 445, 483

O

Oberbeck, V. R., 105
O'Brien, G., 269, 450, 475

O'Dell, S. L., 324-28, 334, 336, 340, 342, 348-50
O'Donnell, W. C., 367
Oegerle, W. R., 302, 450, 459, 462, 473
Oemler, A., 282
Ogden, P. M., 250
Ogelman, H. B., 314
Ogilvie, K. W., 229
Okazaki, A., 128
Oke, J. B., 207, 326, 330, 335, 340, 341
Okuda, H., 275, 277
Olsen, E. H., 143
Olson, E. C., 126-28, 133, 136, 137, 156
Olson, G. L., 447
Onéto, J. L., 145
Oort, J. H., 194, 209, 258, 301
Öpik, E. J., 92, 93, 104
Oppenheimer, M., 248
Ormes, J., 306
Ormes, J. F., 306, 425
Orszag, S. A., 235
Orth, C. D., 306, 309-11
Orton, G. S., 49
Osaki, Y., 19, 379
Osborne, J. L., 307, 308
Osmer, P. S., 203, 209
Oster, L., 269
Osterbrock, D. E., 239, 241, 257, 326, 330, 336
Ostriker, J. P., 25, 187, 221, 233, 234, 251, 255, 291, 304, 305, 307, 308, 364, 377, 428
O'Sullivan, J. D., 193
Owen, F. N., 173, 175, 177-79, 195, 198, 202, 204, 205, 210, 211, 325-28, 333, 338, 339, 342, 477, 479
Owen, T., 49, 53, 55, 56, 58, 60
Owens, A. J., 306
Owens, D. K., 505, 527
Ozel, M. E., 314
Ozernoy, L. M., 542

P

Pacholczyk, A. G., 178, 182, 183, 187, 195, 200
Pacht, E., 324
Pacini, F., 351
Paczynski, B., 381
Padalia, T. D., 126, 130
Padrielli, L., 171, 193, 325, 327
Pagel, B. E. J., 421, 432, 450, 463, 464
Palimaka, J. J., 208
Pallister, W. S., 322
Palmer, H. P., 168, 169
Palmer, P., 265, 273, 275, 278, 282, 283, 415
Papadopoulos, K., 230, 231

AUTHOR INDEX 571

Papanastassiou, D. A., 79
Papousek, J., 128
Pardo, R. C., 434
Paresce, F., 371
Pariyskiy, Yu. N., 194, 551
Parke, E. N., 483
Parker, E. N., 231, 292, 298, 299, 302, 303, 310, 312
Parker, R. A. R., 233, 235, 240
Parma, P., 180
Parsons, S. B., 374, 375
Partridge, R. B., 492, 493, 519, 521-24, 548, 551
Pasachoff, J. M., 471
Paschmann, G., 228
Paul, J. A., 304, 307, 314, 315
Pauliny-Toth, I. I. K., 194-96, 325, 327, 347, 348, 503
Pauls, T., 413
Payne, D. G., 350, 351
Payne-Gaposchkin, C., 277
Pearl, J., 49, 53
Pearson, T. J., 172, 181, 195, 196, 249, 327, 334
Pechernikova, G. V., 91
Pedersen, R. J., 492, 493
Peebles, P. J. E., 540, 541
Peimbert, M., 239, 283, 416, 418-21
Pel, J. W., 16
Pellat, R., 303
Pelyushenko, S. A., 548
Penston, M. J., 336
Penston, M. V., 335, 336
Penzias, A. A., 402, 405, 406, 410-15, 428, 430, 492-94, 521-23, 530, 537, 548
Perley, R. A., 172, 198-200, 202, 204, 205
Perola, G. C., 169, 175-78, 180, 185-87, 189-91, 194, 195, 209, 210, 311, 342
Perrenod, S. C., 543
Perron, C., 310
Persson, S. E., 250, 268, 283
Pesses, M. E., 255
Peters, B., 290, 307
Peterson, B. A., 542
Peterson, B. M., 197, 203, 332
Peterson, C. J., 209
Peterson, J. O., 16, 27, 28
Peterson, V. L., 451, 452
Petrie, R. M., 123, 132, 134, 135
Petschek, A. G., 350
Phillips, J. P., 268, 274, 276, 279
Phillips, M. J., 481
Phillips, M. M., 326, 330, 336
Phillips, T. G., 264, 271, 276, 278, 280
Pica, A. J., 324, 325, 327
Piccinotti, G., 314
Piddington, J., 299

Pierce, W. P., 293, 296
Piirola, V., 479
Pikel'ner, S. B., 235, 241
Pilcher, C. B., 49, 58, 66, 72
Pilkington, J. D. H., 325, 327
Pinkau, K., 315
Pipher, J. L., 45, 508
Pirraglia, J., 49, 53
Plavec, M., 124, 126-28, 130, 136-39, 142, 156
Podosek, F. A., 417, 424
Pollack, J. B., 43, 49, 66, 67, 108
Pollock, A. M. T., 316
Pollock, J. T., 325, 327
Polyachenko, V. L., 84
Ponnamperuma, C., 49, 53
Pooley, G., 326, 327, 334, 342
Pooley, G. G., 172, 173, 180, 187, 325
Popov, M. V., 115
POPPER, D. M., 115-64; 115, 116, 120, 122, 123, 125-30, 132-39, 142, 143, 145, 158-60
Porcas, R. W., 195, 325, 327, 328
Porter, F. C., 272
Potash, R. I., 200
Poteet, W. M., 505
Praderie, F., 440, 441, 446, 447, 450, 482
Prasad, S. S., 248
Pravdo, S. H., 229, 233, 325
Predmore, C. R., 274, 284
Prentice, A. J. R., 80
Preston, G. W., 463
Preuss, E., 195, 325, 327
Price, R. M., 172, 173, 180, 209, 307
Priester, W., 503
Primini, F. A., 371
Prince, T., 310
Pringle, J. E., 204, 351
Prinn, R. G., 53, 54
Priser, J., 365
Prishchep, V. L., 309
Probst, R. G., 149
Probstein, R. F., 227
Ptuskin, V. S., 289, 298, 306, 308, 311
Pucillo, M., 156
Puetter, R. C., 267, 283, 284
Puget, J. L., 548
Pumphrey, W. A., 143
Puplett, E., 510
Purcell, G. H., 195, 327
Purton, C. R., 277, 282-84, 325, 337, 342
Puschell, J. J., 324, 326, 328, 333, 338, 339, 342
Puzanov, V. I., 492, 493
Pye, J. P., 327, 346, 447

R

Rafert, J. B., 139
Rago, J. F., 239
Raisbeck, G. M., 306, 308, 310
Raizer, Yu. P., 222, 224, 227
Rakos, K. D., 147
Ramaty, R., 298, 311
Ramsey, L., 275
Rank, D. M., 49, 58, 60, 266
Rao, K. R., 55
Raphaeli, Y., 551
Rappaport, S. A., 255, 371
Rasmussen, I. L., 290, 307
Rather, J. D. G., 235
Raymond, J. C., 229, 232, 233, 235-37, 240-42, 446, 448, 450, 457, 475
Rayo, J. F., 420
Readhead, A. C. S., 168, 172, 181, 192, 195-97, 202, 204, 327, 334, 349, 352, 355
Reay, N. K., 276
Redman, R., 265, 274, 281
Redman, R. O., 271, 276, 280
Reed, R. A., 49
Rees, M. J., 170, 180, 193, 200, 202, 204, 209, 213, 349-52, 355, 530, 550
Refsdal, S., 18
Reich, W., 166
Reid, M. J., 265, 270, 274, 276, 277, 281, 284
Reid, R. J. O., 316
Reif, K., 166
Reifenstein, E. C., 503
Reimers, D., 278, 279, 381, 473
Reinhardt, M., 185
Rengarajan, T. N., 291
Renzini, A., 278, 280, 382, 441, 447, 459-61
Reynolds, R. J., 250
Rich, A., 324
Richards, P., 543
Richards, P. I., 107
Richards, P. L., 498, 500, 512, 516
Richards, R. S., 168
Richardson, J. A., 371
Richer, H., 365
Richstone, D., 324, 332
Rickett, B. J., 296
Ridgway, S. T., 43, 49, 52, 53, 55, 58, 66, 264-66, 269, 271, 274, 275, 454
Riegler, G., 479
Riegler, G. R., 325
Rieke, G. H., 258, 324-26, 333, 334, 339-41, 345, 350, 351, 355, 505
Riepe, J., 365
Rieu, N. Q., 283
Righini-Cohen, G., 250

Riley, J. M., 169, 171-73, 178, 180, 187, 193, 195
Ringwood, A. E., 80
Roberge, W. G., 245
Roberts, W. W., 221, 251
Robertson, R. M., 450, 455, 460, 466, 472, 474
Robinson, E. L., 364, 374, 378, 379
Robinson, R. D., 481
Robson, E. I., 511, 543
Roche, P. F., 283
Rodgers, A. W., 15, 17, 452
Rodono, M., 450
Rodriguez-Kuiper, E. N., 266
Roesler, F. L., 55, 60, 250
Roffi, G., 325, 327
Rogers, A. E. E., 327, 353
Rogerson, J. B., 402, 429, 447
Roll, P. G., 492, 493, 548
Roman, N. G., 136
Romanishin, W., 185, 367, 382, 393
Romero, H. V., 508
Romney, J. D., 195, 327
Rönnäng, B., 325, 327
Rönnäng, B. O., 194-96
Rood, R. T., 429
Rose, W. K., 278
Rosen, B. R., 277
Rosenbauer, H., 228
Rosner, R., 445, 447, 448, 478, 481
Ross, J. E., 264
Ross, R. R., 16, 22, 27
Rothman, L. S., 499
Rothschild, R. E., 327
Rothschild, R. J., 325
Rowson, B., 168, 169
Rubin, V. C., 209
Rudnick, L., 173, 175-77, 179, 198, 202, 204, 205, 210, 211, 326, 328, 333, 338, 339, 342
Ruskol, E. L., 83
Russell, C. T., 225, 229
Russell, H. N., 122
Russell, R. W., 49, 267, 283, 284
Russell, W. A., 435
Rydbeck, O. E. H., 325, 327
Rydgren, A. E., 119, 152
Rygh, B. O., 463, 464, 466
Ryle, M., 167, 169, 171, 176, 180, 183, 192, 479

S

Sacco, B., 314
Sack, N., 271, 278, 280
Sadik, A. R., 133, 142
Safronov, V. S., 79, 80, 83-86, 89, 91, 94, 98, 99, 101, 103, 105, 107
Sagan, C., 100, 267
Sagdeev, R. Z., 231
Sahade, J., 138, 145
Saito, M., 258
Saito, Y., 258
Salomonovich, A. E., 492, 493
Salpeter, E. E., 234, 240, 247, 268, 269, 274, 278, 280, 368, 370
Salvati, M., 351
Samuelson, R., 49, 53
Sancisi, R., 313
Sandage, A., 16, 21, 22, 25, 36, 38, 341
Sandel, B. R., 240
Sanford, M. T., 272
Sanner, F., 279
Santiago, J. J., 267
Sareyan, J. P., 446
Sarg, K., 475
Sargent, D. G., 277
Sargent, W. L. W., 209, 212, 258, 333, 341, 348, 349, 418
Saslaw, W. C., 193
Sass, C., 370, 371
Sato, H., 538
Sato, S., 275, 277
Saulson, P. R., 522-24, 526, 528, 529
Savage, B. D., 267
Scalo, J. M., 143, 264, 267, 268, 280, 283
Scanlan, R. J., 144
Scarfe, C. D., 16, 128, 134, 135, 145
Scargle, J. D., 269
Scarrotte, S. M., 322
Scarsi, L., 314, 315
Schaack, D., 49
Schallwich, D., 551
Scharmer, G. B., 464
Schatzman, E., 384
Scherb, F., 250
Scheuer, P. A. G., 167, 170, 181, 183, 189, 192, 197, 204, 296, 336, 349, 352, 355
Schiffer, F. H. III, 209, 477, 478
Schild, R. E., 156
Schilizzi, R. T., 189, 192, 194-96, 327
Schlesinger, B. M., 18
Schlickeiser, R., 314
Schmidt, E. G., 15, 21, 25, 28
Schmidt, G. D., 197, 203, 268, 275, 322, 332, 345, 356
Schmidt, K. M., 306
Schmidt, M., 275, 324, 332, 353, 364
Schmidtke, P. C., 446, 448, 450
Schmitt, J. L., 340
Schmitz, F., 441, 447, 455, 458-61
Schmitz, M., 479
Schneeberger, T. J., 450
Schneider, D. P., 131, 156
Schonberner, D., 384
Schoolman, S. A., 119
Schramm, D., 382, 427, 430
Schreier, E. J., 203
Schultz, G. V., 274
Schulz, H., 368, 369
Schwartz, D. A., 185, 190, 203, 325, 327
Schwartz, J., 325, 327
Schwartz, R. D., 239, 242, 243
Schwartz, U. J., 167
Schwarz, J., 185
Schwarzschild, M., 270, 278
Schweizer, F., 209
Sciama, D. W., 349, 352
Sckopke, N., 228
Scott, J. F., 325, 327
Scott, J. S., 172, 178, 186, 187, 189, 195, 198-200, 202, 204, 211, 256, 291, 294, 304
Scott, P. F., 166, 176, 325, 327
Scott, R. L., 325, 327, 331
Scoville, N. Z., 250, 266, 274, 280, 506
Scudder, J. D., 229
Seaquist, E. R., 277, 283
Searle, L., 421
Seaton, M. H., 283
Seaton, M. J., 383
Seiden, P. E., 252
Seielstad, G. A., 327, 349
Sekiya, M., 88, 109
Sellgren, K., 273, 278, 280
Semenova, E. V., 324, 326, 332
Semet, M. P., 79
Serkowski, K., 275, 324, 326
Serlemitsos, P. J., 190, 325, 327, 344, 350, 479
Serov, N. V., 492, 493
Setti, G., 208
Sgro, A. G., 241
Shaffer, B. B., 325, 327
Shaffer, D. B., 195, 325, 327, 342
Shakeshaft, J. R., 492, 493, 503, 548
Shakhovskoy, N. M., 323, 335, 345
Shallis, M. J., 269
Shanny, R., 231
Shapiro, I. I., 327, 353
Shapiro, M. M., 306, 422-24
Shapiro, P. R., 220, 232, 234, 457
Shapiro, S. L., 350, 351
Shaviv, G., 391, 392
Shawl, S. J., 268, 275, 276
Shcheglov, P. V., 234

AUTHOR INDEX 573

Shemansky, D. E., 240
Shen, C. S., 311
Shields, G. A., 268, 283, 350, 353, 419, 421
Shimmins, A. J., 167, 198, 324, 325, 327
Shine, R. A., 451, 452, 455, 456, 461, 463-66, 471, 477
Shipman, H. L., 364, 367-73, 375, 376, 382, 383, 385-87
Shivris, O. N., 194
Shklovsky, I. S., 347
Shmeld, I. K., 253
Shobbrook, R.R., 28, 29, 36, 145
Shortridge, K., 209, 212, 333
Shu, F. H., 251
Shukla, P. G., 314
Shull, J. M., 235, 237-39, 241, 249, 250, 256
Siegmund, W. A., 239
Signer, P., 424
Silberberg, R., 306, 422-24
Silevitch, M. B., 231
Silk, J., 190, 213, 233, 234, 242, 245, 248, 251, 252, 542, 550, 553, 554
Silvaggio, P. M., 61
Silverberg, R. F., 310
Silverglate, P., 264
Simard-Normandin, M., 292
Simkin, S. M., 193, 208
Simon, A. J. B., 172, 177, 179, 181, 189, 210, 211
Simon, M., 250, 306, 349, 505
Simon, N. R., 15, 25, 28, 29
Simon, R. S., 327, 349
Simon, T., 250, 447, 451, 456, 477, 478
Simpson, G. A., 309, 316, 425, 426
Simpson, J. A., 309, 424, 425
Simpson, J. P., 49, 53, 54, 58
Sinn, L. A., 208
Sinton, W. M., 49, 58
Siohan, F., 306
Sion, E. M., 364, 366-68, 370, 388, 392
Siquig, R. A., 18
Sivan, J. P., 420
Sivertsen, S., 508
Skadron, G., 255, 304
Skellern, D. J., 209
Skilling, J., 291, 293, 294, 298, 299, 301, 304
Skumanich, A., 442, 460, 471, 481
Slee, O. B., 168, 169, 325, 327, 342
Slingo, A., 176
Slovak, M. H., 284
Smith, A. G., 325, 327, 331
Smith, B. W., 190, 229, 233, 294, 304, 479

Smith, D., 280
Smith, D. F., 231
Smith, E. O., 324, 326, 327
Smith, E. van P., 466
Smith, G., 421, 432
Smith, G. R., 49, 55
Smith, H., 51, 58, 60
Smith, H. A., 72, 73, 250
Smith, H. E., 203, 324, 326, 327, 334, 342, 406, 421
Smith, L. H., 306
Smith, M. A., 481
Smith, M. D., 187
Smith, R. C., 265
Smith, R. L., 278
Smith, S. E., 277
Smoot, G. F., 306, 310, 311, 495, 505, 522, 523, 527-30
Smythe, C., 442, 460
Sneden, C., 419
Snell, R., 402
Snell, R. L., 269, 271
Snow, T. P., 221, 242, 294, 367, 440, 447
Snyder, L. F., 276
Soboleva, N. S., 194
Socker, D. G., 234
Söderhjelm, S., 136, 156
Sofia, S., 244
Soifer, B. T., 45, 267, 268, 277, 283, 284, 508
Solarz, M., 426
Solinger, A., 225
Solomon, P. M., 506
Solomon, S. C., 105
Soltan, A., 327, 346
Sonneborn, G., 18, 19
Sowell, R., 450
Spalding, J. D., 424-26
Spangler, S. R., 187, 192, 328, 477
Sparks, W. M., 25, 389, 393
Spears, D. L., 269, 272, 274-76, 280
Speed, B., 179
Spencer, J. H., 274, 276, 277
Spencer, R. E., 212
Spenser, P. M., 283
Spiegelhauer, H., 306
Spinrad, H., 60, 324, 326, 327, 376
Spitzer, L., 219, 220, 224, 228, 229, 251
Spruit, H. C., 483
Sreenivasan, S. R., 17, 34
Srivastava, R. K., 126, 129, 130
Staelin, D. H., 492, 493
Stagni, R., 138
Stang, P. D., 311
Stankevich, K. S., 492, 493, 548
Stannard, D., 172, 325, 327, 335, 342

Starikova, G. A., 147
Stark, A. A., 407
Stark, J. P. W., 233
Starrfield, S. G., 364, 378, 379, 389, 393
Stauffer, J., 382
Stawikowski, A., 466
Stecker, F. W., 313, 316, 548
Steigman, G., 219, 429
Stein, R. F., 254, 458, 482, 483
Stein, W. A., 48, 49, 267, 280, 324-28, 333, 334, 336, 338-40, 342, 345, 348-50
Steining, R. F., 505
Stellingwerf, R. F., 17, 18, 27-29, 378
Stencel, R. E., 270, 276, 277, 444, 447, 450, 473, 475
Stenflo, J. O., 482, 483
Stenholm, L. G., 483
Stephens, A. E., 311
Stephens, S. A., 289, 291, 303, 310, 311
Stephenson, C. B., 264
Stern, R., 371
Stevenson, D. J., 393
Stobie, R. S., 15, 16, 21, 22, 25-29, 35
Stocke, J., 210, 211
STOCKMAN, H. S. 321-61; 209, 324, 26, 331, 332, 337-39, 342, 344-46, 350-53, 376
Stokes, R. A., 492-94, 548
Stone, E. C., 314, 424-26
Stone, M. E., 221
Storey, J. W. V., 269, 272, 276
Stothers, R., 18-20, 25, 27, 28, 277
Stotskii, A. A., 194
Stover, R. J., 374, 379
Stoy, R. H., 128, 136
Stramek, R. A., 325, 327, 342
Strand, K. Aa., 146, 147, 150, 151, 153, 365, 368
Strandberg, M. W. P., 498
Strecker, D. W., 43, 49, 53, 54, 58, 66, 269
Strel'nitskii, V. S., 252, 253
Strickland, D. J., 283
Strittmatter, P. A., 21, 25, 26, 35, 219, 324-26, 336, 337, 340, 366, 367, 372, 376, 377, 386-89, 392-94
Strobel, D. F., 54
Strom, K. M., 78, 406
Strom, R. G., 180, 187, 189, 190, 192, 198, 202, 204, 210
Strom, S. E., 78, 406
Strong, A. W., 291, 304, 313, 316
Struve, O., 138, 145
Sturrock, P. A., 169

Suemoto, Z., 471
Sugimoto, D., 20
Sulentic, J. W., 332, 333
Sullivan, W. T. III, 208
Summers, A., 49
SUNYAEV, R. A., 537-60; 252, 489, 538, 541-43, 545-49, 551, 552, 554, 555, 557-59
Sutton, E. C., 269, 272, 279
Sutton, J., 342
Sutton, J. M., 325, 327
Svechnikov, M. N., 115
Swan, G. W., 253
Swanenburg, B. N., 314, 315
Swank, J., 477
Swank, J. H., 325
Swarup, G., 167, 189, 192
Swedlund, J. B., 352
Sweeney, M. A., 382
Swenson, G. W., 276
Syrovatskii, S. I., 289, 313, 348
Szebehely, V., 81, 108
Szeidl, B., 29

T

Tabor, J. E., 20
Tademaru, E., 304
Taira, T., 310
Takeda, H., 538
Takeuchi, J., 105
Takeuti, M., 16, 28
Talbot, R. J. Jr., 209, 241
Talent, D. L., 209, 420
Tammann, G. A., 16, 21, 22, 25, 36, 38
Tananbaum, H., 327, 346
Tapia, S., 268, 273, 275, 324, 326, 327, 339, 369
Tarenghi, M., 189
Tarle, G., 426
Tauber, M. E., 108
Taylor, F. W., 55
Taylor, J. H., 280, 383
Taylor, K., 233
Telesco, C. M., 277
Terashita, Y., 368
Ter Haar, D., 83
Terzian, Y., 264, 283
Testerman, L., 275
Thaddeus, P., 407, 489, 507, 517, 538
Thambyahpillai, T., 312
Theys, J. C., 253, 254
Thielheim, K. O., 314
Thomas, R. N., 440, 441, 444, 465, 482
Thompson, A. R., 168
Thompson, D. J., 314, 316
Thompson, I., 344, 345
Thompson, R. I., 258
Thornton, D. D., 492, 493

Thorsos, T., 314
Thorstensen, J., 326
Thronson, H. A., 277
Tidman, D. A., 222, 223, 230
Tifft, W. G., 325, 327, 418
Timofeeva, G. M., 194
Tinkham, M., 498
Tinsley, B. M., 280, 282, 419, 420, 427, 428, 430
Tisserand, F., 90
Tohline, J. E., 326
Tokunaga, A., 49, 52, 53, 58, 60
Tomasi, P., 325, 327
Tomkin, J., 145, 158, 434
Toombs, R. I., 272
Topka, K., 447, 448
Torres, C. A. O., 476
Torres-Peimbert, S., 239, 283, 420
Toussaint, J., 310
Tovmassian, H., 325, 327, 342
Townes, C. H., 49, 269, 272
Traub, W. A., 517
Trauger, J. T., 55, 60
Treffers, R. R., 49, 55, 66, 70, 250, 251, 267
Tremko, J., 128
Treverton, A. M., 325
Trimble, V., 367, 370, 399, 422, 427
Tripp, D. A., 451, 452
Trivero, P., 508
Truran, J. W., 389
Trussoni, E., 187
Tsao, C. H., 306, 423
Tsuji, T., 265, 269
Tsunemi, H., 234
Tsytovitch, V. N., 296
Tuchman, Y., 271, 278, 280
Tucker, K. D., 251, 407
Tucker, W., 233
Tucker, W. H., 445, 478
Tuggle, R. S., 18, 20, 22, 38
Tümer, T., 314
Tuohy, I. R., 234, 387
Turland, B. D., 194, 195, 198, 203, 204, 332
Turner, B. E., 252, 265, 273, 275, 278, 282, 283, 415
Turner, E. L., 329
Turnshek, D. A., 337
Twigg, L. W., 139
Tyson, J. A., 193, 495

U

Ulfbeck, O., 489, 519
Ulmschneider, P., 440, 441, 446, 447, 449, 455, 458-61, 481, 482
Ulrich, M.-H., 178, 189, 190, 325, 326, 328, 344

Ulrich, R. K., 142, 159, 160, 447, 476
Upgren, A. R., 146
Usher, P. D., 325, 327, 340

V

Vaiana, G. S., 445, 447, 478
Valentijn, E. A., 173, 175-77, 179, 189, 190, 210, 211
Vallée, J. P., 174, 187, 189, 190, 325, 327
Valtonen, M. J., 165, 191, 193, 194, 205
van Altena, W. F., 22, 146
van Ardenne, A., 194-96
van Breugel, W. J. M., 173, 174, 176, 188-90, 198, 201-3, 209
van de Kamp, P., 146, 148, 150, 151
van den Bergh, S., 21, 28, 203, 209, 229, 241, 419
van den Bos, W. H., 147
Vanden Bout, P. A., 265, 272, 274
Van den Heuvel, E. P. J., 383
van der Hucht, K. A., 279, 447, 450, 473-75
van der Kruit, P. C., 165, 169, 177, 190, 313
van der Laan, H., 169, 173, 174, 177, 186, 187, 190
Van Hoosier, M. E., 445, 451, 452, 456, 471
Van Horn, H. M., 364, 367, 373, 375, 390, 391, 396
van Rijsbergen, R., 128
Vapillon, L., 49, 58
Vardya, M. S., 265
Vasilevskis, S., 146
Vasyliunas, V. M., 225
Vauclair, G., 369, 377, 384, 385, 387-90
Vauclair, S., 377, 384, 385, 387, 389, 390
Veeder, G. J., 44, 66, 71, 131, 146, 149, 152
Vemury, S. K., 25
Verma, R. P., 291
Vernazza, J. E., 451, 452, 456, 482
Vernon, F. L., 492, 493
Vetesnik, M., 128
Vickers, D. G., 511, 543
Vidal-Madjar, A., 233
Vigotti, M., 325, 327
Vigroux, L., 421, 432
Visvanathan, N., 322, 324, 326, 330, 331, 335, 338, 339, 345, 346
Vitkevich, V. V., 325, 327
Vittone, A., 284

AUTHOR INDEX

Vogt, R. E., 314, 422, 424-26
Völk, H. J., 294, 296, 304
Vonberg, D. D., 167
von Hoerner, S., 167
von Kap-herr, A., 173, 180, 203
Vrba, F. J., 326, 333, 339, 351, 355

W

Wacker, J. F., 90, 91, 104, 107
Waddington, C. J., 289, 309, 310
Wade, C. M., 169, 171, 197
Waggett, P. C., 198
Wagner, C. E., 230, 231
Walborn, N. R., 138
Walker, G. A. H., 153
Walker, M. F., 325
Walker, R., 365
Walker, R. C., 274, 277, 327, 349
Walker, R. L., 130
Wall, J. V., 324, 325, 327
Wallace, L., 49, 52, 53, 55
Wallerstein, G., 233, 242, 270, 271, 284
Walmsley, C. M., 407, 409
Walter, F. M., 477, 478
Wampler, E. J., 324, 326, 327, 330, 334, 342
Wanner, J. F., 151
WANNIER, P. G., 399-437; 264-66, 271, 276, 278, 280, 405, 407, 410, 411, 413, 415, 417, 430, 432
Ward, W. R., 84, 85, 89
Wardle, J. F. C., 200, 324, 325, 327, 341, 342, 349
Warner, B., 378, 446
Warner, J. W., 326, 333, 339, 340
Warner, P. J., 198
Warren, P. R., 21, 37
Warwick, R. S., 179
Wasserburg, G. J., 79, 435
Wasson, J. T., 105
Watkinson, A., 172, 193
Watson, A. A., 316
Watson, W. D., 245, 247, 249, 402, 403
Wdowczyk, J., 311
Weast, G. J., 364
Weaver, R., 220, 221, 232
Webb, D. F., 418
Webber, J. C., 277
Webber, W. R., 306, 309, 311, 425, 426
Webster, A., 267, 313
Weedman, D. W., 341
Wefel, J. P., 425

Wegner, G., 367, 369-71, 376, 382
Wehinger, P. A., 203
Wehrse, R., 371, 372, 376
Weibel, E. S., 230
Weidemann, V., 364, 368, 369, 377, 380, 382, 383, 392
Weidenbeck, M. E., 424, 425
Weidenschilling, S. J., 83, 85, 89, 100, 106, 107
Weigert, A., 18
Weiler, E. J., 450, 459, 462, 473, 477, 479
Weiler, K. W., 167, 207, 325, 326
Weinberg, S., 540
Weinberger, D., 274
Weisberg, J. M., 477
Weiss, E. W., 156
WEISS, R., 489-535; 496, 500, 502, 505, 508, 517, 522-24, 527, 543
Weiss, W. W., 19
Welch, W. J., 188, 194, 195, 203, 492, 493
Wellington, K. J., 175, 190, 198, 204
Wells, D. C., 209
Wendker, H. J., 269
Wentzel, D. G., 226, 292, 294, 301, 312, 483
Werner, M. W., 268, 273, 278, 280
Wesemael, F., 367, 373, 389, 390
Wesselink, A. J., 16
Wesselius, P., 371, 372
West, F. R., 144
Westbrook, W. E., 268, 275
Westergaard, N. J., 307, 309
Westphal, J. A., 209, 212, 272, 333
WETHERILL, G. W., 77-113; 79, 89, 94, 96-102, 104, 105
Weymann, R., 543, 556-58
Weymann, R. J., 269, 270, 272, 278, 337
Wheeler, J. C., 267, 282, 381
Whipple, F. L., 106
Whitcomb, S. E., 505
White, C., 322
White, N. E., 479
White, O. R., 445, 462, 471, 480
White, S. D. M., 553, 554
Whiteoak, J. B., 187, 188, 325, 327, 346, 407, 409, 411
Whitford, A. E., 136, 139
Whiting, D. L., 273, 278, 280
Whitney, A. R., 327, 353
Whittaker, A. G., 267

Wickramasinghe, D. T., 369, 371-73, 376, 377, 386, 388, 393
Wickramasinghe, N. C., 267
Wielebinski, R., 173, 180, 203, 313, 551
Wielen, R., 147
Wiita, P. J., 170
Wilcken, C. K., 128
Wildt, R., 55, 60
Wilkes, C. T., 190
Wilkes, R. J., 310
Wilkins, D. R., 450
Wilkinson, A., 325
Wilkinson, D. T., 492-94, 521-24, 526, 528, 529, 548
Wilkinson, P. N., 168, 172, 181, 195-97, 334
Williamon, R. M., 128
Williams, I. P., 82
Williams, J. G., 105
Williams, P. J. S., 325, 327
Williams, R., 219
Williams, R. E., 324, 337
Williams, W. L., 324
Williamson, K. D., 508
Williamson, R. M., 326, 333, 339, 351, 355
Willis, A. G., 165, 172, 180, 189, 190, 192, 198-200, 202, 204, 205, 210, 346
Willner, S. P., 267, 275, 283, 284
Wills, B. J., 324-27, 331, 334
Wills, D., 166, 324-27, 331, 334
Wills, R. D., 314, 315
Willson, L. A., 271, 278, 281
Wilson, A. S., 174, 180, 189, 194, 198
Wilson, C. P., 209, 212, 333
Wilson, J. G., 316
Wilson, O. C., 451, 462, 466, 480
Wilson, P. R., 471
Wilson, R., 446, 457
Wilson, R. E., 135, 139, 156
Wilson, R. W., 402, 404, 405, 407, 410, 411, 413, 415, 492-94, 521-23, 530, 537, 548
Wilson, T. L., 407, 409, 413, 429, 503
Wilson, W. J. F., 17, 34
Windram, M. D., 169, 176
Wing, R. F., 264, 446-48, 450
Winnberg, A., 274, 276
Winston, R., 505
Wirtanen, C. A., 326, 335, 338, 340
Wisniewski, W. Z., 127, 144, 149, 156, 326, 333, 339, 351, 447, 455
Witebsky, C., 503

AUTHOR INDEX

Withbroe, G. L., 440, 442, 445, 483
Witteborn, F. C., 43, 66
Wittels, J. J., 327, 353
Witzel, A., 195, 325, 327
Wojslaw, R. S., 274
Wolf, G. W., 447
Wolf, S. C., 463
Wolfendale, A. W., 311, 312, 316
Wolff, R. S., 237, 241, 411, 414
Wollman, E. R., 49, 266, 417
Wolstencroft, R. D., 352
Woltjer, L., 208, 349
Wood, D. B., 122, 129, 130, 132
Wood, F. B., 115, 124, 126-28, 130, 132, 136-39, 142, 156
Wood, H. J., 19
Wood, J. A., 92
Wood, P. R., 271, 278, 280-83, 382
Woodgate, B. E., 234
Woodman, J. H., 56, 60
Woodward, P. R., 80, 251
Woody, D., 543
Woody, D. P., 496, 498, 500, 512, 516
Wooley, R., 16
Woolf, N. J., 279, 345, 352
Wootten, A., 402
Wootten, H. A., 251, 265

Worden, S. P., 442, 450, 452, 455, 456, 460, 461, 466, 472, 474, 481
Worley, C. E., 146-51, 153
Worth, M. D., 146, 149-51
Wray, J. D., 375
Wright, A. E., 325
Wright, K. O., 143, 145
Wright, M. C. H., 167, 194, 195
Wrixon, G., 492, 493
Wyatt, S. P., 268, 271, 280, 281, 389
Wyckoff, S., 203, 264
Wyller, A. A., 150
Wynn-Williams, C. G., 275
Wyse, A. B., 160

Y

Yabushita, S., 202
Yee, H. K. C., 207
Yen, J. L., 168
Yngvesson, S., 281
York, D. G., 233, 367, 402, 429
Yorka, S. B., 264
Young, A., 478
Young, J. S., 309
Young, P. J., 209, 212, 333
Younger, F., 133, 139

Yuan, C., 251
Yushak, S. M., 311

Z

Zahn, J. P., 479
Zamorani, G., 327, 346
Zaninetti, L., 187
Zappala, R. R., 272
Zarnecki, J. C., 234
Zéau, Y., 49, 55, 58
Zebergs, V., 138, 145
ZEL'DOVICH, Ya. B., 537-60; 222, 224, 227, 489, 538, 540-43, 545, 547, 548, 551, 554, 555, 557-59
Zellner, B. H., 275
Ziglina, I. N., 101, 105
Zimmerman, B. A., 427
Zipse, J. E., 310, 311
Zirin, H., 450
Zirker, J. B., 465, 466, 483
Zissell, R., 127
Zotov, N. V., 326
ZUCKERMAN, B., 263-88; 264-66, 271, 273, 275, 278, 279, 282-84, 415
Zverev, Yu. K., 194
Zvyagina, Y. V., 91
Zwaan, C., 483
Zweibel, E. G., 302
Zwicky, F., 341

SUBJECT INDEX

A

Abundance ratios
 in late-type giant stars, 264–66, 277
Accretion disk, 79
Active extragalactic objects, 321–356
 absorption lines in, 323, 331, 337
 accretion disks in, 350–51, 353, 355
 Balmer lines in, 331, 336
 black holes in, 333, 351, 355
 bremsstrahlung in, 350–51
 compact cores of, 322–23, 329, 332, 334–36, 346, 352
 continuum radiation from, 321, 323, 329, 331, 337, 340, 345, 352, 354
 double-lobe radio emission from, 331, 334–35, 339, 346, 349, 351–52, 354–55
 dust grains in, 322, 337, 341, 343–45, 347, 350–52, 356
 ejection of matter in, 332
 emission lines in, 322, 330–31, 334, 336, 340–43, 345, 354–55
 energy release in, 321, 346–47, 351, 353, 355–56
 forbidden-line emission from, 329, 331
 infrared emission from, 322, 333–37, 339, 341, 344–46, 348, 350, 353, 355–56
 jets in, 321–22, 332, 335, 338, 349, 351–55
 luminosities of, 329–30, 338, 341, 343, 351, 353–55
 magnetic field in, 348–50, 352
 models of, 349–56
 nonthermal emission from, 322, 341, 345–46, 350
 optical emission from, 321–23, 329, 331–32, 334–35, 337, 339, 342, 344–46, 348, 350, 352–56
 polarization of, 321–56
  dilution of, 331, 333–34, 346–47, 354
 models of, 322, 347, 350–55
 position angle of, 323–27, 329, 331–33, 335–39, 342, 344–46, 354
 variability of, 323, 329, 331–34, 336–39, 342, 346–47, 350
 wavelength dependence, 335, 337, 339–40, 344–46
 radio emission from, 322–23, 329, 331–32, 334–36, 341–44, 346–50, 352–54, 356
 redshifts of, 323–37, 329, 331, 333–36, 342–43, 345, 354–55
 scattering of radiation in, 322, 337, 343–45, 347, 350–52, 354
 spectra of, 321, 323–27, 329, 331–36, 340–43, 346–48, 350–54, 356
 superluminal expansion in, 334, 342, 347, 349, 352–53
 supermassive objects in, 351
 synchrotron radiation in, 332, 344, 346–50, 352
 thermal emission in, 322, 344–45, 350, 353
 variability of, 321, 329, 332–35, 338–42, 345–49, 351, 353, 355–56
 visual magnitudes of, 323–27, 334, 336, 340, 346
 X-ray emission from, 337, 341, 343, 346, 350, 356
Albedos, planetary, 63–64
Alfvén Mach number, 225, 227–28
Alfvén velocity, 186, 227, 230, 292, 295, 297, 301, 308–9
Alfvén waves, 256, 270, 292, 295, 297
Algol-type binary systems, 155, 157–58
Alpha Herculis, 269, 277, 279
Ambipolar diffusion, 303
Angular resolution, 167–68
Antennas, 491, 494–96, 507, 524
Antenna temperature, 491–95, 503–7, 522–23, 529
AO 0235+164, 333–34, 338–39, 342, 350, 352, 354
AP Librae, 338, 342–43

Arcturus, 448, 450, 452, 472–73, 476–80
Asteroids, 78, 89, 105
 composition of, 69–70, 72
 observations of, 43–44
 surfaces of, 43, 45, 66–69, 71–73
Astrophysics, development of, 1–13
Atmospheric interference
 with background radiation, 490–91, 497–502, 507, 509, 512, 514, 516–17, 521, 524, 531
Atmospheric transmission
 windows, 47–48, 51, 57–58, 61–63

B

Balmer lines, 232–33, 238, 264, 331, 336, 365–66, 369–72, 375, 377, 446, 449–51, 456, 463, 476, 481–82
Bandwidth, 51, 53, 70, 491
Beta Draconis, 468–69, 475
Betelgeuse, 265–66, 268–69, 273, 275, 449, 474–75
Big bang, 399, 414, 427, 489
Binary stars
 Algol-type systems, 155, 157–58
 cannibalism in, 365
 chromospheres of, 476–77
 detached evolved systems, 121, 135, 140–43, 159
 detached main-sequence eclipsing systems, 121–22, 124–31, 133, 155, 157, 159–60, 365
 double-lined eclipsing, 142–43
 eclipsing, 115–16, 119–31, 140, 142, 153–59
 flux ratios in, 132
 light curves of, 116, 121–23, 140, 142
 light ratios in, 122–23, 133
 line intensities in, 123–24, 134
 mass ratios in, 142, 147–49, 155
 OB eclipsing systems, 121, 133–38
 orbits of, 122, 135, 146–47
  inclination of, 132, 140–41, 143–155

577

SUBJECT INDEX

radial velocities of, 122, 142, 145, 153, 155, 158
reflection of light in, 122, 155, 158
resolved spectroscopic, 121, 143–44
semidetached eclipsing, 121, 135, 140, 154–60
spectroscopic, 116, 142–44
tidal distortion in, 122
uncertainties in deriving physical parameters of, 124–25, 134–35, 146–49, 152–55, 159
visual, 115–16, 121, 146–54, 159–60
Blackbody spectrum, 491, 497–98, 500, 514, 516, 519, 527, 545–46
Black holes, 212–13, 333, 351, 355, 542, 548
Blazars, 321–23, 333, 337–38, 341–43, 345–48, 350, 353
BL Lacertae, 329, 336, 338–40, 342–43
BL Lacertae objects, 207, 321–22, 332–33, 338–42, 350, 353–54
Bose-Einstein spectra, 546–47, 549, 556–57, 559
Bremsstrahlung, 313–14, 350–51, 387, 540, 547, 550, 553–54
Brightness temperature, 342, 347–50, 352
B2 1308+326, 333, 338–39, 355
Burnham's nebula, 239–40, 242
BY Draconis stars, 476–77, 479–81

C

Capella, 142, 145, 160, 448, 454, 457, 461
Carbonaceous chondrites, 69, 72
Carbyne, 267
Cassiopeia A, 229, 233, 241, 257, 418–19, 433
Cayrel effect, 441, 444
Centaurus A, 168, 172, 185, 195, 197, 203, 209
Cepheid instability strip see Cepheid variable stars, pulsation instability strip for
Cepheid variable stars, 15–39
beat, 16–20, 27, 36–37
binary-system, 15
blue loops in evolution of, 18, 20, 27, 38
bolometric corrections for, 38–39
bump, 17–20, 26–27, 37
bump phases of, 26
calibrating, 16, 22, 37
Carson opacities in, 20, 22, 25
Christy echoes in, 25–26
color effect in, 38–39
color index of, 22, 38–39
colors of, 21
convection zones in, 27–28
density gradient in, 18–19, 27
double-mode, 16, 18, 26–29, 36–37
envelope structure of, 19, 22, 28, 37
first overtone mode in, 27, 29
fundamental mode in, 16, 19, 26–27, 29
heavy-element content of, 18
helium content of surface layers of, 18–20, 22, 27–29, 34, 36–38
light curve bumps in, 15–16, 18, 25–26, 35
light curve phase in, 16, 25
Los Alamos opacities in, 20, 25
luminosities of, 15–16, 20, 23–24, 30–33, 36–39
magnetic field in, 19–20, 22, 27–29
mass anomaly in, 15–20, 25, 28–29, 34, 37–38
masses of, 15–39
beat, 16, 21, 36, 38
bump, 15, 21–22, 26, 36, 38
double-mode, 27–29
evolution-theory, 15, 17–18, 20–24, 27–29, 35–36, 38
pulsation-theory, 15, 17–18, 21–24, 28, 35–36
Wesselink-radius, 17, 21, 30–33, 34–35, 37
mass loss in, 17–18, 21, 39
mass-luminosity relation for, 18, 21–22
membership in clusters and associations, 17
mode coupling in, 27
models of, 18, 27
homogeneous, 22, 28–29, 34–36, 38
inhomogeneous, 22, 29, 34–36, 38
mode switching in, 29
multiple-mode behavior of, 29, 35
nonradial modes in, 19–20
period-color relations, 38–39
period-luminosity-color relations, 16, 22, 37–39
period–mean density relations, 16–17, 21, 26
period-radius relations, 37, 39
periods of, 15–16, 19, 21, 23–24, 26–33, 37, 39
ratios of, 16–18, 20, 26–29, 37
photometry of, 16, 21, 28
pulsational instability in, 19
pulsational instability strip for, 18, 20, 29, 34, 38
blue edge, 22, 39
red edge, 29, 39
width of, 27, 38
pulsation theory for, 25–26
radii of, 16, 23, 27–28, 30–34, 36–37
resonances in, 26–28
rotation of, 18
second overtone mode in, 16, 26
surface temperatures of, 15, 17–18, 21–25, 28–31, 35, 37–39
triple-mode, 19, 27, 29
turbulence in, 16
velocity curve bumps in, 15, 25–26, 35
Wesselink radii of, 16, 28, 36–37
Ceres, 69
Charge exchange, 227, 233
Chromospheres of stars, 439–83
acoustic waves in, 441, 447, 458–62, 464, 482–83
asymmetries in spectra of, 471–75
in close binary systems, 476–80
continuum radiation in, 440–42, 447, 460, 467, 482
definition of, 440–42, 444
dissipation of energy in, 441–42, 445, 483
EUV radiation from, 443, 445
extent of, 440
flow patterns in, 470–75, 479, 483
heating of, 439–45, 455, 458–60, 464–66, 472, 478, 482–83
hydrodynamic waves in, 458
infrared emission from, 446, 460, 471
inhomogeneities in, 451, 461–62

SUBJECT INDEX 579

magnetic fields in, 442, 445, 460–61, 476–83
mass column density in, 456, 464, 469–70, 477
models of, 439, 441, 449, 451, 454–55, 457–58, 460–61, 466, 468, 477, 482
radiative equilibrium in, 441, 455
radiative loss from, 442, 444–45, 455, 458–62, 481–83
spectral lines of, 439–42, 444–51, 454–78, 480–83
temperatures in, 441–42, 450–53, 456, 458–61, 468, 470, 474, 478, 483
gradient of, 442, 454–55
thermal conduction in, 442–43, 445
turbulence in, 455, 464–65, 468–71
ultraviolet radiation from, 443, 446–47, 460, 479
variability in, 480–81
velocities in, 470, 475, 479
winds in, 444–45, 447, 474–76
X-rays from, 442–43, 445, 447–48, 479
Circumstellar matter, 78, 263–66, 268–71, 273, 275–82, 284, 416–17, 429, 473–75
CIT 6, 264, 266, 279
Comets
formation of, 78
Compton catastrophe, 347, 349
Compton scattering, 313, 347, 350, 352, 538, 540, 542, 544–45, 547–51, 556, 558–59
Condensation temperature, 269
Copernicus satellite, 400, 446–47, 450, 459, 463, 473
COS-B, 289, 314
Cosmic Background Explorer, 531–33
Cosmic background radiation, 489–533, 537–59
anisotropy of, 537–38, 552–53
large-scale, 490, 495, 519–30, 533
small-scale, 519, 537–38
atmospheric interference with, 490–91, 497–502, 507, 509, 512, 514, 516–17, 521, 524, 531
brightness temperature of, 540–53, 554, 557–58

discovery of, 489, 537
isotropy of, 489–91, 519–21, 527, 530
measurements of, 489–533, 538, 547–48, 550–53
high-frequency, 494, 496–98, 500, 502–3, 507–18, 527, 531, 533
low-frequency, 490, 495–97, 500, 503, 506–7, 533
polarization, 489–91, 510–13, 530–31
polarization of, 489–91, 510–13, 530–31, 537, 555–56
spectrum of, 489–96, 503, 507–8, 510, 514–18, 520–21, 533, 537–38, 540–59
distortions of, 489, 515–16, 540–41, 543–50, 555
temperature of, 493, 508–9, 515–16, 518–20, 529
Cosmic rays, 225–26, 242, 244, 255, 289–316, 422, 424, 426
abundances in, 290, 301, 305, 309–10, 399–400, 416, 422–25, 427, 433–34
acceleration of, 291–92, 307, 423
adiabatic losses in, 308
age of, 290, 307, 309, 311
anisotropy of, 389–90, 299, 312, 316
antiprotons in, 311
bubbles of, 299–302
closed model of, 290
collisions of, 293, 296, 297
composition of, 290, 304–9, 311
elemental, 289, 305–6, 308
isotopic, 289, 309
compound diffusion of, 298–99
confinement of, 289, 291–92, 295, 299, 308–9, 316
convection of, 293, 301, 308
Coulomb interactions of, 305, 308
diffusion of, 292–96, 298, 307–9, 312–13
diffusion model of, 290–91, 299–300, 306, 309–11
distribution of, 289, 304, 316
dynamical halo model of, 290, 308–9
electrons in, 289–90, 310–11, 313–14
escape of, 290, 305, 307–9

exclusion from interstellar clouds, 291
extragalactic model, 291
gyration radius of, 291–92, 307
injection spectrum of, 306, 311
interactions with interstellar matter, 289–90, 310–11, 313
intergalactic, 542, 544
leaky-box model, 290, 293, 297, 300, 305–6, 309, 311–12
mean amount of matter traversed, 294, 298, 300, 304–7
mean density of matter traversed, 301
measurements of, 289, 305–6, 311, 316
momentum transfer by, 300–1
nested leaky-box model, 290, 300, 306, 311
number density of, 292, 294, 315
pi mesons in, 313
pitch angle of, 292, 294, 296, 298, 300
positrons in, 289, 311
pressure of, 300, 302–3
primary nuclei in, 290, 305–6
propagation of, 290, 292, 295, 298, 301, 305, 307–9, 312, 422, 424, 426
proton spectrum in, 306, 310
radioactive isotopes in, 310
relativistic, 289, 300, 307, 313
secondary nuclei in, 290, 305–6, 310
solar modulation of, 305, 309, 314
sources of, 289–91, 302, 306
spallation reactions in, 305, 308, 422–24, 426–27
spectrum of, 289–90, 304–7, 314, 316
streaming of, 294–95, 312
transport of, 298
ultraheavy nuclei in, 289
very high energy, 316
Crab nebula, 10, 332, 347
CRL 618, 275, 278, 283
Cross sections, nuclear, 305, 308
Cyclotron radiation, 349
Cygnus A, 168–71, 179–80, 195, 208, 213
Cygnus Loop, 233–34, 240–42

SUBJECT INDEX

D

Deceleration parameter, 538
Degenerate electron gas, 363
Delta Scuti stars, 17, 378, 446, 475
Diffuse galactic background, 332
Diffusion coefficient, 291, 293, 296, 298, 307–8, 313
DI Herculis, 134–35, 138
Dione, 66–67, 70
Doppler cores of spectral lines, 464, 466–68
Doppler effect, 544–45, 550, 554, 556
Doppler redistribution, 465–68
Drift velocity, 292, 303
Dust grains, 77–78, 84–85, 107, 222, 238, 240, 242, 250, 253, 264–273, 275–80, 505–6, 527, 532–33, 541, 550
Dwarf novae, 115

E

Earth
  bow shock of, 228–29, 231
  formation of, 88, 99, 101, 104–5, 108
Einstein X-ray Observatory, 195, 212, 229, 234
Envelopes of late-type stars, 263–84
Epsilon Eridani, 448, 456–57
EQ Virginis, 461, 479–80
Escape velocity, 86–87, 91–94
Extended radio emission, 165–213
Extensive air showers, 316
Extragalactic radio sources, 165–213
  structure of, 165–71, 177–78, 189, 206

F

Faraday polarization, 349
Faraday rotation, 332–33, 349–50
Fermi acceleration, 255–57, 351
Flux-freezing conditions, 224
Fokker-Planck equation, 107
Formation of planets
  see Giant planets, formation of; Terrestrial planets, formation of
Fornax A, 173, 194, 209
Fourier transform spectrometer, 510, 532
Free-fall time, 302

Free-free emission, 503, 504
  See also Bremsstrahlung
FU Orionis stars, 12

G

Galactic ring, 408–9, 412–14, 431–32, 434
Galactic wind, 300–1
Galaxies
  clusters of, 166, 169, 173, 175, 190, 192, 209–11, 538, 544–45, 550–55
  halos of, 175–76
  compact radio cores of, 169–72, 174, 193–97, 200, 202–3, 206–7, 211
  elemental and isotopic abundance ratios in, 406, 432, 435
  elliptical, 165, 170, 173, 194, 208, 329, 336, 343, 355
  evolution of, 1, 7, 10–12
  formation of, 1, 7, 10, 12, 540, 544
  halos of, 322
  jets in, 321–22, 332, 335, 338, 349, 351–55
  nuclei of, 10–11, 169, 181, 194–95, 206, 211–13, 257–58, 321, 323, 332–34, 341, 352, 354, 542, 554
  optical observations of, 170, 172, 182–83, 193, 202–11
  radio emission from, 165–213
  see also Radio galaxies
  Seyfert
    see Seyfert galaxies
  spiral, 165, 194
  spiral arms of, 313
  S0, 194
Galilean satellites, 43, 66–67, 72
Gamma rays, 289, 304, 312–15
Giant planets
  atmospheres of, 43–48, 51–65, 67
  chemical composition of, 44, 47, 53–57, 60–63, 65, 72, 105–6
  evolution of, 44–45
  formation of, 78, 83, 88, 89, 106
  spectra of, 45–46, 49–53, 57–65
Gravitational focusing, 92
Gravitational perturbation, 92–93, 95–97, 100, 103–4
Gravitational radiation, 520

H

H II regions, 10, 219–21, 238–39, 243–44, 252, 296, 389, 419–20, 435, 503–4, 533
HEAO satellites, 229, 367, 387, 447–48, 554
Heavy elements
  production of, 399–400
H-H objects, 238–40, 242–43
Hill sphere, 108
H-R diagram, 18, 277, 378, 383, 446, 448–49, 473
Hubble constant, 323, 330, 538, 554
Hubble time, 543, 547
Hyades, 382–83
Hydromagnetic waves, 226, 292, 294–96, 304, 308
Hydrostatic equilibrium, 302

I

Iapetus, 66–67, 70
Ices, 67, 72
Infrared excess, 78
Infrared spectroscopic observations, 43–74
Instability, 294, 303–3
Intercloud medium, 293–94, 297, 304, 389
Interferometry, 167–69, 510–14, 517, 532
Intergalactic medium, 169, 177–79, 190–91, 193, 197, 204, 211, 540, 550–54
  cooling of, 547, 557
  heating of, 540, 542–44, 552, 557
  ionization of, 540–42
Interplanetary medium, 87, 105, 295
Interstellar medium, 78, 155, 200, 399–435
  absorption and emission in, 4, 400–9, 412–19, 421, 429, 451, 472, 506
  abundance gradients in, 406, 408–9, 415–16, 419–21, 431–32, 435
  abundance variations in, 410, 412, 414, 427, 432, 435
  chemical fractionation in, 400–6, 410, 412, 416–17
  chemical reactions in, 400–2, 406, 414, 417
  clouds in, 399–405, 409–17, 421, 432–55
  collisions of, 221, 235, 252–53, 258

SUBJECT INDEX 581

CNO elements in, 401–3, 406–24, 427–35
cooling rate of, 298
dense (molecular) clouds in, 399–405, 409–17, 421, 432–55
depletion in, 400, 418, 428, 413
deuterium in, 401–3, 414–16, 427–29
diffuse radio emission from, 503–4, 521, 524
enrichment of, 400, 416, 418, 421, 430
evolution of, 399, 406, 416, 427–28, 430, 434–35
FUV radiation in, 237, 244, 247, 252
hydrogen in, 293, 305, 313–15
iron-group elements in, 422–24, 426, 433, 435
isotope abundance ratios in, 399–400, 403–15, 417–20, 422–35
magnetic fields in, 221, 223–28, 230, 232, 240, 254–55, 503
masers in, 252–53
molecules in, 400–1, 403–4
neon in, 424, 433–34
nuclear abundances in, 399–435
models of, 404, 427, 432
uncertainties in, 409–10, 414, 419, 422, 432, 435
polarization in, 329, 337, 346, 352
radiation field in, 264, 275
reddening by, 21, 38
shock waves in, 219–258
silicon and sulfur in, 401, 406–14, 417–20, 423–25, 431–35
tunnels in, 304
ultraviolet radiation in, 234, 247, 403
Invariance principles, 1–4, 11–12
Inverse problems, 1, 5–7, 12
Io, 66, 73
Ion-cyclotron resonance, 295
Ion gyroradius, 230
Ionization, 224, 244, 297
IRC+10216, 264–66, 268, 271–72, 275–76, 280–82, 416–17, 419, 431
IRC+10420, 274, 276–77
IUE satellite, 269, 369, 376, 385, 400, 445–51, 457, 459, 463, 477–78

J

Jacobi parameter, 104
Jeans criterion, 81–82, 84
Jupiter
 formation of, 83, 88–89, 105–6
 infrared observations of, 43–46, 48–58, 60–61, 65, 73
 vertical convection in, 54–55, 61

K

Kelvin-Helmholtz instability, 204
Keplerian motion, 5, 90, 92, 94–95, 99, 106
Kolmogorov spectrum, 296–97
Kompaneets equation, 556, 558
Kraichan spectrum, 297
Kuiper Airborne Observatory, 47, 51, 58, 67, 74

L

Lambda Andromedae, 454, 477, 479
Landau damping, 295–96
Large Magellanic Cloud, 234, 240–41, 243
Larmor frequency, 292
Larmor radius, 228
Late-type stars, 263–84, 446, 466, 473
 carbon in atmospheres of, 265–55
 chemical composition of, 263–65
 chromospheres of, 268–71
 freeze-out model of, 265
 infrared emission from, 263, 265–69, 271–77, 279–80, 282, 284
 isotope abundances in, 263, 266–67
 microwave and radio emission from, 263, 265–66, 269, 271, 275–276, 279–80, 282
 optical observations of, 269–70, 279, 284
 physical properties of, 263
 spectra of, 263–64
Liouville theorem, 304
Lunar occultations, 143–44, 167, 272, 275
Lyman continuum, 2, 236, 444, 451

Lyman lines, 2, 444, 449, 451, 456–57, 463, 474, 481, 541–42

M

M 31, 22
M 87, 168, 173, 179, 186, 194–95, 197, 201, 203, 206, 212, 258, 322, 332–33
Mkn 180, 342–43
Mkn 421, 343, 353
Mkn 501, 338, 341, 343
Mach number, 221, 224, 229–32
Magellanic Clouds, 22
 see also Large Magellanic Cloud; Small Magellanic Cloud
Magnetic field
 in active extragalactic objects, 348–50, 352
 in chromospheres of stars, 442, 445, 460–61, 476–83
 extragalactic, 299
 in interstellar medium, 221, 223–28, 230, 232, 240, 254–55, 289, 291–94, 297, 299, 303–4, 313, 422, 503
 fluctuations of, 292, 295, 298
 inhomogeneities in, 292, 298–99, 304, 312–13
 lines of force, 225, 298–300, 312
 in protoplanets, 80–81, 86
 around stars, 275–76
 turbulence of, 298–99
Magnetic mirrors, 255–56
Magnetoacoustic waves, 297
Magnetosonic waves, 296
Main-sequence stars
 see Stars, main-sequence
Markarian galaxies, 341
Mars
 formation of, 88, 101, 105
 polar caps of, 67
Masers
 circumstellar, 263, 274–84
 interstellar, 252–53
Maunder minima, 481
Maxwellian velocity distribution, 95, 22, 227, 229, 550–51, 556
Maxwell's equations, 223
Mercury
 formation of, 101, 105, 107
Meteorites, 70, 73, 79, 105, 110, 400, 424, 435

SUBJECT INDEX

Michelson interferometry, 143, 510–14, 517, 532
Microturbulence, 302
Microwave background radiation
  see Cosmic background radiation
Milky Way galaxy
  center of, 311, 408, 412–14, 428, 431–32, 434
  corona of, 295, 197, 300–1
  disk of, 291, 293–94, 297, 302, 307–8, 313–14, 408, 420, 432, 504–6, 549
  halo of, 299, 309, 313–14, 316, 421
Minerals
  in planets, satellites, and asteroids, 45, 67, 69, 72
Mira-type stars, 264, 268, 272, 276, 278–79, 281–84
Molecular clouds, 78, 228, 244, 250, 265, 268, 280, 399–405, 409–17, 421, 432–33, 435
Molecules
  in atmospheres of giant planets and their satellites, 44–48, 51–63, 65, 67
  dissociation of, 244–47
  formation of, 221, 244–47, 252
Moon
  formation of, 105, 109
Multiple Mirror Telescope, 379

N

N 49, 239, 241–42
Neptune, 43–44, 46, 61, 63–65
Neutron stars, 363, 383, 426
NGC 315, 179, 201, 204–5
NGC 1068, 343–45, 352
NGC 1265, 173, 175, 191, 198, 204–5, 209
NGC 1275, 195, 197, 209, 258, 323, 329, 338, 342–44, 354–55
NGC 4151, 343-45
NGC 5128, 195, 197, 209
NGC 6251, 168, 195–96, 198, 204, 206
NGC 7027, 266, 278, 283
Nonthermal emission, 368
Novae, 115, 427, 430

O

OB stars, 78
OJ 287, 337–39, 342, 354

Orion nebula, 78, 239, 250, 258, 413, 503
OSO satellites, 289, 314, 462

P

Parker instability, 302–3
Partial redistribution radiative transfer, 451, 455, 465–66, 475
Perseus cluster, 173, 176, 191–92, 197, 323
PHL 5200, 322, 337, 341, 346
Photodesorption, 238
Photodissociation, 264, 275
Photoionization, 219, 232, 235, 237–39, 242–44, 257
Photolysis, 61
Photon diffusion, 545, 547
PKS 0521–36, 329, 331, 343, 354–55
Plages, 454–56, 460–61, 466, 477–78, 482
Planck spectrum, 497, 515, 520, 527, 545–46
Planetary embryos, 87, 100–1, 104–5, 108–9
Planetary nebulae, 8–9, 238–39, 266–68, 275–84, 381–83, 400, 416–21, 427, 429
Planetary nebula stars, 383–84
Planetesimals, 72, 88–89, 92, 99, 105–10
  density of, 81, 98
  mass of, 81–86, 90–94, 98, 100, 107–10
  number of, 101–3
  orbital semimajor axes of, 95, 97, 99–104
  size of, 83, 91, 94, 96, 101, 103, 106–7, 110
Planets
  albedos of, 63–64
  giant
    see Giant planets
  photometry of, 53, 67, 71
  temperatures of, 44–45, 47, 53–54, 57, 61
  terrestrial
    see Terrestrial planets
Plasma instabilities, 221, 230–31, 257
Plasma oscillations, 295
Plasma turbulence, 219, 256–57
Plasmons, 169, 193, 200, 211
Pleiades, 7, 382
Pluto
  infrared observations of, 44, 67, 72
Polarization, 268–69, 275–76, 329, 337, 346, 352

Pre-planetary nebulae, 282–84
Primeval fireball, 489, 519, 537
Procyon, 117, 119–20, 153, 365, 367–68
Protoplanets, 80, 82–83, 86, 92
  giant gaseous, 80, 89
  rotation of, 81
Protosolar nebula, 406, 412, 435
Protostars, 228
Protostellar nebulae, 267, 280
Protosun, 82–83
Pulsars, 115, 280, 383
Pulsation instability strip
  see Cepheid variable stars, pulsation instability strip for

Q

Quasars, 10–11, 165, 169, 171–72, 193–95, 198, 200, 203, 207–9, 235, 258, 321–22, 335–38, 340, 343, 346, 350–56, 540, 542
  see also RADIO SOURCES at end of Index

R

Radiation pressure, 244, 441, 474
Radiative transfer, 1–4, 11–12, 237, 249, 451, 455, 465–66, 475
Radio-continuum emission, 294, 313–14
Radio galaxies, 165–213, 321, 335, 354–55
  ages of, 186–87
  beaming of energy in, 170
  double sources in, 170–73, 193–94, 196, 198, 207, 331, 334–35, 339, 346, 349, 351–55
  dust lanes in, 209
  edge brightening in, 171–72, 177–81, 190, 200, 205, 209, 212
  edge darkening in, 172, 174, 179
  ejection of compact objects from, 193
  Faraday rotation in, 187–88, 203
  head-tail, 169, 173, 194
  hot spots in, 169, 171, 179, 181, 183, 189, 192, 212
  in situ acceleration in, 186–87
  jets in, 170, 172, 189, 196–98, 200–6, 212

SUBJECT INDEX 583

linear size of, 180
lobes of, 169, 170, 181, 183, 192, 202, 211, 331, 334–35, 346, 349, 351–52, 354–55
  magnetic fields in, 170, 182–83, 185–90, 192, 203–5
  minimum energy condition in, 182–83, 186
  models of, 170
  morphology of, 166, 169–70, 177–78, 186, 200, 209–211
  polarization of, 167, 170, 181, 187–90, 192–93, 197, 203–4, 209
  source confinement in, 183–85, 190
  spectra of, 171, 178, 183, 189, 192, 194–97, 203, 206, 212
  synchrotron emission from, 177, 182, 185–87, 193, 202–3, 205
  tailed, 173–78, 189–91, 198, 210–11
  X-ray emission from, 185, 190, 197, 203, 211–12
  See also RADIO SOURCES at end of Index
Radionuclides, 45, 72
Rankine-Hugoniot relations, 223
Rayleigh-Jeans spectrum, 490–91, 494–96, 537, 541, 543, 545–46, 549–51, 558
Rayleigh scattering, 63
Rayleigh-Taylor instability, 251, 254, 303
Receivers
  radio, 491, 494, 496–97, 507, 509–10, 521, 524, 527, 530, 532–33
Red giant stars, 18
Relativistic bulk motion, 321, 334, 342, 349, 351–56
Resolution
  spatial, 45, 53
  spectral, 48–49, 52–53, 57–59, 62–63, 66–67, 71, 73
Resonant scattering, 237
Rhea, 66–67, 70
Roche density, 81
Roche limit, 84, 105
Roche lobe, 142, 157, 159
RR Lyrae stars, 26
RS CVn stars, 140, 447, 450, 456–57, 460–61, 476–79
RX Herculis, 124, 132–33

S

Safronov steady-state velocity, 92–96
SAS-2 satellite, 289, 314
Satellites
  composition of, 67
  infrared observations of, 43–46, 48, 66–67
  surfaces of, 44–45, 66–68
Satellite systems, formation of, 78
Saturn
  formation of, 83, 88–89, 105–6
  infrared observations of, 43–44, 46, 48, 50–51, 57–65, 73
  rings of, 67, 70
Scale height, 291, 294, 303
Scattering of light, 56, 62, 65, 268–9, 273, 275, 278
Scintillations, 166, 192, 296
Seyfert galaxies, 11, 194, 321–22, 340–46, 350, 352, 356
Shock front, 221, 223–24, 228, 237, 245, 254–55
Shock thickness, 227–28, 231
Shock waves
  acceleration of particles in, 255–56
  collision-dominated, 222, 251
  collisionless, 222, 227–32
  cooling behind, 226–27, 233, 236, 245–50
  dissociative, 245–46, 248
  excitation process in, 236
  hydromagnetic, 224
  infrared emission from, 235, 243–33, 248–50
  interstellar, 219–58, 304–5, 307–8
  jump conditions in, 219, 222–26, 229
  laminar, 230
  nondissociative, 248
  nonradiative, 219, 226, 232–34, 241–253
  post-shock relaxation layer in, 221–22, 226
  precursor radiation in, 221–22, 235–37
  preshock ionization in, 236, 240
  propagation of, 219, 221, 233, 254
  radiation from, 221, 225–26, 232, 235, 237–43, 248, 250–51, 253
  radiative, 220, 226, 234–35, 241, 251, 253–54

radio emission from, 250–51
  in stellar atmospheres, 264, 268, 271
  stability of, 253–55
  strength of, 221, 248
  turbulent, 230
Silicates, 67, 72
Sirius, 120, 153, 363, 365, 367, 386–87
SIRTF, 74
Skylab, 375
Small Magellanic Cloud, 240–42
Solar circle, 314–15
Solar faculae, 483
Solar flares, 255, 257, 423–24, 442, 456
Solar nebula, 45, 57, 72, 77, 79–80, 82–83, 88–89, 105–6, 109–10
  density in, 82
  dust layer in, 83–85, 89, 107
  gaseous, 80, 109–10
  gas-free, 80, 89, 96, 98–100, 105, 107
  temperature in, 83, 109
Solar wind, 232, 289, 308, 312, 424
Space Telescope, 213
Spectra
  laboratory measurements of, 47, 56–57, 59–62, 67, 73
Spectral line formation, 57, 61–62
Spectral reflectance, 70, 72
Spica, 138, 145
Spinars, 351
Spiral density wave, 221, 251
Star clusters, 7–9, 365
Stars
  abundance of elements in, 400
  active-chromosphere, 461
  ages of, 160, 480
  AM CVn variables, 374
  angular diameters of, 117, 120, 153, 269, 272, 276, 279, 284
  associations of, 9–10
  BEM flux scale for, 117, 119–20
  binary
    see Binary stars
  birthrate function of, 380, 382–83, 394
  blue helium, 18
  bolometric corrections for, 116–120, 148, 159
  carbon-rich, 264, 266–67, 272, 274–75, 280–82
  Cepheid variable
    see Cepheid variable stars

584 SUBJECT INDEX

chromospheres of
  see Chromospheres of stars
circumstellar envelopes of,
    78, 263–66, 268–71, 273,
    275–82, 284, 416–17,
    429, 473–75
color excess of, 124, 126–31,
    136–37, 139, 156
color index of, 117–21,
    123–33, 136–39, 141–42,
    144–45, 148–52, 156,
    158, 474
composition of, 120, 125,
    160, 439, 465, 480
C/O ratio in, 264, 282–83
coronae of, 268–71, 441–43,
    445, 447–48, 457–58,
    462, 474–75, 480, 483
degenerate
  see White dwarf stars
Delta Scuti, 17, 378, 446,
    475
disk population, 79, 120, 366,
    383, 393
dust grains in atmospheres
    of, 265, 267–70, 272–73,
    275, 277, 280, 283, 447
dwarf, 446–47, 450, 455–56,
    461
early-type, 219, 441, 447–48
effective temperatures of,
    118–21, 124–32, 136–41,
    144, 150–51, 156,
    158–59
ejection of matter from,
    8–11, 268, 271, 274,
    277, 279, 281, 283, 400,
    416, 429–30
evolution of, 1, 7, 10, 12,
    263, 281
flare, 6–7, 12, 450
flares in, 447, 476–79,
    481–82
formation of, 1, 7, 10, 12,
    77–78, 251–52, 280, 380,
    382–83, 549–550
giant, 142–43, 159–60, 430,
    441–42, 444, 447, 456,
    461, 472, 474
halo population, 366–67, 383,
    393
high-velocity, 296
inner atmospheres of, 442–43
late-type
  see Late-type stars
long-period variable, 269,
    271, 281–82
luminosities of, 117, 119–20,
    124–33, 136–37, 139–44,
    149–53, 145–60, 439,
    450, 459, 462–63, 480

M dwarf, 117, 146, 152, 264,
    269–71, 274, 276–79,
    282–84, 365, 368, 446,
    450–51, 456–57, 476
magnetic, 296
magnetic fields in, 275–76
main-sequence, 34, 117–24,
    133, 149, 153, 159, 380,
    442, 445–46, 448, 450,
    455–58, 461, 472,
    475–76, 480, 483
masses of, 115–60, 429
  determination of, 116, 122,
    124, 134–35, 143–44,
    149, 153, 155, 159
mass loss from, 244, 263,
    265–66, 268, 270–73,
    277–82, 400, 406, 416,
    418, 427, 435, 474
metal-poor, 399, 429–30, 434
metal-rich, 399
Mira-type, 264, 268, 272,
    276, 278–79, 281–84
moving groups of, 115
multiple, 283–84
nucleosynthesis in, 266,
    399–400, 412–13, 416,
    419–21, 427–35
  explosive, 427, 433–34
OB, 78
outer atmospheres of, 442–43
oxygen-rich, 264–65, 268,
    272, 274, 280–81
parallaxes of, 115, 121, 142,
    145–54, 160
photometry of, 116–17,
    122–25, 133–35, 140,
    142, 148–49, 158–59
photospheres of, 263–65,
    268–70, 272–73, 276,
    284, 441–44, 447–48,
    455, 458, 460
Population I, 120, 381, 416,
    418
Population II, 416
pre-supernova, 282, 418–19,
    426, 433
pulsation of, 269, 271, 278,
    281
quiet-chromosphere, 461
radiation pressure in, 270–71,
    274, 278, 279–80
radii of, 120–44, 149–53,
    155–60
reddening in spectra of, 135,
    138
red-giant, 18
rotation of, 275, 279, 439,
    455, 477–78
RS CVn, 140, 447, 450,
    456–57, 460–61, 476–79

S-type, 264, 282
semiregular red variables,
    271
speckle observations of,
    142–44, 147
spectra of, 439–78, 480–83
  width-luminosity relation
    in, 462–64, 466,
    469–70
  see also Chromospheres of
    Stars
spectral type of, 117–19,
    121–22, 124–33, 136–41,
    144–56, 158, 160
spectroscopic observations of,
    116–17, 121–23, 125,
    133–35, 140, 142–45,
    155, 160
subdwarf, 364, 373–74, 381,
    383–84
subgiant, 120, 140, 155
supergiant, 142, 441, 444,
    446–47, 457, 466,
    468–70, 474–75
surface gravity of, 120, 124,
    439, 447, 458–60,
    464–66, 468, 483
symbiotic, 283–84
transition region in, 443–46,
    448–49, 457–58, 462,
    473, 477–78
T Tauri
  see T Tauri stars
turnoff masses of, 365
ultraviolet excess in, 125
visual flux of, 120–21, 133,
    148, 152–53, 159
visual observations of,
    115–16, 122–23, 132,
    143, 145–46, 148, 159
white dwarf
  see White dwarf stars
W Ursae Majoris, 115, 138
Stellar winds, 219, 243, 252,
    296
Stokes parameter, 338, 530–31
Sun
  chromosphere of, 439, 445,
    448, 451–53, 455–56,
    458, 462, 470–71,
    477–78
  color index of, 117
  formation of, 77–78, 105
  ultraviolet radiation from,
    88–89
Sunspots, 480
Supermassive objects, 193, 351,
    427
Supernovae, 221, 234, 252,
    281–82, 377, 400, 406,
    416, 419, 427, 429–30, 434

SUBJECT INDEX 585

Supernova remnants, 220–21, 225, 229, 231–35, 238–43, 250, 254–55, 267, 280, 400, 416, 418, 427, 503
Suprathermal particles, 255–57
Synchrotron emission, 177, 182, 185–87, 193, 202–3, 205, 225, 312–13, 332, 344, 346–50, 352, 503, 533

T

Terrestrial planets
  formation of, 77, 79–80, 83–91, 94–98, 100–10
  collisions in, 85–95, 100, 103–105, 107, 110
  eccentricities of orbits in, 79, 90–91, 94, 97, 100–1, 109
  gas drag in, 85, 87, 103, 106–7, 109–10
  gas phase in, 77, 88–89, 106, 110
  gravitational instability in, 77, 80–85, 88–89, 110
  heat sources in, 79, 105
  inclinations of orbits in, 79, 91–92, 94, 100–1, 109
  intersection of orbits in, 107–9
  jet streams in, 104
  orbital angular momentum in, 80, 82
  rotational instability in, 85
  spin angular momentum in, 80, 85
  sticking in, 85–87, 101, 107
  sweeping-up in, 86
  time scale of, 83, 96, 99, 100, 104–6, 108–9
  turbulence in, 83–84
  primordial atmospheres of, 109–10
Tethys, 66–67, 70
Thermal conduction, 223, 225, 227, 229, 297
Thermal instability, 253–54
Thomson scattering, 542, 550, 556
Tidal disruption, 82, 105
Titan
  infrared observations of, 44, 46, 61–65, 67.
Triton, 44, 61
T Tauri, 242–43, 450
T Tauri stars, 9, 12, 240, 242, 252, 447, 457, 476
T Tauri wind, 88

TU Cassiopeae, 16, 28–29
Turbulence
  interstellar, 296–99, 302–3
Twenty-one-centimeter emission, 313
Tycho's supernova remnant, 229, 233, 257

U

Uhuru satellite, 343
Universe
  energy density in, 538–41
  evolution of, 7–8, 537–40
  mass density in, 538–41
  models of, 8, 540
  nuclear reactions in, 540
  perturbations in, 537–38, 545, 548, 555
  recombination in, 541–42
  temperature of, 538, 540–59
  X-ray observations of, 538, 550, 553–54
U Ophiuchi, 134–35, 138
Uranus
  infrared observations of, 43–44, 46, 61–65
UV Ceti stars, 447
UX Arietis, 477–79

V

V 1016 Cygni, 282–84
V 367 Scuti, 16, 28, 35–36
Vega, 120, 447–48
Vela X, 234, 241–42, 258
Venus
  formation of, 88, 101, 104–5
VLA, 167, 172, 199, 201, 212
VLBI observations, 168, 195, 212, 276–77, 284, 329, 332, 334–35, 342, 347, 349, 351, 355
Volatiles in solar system, 45, 67, 81, 110
Voyager spacecraft, 67, 73
VX Sagittarii, 270, 274–75
VY Canis Majoris, 274–75, 277

W

Waves
  damping of, 293–95, 297, 301
Westerbork telescope, 167, 171, 173, 175, 191, 198, 210
White dwarf stars, 153, 160, 284, 363–94
  absolute magnitudes of, 365, 367, 380, 390–94

  accretion of interstellar matter by, 374, 384, 386–90, 393
  ages of, 367
  atmospheric opacities of, 373, 375, 390–91
  Balmer lines in spectra of, 365–66, 369–72, 375, 377
  in binary systems, 365, 367, 374, 379, 393–94
  birthrate of, 380, 382, 384, 391, 393
  black degenerate, 394
  Chandrasekhar mass limit in, 363, 381
  in clusters, 365
  colors of, 364–67, 369, 373–75, 377, 380, 393
  composition of, 366, 368, 370–78, 381, 387–88, 390
  convection in, 370, 373–75, 385–91
  cooling of, 371, 379–80, 389–91, 393–94
  coronae of, 368, 384, 386–87, 389–90
  Coulomb effects in, 391
  crystallization in cores of, 391, 393
  DA-type, 365–73, 377–78, 381, 383–85, 387–88
  densities of, 363, 373
  diffusion in, 373, 384–85, 393
  discovery of, 364
  distances of, 367, 369
  dredging of matter in, 387–88, 390
  EUV radiation from, 367, 371, 373, 386
  evolution of, 374, 379–82, 384–85, 388
  in close binary systems, 365, 374, 381
  gravitational redshift in, 370–71
  helium-atmosphere, 365, 368, 372–75, 384, 386–88
  helium cores in, 381
  helium-rich, 366, 376, 379, 384–86, 389
  hydrogen-atmosphere, 369, 372, 375, 377, 379
  line profiles in spectra of, 367, 375, 377
  luminosities of, 363, 365, 367, 380, 390–94
  luminosity function of, 365, 390–94
  magnetic fields in, 379–80, 383, 389

586   SUBJECT INDEX

mass distribution of, 371
masses of, 364–65, 367–68, 370–71, 374, 379, 381–84, 389, 394
mass loss from, 381–83, 385–87, 389
mass-luminosity relation for, 363
mass-radius relation for, 368, 381
model atmospheres of, 369–75, 377, 384, 386, 391
numbers of, 367
parallaxes of, 364–65, 367, 369, 393
precursors of, 374, 379, 383
progenitors of, 380–83
  mass limit for, 382–83
proper motions of, 364, 392–93
radiative acceleration in, 384–86, 390
radii of, 363–64, 367–68, 371, 374
rotation of, 296, 377, 379, 381, 389
spectra of, 364–73, 378–79, 384, 388, 392

stellar winds in formation of, 381–82, 384
surface gravities of, 363, 367–68, 370, 372–75, 378, 384, 390
surface temperatures of, 364, 366–68, 371–76, 378–80, 384–88, 390, 393
thermal bremsstrahlung in, 387
ultraviolet flux from, 367–69, 371, 373, 376, 384, 386, 389–90
variability of, 377–79
velocities of, 365–67, 370, 379, 388–89
X-ray emission from, 365, 367, 371, 384, 386–87
Wien spectrum, 273, 490, 541, 549–550
Wilson-Bappu relation, 439, 462–63
Wolf-Rayet stars, 445
W Ursae Majoris, 115, 138

XYZ

X-ray emission, 115, 221, 229–34, 241, 294, 365, 367, 371, 384, 386–87

Zeeman effect, 499, 517
Zeta Herculis, 148–49, 152
ZZ Ceti stars, 377–79, 387

RADIO SOURCES

1400+162, 334, 338, 342, 354
3C 31, 172, 198, 202–5
3C 47, 179–80, 188
3C 83.1B, 173, 175, 190–191, 201, 204–5, 209
3C 129, 173, 190, 198, 201–2
3C 219, 188, 198, 202, 204
3C 236, 168, 179–80, 194–96
3C 273, 167, 172, 195, 198, 346
3C 279, 334, 338, 340–42, 346, 353
3C 345, 334–35, 338–42, 346, 353
3C 371, 331, 338, 342–43, 354–44
3C 390.3, 329, 331, 334–36, 338, 342, 354
3C 446, 338, 340–42, 346
3C 449, 172–173, 179, 192, 198–99, 201–2, 204
3C 454.3, 338, 340–42, 346
3C 465, 174, 176, 188, 198, 209

CUMULATIVE INDEXES

CONTRIBUTING AUTHORS, VOLUMES 4–8

A

Aizenman, M. L., 16:215–40
Allen, R. J., 14:417–45;
  16:103–39
Ambartsumian, V. A., 18:1–13
Angel, J. R. P., 16:487–519;
  18:321–61
Audouze, J., 14:43–79

B

Bahcall, J. N., 16:241–64
Bahcall, N. A., 15:505–40
Barshay, S. S., 14:81–94
Baym, G., 17:415–43
Bracewell, R. N., 17:113–34
Brown, R. L., 16:445–85
Burns, J. A., 15:97–126
Burton, W. B., 14:275–306

C

Carbon, D. F., 17:515–49
Carson, T. R., 14:95–117
Cassen, P., 15:97–126
Cassinelli, J. P., 17:275–308
Cesarsky, C. J., 18:289–319
Chaffee, F. H. Jr., 14:23–42
Chapman, C. R., 16:33–75
Chevalier, R. A., 15:175–96
Conti, P. S., 16:371–92
Coroniti, F. V., 15:389–436
Counselman, C. C. III,
  14:197–214
Cox, A. N., 18:15–41
Cox, J. P., 14:247–73

D

De Young, D. S., 14:447–74

E

Elliot, J. L., 17:445–75

F

Faber, S. M., 17:135–87
Ford, W. K. Jr., 17:189–212

G

Gallagher, J. S., 16:171–214;
  17:135–87
Gault, D. E., 15:97–126
Giffard, R. P., 16:521–54
Gorenstein, P., 14:373–416
Gott, J. R. III, 15:235–66
Gursky, H., 15:541–68

H

Habing, H. J., 17:345–85
Hansen, C. J., 16:15–32
Harris, W. E., 17:241–74
Hartmann, W. K., 16:33–75
Heiles, C., 14:1–22
Hoag, A. A., 17:43–71
Hollenbach, D. J., 18:219–62
Howard, R., 15:153–73
Huebner, W. F., 14:143–72

I

Israel, F. P., 17:345–85

K

Kennel, C. F., 15:389–436
Knapp, G. R., 16:445–85
Kraft, R. P., 17:309–43

L

Labeyrie, A., 16:77–102
Larson, H. P., 18:43–75
Lebofsky, M. J., 17:477–511
Leovy, C. B., 17:387–413
Lesh, J. R., 16:215–40
Lewis, J. S., 14:81–94
Liebert, J., 18:363–98
Linsky, J. L., 18:439–88
Lockman, F. J., 16:445–85

M

Manchester, R. N., 15:19–44
Marov, M. Ya., 16:141–69
Mathis, J. S., 17:73–111
McCray, R., 17:213–40
McKee, C. F., 18:219–62
Merrill, K. M., 17:9–41
Miley, G., 18:165–218

N

Nicholls, R. W., 15:197–234
Noyes, R. W., 15:363–87

O

O'Dell, S. L., 14:173–95
Oort, J. H., 15:295–362
Öpik, E. J., 15:1–17

P

Parker, E. N., 15:45–68
Payne-Gaposchkin, C., 16:1–13
Peale, S. J., 14:215–46
Pethick, C., 17:415–43
Pettengill, G. H., 16:265–92
Pipher, J. L., 16:335–69
Podosek, F. A., 16:293–334
Popper, D. M., 18:115–64

R

Racine, R., 17:241–74
Rickett, B. J., 15:479–504
Ridgway, S. T., 17:9–41
Rieke, G. H., 17:477–511
Robinson, E. L., 14:119–42
Rosner, R., 16:393–428

S

Salpeter, E. E., 15:267–93
Savage, B. D., 17:73–111
Schroeder, D. J., 14:23–42
Schwartz, D. A., 15:541–68
Smith, A. G., 17:43–71
Snow, T. P. Jr., 17:213–40
Soifer, B. T., 16:335–69
Starrfield, S., 16:171–214
Steigman, G., 14:339–72

Stein, W. A., 14:173–95
Strittmatter, P. A., 14:173–95, 307–38
Stockman, H. S., 18:321–61
Strom, R. G., 15:97–126
Sunyaev, R. A., 18:537–60
Svalgaard, L., 16:429–43
Swings, P., 17:1–7

T

Taylor, J. H., 15:19–44
Thomas, H.-C., 15:127–51
Tinsley, B. M., 14:43–79
Toomre, A., 15:437–78
Tucker, W. H., 14:373–416
Tyson, J. A., 16:521–54

V

Vaiana, G. S., 16:393–428
van der Kruit, P. C., 14:417–47; 16:103–39

W

Wannier, P. G., 18:399–437
Watson, W. D., 16:585–615
Weedman, D. W., 15:69–95
Weiss, R., 18:489–537
Wetherill, G. W., 18:77–113
Whipple, F. L., 14:143–72
Wilcox, J. M., 16:429–43
Williams, J. G., 16:33–75
Williams, R. E., 14:307–38
Withbroe, G. L., 15:363–87
Woodward, P. R., 16:555–84

Z

Zel'dovich, Ya. B., 18:537–60
Zuckerman, B., 18:263–88

CHAPTER TITLES, VOLUMES 14–18

PREFATORY CHAPTER

About Dogma in Science and Other Recollections of an Astronomer	E. J. Öpik	15:1–17
The Development of our Knowledge of Variable Stars	C. Payne-Gaposchkin	16:1–13
A Few Notes on My Career as an Astrophysicist	P. Swings	17:1–7
On Some Trends in the Development of Astrophysics	V. A. Ambartsumian	18:1–13

SOLAR SYSTEM ASTROPHYSICS

Chemistry of Primitive Solar Material	S. S. Barshay, J. S. Lewis	14:81–94
Physical Processes in Comets	F. L. Whipple, W. F. Huebner	14:143–72
Mercury	D. E. Gault, J. A. Burns, P. Cassen, R. G. Strom	15:97–126
Jupiter's Magnetosphere	C. F. Kennel, F. V. Coroniti	15:389–436
The Asteroids	C. R. Chapman, J. G. Williams, W. K. Hartmann	16:33–75
Physical Properties of the Planets and Satellites from Radar Observations	G. H. Pettengill	16:265–92
Isotopic Structures in Solar System Materials	F. A. Podosek	16:293–334
Results of Venus Missions	M. Ya. Marov	16:141–69
Martian Meteorology	C. B. Leovy	17:387–413
Stellar Occultation Studies of the Solar System	J. L. Elliot	17:445–75
Infrared Spectroscopic Observations of the Outer Planets, Their Satellites, and the Asteroids	H. P. Larson	18:43–75
Formation of Terrestrial Planets	G. W. Wetherill	18:77–113

SOLAR PHYSICS

The Origin of Solar Activity	E. N. Parker	15:45–68
Large-Scale Solar Magnetic Fields	R. Howard	15:153–73
Mass and Energy Flow in the Solar Chromosphere and Corona	G. L. Withbroe, R. W. Noyes	15:363–87
Recent Advances in Coronal Physics	G. S. Vaiana, R. Rosner	16:393–428
A View of Solar Magnetic Fields, the Solar Corona, and the Solar Wind in Three Dimensions	L. Svalgaard, J. M. Wilcox	16:429–43

STELLAR PHYSICS

The Structure of Cataclysmic Variables	E. L. Robinson	14:119–42
Nonradial Oscillations of Stars: Theories and Observations	J. P. Cox	14:247–73
Recent Observations of Pulsars	J. H. Taylor, R. N. Manchester	15:19–44
Secular Stability: Applications to Stellar Structure and Evolution	C. J. Hansen	16:15–32
Theory and Observations of Classical Novae	J. S. Gallagher, S. Starrfield	16:171–214
The Observational Status of β Cephei Stars	J. R. Lesh, M. L. Aizenman	16:215–40
Mass Loss in Early-Type Stars	P. S. Conti	16:371–92
Magnetic White Dwarfs	J. R. P. Angel	16:487–519
Infrared Spectroscopy of Stars	K. M. Merrill, S. T. Ridgway	17:9–41
Stellar Winds	J. P. Cassinelli	17:275–308

CHAPTER TITLES

On the Nonhomogeneity of Metal Abundances in Stars of Globular Clusters and Satellite Subsystems Of the Galaxy	R. P. Kraft	17:309–43
Physics of Neutron Stars	G. Baym, C. Pethick	17:415–43
Model Atmospheres for Intermediate- and Late-Type Stars	D. F. Carbon	17:513–49
The Masses of Cepheids	A. N. Cox	18:15–41
Stellar Masses	D. M. Popper	18:115–64
Envelopes Around Late-Type Giant Stars	B. Zuckerman	18:263–88
White Dwarf Stars	J. Liebert	18:363–98
Stellar Chromospheres	J. L. Linsky	18:439–88

DYNAMICAL ASTRONOMY

Radio Astrometry	C. C. Counselman III	14:197–214
Orbital Resonances in the Solar System	S. J. Peale	14:215–46
Theories of Spiral Structure	A. Toomre	15:437–78

INTERSTELLAR MEDIUM

The Interstellar Magnetic Field	C. Heiles	14:1–22
The Interaction of Supernovae with the Interstellar Medium	R. A. Chevalier	15:175–96
Formation and Destruction of Dust Grains	E. E. Salpeter	15:267–93
Interstellar Scattering and Scintillation of Radio Waves	B. J. Rickett	15:479–504
Radio Recombination Lines	R. L. Brown, F. J. Lockman, G. R. Knapp	16:445–85
Gas Phase Reactions in Astrophysics	W. D. Watson	16:585–615
Observed Properties of Interstellar Dust	B. D. Savage, J. S Mathis	17:73–111
Compact H II Regions and OB Star Formation	H. J. Habing, F. P. Israel	17:345–85
Interstellar Shock Waves	C. F. McKee, D. J. Hollenbach	18:219–62
Cosmic-Ray Confinement in the Galaxy	C. J. Cesarsky	18:289–319
Nuclear Abundances and Evolution of the Interstellar Medium	P. G. Wannier	18:399–437

SMALL STELLAR SYSTEMS

Consequences of Mass Transfer in Close Binary Systems	H.-C. Thomas	15:127–51
Theoretical Models of Star Formation	P. R. Woodward	16:555–84

THE GALAXY

The Morphology of Hydrogen and of Other Tracers in the Galaxy	W. B. Burton	14:275–306
The Galactic Center	J. H. Oort	15:295–362
Cosmic-Ray Confinement in the Galaxy	C. J. Cesarsky	18:289–319

EXTRAGALACTIC ASTRONOMY

Chemical Evolution of Galaxies	J. Audouze, B. M. Tinsley	14:43–79
The Radio Continuum Morphology of Spiral Galaxies	P. C. van der Kruit, R. J. Allen	14:417–45
Extended Extragalactic Radio Sources	D. S. De Young	14:447–74
Seyfert Galaxies	D. W. Weedman	15:69–95
Clusters of Galaxies	N. A. Bahcall	15:505–40
The Kinematics of Spiral and Irregular Galaxies	P. C. van der Kruit, R. J. Allen	16:103–39
Masses and Mass-to-Light Ratios of Galaxies	S. M. Faber, J. S. Gallagher	17:135–87
Globular Clusters in Galaxies	W. E. Harris, R. Racine	17:241–74
Infrared Emission of Extragalactic Sources	G. H. Rieke, M. J. Lebofsky	17:477–511
The Structure of Extended Extragalactic Radio Sources	G. Miley	18:165–218
Optical and Infrared Polarization of Active Extragalactic Objects	J. R. P. Angel, H. S. Stockman	18:321–61

OBSERVATIONAL PHENOMENA

The BL Lacertae Objects	W. A. Stein, S. L. O'Dell, P. A. Strittmatter	14:173–95

The Line Spectra of Quasi-Stellar Objects	P. A. Strittmatter, R. E. Williams	14:307–38
Soft X-Ray Sources	P. Gorenstein, W. H. Tucker	14:373–416
Extragalactic X-Ray Sources	H. Gursky, D. A. Schwartz	15:541–68
Masses of Neutron Stars and Black Holes in X-Ray Binaries	J. N. Bahcall	16:241–64
Infrared Spectroscopic Observations of the Outer Planets, Their Satellites, and the Asteroids	H. P. Larson	18:43–75
Optical and Infrared Polarization of Active Extragalactic Objects	J. R. P. Angel, H. S. Stockman	18:321–61
Measurements of the Cosmic Background Radiation	R. Weiss	18:489–537

GENERAL RELATIVITY AND COSMOLOGY

Observational Tests of Antimatter Cosmologies	G. Steigman	14:339–72
Recent Theories of Galaxy Formation	J. R. Gott III	15:235–66
Gravitational-Wave Astronomy	J. A. Tyson, R. P. Giffard	16:521–54
Measurements of the Cosmic Background Radiation	R. Weiss	18:489–537
Microwave Background Radiation as a Probe of the Contemporary Structure and History of the Universe	R. A. Sunyaev, Ya. B. Zel'dovich	18:537–60

INSTRUMENTATION AND TECHNIQUES

Astronomical Applications of Echelle Spectroscopy	F. H. Chaffee Jr., D. J. Schroeder	14:23–42
Stellar Interferometry Methods	A. Labeyrie	16:77–102
Instrumentation for Infrared Astronomy	B. T. Soifer, J. L. Pipher	16:335–69
Advances in Astronomical Photography at Low Light Levels	A. G. Smith, A. A. Hoag	17:43–71
Computer Image Processing	R. N. Bracewell	17:113–34
Digital Imaging Techniques	W. K. Ford Jr.	17:189–212

PHYSICAL PROCESSES

Stellar Opacity	T. R. Carson	14:95–117
Transition Probability Data for Molecules of Astrophysical Interest	R. W. Nicholls	15:197–234
Interstellar Shock Waves	C. F. McKee, D. J. Hollenbach	18:219–62
Nuclear Abundances and Evolution of the Interstellar Medium	P. G. Wannier	18:399–437

ORDER FORM   ANNUAL REVIEWS INC.

Please list on the order blank on the reverse side the volumes you wish to order and whether you wish a standing order (the latest volume sent to you automatically upon publication each year). Volumes not yet published will be shipped in month and year indicated. Prices subject to change without notice. Out of print volumes subject to special order.

NEW TITLES FOR 1980

ANNUAL REVIEW OF PUBLIC HEALTH   ISSN 0163-7525
  Vol. 1 (avail. May 1980): $17.00 (USA), $17.50 (elsewhere) per copy

ANNUAL REVIEWS REPRINTS: IMMUNOLOGY, 1977–1979   ISBN 0-8243-2502-8
A collection of articles reprinted from recent *Annual Review* series
  Avail. Mar. 1980   Soft cover: $12.00 (USA), $12.50 (elsewhere) per copy

SPECIAL PUBLICATIONS

ANNUAL REVIEWS REPRINTS: CELL MEMBRANES, 1975–1977   ISBN 0-8243-2501-X
A collection of articles reprinted from recent *Annual Review* series
  Published 1978   Soft cover: $12.00 (USA), $12.50 (elsewhere) per copy

THE EXCITEMENT AND FASCINATION OF SCIENCE, VOLUME 1   ISBN 0-8243-1602-9
A collection of autobiographical and philosophical articles by leading scientists
  Published 1965   Clothbound: $6.50 (USA), $7.00 (elsewhere) per copy

THE EXCITEMENT AND FASCINATION OF SCIENCE, VOLUME 2: Reflections by Eminent Scientists
  Published 1978   Hard cover: $12.00 (USA), $12.50 (elsewhere) per copy   ISBN 0-8243-2601-6
                    Soft cover: $10.00 (USA), $10.50 (elsewhere) per copy   ISBN 0-8243-2602-4

THE HISTORY OF ENTOMOLOGY   ISBN 0-8243-2101-7
A special supplement to the *Annual Review of Entomology* series
  Published 1973   Clothbound: $10.00 (USA), $10.50 (elsewhere) per copy

ANNUAL REVIEW SERIES

Annual Review of ANTHROPOLOGY   ISSN 0084-6570
  Vols. 1–8 (1972–79): $17.00 (USA), $17.50 (elsewhere) per copy
  Vol. 9 (avail. Oct. 1980): $20.00 (USA), $21.00 (elsewhere) per copy

Annual Review of ASTRONOMY AND ASTROPHYSICS   ISSN 0066-4146
  Vols. 1–17 (1963–79): $17.00 (USA), $17.50 (elsewhere) per copy
  Vol. 18 (avail. Sept. 1980): $20.00 (USA), $21.00 (elsewhere) per copy

Annual Review of BIOCHEMISTRY   ISSN 0066-4154
  Vols. 28–48 (1959–79): $18.00 (USA), $18.50 (elsewhere) per copy
  Vol. 49 (avail. July 1980): $21.00 (USA), $22.00 (elsewhere) per copy

Annual Review of BIOPHYSICS AND BIOENGINEERING*   ISSN 0084-6589
  Vols. 1–8 (1972–79): $17.00 (USA), $17.50 (elsewhere) per copy
  Vol. 9 (avail. June 1980): $17.00 (USA), $17.50 (elsewhere) per copy

Annual Review of EARTH AND PLANETARY SCIENCES*   ISSN 0084-6597
  Vols. 1–7 (1973–79): $17.00 (USA), $17.50 (elsewhere) per copy
  Vol. 8 (avail. May 1980): $17.00 (USA), $17.50 (elsewhere) per copy

Annual Review of ECOLOGY AND SYSTEMATICS   ISSN 0066-4162
  Vols. 1–10 (1970–79): $17.00 (USA), $17.50 (elsewhere) per copy
  Vol. 11 (avail. Nov. 1980): $20.00 (USA), $21.00 (elsewhere) per copy

Annual Review of ENERGY   ISSN 0362-1626
  Vols. 1–4 (1976–79): $17.00 (USA), $17.50 (elsewhere) per copy
  Vol. 5 (avail. Oct. 1980): $20.00 (USA), $21.00 (elsewhere) per copy

Annual Review of ENTOMOLOGY*   ISSN 0066-4170
  Vols. 7–24 (1962–79): $17.00 (USA), $17.50 (elsewhere) per copy
  Vol. 25 (avail. Jan. 1980): $17.00 (USA), $17.50 (elsewhere) per copy

Annual Review of FLUID MECHANICS*   ISSN 0066-4189
  Vols. 1–11 (1969–79): $17.00 (USA), $17.50 (elsewhere) per copy
  Vol. 12 (avail. Jan. 1980): $17.00 (USA), $17.50 (elsewhere) per copy

Annual Review of GENETICS   ISSN 0066-4197
  Vols. 1–13 (1967–79): $17.00 (USA), $17.50 (elsewhere) per copy
  Vol. 14 (avail. Dec. 1980): $20.00 (USA), $21.00 (elsewhere) per copy

Annual Review of MATERIALS SCIENCE   ISSN 0084-6600
  Vol. 1–9 (1971–79): $17.00 (USA), $17.50 (elsewhere) per copy
  Vol. 10 (avail. Aug. 1980): $20.00 (USA), $21.00 (elsewhere) per copy

(continued on reverse)
*Price will be increased to $20.00 (USA), $21.00 (elsewhere) per copy effective with the 1981 volume.

Annual Review of MEDICINE: Selected Topics in the Clinical Sciences*   ISSN 0066-4219
    Vols. 1–3, 5–15, 17–30 (1950–52, 1954–64, 1966–79): $17.00 (USA), $17.50 (elsewhere) per copy
    Vol. 31 (avail. Apr. 1980): $17.00 (USA), $17.50 (elsewhere) per copy

Annual Review of MICROBIOLOGY   ISSN 0066-4227
    Vols. 15–33 (1961–79): $17.00 (USA), $17.50 (elsewhere) per copy
    Vol. 34 (avail. Oct. 1980): $20.00 (USA), $21.00 (elsewhere) per copy

Annual Review of NEUROSCIENCE*   ISSN 0147-006X
    Vols. 1–2 (1978–79): $17.00 (USA), $17.50 (elsewhere) per copy
    Vol. 3 (avail. Mar. 1980): $17.00 (USA), $17.50 (elsewhere) per copy

Annual Review of NUCLEAR AND PARTICLE SCIENCE   ISSN 0066-4243
    Vols. 10–29 (1960–79): $19.50 (USA), $20.00 (elsewhere) per copy
    Vol. 30 (avail. Dec. 1980): $22.50 (USA), $23.50 (elsewhere) per copy

Annual Review of PHARMACOLOGY AND TOXICOLOGY*   ISSN 0362-1642
    Vols. 1–3, 5–19 (1961–63, 1965–79): $17.00 (USA), $17.50 (elsewhere) per copy
    Vol. 20 (avail. Apr. 1980): $17.00 (USA), $17.50 (elsewhere) per copy

Annual Review of PHYSICAL CHEMISTRY   ISSN 0066-426X
    Vols. 10–21, 23–30 (1959–70, 1972–79): $17.00 (USA), $17.50 (elsewhere) per copy
    Vol. 31 (avail. Nov. 1980): $20.00 (USA), $21.00 (elsewhere) per copy

Annual Review of PHYSIOLOGY*   ISSN 0066-4278
    Vols. 18–41 (1956–79): $17.00 (USA), $17.50 (elsewhere) per copy
    Vol. 42 (avail. Mar. 1980): $17.00 (USA), $17.50 (elsewhere) per copy

Annual Review of PHYTOPATHOLOGY   ISSN 0066-4286
    Vols. 1–17 (1963–79): $17.00 (USA), $17.50 (elsewhere) per copy
    Vol. 18 (avail. Sept. 1980): $20.00 (USA), $21.00 (elsewhere) per copy

Annual Review of PLANT PHYSIOLOGY*   ISSN 0066-4294
    Vols. 10–30 (1959–79): $17.00 (USA), $17.50 (elsewhere) per copy
    Vol. 31 (avail. June 1980): $17.00 (USA), $17.50 (elsewhere) per copy

Annual Review of PSYCHOLOGY*   ISSN 0066-4308
    Vols. 4, 5, 8, 10–30 (1953, 1954, 1957, 1959–79): $17.00 (USA), $17.50 (elsewhere) per copy
    Vol. 31 (avail. Feb. 1980): $17.00 (USA), $17.50 (elsewhere) per copy

Annual Review of SOCIOLOGY   ISSN 0360-0572
    Vols. 1–5 (1975–79): $17.00 (USA), $17.50 (elsewhere) per copy
    Vol. 6 (avail. Aug. 1980): $20.00 (USA), $21.00 (elsewhere) per copy

*Price will be increased to $20.00 (USA), $21.00 (elsewhere) per copy effective with the 1981 volume.

To ANNUAL REVIEWS INC., 4139 El Camino Way, Palo Alto, CA 94306 USA
(Tel. 415-493-4400)

Please enter my order for the following publications:
(Standing orders: indicate which volume you wish order to begin with)

_____, Vol(s). _____ Standing order _____

_____, Vol(s). _____ Standing order _____

_____, Vol(s). _____ Standing order _____

_____, Vol(s). _____ Standing order _____

Amount of remittance enclosed $_____   California residents please add applicable sales tax.
Please bill me ☐   Prices subject to change without notice.

SHIP TO (include institutional purchase order if billing address is different)

Name _____

Address _____

_____

_____ Zip Code _____

Signed _____ Date _____

☐ Please add my name to your mailing list to receive a free copy of the current Prospectus each year.
☐ Send free brochure listing contents of recent back volumes for *Annual Review(s)* of

_____